AF745225

Mechatronic Systems Design

Klaus Janschek

Mechatronic Systems Design

Methods, Models, Concepts

Translation by
Kristof Richmond

Springer

Prof. Dr. techn. Klaus Janschek
Technische Universitaet Dresden
Electrical and Computer Engineering
Institute of Automation
Dresden
Germany
klaus.janschek@tu-dresden.de

Dr. Kristof Richmond
Iowa City, Iowa
USA
kristof@frontierastronautics.com

ISBN 978-3-642-17530-5 e-ISBN 978-3-642-17531-2
DOI 10.1007/978-3-642-17531-2
Springer Heidelberg Dordrecht London New York

Library of Congress Control Number: 2011937832

© 2012 Springer-Verlag Berlin Heidelberg
This work is subject to copyright. All rights are reserved, whether the whole or part of the material is concerned, specifically the rights of translation, reprinting, reuse of illustrations, recitation, broadcasting, reproduction on microfilm or in any other way, and storage in data banks. Duplication of this publication or parts thereof is permitted only under the provisions of the German Copyright Law of September 9, 1965, in its current version, and permission for use must always be obtained from Springer. Violations are liable to prosecution under the German Copyright Law.
The use of general descriptive names, registered names, trademarks, etc. in this publication does not imply, even in the absence of a specific statement, that such names are exempt from the relevant protective laws and regulations and therefore free for general use.

Printed on acid-free paper

Springer is part of Springer Science+Business Media (www.springer.com)

– For Ruth

and

in memory of

Martin Beck –

Preface

This is the translated English edition of the book *"Systementwurf mechatronischer Systeme"* published by Springer in January 2010. For the motivation, background and concept of presentation, the obliging reader is referred to the preface of the German edition, which follows.

Already in the course of preparing the German book and studying the vast literature in this subject area, I recognized that there still exists some room on the international mechatronics stage for the material presented in this monograph. Moreover, the many positive comments of colleagues and students on the German edition encouraged me to think of starting a second round with an English edition.

From the very beginning, it was clear to me that such a project could only be successful with translation support from a native speaker having broad and excellent knowledge of engineering, and particularly mechatronics. Luckily, a previous stay at the Stanford Aerospace Robotics Lab brought me into contact with the best possible partner for this purpose: Dr. *Kristof Richmond.* As a fully bilingual native speaker, a highly qualified Stanford graduate, and an intelligent, critical, and altogether most clever scientific partner, he combines all of the talents which I did not really expect to find in one person. The joint work on this English edition was therefore extremely smooth, mutually enriching, and, in the age of internet and Skype, never bounded by the cross-Atlantic separation between Germany and Iowa. A big thanks to Kristof for this great job.

This English edition covers the contents of the German edition with some minor improvements in presentation, with updated English textbook bibliography, and it gave us the chance to remove errata found in the German edition.

For the acknowledgements in putting together this mass of material, the reader is again referred to the subsequent preface of the German edition. Nevertheless, I must acknowledge two people in the context of this English edition. My first thanks go to my beloved wife *Ruth*, who has also supported and accompanied this second mountain climbing expedition with a fantastic and ever-encouraging mood (though fortunately, this was a much lower summit than the previous one, cf. preface to the German edition).

My thanks and commemoration go also to *Martin Beck*, who tragically died in February 2011 just before finishing his PhD. As one of my most talented PhD students and closest co-worker in my mechatronics courses, he provided major contributions in the critical proofreading of the original material, sound and always critical technical and scientific discussions, and numerous recommendations for improved didactic presentation. His spirit will also remain between the lines of this English edition.

Last, but not least, all the aforementioned efforts would not have resulted in the present book without the valuable support, trust, and the excellent service of the Springer publishing team represented by *Eva Hestermann-Beyerle*.

Dresden, June 2011 *Klaus Janschek*

Preface – German Edition

Motivation

Why another book on mechatronics? And moreover, why such a comprehensive volume with so much descriptive text?

I had answered the first question for myself at the beginning of this project with the justification of *"re-working my apprenticeship"*, whence I derived the motivation for its realization—whose extent could not be guessed at the time. The second question only presented itself in the course of composition, and was answered in cases of doubt by making decisions in favor of more text following the paradigm *"everything need not be hidden between the lines and in the formulae"*.

Now, to the story of "the apprenticeship". This began with me as an *electrical engineering student* at the Graz University of Technology (TU Graz), from which I obtained a very serviceable foundation in mathematics and the natural sciences—as should after all be expected of a university course of study in engineering. A major and then a doctorate in control theory subsequently uncovered to me a view of "systems" and systems-oriented solutions.

My subsequent apprenticeship as a *development engineer* in mechanical engineering and in aerospace led me to application areas which had played practically no role in my studies: complex heterogeneous systems, which today would be called "mechatronic systems". That my entry into this domain was still quite successful is probably due to two things: the broad foundation provided by my university education, and a systems-oriented approach to solving problems.

Alongside fascinating experiences involving challenging new applications, these years of apprenticeship produced an important *realization*: "You must learn to bring the numerous approaches conveyed to you by your education into *suitable combination* with each other!" Finding the correct path to take is, of course, always left to each individual engineer, but the way will be eased by helpful, experienced mentors (of which I was lucky to have a good number). Throughout this process, the thought of "what I would have wished for in my studies as a development engineer" often suggested itself and remained present in my mind.

Now, since 1995, I have had the opportunity in my *"academic apprenticeship"* at the Technische Universität Dresden to pass on to engineering students my experiences regarding the topic "what I would have wished for in my studies as a development engineer" (in the meantime, alongside the classical courses in electrical and mechanical engineering, also as part of the interdisciplinary major of mechatronics). In this way, my personal teaching loop has been closed, or more correctly my *teaching and learning loop*, as *academic teaching* is most tightly bound to *one's own learning*.

This present text came into being from many years of teaching "Modeling and Simulation" and "Mechatronic Systems" as part of the primary curriculum in the above-mentioned courses of study.

In the course of this last apprenticeship, it has however turned out that the desired *knowledge transfer regarding system-oriented problem solving in complex heterogeneous systems* can be only approximately realized within the constraints of a time-limited course. It is easy to convey fundamental methodological and conceptual approaches, as well as their implementation in simple practice examples. The space and time required for a broader and deeper technical treatment is simply not available. Bare, weakly-annotated citation of further scientific and technical works to complement a too-brief syllabus really satisfies neither the student nor the instructor. These reasons finally led to "re-working my apprenticeship", the results of which are presented in this textbook and the basic structure of which is succinctly elucidated below.

Methods, models, concepts

The subtitle of this work is *methods, models, concepts* and arises from the following roots.

Models An awareness of the great importance of *models* in *system development* is based on my own professional experience. Aerospace applications, such as the orbital and attitude control of spacecraft, high-precision pointing, and active vibration isolation for instruments, deal with complex heterogeneous systems. Due to their nature, in today's conception, these represent *mechatronic systems* par excellence. The development and verification of such systems has, for obvious reasons, always been based on models. System verification and reliable projections of behaviors are primarily based on predictive models. *Model-based system development* and *systems design* thus imply *working with models*. Interestingly, in the last years, these model-based development approaches have also established themselves in many terrestrial applications, e.g. in the automotive industry, and now represent the *state of the art* for system development in mechatronically-oriented industries.

Methods In order to be able to trust model-based predictions of behavior, the models and dynamic analyses derived from them must be based on a clean technical and scientific foundation. In the context of systems design, this requires suitable *methods* of *model creation* and of comprehensive *dynamic analysis* of the complete system made up of heterogeneous subsystems. In this context, it is particularly those methods enabling clear, reliable, and simple-to-verify dynamic predictions which are sought after, ranging from feasibility predictions in early project phases up to verifying results from computer-aided design processes (never trust your computer!).

Concepts Systems design—as it includes the term "design"—comprises a most highly *creative* activity. Linked to this are multifarious, intriguing opportunities to exploit available *design degrees of freedom*, to the extent that they and their conditions and boundaries are known. A single monograph can certainly not present a comprehensive view of the material in this sense. This textbook, within the realm of the possible, attempts to present selected and successfully used physical configurations and solution concepts to form the kernels of ideas for one's own solution approaches. Based on the methodologically oriented conception of this text, topics are presented on the basis of mathematical models in order to indicate paths towards *quantifiable evaluation* of different conceptual variants.

> This textbook represents an attempt to place important methodological approaches for the modeling, analysis, and design of mechatronic systems into a common context, and to present them in a systematic and self-contained form.

Acknowledgements

The path is the goal, even if the goal initially appears very clearly formulated. Finding the right path, taking it, and finally also arriving at the original goal, requires—as when climbing a mountain—a trustworthy rope team, to whom, at this point, I wish to pay my heartfelt thanks.

First and foremost, thanks go to my family and particularly my beloved wife Dr.phil. *Ruth Janschek-Schlesinger*. It is not only the period of this summit attempt—and its demands on our personal relationship—which she has accompanied with great understanding and steadfast spiritual support. It is of particular joy to me that our decades-long partnership has also led to mutual professional synergies. Thus, for instance, she was able to quite successfully integrate systems-oriented problem-solving approaches into her art therapy and supervisory duties, and her spontaneous, artistic, boundary-breaking perspective has opened up many new points of view for me.

For intensive technical discussions and valuable encouragement, sincere thanks are due to my colleagues Prof. Dr.-Ing. habil. *Helmut Bischoff*, Prof. Dr.-Ing. Dr.rer.nat. *Kurt Reinschke* (both of TU Dresden), and Dr.-Ing. *Peter Schwarz* (Fraunhofer Institute for Integrated Circuits, Design Automation Division, Dresden).

A manuscript of 800 pages naturally contains more than a few dangerous pitfalls and stumbling blocks. For their careful and knowledgeable proofreading of the manuscript and their well-founded suggestions for corrections, particular thanks are due to my co-workers Dipl.-Ing. *Martin Beck*[1] (who deserves the medal for times read!), PD Dr.-Ing. *Annerose Braune*, Dr.-Ing. *Eckart Giebler*, Dipl.-Ing. *Sylvia Horn*, Dipl.-Ing. *Thomas Kaden*, Dipl.-Ing. *Evelina Koycheva*, Dipl.-Ing. *Arne Sonnenburg*, and Dipl.-Ing. *Edgar Zaunick*.

I also wish to thank the remainder of my lab group for their continuous understanding of the demands on my time due to this project. Great, sincere thanks are especially due Ms. *Petra Möge*, who throughout the past two years has, with commitment and skill, kept my hands and head free in the administrative realm, and thus established an important prerequisite for the project's success.

I very sincerely thank the ladies and gentlemen of *Springer-Verlag* for their extremely cooperative and trusting collaboration and their considerate handling of the planning of contents and schedule.

Dresden, October 2009 *Klaus Janschek*

---

[1] 1978 - 2011

Glossary and List of Abbreviations

ADC	analog-to-digital converter
cf.	compare
CBE	constitutive basic equation(s)
DAC	digital-to-analog converter
DAE	differential algebraic equation(s)
DYMOLA	registered trademark of Dynasim AB
ELM	electromechanical
FEM	finite element model
LABVIEW	registered trademark of National Instruments
LTI	linear time-invariant
LTV	linear time-varying
MAPLE	registered trademark of Waterloo Maple Inc.
MATHEMATICA	registered trademark of Wolfram Research
MATLAB	registered trademark of The MathWorks, Inc.
MBS	multibody system
MEMS	micro-electro-mechanical system(s)
MIMO	multi-input multi-output
MODELICA	registered trademark of the Modelica Association
ODE	ordinary differential equation
OSI	Open Systems Interconnection (OSI reference model)
PID	proportional, integral, derivative
SA	Structured Analysis
SIMULATIONX	registered trademark of ITI GmbH
SIMULINK	registered trademark of The MathWorks, Inc.
UML	Unified Modeling Language
ZOH	zero-order hold

Contents

1 Introduction

> *One must learn to direct one's full attention to the actual occurrence and to coordinate inner tension with contemplative mastery of the difficult practice of water-color painting, which brooks no correction.*
>
> Oskar KOKOSCHKA on the method of his "School of Seeing"[1], 1954.
>
> In (Kokoschka 1975)

In the arena of *systems design* involving complex systems, the engineer stands before challenges similar to those of an artist as envisioned by Oskar KOKOSCHKA: he or she must bring a representation of a real system onto paper with minimal error, even though accurate portrayal requires not simply easy-to-use, but rather unambiguous, clear means of description. For both artist and engineer, the beauty and attraction of this challenge lie in its great *creative freedom*, of which one must take full advantage.

The tension between freedom and constraints, distinguishing the doable and the impossible, and the opportunity to predict using scientific methods the behaviors of a not-yet-existing, real, complex system—all this makes systems design a never-ending, grand adventure.

[1] Oskar KOKOSCHKA, 1886-1980, Austrian expressionist painter, graphic artist, and writer. In 1953, he founded the "School of Seeing", today the International Summer Academy of Fine Arts in Salzburg, Austria.

1.1 Mechatronics, Mechatronic Systems

Origins and meaning The currently well-established, much-used term *"mechatronics"* is an invention of the 70s of the last century, and intuitively hints at a tight interaction between the classic engineering disciplines of mechanics (mechanical engineering) and electronics (electrical engineering). Indeed, from this primordial combination of words came the worldwide beginnings of an engineering science which places *thinking in systems* at the center of *product creation.*

The focus of this practice is competitive, innovative *products*, which optimally fulfill the demands—whether they concern function, quality, economics, or emotions—of stakeholder groups. This type of novel product is created by combining diverse technologies into a functional whole. The process of product creation requires *thinking and understanding at a systems level* above and beyond the individual disciplines.

In this sense, mechatronics is an engineering discipline fundamentally concerned with applications, which, in order to realize ambitious product features, must exploit the most current scientific methods and findings from a variety of disciplines.

Definitions of mechatronics From the plethora of existing definitions for the term "mechatronics", the following two are taken as representative, having, in the opinion of the author, the best connection to the material presented in this book:

> **Mechatronics** describes an interdisciplinary design methodology which solves primarily mechanically oriented product functions through the synergistic spatial and functional integration of mechanical, electronic, and information processing subsystems.
>
> *VDI/VDE Gesellschaft für Mess- und Automatisierungstechnik (GMA) - Society for Measurement and Automatic Control (VDI/VDE GMA), Technical Committee 4.15 "Mechatronics"*

> **Mechatronics** is the synergistic combination of precision mechanical engineering, electronic control and systems thinking in the design of products and manufacturing processes. It covers the integrated design of mechanical parts with an embedded control system and information processing.
>
> *International Federation of Automatic Control (IFAC) – Technical Committee on Mechatronic Systems*

Mechatronic systems: product-oriented perspective Considering the definition of mechatronics, a slight extension of the terminology to *mechatronic systems* immediately suggests itself. What is meant by this, however?

The product-oriented external view of a mechatronic system and its boundaries are depicted in Fig. 1.1 using descriptive methods of the function-based modeling approach of *structured analysis* from (Yourdon 1989)[2]. Alongside the actual mechatronic system (the product), two additional important, interacting agents can be seen: the *user* and the *environment.* The rationale behind the product is clearly reflected in Fig. 1.1 in the *user-based product purpose.* A mechatronic system never exists for its own sake, but rather always has direct interactions with a user, and thus a task-oriented *man-machine interface.* In the simplest case, this consists of a start button to enable a complex task to be carried out completely automatically.

---

[2] In Ch. 2, this modeling viewpoint—useful at the most abstract level—is discussed in detail. In this section, for illustration, several pictorial, non-methodological descriptions are also employed. For a strictly formal yet clear presentation, the reader is referred to Ch. 2.

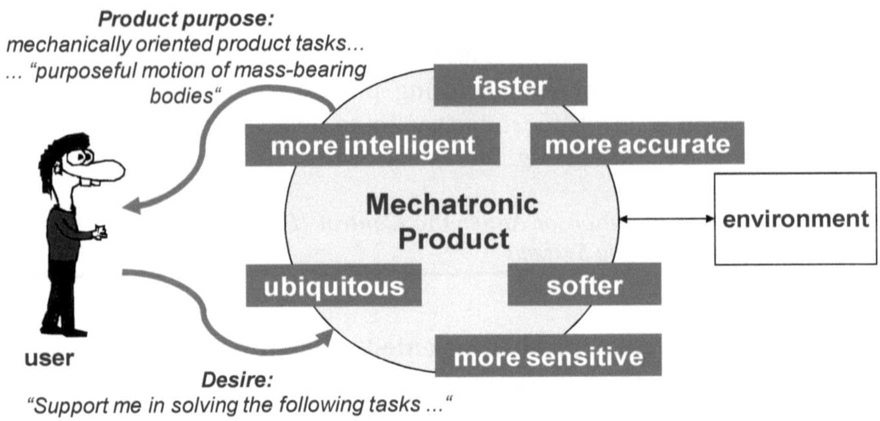

Fig. 1.1. Mechatronic systems as products with exceptional attributes: external perspective

The *generalized product purpose* for mechatronic systems can be characterized in a simplified form as follows[3]:

> A mechatronic product realizes *mechanically oriented* product tasks with the *purposeful* motion of *mass-bearing* bodies and desired, exceptional dynamic properties.

The above characterization can describe practically all products produced under the title of "mechatronic", regardless of whether they be an electronic fuel injection system (activating the injection port), the gripper on a surgical endoscope, the micro-mirrors of a light modulator for video projectors, the autofocus of a photographic camera, or the read head of a computer hard drive (for further examples, see (Isermann 2008) or (Bishop 2007)).

The examples listed above also quite clearly manifest *exceptional product attributes*: for many of these applications, there may be "non-mechatronic" precursors, though with considerably less attractive dynamic

[3] The exceedingly successful implementation of mechatronic development approaches has motivated some actors in the scientific and popular mechatronics community to designate a far wider spectrum of technical systems as "mechatronic". The author objects in the highest degree to this gratuitous misappropriation and dilution of the underlying scientific problem statement and solution approaches. For this reason, in this book, the term "mechatronic system" is limited to the meaning presented above.

properties. In many cases, it is only with a mechatronic solution that innovative, attractive products have been enabled.

The second important agent, the *environment*, primarily determines the operating conditions for a mechatronic system, as shown in Fig. 1.1. The system's purposeful motions must, as a rule, be carried out in the face of disturbing, often uncertain, external influences, placing great demands on its physical implementation (see internal perspective).

Mechatronic systems: function-based internal perspective Realization of the extraordinary dynamic properties of a mechatronic system is based on the *functional structure* sketched in Fig. 1.2 (an interior view of the mechatronic product of Fig. 1.1). This presents a structure for the action flow which is independent of implementation, and in which the interaction of physical system variables is represented at the level of abstract models[4].

The two functions *"generate forces/torques"* (including providing auxiliary energy) and *"generate motion"* describe the motion of a mass-bearing body and the energetic back-effect (mechanical coupling) which is generally present.

It is only in the remaining functions *"measure mechanical states"* and *"process information"* that the actual added value of a mechatronic system can be accessed: the creation of user-oriented, exceptional dynamic behaviors with the *automated implementation* of desired operations using a *closed action chain* (also known as feedback or a control loop).

Mechatronic systems: realization perspective The functions presented in Fig. 1.2 can be physically realized in various ways. In Fig. 1.3, typical *realizing technologies* for the fundamental mechatronic functions are indicated.

---

[4] The close kinship with the block diagrams of control theory is evident. In general, using the verbal/graphical modeling methods of *structured analysis*, intelligible system descriptions can be generated. These include the most important structural and functional properties of the system being described in an easily comprehensible form.

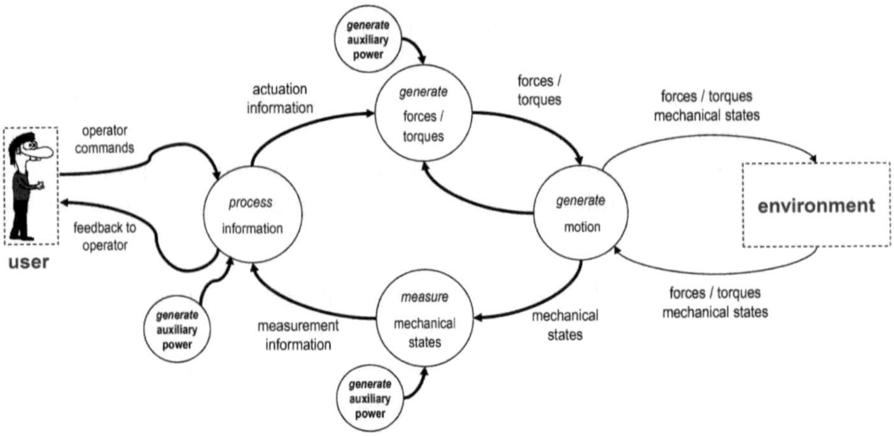

Fig. 1.2. Function-based structure of a mechatronic system: internal perspective (external agents indicated with dashed lines)

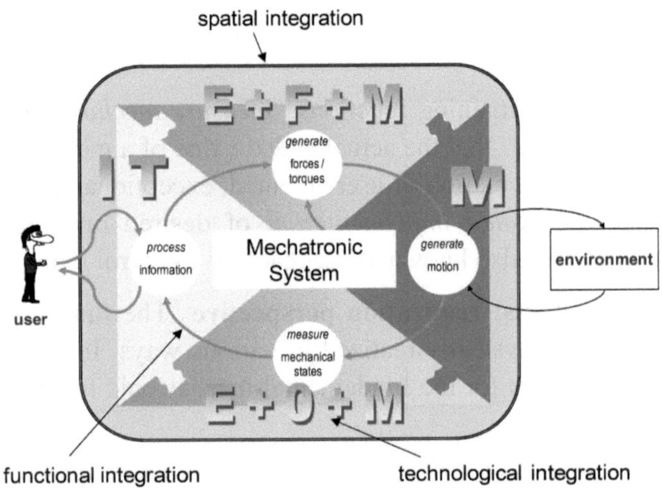

Fig. 1.3. Functional structure of a mechatronic system with underlying realizing technologies: E = electrical (electromagnetic, piezoelectric, etc.); F = fluidic; M = mechanical; O = optical, optoelectronic; IT = embedded information technology

At this level, the previously introduced "synergistic spatial and functional integration of mechanical, electronic, and information processing subsystems" integral to all mechatronic systems clearly reveals itself.

Meanwhile, it becomes clear from the closed action chain that the differing system elements not only require assembly, but must also *interact functionally and technologically* in a well-defined manner (indicated by the puzzle pieces). It is well known that the weakest link of a chain determines its breaking point. A mechatronic system can also only properly complete its product tasks if *all* elements of the closed action chain act in harmony. At the design level, this requires the meshing of system theoretical and physical/technological knowledge from differing scientific and engineering domains (electrical engineering, mechanical engineering, information technology, computer science, physics, etc.).

Key elements of mechatronic systems Mechanically oriented product tasks have been carried out for millennia using machines. This being the case, how do mechatronic products distinguish themselves from "classical" solutions?

The extraordinary attributes of mechatronic systems are based on the *functional* and *spatial integration* of mechanical, electrical, and information-processing subsystems incorporating the following key *conceptual* and *technological* elements:

- closed action chains,
- microelectronics and microsystems,
- new materials, functional materials.

Closed action chains High-grade "purposeful motion" under all possible operating conditions and with ever-present physical uncertainties is only possible using the fundamental concept of a *closed action chain*, or *control loop* (Fig. 1.2). However, the design of a mechatronic system goes far beyond simply putting together a control loop. In particular, in a mechatronic system, all elements of the loop (not simply the control algorithm) are objects of the design, and thus freely modifiable. Despite this—and perhaps precisely for this reason—the key role of *control theory* in systems design can be concisely summarized as follows:

> *"Mechatronics is much more than control, but there is no Mechatronics without control"* (Janschek 2008).

Microelectronics and microsystems It was only with the great technological advances of *microelectronics* starting in the 70s of the 20th century that ever more powerful digital and analog information processing capabilities were able to be packed into ever smaller volumes—following an empirical observation known as *MOORE's Law* (see Fig. 1.4) (Moore 1965), (Schaller 1997).

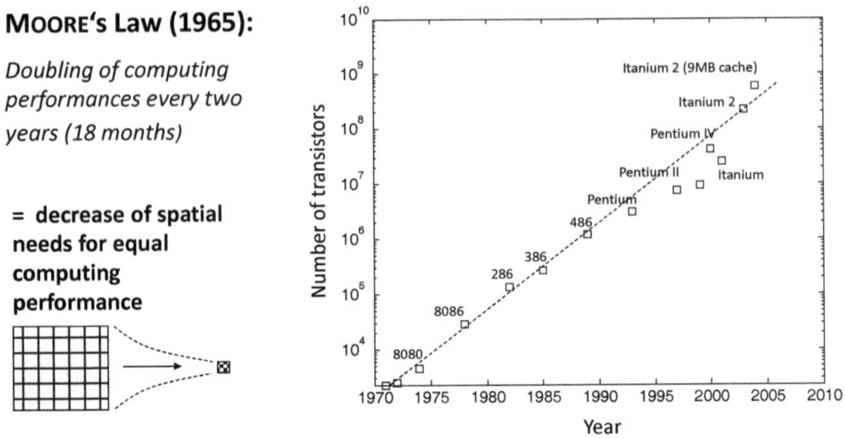

Fig. 1.4. MOORE's Law of microelectronic development[5]

As a consequence, mechatronic functions could be directly incorporated with highly integrated circuits into *microsystems* (Senturia 2001), (Tummala 2004).

As a result, today, there are mechatronic system elements which enable direct measurement at physically relevant locations, "thinking", and action, i.e. nearly perfect realization of functional and spatial integration.

New materials: functional materials Spatial integration which creates compact, moving systems was and continues to be advanced by new materials. As the most important example, take *piezoelectric materials*, which now enable the construction of compact sensors and actuators, and, as a component of *smart structures*, the direct integration of measurement and actuation components into a mechanical structure (see Fig. 1.5) (Preumont 2002), (Srinivasan and McFarland 2001).

Mechatronics: technical community At the international level, two highly respectable journals specialized in the technical area of mechatronics have established themselves:

- *IEEE/ASME Transactions on Mechatronics* (IEEE: Institute of Electrical and Electronics Engineers, ASME: American Society of Mechanical Engineers),
- *IFAC Journal Mechatronics* (IFAC: International Federation of Automatic Control).

5 Processor data from www.Intel.com.

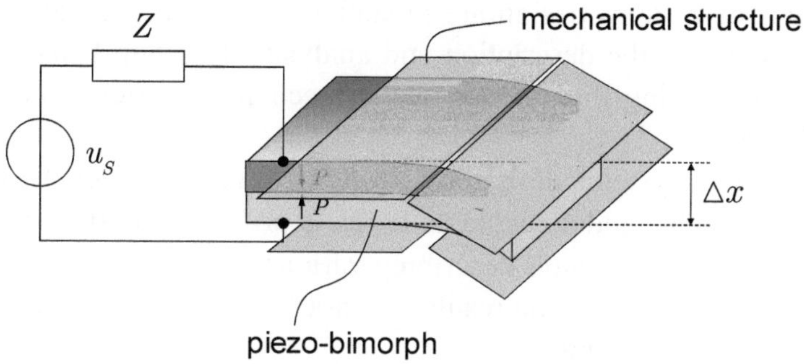

Fig. 1.5. Piezoelectric functional materials (*smart structures*)

These journals reflect the current research into mechatronic questions. To be particularly recommended are their special issues on more focused themes which appear at irregular intervals. The technical societies of the IEEE, ASME, and IFAC also regularly organize international mechatronics conferences.

In the German-speaking countries, the *VDI/VDE Gesellschaft für Mess- und Automatisierungstechnik (Society for Measurement and Automatic Control, VDI/VDE GMA)* is very active in mechatronics with dedicated technical committees and national conferences. The engineering association *Verein Deutscher Ingenieure (VDI)* has issued special *guidelines* for the methodological development of mechatronic systems (VDI 2004).

However, due to the broad nature of mechatronics, many relevant research results are also released in the publications of associated technical communities, e.g. microsystems, automotive engineering, energy engineering (see also the numerous references at the end of each of the subsequent chapters).

1.2 Systems Design

Terminology In this book, the term *systems design* is meant to denote the conception of a technical system (here, a mechatronic system) using *engineering measures*.

The term "conception" as used here denotes the reproducible and consistent *composition* of a system from structured subsystems in such a way that the system can fulfill given *system requirements* to the best extent possible.

The *tools* of systems design are modeling concepts, methods, and procedural models for the description and analysis of system behaviors, e.g. electrical engineering theory, theoretical mechanics, systems theory, and control theory.

The *results* of systems design are *models* ("blueprints") of the solution which are verifiable and assessed as far as is possible, i.e. they are generally abstract dynamic models describing various perspectives on (attributes of) the system. As a rule, the results are not "corporeal" realizations (one exception: rapid prototyping).

Systems design as the central design task The central significance of systems design to the creation of a mechatronic system becomes visible in its functional structure and the meshing of the incorporated realizing technologies (see Fig. 1.6).

The design solution must ensure the functionality of the *complete system*—the harmonious interaction of system elements in a closed action chain taking into account the physical limitations of and conditions for particular realization types. The focus is thus always the *entire system.*

Functional implementation: design verification The behavior of the product is decisively characterized by the fact that all elements of the closed action chain must harmonize with one another and operate according to specifications. The well-known axiom "a chain is only as strong as its weakest link" also proves to be valid here. Thus, systems design is presented with the task of simultaneously delivering a suitable *layout* (design) of all involved system elements and *proving* that the complete (closed) action chain actually operates as specified (*verification* and *validation*, see Sec. 2.1).

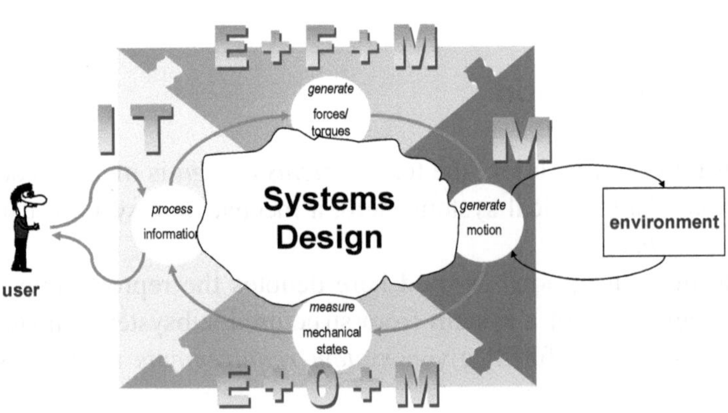

Fig. 1.6. Systems design as the central design task

This last task does not however imply "trying out" the finished system (product), but rather ensuring a high degree of confidence that the functions of the design solution will be correctly implemented even before the actual physical realization. Both tasks—the actual *design* and the *demonstration* of correct implementation—are based on predictive *dynamic models* which describe the true behavior of the mechatronic system with sufficient accuracy.

Dynamic models: multi-domain models All assertions of systems design are based on abstract models of a real mechatronic system. Such models always describe a limited set of attributes of the real system, e.g. static, nonlinear, or linear dynamics. However, they must always account for the complete system with all subsystems, i.e. all relevant interactions and dependences must be made visible in a *common* model, at the same abstract level.

The great challenge in setting up dynamic models lies in the *heterogeneity* of participating system elements possessing quite different physical properties comprising various physical phenomena: *mechanics*, *electromagnetism*, *electrostatics*, *piezoelectricity*, *fluid mechanics*, *optoelectronics*, *digital information processing*, etc. As these heterogeneous subsystems are coupled to and interact with one another in the real system, such attributes must naturally also be correctly represented in a dynamic model. Thus, suitable methods must be available to work with so-called *multi-domain models*, which describe the differing physical phenomena at a common, abstract level.

In <u>Chapter 2: Elements of Modeling</u>, various approaches to multi-domain modeling are discussed, the understanding of which is useful when employing commercial modeling and simulation tools. Some of these approaches also serve as the basis for the dynamic modeling in subsequent chapters.

Model-based behavior analysis Fig. 1.7 presents a simplified model tree for the model-based behavior analysis discussed in this book. The *qualitative system model* serves to structure and functionally describe the external and internal perspectives of a mechatronic system at a highly abstract level.

The principal action structures and mechatronic action loops can already be seen in this model, though as a rule they still lack a concrete physical representation (unless there are predetermined physical limitations or conditions, e.g. the required re-use of a piezo transducer from a previous project as an actuator).

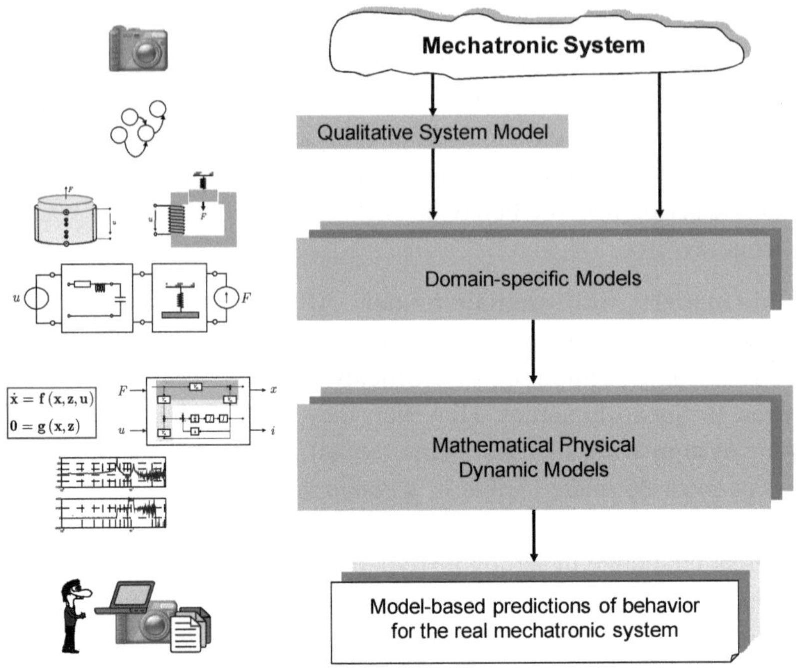

Fig. 1.7. Model-based analysis of system dynamics: model tree

The next design step considers different design variants, each employing particular concrete realizing technologies for the system functions (e.g. an electromagnetic axle bearing, electrostatic measurement of separation, information transfer over a serial data bus). At this point, *physical modeling* comes into play. Predictive, abstract models must be found for the particular realization type (Chapters 4 through 9). This occurs at various levels of abstraction: first using *domain-specific models* (equivalent configurations with lumped parameters, or infinite-dimensional models), and then more concretely with *mathematical physical models* (e.g. differential algebraic models, linearized system models). Using such models, analytical treatment or numerical evaluation can give reliable predictions of the expected behavior of the real mechatronic system.

Mechatronic systems usually place high demands on system performance (e.g. high-precision, highly-dynamic motions). In order to be able to reliably predict performance limits and possible design margins for such a heterogeneous system with its uncertainties and non-deterministic influencing factors, suitable methods for *robust controller design* and dynamic analysis considering random influences are presented in Chapter 10: Control Theoretical Aspects and Chapter 11: Stochastic Dynamic Analysis, respectively.

Analytical vs. computational models Of decisive importance to systems design is a deep understanding of the physical phenomena and their interactions present in a system under consideration. This understanding is first and foremost reflected in the choice of dynamic model. In order to be able to make design decisions in a transparent and reproducible manner, as far as possible, *analytical dynamic models*, in which the relationship between physical parameters and dynamic properties can be explicitly represented, should be used. Experience shows that important design predictions of the *feasibility* or of *critical influences* can already be made using relatively simple system models (design models), as long as the *relevant phenomena* are represented. In this way, clear insight into a complex system is gained by the experienced systems engineer.

As a complement to analytical models, there exist today quite powerful *computer-aided modeling tools* (see Chapter 2: Elements of Modeling). These enable the extremely user-friendly generation of complex models in a short amount of time using domain-specific libraries. However, such a process must be carried out with great care, as it involves the two fundamental difficulties elucidated below.

Complexity of computational models Due to their complexity, the resulting models are, per se, rather opaque, and analytical relationships can only be understood to a limited degree, even if computer algebra programs are employed. On the other hand, when performing purely numerical evaluations, each computational experiment always considers only a *limited set of parameters*—global relationships are difficult to discern.

Correctness and verification of computational models Even model libraries which can be assumed correct still allow for numerous failure modes in the configuration and parameterization of models. Verification of a computer-aided model implementation is also a time-intensive and nontrivial task. Here, the great usefulness of analytical system models comes to the fore. Using these, analytically based predictions of behavior can be produced, which can then serve as references for particular verification experiments using complex computational models. For example, the computer implementation can be configured and parameterized in as equivalent a manner as possible to the analytical model (e.g. zeroing out parasitic effects, defining required conditions) to compare results between the two model approaches. There also exists the possibility of defining certain experiments for which analytical prediction is easily possible (e.g. employing conservation of (angular) momentum in a mechanical system given negligible external excitation).

Design evaluation, design optimization One central task of systems design is to uncover design solutions which conform to the initial requirements. It is the nature of the problem that there exists not *only one single* solution, but rather, as a rule, an infinite number. Which of these, then, is the best?

This question can (at least in principle) be easily answered if the system requirements specify clear, quantitative properties concerning system behavior (i.e. performance metrics). If these properties are quantifiable and thus measureable, then predictive characteristic quantities—*performance metrics* (see Chapter 12: System Budgets)—can also be obtained from the dynamic models. In this way, a design solution can be *evaluated* from an *objective* viewpoint using the model, e.g. with respect to positioning accuracy, rise time, disturbance rejection, or electrical power consumption.

In a further step, by varying the design—i.e. changing implementations (e.g. the technology, structure, configuration, and parameters)—alternative design solutions can be generated. It is then objectively possible to find an optimal design solution with respect to an objective function over the performance metric—this is the process of *design optimization* (see Fig. 1.8).

The central component of design evaluation and optimization consists of predictive models of system behavior, in particular dynamic models. Analytical dynamic models have the advantage of allowing an optimal design solution to be systematically obtained while accounting for additional conditions in a transparent manner (see Chapter 12: System Budgets).

The optimization loop can be closed using manual design variations or with the aid of a computer. Both analytical and computational models offer attractive options for automating *design optimization* for complex models, e.g. (auf der Heide 2005), (auf der Heide et al. 2004).

Systems design in the development process For the development of technical systems, so-called *process models* have established themselves in industry. The motivation for this comes from considerations of quality assurance to guarantee the most efficient allocation of resources (personnel, materials) with minimal development time. These approaches are based on clearly-structured development steps having verifiable results (milestones).

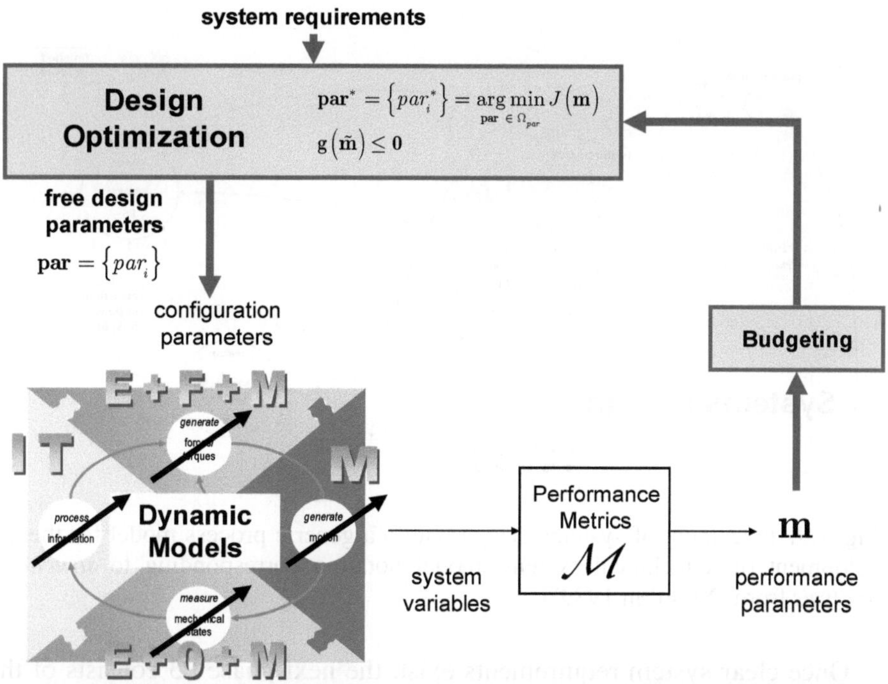

Fig. 1.8. Systems design: design optimization using quantitatively evaluated dynamic models

Fig. 1.9 shows a simplified *generic process model* in the notation of structured analysis (Yourdon 1989) (see also Sec. 2.2). The development activities (development phases) are modeled as functions, the accompanying results are represented by generalized data stores, and the logically causal action flow and interactions between elements are described by arrows.

The beginning of the process is the more or less abstract idea for a new product—here, a mechatronic system. In a first phase P1, the client (possessor of the idea) must develop concrete *user requirements*. This usually occurs in a manner which is not strictly formal.

At the beginning of the technical implementation, in a second phase P2, the user requirements must be translated by the system developer into detailed, technical, and, to the best extent possible, complete, non-contradictory *system requirements*. This step employs increasingly formal or semi-formal descriptive methods, e.g. *structured analysis (SA)* (Yourdon 1989) or the *Unified Modeling Language (UML)* (Vogel-Heuser 2003) (see also Sec. 2.2).

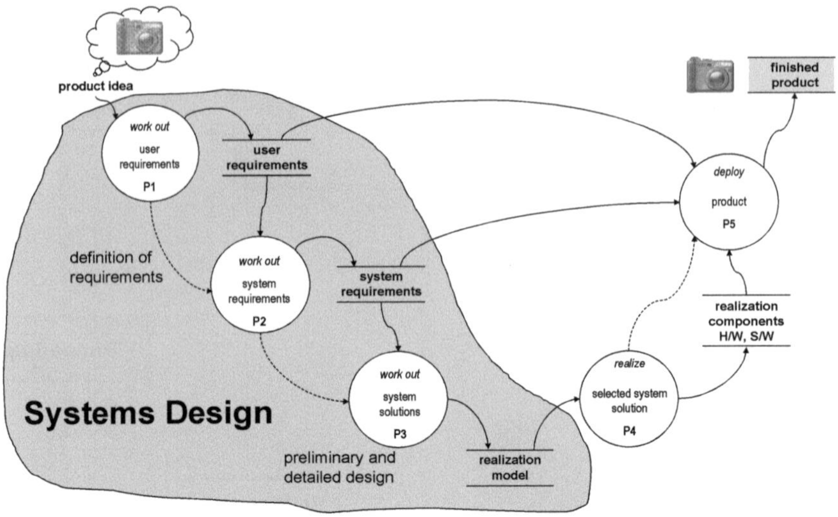

Fig. 1.9. Placement of systems design within a generic process model for the development of a technical system (model notation corresponding to *structured analysis* from (Yourdon 1989))

Once clear system requirements exist, the next phase P3 consists of the concretization of the *technical, functional*, and *technological design*. Starting with a *rough design* comprising different *design variants*, one variant is selected (see discussion of design evaluation and optimization), iteratively refined, and verified based on its model with respect to realizability and conformity to requirements. At the end of phase P3, there is then a detailed *realization model* ("blueprint") with verified and reliable predictions of the system behavior.

In the *realization phase* P4, using the realization model, the manufacture and procurement of system elements (hardware and software) can proceed. In the subsequent phase P5, system integration—including verification, validation, and deployment—is carried out. Delivery of the system is based on the system and user requirements, which laid out the desired concept for the system.

Fig. 1.9 clearly illustrates the activities and central role of systems design. The results of this process, in the form of specifications and "blueprints", serve as the basis for manufacture and verification/validation of the final product.

Of particularly critical importance are the accuracy and the conformance to requirements of the realization model, since the later a design error is recognized, the more costly is the remedy.

V-process model One oft-used process model is the so-called *V-model*, which is also recommended in (VDI 2004) for the development of mechatronic systems, and which underlies the representation in Fig. 1.9. The name results from the similarity of the process structure to the letter V when the development steps are laid out in the manner shown. On the left branch, each step involves a concretization or decomposition of the design, while on the right, each step represents an aggregation or integration of subsystems.

Systems design as a creative composition task The design task for a mechatronic system can be compared to the composition of a piece of music. The design engineer—like a composer—possesses a well-stocked *construction set* containing *methods*—music theory, composition theory—and *technologies*—instruments, ensembles—and has great liberty when creating a "suitable" composition of these building blocks. The process of composition is always a fascinating, deeply creative process, and in systems design, requires skillful play upon the *keyboard of engineering science.* It is the aim of this book to communicate fundamental elements of such "virtuosity".

1.3 Introductory Examples

1.3.1 Telescope with adaptive optics

Problem statement In high-resolution telescopes, the spatial phase distribution of impinging light wavefronts plays a determining role in the achievable resolution. Due to atmospheric turbulence, adjacent light rays experience differing phase delays, resulting in an uneven wavefront at the entry to the telescope (see Fig. 1.10 in the upper section of the light path). If the telescope mirror is constructed as a matrix of movable (controllable) mirrors, the spatial phase delays can be corrected so that an approximately parallel wavefront is formed in the focal plane of the telescope (see Fig. 1.10 in the lower-right section of the light path). To calculate the correction for a wavefront, its phase distribution must be measured with a wavefront sensor, allowing a control loop to calculate corrective signals for the mirror displacements. This principle of optical correction is termed *adaptive optics* (Roddier 2004), (Hardy 1998), (Fedrigo et al. 2005).

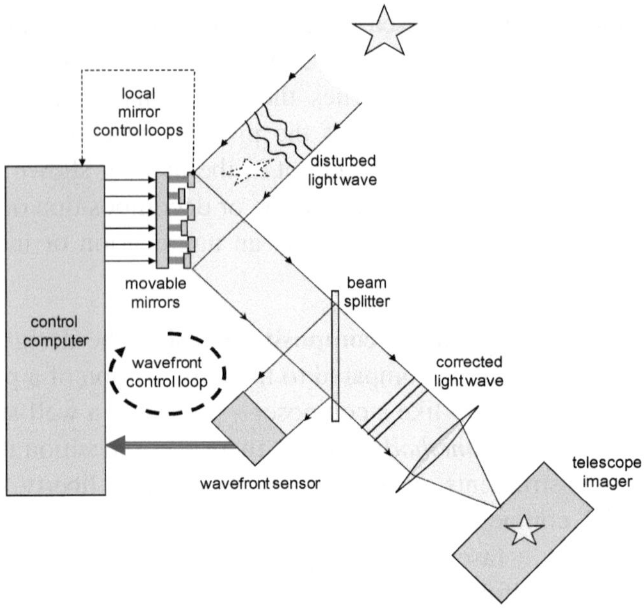

Fig. 1.10. Schematic configuration of a telescope with adaptive optics

Adaptive vs. active optics Adaptive optics should not be confused with the principle of *active optics*. In the latter, mirror elements are also actively displaced, though primarily with the object of compensating geometric deformations due to manufacturing tolerances, environmental factors, and dynamic effects during slewing. This occurs within a typical frequency range ≤ 1 Hz and with relatively large displacements. Adaptive optics, on the other hand, compensates for wavefront errors with considerably higher frequency components > 1 Hz and significantly smaller displacements ≤ 1 mm. From the mechatronics perspective, however, conceptually similar design tasks appear.

Product task Empirically, wavefront sensors (generally SHACK-HARTMANN sensors, see (Roddier 2004)) have a limited bandwidth, so that precision control of the mirror actuators directly in the wavefront control loop is not to be recommended. For this reason, a cascade structure is often employed, with *local mirror displacement loops* (inner control loops), which receive reference inputs from the superordinate wavefront control loop (outer control loop). The remainder of this section will now consider an example of such a local displacement control loop for a moving mirror in the matrix. An overarching problem analysis gives the following

Product requirements for local mirror control:

1. maximum displacement: $\pm\ 1\ \mathrm{mm}$,
2. working frequency range: $0 \leq f \leq 15\ \mathrm{Hz}$,
3. accurate in steady state given constant reference displacements and external force disturbances,
4. positioning accuracy within the operating range: $10\ \mu\mathrm{m}$,
5. mirror mass: $m = 0.1\ \mathrm{kg}$.

Analysis of requirements The requirements listed above have the quality of typical *user requirements* corresponding to phase P1 of the process model in Fig. 1.9, i.e. they are relatively general, not strictly formal, and still leave all design options open. In the next step, detailed *system requirements* must be derived from them (phase P2, Fig. 1.9). In particular, it must specified what exactly is meant by "positioning accuracy: $10\ \mu\mathrm{m}$". Customarily, a suitably weighted sum of all relevant influencing quantities is considered (see <u>*Chapter 12: System Budgets*</u>). Additional precision is also required regarding the externally visible operational behavior, i.e. the interface to the superordinate wavefront control loop, as well as important internal operations (starting, stopping, error handling, etc.). For such a representation, qualitative, semi-formal descriptive methods are quite suitable (see <u>*Chapter 2: Elements of Modeling*</u>).

Rough design solution: discussion of variants On the basis of concrete system requirements, different *variants* for possible design solutions with concrete physical devices can be developed and compared to one another (phase P3, Fig. 1.9, see also design evaluation in Fig. 1.8). This is a deeply creative and engaging process.

Design variant: local mirror displacement control loop Here, *one* possible design variant for the local mirror displacement control loop will be considered more closely. Fig. 1.11 presents a schematic configuration in the form of an illustrative physical model employing a position-controlled actuator and a mechanical linkage to the mirror. The concrete realizing technology for the mirror actuator is still unspecified here (e.g. piezoelectric or electrodynamic). For now, the simplifying assumption is made that the actuator includes internal position control, and thus is able to follow positioning commands with high accuracy, i.e. $y(t) \approx u(t)$. In addition, this allows the motion of the actuator to be decoupled from (have no back-effect from) the motion of the mirror mass.

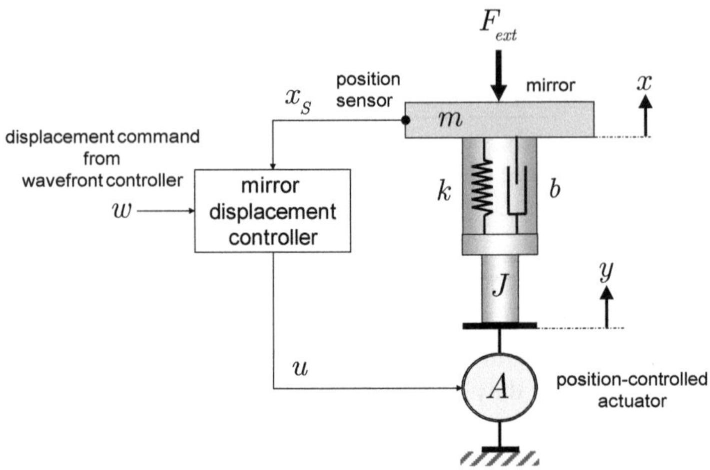

Fig. 1.11. Schematic configuration of a local mirror displacement control loop (rigid mirror body m, position-controlled actuator A, elastic coupling (k, b), joint J

For the mechanical connection, a *joint* J is envisaged, which for the present, is assumed to be *infinitely stiff* (rigid coupling). For the coupling from joint to mirror, a weakly-damped *elastic link* with spring constant k and viscous damping constant $b \ll 1$ is employed. Further, a position sensor is envisaged, with which the mirror position with respect to the baseplate (actuator base) can be measured directly.

Design model: analytical dynamic model Now comes one of the most important and, at the same time, most difficult steps of systems design. From the selected configuration, the *significant dynamic properties* are to be abstracted. However, what are "significant" dynamic properties?

The quality of the model and its particular perspective depend *exclusively* on the *system requirements*. In the case at hand, the specified requirements communicate that a physical model covering the specified operating range and having *lumped parameters* and mechanical *rigid-body models* suffices (see Chapter 4: Multibody Dynamics).

In the general case, due to the ever-present back-effect arising from its energy-based coupling to the mechanical system, *modeling of the actuator* presents a particular challenge. For this purpose, Chapters 2 and 5 present detailed discussions of *general approaches* to *physical modeling* (its foundations and a generic mechatronic transducer), and Chapters 6 through 8 delve into the important and much-used physical phenomena of *electrostatics*, *piezoelectricity*, and *electromagnetism*.

In the current case, modeling the actuator and sensor dynamics is very simple, as, for the time being, ideal dynamics without back-effect are assumed. From this viewpoint, the *mechanical subsystem* can be described with sufficient precision as a damped *single-mass oscillator* (m, k, b) with *displacement excitation* $y(t)$. For the abstract mathematical design model in the time domain, the following linear differential equation is obtained:

$$m\ddot{x} + b\dot{x} + kx = ky + b\dot{y} - F_{ext} \, . \tag{1.1}$$

Taking the LAPLACE transform of Eq. (1.1) gives the following description in the complex plane:

$$X(s) = \frac{1 + \frac{b}{k}s}{1 + \frac{b}{k}s + \frac{m}{k}s^2}\left(Y(s) - \frac{1}{k}F_{ext}(s)\right) = P(s)\cdot\left(Y(s) - \frac{1}{k}F_{ext}(s)\right), \tag{1.2}$$

and the plant transfer function of the *mechanical subsystem*

$$P(s) := \frac{1 + \frac{s}{\omega_{N0}}}{1 + 2d_0\frac{s}{\omega_0} + \frac{s^2}{\omega_0^2}}\,,$$
$$\omega_0 := \sqrt{\frac{k}{m}}\,, \quad d_0 := \frac{b}{2}\sqrt{\frac{1}{mk}}\,, \quad \omega_{N0} := \frac{b}{k}\,. \tag{1.3}$$

Assuming linear dynamics in the actuator, sensor, and the displacement controller, with the corresponding transfer functions $A(s)$, $S(s)$, $H(s)$, results in the signal flow diagram (block diagram) for the design model of the local mirror displacement control loop shown in Fig. 1.12.

Using the *analytical design model* in Fig. 1.12, appropriate methods of analysis can now be applied to obtain reliable performance metrics for design evaluation (see Chapter 11: Stochastic Dynamic Analysis) and control and regulation concepts can be drawn up (see Chapter 10: Control Theoretical Aspects).

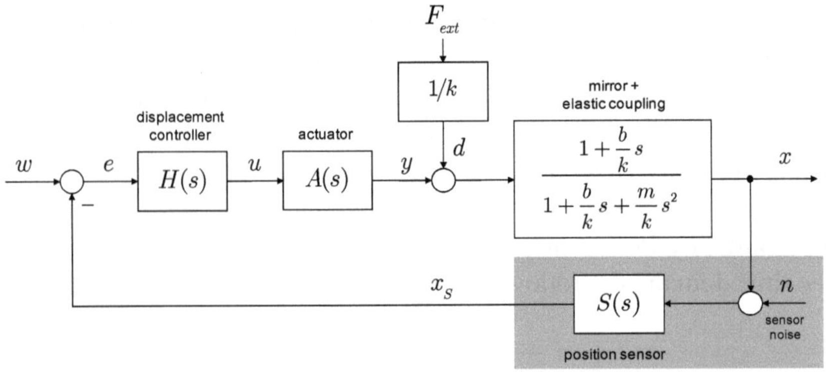

Fig. 1.12. Signal flow diagram for a local mirror displacement control loop

Computational models Alongside the admittedly very simple design model in Fig. 1.12, there exist complementary, more involved models with a higher degree of detail. These are available particularly in later design phases, when more complete knowledge is available. For such models, it is advantageous to employ *computer-aided modeling tools*, which allow for modeling at variously abstract, physical levels, and which support simulation-based experimentation with such models. Fundamental methodological approaches addressing these questions are discussed in *Chapter 2: Elements of Modeling* and in *Chapter 3: Simulation Issues*, enabling insightful use of common computational tools.

Naturally, the simpler design models can also be implemented using such computational tools, to reinforce and verify analytical predictions.

Design variant A: soft mechanical coupling

Mechanical subsystem As a first concretization of the design, consider the mechanical subsystem. Let the *coupling stiffness* $k = 10\ \mathrm{N/m}$ be known from material data. The mechanical damping b can usually be determined only with great difficulty. Here, to ensure particularly robust dynamics, a very low damping $b = 0.01\ \mathrm{Ns/m} \;\hat{=}\; d_0 = 0.005$ will be used (a *worst-case* analysis). Fig. 1.13 plots the magnitude of the frequency response $P(j\omega)$ of the mechanical subsystem, which demonstrates a fundamental, problematic system behavior.

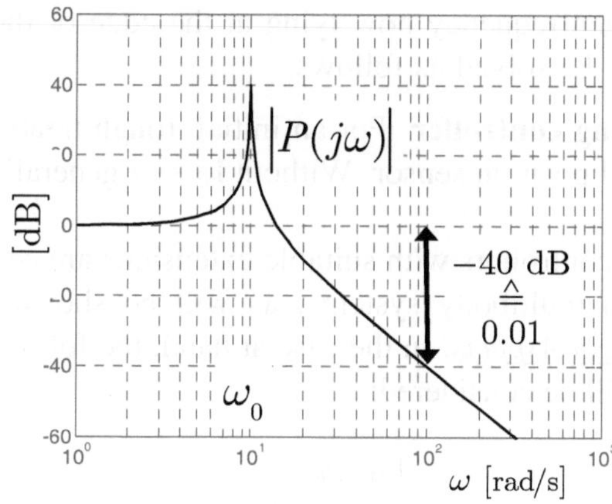

Fig. 1.13. Frequency response of the mechanical subsystem ($k = 10$ N/m)

Conflict between operating range vs. actuator travel The *natural frequency* of the single-mass oscillator with the chosen coupling stiffness lies at $\omega_0 = 10$ rad/s and thus well within the operating frequency range. Troublesome here is less the resonance—this can be quite well handled with a controller—and more the evolution of the magnitude beyond the resonant frequency.

The system requirements specify a mirror travel of $x = \pm$ 1 mm over the entire operating frequency range of $0 \le \omega \le 100$ rad/s. Due to the *magnitude falloff* of -40 dB/dec above the resonant frequency, at an operating frequency of $\omega = 100$ rad/s, an actuator travel $y = \pm$ 100 mm is required (see Fig. 1.13). Obviously, this design variant need not be further pursued. There is a clear design conflict, which cannot be solved at the level of the actuator nor the controller. The physical weak point clearly lies in the excessively soft mechanical coupling and must be corrected there.

Design variant B: medium-stiff mechanical coupling

Mechanical subsystem As an improved mechanical variant, a stiffer coupling mechanism with $k = 1000$ N/m is considered[6]. The new natural frequency is now $\omega_0 = 100$ rad/s. The magnitude plot in Fig. 1.13 shifts one decade to the right, and the mirror can now be moved throughout the entire operating frequency range with a limited actuator travel $y_{max} \approx \pm$ 1...2 mm and the desired displacement. The difficulty pre-

6 Assumption: an even stiffer connection is perhaps uneconomical or technically difficult to implement.

sented by the eigenfrequency now lying at the edge of the operating frequency range can be solved as follows.

PID compensating controller For an initial, rough treatment, assume an ideal actuator and position sensor. Without loss of generality, let $A(s) = 1$ and $S(s) = 1$[7].

Classical PID controllers with suitable extensions are certainly quite applicable, even to multibody systems, as will be shown in Chapter 10: Control Theoretical Aspects. In the case at hand, the following generalized PID controller will be considered:

$$H_{PID}(s) = K_H \frac{1 + 2d_N \dfrac{s}{\omega_N} + \dfrac{s^2}{\omega_N^2}}{s\left(1 + \dfrac{s}{\omega_D}\right)}. \tag{1.4}$$

As an initially completely obvious, but—as will emerge later—*naïve* approach to parameterizing the controller, the controller numerator polynomial is chosen to *compensate* the mechanical *eigenfrequency* of the mechanical subsystem, i.e. $\omega_N = \omega_0$, $d_N = d_0$. According to customary considerations (e.g. the open-loop phase margin (Ogata 2010)), the two remaining free parameters would be chosen to be, for example, $K_H = 250$, $\omega_D = 2\omega_0$. The resulting system dynamics in response to a reference input w are depicted in Fig. 1.14a-c with the curves labeled “1”.

Due to the exact compensation of the eigenfrequency, the frequency response of the open-loop transfer function $L(s) = H(s)A(s)P(s)S(s)$ is smooth (Fig. 1.14a). As desired, the magnitude of the frequency response of the reference input transfer function $T(s) = L(s) / \left(1 + L(s)\right)$ is near unity (0 dB) over the entire operating frequency range (Fig. 1.14b). The time-domain response to reference inputs also appears quite sensible: an error-free steady-state step response (Fig. 1.14c), and harmonic excitation at the maximum operating frequency (Fig. 1.15a). The product requirements numbers 1 to 3 are thus covered as concerns the response to the reference input.

[7] Naturally, the gains have units corresponding to the realizing technology in the actuator and sensor. However, this is of no further concern for the considerations here.

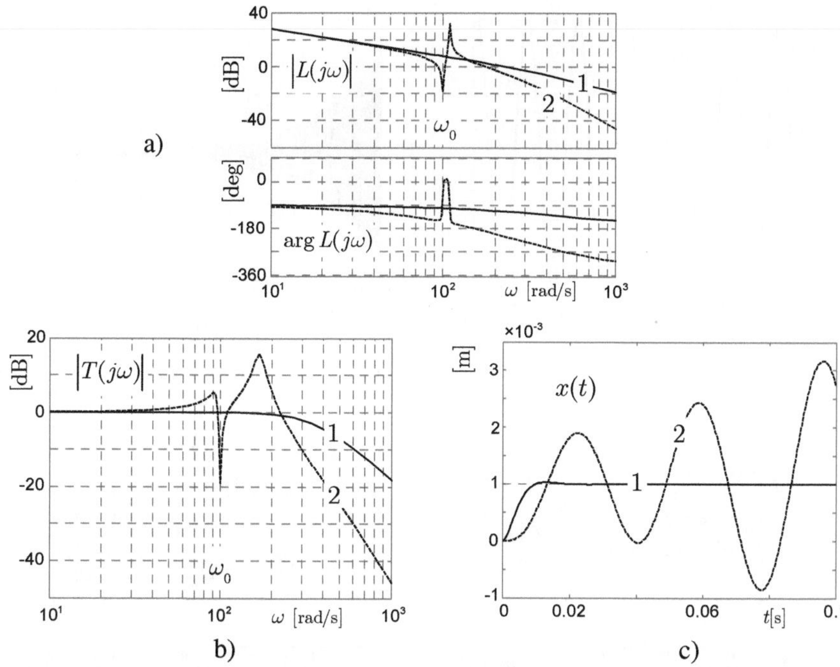

Fig. 1.14. Response characteristics of the mirror displacement control loop with PID compensating controller and $k = 1000$ N/m : a) open-loop BODE diagram $L(j\omega)$, b) closed-loop BODE diagram $T(j\omega)$, c) step response for $w(t) = 10^{-3} \cdot \sigma(t)$ [m] ; curves: 1: exact compensation, 2: incomplete compensation + parasitic dynamics

Deficiencies of the compensating controller Closer analysis of the *disturbance response*, however, discloses clear deficiencies in the compensating controller. Given a *force disturbance* step, the mechanical natural frequency becomes excited and is damped only very slowly by the controller feedback (Fig. 1.15b). The cause for this is precisely the compensating approach. The complex controller zero generates a dip in the gain of the controller frequency response at $\omega = \omega_0$. Thus, near this point, frequency components are fundamentally not at all or only very slightly affected by the controller[8]. The resulting dynamics are completely useless.

[8] The controller filters out these frequency components from reference inputs, so that the natural frequency is never even excited.

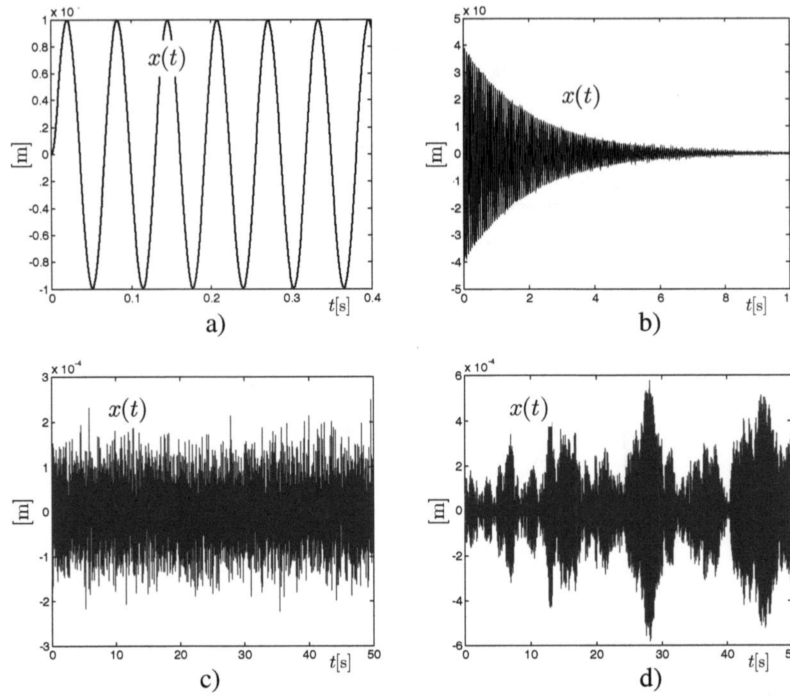

Fig. 1.15. Time-domain response of the mirror displacement control loop with PID compensating controller with exact compensation for different excitations: a) harmonic excitation $w(t) = 10^{-3}\sin(100 \cdot t)$ $[\mathrm{m}]$, b) step disturbance force $F_{ext}(t) = 1 \cdot \sigma(t)$ $[\mathrm{N}]$, c) sensor noise (see main text), d) noise disturbance force (see main text)

A further deficiency can be seen by examining *parameter variations* and additional *parasitic dynamics*, e.g. $A(s) \neq 1$, $S(s) \neq 1$. For the case of a marginally greater coupling stiffness $\tilde{k} = 1.2 \cdot k$ and an additional second-order low-pass element ($d_n = 1$, $\omega_n = 2\omega_0$) in the loop (e.g. due to actuator dynamics), with the same controller parameters, the curves labeled "2" in Fig. 1.14 demonstrate the modified dynamics.

Due to the higher coupling stiffness, the mechanical natural frequency shifts higher, so that now both the controller zero at $\omega = \omega_0$ and the new natural frequency $\tilde{\omega}_0 > \omega_0$ of $L(j\omega)$ are visible in Fig. 1.14a. Closer inspection of the phase plot[9] in Fig. 1.14a additionally reveals a *stability*

[9] This is relatively difficult to see in the BODE diagram. In Ch. 10, an alternative frequency response representation—the NICHOLS diagram—is presented, in which such configurations, which are typical for multibody systems, are much more clearly visible.

problem associated with the parasitic dynamics. The closed control loop now becomes *unstable*, which can also be clearly seen in the step response in Fig. 1.14c. The parameterization of the PID controller as a compensator is thus *unusable* from a practical point of view.

Alternative robust controller design In fact, however, a PID controller with the proper parameterization can achieve useful, robust controller dynamics. Chapter 10: Control Theoretical Aspects conveys approaches for this type of control. Moreover, by employing a specialized representation of the frequency response—the *NICHOLS diagram* (gain-phase plot)—*robust stability regions* can be clearly delineated, enabling simple, manual controller design.

Stochastic dynamic analysis Under real conditions, random disturbances are always present. In the current case, two sources of such disturbances should be noted: noise in the sensors and random excitations due to force disturbances (e.g. vibrations in the environment). The *modeling*, *simulation*, and *dynamic analysis* of *stochastic processes* are discussed in detail in Chapter 11: Stochastic Dynamic Analysis.

For the current case, examples of responses of the system with the compensating PID controller to sensor noise and noisy disturbance forces are shown in Fig. 1.15c and Fig. 1.15d, respectively. These employ the following noise sources:

- sensor noise n: normal distribution, $\mu_n = 0$, $\sigma_n = 0.1\ \mathrm{mm}$;
- noise disturbance force F_{ext}: normal distribution, $\mu_F = 0$, $\sigma_F = 0.1\ \mathrm{N}$;
- both cases: bandwidth $f_b = 100\ \mathrm{Hz}$, first-order low-pass.

The signals shown in Fig. 1.15c,d can in turn be described using statistical parameters. For example, evaluating the time history gives an output standard deviation for $x(t)$ of $\sigma_x = 0.06\ \mathrm{mm}$ for sensor noise (Fig. 1.15c) and $\sigma_x = 0.17\ \mathrm{mm}$ for noisy disturbance forces (Fig. 1.15d). Given a normal distribution of amplitudes, this can also be used to easily approximate the maximum signal magnitudes using their 3σ-values (as can be easily verified in this case).

Such performance predictions of the response to stochastic signals can be found not only using simulation experiments, but also via an analytical treatment—*covariance analysis*. In Chapter 11: Stochastic Dynamic Analysis, relations for this analysis are derived. For lower-order systems, analytical formulae can additionally be determined as functions of the system parameters (the parameters of the transfer function). Thus, even for stochastic inputs and disturbances, simple analytical evaluation of the design is possible (cf. Fig. 1.8).

System budgets: design optimization Up to this point, the important product requirement number 4—"positioning accuracy"—has not been closely considered. This requirement evidently applies to possible deviations of the real mirror position from a desired trajectory during operation. From Figs. 1.12, 1.14d and 1.15, it can be seen that mirror motions are the result of several quite different inputs (*signal sources*), and that during operation, a variety of s*ignal forms* are superimposed. How, then, should "positioning accuracy" be defined and analytically or experimentally determined for such a complex scenario?

The solution lies in the appropriate budgeting of various system responses to characteristic inputs which conform to the requirements. Thus, as a part of the definition of requirements, all relevant operational scenarios should be determined, and characteristic inputs established for each. Then, in the dynamic analysis, individual system responses should be ascertained for each of these, and suitably composed using the principle of superposition. For certain types of statistical uncertainty, this allows the relevant performance metrics—e.g. positioning accuracy—to be sufficiently accurately assessed.

In Chapter 12: Design Evaluation: System Budgets, corresponding theoretical foundations and application-oriented computational procedures are elucidated in detail.

Thus, in the case at hand, all operational scenarios would first have to be specified in detail. However, using the dynamic analyses carried out so far, the critical effect of, for example, sensor and disturbance noise on the positioning accuracy is already apparent. Quantitative metrics obtained from simplified dynamic models already allow the designer to draw important conclusions as to which system elements may require modification. Thus, in this case, both of the assumed influencing variables of sensor noise and disturbance forces are obviously far too large to allow the required positioning accuracy of 10 μm to be achieved. A further design iteration would then require implementing appropriate measures to modify relevant design variables in a balanced manner, e.g. vibration isolation to diminish force disturbances, a lower-noise sensor, reduction of the measurement bandwidth, or modifications in the control algorithm.

Further design aspects

Refining the design The above-discussed aspects enable the determination of critical subsystems within the complete system using only a rough design of manageable complexity. Building upon this base, the design can then be refined, allowing further aspects critical to the realization of the system to be considered. The remainder of this section quickly lays out these refinements with reference to the corresponding book chapters.

Multibody system A mechanical structure always consists of the sum of a number of more or less stiff elements. Depending on the product requirements, these elements can be assumed rigid, elastically coupled, or elastically deformable.

In the current case, joint J was assumed rigid. Assuming instead a finite joint stiffness, the mechanical subsystem must be modeled as an elastically coupled *two-mass system* (see Chapter 4: Multibody Dynamics). This changes the dynamic model: there are now two eigenfrequencies, resulting in possible complications for the closed-loop controller (see Chapter 10: Control Theoretical Aspects).

Sensor/actuator placement: collocation In the case of multibody systems, the placement of sensors and actuators (the measurement and actuation locations) takes on particular importance.

If, in the case at hand, finite joint stiffness is assumed, then the position sensor and the actuation point are located on *different*, elastically-coupled component masses. In this case, the sensor and actuator are termed *non-collocated.* This has profound consequences for the system dynamics, and must also be carefully taken into account during design. Control theoretical measures for doing so are discussed in Chapter 10: Control Theoretical Aspects.

Digital information processing Up to this point, all subsystems have been considered continuous in time. However, mechatronic systems derive their attraction from their capacity for digital information processing. Subsystems implementing this capability have a *discrete-time* nature, which must be suitably considered during analysis of the dynamics of the *closed action chain.* Indeed, properties arising in the digital processing of information, such as *time discretization* (sampling), *value discretization* (quantization), and *time delays*, have a fundamental effect on the dynamics of the system. A manageable procedure for dynamic modeling of such phenomena is presented in Chapter 9: Digital Information Processing, and several suitable control measures are considered in Chapter 10: Control Theoretical Aspects.

Mechatronic transducers The great attraction of mechatronic products is not least a result of their compact physical nature, which is achieved through tight spatial and functional integration. A significant role in this is played by actuators and sensors as the functional and energetic "intermediaries" between the system mechanics and information processing.

In contrast to a purely control theoretical point of view dealing only with subsystems lacking back-effects, in mechatronic systems design it is precisely the physically imposed *energetic back-effects* which are of greatest interest. The fundamental mechanisms for *electromechanical energy exchange* via *electrical* and *electromagnetic fields* allow these phenomena to be used for both sensing and actuation tasks—components implementing these are thus referred to using the general term *transducer*.

In addition to the fundamental electromechanical conversion of energy provided by transducers, knowledge of their internal physical relationships can be used to develop suitable physical measures to implement additional system-level *functional* attributes at the local transducer level. For example, using an *electrical impedance circuit* (in the simplest case, a resistor) at the input terminals of the transducer, *mechanical damping* of an attached structure (multibody system) is possible (a process known as *vibration damping*). Thus, using the least complicated of analog measures, a simpler set up can be created for a superordinate digital controller.

In the current design example, it might thus be considered whether in place of the actuator which has so far been assumed to be position controlled[10], it might not be possible and wise to employ direct *force coupling* using a piezoelectric or electrodynamic actuator. Possibly simpler and cheaper actuators could be employed—an important consideration for a matrix of several hundred mirrors.

The physically oriented modeling and dynamic analysis of mechatronic transducers is given particularly thorough consideration in this book. In

[10] The position-controlled actuator can move the attachment point of the joint coupled to the mirror (the output y) *without back-effect*. The mass properties of the moving structure have no effect on realizing the commanded motion u. Such an actuator is naturally more complex, expensive, voluminous, and requires more auxiliary energy.

Chapter 5: The Generic Mechatronic Transducer, fundamental dynamic models are derived and functional concepts are discussed. These are made concrete in subsequent chapters for selected technological implementations: Chapter 6: Electrostatic Transducers, Chapter 7: Piezoelectric Transducers, and Chapter 8: Electromagnetically-Acting Transducers.

1.3.2 Optomechatronic remote sensing camera

Problem statement From the user's perspective, high-resolution image acquisition systems—here termed *remote sensing* cameras—are desired for aerial and orbital remote sensing. From the engineering perspective, the realization of such cameras requires overcoming a variety of challenges.

Due to the required mobility of such cameras on platforms with limited weight and volume capacities (particularly for spacecraft), solutions should be as *compact* as possible.

High image resolutions, on the other hand, require long focal lengths and large apertures in order to bring as much light as possible onto the image sensor. In compact camera solutions with small apertures, more photons can be brought onto the image sensor with a *longer exposure*. However, this renders image acquisition very sensitive to camera motion during the exposure (Fig. 1.16, upper right).

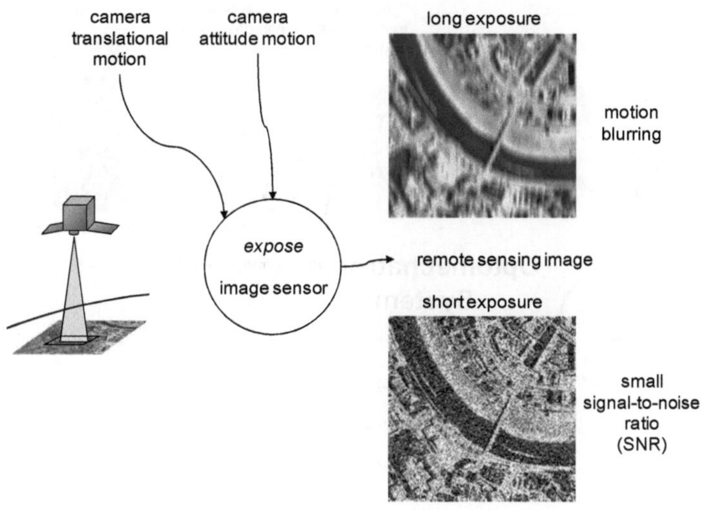

Fig. 1.16. Motion-induced disturbances in remote sensing image acquisition systems

Non-smooth motion profiles are obviously expected for aerial observation platforms (e.g. airplanes or helicopters). An observation satellite is much quieter in comparison. However, due to optical mechanics and the great distance to the observed plane (several hundred kilometers) in such an application, even very small attitude motions have dramatic effects on the image quality. On the other hand, it is easy to see that though shortening the exposure minimizes the effect of motion, due to the reduced number of captured photons, image noise is simultaneously increased (Fig. 1.16, lower left).

All applications mentioned here thus require compact, *motion-compensating imaging systems* enabling undisturbed image acquisition with long exposures.

Optomechatronic solution concept In Fig. 1.17, a function-based descriptive model of an *optomechatronic* system solution is sketched out.

To the extent that the motion of the camera image sensors relative to the observed plane (terrestrial or planetary surface) can be measured, a closed loop can be used to generate corrective signals for a compensating motion of the image sensor such that it remains at rest relative to the observed plane throughout the exposure. The equivalence of the representation in Fig. 1.17 to the generic mechatronic system in Fig. 1.2 is evident.

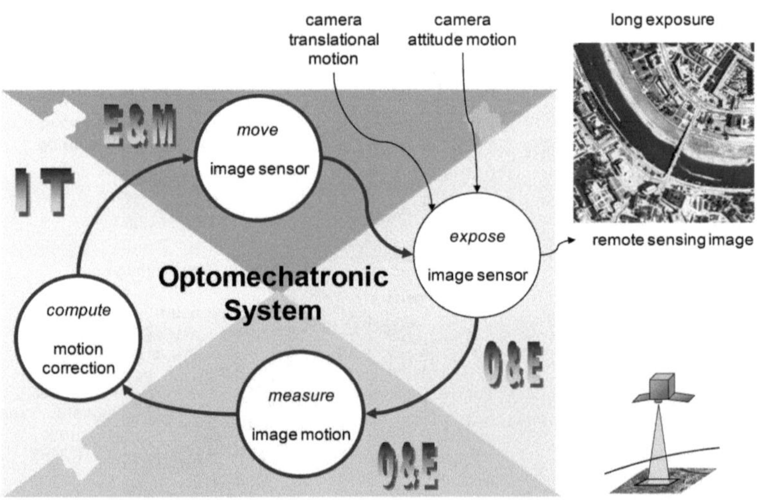

Fig. 1.17. Optomechatronic solution concept for motion compensation of a remote sensing camera

The great challenge in realizing such a concept lies in measuring the motion of the image sensor. In one of the author's workgroups, an opto-electronic measurement procedure based on an *optical correlator* was developed for this purpose and validated in the course of several flight tests (Tchernykh et al. 2004). With the help of such a measurement setup, *in situ* motion measurement in the focal plane of the camera—and thus compact realization of the motion compensation of Fig. 1.17—is made possible. Below, an overview of the results of two design variants is presented.

Design variant 1: Moving mirror

Line scanner For remote sensing tasks, a particularly suitable sensor is a line scanner, due to its relatively simple construction (no shutter is required), capacity for large image widths (by cascading line sensors), and the fact that the scanning motion is realized for free by the moving platform. In the context of an ESA project[11], one of the author's workgroups investigated an optomechatronic concept for a motion-compensating satellite line scanner (Janschek et al. 2005).

Conceptual implementation In Fig. 1.18, a schematic configuration of the motion compensation is sketched. Here, the optical path is directly modified by the corrective motion of the telescope mirror, so that the image acquisition at the line sensor becomes independent of disturbances in the motion of the platform. Such disturbances in the attitude of the satellite arise from *micro-vibrations* in the momentum and reaction wheels for attitude control.

Motion measurement The image motion measurement proceeds by evaluating series of images from a matrix image sensor, which is also placed in the focal plane (*in situ* measurement at the location of the disturbance). From the image series, 2D correlation is used to detect image motion (*image motion tracking*), which is then used to steer the telescope mirror. The correlation-based measurement of image displacements in real time at sub-pixel accuracy is achieved with an optical correlator.

[11] European Space Agency (ESA), ESTEC/Contract No. 17572/03/NL/SFe.

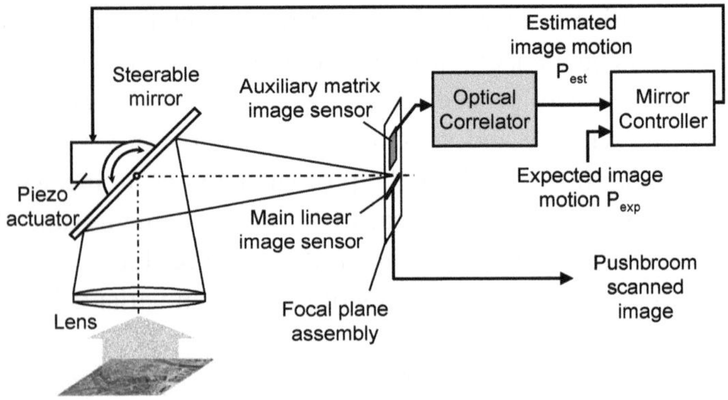

Fig. 1.18. Optomechatronic concept for motion compensation in a remote sensing line scanner: moving mirror (Janschek et al. 2005)

Control concept For the mirror actuator, an *uncontrolled piezo transducer* is envisaged, and the control loop is closed directly via the matrix sensor and the optical correlator (*visual feedback*). Thus, to configure the controller, both the mass-spring system of the piezo-actuator (see Chapter 7: Piezoelectric Transducers) and the delays in image acquisition and processing are determining factors.

For the control algorithm, a *robustly* parameterized *PID controller* with additional filter terms is employed (see Chapter 10: Control Theoretical Aspects). The controller frequency response typical for such applications is depicted in Fig. 1.19a. The performance of the motion compensation is evident from the simulated time histories in Fig. 1.19b.

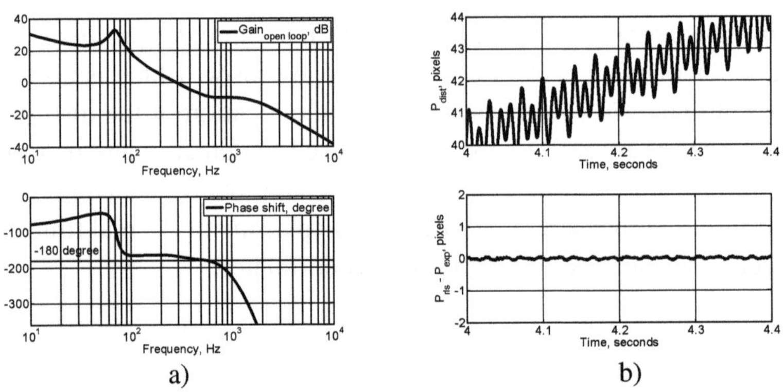

Fig. 1.19. Dynamic analysis of a motion-compensated remote sensing line scanner, (Janschek et al. 2005): a) controller BODE diagram, b) simulated time response of motion compensation, disturbance motion (top), compensated image motion (bottom)

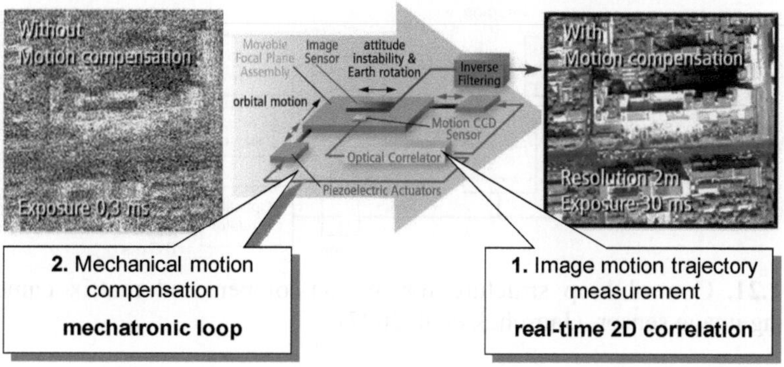

Fig. 1.20. Optomechatronic concept for motion compensation of a remote sensing matrix camera: moving image sensor (Janschek et al. 2007)

Design variant 2: moving image sensor

Matrix camera If no reasonable scan motion is possible, the possibility of employing a camera with a matrix image sensor presents itself (as in commercial digital cameras). Here, the problems portrayed in the introduction regarding motion blur (from long exposure) and noisy images (from short exposure) are quite evident.

Conceptual implementation One of the author's workgroups suggested a solution employing optomechatronic motion compensation using a *moving image sensor* (Janschek et al. 2004), (Janschek et al. 2007), see Fig. 1.20.

The matrix image sensor is mounted on an *X-Y piezo platform*, which is controlled using a correlation-based image motion measurement (same principle as in design variant 1) such that the image sensor remains at rest relative to the observed plane throughout the exposure.

Control concept As this concept for motion compensation requires larger travel for the image sensor (up to several mm), a piezo transducer[12] cannot be employed here. Rather, a so-called piezo ultrasonic motor is used, in which high-frequency excitations move a rotor in micro-steps, allowing relatively large feed rates and travel to be achieved (see Chapter 7: Piezoelectric Transducers, Sec. 7.5).

---

[12] These only allow for operating displacements in the μm range.

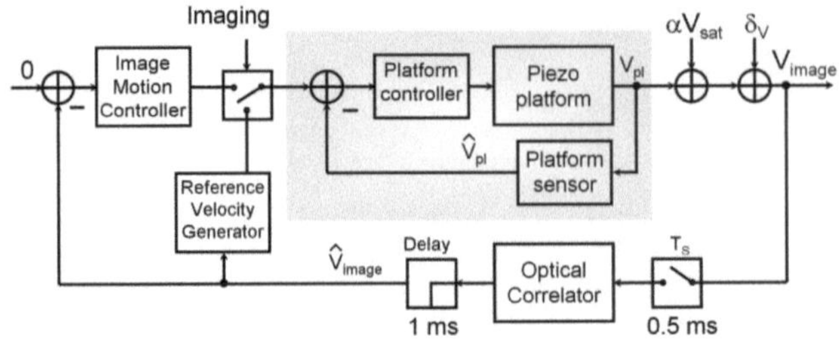

Fig. 1.21. Control loop structure for motion-compensated matrix camera with moving image sensor (Janschek et al. 2007)

This micro-step drive does not provide accurate positioning, so that a *cascaded controller* as in Fig. 1.21 is proposed. For each of the two orthogonal axes X and Y, an inner control loop—the *platform control loop*—adjusts the platform position locally to a reference value generated by the outer control loop—the *imaging control loop*. Due to its position control, the elastomechanical properties of the piezo platform are no longer visible to the outer control loop, so that the control accuracy is primarily determined by the delay for image processing.

Fig. 1.22 presents typical time evolutions; a laboratory demonstration and a hardware-in-the-loop test stand are shown in Fig. 1.23.

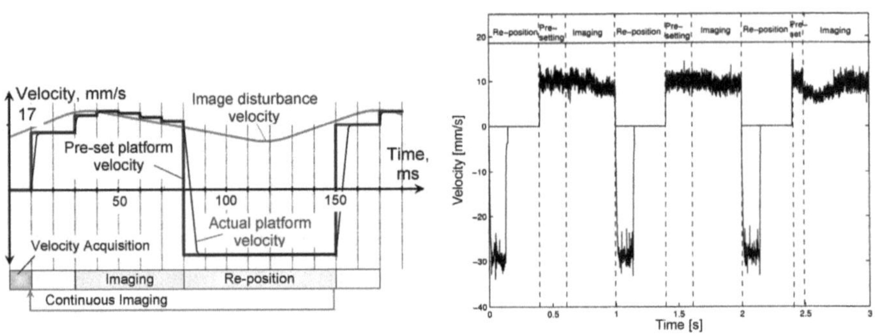

a) b)

Fig. 1.22. Time evolution of motion for a motion-compensated matrix camera with moving image sensor (Janschek et al. 2007): a) motion profile for the cycle "imaging, repositioning image sensor", b) time evolution for hardware-in-the-loop test

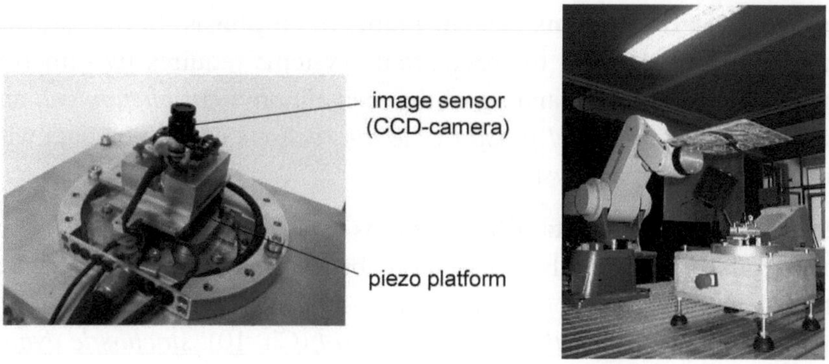

a) b)

Fig. 1.23. Laboratory demonstration of a motion-compensated matrix camera with moving image sensor (Janschek et al. 2007): a) camera configuration, b) hardware-in-the-loop test stand

1.4 About This Book

Methods, models, concepts

This *textbook* imparts fundamental knowledge concerning the *systems oriented* treatment of mechatronic systems. It presents *methods* appropriate to design tasks (e.g. modeling, dynamic analysis, and configuration), develops representative *dynamic models* to aid in understanding applicable physical phenomena, and discusses illustrative mechatronic *solution concepts* for selected examples from practice.

What is covered?

Subject canon The tasks of systems design for *heterogeneous* systems—here, the subject is *mechatronic systems*—are very broad in their nature. As a result, the preparation of a textbook on this topic presents the challenge of representing the *volume* of technical subject matter spanning the dimensions of (domain *multiplicity*) × (method *depth*) × (application *multiplicity*) with a limited, subjectively essential subject cannon, in a readable form, and in the manageable volume of a textbook. In the process, certain points must be naturally be emphasized, and certain omissions must be accepted, as will be described in somewhat more detail below.

Competencies for systems design: points of emphasis In the estimation of the author, the design of mechatronic systems requires two important competencies: an understanding of the relevant *physical phenomena* and a *quantitative, systems-level grasp* of the *interactions* of all elements within a concrete mechatronic system.

Method canon: understanding systems An *understanding of systems* comes from dynamic analysis based on abstract mathematical models, and is thus largely *domain-* and *application-neutral*. Thus, the chapters on *modeling* (Ch. 2), *simulation* (Ch. 3), *control* (Ch. 10), *stochastic dynamic analysis* (Ch. 11), and *system budgets* (Ch. 12) present a generally useful *method canon*, though with a focus on *generic modeling particularities* of mechatronic systems, e.g. multi-domain modeling and multibody dynamics. This results in a quite novel perspective as compared to representations in more general textbooks.

Physical/technical domains Which physical phenomena take on which role in mechatronic systems can be seen quite well in the function-based representation in Fig. 1.3. For any realization, there is always a *mechanical structure* and *embedded information processing*. To this extent, these two physical/technical domains always require in-depth physical modeling (*Chapter 4: Multibody Dynamics*, *Chapter 9: Digital Information Processing*).

The *multiplicity of domains* in mechatronic systems becomes particularly obvious in the technologies realizing the functions *"generate forces/torques"* and *"measure mechanical states"*. Here, a formidably wide spectrum of useable and used physical phenomena (electrostatics, piezoelectricity, etc.) presents itself. Given non-negligible power coupling, many of these can be equally employed for *both* functional realizations—the resulting components are thus termed (mechatronic) *transducers* or *transducer technologies*.

For didactic reasons, this multiplicity of domains had to be trimmed in the following manner for the presentation of this textbook.

Generic mechatronic transducer In many presentations of mechatronics, in the estimation of the author, the focus is shifted much too early to specific realization domains or specific sensor and actuator technologies. The fact that this is neither necessary nor sensible from either didactic or scientific/methodological points of view is demonstrated in the central Chapter 5: Functional Realization: Generic Mechatronic Transducer.

In that chapter, using a generic transducer model, general *commonalities* related to power coupling and transfer characteristics of transducers independent of any concrete physical conversion phenomenon are discussed: force generation, electrical properties, causal structures, and dynamic models. As a basis for modeling, three methods from Chapter 2: Elements of Modeling are favorably combined: starting from basic constitutive equations (i.e. natural laws or first principles), this proceeds with (i) *energy-based modeling* employing EULER-LAGRANGE equations (nonlinear and linearized constitutive transducer equations), from which is derived (ii) a particular *two-port parameterization* with electrical and mechanical terminals (including considerations of voltage or current sources, lossy transducers, and rigid-body or multibody loads), and finally (iii) a control-theoretical, signal-based model representation in the form of a transfer matrix.

These models enable a general discussion of generic dynamic properties such as eigenfrequencies, transducer stiffness, transfer functions (for both sensor and actuator functions), and the electromechanical coupling factor, maintaining completely transparent relationships to the physical parameters of the underlying basic constitutive equations.

Interestingly, using this generic transducer model, general conceptual solution approaches such as impedance feedback, vibration damping (mechatronic resonators), and energy generation (mechatronic vibrating generators or energy harvesting) can also be discussed in an exceptionally domain-neutral manner.

Thus, this generic mechatronic transducer forms the *methodological fixture* and the *model framework* for a general understanding of the principles of *power-coupled* transducers, and for the detailed representation of physical transducer principles in subsequent chapters. In its breadth of presentation and thorough application to models, to the best knowledge of the author, this representation is completely novel.

Physical transducer principles Of particular interest from a mechatronic point of view are *power-conserving* physical phenomena, as these allow special synergies to be exploited in functional realizations. For this reason, three power-conserving physical transducer principles with wide *industrial* *application* are considered in detail in their own chapters as concrete implementations of the generic mechatronic transducer:

- *Chapter 6: Electrostatic Transducers*,
- *Chapter 7: Piezoelectric Transducers*,
- *Chapter 8: Electromagnetically-Acting Transducers* (electromagnetic and electrodynamic transducers).

The discussion of transducer principles proceeds in complete harmony with the generic models of *Chapter 5: The Generic Mechatronic Transducer*, though with domain-specific *physical and technical concretization* starting with the fundamental constitutive laws through to structural implementational types. In the process, for each transducer principle, an attempt is made to consider the current state of technology as concerns implementations, and to highlight and represent in closed form generalized structural and functional properties, e.g. lateral vs. transverse degrees of freedom for armature and stator elements, differential configurations, or comb structures.

The presented breadth, model detail, and uniformity of representation is, to the best knowledge of the author, thoroughly distinct from the material presented in other textbooks, where often only particular physical principles are delved into, or particular model representations less-suited to control-oriented dynamic analysis are employed.

Multiplicity of applications The spectrum of mechatronic products grows on a nearly daily basis, whether through actual innovations or incremental improvements as part of ever-shorter innovation cycles. Presenting this spectrum is hardly the place of a textbook. The rate of obsolescence would be too great and the potential for generalization would be too low. For these reasons, it is here that the greatest compromises in content are made, and omissions are consciously hazarded. For an overview of mechatronic applications, the interested reader is thus referred to more encyclopedic works or contemporary internet searches.

What is omitted The careful reader and the experienced systems engineer—particularly one at home in the world of mechatronics—may miss a few important elements of systems design in the canon of subject matter presented so far. This impression is shared by the author to the greatest degree—he is fully aware of the *omissions*, which have been accepted (with much gnashing of teeth) as a practical matter.

Important from the perspective of *design methodology* are certainly more current methods for and approaches to stricter formal description of systems in early development phases, e.g. verifiable system models based on UML (Vogel-Heuser 2003), and extended to *heterogeneous hardware-software systems*, e.g. (Koycheva and Janschek 2009).

From the point of view of applications, a further important aspect of design coming from automotive applications concerns *safety-critical* system attributes, generally termed *Reliability, Availability, Maintainability and Safety* (*RAMS*). Here, experiences in aerospace demonstrate that properties of the system related to safety and reliability must not be designed decoupled from nominal system functions, but rather must be considered as an integral component of system functionality. Thus, such properties should be suitably represented in dynamic models (using quantitative mathematical models), and conceptual approaches for design and realization thereof should be established (e.g. redundancy), see (Bertsche et al. 2009), (Isermann 2003).

For actuator implementations, certainly *fluid mechanics* (*pneumatics, hydraulics*) is of great importance in macro-mechatronic applications (Houghtalen et al. 2010), (Manring 2005). An especially interesting area is the so-called *servo-based* approaches. These are understood to indicate a fluidic actuator which—from the control side—can be electrically driven without back-effect (termed a servo valve). Such a device is not truly a transducer, as power flows only in one direction. Despite this, is can be shown that the model framework of the generic transducer can also be applied to this case.

In addition to the power-conserving physical phenomena described in detail in this book, there exists a plethora of other interesting action principles, collectively termed *unconventional actuators*—e.g. magnetostriction (piezomagnetism, (Lenk et al. 2011)), electrorheological fluids, shape memory alloys, bi-metals (Janocha 2004)—which however, compared to the principals presented in this book, have not (yet) attained great industrial importance.

One scientific domain important for realizations is *technical optics* and *optoelectronics* (Hecht 2001). In addition to many demanding optical applications (see introductory examples in Sec. 1.3), optical principles are used in modern optoelectronic elements and microsystems as touch-less measurement devices for motion variables (e.g. lasers, image sensors).

In the original concept for this book, it was envisaged that all topics listed here would be included. In the course of its composition, however, it became increasingly apparent that given a uniform scientific and methodological depth of treatment, the material would far surpass the page count agreed upon with the publisher (see introductory remarks to this section). For this reason, the author made the *"decision to omit"* with the justifications elucidated in this section[13]. For the topics missing here, the interested reader is referred for her initial contact to the specified literature.

How to use this book

The *navigation chart* depicted in Fig. 1.24 is meant to simplify use of this book.

Chapters of key importance are indicated with a ★.

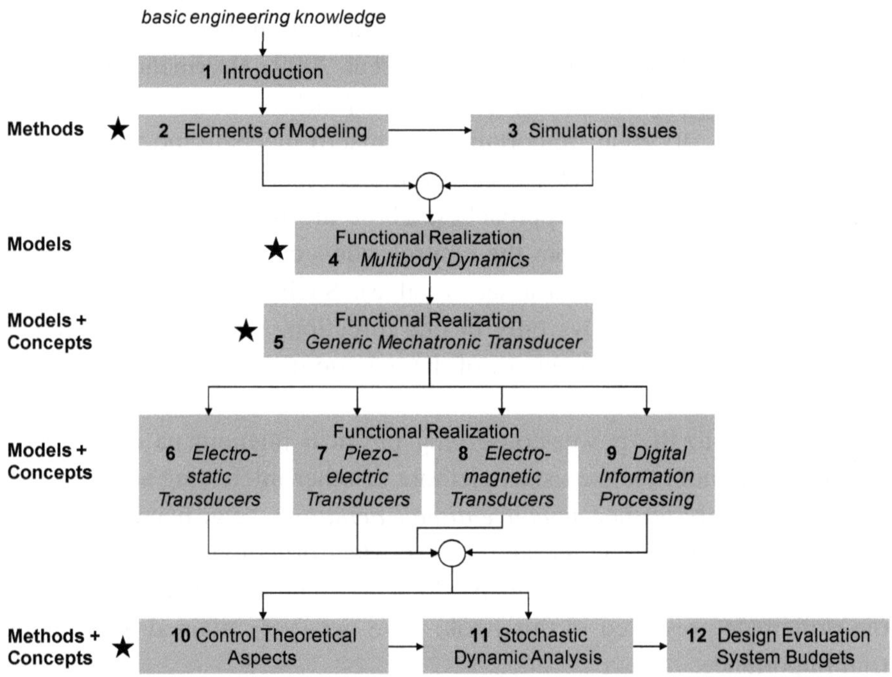

Fig. 1.24. Navigation chart for the book chapters

---

13 If, in the future, demand for supplementary material presents itself, the author and publisher will certainly find a way to accommodate it.

For a novitiate to this field, the following order of book chapters is recommended:

- Chapter 2: Elements of Modeling,
- Chapter 4: Functional Realization: Multibody Dynamics,
- Chapter 5: Functional Realization: Generic Mechatronic Transducer,
- Chapter 10: Control Theoretical Aspects,
- Chapter 11: Stochastic Dynamic Analysis,
- Chapter 12: Design Evaluation: System Budgets.

The remaining chapters can subsequently be read and worked through independently depending on interest or to address particular questions.

Audience

Required knowledge Basic knowledge of electrical engineering, mechanics, embedded information processing, systems theory, and control theory.

Targeted groups

- *students* at the *masters* and *doctoral levels* in electrical engineering, mechanical engineering, mechatronics, or computer science;
- *engineers* in mechatronic-oriented fields of industry: automotive engineering, railroad engineering, nautical engineering, aerospace, electrical drivetrain engineering, automation, machine tools, robotics, medical devices, microsystems.

Nomenclature and terminology

Polynomial terms in transfer functions In many cases, the following shorthand for linear and quadratic polynomial terms is employed:

$$\left[\omega_i\right] := 1 + \frac{s}{\omega_i}$$

$$\left\{d_i, \omega_i\right\} := 1 + 2d_i\frac{s}{\omega_i} + \frac{s^2}{\omega_i^{\,2}}, \qquad \left\{\omega_i\right\} := 1 + \frac{s^2}{\omega_i^{\,2}}$$

Scalars, vectors, matrices In contrast to scalar variables ($x, y, z, F, \ldots$), *vectors* and *matrices* are printed in *boldface* ($\mathbf{x}, \mathbf{F}, \ldots$). Whether lower or upper case is used for these can be determined from the context (as a rule, vectors are lower case).

Bibliography for Chapter 1

auf der Heide, K. (2005). *Integriertes Energie- und Drallmanagement von Satelliten mittels zweiachsig schwenkbarer Solargeneratoren.* Dissertation. Fakultät Maschinenwesen, Technische Universität Dresden.

auf der Heide, K., K. Janschek and A. Tkocz (2004). Synergy in Power and Momentum Management for Spacecraft using Double Gimbaled Solar Arrays. *16th IFAC Symposium on Automatic Control in Aerospace 2004* St.Petersburg, Russia. pp.290-295.

Bertsche, B., P. Göhner, U. Jensen, W. Schinköthe, et al. (2009). *Zuverlässigkeit mechatronischer Systeme - Grundlagen und Bewertung in frühen Entwicklungsphasen.* Springer.

Bishop, R. H., Ed. (2007). *The Mechatronics Handbook*, CRC Press.

Fedrigo, E., M. Kasper, L. Ivanescu and H. Bonnet (2005). "Real-time Control of ESO Adaptive Optics Systems." *at-Automatisierungstechnik* **53**(10): 470-483.

Hardy, J. W. (1998). *Adaptive Optics for Astronomical Telescopes.* Oxford University Press.

Hecht, E. (2001). *Optics*. Addison Wesley.

Houghtalen, R. J., A. O. Akan and N. H. C. Hwang (2010). *Fundamentals of Hydraulic Engineering Systems.* Prentice Hall.

Isermann, R. (2003). *Mechatronic Systems - Fundamentals.* Springer.

Isermann, R. (2008). "Mechatronic systems - Innovative products with embedded control." *Control Engineering Practice* **16**(1): 14-29.

Janocha, H., Ed. (2004). *Actuators*, Springer-Verlag Berlin Heidelberg.

Janschek, K. (2008). "Optimized system performances through balanced control strategies (Editorial)." *Mechatronics* **18**(5-6): 262-263.

Janschek, K., V. Tchernykh and S. Dyblenko (2004). Opto-Mechatronic Image Stabilization for a Compact Space Camera. *3rd IFAC Conference on Mechatronic Systems - Mechatronics 2004, 6-8 September 2004*, Sydney, Australia. pp.547-552 **Conference Best Paper Award**.

Janschek, K., V. Tchernykh and S. Dyblenko (2007). "Performance analysis of opto-mechatronic image stabilization for a compact space camera." *Control Engineering Practice* **15**(3 SPEC. ISS.): 333-347.

Janschek, K., V. Tchernykh, S. Dyblenko and G. Flandin (2005). "A Visual Feedback Approach for Focal Plane Stabilization of a High Resolution Space Camera." *at-Automatisierungstechnik* **53**(10): 484-492.

Kokoschka, O. (1975). *Aufsätze, Vorträge, Essays zur Kunst.* Hans Christian Verlag.

Koycheva, E. and K. Janschek (2009). "Leistungsanalyse von Systementwürfen mit UML und Generalisierten Netzen - Ein Framework zur frühen Qualitätssicherung." *atp - Automatisierungstechnische Praxis* **50**(8): 62-69.

Lenk, A., R. G. Ballas, R. Werthschützky and G. Pfeifer (2011). *Electromechanical Systems in Microtechnology and Mechatronics.* Springer.

Manring, N. (2005). *Hydraulic Control Systems.* Wiley.

Moore, G. E. (1965). "Cramming more components onto integrated circuits." *Electronics Magazine* **38**(8).

Ogata, K. (2010). *Modern Control Engineering.* Prentice Hall.

Preumont, A. (2002). *Vibration Control of Active Structures - An Introduction.* Kluwer Academic Publishers.

Roddier, F. (2004). *Adaptive Optics in Astronomy.* Cambridge University Press.

Schaller, R. R. (1997). "Moore's law: past, present and future." *Spectrum, IEEE* **34**(6): 52-59.

Senturia, S. D. (2001). *Microsystem Design.* Kluwer Academic Publishers.

Srinivasan, A. V. and D. M. McFarland (2001). *Smart Structures: Analysis and Design* Cambridge University Press.

Tchernykh, V., S. Dyblenko, K. Janschek, K. Seifart, et al. (2004). Airborne test results for a smart pushbroom imaging system with optoelectronic image correction. *Proc. SPIE - Sensors, Systems and Next-Generation Satellites Vii*, SPIE. pp.550-559.

Tummala, R. R. (2004). "SOP: what is it and why? A new microsystem-integration technology paradigm-Moore's law for system integration of miniaturized convergent systems of the next decade." *Advanced Packaging, IEEE Transactions on* **27**(2): 241-249.

VDI (2004). Entwicklungsmethodik für mechatronische Systeme (in German) - Design methodology for mechatronic systems (in English). V. D. I. (VDI). Berlin-Wien-Zürich, Beuth Verlag GmbH. **VDI 2206**.

Vogel-Heuser, B. (2003). *Systems Software Engineering*. München. Oldenbourg.

Yourdon, E. (1989). *Modern Structured Analysis*. Yourdon Press.

[illegible]

2 Elements of Modeling

Background Abstract *dynamic models* play a central role in the design process for mechatronic systems. The *dynamic* behaviors and desired (or undesired) interactions of system components fundamentally define favorable (and unfavorable) product properties. The primary challenge in modeling mechatronic systems lies in their *multi-domain nature*. To the extent that *heterogeneous* physical components are interconnected in a *homogeneously* operating functional unit, models of the components must naturally also be expressed in a *domain-independent abstract structure*. Naturally, when creating an abstract model, relevant physical and dynamic properties must be correctly depicted, and the assignment of real component properties to model parameters should remain sufficiently transparent.

Contents of Chapter 2 In this chapter, a number of selected, fundamental *methodological* approaches to the *modeling* of mechatronic systems with *lumped system elements* are discussed, which, in the estimation of the author, belong to the basic tool chest of the system engineer, and which go beyond the standard subject matter of individual disciplines. *Structured analysis* delivers *qualitative* system models with a clear functional structure. For *quantitative* physical multi-domain modeling, the *LAGRANGE formalism* in a generalized formulation (an energy-based method) and a generalized *network-based approach* (a power-flow-based method) are presented, and their strengths and weaknesses are discussed. Particular attention is paid to the *modularization* of models, to which *multi-port models* are particularly suited. In addition, these models form the basis for modern *object-oriented* modeling tools. The resulting mathematical models are generally systems of *differential-algebraic equations*, the non-trivial handling of which is discussed in detail. Discontinuous behaviors of mechatronic systems can be represented using *hybrid* descriptions, for which *net-state models* are introduced as a practical standard structure. To round out the toolbox of methods, a summary review of *linearization* of nonlinear system models is included. A methodological and practical approach to the *experimental* determination of *frequency responses* concludes the theoretical concepts of modeling. ■

2.1 Systems Engineering Context

System models Abstract, mathematical models of a mechatronic system as a depiction of the real world play a central role in systems design. As a rule, these models are developed and manipulated long before the actual components of the system are available. It is on the basis of such abstract models that robust predictions of the capabilities of the (possibly yet to be produced) real system must already be made.

Experimenting with models: simulation In this context, the construction of these models—the *modeling task*—is a key task of systems design which should be carried out with the greatest degree of diligence and care. Ultimately, product capabilities of interest are always stated in terms of statistics concerning system performance at a *specified time* or over a defined *time interval* (e.g. the nominal operating period). Such time-based performance properties can be determined via *simulation*, that is, through experiments on available models. As a part of this process, each experiment should be set up in a clearly verifiable manner with an *experiment frame* ε (the experimental conditions) and assessable system responses $y(t)$. Such experiments can be performed on the real system (ε_S, y_S) or a model of the real system (ε_M, y_M) via simulation, see Fig. 2.1.

To perform a simulation experiment, the mathematical model must be "animated" in such a way that it becomes possible to calculate the time history of all outputs that are of interest (Fig. 2.2).

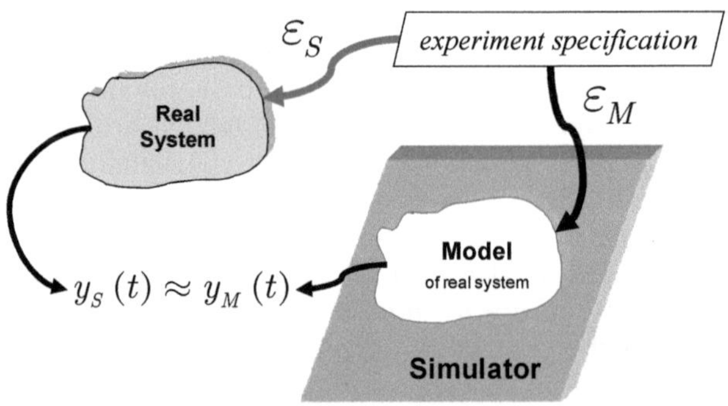

Fig. 2.1. Simulation as experimentation on models

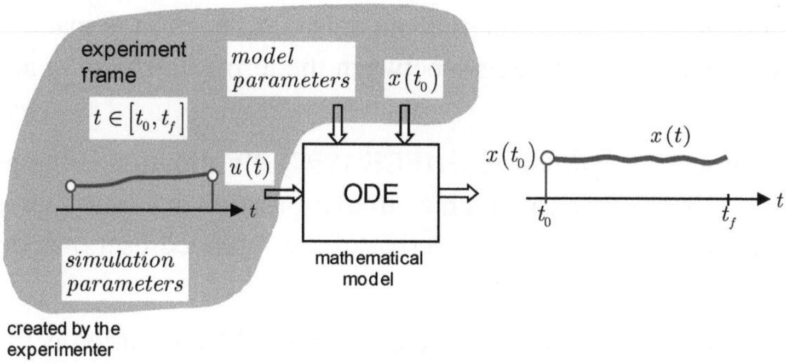

Fig. 2.2. Simulation experiment on an abstract mathematical model of an illustrative system with one input $u(t)$, one output $y(t)$, and internal states $\mathbf{x}(t)$ with initial values $\mathbf{x}(t_0)$

In general, this proceeds via the solution of a system of differential equations using appropriate numerical methods (numerical integration algorithms) which are implemented on a computational platform (digital *simulator*, Fig. 2.1). The methods used to implement simulation models inside of simulators are termed *simulation techniques.*

Validation vs. verification

Simulation experiments By performing a simulation experiment, it becomes possible to predict the behavior $y_S^i(t)$ of the real system for *one* specific experiment ε_S^i, using the simulation solution $y_M^i(t)$ as the result of *one* equivalent simulation experiment ε_M^i. The following must be kept in mind in this context: the comparability of $y_S^i(t)$ and $y_M^i(t)$ depends both on the model accuracy and on the concrete computational implementation of the mathematical model. The predictive capability of the simulation results should thus always be *critically* scrutinized: "Does my model consider all properties important to me?", "How are my model equations actually implemented in the simulator?", "Which method-dependent approximation errors are entailed by the solution algorithms employed?", "What numerical errors result from the concrete implementation on the chosen computational platform?". All of these questions *predictably* influence the accuracy of the simulation, i.e. the best possible equivalence $y_S^i(t) \approx y_M^i(t)$ under the chosen simulation boundary conditions.

In confirming the correctness of models, a distinction is made—depending on the types of the models—between the terms *verification* and *validation*[1].

Definition 2.1. *validation* – (IEEE 1997) "Validation is the process of determining the degree to which a simulation is an accurate representation of the real world from the perspective of the intended use(s) as defined by the requirements."

Definition 2.2. *verification* – (IEEE 1997) "Verification is the process of determining that an implementation of a simulation accurately represents the developers conceptual description and specifications."

In summary, the definitions can be expressed as follows:

- *validation* = "Have I created the *correct model*?"
- *verification* = "Have I *correctly created* (implemented) the model?"

The three fundamental steps of verification and validation (V&V) in the context of model creation are further discussed below.

Experimental model validation If sufficiently meaningful results regarding the output $y_S^i(t)$ of the real system are available for comparison, *experimental validation* of *mathematical models* is possible based on simulation (Fig. 2.3). Non-trivial questions in this context include: "Where do the comparison results come from, given an as yet non-existent system?", "How many simulation experiments are really sufficient for validation?".

Verification of simulation models For model validation, an implied premise is that the mathematical models, including the accompanying experiment frame, have been correctly implemented in the employed simulation model. Assessing the correctness of the implementation is termed *verification* of the *simulation model* (Fig. 2.3). Verification proceeds by comparing simulation results $y_M^i(t)$ with significant reference data $y_S^i(t)$. Predictions of system behavior which can be obtained through *analytical* consideration of the underlying mathematical models are particularly well-suited to this purpose, e.g. steady-state values from limiting values of a

1 The terms "verification" and "validation" are unfortunately understood quite differently in different technical communities. The interpretations and definitions presented here follow international standards which have generally proven valuable in industrial fields designing complex (mechatronic) systems, e.g. aerospace, for many years (ESA 1995), (IEEE 1998).

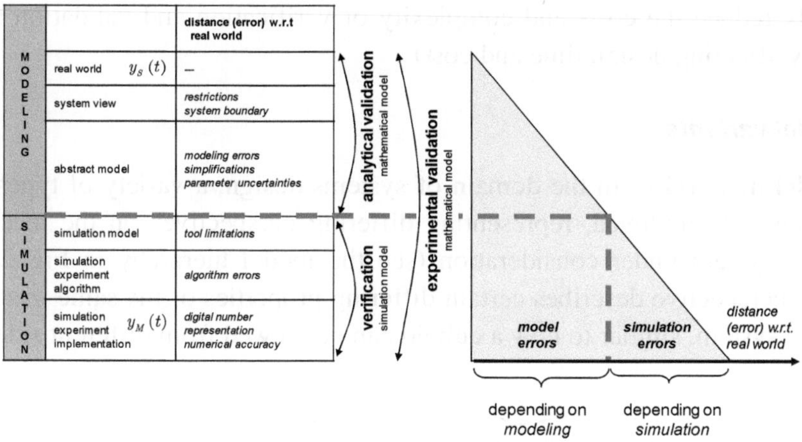

Fig. 2.3. Verification & validation vs. modeling & simulation

LAPLACE transform, oscillatory dynamics under harmonic excitation from the frequency response, angular momentum from conservation principles in mechanical systems. For any such analytical prediction, appropriate test cases (experimental frames) should be created. The deeper the theoretical understanding of the system incorporated at this point, the greater the chance of generating an accurate simulation model[2].

Analytical model validation Analytical predictions of system behavior bring forth the possibility of *analytical validation* of mathematical system models, i.e. a direct comparison with real system data. This can happen independently of, or complementary to, experimental validation (Fig. 2.3).

A further strength of *analytically*-based predictions is that, in the best case, properties with general validity can be derived for a large class of experiment frames.

A good example is provided by *stability* predictions for mechatronic systems. Experimental confirmation of system linearity and time-invariance within a few characteristic experiment frames permits general predictions for arbitrary inputs in terms of *bounded-input bounded-output* (*BIBO*) stability within a particular operational regime. Conversely, *one single* simulation experiment only allows inference about that *specific, unique* experimental frame. Using analytical predictions can thus signifi-

---

2 The converse holds equally, i.e. carelessness and ignorance are the enemies of reliable systems design!

cantly reduce the costs and complexity of verification and validation (directly affecting design time and cost).

Model variants

Model hierarchy In the domain of systems design, a variety of types of models are employed, representing differing perspectives on the mechatronic system under consideration (see the model hierarchy in Fig. 2.4). Each perspective describes certain differing properties of the same mechatronic system, similar to how a cuboid can be viewed from different sides.

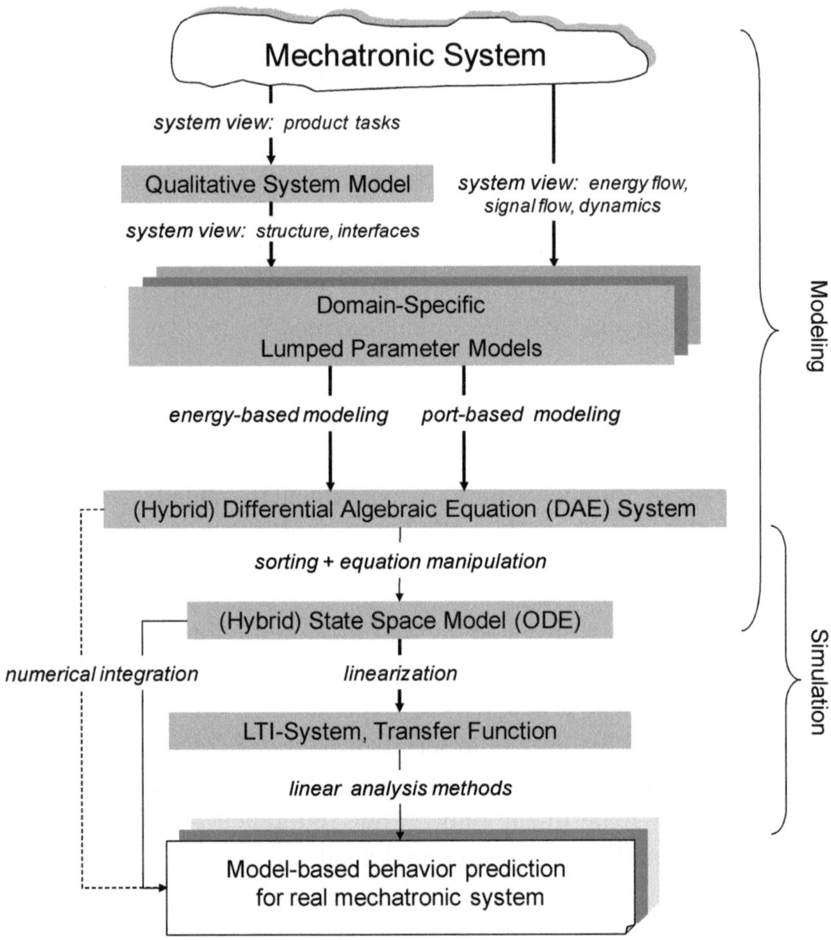

Fig. 2.4. Model hierarchy for the design of mechatronic systems

Qualitative system model At the most abstract level, a system can be described using purely *qualitative attributes*, resulting in a qualitative system model (Fig. 2.4. upper left, see also introductory example Fig. 1.17). In a mechatronic system, important aspects of the model include defining the behavior of the system with respect to the environment (user) and assigning product tasks to realizing "functions" (in the sense of "accomplish task"). At the same time, a preliminary functional system structure and important functional interfaces should be defined (see Sec. 2.2).

Quantitative models For quantitative predictions of system behavior, mathematical models suitable for computation must be generated. To accomplish this, attention should be paid to energy flow, signal flow, and the dynamics within and between the functions defined in the qualitative model. The challenge in mechatronic systems lies in the variety of physical domains involved (electrical, mechanical, hydraulic, thermal, etc.). As a result, a broad technical understanding of the different domains must be present to form a firm basis for modeling.

When creating a model, attention must be paid to the fact that interactions between system elements from different physical domains always take place via energy flows with power back-effects. The resulting *heterogeneously*-coupled *system behavior* should be captured in *domain-independent models*.

Modeling paradigms For modeling in multiple physical domains, there exist only a few suitable modeling methodologies. This book considers system descriptions with *lumped system elements*, and two methodological approaches particularly well-suited to this formulation are presented in more detail. First, *energy-based* modeling using the *LAGRANGE formalism* is well-suited to smaller systems with nonlinear equations, and can be easily worked out by hand. *Multi-port* modeling methods, on the other hand, are particularly appropriate for very large systems and for computer modeling. Combined with *object-oriented* concepts, multi-port methods nowadays offer an attractive array of efficient computational tools for multi-domain modeling and simulation.

Mathematical system model The end results of the various modeling methodologies are so-called *differential-algebraic equations* (*DAE*). These are systems of (generally nonlinear) differential and algebraic equations involving relevant system variables (Fig. 2.4 center). They are the final, actual mathematical model formulation—a DAE system presents a domain-independent mathematical model in which all physical phenomena are represented. Predictions of system behavior are obtained via appropriate experimentation on the DAE system model. Solving such DAE systems in the general case is, however, (very) difficult, so that a DAE system model must generally be manipulated so as to enable computations based on it (e.g. converting it to state-space representation, possibly linearization). Note that certain special phenomena—such as switching operations, mechanical contact problems, stiction, and discrete-time (primarily in the context of information processing) and discrete-event phenomena—should be modeled by extending the model into a *hybrid system*. Working with DAEs and hybrid systems is elaborated upon in the following sections.

Model variants

Model purpose, model accuracy One significant task in systems design is to incorporate the "correct" abstraction and simplification of the real physical behavior of a system under examination into a mathematical model. There are no set rules for this process; rather, it requires experience and engineering judgment. In the end, the idea is to separate significant aspects from the negligible. For example, the question of whether to consider parasitic effects such as electrical line resistance or mechanical friction in the model always depends on the purpose of the model. For this reason, in the context of design, there will generally be several mathematical models with *differing model accuracy* and *fidelity* for a single mechatronic system.

Low-fidelity models For controller design, stochastic performance predictions, or a rough design (feasibility studies), *simplified* or *low-fidelity models* are typically employed. This type of model is generally a linear time-invariant model (LTI system) of low order. In the case of multibody systems, these consider only a select subset of the mechanical structural frequencies (eigenmodes). Implementing such models to produce output time histories in simulation presents hardly any problems using standard algorithms, and is thus not further discussed here.

High-fidelity models On the other hand, for design verification and validation (e.g. sensitivity analysis with differing experiment frames, statistical analysis via Monte-Carlo simulation), *detailed, high-fidelity models* with the smallest possible modeling errors (Fig. 2.3) should be used to the greatest extent possible. Such models generally take into account all relevant nonlinearities, broadband dynamic system behavior, and, in particular, high-frequency structural modes. This results in models of (very) high order and considerable complexity. Additional issues in the treatment of models, e.g. the modularity of simulation models, play a key role here. *High-fidelity* models pose some of the greatest challenges to simulation technology, and can often only be *accurately* simulated by taking special measures. For this reason, Ch. 3 is devoted to certain specialized methods and solution approaches for simulating mechatronic systems.

2.2 System Modeling with Structured Analysis

Goal This section introduces several fundamental elements of *systems analysis and modeling*. These formal tools permit the *structuring* of a system in such a way that it becomes manageable and that its *inner relationships* become visible to the engineer. Exposing system structures is a task which is generally performed without requiring mathematical formulae, but which should nonetheless not be underestimated.

Qualitative system models Due to the complexity of some systems, developing a system model is generally a non-trivial task with an uncertain outcome (there are infinite solutions, depending on which system properties are considered). Though *qualitative system models* do not allow for numerical computation, they do enable—in addition to the recognition of fundamental system relations (causality loops, dynamics)—the preliminary establishment of a foundation for quantitative (mathematical) models by setting up clearly-delineated and manageable subsystems. On the basis of such structured system models, design variants of the device can be developed. Only once such device variants are available is it even possible to engage in the creation of a physical (i.e. quantitative) model. Indeed, it is only at this point that the candidate devices can start to be considered (for example, electromechanical vs. servo-hydraulic actuation). It is only in textbooks that standard problem formulations of the type "given:, find:" exist.

Top-down modeling The workflow described above belongs to the so-called *top-down* system modeling paradigm, and always takes place at the beginning of the product design process. During the definition of requirements, the description of properties of the product under development is made more and more detailed following to the above procedure. Today, techniques used during this process often come from the realm of software development and can be divided into *function-oriented* and *object-oriented* modeling methods (Vogel-Heuser 2003).

Function-oriented models *Function-oriented* modeling methods offer a natural approach for the design of *mechatronic systems*, as they are centered on workflows and input/output relations in a conventional way. Thus, following a few fundamental definitions, several elements of modeling via *structured analysis* (*SA*)—which are practical for qualitative modeling due to their intuitive simplicity—are presented below.

Fig. 2.5. Colorful definition of the term "system" (adopted from (Yourdon 1989) with kind permission of Ed Yourdon)

2.2.1 Definitions

Definition 2.3. *System* – (Cellier 1991) "A system is characterized by the fact that we can say what belongs to it and what does not, and by the fact that we can specify how it interacts with its environment. System definitions can furthermore be hierarchical. We can take the piece from before, cut out a yet smaller part of it, and we have a new 'system'." See Fig. 2.5.

Definition 2.4. *System* – (Schnieder 1999) "A system is marked by the presence of certain properties and is characterized by the following four *axioms*:

- **Principle of Structure** The system consists of a quantity of parts, which have mutual relationships to each other and the (system) environment. The system has reciprocal influences with its environment via physical quantities describing the energy, mass and information state of the system.
- **Principle of Decomposition** The system consists of a quantity of parts, which can further be decomposed into a number of mutually influencing sub-parts. When examined in detail, the sub-parts in turn exhibit a certain complexity or general system characteristics.

- **Principle of Causality** The system consist of a quantity of parts, whose mutual relationships and own variations are clearly defined in themselves. Following a causal interrelationship, later states can only depend on previous ones. Causality is understood as the logic of events.
- **Principle of Temporality** The system consists of a quantity of parts, whose structure and state to a greater or lesser extent determine changes occurring over time. Temporality is the sequence of events and variations over time."

These two definitions are, in the end, equivalent, though the second definition also brings into play the temporal aspect important for physical systems. In the context of model creation, the principle of causality is unfortunately sometimes cited in the wrong setting; Sec. 2.3.8 further clarifies this terminology.

2.2.2 Ordering principles

Complex systems One significant task of system modeling is the ordering of the various components of a system. Complex systems are understood to be those consisting of a great number of components. Experience shows that humans are limited in the number of elements (graphic symbols) they are capable of simultaneously recognizing and understanding the contents (the semantics) of in a depiction. The important design principles of *structuring*, *decomposition*, *aggregation*, and *hierarchy* allow the designer to limit the number of system components under consideration at any time in the face of increasing model detail; these principles are described below.

Structuring
- establishing the relationships between entities in a system according to given criteria,
- deconstructing a given system according to given criteria so that its relationships become recognizable.

Decomposition
- "breaking into fundamental elements",
- systems are *decomposed* into subsystems,
- more details are revealed in an existing model,
- refinement of a structure.

Aggregation[3]
- "layering together of individual elements",
- subsystems are *aggregated* into a system.

---

[3] Aggregation is the opposite of decomposition.

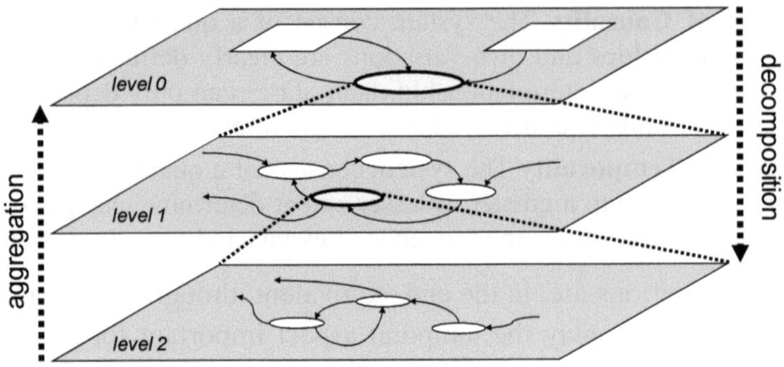

Fig. 2.6. Hierarchical levels of a system model: *decomposition* from level 0 → level 1, *aggregation* from level 1 → level 0

Hierarchy
- "(pyramidal) ranking",
- system hierarchy: system definitions are *hierarchical*, i.e. one piece of the system can be extracted from the whole and be considered as a new system (see subsystems),
- hierarchical level: a particular level of consideration of a system, generally representing a subsystem,
- top level: a global view of the system
- lower level: a detailed view of the system (view of the interior).

2.2.3 Modeling elements of structured analysis

Introductory considerations The method of *structured analysis* (*SA*) known from the field of software development offers a very natural approach for system modeling for the design of mechatronic systems.

This section presents a simplified variation of the YOURDON SA approach (Yourdon 1989). In the experience of the author, the modeling elements presented here completely suffice for the creation of manageable functional descriptions of mechatronic products in a short time and a clear manner, without requiring extensive methodological knowledge or specialized tools[4].

[4] The practical procedure presented here has been successfully employed in the author's teaching for over 10 years as a requirement in all student projects (final projects and master's theses).

An expanded formulation—*structured analysis with real-time extension* (*SA/RT*) (Hatley and Pirbhai 1987; Hatley and Pirbhai 1993), (Vogel-Heuser 2003)—permits a strictly formal specification of temporal-causal system dynamics and is recommended for larger (more complex) tasks.

Function-oriented modeling Using structured analysis results in a primarily *function-oriented model* as it takes as a starting point product functions and then considers their *logically causal* interconnectedness via data and signals. Note further, that for a complete system description, structured analysis additionally models *temporally causal* relationships using state machines.

Model elements

The most important elements of function-oriented modeling are shown in Table 2.1.

Table 2.1. Model elements for structured analysis

Symbol	Property	Description
	***data flow***	A *data flow* represents the transport of abstract data between processes (functions)
	***control flow***	A *control flow* represents the transport of abstract *control data* (events) between processes (functions)
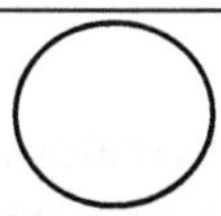	***process, function***	A *process (function)* transforms input data to output data using specified rules; e.g. *measure* speed (*predicate* + object)
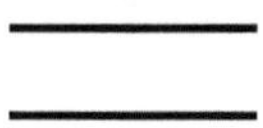	***store***	A data *store* preserves data elements for a certain duration of time (non-volatile)
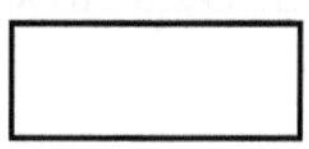	***terminator***	A *terminator* represents an entity (function, device, person) which exists <u>outside</u> the system under consideration and exchanges data with it

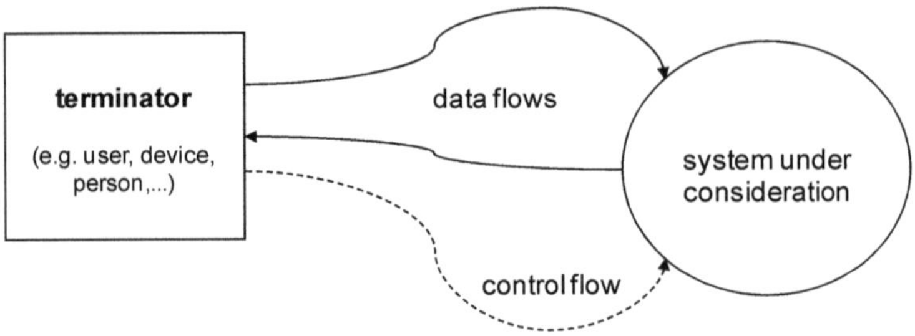

Fig. 2.7. System delineation via a data context diagram

Context diagram

Data context diagram (DCD) Context diagrams describe the system under development from the point of view of a user. The purpose of the system is summarized as a single system process. This process converts inputs from terminators/end users into outputs to terminators/end users (Fig. 2.7). The context diagram describes the interaction of the system with its environment.

Data flow diagram

Data flow diagram (DFD) Data flow diagrams are the primary tool for determining the functional properties of a system. They make structures of the system concrete by defining component functions (processes) which are tied together with data flows. A DFD contains processes, data flows and data storage locations, but no terminators (Fig. 2.8).

Process A process (also called a function, activity, or task) creates an output from an input by performing an operation. Processes have names and numbers.

Data flow Data flows represent all possible types of generalized information (signals, action flows) and can be further decomposed. Data flows can be binary, digital, or analog.

Leveling The decomposition of a parent DFD into child DFDs with an increased level of detail. The decomposition levels can have various depths.

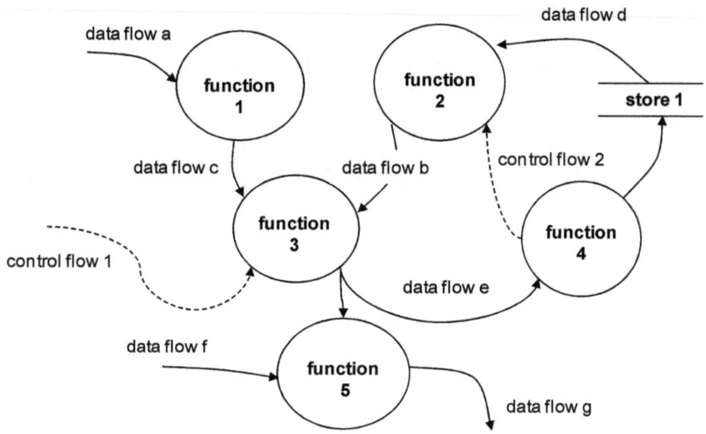

Fig. 2.8. Functional decomposition via a data flow diagram

Balancing A test for inconsistent data flows. Input and output data flows from parent and child processes must match up in a consistent manner. Data flows without a source or sink create inconsistencies which can be tested for with automated or manual procedures. For each level, preferably *five to seven* (at most ten) processes (functions) should appear (a greater number becomes unwieldy and encumbers any clear overview for the designer).

Process specification (PSPEC) A process is continuously decomposed until a short and unambiguous component description becomes possible. Possible methods for describing the process include anything elucidating its content, e.g. tables, prose description, equations, control theory transfer functions. PSPECs can appear at all levels of refinement (Fig. 2.9).

PSPEC function x.y

- textual specification
- pseudocode
- mathematical equations, etc.

Fig. 2.9. Function definition via process specification

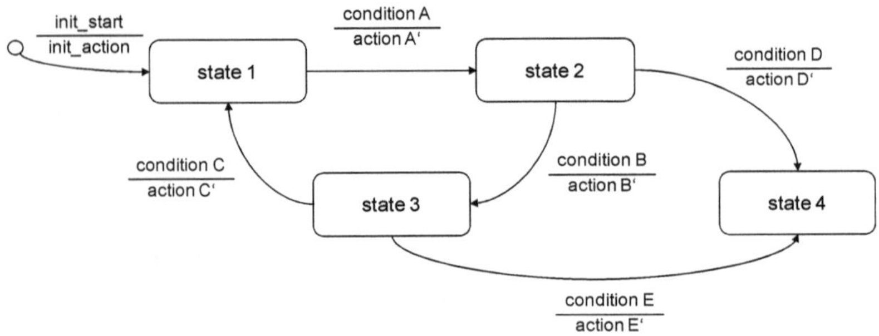

Fig. 2.10. Specification of the temporal/logical implementation of functions via a state transition diagram (here in MEALY machine notation)

Control specification

State transition diagram (STD) Control specifications describe the processing of control flows. Usually, these flows trigger state transitions or are combined with other signals to form new control signals. Typical means of description include state transition diagrams, decision tables, or prose descriptions. In general, each process has *its own* control specification. Fig. 2.10 shows an example of a state transitions diagram in the form of a *MEALY machine* (Litz 2005). When *State 1* is active, fulfillment of *Condition A* results in the activation of *State 2* (with simultaneous deactivation of *State 1*) and the carrying out of *Action A'*.

Data handling

Data dictionary (DD) Every data and control flow—as well as all storage locations—must be defined in a data dictionary. Flows are either primitives or non-primitives, the latter consist of groupings of primitives. A dictionary is usually laid out in a computer-readable format, and can be formulated in written form, table form, or as a database.

Architecture diagram

Implementation structure An architecture diagram describes an implementation structure which realizes the functional relationships indicated in a DFD (device elements and their linkages, the *device architecture*, see Fig. 2.11). In addition, the distribution of functions among device elements and the distribution of data flows among device interfaces are documented. This gives a *semantic* description of devices, i.e. "which task(s) should the device fulfill". Further, clear justification for the existence of any particular device in the system becomes evident within a device architecture.

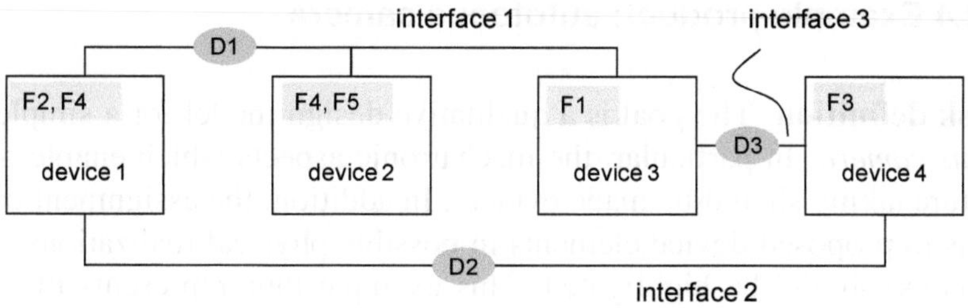

Fig. 2.11. Assignment of functions and data flows to device components in an architecture diagram

System model

A complete qualitative *structured analysis system model* is presented Fig. 2.12. In a concise, semi-formal form, this model describes different views of the system (logically causal = functional vs. temporally causal = dynamic), possesses different hierarchical levels (context, DFD level 1, DFD level 2…), and contains one or more possible device architectures (design variants) along with the assignment of device elements to their functions.

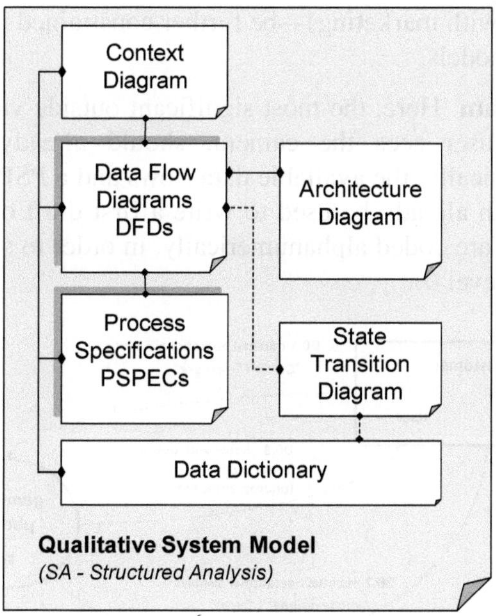

Fig. 2.12. Complete system model obtained from structured analysis

2.2.4 Example product: autofocus camera

Task definition The goal is a qualitative design model for a simple *autofocus camera.* In particular, the mechatronic aspects which enable optimal picture taking should be made evident. In addition, the assignment of functions to proposed device elements in possible physical realizations (design variants) should be highlighted. This example thus represents the typical first steps in the design of a new product.

Written product specification "The autofocus camera should be very user-friendly (target user is a photographic layperson) and have as few operational and display functions as possible. Sharp pictures should be produced without special manual intervention (industry-standard autofocus functionality). The camera should work with standard rolls of film (black-and-white, color). It should be possible to use standard rechargeable batteries. The camera should fit into the low-cost market, preferably be light weight, and enable as long an operational time as possible for every full battery charge."

Typically, user requirements (e.g. from marketing) are given in purely verbal form, and at this point, still allow for great freedom in the design. To begin the design process, this freedom should—in cooperation with the customer (e.g. with marketing)—be further constrained using formal qualitative system models.

Context diagram Here, the most significant outside view of the product, i.e. how the user sees the camera, should already be recognizable (Fig. 2.13). Basically, the available data flows and a PSPEC of the Primary Function F0 can already be used to write a first draft of the user manual. The data flows are coded alphanumerically, in order to simplify later reference (D*0*.x = Level *0*).

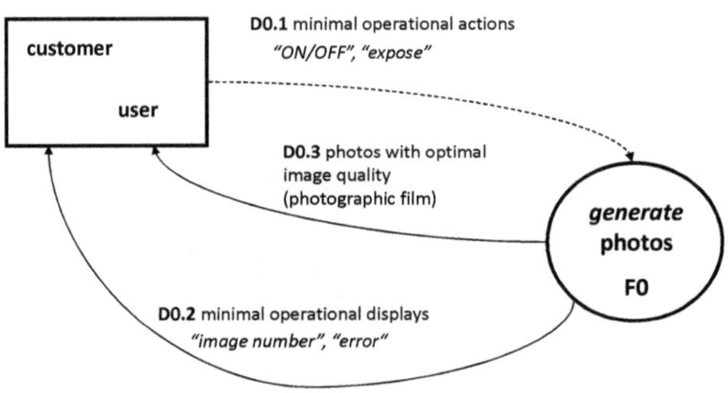

Fig. 2.13. Level 0: data context diagram for an autofocus camera

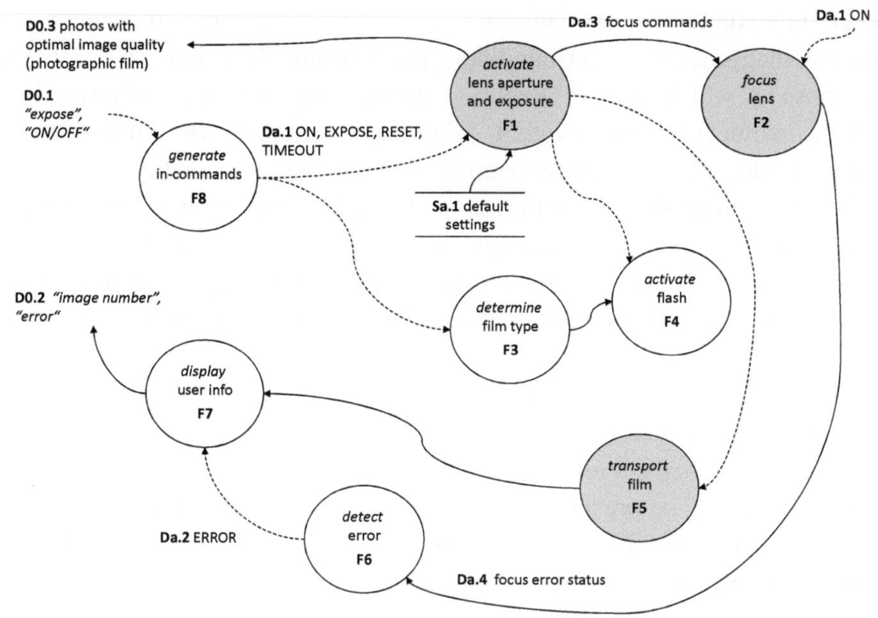

Fig. 2.14. Level a: data flow diagram for Primary Function F0 *generate_photos*; mechatronic functions in gray (labels excerpted)

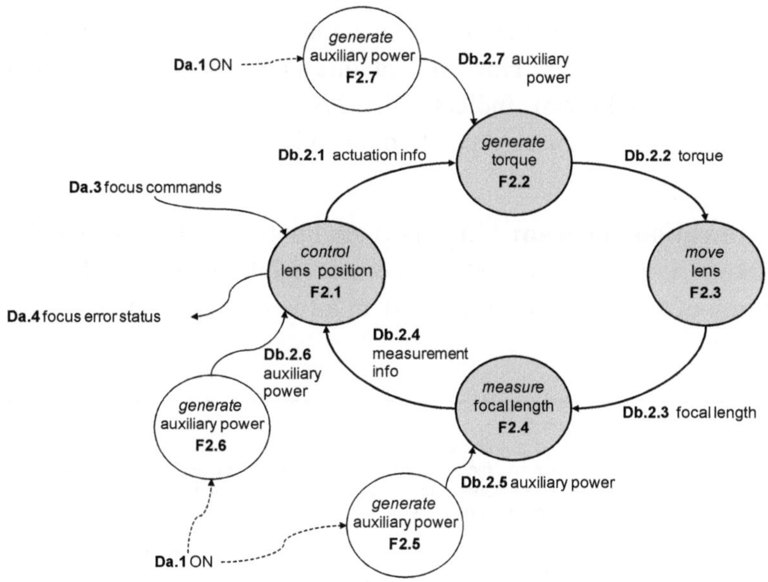

Fig. 2.15. Level b: data flow diagram for function F2 *focus_lens*; mechatronic functions in gray

Data flow diagrams Already at Level *a* (Fig. 2.14)—the *first* below the context diagram—the first candidate mechatronic functions (F1, F2, F5, i.e. those in which it is obvious that masses must be made to move in a precise manner) become evident. With eight functions, the DFD remains just readable; a larger number would render it rather unwieldy.

Of course, this decomposition is anything but unique. It is the result of a subjective perspective on the system by the design engineer, and should always be coordinated with the customer. The textual representation of data flows and functions in this model obviously eases such discussions across professional boundaries (all the way down to the layperson). The functions and data flows are coded alphanumerically, in order to simplify later references (D*a*.x = Level *a*, F*i* = Function i).

Function F7 *focus_lens* is further decomposed at the *second* level, Level *b* (Fig. 2.15). The presence of a closed functional chain is already clearly discernable here. The coding of functions and data flows is as before (D*b*.x = Level *b*, F2.*j* = Subfunction *j* of F2).

PSPECs Though the functional descriptions of the DFD elements are already readily interpretable in clear text, it is advisable to further specify the contents of the functions.

In the present case (Fig. 2.16), Function F2 *focus_lens* has a detailed written specification, which is supplemented with important numerical performance parameters. However, at this point, concrete values for the parameters are unknown, indicated by *TBD* = *to be defined.* Such concrete values must then be specified in subsequent steps of the design process.

State transition diagram In this diagram (Fig. 2.17), a model of the flow between operational modes becomes visible. Note that the transition conditions of the state machine must involve only available control flow signals.

PSPEC F2 *focus_lens*
The lens shall be moved automatically such that sharp images are always produced. It shall be possible to select an object in front of the lens as the distance reference.
The focal length shall be positioned with an accuracy of *TBD* [mm].
The automatic positioning shall complete in not more than *TBD* [sec].

Fig. 2.16. PSPEC for Subfunction F2 *focus_lens* (*TBD* = *to be defined*)

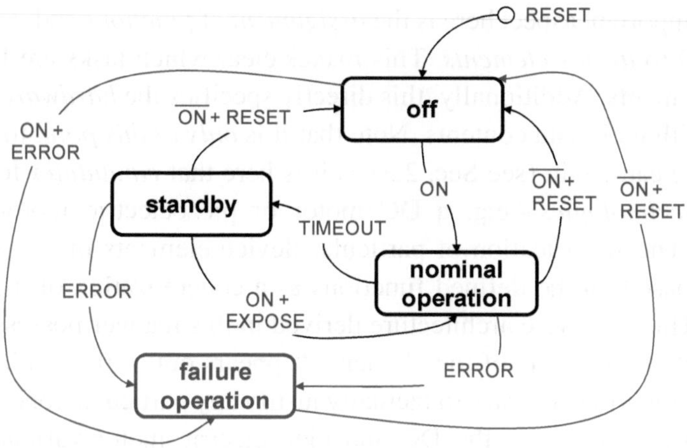

Fig. 2.17. State transition diagram for Primary Function F0 *generate_photos* (specification of the actions of the MEALY machine definition shown in Fig. 2.10 are omitted here for clarity)

Level	Data flow	Type	Description
0	**D0.1** Minimal user intervention	C	operation by user
	D0.2 Minimal operation feedback	D	feedback to user
	D0.3 Optimal picture quality	D	output: exposed, safely rewound roll of film
a	**Da.1** ON, SHUTTER, RESET, TIMEOUT	C	internal control signals
	Da.2 ERROR	C	internal control signal
	Da.3 Focus commands	D	default system parameters
:	:	:	
b	**Db.2.1** Set point information	D	computed set point signal
	Db.2.2 Torque	D	rotational moment on lens
	Db.2.3 Focal length	D	physical focal length
	Db.2.4 Measurement information	D	measured focal length
	Db.2.5 Auxiliary energy	D	auxiliary energy for measurement
:	:	:	

Fig. 2.18. Data dictionary for Primary Function F0 *generate_photos* (excerpt)

Data dictionary In this concise representation (Fig. 2.18), all data flows are described in more detail (D = data flow, C = control flow).

Design variants Fig. 2.19 and Fig. 2.20 show *two* different *design variants* (alternatives) for Function F2 *focus_lens*. In the two cases, the same functional characteristics are realized using different technologies. Conceptualization of the physical implementation is completely up to the design engineer. This fact can be used to justify, for example, the design decision to use two microcontrollers.

The important aspect here is the *assignment* of *functions* and *data flows* of the DFD to *device elements*. This makes clear which tasks are fulfilled by these elements. Additionally, this directly specifies the *hardware interfaces* along with their data contents. Note that it is *only at this point* that *physical modeling* can begin (see Sec. 2.3) as it is here that *candidates* for *concrete device technologies*—e.g. a DC motor or piezoelectric motor—become known. The specification of particular device elements in this example is clearly based on the defined functions as a concretization of user requirements. Thus, a device architecture derived in this manner possesses a clear, justifiable basis; none of the elements "appears out of thin air". Naturally, though, it is still necessary to factually justify any particular concrete choice. In this particular case, the DC and piezoelectric motor variants result in completely *different device properties*, e.g. device complexity (fewer elements for piezoelectric), power requirements (high-voltage for piezoelectric), motion properties, mass, cost, etc. The final choice can only take place following analysis of all performance aspects and device capabilities. However, the starting point for all of these further investigations is the qualitative functional system model.

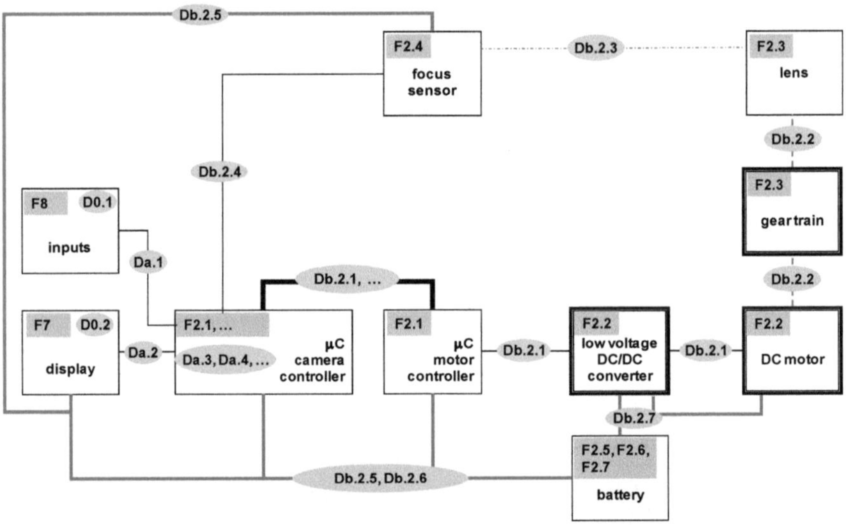

Fig. 2.19. Design Variant A for Function F2 *focus_lens* with a *DC motor and gear train*; architecture diagram with assignment of functions/data flows ↔ device technology

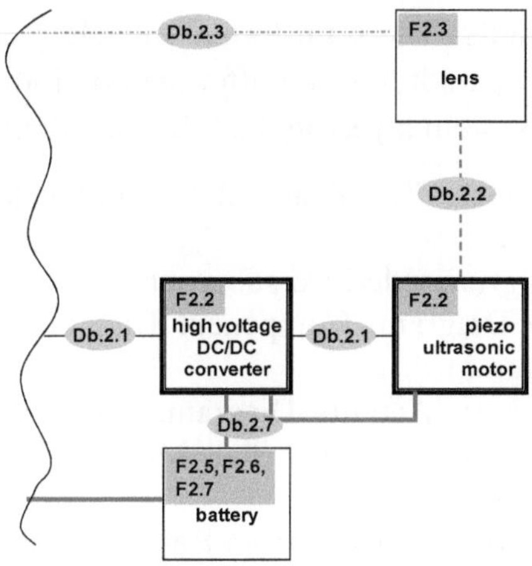

Fig. 2.20. Design Variant B for Function F2 *focus_lens* with a *direct drive piezoelectric ultrasonic motor*; architecture diagram (detail) with assignment of functions/data flows ↔ device technology

Nomenclature Structured analysis does not have a standardized nomenclature; at most in the context of computer-aided tools are there strict syntax rules which must be followed. In this example, a very simple and pragmatic nomenclature was employed, which has demonstrated its usefulness in the author's many years of experience in industry and teaching. The most important syntactic element has proven to be *alphanumeric coding*, which enables simple, concise, and unambiguous naming of model elements. This example also demonstrates quite nicely that structured analysis can be easily applied *without specialized tools*—commonly used word processing tools are thoroughly sufficient.

2.2.5 Alternative modeling methods

Object-oriented system modeling with UML

Unified Modeling Language (UML) In addition to the above-mentioned *structured analysis with real-time extensions (SA/RT)* following *Hatley* and *Pirbhai*, the only real alternative—due to its engineering-oriented approach and wide international acceptance—for *comprehensive* system modeling in the *object-oriented* paradigm is the *Unified Modeling Language (UML)* (Oestereich 2006), (Vogel-Heuser 2003).

UML model elements The semi-formal modeling elements of UML follow *object-oriented* paradigms and offer the possibility of depicting *structural* and *dynamic* system layers in the following diagram types:

- *Structure Diagrams*: Class Diagram, Object Diagram, Profile Diagram, Package Diagram,
- *Architecture Diagrams* (derived subgroup of structure diagrams): Composite Structure Diagram, Component Diagram, Subsystem Diagram, Deployment Diagram,
- *Behavior Diagrams*: Activity Diagram, Use Case Diagram, State Machine Diagram (a special type of State Machine Diagram is the so-called Protocol State Machine),
- *Interaction Diagrams* (derived from Behavior Diagrams): Sequence Diagram, Communication Diagram, Timing Diagram, Interaction Overview Diagram.

UML is particularly well-suited to modeling and specification of *complex systems* and is primarily employed in software development. Since 1997, *UML* has been an *international standard*[5] and is supported with an excellent selection of computer-aided tools (primarily for software design, though equally suited to general systems design). Newer extensions permit considerably improved options to specify requirements, system blocks, and allocations (see SysML[6]), as well as timing properties and performance parameters, see *Profile for Modeling and Analysis of Real-Time and Embedded systems (MARTE)*[7].

Model verification UML itself is—as are SA and SA/RT—a *semi-formal* definition, whose model descriptions are not strictly formally (analytically) verifiable, nor do they lead directly to executable models (simulation models). Though computer-based UML models can be checked for syntactic integrity with automated tools, no semantic integrity tests of any substance are possible. Thus, while such system models provide structured information about the system, they can still exhibit arbitrary inconsistencies in the logical and temporal flow. In the case of quantitative models (complex nonlinear dynamic systems, see Sec. 2.3), such a lack of comprehensive analytical verification can be circumvented by performing simulation experiments. This type of procedure has recently been more intensively stud-

5 www.omg.org

6 http://www.sysml.org/, September 2009

7 http://www.omgmarte.org/Specification.htm, September 2009

ied for these more qualitative models, giving rise to so-called *executable UML models*, e.g. (Koycheva and Janschek 2007), so that in the future, the use of UML system models in simulations may become more feasible.

Application From the viewpoint of mechatronics, UML certainly places greater demands on the formal abstractive capabilities of the design engineer than does structured analysis. The function-oriented mindset of SA coincides very nicely with control theoretical viewpoints, and is thus directly usable even without further training. In order to truly be able to fully apply UML, however, training in object-oriented paradigms and practical design exercises are absolutely recommended. Familiarity with the modeling syntax and the mindset of object-oriented methods is necessary in order to meet formal requirements in the models. Thus, for an *introduction* to systematic and structured system modeling, structured analysis is to be preferred, and, in the author's experience, when applied to problems with low to moderate complexity, this latter method quickly gives very useable results.

Model-based systems design

An alternative approach—which is increasingly experiencing great popularity among mechatronics users (particularly in the automotive industry)—is so-called *model-based systems design* (Rau 2002), (Conrad et al. 2005), (Short and Pont 2008). In general, this approach does *not* encompass qualitative system modeling in the above sense, but rather quantitative *hybrid* system models, with mixed *discrete event* and *continuous* dynamics (see Sec. 2.5). For the modeling of flow-oriented properties, generalized state machines are used in combination with descriptive languages employing block diagrams (e.g. STATEFLOW / SIMULINK (Angermann et al. 2005)). As a result of the hierarchical structure of these tools, structural characteristics can be depicted with hierarchical subsystems (SIMULINK) and can be coupled with flow-oriented model components (STATEFLOW). Given consistent modeling, the descriptive elements of structured analysis (the data flow diagram and the state transition diagram) can, to a certain extent, be transformed into executable models. They thus result in executable specifications, which can be continually refined throughout the design process. For example, purely written PSPECs can be used in the form of text comments in SIMULINK blocks, and can be replaced step by step with mathematical functions (transfer functions, state machines, etc.) during the course of the design process in an *evolution of system models.* A particularly attractive aspect of this latter concept is the possibility of directly generating code and immediately testing models in the context of *rapid prototyping*.

2.3 Modeling Paradigms for Mechatronic Systems

Modeling goals The physical modeling of mechatronic systems is predominantly characterized by the multidisciplinary (multi-domain) character of the different component systems. Naturally, the goal is a complete abstract model, representing overall system behavior—a *domain-independent model.* In this and following sections, this will be accomplished by transferring and simplifying fundamental physical model equations (differential equations, algebraic equations) into linear time-invariant (LTI) models in a frequency-domain representation (transfer functions). Using such models, a series of significant analyses of system behavior can be efficiently worked through using established (commercial) computational tools. However, to deal with more complex *high-fidelity* models, further modeling must be undertaken.

Modeling approaches Starting from the view of a system using *lumped system elements* and generally valid statements of energy conservation, there are fundamentally two modeling approaches which emerge (Fig. 2.21):

- *energy-based modeling* employing *scalar energy functions* (LAGRANGE formalism, HAMILTON's equations)
- *multi-port modeling* employing *component-based* system models with *power-conserving network rules* (KIRCHHOFF networks, bond graphs).

In both modeling approaches, the aspects of *back-effect* arising from mutual power exchanges between interacting system components is taken into account in different ways. In the case of *unidirectional* KIRCHHOFF networks, modeling can be simplified using *signal-coupled networks* (e.g. control system signal-oriented diagrams). The port-HAMILTONian formulation—an interesting, relatively new approach—combines properties of energy-based and multi-port modeling and enables a specific type of mathematical model which is particularly useful for nonlinear controller design.

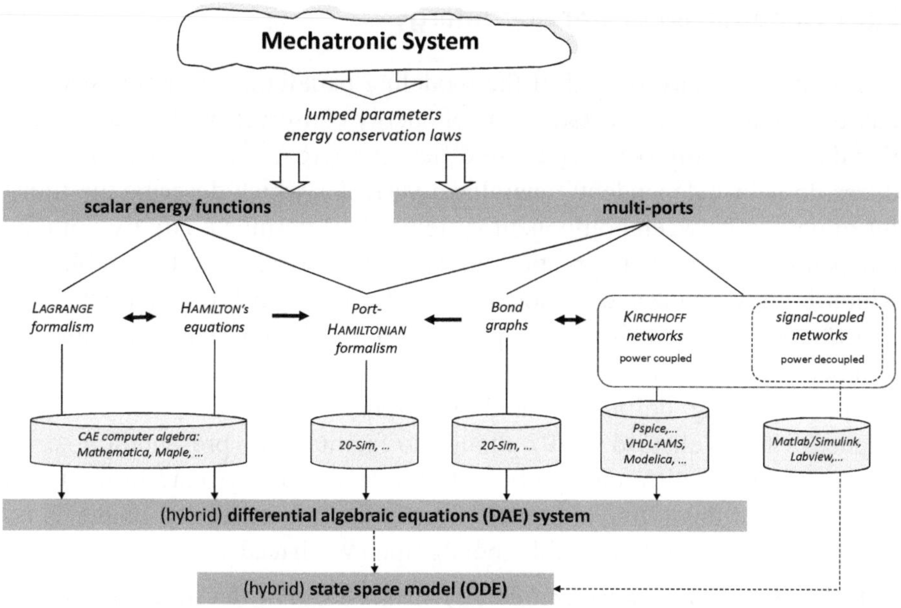

Fig. 2.21. Paradigms and commercial computational tools for multi-domain modeling of mechatronic systems using lumped system elements

Common model basis: DAE systems All of the modeling paradigms presented here accomplish the task of describing the physical dynamics of a *coupled heterogeneous* mechatronic system using a *domain-independent* mathematical model in the form of a system of differential-algebraic equations (DAE system) or a state space model. This mathematical model is then the starting point for all further analyses of system behavior.

A thorough discussion of these approaches is far beyond the scope of this book; for such, respective standards and specialized references should be consulted. However, in order to sharpen understanding of the scope of the problem and to ease classification of simulation approaches and tools, the remainder of this section presents a concise summary of the paradigms introduced above for multi-domain modeling of mechatronic systems. Due to their foundational importance and ease of application, the LAGRANGE formalism and the KIRCHHOFF network approach are described in somewhat more detail. In addition, these two approaches are applied in subsequent chapters for modeling physical phenomena implementing mechatronic system functions (Chs. 5 through 8).

2.3.1 Generalized power and energy

Axiomatic foundation All of the modeling paradigms presented here are based on generally valid statements of energy conservation. In order to clarify the relationship between these different approaches, this section introduces domain-independent generalized variables which describe the transfer of energy between component systems. This permits a largely domain-independent, generic, axiomatic foundation for modeling to be established, which is only connected to domain-specific physical laws—the so-called *constitutive equations*—at very precise points to realize a functional instrument for computation.

The following definitions and descriptions are conceptually largely aligned with (Wellstead 1979), though to facilitate the presentation sometimes employing different symbols. For a deeper background in modeling approaches utilizing this axiomatic foundation, the interested reader is referred to the eminently readable monograph (Wellstead 1979).

Definition 2.5. *Generalized energy variables.* For the modeling of mechatronic systems, the following generalized energy variables are defined:

- Generalized *potential, effort* e
- Generalized *velocity, flow* f
- Generalized *momentum*

$$p(t) := \int_{t_0}^{t} e(\tau) \cdot d\tau + p(t_0) \quad \text{or} \quad dp = e \cdot dt\,,\ e = \dot{p}$$

- Generalized *coordinate, displacement*

$$q(t) := \int_{t_0}^{t} f(\tau) \cdot d\tau + q(t_0) \quad \text{or} \quad dq = f \cdot dt\,,\ \ f = \dot{q}$$

- Generalized power

$$P(t) := \frac{dE(t)}{dt} := f(t) \cdot e(t) = \frac{dq}{dt}\frac{dp}{dt}$$

- Generalized energy

$$dE(t) = P(t) \cdot dt = f(t) \cdot e(t) \cdot dt = \frac{dq}{dt} e \cdot dt = e \cdot dq = f \frac{dp}{dt} dt = f \cdot dp$$

$$\Rightarrow \quad E(t) = \int_{t_0}^{t} f(\tau) \cdot e(\tau)\, d\tau$$

- Generalized potential energy

$$V(q) = \int_{q_0}^{q} e(q) \cdot dq \tag{2.1}$$

- Generalized potential co-energy

$$V^*(e) = \int_{e_0}^{e} q(e) \cdot de$$

- Generalized kinetic energy

$$T(p) = \int_{p_0}^{p} f(p) \cdot dp$$

- Generalized kinetic co-energy

$$T^*(f) = \int_{f_0}^{f} p(f) \cdot df \tag{2.2}$$

A summary representation of the relationships created by these definitions is presented in Fig. 2.22 and Fig. 2.23.

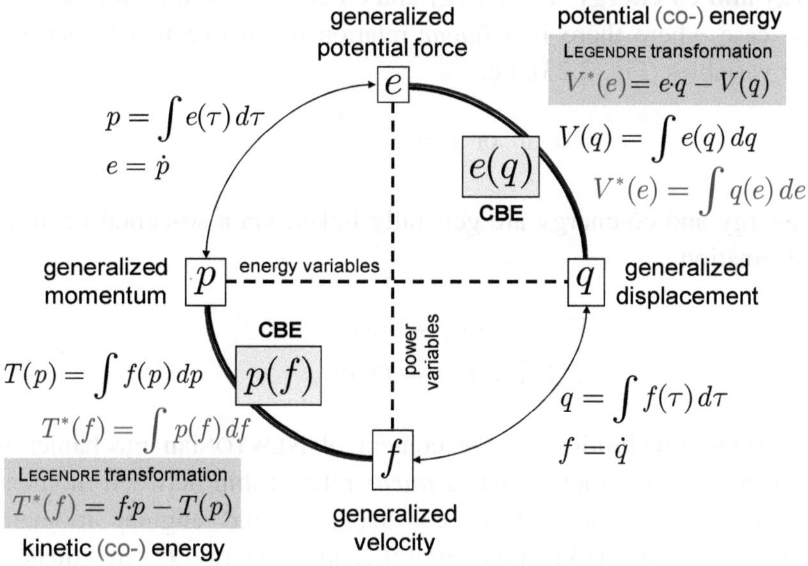

Fig. 2.22. Generalized energy variables: Definitional Relationships I (CBE = constitutive basic equations)

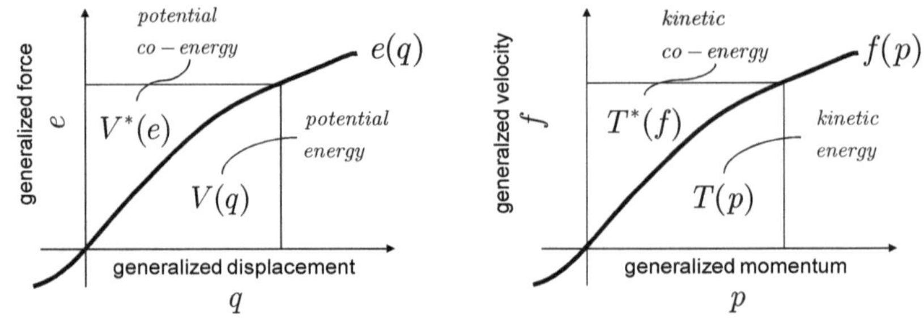

Fig. 2.23. Generalized energy variables: Definitional Relationships II

Conjugate variables The tuples (p, q) and (e, f) are respectively termed *conjugate energy variables* and *conjugate power variables*.

Constitutive equations Among the equations involving the generalized energy variables, the so-called *constitutive equations* $e = e(q)$ and its inverse relation $q = q(e)$, as well as $\dot{q} = \dot{q}(p)$ and $p = p(\dot{q})$, represent the *domain-specific physical laws*, which relate energy variables to one another.

For the majority of applications, the generalized displacement coordinates q and $\dot{q}$ along with the constitutive equations $e = e(q)$ and $p = p(\dot{q})$ offer a convenient starting point for model creation.

Energy and co-energy The energy and co-energy variables are only equal in the case where there is a *linear* relationship between the energy and power variables (Fig. 2.23), i.e.

$$e = \alpha \cdot q \quad \text{or} \quad f = \dot{q} = \frac{1}{\beta} \cdot p \, .$$

The energy and co-energy are generally linked via a so-called LEGENDRE transformation

$$V^*(e) = e \cdot q - V(q),$$
$$T^*(f) = f \cdot p - T(p) \, .$$

NEWTONian mechanics In the context of NEWTONian mechanics with lumped parameters, there exists a *linear* relationship between momentum and velocity, i.e. momentum $p = M(x) \cdot v$, and angular momentum $h = I(\theta) \cdot \omega$. Thus the kinetic energy T and co-energy T^* are equal.

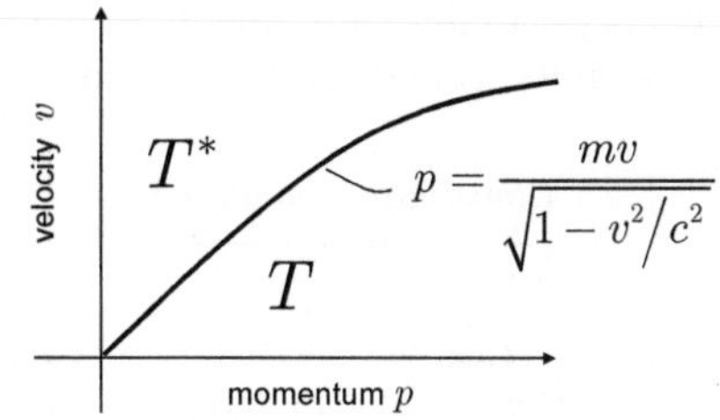

Fig. 2.24. Mechanical kinetic (co-)energy in the presence of relativistic effects

It is only with *relativistic* effects that kinetic energy T and co-energy T^* begin to differ (Fig. 2.24) due to

$$p = \frac{mv}{\sqrt{1 - v^2/c^2}}, \quad \text{c = speed of light.}$$

For the applications considered in this book, however, relativistic effects do not play a role.

There is however one more important aspect to consider. Typically when working in the context of NEWTONian mechanics, the kinetic *co-energy* is employed

$$T^*(\dot{q}) = \int_0^{\dot{q}} p(\dot{q}) \cdot d\dot{q} = \int_0^{\dot{q}} m \cdot \dot{q} \cdot d\dot{q} = \frac{1}{2} m\dot{q}^2 = \frac{1}{2} mv^2 .$$

Due to the equality of T and T^*, this is not an issue in most applications (as long as NEWTONian conditions are maintained!). Care should however be taken when applying the LAGRANGE formalism (see Sec. 2.3), as there the distinction between the energy and co-energy functions must be carefully maintained.

Domain-specific relations

Table 2.2 shows examples of a few selected physical domains with their domain-specific energy variables and the corresponding constitutive equations. Corresponding elementary mechanical and electrical system configurations with linear *energy storage elements* are depicted in Fig. 2.25.

Table 2.2. Power and energy variables for selected physical domains and linear energy storage elements (SI units, generalized coordinates = domain-specific generalized displacements)

Constitutive Equations (spanning Gen. Momentum and Gen. Restoring Force)

System Type	Gen. Coordinate (displacement) q	Gen. Velocity (flow) $\dot{q}$	Gen. Momentum $p(\dot{q})$	Gen. Restoring Force (effort) $e(q)$	Other Gen. Force (excitation) $e(q)$	Kinetic Co-energy $T^*(q,\dot{q})$	Potential Energy $V(q)$
Mechanical Translation	Position [m] x	Velocity [m/s] $\dot{x} = v$	Momentum [Ns] $p = M\dot{x}$	Force [N] $F = Kx$	Force [N] F	$\frac{1}{2}M\dot{x}^2$	$\frac{1}{2}Kx^2$
Mechanical Rotation	Angle [rad] θ	Angular Velocity [rad/s] $\dot{\theta} = \omega$	Angular Momentum [Nms] $h = J\dot{\theta}$	Torque [Nm] $\tau = K\theta$	Torque [Nm] τ	$\frac{1}{2}J\dot{\theta}^2$	$\frac{1}{2}K\theta^2$
Electrical	Charge [C=As] q	Current [A=C/s] $\dot{q} = i$	Flux Linkage [Vs] $\psi = L\dot{q}$	Voltage [V] $u = \frac{1}{C}q$	Voltage Source [V] u	$\frac{1}{2}L\dot{q}^2$	$\frac{1}{2C}q^2$
Hydraulic	Volume [m³] V	Volume Flow Rate [m³/s] Q	Pressure Impulse [Ns/m²] $p_P = I_h Q$	Pressure [N/m²] $P = \frac{1}{C_h}V$	Pressure [N/m²] P	$\frac{1}{2}I_h Q^2$	$\frac{1}{2C_h}V^2$

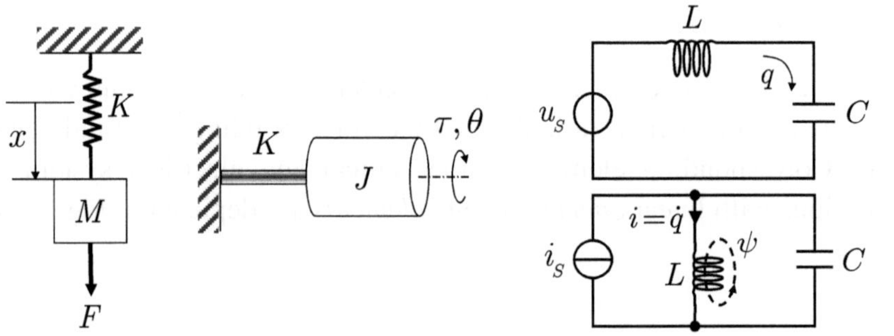

Fig. 2.25. Elementary physical systems (mechanical, electrical)

Table 2.3. Network-based power and energy variables for mechanical systems (SI units)

System Type	Integrated Flow Variable q	Flow Variable f	Integrated Effort Variable $p(f)$	Effort Variable $e(q)$
			Constitutive Equations	
Mechanical Translation	Momentum [Ns] p	Force [N] F	Position [m] $x = \frac{1}{K} F$	Velocity [m/s] $v = \frac{1}{M} p$
Mechanical Rotation	Angular Momentum [Nms] h	Torque [Nm] τ	Angle [rad] $\theta = \frac{1}{K} \tau$	Angular Velocity [rad/s] $\omega = \frac{1}{J} h$

Mechanical power variable in KIRCHHOFF networks Interestingly, in the domain of KIRCHHOFF network models for mechanical systems, slightly different definitions are often used for the flow and effort variables (Lenk et al. 2011), (Reinschke and Schwarz 1976). In this formulation, the conjugate power variables are switched $e \rightleftarrows f$ while the remainder of the definitional relations are retained (see Table 2.3).

Note that under this alternate assignment of variables, the power $P = e \cdot f$ does not change. However, as is easy to verify, the definitional assignment of potential and kinetic (co-)energy does change. For this reason, when applying the energy-based LAGRANGian formulation, it is better to use the power and energy variables defined in Table 2.2 (see Sec. 2.3.2).

2.3.2 Energy-based modeling: LAGRANGE formalism

Background An exceedingly elegant method for domain-independent modeling is offered by the *LAGRANGE formalism* (Schultz and Melsa 1967), (Goldstein et al. 2001), (Wellstead 1979), based on the calculus of variations. Using the generalized energy and power variables (Table 2.2), not only can the equations of motion for *mechanical* systems be derived (as is commonly taught in courses on mechanics), but in a very natural way, the dynamic equations for *heterogeneous physical* systems—in particular for *mechatronic* systems—can also be obtained.

A basic knowledge of the LAGRANGE formalism is assumed here; interested readers are directed to one of the above-cited monographs to refresh or update their knowledge. The remainder of this section describes the most important steps and requirements for modeling, and demonstrates them with an example.

The LAGRANGE formalism is further used in Chapter 5: Generic Mechatronic Transducer as a fundamental basis for deriving a generalized dynamic transducer model.

Generalized coordinates One important modeling decision is the choice of domain-specific (physical) generalized coordinates. Candidates include two of the energy variables: the generalized *displacement* q or the generalized *momentum* p (see Fig. 2.22). Both coordinates are equally valid; note however the differing definitions of the energy functions depending on the choice of variables (energy vs. co-energy).

In the case of *mechanical* systems, the mechanical *displacement* x is commonly chosen as the generalized coordinate, so that the usual kinetic *mechanical co-energy* $T^*(x,\dot{x})$ [8] and potential mechanical energy $V(x)$ are employed, see Table 2.2.

In the case of *electrical* systems, there is a choice of whether to choose the electrical displacement (*charge*) q_{el} or the electrical momentum (*flux linkage*) ψ as the generalized coordinate. A detailed discussion of this aspect of modeling is undertaken in Chapter 5: Generic Mechatronic Transducer, so that this topic is not further pursued here.

For compatibility with mechanical systems, the following discussion considers the generalized coordinate to be the respective *domain-specific displacement coordinate*, in other words, *electric charge* q_{el} for electrical components and *volume* V_{fluid} for hydraulic components. Under these assumptions, it is thus always the kinetic co-energy function and the potential energy function which apply.

8 In certain cases, the kinetic energy depends not only on the generalized velocity, but also on the generalized displacement (position), e.g. in a multi-link manipulator, the mass matrix depends on the joint angles.

Configuration space: redundant coordinates If every energy storage element $\nu = 1,\ldots,N$ is assigned a generalized coordinate $\tilde{q}_\nu$ according to Table 2.2, the result is a vector of generalized coordinates

$$\tilde{\mathbf{q}} = (\tilde{q}_1, \tilde{q}_2, \ldots, \tilde{q}_N)^T \tag{2.3}$$

defining the *configuration space* of the system under consideration. Since these coordinates are generally not independent due to configuration constraints, they are termed *redundant* coordinates.

Holonomic constraints: minimal coordinates In general, energy storage elements are coupled via constraints. These manifest themselves as dependencies between the generalized coordinates, which, in the simplest case, can be described via *algebraic* relations between the coordinates. These are termed *holonomic* constraints[9].

With N_C such *holonomic constraints*

$$h_j\left(\tilde{q}_1, \tilde{q}_2, \ldots, \tilde{q}_N\right) = 0\,, \quad j = 1,\ldots,N_C \tag{2.4}$$

elimination of N_C components of the redundant coordinates is possible (by expressing them in terms of other coordinates) resulting in a set of *independent* coordinates, the *minimal coordinates*

$$\mathbf{q} = (q_1, q_2, \ldots, q_{N_{FG}})^T\,,\ N_{DOF} = N - N_C \tag{2.5}$$

where N_{DOF} is termed the number of *degrees of freedom* of the system.

Nonholonomic constraints If the configuration constraints between redundant generalized coordinates cannot be described via algebraic equations, they are termed *nonholonomic*[10] constraints. Examples include non-integrable differential equations (e.g. a rolling wheel) or inequality constraints.

Differential constraints: PFAFFian form The manipulation of holonomic and nonholonomic constraints is simplified by examining their *differential* form, i.e. the relationship between derivatives of the generalized coordi-

---

9 A more detailed discussion of *holonomic* and *nonholonomic* conditions is included in Chapter 4 (Sec. 4.3.3) and is thus not further pursued here.

10 See Sec. 4.3.3.

nates. This results in first-order differential equations in a particular form—the *PFAFFian form*:

$$\sum_{i=1}^{N} a_{ji}\left(\tilde{\mathbf{q}}\right)\cdot\dot{\tilde{q}}_i = 0,\ j = 1,\ldots,N_C\,. \tag{2.6}$$

If a differential constraint following Eq. (2.6) has been derived from a holonomic constraint (Eq. (2.4)), it then holds that

$$a_{ji} := \frac{\partial h_j\left(\tilde{\mathbf{q}},t\right)}{\partial \tilde{q}_i},\ i = 1,\ldots,N;\ j = 1,\ldots,N_C\,. \tag{2.7}$$

EULER-LAGRANGE equations of the second kind Under the condition that the chosen *minimal coordinates* are not only *independent*, but also unconstrained, component-wise construction of the kinetic co-energy $T^*(\mathbf{q},\dot{\mathbf{q}})$ and the potential energy $V(\mathbf{q})$ enables construction of *EULER-LAGRANGE equations of the second kind:*

$$\frac{d}{dt}\frac{\partial L(\mathbf{q},\dot{\mathbf{q}})}{\partial \dot{q}_i} - \frac{\partial L(\mathbf{q},\dot{\mathbf{q}})}{\partial q_i} = f_i - D_i,\ i = 1,\ldots,N_{DOF} \tag{2.8}$$

with the *LAGRANGian*[11]

$$L(\mathbf{q},\dot{\mathbf{q}}) := T^*(\mathbf{q},\dot{\mathbf{q}}) - V(\mathbf{q})\,. \tag{2.9}$$

The LAGRANGian (2.9) contains, at a consistent, abstract level, all *conservative* dynamic components of the modeled heterogeneous physical (mechatronic) system (see Table 2.2). Domain-specific generalized *forcing* terms f_i (see Table 2.2, column 6) and *dissipative* forces D_i can be accounted for on an equally consistent, abstract level within the EULER-LAGRANGE equations (2.8) (right-hand side) (Schultz and Melsa 1967), (Goldstein et al. 2001).

---

[11] Note: the often used form of the LAGRANGE function "kinetic co-energy minus potential energy" is only valid when displacement coordinates are chosen as the generalized coordinates (this is very practical for modeling of mechanical systems). In general, "LAGRANGE function = total co-energy minus total energy".

The system of equations (2.8) leads to a set of nonlinear second-order differential equations in terms of the generalized minimal coordinates $\mathbf{q} = (q_1, q_2, \ldots, q_{N_{DOF}})^T$. Using the state definition $x_j := q_i$, $x_{j+1} := \dot{q}_i$ and $u_i := f_i$, this can be easily transformed into a tractable *state space model* of the form

$$\dot{\mathbf{x}} = \mathbf{f}\left(\mathbf{x}, \mathbf{u}, t\right). \tag{2.10}$$

EULER-LAGRANGE equations of the first kind In the case of *nonholonomic* constraints between redundant generalized coordinates $\tilde{\mathbf{q}} = (\tilde{q}_1, \tilde{q}_2, \ldots, \tilde{q}_N)^T$ or when *constraint forces* are called upon to maintain system constraints, the EULER-LAGRANGE equations can be extended with LAGRANGE multipliers λ_j to form *EULER-LAGRANGE equations of the first kind*

$$\frac{d}{dt}\frac{\partial L\left(\tilde{\mathbf{q}}, \dot{\tilde{\mathbf{q}}}\right)}{\partial \dot{\tilde{q}}_i} - \frac{\partial L\left(\tilde{\mathbf{q}}, \dot{\tilde{\mathbf{q}}}\right)}{\partial \tilde{q}_i} = f_i - D_i + \sum_{j=1}^{N_C} \lambda_j a_{ji}\left(\tilde{\mathbf{q}}\right), \; i = 1, \ldots, N \tag{2.11}$$

with differential-algebraic *constraints* (holonomic and/or nonholonomic, the PFAFFian form) following Eq. (2.6).

The second-order differential system of equations (2.11) and the first-order differential system of equations (2.6) give a total of $\left(N + N_C\right)$ equations for computation of the $\left(N + N_C\right)$ unknowns $\tilde{q}_1, \ldots, \tilde{q}_N$ and $\lambda_1, \ldots, \lambda_{N_C}$.

Here, the LAGRANGE multipliers $\lambda_j(t)$ represent the *constraint forces* required to maintain the constraints in Eq. (2.6) (Goldstein et al. 2001). To this extent, the EULER-LAGRANGE equations of the first kind also commend themselves in the case of holonomic constraints if the constraint forces are of interest.

In the general case, using the state definition $x_j := \tilde{q}_i$, $x_{j+1} := \dot{\tilde{q}}_i$ or the algebraic variables $z_j := \lambda_j$ results in a system of *differential-algebraic equations* (*DAE*) in the form

$$\dot{\mathbf{x}} = \mathbf{f}\left(\mathbf{x}, \mathbf{z}, \mathbf{u}, t\right) \tag{2.12}$$

$$\mathbf{0} = \mathbf{g}\left(\mathbf{x}, \mathbf{z}, t\right). \tag{2.13}$$

The dimension of the state vector $\mathbf{x}$ can vary from N to $2N$.

Example 2.1 *Electrostatic saddle bearing: LAGRANGE formalism.*

System configuration Gyroscopes are used to measure rotation rates of moving objects (airplanes, satellites, etc.). An electrostatic suspension in a gyroscope enables friction-free rotation of the inertial mass spinning at a constant rate. A full two-sided electrostatic suspension (electrostatic bearing) is treated in detail in Ch. 6 (Sec. 6.6.4, Example 6.2).

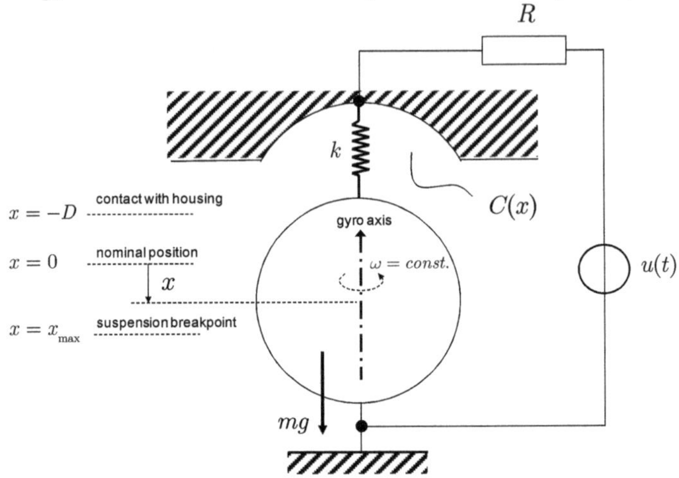

Fig. 2.26. Simplified configuration for an electrostatic saddle bearing (e.g. a test assembly for functional tests of bearings for an electrostatic gyro)

In this example, however, the *equations of motion* of the *elastically suspended* inertial mass m are to be derived for the greatly *simplified model* of an *electrostatic saddle bearing* shown in Fig. 2.26 (from (Schultz and Melsa 1967)) with help of the LAGRANGE formalism. This configuration can be considered to be a test assembly for functional tests of a full electrostatic bearing. Here, instead of the lower electrostatic bearing, an elastic suspension is inserted in order to prevent the sphere from sticking to the upper housing.

Model validity A few remarks regarding restrictions on the model validity: in this example, only the vertical motion of the sphere is of interest. The vertical translational motion is decoupled from the rotational motion of the sphere. The rotational energy of the sphere is a conserved variable (i.e. a constant physical quantity), and does not affect the translational motion. Thus, the sphere rotation can be ignored when creating the model.

Two *nonholonomic* constraints are apparent:

$$x < x_{\max} \text{ : when violated } \rightarrow \text{ detachment of suspension}$$
$$x > -D \text{ : when violated } \rightarrow \text{ contact with housing.}$$

In the remainder of this example, it is assumed that the sphere only moves within its free range of motion. Thus, the nonholonomic constraints do not bind, and all relevant generalized coordinates remain *unconstrained* (an important prerequisite for EULER-LAGRANGE equations of the second kind).

Model creation To start, a generalized coordinate is chosen for each of the three energy storage elements (see Table 2.2):

$$\begin{aligned} C: &\quad q_1 := \text{charge } Q \\ k: &\quad q_2 := \text{displacement } x \\ m: &\quad q_3 := \text{displacement } x \end{aligned}$$

Due to the *holonomic* condition $q_2 = q_3$, there remain only two degrees of freedom with *independent, unconstrained* (in the sense of the above assumption!) generalized coordinates q_1, q_2.

The energy functions obtained for this case are (see also Sec. 5.3.1):

$$V(\mathbf{q}) = \frac{1}{2C(q_2)} q_1^2 + \frac{1}{2} k q_2^2 - m g q_2,$$

$$T^*(\dot{\mathbf{q}}) = \frac{1}{2} m \dot{q}_2^2,$$

giving the LAGRANGian $L(\mathbf{q}, \dot{\mathbf{q}}) := T^*(\dot{\mathbf{q}}) - V(\mathbf{q})$.

An approximation for the computation of the gap-dependent capacitance $C(x)$ is a simplified *plate capacitor* where

$$C(q_2) = \frac{A\varepsilon_0}{D + q_2}$$

A ... capacitor area
ε_0 ... permittivity of free space
D ... see nonholonomic condition.

The only *dissipative* term which requires consideration is the resistive contribution $R\dot{q}_1$ for the electrical coordinate q_1. The generalized *forcing* term is to be found in the voltage source $u(t)$ (in the q_1-direction).

By evaluating the EULER-LAGRANGE equations of the second kind (2.8), the (nonlinear) *equations of motion* are then obtained in *problem coordinates*:

$$\begin{aligned} m\ddot{x} + kx + \frac{Q^2}{2\varepsilon_0 A} &= mg \ , \\ R\dot{q}_C + \frac{D + x}{\varepsilon_0 A} Q &= u \ . \end{aligned} \tag{2.14}$$

Using the *state definition* $x_1 := x,\ x_2 := \dot{x},\ x_3 := Q$, in Eq. (2.14), the *state space model* is then

$$\begin{aligned}
\dot{x}_1 &= x_2 \\
\dot{x}_2 &= -\frac{k}{m}x_1 - \frac{1}{2\varepsilon_0 Am}x_3^{\,2} + g \\
\dot{x}_3 &= -\frac{D}{\varepsilon_0 AR}x_3 - \frac{1}{\varepsilon_0 AR}x_1 x_3 + \frac{1}{R}u\ .
\end{aligned} \tag{2.15}$$

Discussion As is apparent, Eq. (2.15) represents an ordinary state space model without algebraic variables or equations. This is not surprising, as a set of independent generalized coordinates q_1, q_2 without constraints was already present.

One other particularity is worth noting in this example: as is apparent, the EULER-LAGRANGE equations—despite representing a system of second-order differential equations with *two degrees of freedom*—do not (as in the case of purely mechanical systems) result in a state space model with *four* states. Rather, only *three* states appear in Eq. (2.15). This is due to the fact that in the electrical subsystem including the capacitor, only one potential energy storage element is present, so that the electrical coordinate makes no contribution to the total kinetic energy. As a result, the typically-expected second time derivative of the electrical coordinate—which would require an additional state variable—is missing. ■

Ease of use The charm of the LAGRANGian formulation lies in both the universal applicability of its definition of energy functions and in its automatable, formal procedure. The actual creative modeling task consists of establishing generalized coordinates and constraints, and the energy functions. Derivation of the systems of equations in (2.8) and (2.11) can be transferred to a computer algebra program. The LAGRANGE formalism is particularly well-suited to problem statements of sufficiently *low order* with *smooth nonlinearities*. For higher-order systems, however, the analytic derivation of EULER-LAGRANGE equations very quickly becomes intractable.

Multi-domain properties Example 2.1 nicely demonstrates the strengths of the LAGRANGE formalism. The simplicity of combining differing domains at the model level is noteworthy. Starting from the elementary constitutive relations for each energy storage element in the system, and the geometric and algebraic dependencies between coordinates in the problem, the derivation of the EULER-LAGRANGE equations so-to-speak automatically reproduces the multi-domain *interactions* between the electrical and mechanical component systems, i.e., in this case, the *charge-dependent* and *gap-dependent* mechanical *COULOMB force*

$$F_{Coul}(t) = \frac{Q(t)^2}{2\varepsilon_0 A} \tag{2.16}$$

and the *gap-dependent capacitor voltage*

$$u_C(t) = \frac{D + x(t)}{\varepsilon_0 A} Q(t) . \tag{2.17}$$

2.3.3 Energy-based modeling: HAMILTON's equations

Special state definition: canonical momentum Transforming the system of second-order differential EULER-LAGRANGE equations into an equivalent system of first-order differential equations can be effected via a very specialized *state definition*, motivated by analytical mechanics (Reinschke 2006).

Define as the state variable the existing generalized position coordinates q_i as well as a set of new variables, their *conjugate* coordinates p_i:

$$p_i := \frac{\partial L(\mathbf{q}, \dot{\mathbf{q}})}{\partial \dot{q}_i}, \; i = 1, \ldots, N_{DOF} . \tag{2.18}$$

The variables p_i are also termed the *canonical momentum* variables[12]. It is easy to see that, from a physical point of view, the p_i actually correspond to generalized momenta by simply evaluating Eq. (2.18) while accounting for $L = T^* - V$ and relation (2.2) with $f = \dot{q}$.

Due to Eq. (2.8) (and assuming an autonomous system), the canonical momentum variables p_i satisfy the following system of first-order equations

$$\dot{p}_i = \frac{\partial L(\mathbf{q}, \dot{\mathbf{q}})}{\partial q_i}, \; i = 1, \ldots, N_{DOF} .$$

12 In physics, the term “canonical” has the meaning of *natural* or *regular*.

Employing a manipulation with various justifications (Goldstein et al. 2001), (Reinschke 2006), and with help of Eq. (2.18), the scalar *HAMILTONian* $\mathcal{H}(\mathbf{q},\mathbf{p})$ can be introduced and derived from the LAGRANGian via the following LEGENDRE transformation

$$\mathcal{H}(\mathbf{q},\mathbf{p}) := \mathbf{p}^T\dot{\mathbf{q}} - L(\mathbf{q},\dot{\mathbf{q}}). \tag{2.19}$$

With the resulting HAMILTONian and the canonical momentum, the EULER-LAGRANGE equations, after a few intermediate steps, can be written in the following, equivalent form:

$$\dot{q}_i = \frac{\partial\mathcal{H}(\mathbf{q},\mathbf{p})}{\partial p_i} \quad \dot{p}_i = -\frac{\partial\mathcal{H}(\mathbf{q},\mathbf{p})}{\partial q_i},\ i = 1,\dots,N_{DOF}. \tag{2.20}$$

The $2N_{DOF}$ differential equations (2.20) are called *HAMILTON's equations* or also the *canonical equations of motion* (due to their formal simplicity and symmetric structure). The $2N_{DOF}$ state variables q_i, p_i are termed the *canonical variables*. In general, systems which can be described via a function $\mathcal{H}(\mathbf{q},\mathbf{p})$ and a system of differential equations (2.20) are called *HAMILTONian systems*.

In any concrete case, the computation of the HAMILTONian (2.19) presents a certain obstacle. Note that it is a function of $\mathbf{q}$ and $\mathbf{p}$. However, the right-hand side of Eq. (2.19) still contains the generalized velocities $\dot{\mathbf{q}}$. In order to eliminate the velocities, Eq. (2.18) is evaluated, giving $\dot{q}_i = \dot{q}_i(\mathbf{q},\mathbf{p})$.

Special case In the *special case* of scleronomic (time-independent), holonomic constraints, inertial coordinates, and conservative forces, this laborious computation can, however, be avoided, and the fact that in this case (only with the indicated prerequisite conditions!), the HAMILTONian describes the *total energy* of the system under consideration (Goldstein et al. 2001) can be used to advantage, i.e.

$$\mathcal{H} = T^* + V = E_{tot}.$$

Significance The HAMILTON's equations (2.20) are completely equivalent to the LAGRANGian equations. Thus, all properties discussed there hold here as well. The HAMILTONian procedure seems, in the general case, somewhat less tractable (and is also much less taught). However, it offers certain advantages in terms of a geometric interpretation of the motion in the coordinate/momentum state space (also called the *phase space*). The HAMILTONian approach in the elementary form presented here has no noteworthy significance in the modeling of mechatronic systems. However, it forms the basis for the port-HAMILTONian formalism presented in Sec. 2.3.6[13].

2.3.4 Multi-port modeling: KIRCHHOFF networks

Motivation The abstract depiction of physical systems as *networks* permits an additional, very powerful approach to multi-domain modeling for mechatronic systems. Methods and procedures for network modeling and analysis are especially developed in the field of electronics, and have also been made broadly useful in industry with powerful computational tools for highly complex applications in the field of microelectronics. It is noteworthy that mechanical, acoustic, fluidic, and thermal systems can also be neatly abstracted as networks; generalized KIRCHHOFF's Laws suffice for their modeling and analysis. Such networks are termed *KIRCHHOFF networks*. Due to the existence of well-developed methods and tools for electrical networks, it is convenient to describe network models in the other domains as equivalent electrical networks. These are then called *"analogous electrical"* networks (Lenk et al. 2011). By finding these analogs, it is thus possible to depict networks from a variety of domains at a common, abstract level in a unified system model.

Lumped element networks The following discussion examines networks having *lumped elements*. Such spatially and functionally delineated elements can mutually exchange energy with other coupled elements (the *network*) via interfaces (*terminals* or *poles*, with a pair of terminals forming a *port*). Thus network elements are termed two-terminal elements (bipoles), three-terminal (tripoles), four-terminal (quadripoles), or one-port or multi-port (Fig. 2.27). Spatially lumped implies that spatial delay effects (wave propagation) play no role at the element level.

---

[13] In the field of mechanics, however, HAMILTON's equations are of great importance to statistical mechanics and quantum mechanics, see (Goldstein 2001).

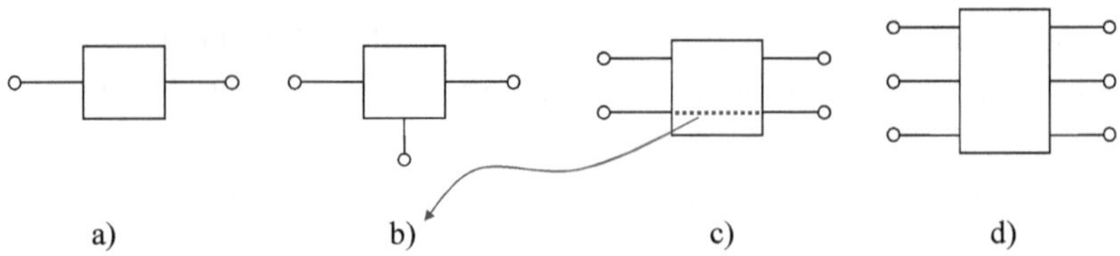

Fig. 2.27. Lumped network elements: a) two-terminal = one-port, b) three-terminal, c) four-terminal = two-port, d) multi-port

Effort and flow variables The behavior of an element can be described using two generic terminal variables (Fig. 2.28):

- *Flow f.* Describes variables which transit through the network element, i.e. flow into and out of the element (e.g. electric current). Such a variable can be measured at any single point (terminal) in the network.
- *Effort e.* Describes variables which are applied between terminals of the network element (e.g. electric voltage). Such a variable can be measured between two different points (terminals) in the network.

In the international literature, the synonymous terms for flow and effort variables shown in Table 2.4 have become commonplace.

Domain-specific flow and effort variables A straightforward assignment of flow and effort variables to concrete physical quantities is possible via simple energy considerations. The resulting *power flow* (= energy flow per unit time) can be unambiguously described for each network element by the product of two system variables—termed *conjugate variables.* If these are chosen to be flow and effort variables, then the domain-specific

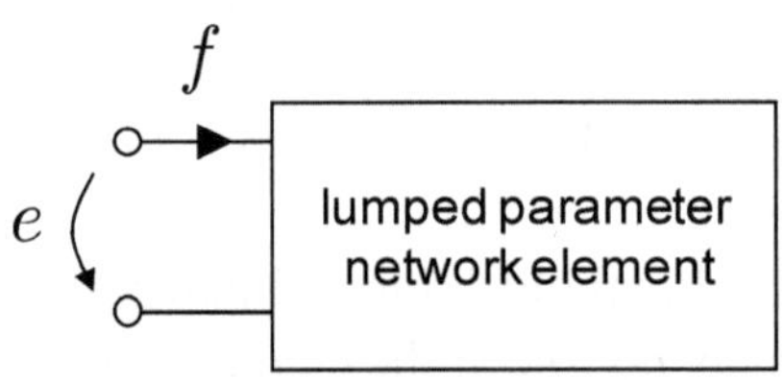

Fig. 2.28. Effort variable e and flow variable f for a network element

Table 2.4. Different terms for flow and effort variables

effort variable "e"	*flow* variable "f"
across variable	through variable
potential	

choice of these variables is, in effect, unrestricted (and thus unfortunately non-unique and somewhat ambiguous), as it is only the *product = effort x flow* which must have the physical units of power.

A few examples of domain-specific power variables can be found in Tables 2.2 and 2.3. Note the different conventions for mechanical systems: *e:=force, f:= velocity* vs. *e:= velocity, f:= force.* In both cases, the product is power, but with quite differing physical interpretations of flow and effort variables. However, this is not the only ambiguity resulting from the usage of conjugate variables, as discussed below.

For reasons of convenience, flow and effort can also be chosen deviating from the power convention presented above. For some mechanical systems, e.g. servo drives, it is the axis position which is of primary interest. In this case, without further consideration, *f:= force* and *e:= position* can be chosen. In this case, the product is energy, i.e. the work expended. Regardless of the specific choice of flow and effort variables, the product should always result in an energy-related quantity.

Circuit laws Combining elements into a network proceeds by considering the following elementary conservation laws (Reinschke and Schwarz 1976), (Fig. 2.29):

- *Junction rule for flow variables* (KIRCHHOFF's node rule)
  The algebraic sum of all flow variables (inflows and outflows) at any arbitrary network node is zero at every point in time.
- *Loop rule for effort variables* (KIRCHHOFF's mesh rule)
  The algebraic sum of all effort variable differences around any arbitrary closed network path is zero at every point in time.

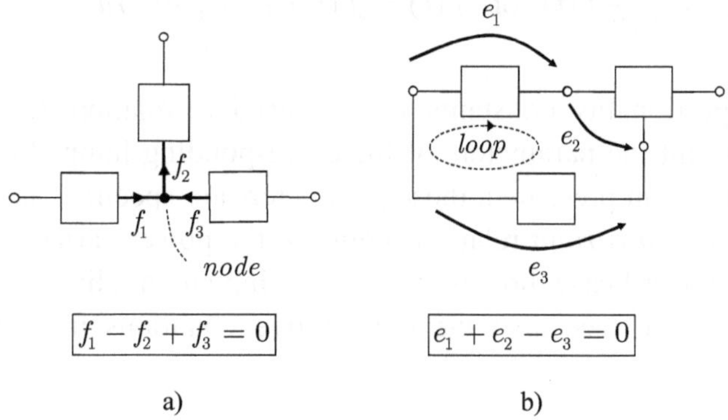

Fig. 2.29. Generalized KIRCHHOFF's laws: a) junction rule, b) loop rule

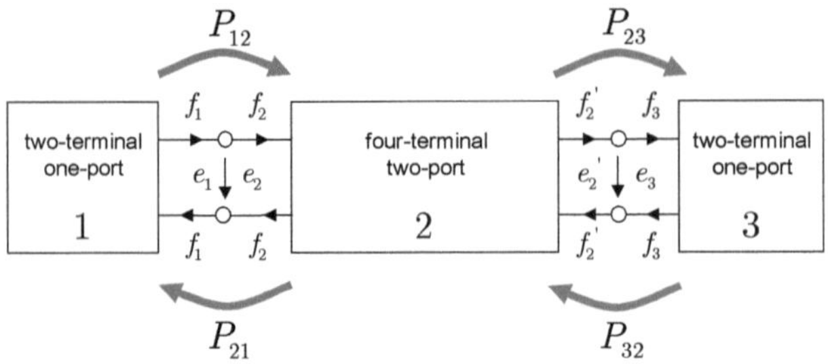

Fig. 2.30. Port-based network

Port-based networks So-called *two-terminal* or *one-port elements* and *four-terminal* or *two-port elements* are the most important network elements. Networks resulting from the interconnection of such elements are called *port-based networks* (Fig. 2.30). When creating networks, it should be ensured that the same types of flow and effort variables connect at circuit nodes.

Elementary two-terminal element: the one-port Characteristic *constitutive* relations between effort and flow can be described using the following elementary two-terminal or one-port network elements:

- *Load*

$$e(t) = \alpha \cdot f(t) \tag{2.21}$$

- *Flow accumulator*

$$f(t) = \beta \frac{d}{dt} e(t) \quad \text{or} \quad e(t) = e(t_0) + \frac{1}{\beta} \int_{t_0}^{t} f(\tau)\, d\tau \tag{2.22}$$

- *Effort accumulator*

$$e(t) = \gamma \frac{d}{dt} f(t) \quad \text{or} \quad f(t) = f(t_0) + \frac{1}{\gamma} \int_{t_0}^{t} e(\tau)\, d\tau\,. \tag{2.23}$$

The proportionality constants α, β, γ used in relations (2.21) through (2.23) represent the parameters of the corresponding lumped network elements. In the form presented, these parameters are specified as constant, so that *linear time-invariant* relations between the power variables *e* and *f* result. In the general case, however, time varying and nonlinear relationships are also possible (see also the constitutive equations in Sec. 2.3.1 and Fig. 2.23).

Table 2.5. Elementary two-terminal (one-port) network elements (general, electric, mechanical)

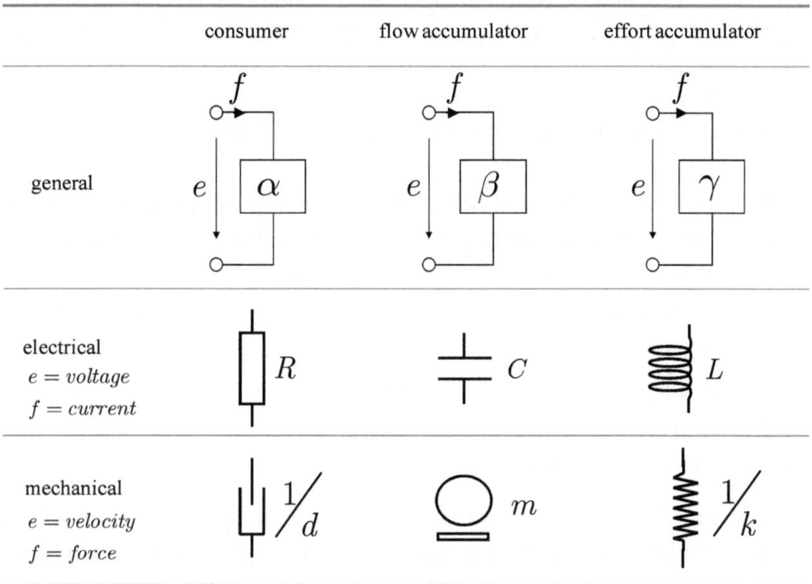

Analogous relations In any particular domain, the parameters α, β, γ are typically assigned individual names and symbols (Tables 2.2, 2.3). To the great chagrin of the engineer, however, this assignment is rather inconsistent, particularly the assignments for *mechanical* systems. In the field of *network theory* (Thomas et al. 2009), (Reinschke and Schwarz 1976), (Lenk et al. 2011), the assignment of effort and flow shown in Table 2.3 has become the norm. The reader can verify that as a result, the assignment of electrical analogs shown in Table 2.5 and Table 2.6 is then appropriate for mechanical network parameters.

Table 2.6. Analogous electrical assignment of mechanical network parameters

Flow *f*	Force or torque F or τ
Effort *e*	Velocity or angular velocity v or ω
Load *R*	1/friction (damping) coefficient $h = 1/b$
Capacitance *C*	Mass or moment of inertia m or J
Inductance *L*	Compliance (= 1/stiffness) $n = 1/k$

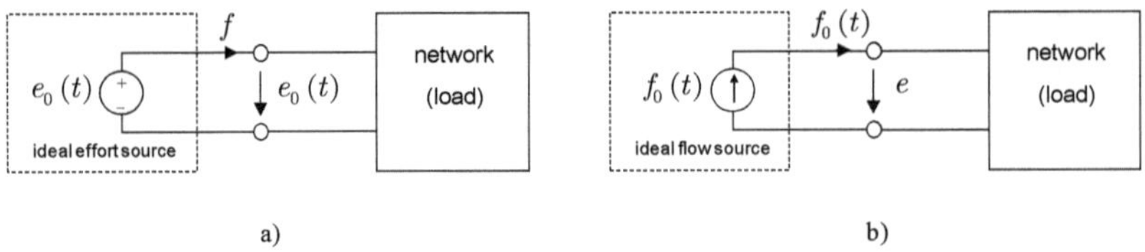

Fig. 2.31. Ideal network sources: a) potential source, b) flow source

Independence of power variables One hallmark of port-based networks is that only one of the two conjugate power variables can be independently controlled at any one time. An important example is seen in ideal (lossless) sources for effort and flow (Fig. 2.31). In the case of a *potential source* (source for effort variables), the output effort $e_0(t)$ can be controlled independently of the applied load, whereas the flow $f(t)$ is load-dependent. The equivalent holds for *flow sources*, where the flow $f_0(t)$ can be controlled independently of the load.

Transducers The domain-specific two-terminal network elements introduced so far can only be connected within a single physical domain, i.e. only the same types of effort and flow variables can be coupled to the network port. In order to model the interaction of variables from different domains—e.g. electrical-mechanical, mechanical-hydraulic, translation-rotation—the *four-terminal* or *two-port* transducer elements shown in Fig. 2.30 (element 2) are required[14].

Transformer vs. gyrator These two transducer elements describe a generalized power flow between different ports. In particular, differing physical domains can be present at the two ports, e.g. $(e_1, f_1) \triangleq$ electrical and $(e_2, f_2) \triangleq$ mechanical. Fig. 2.32 shows two elementary *two-port elements* (*quadripoles*): an ideal *transformer* (*transducer*) and an ideal *gyrator*.

The two-port elements represent a *lossless* transfer of power, i.e. regardless of the particular power variables, $e_1 \cdot f_1 = e_2 \cdot f_2$.

In the case of a *transformer*, the transducer constant *n* determines the relation between pairs of the same power variables (e_1, e_2) or (f_1, f_2) of the two ports. In the case of a *gyrator*, the transducer constant *r* is used to relate pairs of differing power variables (e_1, f_2) or (e_2, f_1) of the two ports.

14 Using so-called *multi-ports*, the simultaneous interactions between more than two domains can be described in an expanded form.

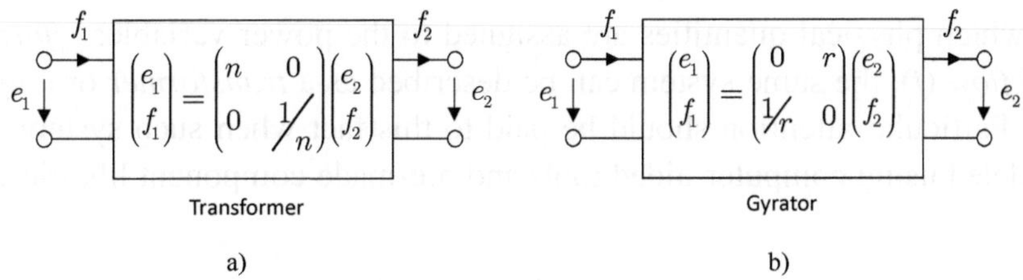

Fig. 2.32. Ideal transducers (lossless): a) transformer, b) gyrator

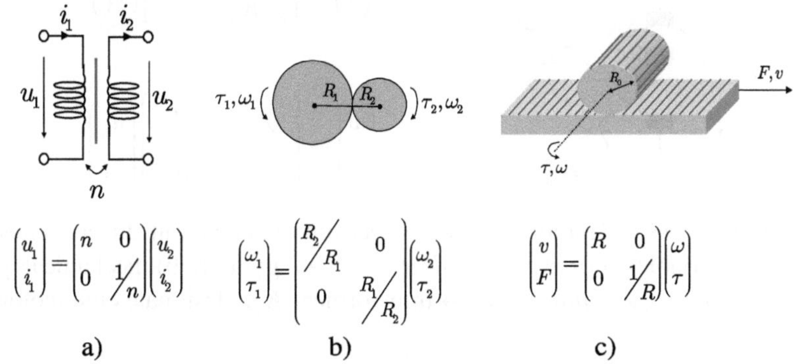

Fig. 2.33. Elementary mechatronic transformer transducers (ideal, lossless): a) electrical transformer, b) mechanical transmission, c) rack and pinion

Fig. 2.34. Elementary mechatronic gyrator transducer (ideal, lossless): hydraulic transducer

The physical units of the transducer constants n, r are determined by the domain-specific power variables. Fig. 2.33 shows physical examples of *transformer* type transducers and Fig. 2.34 shows a physical example of a *gyrator* type transducer.

In the matrix form shown, the transfer matrix represents the so-called chain matrix of two-port network theory.

Mechanical transducers: variable definitions Recall that, due to the arbitrary assignment of domain-specific power variables, the description of physical transducer implementations is not unique. Particular attention is due in the case of *mechanical transducers*. Fig. 2.35 shows a lossless DC motor as an example of an ideal electromechanical transducer. Depending on which physical quantities are assigned to the power variables *effort* (e) and *flow* (f), the same system can be described as a *transformer* or a *gyrator*. Particular attention should be paid to this fact when such systems are modeled using computer-aided tools and pre-made component libraries.

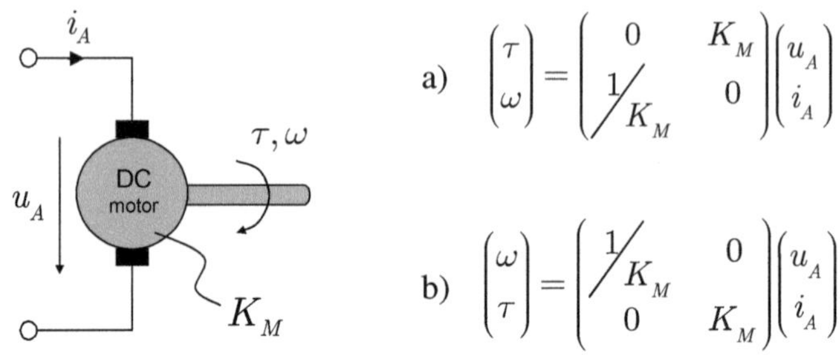

Fig. 2.35. Lossless DC motor as an ideal electromechanical transducer: a) mechanical power variables $effort := \tau$, $flow := \omega$ $\rightarrow$ gyrator, b) mechanical power variables $effort := \omega$, $flow := \tau$ $\rightarrow$ transformer, K_M designates the motor constant

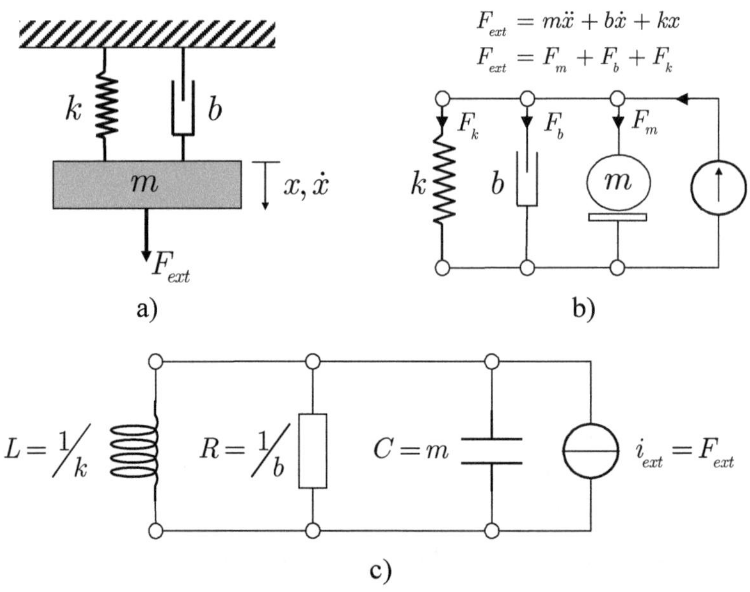

Fig. 2.36. Example of topological rules of construction: a) physical configuration of a mechanical system, b) mechanical network, c) analogous electrical network

Topological rules of construction Starting with the physical topological structure of network elements, an abstract topological network model can be constructed as an undirected graph including standardized network elements (e.g. Table 2.5) using the following elementary rules (Fig. 2.36):

- *Common potential:* Elements with a common potential are connected *in parallel* in the abstract network graph,
- *Common flow:* Elements with a common flow are connected *in series* in the abstract network graph.

Connection rule for inertial masses One pole of the mass symbol (the one with the *bar*) must always be connected to an *inertial system* (corresponding to "ground" in an electrical network).

This is due to the applicability of NEWTON's second law of motion and thus the inertia $m\ddot{x}$ relative to inertial space. Note that in many cases, it suffices to consider a *non-accelerating* reference coordinate system as a *"virtual" inertial* coordinate system, e.g. a coordinate system attached to the Earth and with a constant inertial orientation, or a coordinate system attached to a vehicle if the vehicle is moving with constant velocity (i.e. is not accelerating) (see Sec. 4.3.2, Example 4.1).

> *KIRCHHOFF networks* are defined by
>
> - flow variables f
> - effort variables e
> - a conservation rule for flow (junction rule, node rule)
> - a conservation rule for effort (loop rule, mesh rule)
> - constitutive equations $f = f(e)$ or $e = e(f)$

Example 2.2 *Electrostatic saddle bearing – multi-port model.*

System configuration A multi-port system model is to be derived for the electrostatic saddle bearing already described in Example 2.1 (Fig. 2.26).

From the physical configuration, a locally limited electrical and a mechanical network can be directly extracted. These two networks are depicted as (for the time being) separate networks in Fig. 2.37.

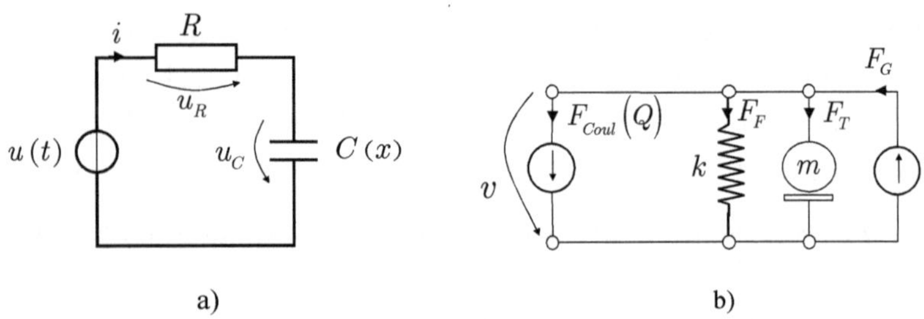

Fig. 2.37. Electrostatic saddle bearing: separate domain-specific networks: a) electrical network, b) mechanical network

Mathematical model: DAE system For each of these networks, the elementary circuit rules can now be applied, and the constitutive equations for the network elements can be constructed.

(a) Electrical network
Loop rule for effort variables (electrical voltages, KIRCHHOFF's mesh rule)

$$\sum_i u_i = 0: \quad u = u_R + u_C \tag{2.24}$$

Constitutive equations

$$\begin{aligned} i &= \dot{Q} \\ u_R &= R \cdot \dot{Q} \\ u_C &= \frac{1}{C(x)} Q = \frac{D + x}{\varepsilon_0 A} Q \end{aligned} \tag{2.25}$$

(b) Mechanical network
Junction rule for flow variables (forces, NEWTON's third law)

$$\sum_i F_i = 0: \quad F_T = F_G - F_F - F_{Coul} \tag{2.26}$$

Constitutive equations

$$\begin{aligned} F_T &= m\ddot{x} \\ F_G &= mg \quad F_{Coul}(Q) = \frac{1}{2\varepsilon_0 A} Q^2 \\ F_F &= kx \end{aligned} \tag{2.27}$$

The system of equations (2.24) through (2.27) fully describes the system behavior already modeled in Example 2.1, though here it is in the form of a general DAE system integrating all involved physical domains (a domain-independent system model). Via suitable manipulation, this DAE system can, without much difficulty, be transformed into the state space form (2.15) (this is left as an exercise to the reader).

Thus, network-based modeling results in equations equivalent to those produced following the LAGRANGE formalism (which should not surprise the astute reader).

Nonlinear network representation (multi-domain) Network-based modeling offers additional advantages which have so far not been called upon in this exposition. Though the DAE system (2.24) through (2.27) represents a complete computational specification of the system behavior, it obscures topological relationships between the networks. The electromechanical coupling between the two physical domains occurs via the charge-dependent COULOMB force and the displacement-dependent capacitance. This coupling, which is only indirectly apparent in the DAE system and in Fig. 2.37, can, however, be made more visible in an extended network representation, where a *two-port element* (transducer) is introduced as a *coupling element* between the two networks (Fig. 2.38). Due to the nonlinear characteristics of the coupling, however, none of the simple transducer elements introduced above can be employed.

On the mechanical side, the configuration can also be described using domain-specific mechanical network elements. The resulting network is then directly coupled, clearly exposing the physical system topology and the coupling mechanism, but possesses a ***heterogeneous*** structure as relates to the *physical domains*. It is self-evident that applying the junction rule or mesh rule will lead to the same DAE equations.

Linear network representation The strengths of a network representation can be particularly well exploited in linear networks, e.g. the two-port representation, and graphical and analytical network analysis (Reinschke and Schwarz 1976). A linear description is possible in the present case, for example, if *small deviations* about a resting position of the system are considered. The linearization of system models is considered more closely in Sec. 2.6, so that only the results of this process are presented here.

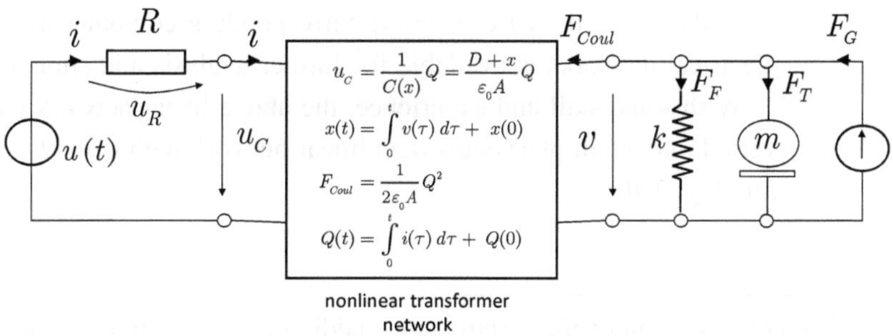

Fig. 2.38. Electrostatic saddle bearing: coupled domain-specific networks (nonlinear)

Resting position (compensation of the weight of the sphere with a charge Q^0 or equivalently a voltage u^0):

$$x^0 = \dot{x}^0 = 0\,,\; Q^0 = \sqrt{2\varepsilon_0 A m g}\,,\; u^0 = D\sqrt{\frac{2mg}{\varepsilon_0 A}}\,. \tag{2.28}$$

Linearized constitutive equations of the transducer ($\Delta i, \Delta u_C, \Delta F, \Delta v$ are displacements about the resting position):

$$\begin{aligned} \Delta F &= \sqrt{\frac{2mg}{\varepsilon_0 A}}\Delta Q \\ \Delta u_C &= \sqrt{\frac{2mg}{\varepsilon_0 A}}\Delta x + \frac{1}{C_0}\Delta Q \end{aligned} \quad \text{where } C_0 := \frac{\varepsilon_0 A}{D}\,. \tag{2.29}$$

Linear analogous electrical network For the network representation (Fig. 2.38), consistent network elements are now to be used—for example, the *electrical* elements from Table 2.5. For the mechanical elements, the corresponding analogs in Table 2.6 are to be used. An appropriate representation of the transducer network is somewhat trickier. Indeed, the constitutive relations (2.29) do not relate $\Delta F \leftrightarrow \Delta i$ as is needed, but rather $\Delta F \leftrightarrow \Delta Q = \int \Delta i \, d\tau$. This relationship can be represented in a symbolic, elegant manner via the LAPLACE transform of the integral. This results in an algebraic relationship $\Delta F \leftrightarrow \left(1/s\right)\Delta i$ and further, using a "complex" transducer constant N/s, a *transformer* transducer network (see Fig. 2.39).

The resulting network now exhibits a *homogeneous* structure in the *physical domains*. It can either be interpreted as a *generalized* network with abstract generalized components, or as an *analogous electrical* network with *electrical* network elements. The latter interpretation is particularly apt in the case where a corresponding computer-aided tool for electrical networks is available for further analysis and simulation tasks.

With some skill and experience, the above linear network can be converted into a (simpler) equivalent linear network with a real *lossless gyrator* (Fig. 2.40)[15].

[15] The configuration of the electrostatic saddle bearing examined here is equivalent to an *electrostatic plate transducer*, as is discussed in depth in, e.g., (Lenk et al. 2011) or (Senturia 2001). Further discussion of the system behavior (negative k_C !) follows in Ch. 6.

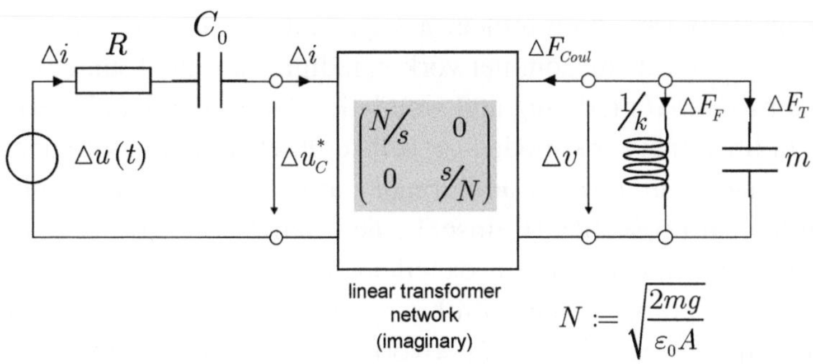

Fig. 2.39. Electrostatic saddle bearing: analogous electrical linear network (small deviations about equilibrium position); Variant 1 with imaginary *transformer network*

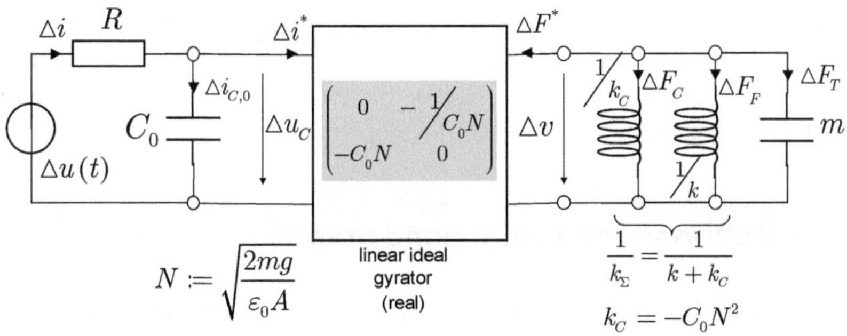

Fig. 2.40. Electrostatic saddle bearing: analogous electrical linear network (small deviations about equilibrium position); Variant 2 with real *gyrator network*

■

Ease of use and multi-domain properties The example discussed above very nicely demonstrates the strengths and weaknesses of network-based modeling. From the network topology, elementary circuit rules (junction rule, mesh rule), and constitutive equations with (lumped) network parameters, a mathematical model in the form of a (high-dimensional) DAE system can be systematically constructed. Via appropriate (automatable) manipulation, this can then be brought into a more concise form, e.g. a state-space representation (though in some cases, due to certain restrictions, this is only possible to a limited extent, see Sec. 2.4).

Among the advantages is certainly also the structure of the network model, which replicates the physical topology. This presents an excellent opportunity to *modularize* physical models by partitioning them into component networks, with a clear assignment of components to physical subsystems. For this reason, network paradigms are the foundation for modern *object-oriented* modeling and simulation tools (see Sec. 2.3.9).

For linear network models, a well-researched methodological mechanism for *network analysis* is additionally available (e.g. while also exploiting topological properties (Reinschke and Schwarz 1976)). For electrical networks in particular, there is excellent support in the form of available tools, with which complex mechatronic systems can also be very efficiently modeled as *analogous electrical* networks in a domain-independent manner.

It must however be counted as a disadvantage that it is not always completely trivial to employ domain-independent analogs in a proper manner. This is made particularly difficult by the inconsistent physical assignment of conjugate power variables (especially for mechanical elements!). In addition, it is occasionally tricky—and thus creates a barrier for access to the "network world" for non-electrical engineers—to recast components into standard structures (see Fig. 2.40), for example to be able to use component libraries for computer-aided tools.

2.3.5 Multi-port modeling: bond graphs

Power bonds One particular network-oriented modeling approach is offered by so-called *bond graphs*. This approach has gained popularity lately—in particular in the field of mechatronics—though often its close relationship to the significantly older network modeling approach is obscured. Bond graphs employ a more concise, graph-oriented representation of network elements and their links via power variables (*power bonds*, see Fig. 2.41, left).

Graph-oriented model In place of two directed edges (arrows) for flow and effort, only a single weighted, directed edge is employed, on which symbolic identification of the flow and effort and their defined direction of effect are indicated in a concise form. The network parameters appear as weighted end nodes in the bond graph. Further, specialized nodes (*junctions*) are defined. Via a suitable assignment of definitions, a 1:1 depiction of heterogeneous networks as bond graphs on a homogeneous domain is more or less straightforwardly possible.

In order to give an impression of this type of modeling, Fig. 2.41 shows the bond graph model (Geitner 2008) equivalent to the network model in Fig. 2.38 of the electrostatic saddle bearing of Fig. 2.26.

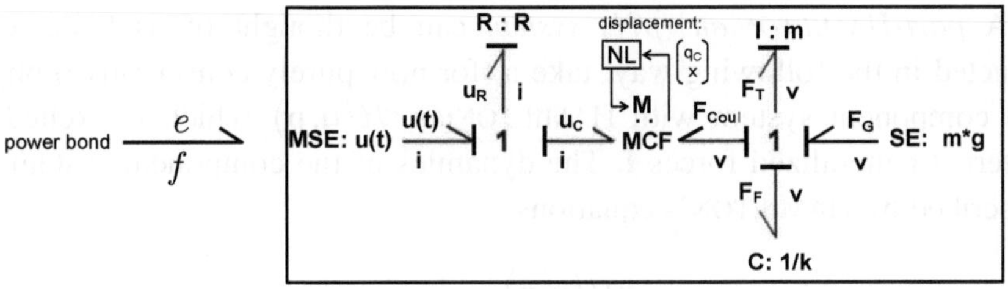

Fig. 2.41. Bond graph model of the electrostatic saddle bearing (Fig. 2.26), equivalent model to the network model in Fig. 2.38, from (Geitner 2008)

Ease of use Depending on the degree of familiarity with the various modeling formulations, one or the other of these representations may seem clearer to the user. Since modeling with bond graphs does not offer anything methodologically new as compared to network modeling, this approach will not be further discussed here. The interested reader is directed to the copious literature on this subject, e.g. (Paynter 1961), (Cellier 1991), (Damic and Montgomery 2003), (Karnopp et al. 2006).

Notation for power variables One remark relevant to applications will however not be dispensed with, in case the reader wishes to adapt *mechanical bond graph models* from the literature for her own purposes. In contrast to the network world, in the bond graph community, it is customary to use the physical assignments for power variables listed in Table 2.2, i.e. *effort:= force/torque*, *flow:= velocity/angular velocity* (though here too, the exception unfortunately proves the rule!). As a consequence, the assignment of network parameters also changes. Thus, when working with bond graph models, careful attention should be paid to the assignment of domain-specific quantities.

2.3.6 Energy / multi-port modeling: port-HAMILTONian systems

Foundations The advantages of universal scalar energy functions and the excellent modularizability of KIRCHHOFF networks are combined in the still rather young modeling approach of *port-HAMILTONian systems* (Maschke and van der Schaft 1992), (Cervera et al. 2007), (Duindam et al. 2009).

A *port-HAMILTONian (p$\mathcal{H}$) system* can be thought of as being constructed in the following way: take a (for now purely conservative) physical component system with HAMILTONian $\mathcal{H}(\mathbf{q},\mathbf{p})$ which is excited by external generalized forces **f**. The dynamics of the component system are described by HAMILTON's equations

$$\begin{aligned} \dot{\mathbf{q}} &= \frac{\partial \mathcal{H}(\mathbf{q},\mathbf{p})}{\partial \mathbf{p}}, \\ \dot{\mathbf{p}} &= -\frac{\partial \mathcal{H}(\mathbf{q},\mathbf{p})}{\partial \mathbf{q}} + \mathbf{f}. \end{aligned} \tag{2.30}$$

The power used can be described as the product of conjugated power variables, i.e. in a concrete case, by the scalar product

$$P = \dot{\mathbf{q}}^T \mathbf{f}\,.$$

Power port If Eq. (2.30) is interpreted as a network element with power variables $\mathbf{u} = \mathbf{f}$ and $\mathbf{y} = \dot{\mathbf{q}}$ at its terminals[16], then the system effects on the outside (at the *interface*) are described by the *vector* terminal variables $\langle \mathbf{u}, \mathbf{y} \rangle$ and its internal structure (*functionality*) is described by the *scalar* HAMILTONian $\mathcal{H}(\mathbf{q},\mathbf{p})$ (Fig. 2.42).

It is easy to verify that the following energy conservation relation holds:

$$\frac{d}{dt}\mathcal{H} = \frac{\partial^T \mathcal{H}}{\partial \mathbf{q}}\dot{\mathbf{q}} + \frac{\partial^T \mathcal{H}}{\partial \mathbf{p}}\dot{\mathbf{p}} = \frac{\partial^T \mathcal{H}}{\partial \mathbf{p}}\mathbf{u} = \dot{\mathbf{q}}^T \mathbf{u} = \mathbf{y}^T \mathbf{u}\,,$$

i.e. the increase in energy of the system equals the power applied to the port $\langle \mathbf{u}, \mathbf{y} \rangle$.

In Fig. 2.42, a concise, alternative representation of a generalized network element with a *power port* is recognizable. This can be linked in the normal manner with other components according to the circuit rules of KIRCHHOFF networks. It can be shown that connecting p$\mathcal{H}$ systems results in a p$\mathcal{H}$ system (van der Schaft and Maschke 1995).

[16] The signals **u** and **y** are termed *collocated* in- and outputs of the p$\mathcal{H}$ system.

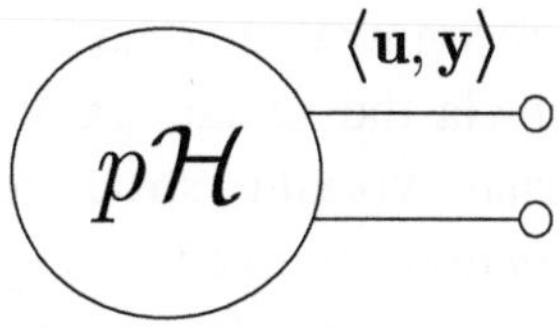

Fig. 2.42. Port-HAMILTONian system with power port

Generalized p$\mathcal{H}$ system HAMILTON's equations (2.30) can be represented in the following way:

$$\begin{aligned} \dot{\mathbf{x}} &= \mathbf{J}\frac{\partial \mathcal{H}(\mathbf{x})}{\partial \mathbf{x}} + \mathbf{G}\mathbf{u} \\ \mathbf{y} &= \mathbf{G}^T \frac{\partial \mathcal{H}(\mathbf{x})}{\partial \mathbf{x}} \end{aligned} \quad \text{where} \quad \begin{aligned} \mathbf{x} &= \begin{pmatrix} \mathbf{q} & \mathbf{p} \end{pmatrix}^T \\ \mathbf{J} &= \begin{pmatrix} \mathbf{0} & \mathbf{E} \\ \text{-}\mathbf{E} & \mathbf{0} \end{pmatrix}, \quad \mathbf{G} = \begin{pmatrix} \mathbf{0} \\ \mathbf{E} \end{pmatrix}, \quad \mathbf{J} = -\mathbf{J}^T, \end{aligned} \tag{2.31}$$

and the skew-symmetric matrix **J** describes the *internal structure* of the system (in the sense of a generalized geometric structure). This matrix is also called the POISSON structure matrix, and the system thus described (2.31) is thus said to have POISSON structure (van der Schaft and Maschke 1995).

The mathematical description introduced in Eq. (2.31) can be further generalized and extended with the introduction of dissipative elements, giving a *generalized port-HAMILTONian system* with dissipation:

$$p\mathcal{H} := \left\{ \begin{aligned} \dot{\mathbf{x}} &= [\mathbf{J}(\mathbf{x}) - \mathbf{R}(\mathbf{x})]\frac{\partial \mathcal{H}(\mathbf{x})}{\partial \mathbf{x}} + \mathbf{G}(\mathbf{x})\mathbf{u} \\ \mathbf{y} &= \mathbf{G}^T(\mathbf{x})\frac{\partial \mathcal{H}(\mathbf{x})}{\partial \mathbf{x}} \end{aligned} \right. \tag{2.32}$$

$$\text{where} \quad \begin{aligned} \mathbf{J}(\mathbf{x}) &= -\mathbf{J}^T(\mathbf{x}) \\ \mathbf{R}(\mathbf{x}) &\geq 0 \end{aligned},$$

i.e. the structure matrix **J** is skew-symmetric and the dissipation matrix **R** is symmetric positive-definite.

Significance From this intentionally short presentation of the port-HAMILTONian formulation it should have become clear where its strengths lie with respect to model creation. Via the power ports, the p$\mathcal{H}$ formulation supports modular modeling at the physical level and permits the direct linking of physical structures (equivalent to KIRCHHOFF networks and bond graphs). An additional recognizable advantage is the concise description of internal functionality via the scalar HAMILTONian, as well as the clear depiction of the internal working structure of the system in the POISSON structure and dissipation matrices. The greatest challenge, however, consists of the construction a suitable HAMILTONian.

As was mentioned in the beginning, the port-HAMILTONian formulation is still relatively young and is the subject of current research. In this context, it should be mentioned that the significance of the p$\mathcal{H}$ formulation appreciably exceeds the narrower task of modeling. For example, the formal description of a system as a POISSON structure shown in Eq. (2.32) presents valuable properties which directly support the construction of nonlinear control algorithms and stability proofs (Ortega et al. 2002), (Kugi and Schlacher 2002), (Kugi and Schlacher 2001), (Fuchshumer et al. 2003). One further focus of current research concerns the extension to infinite-dimensional system descriptions (via partial differential equations) (van der Schaft and Maschke 2002).

2.3.7 Signal-coupled networks

One important set of questions to be asked in the context of model creation deals with the modularization of dynamic models. This is particularly of interest in the case when the models are to be used libraries as part of computer-aided simulation tools.

The following questions are appropriate in this context:

- When and under what conditions can models of physical (component) systems be directly connected to each other?
- Why are signal-oriented models of physical systems only conditionally modularizable?

Both questions are fundamentally connected to the problem of *back-effects* between systems.

Power flow The physical coupling between system components always occurs via a power flow (Fig. 2.43). This means that for system models, at every interface, the corresponding conjugate power variables effort e and flow f should be examined: it is already known that the transferred energy is equal to the product of the power variables. In the general case, inter-

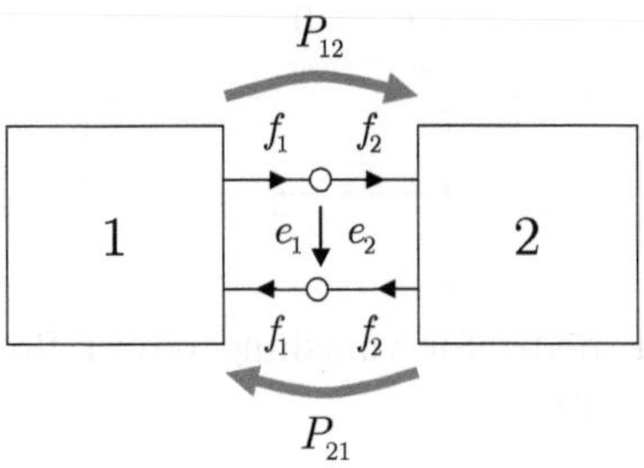

Fig. 2.43. Coupling of physical system models

faces between system models must thus always contain a *pair* of *conjugate power variables* (e, f). One additional condition which must be met is that *both* interface variables must be compatible with respect to the *physical domains* considered and the *physical units* used, i.e. $[e_1] = [e_2]$ and $[f_1] = [f_2]$ in Fig. 2.43. Thus, for example, only electrical power variables from the same physical domain (volt, ampere) may be connected to one another; the direct coupling of electrical and mechanical interface variables is, in contrast, trivially impossible.

Power back-effect Why must both power variables always be taken into account?

In a one-port element, at most one of the two power variables can be freely controlled, the other power variable is determined by the coupling. If, for example, in the connected system in Fig. 2.43, the effort e_1 can be arbitrarily set via an ideal potential source, the arising flow f_1—and as a consequence, due to the coupling conditions, also the flow f_2—depends on the structure of System 2, which serves as the load for System 1. From the perspective of the power flow P_{12}, in this sense there is a *back-effect* from System 2 to System 1. The flow f_1 can thus only be computed if the inner structure or the behavior at the terminals of System 2 is known.

> System coupling via *power flows* is always subject to power *back-effects.*

Signal coupling: absence of back-effects How, then, do models behave in which no appreciable power flow occurs?

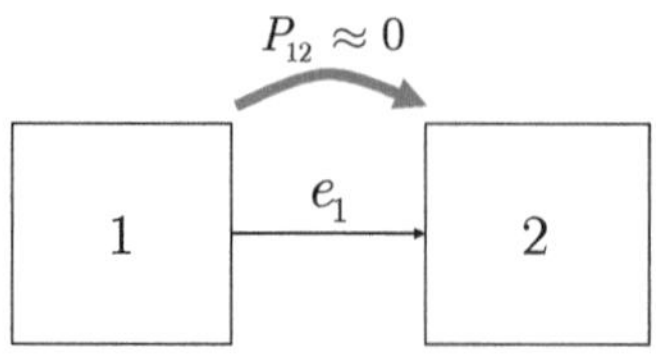

Fig. 2.44. Signal-coupled model for vanishing power flow between Systems 1 and 2 (back-effect-free linking)

This is the case, for example, when one of the two power variables possesses a very small value. In a system configuration as shown in Fig. 2.43, given a very large input resistance in System 2, only vanishingly small flow f_1 (or f_2) appears. The transferred power $P_{12} = e_1 \cdot f_1$ is thus approximately zero. Thus, the interface between Systems 1 and 2 becomes dependent on *only one single* interface variable, in this case the effort e_1. The particular characteristics of System 2 have no influence whatsoever on the effort e_1. In this context, the relationship between the two Systems 1 and 2 is called *back-effect-free* and the connection is termed *signal-coupled* using a single interface variable (the *signal*).

Definition 2.6. *Signal coupling* – A signal flow between two physical systems implies an information flow with negligible power flow, i.e. one of the two conjugate power variables at the system interface is approximately zero. The non-zero power variable is termed the connecting *signal* between the two systems. Under these conditions, the interface between two systems is described solely by the connecting signal, the two systems are said to be *signal-coupled.*

Signal-coupled model The type of back-effect-free action relationship between Systems 1 and 2 in Fig. 2.43 is commonly depicted in the form of a *signal-coupled model* (Fig. 2.44). Well-know examples of signal-coupled models are, e.g., *control system block diagrams.*

> In signal-coupled networks (in general), all *flows equal zero*; the *efforts* (measured with respect to a common reference) serve as *signal quantities* and alone describe the flow of action.
>
> In the case of system connection via *signals*, *back-effect-free* interaction of the components is always presumed.

Example 2.3 *Back-effect in an RC network.*

System configuration Fig. 2.45 shows an RC network whose input is excited via an ideal potential source u_Q and whose output is connected to a one-port with impedance Z_L as a load.

Model creation The mathematical model of the RC network is formally given in Eqs. (2.24), (2.25) (assuming a constant capacitance C). Given the constitutive load relation (LAPLACE-transformed variables)

$$U_L(s) = Z_L \cdot I_L(s),$$

it follows for the input interface variables that[17]

$$U_S(s) = \left(1 + \frac{R}{Z_L} + RCs\right) U_L(s),$$

$$I_S(s) = \left(\frac{1}{Z_L} + Cs\right) U_L(s),$$

and for the *load-dependent* input impedance

$$Z_S(s) = \frac{U_S(s)}{I_S(s)} = \frac{1 + \dfrac{R}{Z_L} + RCs}{\dfrac{1}{Z_L} + Cs}. \tag{2.33}$$

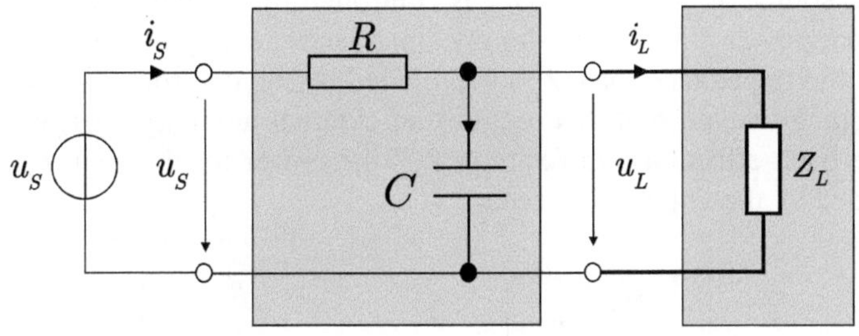

Fig. 2.45. Loaded RC network

[17] A model with equivalent computational structure taking the form of *linear differential equations* can be obtained in the well-known manner using the inverse LAPLACE transform.

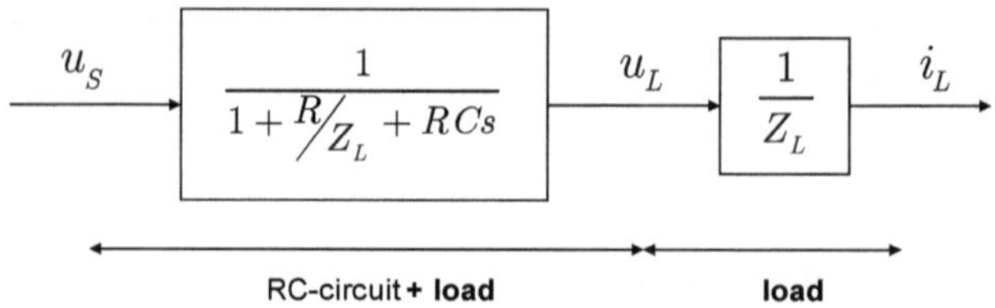

Fig. 2.46. RC network with back-effect as a signal-oriented block diagram obscuring the physical system topology

Signal-coupled model Even in the signal-oriented block-diagram representation (using transfer functions), representing the dependence on the load in the system model is ungainly. It is certainly possible to represent the impedance load in its own block. However, the model of the RC network itself also depends on the impedance load Z_L. In this case, then, using a *signal-oriented* graphical model (Fig. 2.46) hides the original physical structure, and thus a modular signal-coupled model at the level of physical components is not possible.

■

Example 2.4 *Back-effect-free connection for an RC network.*

System configuration Fig. 2.47 depicts the same RC network as in Example 2.3, though here, the impedance load is connected via a so-called *isolation amplifier*. The isolation amplifier possesses a high-resistance input stage (input current $i_{in} \approx 0$). At the output, a low-resistance voltage source $u_{out}(u_{in}) = V \cdot u_{in}$ is controlled by the input voltage u_{in}. The input power draw is vanishingly small, and the output power (determined by the impedance load Z_L) is provided by the controlled voltage source. Note, however, that this requires an external auxiliary energy source. In the back-effect case in Example 2.3, the power for the load Z_L was provided by the voltage source u_S.

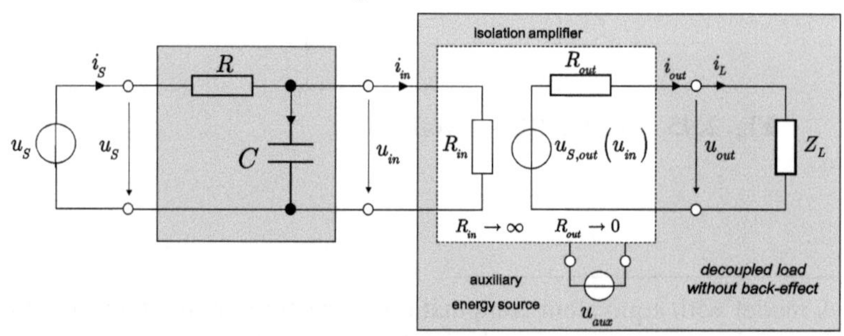

Fig. 2.47. Loaded RC network with isolation amplifier

Model creation The mathematical model of the network is once again formally given in Eqs. (2.24), (2.25), and, since $i_{in} \approx 0$, it holds that

$$\begin{pmatrix} U_S(s) \\ I_S(s) \end{pmatrix} = \begin{pmatrix} 1+RCs & R \\ Cs & 1 \end{pmatrix} \begin{pmatrix} U_{in}(s) \\ 0 \end{pmatrix}.$$

Noting that $u_{out}(u_{in}) = V \cdot u_{in}$, it follows that

$$\begin{aligned} U_{out}(s) &= \frac{V}{1+RCs} U_S(s), \\ I_S(s) &= \frac{Cs}{1+RCs} U_S(s). \end{aligned} \tag{2.34}$$

In this case, the impedance of the networks at the exciting input is independent of the load (compare Eqs. (2.33), (2.34)). The voltage u_{out} at the load can be controlled without back-effect (independent of the actual load) by the voltage source $u_{out}\left(u_{in}\right)$.

Signal-coupled model This state of affairs can also be discerned from the signal-oriented block diagram (transfer function, Fig. 2.48). In addition, the block diagram allows the separation of the RC network and the load via the isolation amplifier to be distinguished, reflecting the physical topology. In this case, then, the original physical structure of the system can be easily reproduced in the signal-oriented graphical model. It is only under the conditions given here that *signal-oriented* modeling at the level of physical components is possible.

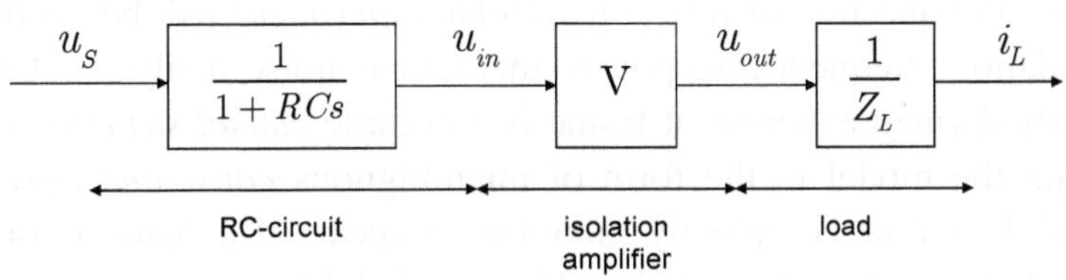

Fig. 2.48. Back-effect-free RC network as a signal-oriented block diagram with clear replication of the physical system topology

■

Further examples of back-effect-free system interfaces—and thus signal-oriented models—are *measurement amplifiers*, *power stages* for actuators, *information processing* systems, and *analog* and *digital signal processing* elements. The latter can be used with familiar block diagrams from control theory (transfer functions) in the accustomed manner.

2.3.8 Model causality

Causal vs. acausal models Within the context of modeling of physical systems, the terms *causal* and *acausal model* often appear, usually in connection with certain modeling paradigms or simulation tools. This section attempts to place these imprecise and sometime incorrectly used designations into a consistent framework, and to explain their meanings.

Causality in system theory In the context of the design of physical systems, the term *causality* should be used in accord with the established meaning in system theory.

Definition 2.7. *Causality:* A system with input $u(t)$ and output $y(t)$ [18] is called *causal* if at any time t_1, the output $y(t_1)$ is only affected by the evolution of the input $u(t)$ up to time t_1 (Fig. 2.49). A system is called *acausal* if it does not meet the above condition.

At its core, this definition states that in a realizable, physical system, no future inputs can affect the current output. In this sense, naturally all mechatronic systems are *causal by definition*, and the same should consequently also be expected of their dynamic models[19].

Causal structure Given the above definition, it is quite clear that system dynamics must exhibit a *cause-and-effect* behavior (input-output, excitation-response). In an unconnected multi-port model (with open terminals), any particular system variable can be defined as a cause or an effect by asking the question of how a particular *experiment* will be conducted (e.g. simulation, frequency response, theoretical noise analysis). Only with a clearly defined experiment frame is a definite *causal structure* established within the model in the form of unambiguous *cause-and-effect relationships*. It is thus completely incorrect to speak of a "causal" or "acausal" model given an unconnected multi-port model[20].

---

[18] The limited formulation here of single-input single-output (SISO) systems can easily be extended to systems with an arbitrary number of inputs and outputs.

[19] One exception, however, is so-called *prediction models,* which are *acausal* in the above sense.

[20] Models referred to as "acausal" (e.g. networks, bond graphs, object-oriented models) are often pointed to as having the most beneficial of properties. However, such a paradoxical labeling unnecessarily obscures the actual strengths of such models.

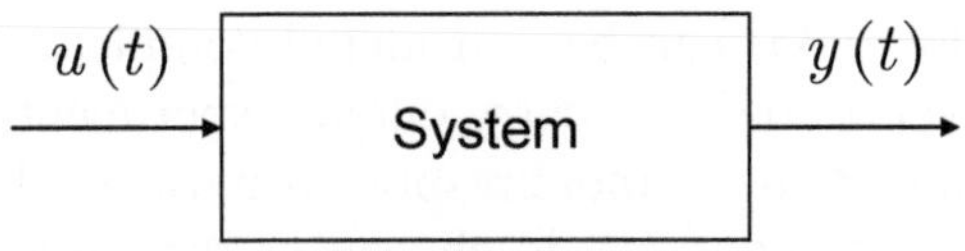

Fig. 2.49. Causal structure (cause → effect) of a system

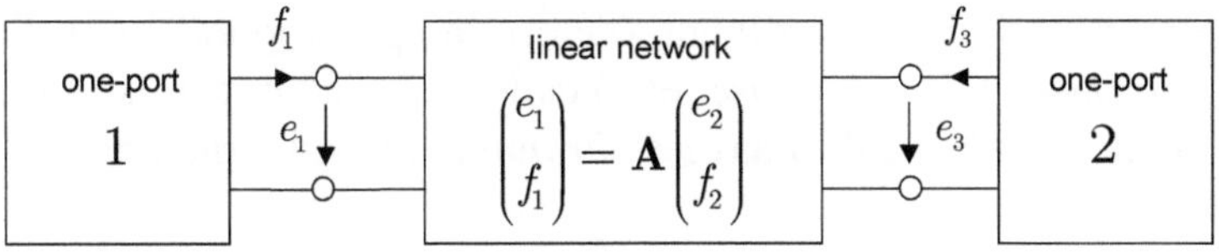

Fig. 2.50. Linear network with indeterminate causal structure

What is actually meant by such statements is whether the experiment-dependent cause-and-effect relationship—and thus the casual computational structure—of an abstract model is determined or still open. To be correct, it is thus better to speak of models with *determinate* or *indeterminate causal structure*.

Causal structures in network models Without more detailed specification of the input or output connections, network models generally have an *indeterminate causal structure*. Fig. 2.50 shows a linear network model, described by its chain matrix $\mathbf{A}$[21]

$$\begin{pmatrix} e_1 \\ f_1 \end{pmatrix} = \mathbf{A} \begin{pmatrix} e_2 \\ f_2 \end{pmatrix} = \begin{pmatrix} a_{11} & a_{12} \\ a_{21} & a_{22} \end{pmatrix} \begin{pmatrix} e_2 \\ f_2 \end{pmatrix} . \tag{2.35}$$

The actual model equations are described by the matrix $\mathbf{A}$. The behavior of the network in response to external "excitations" depends on the external connections of the network to One-ports 1 and 2. Each of these one-ports can represent a passive load or an active network, e.g. a source.

The input, or cause, of a causal structure can only be an independent source, i.e. either a potential or flow source. The description given in Eq. (2.35) does not, however, answer the questions of (a) which of the two one-ports will act as the source and (b) which type of source is being considered, i.e.

$$\begin{array}{ll} \text{potential source} & e_i(t), \;\; i = 1,2 \\ \text{flow source} & f_i(t), \;\; i = 1,2 . \end{array}$$

---

[21] In general, complex elements $a_{ij}(s)$ can also be incorporated, i.e. arbitrary passive (generalized) network elements R, L, C can be included in the circuit.

Thus, the model (2.35) can be considered complete, yet *causally indeterminate*. However, in order to carry out an experiment, a causal structure with an excitation and one or more responses must be identified. If, for example, a potential source $e_1(t)$ is chosen as the excitation (One-port 1) and the effort $e_2(t)$ is taken as the output with an open circuit (One-port 2 has infinite input resistance), then an *unambiguous casual structure* has been defined and the experiment (determining temporal evolution, frequency response, noise response, etc.) can be carried out. It is in this sense that the model can be said to have *determinate causal structure*.

Example 2.5 *Causal structures for an RC network.*

Indeterminate external connection Consider the RC network with *indeterminate external connections* shown in Fig. 2.51. A mathematical dynamic model can be directly constructed by applying the modeling methodology introduced in Sec. 2.3.3 (using the junction rule mesh rule, and constitutive equations with lumped, constant parameters). Using LAPLACE-transformed system variables and the complex chain matrix, this model can be concisely represented as

$$\begin{pmatrix} U_1(s) \\ I_1(s) \end{pmatrix} = \begin{pmatrix} 1+RCs & R \\ Cs & 1 \end{pmatrix} \begin{pmatrix} U_2(s) \\ I_2(s) \end{pmatrix} = \mathbf{A}(s) \begin{pmatrix} U_2(s) \\ I_2(s) \end{pmatrix}. \tag{2.36}$$

As is readily apparent, cause-and-effect relationships in Eq. (2.36) can be seen in either direction. Without more detailed specification of the external connections—i.e. constraints on the input and output power variables u_1, i_1, u_2, i_2—the *causal structure* of the model (2.36) remains *indeterminate*.

Load For the first concrete experiment, let One-port 2 be further specified: let it be a passive network with infinite input resistance, i.e. no current should be able to flow into One-port 2 (an open circuit at terminal 2 of the RC network).

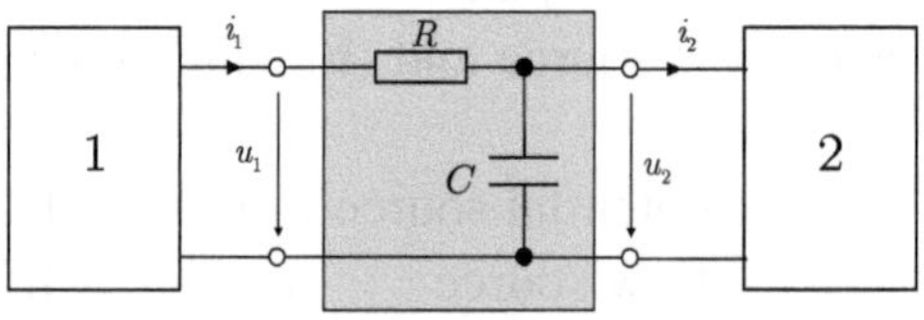

Fig. 2.51. RC network with indeterminate external connections

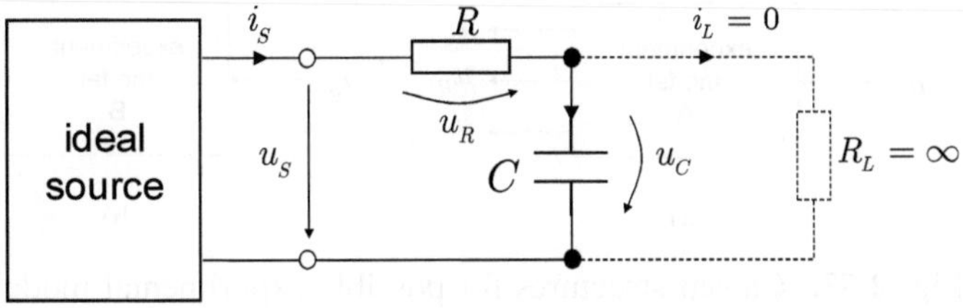

Fig. 2.52. Load-free RC network with indeterminate source

Source This makes it clear that One-port 1 must be regarded as a source (Fig. 2.52). However, the causal structure has still in no way been unambiguously determined. The question of which of the two power variables at the source are to be independently forced—serving as the *excitation* for the network (or *input of the causal structure*)—still remains open. Which of the system variables should be regarded as *response variables* (or *outputs of the causal structure*) also remains to be defined.

A mathematical model of the load-free RC network with an indeterminate source—in the form of a DAE system—thus has the following form:

$$\begin{aligned} u_S - u_R - u_C &= 0, \\ u_R &= R i_S, \\ \dot{u}_C &= \frac{1}{C} i_S, \\ u_S(i_S) &= 0. \end{aligned} \tag{2.37}$$

As can be easily recognized, a self-contained, complete description of the dynamics of this system is available from the model in Eq. (2.37). All that remains to be determined is the algebraic relation between source voltage and source current.

Experiments In order to be able to investigate concrete system behaviors (experiments), suitable, independently controllable excitations must be defined. For this purpose, in the present case, the only possible candidates are the two source power variables .

Case A: *ideal voltage source* → excitation $u_S = u_S(t)$, no constraints on $i_S(t)$, the output is considered to be all time-varying system variables (Fig. 2.53a).

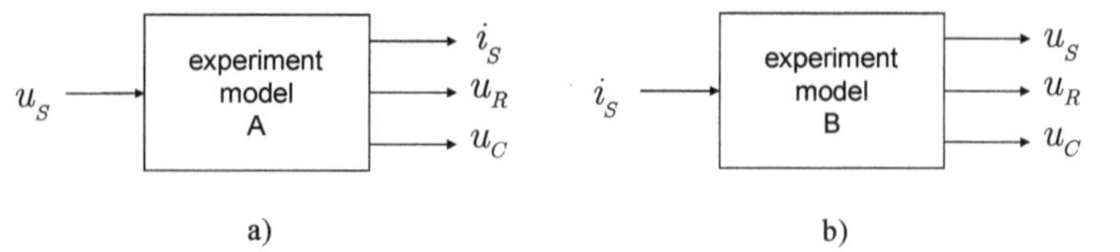

Fig. 2.53. Causal structures for possible experimental models of the RC network: a) voltage excitation, b) current excitation

Mathematical model for *Experiment A* (via simple manipulation of the model in Eq. (2.37)):

$$\begin{aligned} \dot{u}_C &= -\frac{1}{RC} u_C + \frac{1}{RC} u_S \,, \\ i_S &= -\frac{1}{R} u_C + \frac{1}{R} u_S \,, \\ u_R &= u_S - u_C \,. \end{aligned} \tag{2.38}$$

Case B: *ideal current source* $\rightarrow$ excitation $i_S = i_S(t)$, no constraints on $u_S(t)$, the output is considered to be all time-varying system variables (Fig. 2.53b).

Mathematical model for *Experiment B* (also via simple manipulation of the model in Eq. (2.37)):

$$\begin{aligned} u_S &= Ri_S + \frac{1}{C}\int i_S(\tau)\, d\tau \,, \\ u_R &= Ri_S \,, \\ \dot{u}_C &= \frac{1}{C} i_S \,. \end{aligned} \tag{2.39}$$

Now both experiments exhibit clear causal structures in their models (2.38), (2.39), which additionally present a causal computational structure, i.e. all unknown quantities (outputs, system responses) can be computed explicitly using the DAE systems (2.38) and (2.39) given known input functions $u_S(t)$ or $i_S(t)$ (this is a determining property for simulation experiments).

■

Computational causality A different causality concept is connected with the technical solution of a DAE system for simulation—that is, carrying out a (simulation) experiment. The term *computational causality* describes a suitable *structure* and *sequencing* of the equations of the DAE model permitting the *sequential numerical* solution of the DAE system.

For example, the system of equations (2.37)—even given a defined source—does not permit sequential computation. In contrast, the system of equations (2.38) presents a causal computational structure. In the first equation, given the source voltage u_Q, the capacitor voltage u_C can be computed. In the next step, using u_Q and u_C, the source current i_Q is obtained, and in the third step, the voltage across the resistor is determined (steps 2 and 3 can also be interchanged). In modern simulation tools, this manipulation of equations is automated using symbolic computation.

An extended interpretation of computational causality additionally requires *explicit* representation of the equations in the form $a = f(b, c)$ with known variables *b* and *c*, as opposed to an implicit representation $a = f(a, b, c)$. Implicit equations actually pose no extraordinary problems for modern simulation tools, so that nowadays, explicit representation of the equations is not a particular requirement.

Causal structure vs. modeling approaches

Multi-ports It should be noted that, in a way, multi-port modeling automatically results in models in the form of DAE systems with indeterminate causal structure. This in turn means that the causal structure only needs to be established at the time of experimentation by fixing the assignment of inputs and outputs. The DAE model is thus universally reusable once generated (see Eq. (2.37)). Of course, *before* carrying out the actual experiments, the equations must be suitably manipulated to achieve a causal computational structure. This task can, however, be delegated to a computer-aided tool, just as can be done for determining the DAE system equations from the network topology. This approach is followed by modern equation-oriented modeling and simulation tools, e.g. object-oriented modeling with MODELICA (Tiller 2001) and network-based modeling with VHDL-AMS (Schwarz et al. 2001).

LAGRANGE formalism In comparison, in the modeling approach using the LAGRANGE formalism, specification of the causal structure must occur significantly earlier in the modeling process. The generalized coordinates must already be established during the process of setting up the LAGRAN-Gian equations. As previously mentioned, these must be independent and unconstrained when dealing with *EULER-LAGRANGE* equations of the second kind. For example, in Example 2.5, given a voltage excitation, charge would be used as a generalized coordinate; on the other hand, given a current excitation, flux linkage would be used (see also Table 2.2). Furthermore, the forcing function must be defined in the form of an external generalized force or potential function. As a result, each input case directly gives rise to its own particular model with a determinate causal structure—though naturally, this is ultimately identical to the equivalent network-based model (see Eqs. (2.38), (2.39)).

Evaluation Given the perspectives presented here, the network-based modeling approach is justifiably favored within the technical community. Due particularly to the substantial available tool set, it is eminently suited to (very) large, complex systems. Certainly, it should also be pointed out again that the advantages of multi-port physical modeling come from its equation-oriented description with indeterminate causal structure, and not, mistakenly, from "acausal" models.

2.3.9 Modular modeling of mechatronic systems

DAE systems as dynamic models When speaking of the modular modeling of mechatronic systems, it is first and foremost desirable that the *physical topology* of real systems be equivalently replicated in their dynamic models. Such computable dynamic models (see also Fig. 2.4) are given in a general form by a system of *differential-algebraic equations* (*DAE system*)[22] (here in semi-explicit form, see Sec. 2.4):

$$\dot{\mathbf{x}} = \mathbf{f}(\mathbf{x}, \mathbf{z}, \mathbf{u}, t), \tag{2.40}$$

$$\mathbf{0} = \mathbf{g}(\mathbf{x}, \mathbf{z}, t). \tag{2.41}$$

As a typical example of a mechatronic system, consider the *servo-controlled drive shaft* (e.g. in a robot or machine tool) depicted in Fig. 2.54.

[22] This applies in any case for most physical system components which can be described via *continuous-time models*. Switching, mechanical contact issues, stick-slip behavior, and discrete-time and discrete-event phenomena (mostly in the context of information processing) can be described using specialized model extensions termed *hybrid models* (see Sec. 2.5).

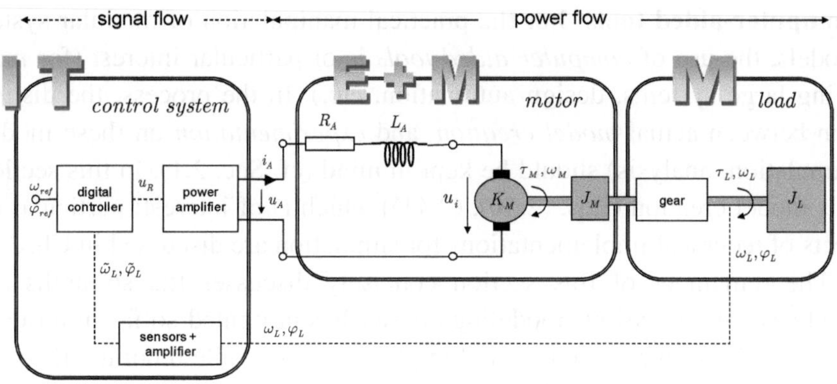

Fig. 2.54. Example of a multi-domain model: servo-controlled drive shaft (e.g. robot, machine tool)

Requirements A useable *modular dynamic model* for a mechatronic system should fulfill the following general *requirements*:

1. The *physical system topology* should remain visible in the system of equations (2.40), (2.41), e.g. the armature inductance should have a physical relation to the armature resistance and motor, the transmission should have a physical relation to the motor and load axles.
2. Models of individual system elements should be *interchangeable* while maintaining the power relation at interfaces (i.e., power-conserving), e.g. a rigid transmission interchangeable with an elastic transmission, a rigid load interchangeable with a multibody system.
3. Models of system elements should support *hierarchical* operations (decomposition, aggregation), e.g. aggregate motor model → new model "*Motor*", decomposed transmission model → detailed component model.
4. System components from *different physical domains* (multi-domain components) should be described using *consistent models*; this requirement is automatically met at the computational level using a DAE system (2.40), (2.41), but is equally desirable for abstract model precursors (graph-oriented models).
5. *Coupling* should be possible for *power-flow-* and *signal-oriented* models, e.g. control → motor, load → control.

Requirements 2 and 3 generally suggest the use of *model libraries*; Requirement 4 can be implemented with *domain-specific* model libraries.

Computer-aided tools For the practical manipulation of modular system models, the use of *computer-aided tools* is of particular interest (for managing large systems, design automation, etc.). In the process, the distinction between actual *model creation*, and *experimentation* on these models (simulation, analysis) should be kept in mind (cf. Sec. 2.1). In this section, it is model creation (Eqs. (2.40), (2.41)) which is of interest; particular aspects of technical implementations for simulation are discussed in Ch. 3.

The remainder of this section concisely discusses the strengths and weakness of the various modeling approaches presented so far in terms of their suitability for computational implementation while retaining the form of modular physical models.

Energy-based modular modeling

Strengths and weaknesses The LAGRANGE formalism presented in Sec. 2.3.2 is actually not suited to modular modeling. Requirements 2 and 4 can be implemented with uncoupled coordinates at the energy level using suitable model libraries. In the same way, conversion into a DAE system can, in principle, be automated using computer algebra systems (for differentiation of the LAGRANGian). System coupling is represented by equating coordinates in the problem. However, the coupling of power flows behind these coordinates is only indirectly visible via differentiation of the LAGRANGian, and is thus not evident in the DAE system. This is likely the reason that no multi-domain tools employing this paradigm have penetrated the market.

Multi-port modular modeling

Strengths and weaknesses From the material in the preceding sections, it should already be clear that essentially only *multi-port* modeling approaches can fulfill all of the modularity requirements presented above. They inherently exhibit all necessary properties, such as *power-conserving interfaces*, *open causal structures*, *multi-domain properties* via the analogous substitutions, and *generalized network elements* (see Examples 2.2 through 2.4).

The *hierarchy requirements* can also—at least in the case of linear models—easily be met using concise *two-port descriptions* with complex elements (see Eq. (2.36)).

At this point, the particular assignment of power variables to mechanical quantities used in network theory can finally be clearly justified (see Table 2.3).

The assignment is as follows

flow := *force or torque*

effort := *velocity or angular velocity.*

For example, examining the *rigid coupling* between motor axle and transmission (τ_M, ω_M) in the servo-controlled drive shaft in Fig. 2.54 from a physical point of view shows that the angular velocities—not the torques—must be identical at the interface. However, following the rules of network theory, such a condition is only possible for effort variables, from which it follows that the *(angular) velocity* must be used as the effort variable (see Example 2.7).

Note that in the bond graph world, the opposite assignment (that in Table 2.2) is typically used, complicating the representation of rigid couplings (see Example 2.6).

Exploiting the advantageous properties of port-based modeling, a series of powerful modeling and simulation tools have been developed, which in particular support *domain-specific model libraries* at the physical level, and which appreciably lighten the burden of the modeling task, e.g. (Schwarz et al. 2001), (Geitner 2006).

Example 2.6 *Port-based model (bond graph) of a drive shaft.*

System configuration Consider the servo-controlled drive shaft in Fig. 2.54. Of particular interest is the rigid coupling between the motor arbor and the transmission to the load. This physical interface is to be replicated with as little modification as possible in a modular dynamic model.

Bond-graph-based dynamic model As laid out in the preceding paragraphs, most bond graph tools employ forces/torques as mechanical effort variables. As explained above, this convention brings with it certain complication for the case of rigidly-coupled arbors. Fig. 2.55a shows a bond graph model (Geitner 2008) in which such a rigid coupling is in fact directly implemented, though with an accompanying loss of correspondence between physical and model modules. The coupling is only realizable by accounting for the combined moment of inertia $J_{ges} = J_M + J_L$ at the driving input (the motor).

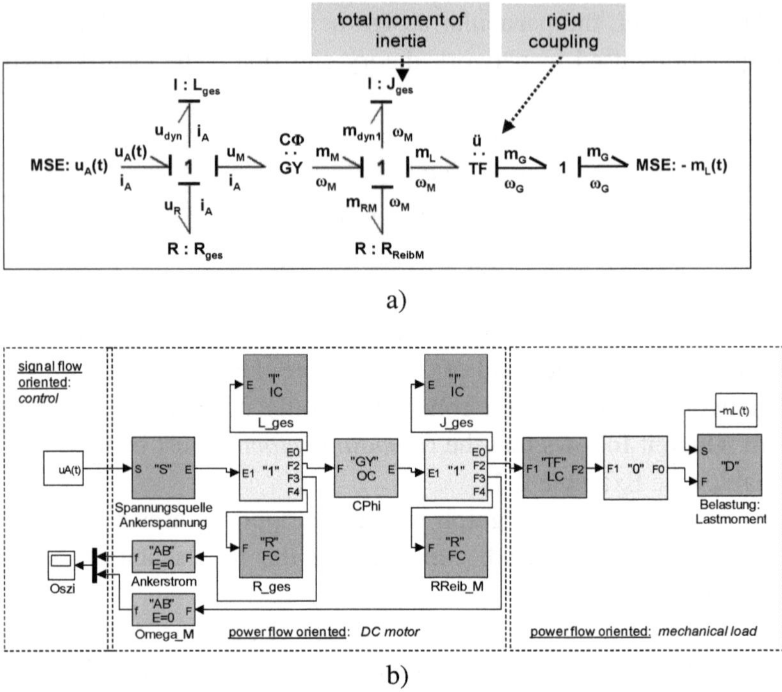

Fig. 2.55. Bond graph model for the servo-controlled drive shaft in Fig. 2.54 with a *rigid* motor-transmission coupling: a) bond graph, b) bond graph computer model embedded in SIMULINK, from (Geitner 2008)

In contrast, Fig. 2.56a shows a modular bond graph model which strictly maintains correspondence to physical modules, so that the motor and load each retain their individual inertia parameters (Geitner 2008). Admittedly, in order to maintain consistency at the interfaces—i.e. the coupling of the velocities as flows—a virtual *elastic coupling* employing a high-stiffness spring had to be introduced. In this way, relative motions between the two arbors are allowed, though they remain small only with very high spring stiffness. In addition to this thoroughly problematic *model falsification*[23], a new problem has been introduced which impacts the simulation of the system. The new model is a stiff differential system of equations and is hence more difficult to solve numerically when performing simulation experiments (see Ch. 3).

[23] It is a question of the system point of view—i.e. of what the modeling objective is—whether this model modification is treated as a *falsification* of the rigid coupling assumption, or as an *enhancement* of the modeling of physical system properties (strictly speaking, there are no real "rigid" bodies).

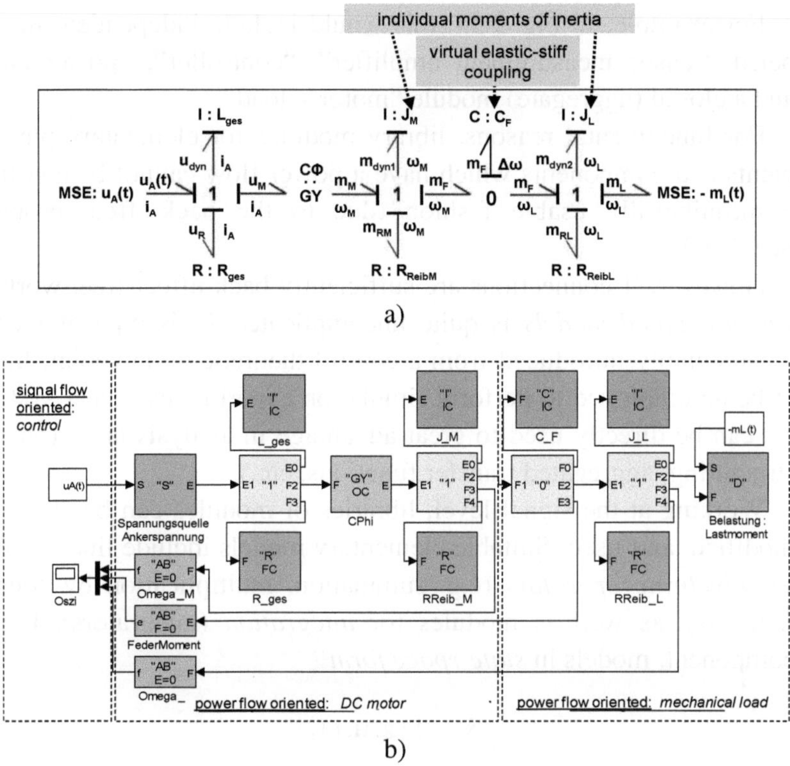

Fig. 2.56. Bond graph model for the servo-controlled drive shaft in Fig. 2.54 with a *virtual stiff elastic* motor-transmission coupling: a) bond graph, b) bond graph computer model embedded in SIMULINK, from (Geitner 2008)

Bond-graph-based computer models are shown in Fig. 2.55b and Fig. 2.56b for both cases considered (embedded here in the computer-based tool SIMULINK, (Geitner 2006)).

■

Signal-oriented modular modeling

Strengths and weaknesses Modularization using signal-oriented models—while meeting the Requirements 1 through 4 established above—is only meaningfully possible for *functional physical device groups* having *back-effect-free* coupling.

For example, in Fig. 2.54, this would include independent modules labeled “sensor measurement amplifier”, “controller”, “power amplifier”, and a closed (aggregate) module “motor + load”.

For fundamental reasons, library modules for elementary physical elements and components which have a power flow cannot be constructed in a meaningfully usable fashion (due to the back-effect problem, see Sec. 2.3.5).

However, if connections are sufficiently back-effect-free, working with *signal-coupled models* is quite uncomplicated. This type of modeling is most suitably introduced from a control theoretical viewpoint. In addition to being employed to perform simulation experiments, linear system models can be directly used to great advantage in analysis tasks (control loop structuring, aggregated transfer functions, etc.).

Working at the signal level, libraries of modules can be easily created, modified, and used. Suitable elementary models include linear and nonlinear *algebraic operators* (e.g. summation, multiplication by constant parameters), as well as modules for *integration* (integrators). Using such component, models in *state-space form*[24]

$$\dot{\mathbf{x}} = \mathbf{f}(\mathbf{x}, \mathbf{u}, t), \tag{2.42}$$

$$\mathbf{y} = \mathbf{g}(\mathbf{x}, \mathbf{u}, t), \tag{2.43}$$

are easily modularized.

Note however, that connecting several state-space models (2.42), (2.43) is only straightforward if no *algebraic loops* appear (corresponding to a index-0 DAE system, see Sec. 2.4).

Due to this methodologically simpler process, computer-aided tools for signal-oriented modeling, simulation, and analysis (MATLAB/SIMULINK, LABVIEW) are significantly more widely distributed in practice than those which are network based. Thus, in many cases, a signal-oriented modeling approach is preferred intuitively, despite the fact that this entails considerable *limitations* on physical modular models incorporated into libraries.

---

[24] Here, as is usual, $\mathbf{u}$ represents the vector of inputs, $\mathbf{y}$ the vector of outputs, and $\mathbf{x}$ the state vector (see also Sec. 2.6).

A characteristic example highlighting the inherent obstacles to modularizability was already presented in Sec. 2.3.7 (see Example 2.3). The problem of rigid coupling in mechanical systems introduced there is thoroughly discussed in Sec. 2.4 (DAE systems, index-3 systems). For the case of signal-oriented modeling (as with the bond graph model), only two solutions come into question, neither of which is fully satisfactory: a *common* model including both masses (the correspondence to physical objects is lost), or *elastic coupling* with sufficient stiffness (resulting in a stiff system of differential equations).

Object-oriented modeling of physical systems

Dual meaning of "object-oriented" The advantageous aspects of *network-based* modeling discussed above can be efficiently combined with *object-oriented* concepts from software design to create computer-based modeling implementations. The resulting paradigm—termed *object-oriented modeling of physical systems*—employs the attribute "object-oriented" in two ways:

- object-oriented in the *conceptual* sense: using *encapsulated dynamic models* of physical components (e.g. capacitors, transmissions, masses, motors) while maintaining power-based interrelationships,
- object-oriented in the *software* sense: using *encapsulated software entities* with classic object-oriented software properties (polymorphism, inheritance, hierarchical classes, etc.).

Object-oriented computer tools *Computer-based tools* created following the object-oriented physical modeling paradigm enable—alongside a *user-friendly model representation*—automated model creation and computer-based simulation and analysis (set point computation, linearization, frequency response calculation, etc.). Fundamentally, this can be viewed as creating a class of generalized network analyzers having a multi-domain man-machine interface.

Fig. 2.57 depicts the generalized model hierarchy for object-oriented physical modeling, cf. the generalized model hierarchy in Fig. 2.4.

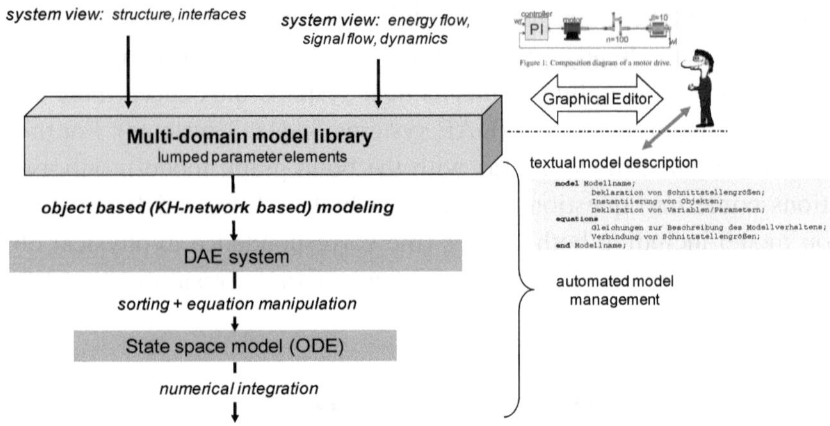

Fig. 2.57. Model hierarchy for object-oriented physical modeling

Model description languages At the core of object-oriented modeling approaches lie *textual model description languages.* It is here that the concepts of network modeling and of object-oriented software design are united, including:

- the mathematical description of real processes in encapsulated objects,
- models implemented as (object) classes;
- hierarchical class structures: the parent-child concept ⇒ specialization at lower levels of the hierarchy;
- inheritance of properties from superordinate classes in the hierarchy;
- polymorphism: inherited properties can be locally overridden or, in the case of multiple definitions, the property of the closest ancestor applies;
- the use of models in the form of (object) instances (predefined in the form of classes);
- simplified reuse of models and components.

MODELICA An important example worth mentioning is the language MODELICA[25] (Fritzson 2011), (Tiller 2001), which offers several specialized attributes:

- a model-specific rather than mathematically-oriented description,
- application neutrality,
- is a quasi-standard for object-oriented modeling languages,

---

[25] www.modelica.org

- supports mixed continuous-discrete systems (hybrid systems),
- offers tracking and accounting of physical units.

The translation of textual models into a DAE system along with subsequent model manipulation in computer-based tools takes place in a largely automated fashion using powerful *transformation algorithms* (Otter 1999; Otter and Bachmann 1999; Otter and Bachmann 1999).

Graphical editors Though not methodologically relevant for modeling, an additional helpful (and attractive!) user option offered by several tools is a powerful *graphical editor*, where—using domain-specific *object diagrams*—heterogeneous (e.g. mechatronic) systems can be graphically composed and automatically translated into the textual description language.

Advantages for the user The particular *advantages* of object-oriented modeling tools lie in their accessibility for the user (particularly if a graphical editor is available), the straightforward reusability of previously-constructed models (evolving model libraries), inherently physically accurate model composition using the network approach, and efficient, automated management of models.

A superficial consideration of the model hierarchy shown in Fig. 2.57 would give the impression that a user has only to open the right toolkit, arrange and connect the components in a physically relevant manner and voila! an accurate system model is generated.

It is, however, precisely in such simplistic user interventions (lacking the requisite technical understanding), in combination with concealed, automated model management processes, that, in the experience of the author, lie potential dangers which should not be underestimated.

Problems during use Even if it is assumed that computer-based tools contain validated software (but what software is really error-free?), sufficient opportunity still exists to create an incorrect or fundamentally numerically ill-conditioned model, which will hit (often mostly hidden) snags only during the course of experimentation (simulation). It should be noted that *all* possible sources of difficulty discussed in Chapter 3: Simulation Issues can also appear precisely in object-oriented models. In many cases, these problems can be countered with a suitable parameterization of the simulation algorithms; in other cases, however, they can only be overcome with changes to the model. Knowledge of these aspects of modeling is indispensible even—and precisely—when using object-oriented computational tools.

Example 2.7 *Object-oriented model of a drive shaft.*

System configuration Consider again the servo-controlled drive shaft in Fig. 2.54. In the example models so far (bond graphs, signal-oriented), the rigid coupling between the motor arbor and the transmission to the load could only be represented in a physical, modular model in a very limited manner. The following modeling approach using the modeling language MODELICA permits *freely configurable*, *port-based* model interfaces, and thus enables the much more straightforward generation of general physical, modular models and model libraries.

MODELICA dynamic model The object-oriented dynamic model presented below is based on (Schwarz and Zaiczek 2008) and is characterized by:

- on the mechanical side: a port for torque and angular rotation,
- on the electrical side: two terminals, which combine to form an electrical port,
- a "ground" (at the "top-level schematic") which also should not be omitted from hand-drawn system descriptions,
- the load: damping (with value "d") proportional to the angular velocity,
- *"protected"* variables which cannot be used outside of their scope of validity (encapsulation, a typifying property of object orientation).

The models are annotated in such a way as to be understandable to the unpracticed eye, so that the model hierarchy and the interface concept should be self-evident to the knowledgeable reader. In order to simulate these models using a computational tool, they must first be suitably prepared (e.g. modules must be arranged as *files* in *directories* or compiled into *packages*). These latter issues are however questions of implementation, exceeding the scope of modeling methodology of interest here. For a detailed understanding of the modeling language MODELICA, the interested reader is referred to e.g. (Tiller 2001).

Rigid mechanical coupling The strengths of the *port-based* approach of a modeling language like MODELICA now become apparent. Due to the *user-specific* choice of flow and effort, suitable, consistent variables can be defined at the mechanical ports of the drive and load shafts for the case presented here, directly enabling their rigid coupling. In this example, the most obvious choice is to select motion variables as the effort, that is, the *angular velocity* or the *rotation angle* of the shaft. Both quantities are equally permissible (see remarks in Sec. 2.3.4 in the paragraph "Domain-specific flow and effort variables") and must be—both as efforts and as physical quantities—identically equal for rigid arbor coupling. For convenience, for servo control (as is the case here), the *rotation angle* will be chosen as the effort variable. Naturally, constitutive equations should be matched to the choice of port variables.

MODELICA program module (complete model, from (Schwarz and Zaiczek 2008))

```
model Motor_Multiport_complete "consists of motor + voltage source + load +
ground"
// Description of the overall system, "top-most" model level
// Motor connected to voltage source and load
// List of all models used
  Motor_MultiportModel MM;
  SourceOnePort source;
  Load_damping load;
  Ground ground;
// Linking of models
equation
  connect( source.p, MM.p );
  connect( source.n, MM.n );
  connect( source.n, erde.p );
  connect( MM.portM, load.port );
end Motor_Multiport_complete;
```

```
model Motor_MultiportModel "motor model as reusable multi-port"
//Model parameters: normalized, dimensionless
  parameter Real RA = 1;
  parameter Real LA = 1;
  parameter Real JM = 2;
  parameter Real KM = 3;
  //electrical and mechanical terminals
  Pole p              " together form  " ;
  Pole n              " an electral port " ;
  MechanicalPort portM;
  //"protected" encapsulates internal data (object-oriented!)
  protected
  Real tauE;
  Real uA;
  Real iA;
  Real ui;
  Real tauM;
  Real omM;
  Real omE;
  Real phiE;

equation
  //internal model
  uA = RA*iA + LA*der(iA) + ui;
  tauM = tauE + JM*der(omM);
  tauM = KM*iA;
  ui = KM*omM;
  omE = omM;
  der(phiE)=omE;

  // communication with the environment, interface
  p.v-n.v = uA;
  p.i = iA;
  p.i+n.i = 0                "electrical port condition";
  portM.tau = tauE;
  portM.phi = phiE;
end Motor_MultiportModel;
```

```
connector MechanicalPort
  Real phi;
  flow Real tau;
end MechanicalPort;
```

```
model Load_Damping
  MechanicalPort port;
  parameter Real d=1;
equation
  port.tau = -d*der(port.phi);
end Load_Damping;
```

```
model SourceOnePort
  parameter Real val=1.0;
  Pole p;
  Pole n;
equation
  p.v-n.v = if time >= 0 then val else 0.0  "step function as example" ;
  p.i+n.i = 0 "electrical port condition";
end SourceOnePort;
```

```
model Ground "required for definition of reference node"
  Pole p;
equation
  p.v=0;
end Ground;
```

```
connector Pole
  Real v "potential at terminal";
  flow Real i "current flow into the terminal";
end Pole;
```

Dynamic model as object diagram The benefits of an object-oriented modeling language like MODELICA can be taken advantage of in user-friendly computer-aided modeling and simulation tools, such as DYMOLA[26] and SIMULATIONX[27]. These also employ so-called *object diagrams*, easing model creation (for the possible hazards, see above).

Fig. 2.58 shows such an object diagram created with the modeling tool DYMOLA (Schwarz and Zaiczek 2008). This diagram uses basic blocks from the MODELICA standard library. Here, no load has been connected on the mechanical side (free running motors).

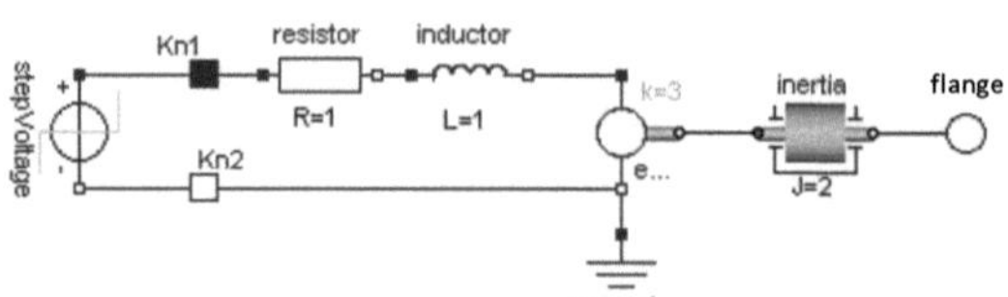

Fig. 2.58. DYMOLA object diagram for the motor of the servo-controlled drive shaft in Fig. 2.53, from (Schwarz and Zaiczek 2008)

[26] www.dymola.com

[27] www.simulationx.com

The object diagram has a *textual* description generated automatically by the simulator, but which could also be entered manually (cf. the tool architecture in Fig. 2.57). For each model component (`Resistor`, `Inductor`, `EMF`, `Inertia`, etc.), the encapsulated program modules (see above) and functional and interface reference descriptions (e.g. the meanings of the parameters `R`, `L`, `k`, `J`) are stored in a model library. An example of the automatically-generated text description is shown in excerpt below. The functional description of the model modules is equivalent to the MODELICA models presented above.

DYMOLA / MODELICA Structural description (excerpt)

```
model Motor_with_voltage_step
// List of all components
  Modelica.Electrical.Analog.Basic.Resistor resistor(R=1) ;
  Modelica.Electrical.Analog.Basic.Inductor inductor(L=1) ;
  Modelica.Electrical.Analog.Basic.EMF eMF(k=3) ;
  Modelica.Mechanics.Rotational.Inertia inertia(J=2) ;
  Modelica.Electrical.Analog.Interfaces.NegativePin Kn2 ;
  Modelica.Electrical.Analog.Interfaces.PositivePin Kn1 ;
  Modelica.Mechanics.Rotational.Interfaces.Flange_b Flange ;
  Modelica.Electrical.Analog.Sources.StepVoltage stepVoltage ;
  Modelica.Electrical.Analog.Basic.Ground ground ;
// component connections
equation
  connect(resistor.n, inductor.p) ;
  connect(inertia.flange_a, eMF.flange_b) ;
  connect(eMF.n, Kn2) ;
  connect(inertia.flange_b, Flange) ;
  connect(inductor.n, eMF.p) ;
  connect(resistor.p, Kn1) ;
  connect(stepVoltage.p, Kn1) ;
  connect(stepVoltage.n, Kn2) ;
  connect(eMF.n, ground.p) ;
end Motor_with_voltage_step;
```

■

2.4 Systems of Differential-Algebraic Equations

2.4.1 Introduction to DAE systems

DAE systems In the course of modeling the dynamics of mechatronic systems, *systems of differential equations* appear which are subject to *algebraic constraints.* Multi-port modeling using KIRCHHOFF networks discussed in Sec. 2.3.4 naturally leads to this form of model, as does the application of the EULER-LAGRANGE equations of the first kind given holonomic and nonholonomic constraints[28].

[28] These models consist of a system of second-order differential equations and can easily be transformed into the standard form of first-order differential equations used here.

Such systems of equations are called *differential-algebraic equations* (*DAEs*). As will be shown in Ch. 3, the numeric solution of such models requires the application of special methods.

For reasons of space, only a short introduction to the important terms and relationships fundamental to understanding DAE systems as applied to system models can be given here. For a deeper understanding, the reader is referred to the very readable and extensive monographs (Cellier and Kofman 2006) and (Brenan et al. 1996).

Representations

- *implicit* DAE system

  $\mathbf{0} = \tilde{\mathbf{f}}\left(\dot{\mathbf{x}}, \mathbf{x}, \mathbf{z}, \mathbf{u}, t\right)$

$\mathbf{x}(t) \in \mathbb{R}^n$	state variables, appear with differentials
$\mathbf{z}(t) \in \mathbb{R}^m$	algebraic variables
$\mathbf{u}(t) \in \mathbb{R}^r$	inputs (externally imposed)

$\tilde{\mathbf{f}} : \mathbb{R}^n \times \mathbb{R}^n \times \mathbb{R}^m \times \mathbb{R}^r \times \mathbb{R} \to \mathbb{R}^{n+m}$ ... set of differential and algebraic equations

Note: a total of $(n + m)$ equations is required for the $(n + m)$ unknowns $\mathbf{x}(t)$, $\mathbf{z}(t)$.

- *semi-explicit* DAE system

$$\dot{\mathbf{x}} = \mathbf{f}\left(\mathbf{x}, \mathbf{z}, \mathbf{u}, t\right) \tag{2.44}$$

$$\mathbf{0} = \mathbf{g}\left(\mathbf{x}, \mathbf{z}, t\right) \tag{2.45}$$

$$\mathbf{f} : \mathbb{R}^n \times \mathbb{R}^m \times \mathbb{R}^r \times \mathbb{R} \to \mathbb{R}^n, \ \mathbf{g} : \mathbb{R}^n \times \mathbb{R}^m \times \mathbb{R} \to \mathbb{R}^m$$

Index of a DAE system

Difficulty of solving a DAE The classification of DAE systems commonly takes place using a special measure, called the *index* i of the DAE system. In a certain manner, this measure describes the difficulty of solving the DAE system. A system of ordinary differential equations has index $i = 0$, DAE systems have $i > 0$. A DAE system is called *high index* if $i \geq 2$.

However, there are varying definitions of the index, so that in some cases, different orderings are possible. In this book, the most commonly-used definition will be referenced: the *differential index*.

Definition 2.8. *Differential index*: The *differential index* is the minimum number of differentiations required for the equations of a DAE system to arrive at a system of explicit ordinary differential equations.

2.4.2 DAE index tests

Goal: Determination of the *differential index*, i.e. the minimal number of differentiations d/dt of Eq. (2.45) required so that—given Eq. (2.44)—a system of explicit ordinary differential equations ensues.

Index-1 systems

$$\frac{d}{dt}\mathbf{g}(\mathbf{x},\mathbf{z}) = \frac{\partial \mathbf{g}}{\partial \mathbf{x}}\dot{\mathbf{x}} + \frac{\partial \mathbf{g}}{\partial \mathbf{z}}\dot{\mathbf{z}} = \mathbf{0}$$

$$\text{substituting Eq. (2.44):} \quad \frac{\partial \mathbf{g}}{\partial \mathbf{x}}\mathbf{f} + \frac{\partial \mathbf{g}}{\partial \mathbf{z}}\dot{\mathbf{z}} = \mathbf{0}$$

- *Index-1 condition*

$$\boxed{\det\left(\frac{\partial \mathbf{g}}{\partial \mathbf{z}}\right) = \det\begin{pmatrix} \frac{\partial g_1}{\partial z_1} & \cdots & \frac{\partial g_1}{\partial z_m} \\ \vdots & \ddots & \vdots \\ \frac{\partial g_m}{\partial z_1} & \cdots & \frac{\partial g_m}{\partial z_m} \end{pmatrix} \neq 0} \tag{2.46}$$

If Eq. (2.46) holds, then the algebraic constraint can be written as a system of first-order differential equations

$$\dot{\mathbf{z}}(\mathbf{x},\mathbf{z},\mathbf{u}) = -\left(\frac{\partial \mathbf{g}}{\partial \mathbf{z}}\right)^{-1}\frac{\partial \mathbf{g}}{\partial \mathbf{x}}\mathbf{f}\,.$$

Example 2.8 *RC network.*

System configuration Consider an unloaded RC network following Fig. 2.52, with an ideal (lossless) voltage source $u_S(t)$ and constant components R, C.

Modeling as a KIRCHHOFF network gives the mathematical model

$$\text{Conservation equation:} \quad u_S - u_R - u_C = 0$$

$$\text{Constitutive equations:} \quad \begin{aligned} u_R &= R i_S \\ \dot{u}_C &= \frac{1}{C} i_S \end{aligned}$$

and, along with the definitional relations

$$x := u_C, \; z_1 := i, \; z_2 := u_R, \; u := u_S,$$

this gives the DAE system in semi-explicit standard form:

$$\begin{aligned} \dot{x} &= f(\mathbf{z}) = \frac{1}{C} z_1 \\ 0 &= g_1(\mathbf{z}) = R z_1 - z_2 \\ 0 &= g_2(x, \mathbf{z}, u) = x + z_2 - u . \end{aligned}$$

The index-1 condition

$$\det\left(\frac{\partial \mathbf{g}}{\partial \mathbf{z}}\right) = \det\begin{pmatrix} \dfrac{\partial g_1}{\partial z_1} & \dfrac{\partial g_1}{\partial z_2} \\ \dfrac{\partial g_2}{\partial z_1} & \dfrac{\partial g_2}{\partial z_2} \end{pmatrix} = \det\begin{pmatrix} R & -1 \\ 0 & 1 \end{pmatrix} = R \neq 0$$

holds for all $R > 0$, so that this represents a DAE system with differential index $i = 1$.

■

***Index-2 systems*[29]**

$$\begin{aligned}\dot{\mathbf{x}} &= \mathbf{f}(\mathbf{x},\mathbf{z},\mathbf{u})\\ \mathbf{0} &= \mathbf{g}(\mathbf{x})\end{aligned} \tag{2.47}$$

$$\frac{d}{dt}\left[\frac{d}{dt}\mathbf{g}(\mathbf{x})\right] = \frac{d}{dt}\left[\frac{\partial \mathbf{g}}{\partial \mathbf{x}}\dot{\mathbf{x}}\right] = \frac{d}{dt}\left[\frac{\partial \mathbf{g}}{\partial \mathbf{x}}\mathbf{f}(\mathbf{x},\mathbf{z},\mathbf{u})\right] = \mathbf{0} \Rightarrow \ldots + \frac{\partial \mathbf{g}}{\partial \mathbf{x}}\frac{\partial \mathbf{f}}{\partial \mathbf{z}}\dot{\mathbf{z}} = \mathbf{0}$$

- *Index-2 condition*[30]

$$\det\left(\frac{\partial \mathbf{g}}{\partial \mathbf{x}}\frac{\partial \mathbf{f}}{\partial \mathbf{z}}\right) = \det\left(\begin{pmatrix}\frac{\partial g_1}{\partial x_1} & \cdots & \frac{\partial g_1}{\partial x_n}\\ \vdots & \ddots & \vdots\\ \frac{\partial g_m}{\partial x_1} & \cdots & \frac{\partial g_m}{\partial x_n}\end{pmatrix}\begin{pmatrix}\frac{\partial f_1}{\partial z_1} & \cdots & \frac{\partial f_1}{\partial z_m}\\ \vdots & \ddots & \vdots\\ \frac{\partial f_n}{\partial z_1} & \cdots & \frac{\partial f_n}{\partial z_m}\end{pmatrix}\right) \neq 0 \tag{2.48}$$

If Eq. (2.48) holds, then the algebraic constraint (2.47) can again be written as a system of first-order differential equations: $\dot{\mathbf{z}} = \dot{\mathbf{z}}(\mathbf{x},\mathbf{z},\mathbf{u})$.

Example 2.19 *Linear index-2 system.*

Problem statement Determine the differential index of the following DAE system ($\phi(t)$ is a given function of time):

$$\dot{x} = f(x,z) = (z-x)a\,,$$

$$0 = g(x,t) = x - \phi(t)\,.$$

Solution

Test index-1 condition: $\frac{\partial g}{\partial z} = 0$ condition violated ➔ index > 1.

Test index-2 condition: $\frac{\partial g}{\partial x}\frac{\partial f}{\partial z} = 1\cdot a \neq 0$ ➔ index $i = 2$.

■

---

29 To simplify the presentation, only index-2 candidates are considered for which the index-1 condition Eq. (2.46) is *not* met, i.e. the algebraic constraints are *independent* of the algebraic variables.

30 Note: This condition only holds for the case $\mathbf{g} = \mathbf{g}(\mathbf{x})$, i.e. *independence* from the algebraic variables $\mathbf{z}$.

Index-3 systems

Negative index test If Eq. (2.48) does not hold, then the DAE system possesses an index $i > 2$. In most cases, this means it is an index-3 system. This should be considered a very difficult problem, as will be elucidated in the following typical example.

Example 2.10 *Rigidly-coupled two-mass system.*

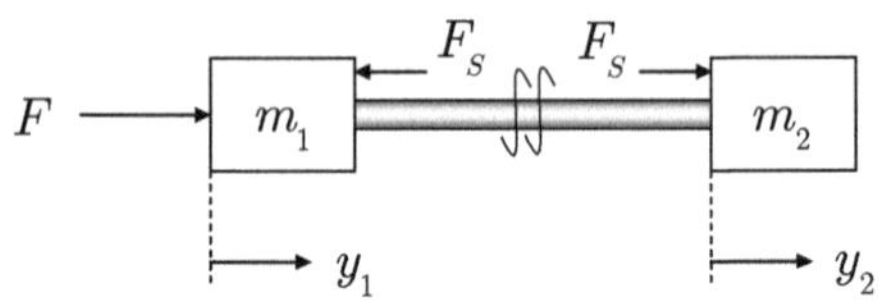

Fig. 2.59. Rigidly-coupled two-mass system

System configuration Let two masses be coupled via a rigid connection (Fig. 2.59). Find the equations of motion and the constraining force in the connecting rod, as well as the classification of the resulting DAE system.

Model Using LAGRANGian equations of the first kind gives the *equations of motion*[31]

$$m_1 \cdot \ddot{y}_1 = F - F_S\,,$$
$$m_2 \cdot \ddot{y}_2 = F_S\,,$$

and the *holonomic* constraint (rigid coupling)

$$y_2 - y_1 = 0\,.$$

Using the definitions $x_1 := y_1$, $x_2 := \dot{y}_1$, $x_3 := y_2$, $x_4 := \dot{y}_2$, $u := F, z := F_S$, gives the equivalent DAE system in *semi-explicit* form:

$$\begin{pmatrix} \dot{x}_1 \\ \dot{x}_2 \\ \dot{x}_3 \\ \dot{x}_4 \end{pmatrix} = \begin{pmatrix} x_2 \\ -\dfrac{1}{m_1} z + \dfrac{1}{m_1} u \\ x_4 \\ \dfrac{1}{m_2} z \end{pmatrix} = \mathbf{f}\left(\mathbf{x}, z, u\right), \tag{2.49}$$

[31] The constraint force F_S represents the LAGRANGE multiplier in Eq. (2.11).

$$0 = x_3 - x_1 = g(\mathbf{x})\,. \tag{2.50}$$

Note: $m = 1$, i.e. there is *one* algebraic variable and *one* algebraic equation.

Index test

Index-1 condition $\dfrac{\partial g}{\partial z} = 0$ ➔ violated

Index-2 condition $\dfrac{\partial g}{\partial \mathbf{x}}\dfrac{\partial \mathbf{f}}{\partial z} = \begin{pmatrix} -1 & 0 & 1 & 0 \end{pmatrix}\begin{pmatrix} 0 \\ -1/m_1 \\ 0 \\ 1/m_2 \end{pmatrix} = 0$ ➔ Index > 2 !

Index determination What is the index of the DAE system (2.49), (2.50)?

Solution: Applying the definition of the differential index, the minimal number of differentiations d/dt of Eq. (2.50) which—given Eq. (2.49)—result in a system of explicit ordinary differential equations can be found.

(1) $\dfrac{d}{dt}$ Eq. (2.50) ➔ $\dot{x}_3 - \dot{x}_1 = 0$

with Eq. (2.49): $$x_4 - x_2 = 0 \tag{2.51}$$

(2) $\dfrac{d}{dt}$ Eq. (2.51) ➔ $\dot{x}_4 - \dot{x}_2 = 0$

with Eq. (2.49): $$\left(\frac{1}{m_1} + \frac{1}{m_2}\right) z - \frac{1}{m_1} u = 0 \tag{2.52}$$

(3) $\dfrac{d}{dt}$ Eq. (2.52) ➔ $\left(\dfrac{1}{m_1} + \dfrac{1}{m_2}\right) \dot{z} - \dfrac{1}{m_1} \dot{u} = 0$

ODE $$\boxed{\dot{z} = \frac{m_2}{m_1 + m_2}\dot{u}} \tag{2.53}$$

➔ Index $i = 3$ since the original algebraic condition (2.50) had to be differentiated three times in order to arrive at an ODE (2.53) in the algebraic variable z.

This state of affairs applies *in general* for mechanical systems with *rigid coupling*.

■

2.4.3 DAE index reduction

Goal Solution of a DAE system (2.44), (2.45) via *explicit* numerical integration (an *ordinary differential equation* (*ODE*) *solver*, e.g. RUNGE-KUTTA).

Solution Using *index reduction*, the DAE system is re-formed into a system of ordinary differential equations, i.e. the algebraic variables also become defined by a system of differential equations. One fundamental hurdle in the process is the determination of consistent initial values $\mathbf{x}(0), \mathbf{z}(0)$ for the state and algebraic variables.

Index reduction by a factor k

- *differentiate* the algebraic equations (2.45) k times: $\frac{d^k}{dt^k}(\mathbf{g}(\mathbf{x}, \mathbf{z}, t)) = \mathbf{0}$, i.e. since $\mathbf{g} : \mathbb{R}^n \times \mathbb{R}^m \times \mathbb{R} \to \mathbb{R}^m$, there are m new differential equations
- k statements of consistent *initial values* from the nonlinear system of equations:

$$\mathbf{g}(\mathbf{x}(0), \mathbf{z}(0), t = 0) = \mathbf{0} \quad \ldots \text{ algebraic equations}$$

$$\left.\begin{aligned} \frac{d}{dt}\Big[\mathbf{g}(\mathbf{x}, \mathbf{z}, t)\Big]_{\mathbf{x}=\mathbf{x}(0),\mathbf{z}=\mathbf{z}(0),t=0} &= \mathbf{0} \\ &\vdots \\ \frac{d^{k-1}}{dt^{k-1}}\Big[\mathbf{g}(\mathbf{x}, \mathbf{z}, t)\Big]_{\mathbf{x}=\mathbf{x}(0),\mathbf{z}=\mathbf{z}(0),t=0} &= \mathbf{0} \end{aligned}\right\} \ldots (k-1) \text{ derivatives}$$

- $\mathbf{x}(0), \mathbf{z}(0)$ must satisfy the above system of equations, i.e. they generally can not be chosen independently of one another.

Further index reduction procedures Index reduction of DAE systems is the key to successfully employing object-oriented physical models. As will be shown in Ch. 3, only DAE systems with sufficiently low index can be solved in a numerically stable, usable form. For this reason, index reduction should always be attempted for DAE systems with a higher index to bring them into a tractable form. In addition to the further readings presented above, the interested reader is directed to three core original publications: (Pantelides 1988), (Cellier and Elmqvist 1993), (Mattsson and Söderlind 1993).

Example 2.11 *Index reduction for a rigidly-coupled two-mass system.*

Algebraic conditions: consistent initial conditions The calculations in Example 2.10 lead to the DAE system (2.49), (2.50). Derived from the thrice-differentiated algebraic constraints, the algebraic conditions below hold for the entire trajectory (cf. Eqs. (2.51) through (2.53)) and thus for $t = 0$, from which follow *consistent* initial conditions:

$$\left.\begin{aligned} x_1(t) &\overset{!}{=} x_3(t) \\ x_2(t) &\overset{!}{=} x_4(t) \\ z(t) &\overset{!}{=} \frac{m_2}{m_1 + m_2} u(t) \end{aligned}\right\} \forall t \Rightarrow t = 0 : \boxed{\begin{aligned} x_1(0) &\overset{!}{=} x_3(0) \\ x_2(0) &\overset{!}{=} x_4(0) \\ z(0) &\overset{!}{=} \frac{m_2}{m_1 + m_2} u(0) \end{aligned}} \tag{2.54}$$

Given the initial conditions (2.54) and the differential equations (2.49), (2.53), a causal block diagram simulation model can now be constructed, where the differentiation of the input value (Fig. 2.60a) can be eliminated by integrating Eq. (2.53) (Fig. 2.60b).

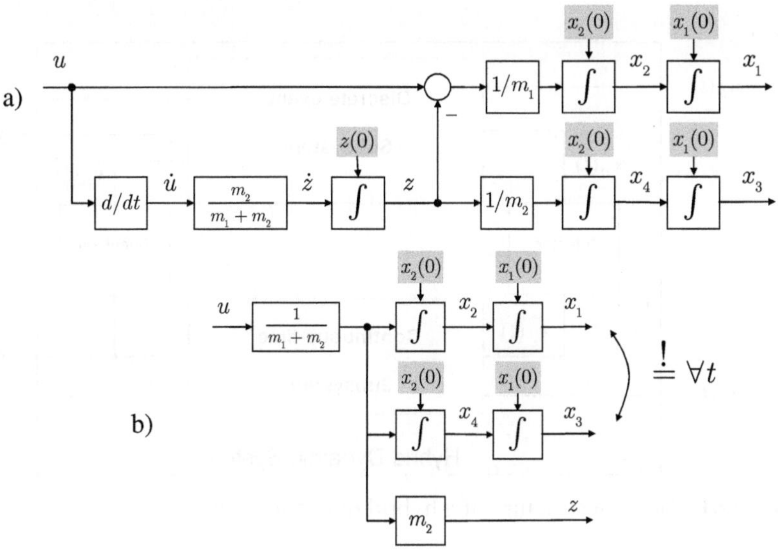

Fig. 2.60. Signal-oriented simulation model of the rigidly-coupled two-mass system: a) with integrator for the algebraic variable z, b) equivalent representation with integrator for the algebraic variable z eliminated

■

2.5 Hybrid Systems

2.5.1 General structure of a hybrid system

Hybrid systems: terminology In many technical systems, especially if the boundaries of the system are sufficiently broadly defined (see Sec. 2.1) or if the system dynamics are modeled with sufficient accuracy, the simultaneous appearance of continuous-time and discrete-event phenomena will be observed. Such a system model is then termed a *hybrid dynamic* system, referred to concisely as a "*hybrid* system" below[32].

For example, in mechatronic systems, a hybrid description is called for when, during the course of operation, the governing mathematical model changes due to a deterministic external influence (e.g. user intervention) or state-dependent conditions (e.g. mechanical contact effects, state-dependent parameter changes). Such hybrid dynamics resulting from continuous-time and discrete-event changes—and particularly their resulting interactions—must be appropriately represented in a *combined* system model (Engell et al. 2002), (Buss 2002).

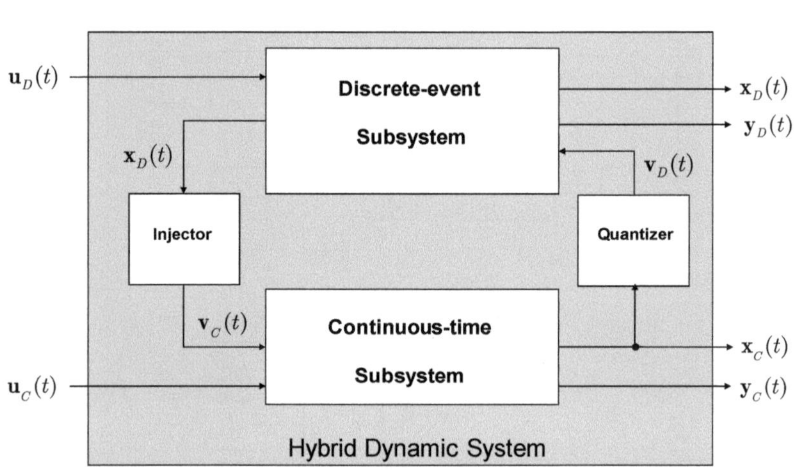

Fig. 2.61. General structure of a hybrid dynamic system

[32] Often mixed *continuous-time, discrete-time* models are termed *hybrid.* At the simulation implementation level, such models should be considered special cases of hybrid discrete-event, continuous-time systems, in which the events occur at well-defined, predetermined, periodic times. The principle of energy-based adaptive step size (Sec. 3.7) can be correspondingly applied in a greatly simplified form.

Hybrid system structure The general structure of a *hybrid discrete-event, continuous-time* system (or *hybrid* system) is represented in Fig. 2.60 (Lunze 2002). A hybrid system comprises the following components and system variables:

$\mathbf{x}_C, \mathbf{u}_C, \mathbf{y}_C$	continuous states/inputs/outputs
$\mathbf{x}_D, \mathbf{u}_D, \mathbf{y}_D$	discrete states/inputs/outputs
Continuous-time subsystem	the principal, continuous-time system dynamics
Discrete-event subsystem	varying operating states of the system
Injector	unique map from a discrete-valued signal defined over a finite set of symbols to a real-valued signal
Quantizer	discrete-valued map of a real-valued signal

Research in this area is still relatively young, and as yet no generally-valid, easily-applicable methods particular to the design and analysis of hybrid systems have been established. The situation is somewhat better in the area of methodological support and toolset implementation for modeling and simulation. This is of rather more practical significance, as such tools enable—at least at the model level—the prediction of complex hybrid system dynamics via simulation (design verification). The remainder of this section introduces the most significant hybrid phenomena in mathematical notation (Lunze 2002); presents a specialized, practical approach—*net-state models* (Nenninger et al. 1999)—and discusses particulars of the implementation of simulations.

2.5.2 Hybrid phenomena

Hybrid system Consider as given the following description of a *continuous subsystem*[33]

$$\dot{\mathbf{x}} = \mathbf{f}^1(\mathbf{x}, \mathbf{u}, t) \text{ where } \mathbf{x} \in M_1, \tag{2.55}$$

$$\dot{\mathbf{x}} = \mathbf{f}^2(\mathbf{x}, \mathbf{u}, t) \text{ where } \mathbf{x} \in M_2, \tag{2.56}$$

$$\dot{\mathbf{x}} = \mathbf{f}^\delta \delta\left(m^\delta(\mathbf{x}, \mathbf{u}, t)\right), \tag{2.57}$$

---

[33] This representation can also be directly applied to DAE systems, i.e. the conditions presented for the vector field of the state variable derivatives (= right-hand side of the ODE) hold equivalently for the vector field of the algebraic constraints.

with *discontinuity hypersurface*

$$m(\mathbf{x}, \mathbf{u}, t) = 0. \tag{2.58}$$

Discontinuous vector field One characteristic of hybrid systems is the existence of discontinuities in the vector field of derivatives of the state variables (= right-hand side of the ordinary differential equations (2.55) through (2.57)). In such cases, the smoothness of the vector field is violated, i.e. there exist states $\mathbf{x}_s$, for which *no* LIPSCHITZ condition of the type

$$\left\|\mathbf{f}^i(\mathbf{x}_s, \mathbf{u}, t) - \mathbf{f}^j(\tilde{\mathbf{x}}, \mathbf{u}, t)\right\| \le L \cdot \left\|\mathbf{x}_s - \tilde{\mathbf{x}}\right\|, \quad i, j \in \{1, 2, \delta\}$$
$$\|\bullet\| \text{ arbitrary vector norm, } L \in \mathbb{R}^+ \text{ finite} \tag{2.59}$$

exists (holds similarly for $\mathbf{u}$ if its components $u_j(t)$ exhibit step-wise changes).

Sets of states The sets M_1, M_2 describe regions in the n-dimensional state space $\mathbb{R}^n$ in which the corresponding vector fields $\mathbf{f}^1(\mathbf{x}, \mathbf{u}, t)$, $\mathbf{f}^2(\mathbf{x}, \mathbf{u}, t)$ define the dynamics of the continuous subsystem. The boundary between M_1 and M_2 is described by the *discontinuity hypersurface* (2.58). Systems with a description following Eq. (2.55), (2.56) are distinguished by the fact that the *states* themselves remain *continuous* at the discontinuity hypersurfaces, whereas in the case of a system following Eq. (2.57) so-called *state discontinuities* appear (Fig. 2.62).

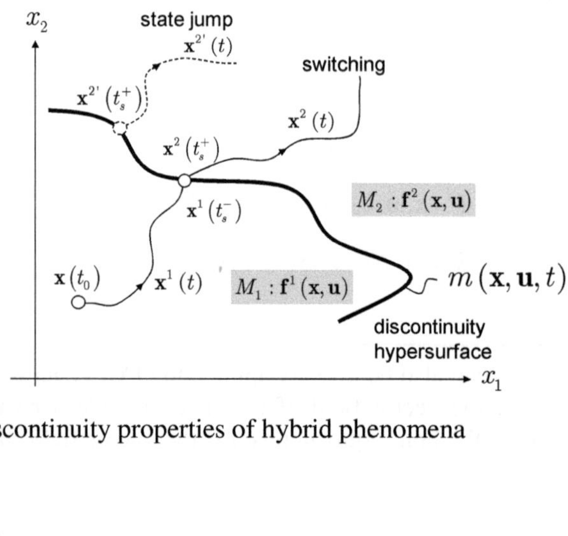

Fig. 2.62. Discontinuity properties of hybrid phenomena

Hybrid dynamics of systems can be systematically described by the following four *hybrid phenomena*. In concrete cases, an arbitrary combination of these phenomena can be present.

Autonomous switching

Entry of the trajectory of the continuous system into a particular subset of the continuous state space which induces a change in the discrete state, which in turn modifies the continuous system dynamics (⇒ Eqs. (2.55), (2.56), (2.58); states remain continuous).

Controlled switching

An *external* intervention triggers a change in the discrete state, which in turn modifies the continuous system dynamics (⇒ Eqs. (2.55), (2.56), (2.58), discontinuous change in $\mathbf{u}(t)$; states remain continuous).

Autonomous state jumps

Entry of the trajectory of the continuous system into a particular subset of the continuous state space which induces a change in the discrete state, which directly triggers a jump in the continuous state (or subset of the state variables) (⇒ Eq. (2.57)).

This behavior can be illustrated with a *first-order system*:

$$\left.\begin{aligned} \dot{x}(t) &= a \cdot \delta\left(x(t) - x_s\right) \\ x(t_s) &= x_s \end{aligned}\right\}$$

When the threshold x_s is exceeded, the state immediately jumps by an amount a, i.e. the derivative becomes infinite, as modeled by the DIRAC delta function[34] $\delta(\bullet)$.

$$\Rightarrow \quad x(t_s + 0) = x(t_s - 0) + a \quad \text{and} \quad x(t_s) = x(t_s^-) + a = x_s + a\,.$$

[34] More precisely, this is a DIRAC distribution, as the argument is a function of time.

Controlled state jumps

An external intervention triggers a change in the discrete state, which in turn directly triggers a jump in the continuous state (⇒ Eq. (2.57)).

To illustrate, once again consider a *first-order system*:

$$\left.\begin{aligned} \dot{x}(t) &= a \cdot \delta\left(u(t) - u_s\right) \\ u(t_s) &= u_s \end{aligned}\right\}$$

$$\Rightarrow \quad x(t_s + 0) = x(t_s - 0) + a \quad \text{and} \quad x(t_s) = x(t_s^-) + a$$

Note: $u(t)$ is continuous, i.e. does not have a jump/DIRAC character.

Switching occurs most commonly (e.g. structure-variable control), *discontinuities* are more rare, e.g. in the case of collisions.

2.5.3 Net-state models

Model structure The general structure of a *net-state model* (*NSM*) is shown in Fig. 2.63. Here, the discrete-event (DE) subsystem is realized as an *interpreted PETRI net* (*IPN*) (with inputs and outputs) (Litz 2005). When combined with a continuity condition on states in the continuous-time subsystem (*extended state space model* (*ESM*)), all above-mentioned hybrid phenomena can be modeled using a net-state model (Nenninger et al. 1999).

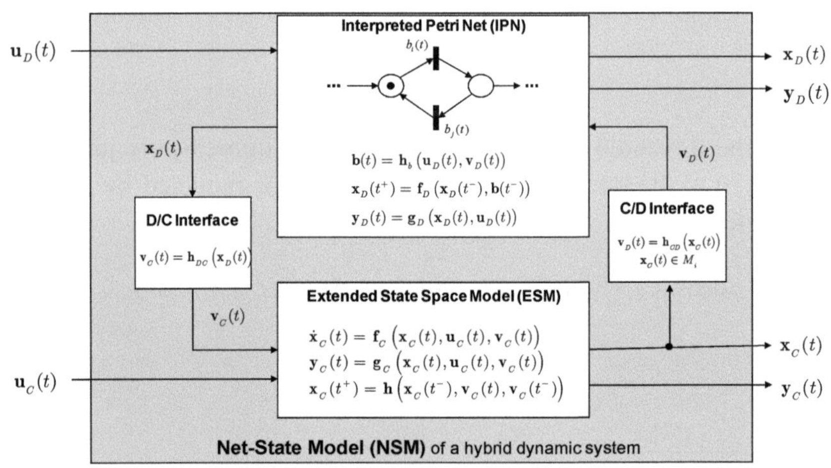

Fig. 2.63. General net-state model (NSM)

Properties of net-state models

- *Discrete-event subsystem*
  By using interpreted PETRI nets (IPNs), a great diversity of models is achievable, as both purely sequential and parallel processes can be modeled (state machines vs. synchronization graphs); interpretation is conventional: switching conditions at the transitions represent inputs to the DE subsystem, and output places represent outputs of the DE subsystem.
- *Hybrid model state*

  $$\mathbf{x}_H(t) = \left(\mathbf{x}_D(t), \mathbf{x}_C(t)\right)^T .$$

- *Discrete model state*
  Describes the current marking vector of the interpreted PETRI net (IPN token distribution).
- Dynamics of the discrete model variables $\mathbf{x}_D(t)$, $\mathbf{u}_D(t)$
  Piecewise-constant, until the occurrence of a switching event at time t^-.
- *D/C interface, injection*
  Dynamics of the continuous model variable $\mathbf{v}_C(t)$; piecewise-constant, until the occurrence of a switching event at time t^-.
- *Firing requirement for transition j*
  If the transition j is activated according to the marking, and the Boolean switching expression $b_j(t)$ is true, then transition j fires immediately; if several parallel transitions are ready to fire, then all fire simultaneously.
- Reinitialization of the continuous model state $\mathbf{x}_C$
  Upon an external/autonomous jump in the state, the continuous state vector $\mathbf{x}_C(t^+)$ is updated according to the discontinuity equation.
- *C/D interface, quantization*
  Threshold violation by the continuous model state $\mathbf{x}_C$. When the continuous state vector $\mathbf{x}_C(t)$ enters the set Φ_i, the corresponding internal model variable $v_{Di}(t)$ is set to 1 (i.e. $v_{Di}(t) = 0$ for $\mathbf{x}_C(t) \notin M_i$).

Modeling the discrete-event subsystem A detailed description of discrete-event modeling paradigms is beyond the scope of this book. For the reader unacquainted with this area of modeling, the very readable monograph (Litz 2005) is recommended, where, in particular, the *interpreted PETRI net* employed here is described in detail.

PETRI nets are not required to describe net-state models. However, in comparison to sequential *automaton models*, they do appear more suitable, since parallel PETRI nets produce a clearer DE system structure when several discrete system variables are present. Note, however, that for a comprehensive analysis of the dynamics of a DE subsystem, the entire *reachability graph* of the PETRI net must be examined. This graph is itself a sequential automaton (with a potentially very large number of states). In this respect, there is no difference in the amount of effort required for these two representations. The use of one over the other is rather a matter of taste for the designer.

Computational implementation with STATECHARTS The implementation of net-state models for computation requires a common platform for continuous-time and discrete-event models. For both automaton models and PETRI net models, there exists only a limited number of publicly-available simulation platforms which allow for the incorporation of continuous-time models. A practicable solution is offered by the *STATECHARTS* modeling paradigm (Harel 1987), with which hierarchical and parallel automaton structures can be very efficiently modeled. However, when representing parallel, structurally limited PETRI net models using STATECHARTS, particular transformation limitations must be observed (Schnabel et al. 1999).

The attraction of STATECHARTS for discrete-event modeling in hybrid systems is not least due to the fact that this modeling paradigm has been successfully implemented in commercial simulation tools[35] so that easy-to-use computational platforms for efficient simulation of hybrid systems are available.

Example 2.12 *Single-joint manipulator with collision.*

System configuration An elastically suspended single-joint manipulator (massless arm of length l, end effector mass m, spring constant k, motor torque τ, equilibrium position at $\theta = 0$) moves horizontally on a level surface with position-dependent, velocity-proportional coefficients of friction μ_1, μ_2 (Fig. 2.64). Along the x-axis shown, collisions with a hard boundary are possible (assuming elastic collisions).
For this system, a *net-state model* is to be created.

---

[35] E.g. STATEFLOW, a component of MATLAB / SIMULINK.

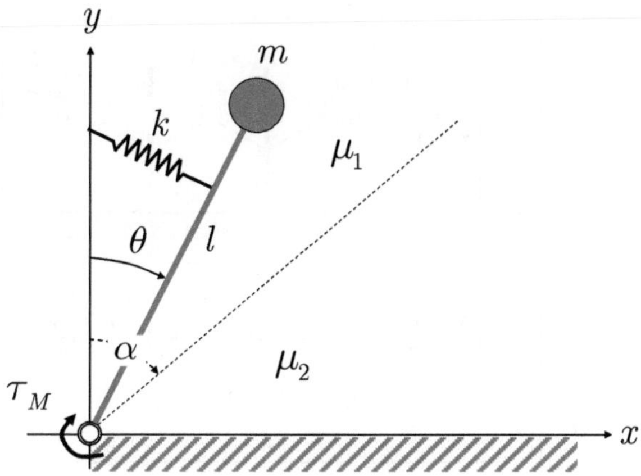

Fig. 2.64. Single-joint manipulator with collision

Model creation In this system, two *hybrid phenomena* appear:

- *autonomous switching*: due to the position-dependent friction μ_1, μ_2, the vector field of the equations of motion changes when the threshold on the arm angle (state) $\theta > \alpha$ is exceeded,
- *autonomous state jump*: in the case of contact at $\theta = 90°$ under the assumption of an elastic collision, the state θ remains continuous; however, there is a jump in the state $\dot{\theta}$: the angular velocity changes sign and its magnitude decreases according to the *coefficient of restitution* ρ, $0 \leq \rho \leq 1$.

In the state plane $\theta, \dot{\theta}$, three disjoint sets of states $\mathcal{M}_1, \mathcal{M}_2, \mathcal{M}_3$ can be defined. These sets are in turn assigned (binary) variables $v_{Di} \in \{0,1\}$, $i = 1,2,3$ in the C/D interface.

$$v_{Di} \begin{cases} 1, & \mathbf{x}_C \in \mathcal{M}_i \\ 0, & \mathbf{x}_C \notin \mathcal{M}_i \end{cases}, i = 1,2,3 \quad \begin{aligned} v_{D1} &: \quad \mathcal{M}_1 = \{\theta \,|\, 0 \leq \theta < \alpha\} \\ v_{D2} &: \quad \mathcal{M}_2 = \{\theta \,|\, \alpha \leq \theta < 90°\} \\ v_{D3} &: \quad \mathcal{M}_3 = \{\theta \,|\, \theta \geq 90°\} \end{aligned}$$

The alternation of the trajectory $\mathbf{x}_C(t)$ between the subsets can be described via a simple, structurally constrained PETRI net of the *state machine* type (Fig. 2.65). The discrete states $x_{Di} \in \{0,1\}$, $i = 1,2,3$ describe the current subset occupied by $\mathbf{x}_C(t)$.

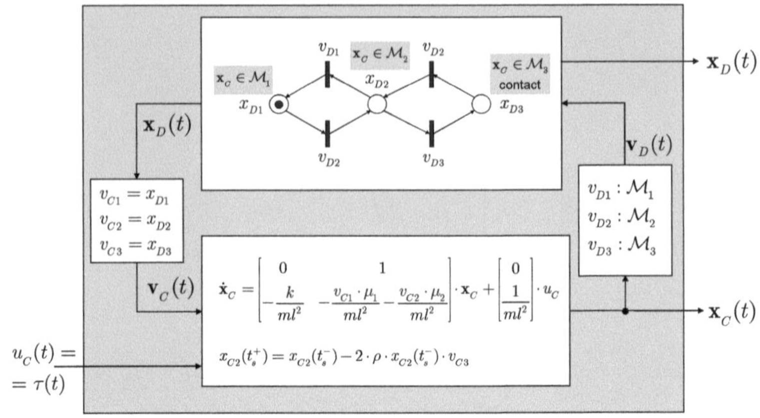

Fig. 2.65. Net-state model for single-joint manipulator with collision

In the D/C interface, the discrete states are used to generate continuous-time switching variables $v_{Ci}(t)$, for which

$$v_{Ci}(t) = \begin{cases} 0, & x_{Di} = 0 \\ 1, & x_{Di} = 1 \end{cases}$$

holds.

Using the switching variables $v_{Ci}(t)$, the vector field in the state model can then be modified in a timely manner or the state discontinuity can be modeled. The complete net-state model is presented in Fig. 2.65.

■

2.6 Linear System Models

Linear dynamic analysis As shown in the previous sections, the class of mechatronic systems with lumped elements considered in this book can be described in a general form by a system of *nonlinear* differential-algebraic equations. The methods used throughout this book for dynamic analysis and controller design are, however, based on *linear time invariant* (*LTI*) models. This section thus presents a primer on *local linearization* (or *JACOBI linearization*) of nonlinear dynamic systems. This type of linearization is well known in many technical disciplines and is discussed here to complete the methodological toolbox.

Local linearization As a precondition for local linearization, the dynamics of the nonlinear system in the neighborhood of certain solutions (trajectories) are examined. Even with a completely known (nonlinear) model, the result is always an *approximation* of the actual system behavior in the vicinity of a certain trajectory and it is no longer representative if "larger" deviations from the reference solution are considered. Results based on such linearized models should thus always be evaluated with all due caution, and the observance of "sufficiently small deviations" should be carefully verified in every particular case.

Exact linearization Attention of the interested reader is directed to one extended form of linearization, which has in the last two decades significantly extended control system theory in particular. In *exact linearization, input-output linearization*, or *feedback linearization*—e.g. (Isidori 2006)—a nonlinear transformation (e.g. feedback) is inserted in the system to generate a "linear" system (i.e. the system is not simply "linearized", but in fact "linear") to which linear control laws can then be applied (for examples in robot controllers see (Siciliano et al. 2009)). In such cases, then, given exact knowledge of the original nonlinear system, an *exactly linear system* is generated. However, since this type of linear model is of only limited use for general analysis of system dynamics, it is *local linearization* which is pursued below.

2.6.1 Local linearization of nonlinear state space models

System description Consider the following nonlinear state space model as a special case of a DAE system:

$$\dot{\mathbf{x}} = \mathbf{f}(\mathbf{x}, \mathbf{u}), \tag{2.60}$$

$$\mathbf{y} = \mathbf{g}(\mathbf{x}, \mathbf{u}), \tag{2.61}$$

$$\mathbf{f} : \mathbb{R}^n \times \mathbb{R}^r \to \mathbb{R}^n, \ \mathbf{g} : \mathbb{R}^n \times \mathbb{R}^r \to \mathbb{R}^m .$$

For a given input $\mathbf{u}_*(t)$ and the resulting solution $\mathbf{x}_*(t)$ of the differential equation (2.60), it holds by definition that

$$\begin{aligned} \dot{\mathbf{x}}_* &= \mathbf{f}(\mathbf{x}_*, \mathbf{u}_*) \\ \mathbf{y}_* &= \mathbf{g}(\mathbf{x}_*, \mathbf{u}_*) \end{aligned} . \tag{2.62}$$

Now, considering (arbitrary) deviations $\mathbf{x}(t) = \mathbf{x}_*(t) + \Delta\mathbf{x}(t)$ from the *reference solution* resulting from input deviations $\mathbf{u}(t) = \mathbf{u}_*(t) + \Delta\mathbf{u}(t)$, it then follows from Eqs. (2.60), (2.61) that

$$\begin{aligned} \dot{\mathbf{x}}_* + \Delta\dot{\mathbf{x}} &= \mathbf{f}\left(\mathbf{x}_* + \Delta\mathbf{x}, \mathbf{u}_* + \Delta\mathbf{u}\right) \\ \mathbf{y}_* + \Delta\mathbf{y} &= \mathbf{g}\left(\mathbf{x}_* + \Delta\mathbf{x}, \mathbf{u}_* + \Delta\mathbf{u}\right) \end{aligned} \qquad (2.63)$$

where the substitution $\mathbf{y} = \mathbf{y}_* + \Delta\mathbf{y}$ was applied to the output vector.

Linear approximation Replacing the vector fields **f** and **g** in Eq. (2.63) by their *linear approximations* (Taylor expansion) and considering *small deviations* $\Delta\mathbf{x}$, $\Delta\mathbf{u}$, higher-order terms can be neglected, so that

$$\begin{aligned} \dot{\mathbf{x}}_* + \Delta\dot{\mathbf{x}} &= \mathbf{f}\left(\mathbf{x}_*, \mathbf{u}_*\right) + \mathbf{A}\Delta\mathbf{x} + \mathbf{B}\Delta\mathbf{u} \\ \mathbf{y}_* + \Delta\mathbf{y} &= \mathbf{g}\left(\mathbf{x}_*, \mathbf{u}_*\right) + \mathbf{C}\Delta\mathbf{x} + \mathbf{D}\Delta\mathbf{u} \end{aligned} \qquad (2.64)$$

where

$$\begin{aligned} A_{ij} &= \frac{\partial f_i}{\partial x_j}\left(x_*, u_*\right), \quad \mathbf{A} \in \mathbb{R}^{n\times n} \quad C_{ij} = \frac{\partial g_i}{\partial x_j}\left(x_*, u_*\right), \quad \mathbf{C} \in \mathbb{R}^{m\times n} \\ B_{ij} &= \frac{\partial f_i}{\partial u_j}\left(x_*, u_*\right), \quad \mathbf{B} \in \mathbb{R}^{n\times r}, \quad D_{ij} = \frac{\partial g_i}{\partial u_j}\left(x_*, u_*\right), \quad \mathbf{D} \in \mathbb{R}^{m\times r}. \end{aligned} \qquad (2.65)$$

Due to (2.62), Eq. (2.64) can be simplified to give the standard representation of a *linear state space model*

$$\begin{aligned} \Delta\dot{\mathbf{x}} &= \mathbf{A}\Delta\mathbf{x} + \mathbf{B}\Delta\mathbf{u} \\ \Delta\mathbf{y} &= \mathbf{C}\Delta\mathbf{x} + \mathbf{D}\Delta\mathbf{u} \end{aligned}. \qquad (2.66)$$

The system matrices (2.65) are the well-known *JACOBIAN matrices* of the vector fields **f** and **g**, e.g.

$$\mathbf{A} := \frac{\partial \mathbf{f}}{\partial \mathbf{x}}\left(\mathbf{x}_*, \mathbf{u}_*\right) = \begin{bmatrix} \frac{\partial f_1}{\partial x_1}\left(\mathbf{x}_*, \mathbf{u}_*\right) & \cdots & \frac{\partial f_1}{\partial x_n}\left(\mathbf{x}_*, \mathbf{u}_*\right) \\ \vdots & & \vdots \\ \frac{\partial f_n}{\partial x_1}\left(\mathbf{x}_*, \mathbf{u}_*\right) & \cdots & \frac{\partial f_n}{\partial x_n}\left(\mathbf{x}_*, \mathbf{u}_*\right) \end{bmatrix}. \qquad (2.67)$$

Thus, the state space system (2.66) is called the *local linearization* or *JACOBI linearization* of (2.60), (2.61).

Eq. (2.67) predicts that at every time $\tilde{t}$, the partial derivatives of the respective vector fields should be calculated and subsequently the variables x_i, u_j should be replaced by the corresponding values $x_{i*}(\tilde{t})$, $u_{j*}(\tilde{t})$ from the reference solution. This results in the following two characteristic cases.

Linearization about an equilibrium For constant inputs $\mathbf{u}_*(t) = \mathbf{u}_{*0} = const.$, the *equilibria* $\mathbf{x}_{*0} = const.$ of the system in Eq. (2.60) are computed from the algebraic system of equations

$$\mathbf{0} = \mathbf{f}\left(\mathbf{x}_{*0}, \mathbf{u}_{*0}\right). \tag{2.68}$$

With the (constant) solutions $\mathbf{u}_{*0}, \mathbf{x}_{*0}$ of Eq. (2.68), *constant* system matrices (Eq. (2.65)) follow, and thus a linear *time-invariant* (LTI) state space model

$$\begin{aligned} \Delta\dot{\mathbf{x}} &= \mathbf{A}_0 \Delta\mathbf{x} + \mathbf{B}_0 \Delta\mathbf{u}, \\ \Delta\mathbf{y} &= \mathbf{C}_0 \Delta\mathbf{x} + \mathbf{D}_0 \Delta\mathbf{u}. \end{aligned} \tag{2.69}$$

The tuple $\left(\mathbf{x}_{*0}, \mathbf{u}_{*0}\right)$ is also called the *operating point* of the system so that Eq. (2.69) is referred to as a *linearization* about the operating point $\left(\mathbf{x}_{*0}, \mathbf{u}_{*0}\right)$.

Linearization about a trajectory For general, non-constant inputs $\mathbf{u}_*(t)$, the time-varying *solution trajectory* $\mathbf{x}_*(t)$ must also be incorporated into the computation of the system matrices (2.65)[36], resulting in time-varying system matrices and, overall, a *linear time-varying (LTV) state space system*

$$\begin{aligned} \Delta\dot{\mathbf{x}} &= \mathbf{A}(t) \cdot \Delta\mathbf{x} + \mathbf{B}(t) \cdot \Delta\mathbf{u}, \\ \Delta\mathbf{y} &= \mathbf{C}(t) \cdot \Delta\mathbf{x} + \mathbf{D}(t) \cdot \Delta\mathbf{u}. \end{aligned} \tag{2.70}$$

As usual, the eigenvalues of the system matrix $\mathbf{A}$ determine the dynamics and stability of the locally linearized system.

[36] This solution trajectory comes from the solution of the nonlinear system model (2.60). Only in exceptional cases is this analytically possible. For simulation experiments, the value for $\mathbf{x}_*(t)$ is taken to be the approximate value from the numerical solution. For stochastic dynamic analysis (Ch. 11), pre-calculated (numerical) nominal trajectories can be used.

Example 2.13 *Particle motion with viscous friction.*

System configuration Consider a point mass *m* moving weightlessly in a plane subject to viscous friction (coefficient of friction μ) and an applied force $\vec{F}$. The current distance *r* from the point mass to the origin can be measured with a suitable measuring device (*range measurements*) (Fig. 2.66).

Find a linear state space model given a reference trajectory $\vec{s}_*(t), \vec{v}_*(t), \vec{F}_*(t)$.

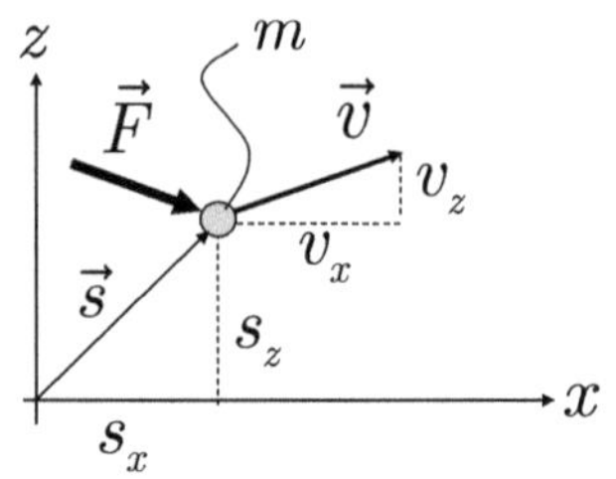

Fig. 2.66. Linearization of a nonlinear state space model

Model creation

- Equation of motion for the point mass:

$$m\dot{\vec{v}} = \vec{F} - \mu|\vec{v}|^2 \frac{\vec{v}}{|\vec{v}|}$$

- Measurement equation:

$$r = |\vec{s}|$$

- Nonlinear state space model:

$$x_1 := s_x,\ x_2 := v_x,\ x_3 := s_z,\ x_4 := v_z$$

$$u_1 := F_x,\ u_2 := F_z, \quad y := r$$

$$\dot{x}_1 = x_2$$

$$\dot{x}_2 = -\frac{\mu}{m} x_2 \sqrt{x_2^2 + x_4^2} + \frac{1}{m} u_1$$

$$\dot{x}_3 = x_4$$

$$\dot{x}_4 = -\frac{\mu}{m} x_4 \sqrt{x_2^2 + x_4^2} + \frac{1}{m} u_2$$

$$y = \sqrt{x_1^2 + x_3^2}$$

- Linearized state space model (time-varying due to $x_{i*}(t)$):

$$\Delta\dot{x}_1 = \Delta x_2$$

$$\Delta\dot{x}_2 = -\frac{\mu}{m}\frac{2x_{2*}^2 + x_{4*}^2}{\sqrt{x_{2*}^2 + x_{4*}^2}}\Delta x_2 - \frac{\mu}{m}\frac{2x_{2*}x_{4*}}{\sqrt{x_{2*}^2 + x_{4*}^2}}\Delta x_4 + \frac{1}{m}\Delta u_1$$

$$\Delta\dot{x}_3 = \Delta x_4$$

$$\Delta\dot{x}_4 = -\frac{\mu}{m}\frac{2x_{2*}x_{4*}}{\sqrt{x_{2*}^2 + x_{4*}^2}}\Delta x_2 - \frac{\mu}{m}\frac{2x_{4*}^2 + x_{2*}^2}{\sqrt{x_{2*}^2 + x_{4*}^2}}\Delta x_4 + \frac{1}{m}\Delta u_2$$

$$\Delta y = \frac{2x_{1*}}{\sqrt{x_{1*}^2 + x_{3*}^2}}\Delta x_1 + \frac{2x_{3*}}{\sqrt{x_{1*}^2 + x_{3*}^2}}\Delta x_3$$

■

2.6.2 Local linearization of nonlinear DAE systems

Consider the *semi-explicit* time-invariant DAE system

$$\begin{aligned}\dot{\mathbf{x}} &= \mathbf{f}(\mathbf{x},\mathbf{z},\mathbf{u}),\\ \mathbf{0} &= \mathbf{g}(\mathbf{x},\mathbf{z}),\end{aligned} \tag{2.71}$$

$$\mathbf{f}: \mathbb{R}^n \times \mathbb{R}^m \times \mathbb{R}^r \to \mathbb{R}^n,\ \mathbf{g}: \mathbb{R}^n \times \mathbb{R}^m \to \mathbb{R}^m.$$

Generalizing the results in Sec. 2.6.1 for small deviations $\Delta\mathbf{x}$, $\Delta\mathbf{z}$, $\Delta\mathbf{u}$ from a reference solution $\mathbf{x}_*, \mathbf{z}_*, \mathbf{u}_*$ (i.e. $\mathbf{x} = \mathbf{x}_* + \Delta\mathbf{x}$, $\mathbf{z} = \mathbf{z}_* + \Delta\mathbf{z}$, $\mathbf{u} = \mathbf{u}_* + \Delta\mathbf{u}$) gives the *locally linearized DAE system*

$$\Delta\dot{\mathbf{x}} = \mathbf{A}\Delta\mathbf{x} + \mathbf{B}\Delta\mathbf{u} + \mathbf{F}\Delta\mathbf{z}$$

$$\mathbf{0} = \mathbf{G}\Delta\mathbf{x} + \mathbf{H}\Delta\mathbf{z}$$

with system matrices

$$A_{ij} = \frac{\partial f_i}{\partial x_j}(x_*, z_*, u_*),\quad \mathbf{A} \in \mathbb{R}^{n\times n}$$

$$B_{ij} = \frac{\partial f_i}{\partial u_j}(x_*, z_*, u_*),\quad \mathbf{B} \in \mathbb{R}^{n\times r},$$

$$F_{ij} = \frac{\partial f_i}{\partial z_j}(x_*, z_*, u_*),\quad \mathbf{F} \in \mathbb{R}^{n\times m}$$

$$G_{ij} = \frac{\partial g_i}{\partial x_j}(x_*, z_*),\quad \mathbf{G} \in \mathbb{R}^{m\times n}$$

$$H_{ij} = \frac{\partial g_i}{\partial z_j}(x_*, z_*),\quad \mathbf{H} \in \mathbb{R}^{m\times m}.$$

Note that depending on the index of the DAE system, the matrices **G** and **H** can also be singular.

2.6.3 LTI systems: transfer function, frequency response

Frequency domain representation Starting with mathematical models of a mechatronic system in DAE form (2.71) or state space form (2.60), (2.61), a coupled dynamic analysis is particularly clear if undertaken with *linearized models* in the *frequency domain.* The most important tool presented in this book for evaluating system dynamics is thus the use of *transfer functions* of the system elements under consideration. As usual, the linear (LTI) state space model (2.69) can be used to compute the transfer function via the *LAPLACE transform* (Ogata 1992).

As an important, representative case, the rest of this section considers the *single-input single-output* (*SISO*) LTI system

$$\begin{aligned} \dot{\mathbf{x}} &= \mathbf{A} \cdot \mathbf{x} + \mathbf{b} \cdot u \qquad && \mathbf{x}, \mathbf{b}, \mathbf{c} \in \mathbb{R}^n, \ \mathbf{A} \in \mathbb{R}^{n \times n} \\ y &= \mathbf{c}^T \cdot \mathbf{x} + d \cdot u && d, u, y \in \mathbb{R} \end{aligned} \quad . \tag{2.72}$$

Transfer function

Procedure By applying the LAPLACE transform to Eq. (2.72) and with straightforward manipulation of the resulting algebraic equations in the complex variable *s*, the *transfer function G*(*s*) is obtained:

$$G(s) = \frac{Y(s)}{U(s)} = \mathbf{c}^T (s\mathbf{E} - \mathbf{A})^{-1} \mathbf{b} + d, \tag{2.73}$$

where $U(s)$ and $Y(s)$ represent the LAPLACE transforms of the input $u(t)$ and the output $y(t)$, respectively (Fig. 2.67, transform direction from left to right).

The *transfer function G*(*s*) is a rational function of the form

$$\begin{aligned} G(s) = \frac{Y(s)}{U(s)} &= \frac{b_m \cdot s^m + \cdots + b_1 \cdot s + b_0}{s^n + a_{n-1} \cdot s^{n-1} + \cdots + a_1 \cdot s + a_0} \\ &= \frac{N_G(s)}{D_G(s)} = b_m \frac{\prod_{j=1}^{m} (s - n_j)}{\prod_{i=1}^{n} (s - p_i)} \end{aligned} \quad , \quad m \le n \,, \tag{2.74}$$

with *numerator polynomial* $N_G(s)$, *denominator polynomial* $D_G(s)$, *zeroes* n_i, and *poles* p_j. If the degree *m* of the numerator equals the degree *n* of the denominator, there is a direct feedthrough of the input *u* to the output *y*—equivalent to $d \neq 0$ in Eq. (2.72).

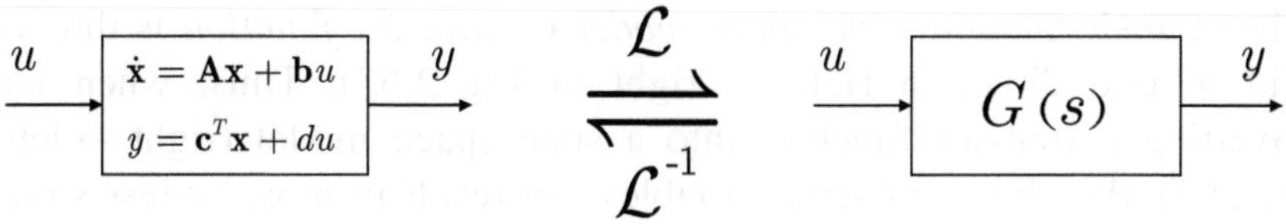

Fig. 2.67. LTI system: transfer function (single-input single-output (SISO))

Poles and zeroes The dynamic behavior, and in particular the stability, of the system is determined by the *poles* of the transfer function $G(s)$, which as a rule are identical to the *eigenvalues* of the system matrix **A**[37]. The *zeroes* of the transfer function depend on the *input vector* **b** and the *output vector* **c** of the state space model in a complicated manner which can be expressed analytically only to a limited extent[38].

Inverse transformation to state space model Given a transfer function G(s) (2.74), an infinite number of state space representations is possible (Fig. 2.67, transform direction from right to left). One possible form is the so-called *control canonical form* (Ogata 2010): (for $m = n$)

$$\begin{pmatrix} \dot{x}_1 \\ \dot{x}_2 \\ \vdots \\ \dot{x}_n \end{pmatrix} = \begin{pmatrix} 0 & 1 & 0 & \cdots & 0 \\ 0 & 0 & 1 & \cdots & 0 \\ \vdots & & & \ddots & \\ 0 & 0 & 0 & \cdots & 1 \\ -a_0 & -a_1 & -a_2 & \cdots & -a_{n-1} \end{pmatrix} \cdot \begin{pmatrix} x_1 \\ x_2 \\ \vdots \\ x_n \end{pmatrix} + \begin{pmatrix} 0 \\ 0 \\ \vdots \\ 1 \end{pmatrix} \cdot u, \tag{2.75}$$

$$y = \begin{pmatrix} (b_0 - b_n a_0) & (b_1 - b_n a_1) & \cdots & (b_{n-1} - b_n a_{n-1}) \end{pmatrix} \cdot \begin{pmatrix} x_1 \\ x_2 \\ \vdots \\ x_n \end{pmatrix} + b_n u .$$

---

37 This is *never* the case when the state space model (2.72) is *uncontrollable* or *unobservable*. In these cases, certain pole and zero terms cancel in Eq. (2.74), i.e. the number of poles in the transfer function is then *smaller* than the number of eigenvalues of **A**.

38 If the state space system (2.73) describes the "plant" in a mechatronic system—e.g. a mechanical structure (multibody system)—then the zeroes of the transfer function between the actuator and sensor depend significantly on the locations of the actuator and sensor in the multibody system. This fundamental behavior is discussed in detail in Ch. 4.

The transformation *state space model* ↔ *transfer function* is thus only valid in one direction (left → right in Fig. 2.67). Thus, when back-converting a transfer function into a state space model (right → left in Fig. 2.67), the choice of state variables can result in more or less suitable state representations for numerical integration (see Ch. 3).

Frequency response

The transfer function $G(s)$ is used to derive the *frequency response* $G(j\omega)$ in the usual manner (Ogata 2010):

$$G(j\omega) = G(s)\big|_{s=j\omega} = \left|G(j\omega)\right| \cdot \mathrm{e}^{j \cdot \arg G(j\omega)} . \tag{2.76}$$

The frequency response (2.76) represents a *non-parametric* description of the LTI system (2.72) in the form of the frequency-dependent *amplitude response* $\left|G(j\omega)\right|$ and *phase response* $\arg G(j\omega)$, and can also be very efficiently determined *experimentally* (see Sec. 2.7).

Model properties Such a non-parametric description is particularly useful when the LTI system exhibits a high system order and contains transcendental components (delay terms in the form of exponential functions in $j\omega$). In such cases, greater model complexity is simply reflected in a greater degree of detail in $G(j\omega)$. As will be shown below, the dynamics of *multibody systems* and the influences of significant physical parameters are, in particular, very clearly revealed in the frequency response.

Graphical representation The ordinary representation of the frequency response in the form of a *frequency response locus* $G(j\omega)$ in the complex plane proves to be rather impractical when working with more complex systems. For this reason, this book prefers the well-known representation in the form of *logarithmic frequency curves* or *BODE diagrams* (Ogata 2010). This representation depicts the amplitude and phase curves as a function of frequency in separate plots. By using a semi-logarithmic representation, construction can be simplified, so that even hand sketches are possible, which can then be used for quick checks (in the sense of "verification", see Sec. 2.1) on computer-generated results.

For more specialized applications in controller design, an additional representation of the frequency response in the phase-amplitude plane—the so-called NICHOLS plot—is of particular use. Its use is explained in detail in the context of robust controller design using the NYQUIST criterion in the *intersection form* in Ch. 10.

Controller design In addition to advantageous properties for modeling, the frequency response approach also offers excellent opportunities for *controller design*. There exists an extensive methodological toolbox for control based on the frequency response which can answer questions of closed-loop *stability* (NYQUIST criterion) and synthesize *robust control algorithms*. Practical methodological approaches for both of these areas are presented in detail for mechatronic systems in Ch. 10.

Example 2.14 *Two-mass oscillator with force excitation.*

System configuration and model creation Any of the methods for physical modeling shown in Sec. 2.3 leads to the following *equations of motion* for the coupled two-mass system in Fig. 2.68 (with assumed negligible damping):

$$\begin{aligned} m\ddot{y}_1 + 2ky_1 - ky_2 &= F\,, \\ m\ddot{y}_2 - ky_1 + 2ky_2 &= 0\,. \end{aligned} \tag{2.77}$$

Using the state definition $x_1 := y_1$, $x_2 := y_2$, $x_3 := \dot{y}_1$, $x_4 := \dot{y}_2$ and $u := F$, the *state space model* is

$$\begin{aligned} \dot{\mathbf{x}} &= \begin{pmatrix} 0 & 0 & 1 & 0 \\ 0 & 0 & 0 & 1 \\ -\dfrac{2k}{m} & \dfrac{k}{m} & 0 & 0 \\ \dfrac{k}{m} & -\dfrac{2k}{m} & 0 & 0 \end{pmatrix} \mathbf{x} + \begin{pmatrix} 0 \\ 0 \\ \dfrac{1}{m} \\ 0 \end{pmatrix} u\,, \\ y &= \begin{pmatrix} 1 & 0 & 0 & 0 \end{pmatrix} \mathbf{x}\,. \end{aligned} \tag{2.78}$$

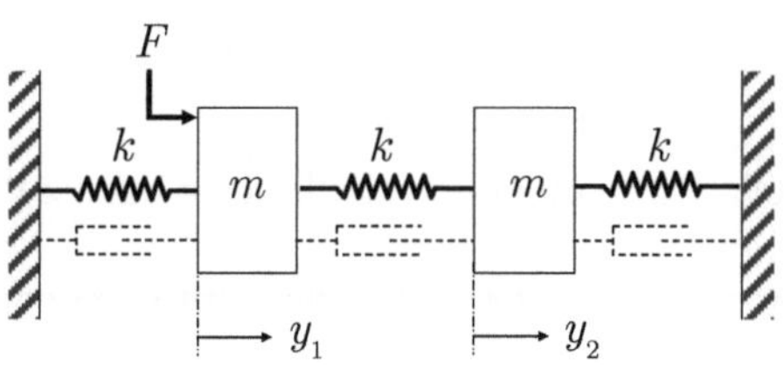

Fig. 2.68. Two-mass oscillator with very low damping

Via a LAPLACE transform of Eq. (2.77), or using relation (2.73), Eq. (2.78) yields the *transfer function* between the excitation force F and the mass displacement y_1:

$$G(s) = \frac{Y_1(s)}{F(s)} = V \frac{1 + \frac{s^2}{\omega_Z^2}}{\left(1 + \frac{s^2}{\omega_{P1}^2}\right)\left(1 + \frac{s^2}{\omega_{P2}^2}\right)}, \tag{2.79}$$

$$\omega_{P1} = \sqrt{\frac{k}{m}}, \quad \omega_{P2} = \sqrt{\frac{3k}{m}}, \quad \omega_Z = \sqrt{\frac{2k}{m}}, \quad V = \frac{2}{3k}.$$

The values ω_{P1}, ω_{P2} are termed the *natural frequencies* of the multi-body system. For a more detailed physical interpretation of these natural frequencies, as well as the numerator term containing ω_Z (antiresonant frequency), refer to Ch. 4.

The *amplitude response* component of the frequency response $G(j\omega)$ is shown in Fig. 2.69 in the form of a BODE diagram (in the graphical representation, a finite, very small damping is assumed). The natural frequencies and the anti-resonant frequency are clearly discernable.

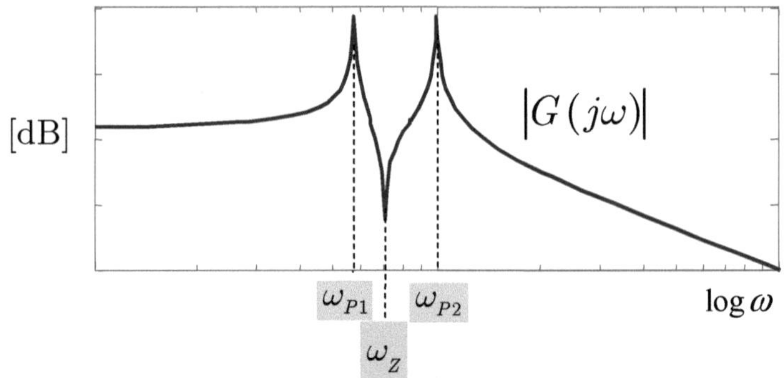

Fig. 2.69. BODE diagram (amplitude response) of the two-mass oscillator

■

Elastic structures: harmonic oscillators Example 2.14 exemplifies typical dynamics of mechatronic systems. In the task "purposeful motion of mass-bearing bodies", *elastically*-connected *multibody systems* (*MBS*) are, as a rule, involved. Either the mechanical structure is elastically joined, or force application happens not via rigid, but rather elastic structures. These particular dynamics—in the form of *resonant natural frequencies* (eigenmodes) of the elastic multibody systems—are significantly easier to recognize in the transfer function (*complex conjugate pairs* of poles) than in the state space model. The natural frequencies are particularly recognizable in the amplitude responses of the BODE diagram (as spikes). If the possibility of designing controllers based on the frequency response of the *open* loop is taken into account, BODE diagrams of the (open) loop consisting of the *actuator, mechanical structure (MBS), and sensor* represent central analysis and design aids.

Notation For a concise and expressive representation of transfer functions of mechatronic systems, later chapters will, wherever useful, employ the following *shorthand* for linear and quadratic factors appearing there:

$$\boxed{\begin{aligned} \left[\omega_i\right] &:= 1+\frac{s}{\omega_i} \\ \left\{d_i,\omega_i\right\} &:= 1+2d_i\frac{s}{\omega_i}+\frac{s^2}{\omega_i^2} \\ \left\{\omega_i\right\} &:= 1+\frac{s^2}{\omega_i^2} \end{aligned}}. \tag{2.80}$$

In particular, the representation in Eq. (2.80) enables the concise specification of eigenmodes with *natural frequency* ω_i and, if applicable, damping d_i.

The transfer function (2.79) in Example 2.14 would thus be represented as follows

$$G(s) = K\frac{\left\{\omega_Z\right\}}{\left\{\omega_{P1}\right\}\left\{\omega_{P2}\right\}}.$$

2.7 Experimental Determination of the Frequency Response

2.7.1 General considerations

As a complement to the creation of system models by theoretical and analytical means, experimental modeling methods and procedures present significant advantages. For models derived from physical experiments, the following distinctions regarding the form of model are made:

- *parametric* models, e.g. transfer functions with a finite number of parameters ➔ *parameter estimation*,
- *non-parametric* models, e.g. impulse response, frequency response ➔ *signal-oriented estimation.*

Parametric models Parameter estimation is well-suited to determining models of sufficiently *low* order, and possibly even the direct estimation of physical parameters, while also accounting for the most *salient* static and dynamic system properties. However, mechatronic systems with pronounced multibody system properties (many resonant frequencies) and complex dynamics generally require a very high order model.

Non-parametric models: frequency response As is shown in following chapters, properties of complex mechatronic systems can be very efficiently and comprehensively described with transfer functions or *frequency responses*. Using the procedures presented in Ch. 10 for *robust controller design* based on frequency responses, it is in principle possible to carry out controller design without knowledge of the physical system parameters and using only knowledge of the complete dynamic response between actuator and sensor (including delays!). Since, as is shown below, the measurement of the frequency response can be carried out very efficiently, experimentally-determined *non-parametric* models in the form of frequency responses offer themselves as ideal complements to theoretically-based models for mechatronic systems.

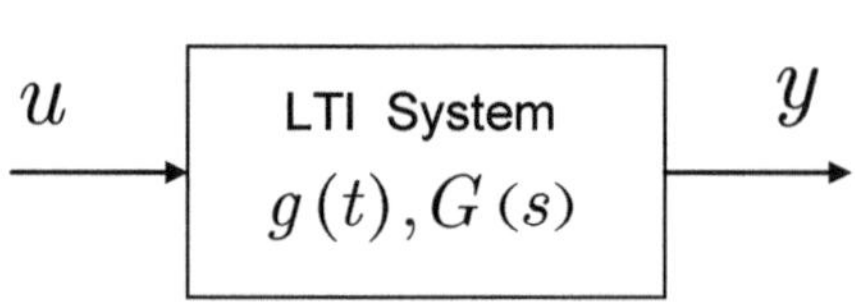

Fig. 2.70. Linear time-invariant system

2.7.2 Methodological approach

System configuration To facilitate further discussion, consider the linear time-invariant system shown in Fig. 2.70. By definition, the frequency response

$$G(jw) = \frac{Y(j\omega)}{U(j\omega)}$$

represents the relationship between the LAPLACE or FOURIER transforms of the input and output of the systems. A convenient arrangement for the experimental determination of $G(j\omega)$ for an example mechatronic system is shown in Fig. 2.71.

The following methodological approaches have proven valuable in practice, and are correspondingly supported with industrial devices or are easily implemented with basic signal processing (e.g. in MATLAB).

Harmonic excitation

Signal generator:	$u(t) = U_0 \sin \omega t$, $\omega \in [\omega_{min}, \omega_{max}]$
Signal evaluation:	after transients have dissipated, the amplitude ratio $\lvert Y(j\omega)\rvert / \lvert U(j\omega)\rvert$ and the phase offset $\arg Y(j\omega) - \arg U(j\omega)$ are determined
Advantage:	simple signal processing
Disadvantage:	wait time for transients; large amplitudes when MBS eigenmodes are excited ➔ danger of mechanical failure!

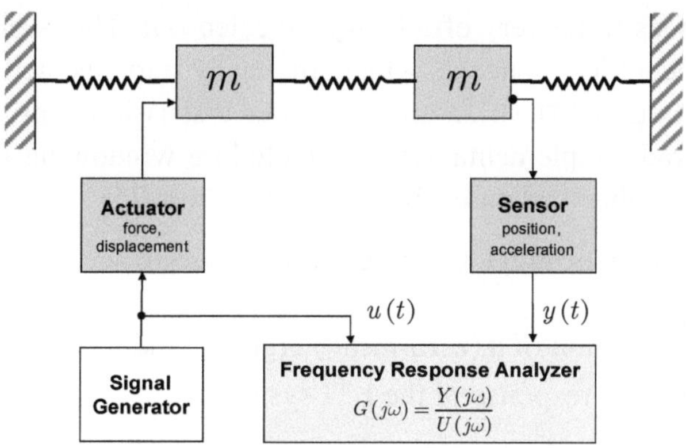

Fig. 2.71. Configuration for measurement of frequency response

Impulse excitation

Signal generator:	$u(t) = U_0 \cdot \delta(t)$
Signal evaluation:	The impulse response $g(t)$ is measured directly, i.e. $Y(s) = L\{g(t)\} \cdot L\{u(t)\} = G(s) \cdot U_0$, $G(j\omega)$ can then be easily computed via FFT
Advantage:	small excitation energy at MBS resonant frequencies
Disadvantage:	complicated signal processing (FFT); only direct force excitation of the MBS is possible

Noise excitation

Signal generator:	$u(t) = random\ signal$
Signal evaluation:	The impulse response $g(t)$ is measured directly, i.e. $Y(s) = L\{g(t)\} \cdot L\{u(t)\} = G(s) \cdot U_0$; $G(j\omega)$ can be easily computed from the series $(g(kT_A))$ via FFT
Advantage:	small excitation energy at MBS resonant frequencies; very robust to signal disturbances; direct force and displacement excitation of the MBS possible (given suitable actuators)
Disadvantage:	complicated signal processing (FFT)

2.7.3 Frequency responses measurement via noise excitation

Introductory remarks Frequency response measurement using noise excitation has proven itself particularly suitable in practical situations. In particular, the use of the correlation function allows the effects of random signal disturbances to be very efficiently canceled out. This section sketches significant theoretical concepts which are easily implemented in a computational tool (e.g. MATLAB), a more rigorous discussion is given in Ch. 11. For a high-grade implementation, e.g. including windowing functions, refer to the applicable literature (Rabiner and Gold 1975).

Computational procedure Consider an LTI system according to Fig. 2.70 with

- $u(t)$ ... Realization of a (zero-mean) ergodic random process,
- $g(t)$ ... Impulse response of the LTI system.

The output $y(t)$ is then computed via the convolution integral

$$y(t) = g(t) * u(t) = \int_{-\infty}^{\infty} g(\tau)\, u\,(t-\tau)\, d\tau\,. \tag{2.81}$$

The FOURIER transform of a signal $x(t),\;\; x \in \{\, u, y, g\}$ is, by definition,

$$X\big(j\omega\big) := \int_{-\infty}^{\infty} x(t)\, e^{-j\omega t}\, dt\,.$$

For the LTI system in Fig. 2.70, it holds that

$$Y\,(j\omega) = G\,(j\omega)\, U\,(j\omega)\,.$$

The *cross-correlation* $r_{uy}(\tau)$ of the signals $y(t),\, u(t)$ is defined as

$$r_{uy}\big(\tau\big) := \lim_{T_0\to\infty} \frac{1}{2T_0} \int_{-T_0}^{T_0} u(t)\, y(t+\tau)\, dt\,. \tag{2.82}$$

Replacing $y(t)$ in Eq. (2.82) by Eq. (2.81),

$$r_{uy}\,(\tau) := \lim_{T_0\to\infty} \frac{1}{2T_0} \int_{-T_0}^{T_0} u(t) \left[\int_{-\infty}^{\infty} g\big(\lambda\big)\, u\,\big(t+\tau-\lambda\big)\, d\lambda\right] dt\,,$$

and changing the order of integration

$$r_{uy}(\tau) = \int_{-\infty}^{\infty} g(\lambda) \left[\lim_{T_0\to\infty} \frac{1}{2T_0} \int_{-T_0}^{T_0} u(t)\, u(t+\tau-\lambda)\, dt\right] d\lambda\,. \tag{2.83}$$

For the autocorrelation $r_{uu}(\tau)$ of the input $u(t)$ (see also Eq. (2.82)), a similar procedure gives

$$r_{uu}\big(\tau-\lambda\big) = \lim_{T_0\to\infty} \frac{1}{2T_0} \int_{-T_0}^{T_0} u(t)\, u(t+\tau-\lambda)\, dt\,. \tag{2.84}$$

From Eqs. (2.83) and (2.84), the convolution integral for the correlations is then

$$r_{uy}(\tau) = \int_{-\infty}^{\infty} g(\lambda)\, r_{uu}\left(\tau - \lambda\right) d\lambda = g(\tau) * r_{uu}(\tau). \tag{2.85}$$

The FOURIER transform of Eq. (2.85) gives

$$S_{uy}\left(j\omega\right) = G\left(j\omega\right) S_{uu}\left(j\omega\right), \tag{2.86}$$

where

$$S_{uy}\left(j\omega\right) = \int_{-\infty}^{\infty} r_{uy}\left(\tau\right) e^{-j\omega\tau}\, d\tau \quad \text{cross spectral density,} \tag{2.87}$$

$$S_{uu}\left(j\omega\right) = \int_{-\infty}^{\infty} r_{uu}\left(\tau\right) e^{-j\omega\tau}\, d\tau \quad \text{power spectral density.} \tag{2.88}$$

The frequency response $G(j\omega)$ can thus be determined via the power spectral densities (2.87) and (2.88), along with Eq. (2.86), as

$$\boxed{G\left(j\omega\right) = \frac{S_{uy}\left(j\omega\right)}{S_{uu}\left(j\omega\right)}}. \tag{2.89}$$

Measurement of the power spectral densities (2.87), (2.88) is easily achieved. For example, MATLAB makes available the following pre-made functions for processing signal sequences $\left(u(kT_a)\right)$, $\left(y(kT_a)\right)$:

- `psd` *power spectral density estimate,*
- `csd` *cross spectral density estimate.*

Measurement noise Under small signal disturbances (measurement noise), averaging even a few individual measurement sequences enables a representative estimate of the frequency responses (see Example 2.15). In particular, MBS eigenmodes and complex zeroes (the collocation problem), as well as phase lag due to delays and low-pass elements, can be elegantly ascertained with this method. Basically, the measured frequency response (following possible smoothing of "outliers") can be employed directly in robust controller design using a computational tool (e.g. MATLAB), as presented in Ch. 10.

Example 2.15 *Experimental frequency response determination for two-mass oscillator with force excitation.*

System configuration For the multibody system shown in Fig. 2.68, noise excitation is to be used to experimentally determine the frequency response $G(j\omega) = Y_1(j\omega) / F(j\omega)$ and the mechanical parameters (m, k).

Experiment frame To model the excitation with a wideband noise source, a second-order shaping filter with $\omega_n = 10$ rad/s, $d_n = 1$ is assumed (Fig. 2.72). Following the dissipation of transients, the measurements of u and y are sampled with a sampling time of $T_a = 0.1$ s and are stored in blocks of 1024 values. Ten frequency series calculated according to Eq. (2.89) are then averaged and output as the *estimated* frequency response.

Discussion The results of frequency response estimation are shown in Fig. 2.73 for noise-free and noisy measurements. Even assuming measurement disturbances, the resonant and anti-resonant frequencies are clearly visible. However, in the noisy case, due to the reduction in the signal-to-noise ratio, the measured response is no longer useful at high frequencies (there is only a small output signal due to the amplitude falloff of -40 dB/decade).

Using the measured resonant and antiresonant frequencies and the mathematical model in Eq. (2.79), the masses and spring constants of the system can be estimated (as a check for the interested reader: $m \approx 10$ kg, $k \approx 400$ N/m). Due to measurement uncertainty, the damping can be only relatively imprecisely estimated (overshoot at resonance, undershoot in the anti-resonance), as is usual in practice. In any case, an experimentally determined frequency response model as shown in Fig. 2.73b gives direct access to *robust controller design*, as discussed in more detail in Ch. 10.

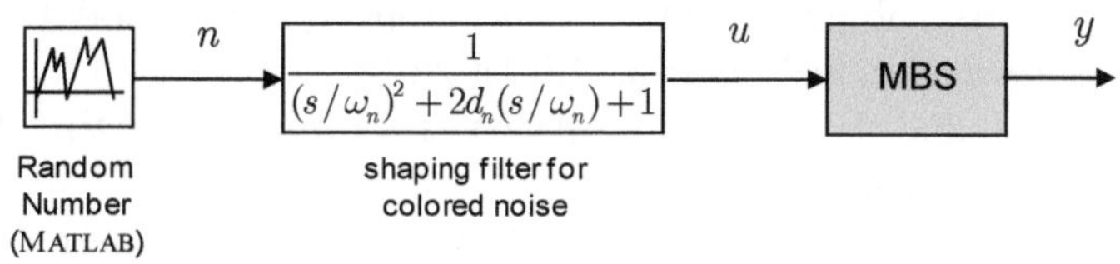

Fig. 2.72. Signal model for frequency response measurement

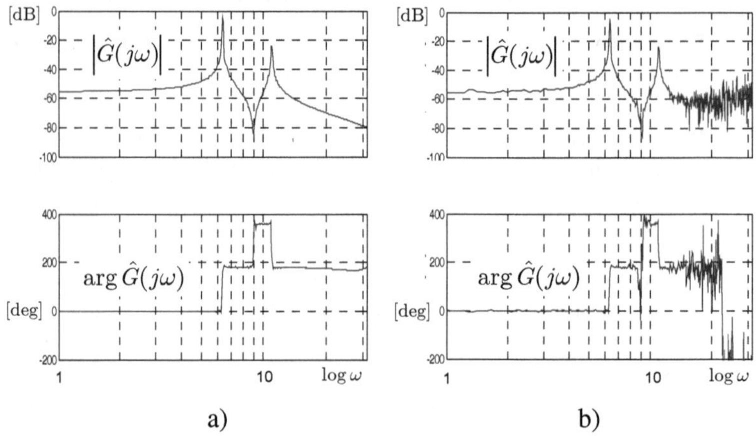

Fig. 2.73. Estimated frequency responses (BODE diagrams) for two-mass oscillator, averaged values from 10 measured frequency series:
a) without measurement noise, b) with measurement noise

■

Bibliography for Chapter 2

Angermann, A., M. Beuschel, M. Rau and U. Wohlfarth (2005). *Matlab-Simulink-Stateflow. Grundlagen, Toolboxen, Beispiele.* München. Oldenbourg Wissenschaftsverlag.

Brenan, K. E., S. L. Campbell and L. R. Petzold (1996). *Numerical Solution of Initial-Value Problems in Differential-Algebraic Equations*. SIAM.

Buss, M. (2002). *Methoden zur Regelung Hybrider Dynamischer Systeme.* Fortschritt-Berichte, VDI Reihe 8, Nr. 970.

Cellier, F. E. (1991). *Continuous System Modeling*. Springer.

Cellier, F. E. and H. Elmqvist (1993). "Automated formula manipulation supports object-oriented continuous-system modelling." *IEEE Control System Magazine* **13**(2): 28-38.

Cellier, F. E. and E. Kofman (2006). *Continuous System Simulation.* Berlin. Springer.

Cervera, J., A. J. van der Schaft and A. Banos (2007). "Interconnection of port-Hamiltonian systems and composition of Dirac structures." *Automatica* **43**(2): 212-225.

Conrad, M., I. Fey and S. Sadeghipour (2005). "Systematic Model-Based Testing of Embedded Automotive Software " *Electronic Notes in Theoretical Computer Science* **111**: 13-26

Damic, V. and J. Montgomery (2003). *Mechatronics by Bond Graphs*. Springer.

Duindam, V., A. Macchelli, S. Stramigioli and H. Bruyninckx, Eds. (2009). *Modeling and Control of Complex Physical Systems - The Port-Hamiltonian Approach*, Springer.

Engell, S., G. Frehse and E. Schnieder, Eds. (2002). *Modelling, analysis, and design of hybrid systems*. Lecture notes in control and information sciences, Springer.

Fritzson, P. (2011). *Introduction to Modeling and Simulation of Technical and Physical Systems with Modelica*. John Wiley & Sons Inc.

Fuchshumer, S., G. Grabmair, K. Schlacher and G. Keintzel (2003). "Automatisierungstechnik in der Mechatronik — zwei Beispiele aus der Stahlindustrie " *e&I* **120**(5): 164-171.

Geitner, G. H. (2006). Power Flow Diagrams Using a Bond Graph Library under Simulink. *Proc. of 32nd Annual Conference on IEEE Industrial Electronics, IECON 2006-*. pp.5282-5288.

Geitner, G. H. (2008). Bondgraphen-Modelle für ausgewählte mechatronische Anschauungsbeispiele. Persönliche Kommunikation, Elektro-technisches Institut, Technische Universität Dresden.

Goldstein, H., C. P. Poole and J. L. Safko (2001). *Classical Mechanics*. Addison Wesley.

Harel, D. (1987). "Statecharts - A Visual Formalism for Complex Systems." *Science of Computer Programming* **8**: 231-274.

Hatley, D. J. and I. A. Pirbhai (1987). *Strategies for Real-Time System Specification*. New York, NY. Dorset House.

Hatley, D. J. and I. A. Pirbhai (1993). *Strategien für die Echtzeitpro-grammierung*. München, Wien. Hanser.

IEEE (1997). IEEE Trial-Use Recommended Practice for Distributed Interactive Simulation -Verification, Validation, and Accreditation. IEEE Std 1278.4-1997. I. C. Society.

Isidori, A. (2006). *Nonlinear Control Systems*. Springer.

Karnopp, D. C., D. L. Margolis and R. C. Rosenberg (2006). *System dynamics: modeling and simulation of mechatronic systems*. John Wiley & Sons, Inc.

Koycheva, E. and K. Janschek (2007)._Performance analysis of system models with UML and Generalized Nets. *EUROSIM 2007, 6th EUROSIM Congress on Modelling and Simulation*, Ljubljana, Slovenia

Kugi, A. and K. Schlacher (2001). "Dissipativit&ts- und passivitätsbasierte Regelung nichtlinearer mechatronischer Systeme." *e&i* **120**(1): 40-48.

Kugi, A. and K. Schlacher (2002). "Analyse und Synthese nichtlinearer dissipativer Systeme: Ein Überblick (Teil 2)." *at - Automatisierungs-technik* **50**(3): 103-111.

Lenk, A., R. G. Ballas, R. Werthschützky and G. Pfeifer (2011). *Electromechanical Systems in Microtechnology and Mechatronics.* Springer.

Litz, L. (2005). *Grundlagen der Automatisierungstechnik.* Oldenbourg Verlag München Wien.

Lunze, J. (2002). What Is a Hybrid System? *Modelling, Analysis, and Design of Hybrid Systems.* S. Engell, G. Frehse and E. Schnieder. Springer: 3-14.

Maschke, B. M. and A. J. van der Schaft (1992). Port-controlled Hamiltonian systems: Modelling origins and system theoretic properties. *IFAC Symposium on Nonlinear Control Systems Design (NOLCOS) 1992*, Bordeaux, France. pp.359-365.

Mattsson, S. E. and G. Söderlind (1993). "Index Reduction in Differential-Algebraic Equations Using Dummy Derivatives." *SIAM Journal on Scientific Computing* **14**(677-692).

Nenninger, G., M. Schnabel and V. Krebs (1999). "Modellierung, Simulation und Analyse hybrider dynamischer Systeme mit Netz-Zustands-Modellen." *at-Automatisierungstechnik* **47**(3): 118-126.

Oestereich, B. (2006). *Analyse und Design mit der UML 2.1 - Objektorientierte Softwareentwicklung.* Oldenbourg Wissenschaftsverlag.

Ogata, K. (1992). *System Dynamics.* Prentice Hall.

Ogata, K. (2010). *Modern Control Engineering.* Prentice Hall.

Ortega, R., A. J. van der Schaft, B. M. Maschke and G. Escobar (2002). "Interconnection and damping assignment passivity-based control of port-controlled Hamiltonian systems." *automatica* **38**: 585-596.

Otter, M. (1999). "Objektorientierte Modellierung Physikalischer Systeme, Teil 4." *at-Automatisierungstechnik* **47**(4): A13-A16.

Otter, M. and B. Bachmann (1999). "Objektorientierte Modellierung Physikalischer Systeme, Teil 5." *at-Automatisierungstechnik* **47**(5): A17-A20.

Otter, M. and B. Bachmann (1999). "Objektorientierte Modellierung Physikalischer Systeme, Teil 6." *at-Automatisierungstechnik* **47**(6): A21-A24.

Pantelides, C. C. (1988). "The consistent initialization of differential-algebraic systems." *SIAM Journal of Scientific and Statistical Computing* **9**: 213-231.

Paynter, H. M. (1961). *Analysis and Design of Engineering Systems.* MIT Press, Cambridge, Mass.

Rabiner, L. R. and B. Gold (1975). *Theory and Application of Digital Signal Processing*. Englewood Cliffs, New Jersey. Prentice Hall.

Rau, A. (2002). *Model-Based Development of Embedded Automotive Control Systems*. Dissertation, Universität Tübingen.

Reinschke, K. (2006). *Lineare Regelungs- und Steuerungstheorie*. Springer.

Reinschke, K. and P. Schwarz (1976). *Verfahren zur rechnergestützten Analyse linearer Netzwerke*. Akademie Verlag Berlin.

Schnabel, M., G. Nenninger and V. Krebs (1999). "Konvertierung sicherer Petri-Netze in Statecharts." *at - Automatisierungstechnik* **47**(12): 571-580.

Schnieder, E. (1999). *Methoden der Automatisierung. Beschreibungsmittel, Modellkonzepte und Werkzeuge für Automatisierungssysteme*. Braunschweig, Wiesbaden. Vieweg.

Schultz, D. G. and J. L. Melsa (1967). *State functions and linear control systems*. McGraw-Hill Book Company.

Schwarz, P., C. Clauß, J. Haase and A. Schneider (2001)._VHDL-AMS und Modelica - ein Vergleich zweier Modellierungssprachen. *15. Symposium Simulationstechnik ASIM 2001*, Paderborn. pp.85-94.

Schwarz, P. and T. Zaiczek (2008). Torbasierte Rechnermodelle für ausgewählte mechatronische Anschauungsbeispiele. Persönliche Kommunikation, Fraunhofer Institut Integrierte Schaltungen, Institutsteil Entwurfsautomatisierung, Dresden.

Short, M. and M. J. Pont (2008). "Assessment of high-integrity embedded automotive control systems using hardware in the loop simulation." *Journal of Systems and Software* **81**(7): 1163-1183.

Siciliano, B., L. Sciavicco, L. Villani and G. Oriolo (2009). *Robotics: Modelling, Planning and Control*. Springer.

Thomas, R. E., A. J. Rosa and G. J. Toussaint (2009). *The Analysis and Design of Linear Circuits*. John Wiley and Sons, Inc.

Tiller, M. M. (2001). *Introduction to Physical Modeling with Modelica*. Kluwer Academic Publishers.

van der Schaft, A. J. and B. M. Maschke (1995). "The Hamiltonian formulation of energy conserving physical systems with external ports." *Archiv für Elektronik und Übertragungstechnik* **49**: 362-371.

van der Schaft, A. J. and B. M. Maschke (2002). "Hamiltonian representation of distributed parameter systems with boundary energy flow." *Journal of Geometry and Physics* **42**: 166-194.

Vogel-Heuser, B. (2003). *Systems Software Engineering*. München. Oldenbourg.

Wellstead, P. E. (1979). *Introduction to Physical System Modelling*. London. Academic Press Ltd.

Yourdon, E. (1989). *Modern Structured Analysis*. Yourdon Press.

3 Simulation Issues

Background Experimenting with dynamic system models is one of the standard tasks of systems design, and the results of such simulations provide the basis for far-reaching design decisions. Nowadays, (commercial) computer-aided tools generally provide the platform for modeling and simulation, and pre-existing model libraries are thus often employed. However, as a result, there is often a dangerous gap in understanding between the "computerized" simulation model and the (in the extreme case, naïve) user during this extremely important phase of design. In unfavorable cases, this can easily lead to flawed simulation results. Thus, knowledge of the particularities of *simulation implementations* and solution approaches is an essential area of competence for systems engineers, even and especially when employing modern simulation tools. Only when armed with such knowledge is it at all possible to recognize potential problems and to alleviate them with appropriate measures, whether by modifying the model or by employing a targeted choice and parameterization of available simulator functions: *"using the tool with appreciation and awareness"*.

Contents of Chapter 3 This chapter discusses selected aspects of the *implementation of mathematical models* for *simulation experiments* and particular problems and solution approaches related to models of *mechatronic systems*. To this extent, basic knowledge on the part of the reader regarding numerical integration and general simulation methods is assumed. Following a brief discussion of *numerical stability*, the fundamental effects of integration step size, and the properties of different integration methods, this chapter presents typical problems in and solution approaches to the simulation of multibody systems which are characterized by *stiff* system configurations with clearly differentiable eigenmodes and weak-to-*undamped eigenmodes*. For *linear, high-order* multibody models (such as those produced using finite element methods (FEM)), a highly efficient and precise integration procedure using the *state transition matrix* is presented. The non-trivial numerical integration of *systems of differential-algebraic equations (DAE systems)* and the handling of *hybrid phenomena* are elucidated using fundamental concepts. A concluding example demonstrates the closed-form modeling of a DAE system and its implementation in a simulation.

■

3.1 Systems Engineering Context

Modeling vs. simulation Systems design taken as "working with models" comprises two closely-knit tasks: constructing models (model creation) and experimenting with models (simulation). It should be clear from Fig. 2.3 that the predictive capability of a simulation result—i.e. how representative it is of the real system behavior—depends on the sum of modeling errors and simulation errors. The choice of a particular type of model determines the difficulty of the simulation task, and in turn, the resulting simulation errors. A concise model in the form of a system of ordinary differential equations in minimal coordinates can be implemented and computed with much less effort than a highly-redundant DAE system obtained from object-oriented modeling. Thus, the effort required to construct a model and the desired model accuracy must always be weighed against the effort required to carry out simulations.

Computer-aided simulation Along with the computer-aided construction of models, modern design tools enable nearly effortless execution of simulation experiments. From the user's point of view, this ease is certainly desirable. However, it hides great dangers if the models under consideration possess certain unfavorable properties. Despite the fact that good computer-based tools have a number of built-in sanity checks, a flawed parameterization of the solution algorithms can lead to completely false simulation results. In particularly malignant cases (e.g. with complex models) such errors are difficult to detect. Computer-based tools usually only examine the syntax and parameters of models, and the parameters of an experiment. As a matter of principle, the *semantics* of models must remain unmonitored, and thus represent a latent source of errors.

Knowledgeable use of simulation tools This chapter particularly concerns itself with the semantics of *numerical* solution procedures for *systems of ordinary differential equations* and *DAE systems*, i.e. solution algorithms (numerical integration procedures) and the significance of their more important parameters (step size, order, etc.). This background is intended to enable knowledgeable selection and use of common methods as they are implemented in today's commercial computer-based tools.

Prerequisites It is assumed that the reader is familiar with the fundamental concepts of *numerical integration* (e.g. explicit vs. implicit methods, single-step vs. multi-step methods, the RUNGE-KUTTA method, adaptive step sizes with error monitoring, etc.). If needed as an update or refresher, relevant literature from the field of *numerical analysis* is recommended, e.g. (Faires and Burden 2002). Procedures more immediately tailored to the simulation of dynamic systems can be found in the monograph (Cellier and Kofman 2006).

3.2 Elements of Numerical Integration

3.2.1 Numerical integration of differential equations

Simulation experiments To carry out a computer-aided simulation experiment, an approximate solution for system variables of interest must be calculated based on the underlying mathematical model. It is then often said: "The mathematical model is being simulated."

For this purpose, first consider the following ordinary nonlinear *state space model* with a single input $u(t)$ and a single output $y(t)$ (Fig. 3.1)[1]:

$$\dot{\mathbf{x}} = \tilde{\mathbf{f}}\left(\mathbf{x}, u, t\right), \tag{3.1}$$

$$y = \mathbf{g}\left(\mathbf{x}, u, t\right). \tag{3.2}$$

When *simulating* this system, it is generally the time evolution of the solution $\mathbf{x}(t)$ or $y(t)$ over a finite time interval $\left[t_0, t_f\right]$ which is of interest. In such a case, it can be assumed that the evolution of the input $u(t)$ over $\left[t_0, t_f\right]$ is known.

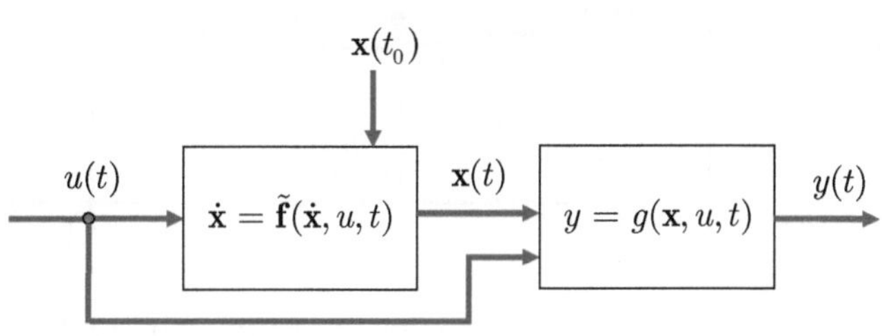

Fig. 3.1. State space model of a single-input single-output (SISO) dynamic system

1 This model represents a DAE system of index 0 (cf. Sec. 2.4). The solution of DAE systems of higher index is discussed in Sec. 3.6.

To compute the output $y(t)$ using Eq. (3.2), all that is required is to determine $\mathbf{x}(t)$ as the solution of the system of *n* first-order differential equations (3.1) within the time interval $\left[t_0, t_f\right]$.

Given the above assumption, the following fundamental problem can be formulated—the *numerical integration of differential equations*: find an approximation $\hat{\mathbf{x}}(t)$ of the time evolution of the solution $\mathbf{x}(t)$ of the system of ordinary differential equations[2]

$$\dot{\mathbf{x}} = \mathbf{f}(\mathbf{x}, t), \quad \mathbf{x}(t_0) = \mathbf{x}_0 \in \mathbb{R}^n . \tag{3.3}$$

Single-step methods: explicit vs. implicit An approximate solution for Eq. (3.3) is obtained for a finite number of values $\hat{\mathbf{x}}(t_k)$ using a difference approximation to the differential equation (3.3) or the corresponding integral equation. Then, to compute a new approximate value $\hat{\mathbf{x}}(t_{k+1})$ considering only the last-computed value $\hat{\mathbf{x}}(t_k)$, the following general recursion is obtained for so-called *single-step methods*:

$$\hat{\mathbf{x}}(t_{k+1}) = \hat{\mathbf{x}}(t_k) + h\varphi\left(\hat{\mathbf{x}}(t_k), \hat{\mathbf{x}}(t_{k+1}), t_k, h\right), \tag{3.4}$$

with *increment function* $\varphi(\cdot)$ and *step size* h. If the increment function does not depend on $\hat{\mathbf{x}}(t_{k+1})$, the method is called *explicit* (e.g. EULER, RUNGE-KUTTA), otherwise, it is termed *implicit* (e.g. the trapezoid method) (Faires and Burden 2002).

The particular choice of increment function $\varphi(\cdot)$ and step size h determines the accuracy of the approximation (Fig. 3.2).

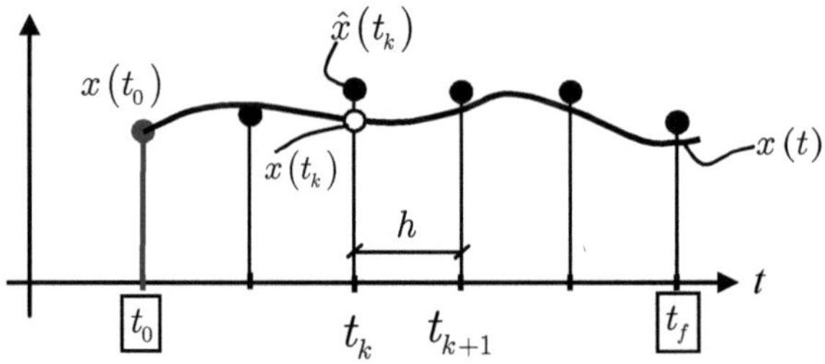

Fig. 3.2. Numerical integration: approximation $\hat{x}$ of a differential equation

---

2 A smooth vector field $\mathbf{f}(\cdot)$ is assumed. If discontinuities exist in $\mathbf{f}(\cdot)$ (e.g. steps in the input excitation functions or the state variables $\mathbf{x}(t)$), special provision must be made, see Sec. 3.7.

3.2.2 Concepts of stability

Definition 3.1. *Local discretization error*: The *local discretization error* (LDE) of an explicit single-step method[3] at time t_{k+1} is understood to be the value

$$\mathbf{d}_{k+1} := \underbrace{\left\{\mathbf{x}(t_{k+1}) - \mathbf{x}(t_k)\right\}}_{\substack{\text{Single-step change} \\ \text{in the \underline{true} solution}}} - \underbrace{h \cdot \varphi\left(\mathbf{x}(t_k), t_k, h\right)}_{\substack{\text{Single-step change} \\ \text{from application of the} \\ \text{integration algorithm} \\ \left(\text{to the \underline{true} } \mathbf{x}(t_k)\right)}} .$$

The LDE $\mathbf{d}_{k+1}$ represents the deviation of the integration method from the *true* solution $\mathbf{x}(t_k)$ for *a single step*. The LDE is thus a measure of how nearly the solution in Eq. (3.4) approximates the true solution $\mathbf{x}(t_k)$.

Definition 3.2. *Global discretization error*: The *global discretization error* (GDE) at a fixed time t_k is understood to be the value

$$\mathbf{g}_k := \mathbf{x}(t_k) - \hat{\mathbf{x}}(t_k) .$$

The GDE $\underline{g_i}$ thus represents the deviation of the approximate solution $\hat{\mathbf{x}}(t_k)$ from the true solution $\mathbf{x}(t_k)$, and in particular includes the accumulated errors (LDE and GDE) of all previous steps $j = 0, 1, \ldots, (k-1)$.

Definition 3.3. *Consistency*: A numerical integration method solving an initial value problem is called *consistent* if the sum of local discretization errors R_{LDE} approaches zero, given a step size that also approaches zero:

$$\lim_{h \to 0} \left(\frac{1}{h} R_{\text{LDE}} \right) = 0 .$$

Definition 3.4. *Convergence*: A numerical integration method solving an initial value problem is called *convergent* if the global discretization error approaches zero over the entire integration interval as the step size approaches zero:

$$\lim_{h \to 0} \left(\hat{x}_k - x_k \right) = \lim_{h \to 0} g_k = 0 \quad \forall k, \quad \text{i.e. } t \in \left[t_0, t_f\right].$$

[3] For other methods (e.g. implicit or multi-step methods), the LDE is defined analogously.

Stability

The following types of stability should be distinguished:

- *Inherent stability of the system model*
  Stability concepts employed are, for example, input/output stability (e.g. BIBO stability), or (asymptotic) state stability (Ogata 2010). The system is termed (inherently) *stable* if the system model is stable in the sense stated above.
- *Numerical stability of the integration algorithm*
  A numerical integration method solving an initial value problem is called *numerically stable*, if "small errors" in the integrated values $\hat{x}_k$ also engender only "small errors" in computing subsequent steps $\hat{x}_{k+1}$ (i.e. there is sufficient error damping) (Faires and Burden 2002).

Given the above definitions, the following elementary theorem ensues:

> **Theorem 3.1.** A numerical integration method is *convergent* if and only if it is *consistent* and numerically *stable*.

The properties of convergence, consistency, and numerical stability are thus intimately interrelated, and represent fundamental properties for the execution of simulation experiments. Though commercial simulation tools generally include an ample quantity of *consistent* integration methods as built-in functions (only such would even make sense!), it is *not* thus a foregone conclusion that a *convergent* approximating solution will be obtained (though nothing less would be expected from a sensible simulation experiment!). In accordance with Statement 3.1, numerical stability is also required, and this in turn fundamentally depends on the step size h, which—as a freely-selectable simulation parameter—can also be set arbitrarily incorrectly if there is a corresponding lack of understanding of its action and effect (see Sec. 3.2.3).

It is quite clear that to achieve higher accuracy in the approximate solution, the step size h should be chosen to be as *small* as possible. On the other hand, for a fixed integration interval, this increases the computational requirements (a greater number of recursions), so that for faster computing speed, it is rather the largest possible step size which is desirable. In any particular case, there is thus always a tradeoff to be made between accuracy and computational load in the choice of the step size h.

3.2.3 Numerical stability

Linear test system: initial value problem A numerical integration process can be expressed as a discrete-time dynamic system represented by a system of nonlinear difference equations. This allows well-known stability concepts and criteria to be called upon for the analysis of its numerical stability.

For the purpose of discussion, consider the following *linear* (inherently stable) *test initial value problem*:

$$\dot{x} = \lambda x, \quad \text{where} \quad x(0) = x_0, \ \lambda < 0 . \tag{3.5}$$

For the *EULER method*

$$\hat{x}_{k+1} = \hat{x}_k + h \cdot f(\hat{x}_k),$$

so that, along with Eq. (3.5), the linear first order difference equation

$$\hat{x}_{k+1} = (1 + h \cdot \lambda)\hat{x}_k \tag{3.6}$$

follows. The general solution of Eq. (3.6) is:

$$\hat{x}_{k+1} = (1 + h \cdot \lambda)^{k+1} \cdot x_0 . \tag{3.7}$$

Numerical stability then results when the series of approximation values $(\hat{x}_k) = (x_0, \hat{x}_1, \hat{x}_2, \ldots)$ from Eq. (3.7) converges to the stationary final value of the true solution $x_\infty = 0$ for $k \to \infty$, i.e. if the *numerical stability condition*

$$|1 + h \cdot \lambda| < 1 \tag{3.8}$$

holds. The condition in Eq. (3.8) corresponds to the well-known discrete-time stability criterion "magnitude of the eigenvalue is less than 1" for the linear difference equation (3.6), see (Franklin et al. 1998).

Inherently stable system In an *inherently stable* system ($\lambda < 0$), the numerical stability condition (3.8) is satisfied if and only if:

$$h\lambda < 0 \ \text{ and } \ h < \frac{2}{-\lambda} = h_{crit}, \tag{3.9}$$

so that the allowable step size is bounded *above* by h_{crit}.

Example 3.1 *Explicit* Euler *integration.*

Fig. 3.3 demonstrates the effect of the integration step size h on the simulation solution. For $h \geq 2$ (with $\lambda = -1$), the numerical integration is unstable, see Eq. (3.9).

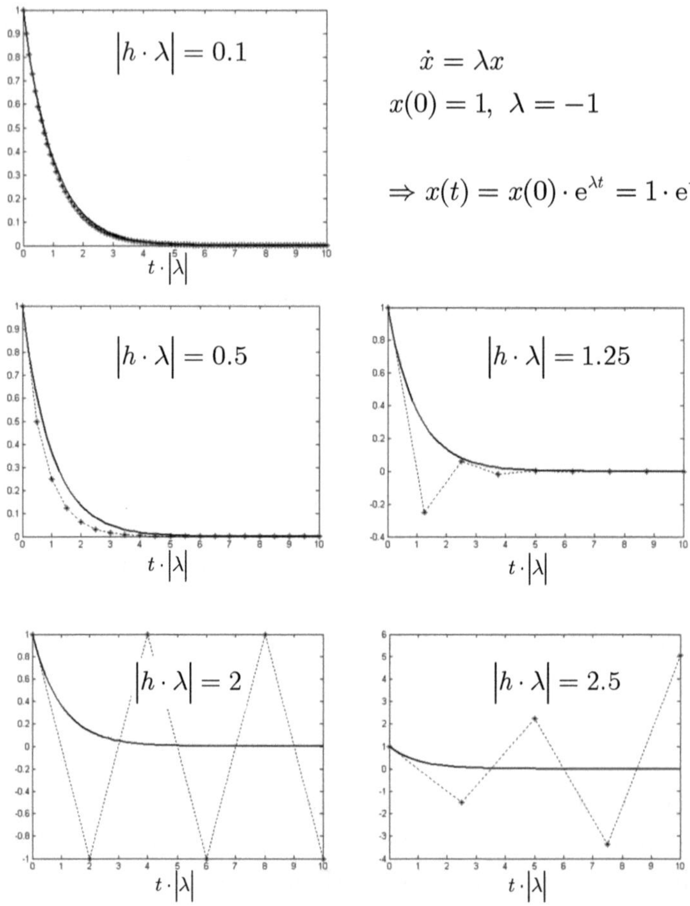

Fig. 3.3. Numerical stability as a function of step size

■

Absolute numerical stability

Definition 3.5. *Absolute stability*: For a single-step method applied to the *test initial value problem*

$$\dot{x} = \lambda x\,,\ x(t_0) = x_0\,, \tag{3.10}$$

leading to the recursion

$$\hat{x}_{k+1} = \varphi_R\,(h\lambda)\,\hat{x}_k\,, \tag{3.11}$$

the set $B := \left\{\mu \in \mathbb{C},\ \left|\varphi_R\,(\mu)\right| < 1\right\},\ \mu = h\lambda$, is called the *region of absolute stability.*

The integration step size h should thus always be chosen such that, for $\mathrm{Re}\,(\lambda) < 0$, it always holds that $(h \cdot \lambda) \in B$.

When applied to the test linear initial value problem (3.5), different explicit and implicit single-step methods with integration accuracies of *order* p exhibit the characteristic polynomials in $\mu = h\lambda$ shown in Table 3.1.

Table 3.1. Regions of absolute stability for single-step methods

Explicit single-step methods Order p	$\varphi_R(\mu)\,,\ \mu = h\lambda$
p = 1: Forward EULER (EUL)	$1 + \mu$
p = 2: Midpoint method. (MIP)	$1 + \mu + \frac{1}{2}\mu^2$
p = 3: Third-order RUNGE-KUTTA (RK3)	$1 + \mu + \frac{1}{2}\mu^2 + \frac{1}{3!}\mu^3$
p = 4: Fourth-order RUNGE-KUTTA (RK4)	$1 + \mu + \frac{1}{2}\mu^2 + \frac{1}{3!}\mu^3 + \frac{1}{4!}\mu^4$
p = 5: Fifth-order RUNGE-KUTTA (RK5)	$1 + \mu + \frac{1}{2}\mu^2 + \frac{1}{3!}\mu^3 + \frac{1}{4!}\mu^4 + \frac{1}{5!}\mu^5$
$p \to \infty$: State transition matrix (LIN)	e^{μ}
Implicit single-step methods Order p	$\varphi_R(\mu)\,,\ \mu = h\lambda$
p = 2: Trapezoid method (TRA)	$\dfrac{1 + \frac{1}{2}\mu}{1 - \frac{1}{2}\mu}$

For RUNGE-KUTTA methods, applicable stability regions in the complex plane (corresponding to complex eigenvalues) are shown in Fig. 3.4; the stability intervals for real eigenvalues are given in Table 3.2. It can be shown that the recursion coefficients of $\varphi_R(\mu)$ are the first p terms (where p is the order of the integration accuracy of the respective integration algorithm) of the Taylor expansion of the exponential; in the limit $p \to \infty$, $\varphi_R(\mu)$ is the exponential function itself.

It is particularly worth noting the numerical stability regions for *explicit* integration methods with $p \to \infty$ (state transition matrix, LIN) and for *implicit* integration methods (for example the trapezoid method, TRA, Table 3.1, bottom). In both cases, the region of absolute stability is the entire open left half-plane. For inherently stable systems ($\lambda < 0$), any arbitrary $h > 0$ thus yields a *stable* integration algorithm (a decisive advantage for *stiff* systems of ordinary differential equations, see Sec. 3.3).

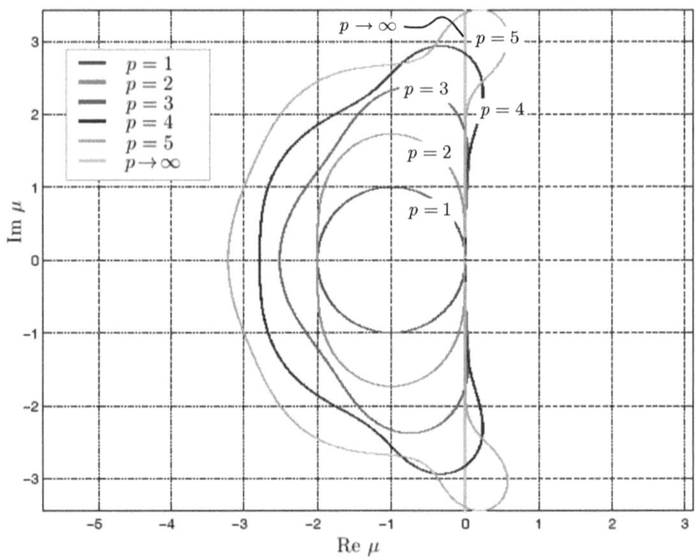

Fig. 3.4. Stability regions for RUNGE-KUTTA methods, from (Potthoff 2003)

Table 3.2. Real stability intervals for RUNGE-KUTTA methods (cf. Fig. 3.4)

Order p	1	2	3	4	5
Interval for real $h\lambda$	[-2.0, 0]	[-2.0, 0]	[-2.51, 0]	[-2.78, 0]	[-3.21, 0]

3.3 Stiff Systems

Technical background Complex models of real engineering systems are generally characterized by widely disparate time constants among component subsystems. Such arrangements are called *stiff systems*.

This concept can be illustrated using a multibody system. Fig. 3.5 depicts a two-mass system with differing spring constants. The *stiffer* spring (k_2) results in a resonant frequency ω_2 with greater magnitude, corresponding to the imaginary eigenvalues $\lambda_2 = \pm j\omega_2$. Equivalently, large-magnitude real eigenvalues (or small time constants) are also termed "stiff".

Local linearization: the JACOBIan In the case of nonlinear system descriptions of the form

$$\dot{\mathbf{x}}(t) = \mathbf{f}(\mathbf{x}, t),$$

quantitative statements about the system dynamics at a *fixed time* t_k (and approximated over the interval $[t_k, t_k + h]$) can be derived via local linearization of the trajectory about $\mathbf{x}(t_k)$. For this purpose, consider the *JACOBIan* $\mathbf{J}$:

$$\mathbf{J}(t_k) = \left.\begin{bmatrix} \frac{\partial f_1}{\partial x_1} & \frac{\partial f_1}{\partial x_2} & \cdots & \frac{\partial f_1}{\partial x_n} \\ \frac{\partial f_2}{\partial x_1} & \frac{\partial f_2}{\partial x_2} & \cdots & \frac{\partial f_2}{\partial x_n} \\ \vdots & \vdots & & \vdots \\ \frac{\partial f_n}{\partial x_1} & \frac{\partial f_n}{\partial x_2} & \cdots & \frac{\partial f_n}{\partial x_n} \end{bmatrix}\right|_{t=t_k,\ \mathbf{x}=\mathbf{x}(t_k)} .$$

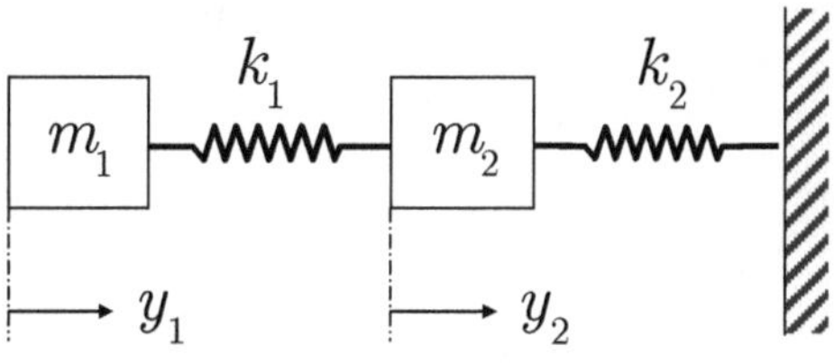

Fig. 3.5. Stiff multibody system with $k_2 \gg k_1$

Stiffness The eigenvalues $\lambda_j\ (j=1,\ldots,n)$ of $\mathbf{J}(t_k)$ describe the dynamics of the system at time t_k. The *stiffness* S of the system is characterized by the following quotient:

$$S=\frac{\max\left(|\,\lambda_j\,|\right)}{\min\left(|\,\lambda_j\,|\right)}\ ,\qquad j=1,\ldots,n\,;$$

where $S\gg 1$ for stiff systems.

Linear test system: method comparison For this set of comparisons, consider the following linear system:

$$\dot{\mathbf{x}}=\mathbf{A}\mathbf{x}\ ,\quad \mathbf{x}\left(0\right)=\mathbf{x}_0\,.$$

- *Midpoint method* – MIP (order $p = 2$):

$$\begin{aligned}
\mathbf{k}_1 &= \mathbf{A}\hat{\mathbf{x}}_i\,,\\
\mathbf{k}_2 &= \mathbf{A}\left(\hat{\mathbf{x}}_k+\frac{h}{2}\,\mathbf{A}\,\hat{\mathbf{x}}_k\right),\\
\hat{\mathbf{x}}_{k+1} &= \hat{\mathbf{x}}_k+h\left(\mathbf{A}\,\hat{\mathbf{x}}_k+\frac{h}{2}\,\mathbf{A}^2\,\hat{\mathbf{x}}_k\right)=\left(\mathbf{I}+h\,\mathbf{A}+\frac{h^2}{2}\mathbf{A}^2\right)\hat{\mathbf{x}}_k=\mathbf{\Phi}_{MIP}\,\hat{\mathbf{x}}_k\,.
\end{aligned}$$

- *Trapezoid method* – TRA (order $p = 2$):

$$\begin{aligned}
\hat{\mathbf{x}}_{k+1} &= \hat{\mathbf{x}}_k+\frac{h}{2}\left(\mathbf{A}\hat{\mathbf{x}}_k+\mathbf{A}\hat{\mathbf{x}}_{k+1}\right),\\
\hat{\mathbf{x}}_{k+1} &= \left(\mathbf{I}-\frac{h}{2}\mathbf{A}\right)^{-1}\cdot\left(\mathbf{I}+\frac{h}{2}\,\mathbf{A}\right)\hat{\mathbf{x}}_k=\mathbf{\Phi}_{TRA}\cdot\hat{\mathbf{x}}_k\,.
\end{aligned}$$

In the special case of a diagonal $\mathbf{A}$ (possible if using modal transformations), a *second-order system* results in:

$$\mathbf{A}=\begin{bmatrix}\lambda_1 & 0\\ 0 & \lambda_2\end{bmatrix},\quad \lambda_1\ll\lambda_2 \text{ are eigenvalues of the system,}$$

$$\mathbf{\Phi}_{MIP}=\begin{bmatrix}1+h\lambda_1+\frac{h}{2}\lambda_1^2 & 0\\ 0 & 1+h\lambda_2+\frac{h}{2}\lambda_2^2\end{bmatrix},$$

$$\mathbf{\Phi}_{TRA} = \begin{bmatrix} \dfrac{1+\dfrac{h}{2}\lambda_1}{1-\dfrac{h}{2}\lambda_1} & 0 \\ 0 & \dfrac{1+\dfrac{h}{2}\lambda_2}{1-\dfrac{h}{2}\lambda_2} \end{bmatrix}.$$

To ensure numerical stability, the step size h should be chosen such that all diagonal elements of the matrices $\mathbf{\Phi}_{MIP}$, $\mathbf{\Phi}_{TRA}$ have magnitude less than 1 (cf. *absolute stability*).

For the *explicit MIP method*, the largest-magnitude eigenvalue λ_2 ("fastest" component of the solution) thus determines the step size. In stiff systems, the remaining "slow" components (λ_1) must also be integrated using this "small" step size. Clearly this happens at the expense of computational speed.

The advantage of the *implicit TRA method* becomes clear here, as a choice of larger step size—though increasing the error for the fast component λ_1— on the whole maintains guaranteed absolute stability of the integration.

Example 3.2 *Stiff second-order system.*

$$G(s) = \frac{Y(s)}{U(s)} = \frac{1}{\left(1+\dfrac{s}{a_1}\right)\left(1+\dfrac{s}{a_2}\right)} = \frac{a_1 a_2}{s^2 + \left(a_1 + a_2\right)s + a_1 a_2}.$$

State space model

$$\dot{\mathbf{x}} = \begin{pmatrix} 0 & 1 \\ -a_1 a_2 & -\left(a_1 + a_2\right) \end{pmatrix}\mathbf{x} + \begin{pmatrix} 0 \\ 1 \end{pmatrix} u$$

$$y = \begin{pmatrix} a_1 a_2 & 0 \end{pmatrix}\mathbf{x}.$$

Eigenvalues $\quad \lambda_1 = -a_1,\ \lambda_2 = -a_2.$

Permissible step size for an explicit single-step method with order $p = 1$ (e.g. EUL) or $p = 2$ (e.g. MIP),
i.e. $\mu = h\lambda \in [-2.0, 0]$.

Numerical example

$$a_1 = 1 \quad \Rightarrow \quad h_{1,\max} \le 2$$
$$a_2 = 0.01 \quad \Rightarrow \quad h_{2,\max} \le 200$$

➔ maximum permissible step size: $\underline{\underline{h_{\max}}} = \min\left(h_{1,\max}, h_{2,\max}\right) = \underline{\underline{2}}$!

■

Avoiding long computation times In order to avoid disproportionately long computation times, the following approaches can be employed:

- Use of an absolutely stable *implicit* integration method (e.g. TRA). Advantage: arbitrarily large step sizes are possible; disadvantage: step-size-dependent error.
- *Linear* simulation using the *state transition matrix* (see Sec. 3.5). Advantage: high accuracy regardless of step size; disadvantage: only possible for linear systems, limitations on the temporal behavior of the inputs (piecewise-constant, input models).
- *Model modification* in which "stiff" components are approximated by proportional gain or nonlinear transfer elements.

Model modification: model reduction The final possibility presented above—that of model modification—should always be considered as a serious alternative. However, such a modification should also always be critically scrutinized to determine whether it is truly allowable, i.e. whether it retains the desired degree of representation of the system dynamics. In many cases, well-damped low-pass elements can in fact be approximated by gain elements within a matching frequency band (within the controller bandwidth). This type of element substitution is termed *model reduction* (Fig. 3.6, e.g. fast actuators, broadband measurement amplifiers).

High-frequency eigenmodes of multibody systems Significantly greater care must be taken, however, when dealing with high-frequency eigenmodes of multibody systems (MBS). In many cases, such eigenmodes are neglected in the controller design (the *spillover* effect, see Sec. 10.3). However, it is precisely the simulation which should demonstrate that such

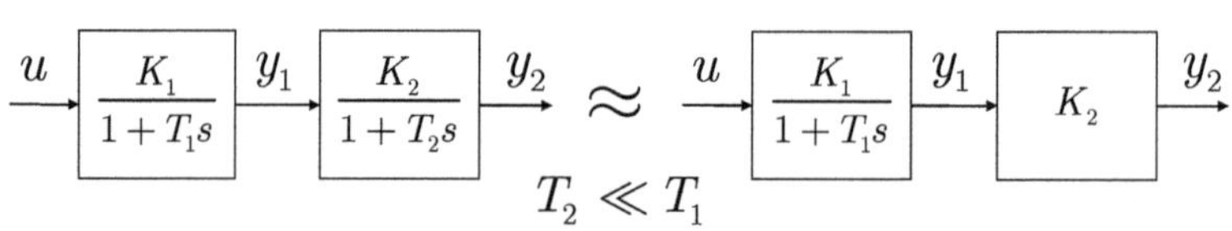

Fig. 3.6. Model modification in the case of broadband low-pass elements

model uncertainties have no undesired effects on the completed control loop and its robust stability. To this extent, it is downright unacceptable to suppress these high-frequency eigenmodes in the simulation. For such cases, simulation using the state transition matrix—allowable due to the generally linear description of MBSs (e.g. via the finite element method)—offers an ideal solution alternative (see Sec. 3.5).

3.4 Weakly-Damped Systems

Eigenmodes of multibody systems One oft-overlooked difficulty appears in the course of simulating weakly-damped (or in the extreme case, undamped) oscillatory systems (harmonic oscillators). In particular, for multibody systems, high-frequency eigenmodes with *vanishingly small damping* are examined during sensitivity and robustness investigations to verify robust stability.

Harmonic oscillator Fig. 3.7 shows two typical cases for a harmonic oscillator (undamped oscillations). In this example, the concept of absolute stability suggests that for the conjugate complex eigenvalues $\lambda_2 = \pm j\omega_2$, the step size h should be reduced until $\mu_2 = \pm jh\omega_2$ lies within the region of absolute stability. As this is already the case for the conjugate complex eigenvalues $\lambda_1 = \pm j\omega_1$, the question presents itself: does a simulation with $\mu_1 = \pm jh\omega_1$ deliver the correct result?

The answer is: no! The true solution for this harmonic oscillator is $y(t) = y(0) \cdot \sin(\omega_1 t + \varphi_0)$, i.e. an *undamped* harmonic oscillation. However, the region of absolute stability guarantees that the approximate discrete solution for $y(t)$ using Eq. (3.11) will be a *decaying* series, as the eigenvalues of the difference equation will have magnitude less than one.

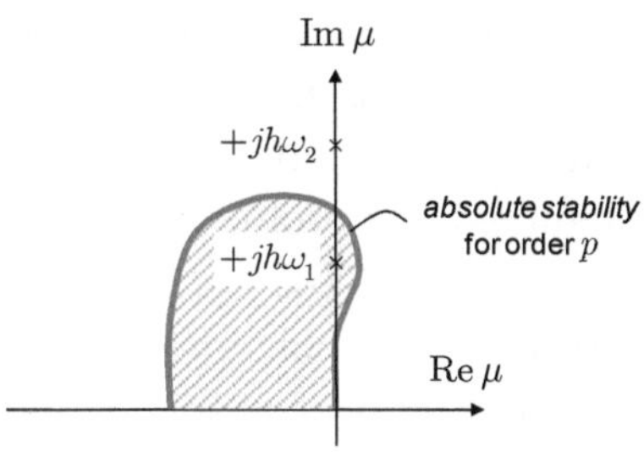

Fig. 3.7. Undamped eigenmodes under an explicit single-step method (only top half-plane shown)

Problem of explicit single-step methods This means that the marginally-stable case of a harmonic oscillator fundamentally *cannot* be correctly solved using an explicit single-step method (nor can general linear models with unstable eigenvalues). Steady-state oscillatory dynamics cannot be correctly replicated, as damping appears due to the integration process[4]. Only for the case with order $p \to \infty$ (i.e. the state transition matrix method, LIN) does a true steady oscillation result. This once again demonstrates the advantages of the state transition matrix method for multibody systems (see Sec. 3.5).

Error equation This type of undesirable behavior for methods of finite orders p can be easily verified with an *error equation* (Potthoff 2003). Comparing the true solution $x(t) = x(0) \cdot e^{\lambda t}$ of the test initial value problem (3.10) and its approximated solution $\hat{x}_{k+1} = \varphi_R\left(h\lambda\right)\hat{x}_k$ (Eq. (3.11)), the error equation for the *global discretization error* (*GDE*) is obtained:

$$\Delta x_{k+1} = x\left((k+1)\cdot\mu\right) - \hat{x}_{k+1} = \left(e^{(k+1)\cdot\mu} - {\varphi_R}^k\left(\mu\right)\right)\cdot x\left(0\right).$$

For example, for $p = 1$, the error equation is

$$\Delta x_{k+1} = \left(e^{(k+1)\cdot\mu} - \left(1+\mu\right)^k\right)\cdot x\left(0\right).$$

Under *unstable* dynamics, i.e. $\mathrm{Re}\,\mu > 0$, a divergent error equation for the *global discretization error* (*GDE*) results:

$$\lim_{k\to\infty}\left[\left(1+\mu+\frac{1}{2!}\mu^2+\ldots\right)^{k+1} - \left(1+\mu\right)^k\right] = \infty \text{ for } \mu > 0.$$

For other methods and method orders, the dynamics can be investigated in an equivalent manner.

What possibilities thus exist to nevertheless correctly implement a simulation of this case of great practical importance?

Solution possibilities For *linear* models, the previously mentioned method using the *state transition matrix* is available (see Sec. 3.5). If a *nonlinear* simulation—and thus a generally-applicable numerical integration method—is absolutely required, the problem can only be handled by limiting the global discretization error $GDE = \mathcal{O}\left(h^p\right)$. This requires selecting a sufficiently small step size. "Sufficiently small" not only means that $\mu = h\lambda$ must be within the region of absolute stability, but additionally

[4] This holds in general for unstable systems, i.e. those in which $\mathrm{Re}\,\lambda_i > 0$.

that $\mu = h\lambda$ must be placed near the origin, implying, in turn, a significant increase in the required computation time.

Example 3.3 *Harmonic oscillator.*

An undamped mass-spring system (m, k) represents a harmonic oscillator with resonant frequency $\omega_0 = \sqrt{k/m}$. In the absence of external excitation, this system is described by the following autonomous differential equation:

$$\dot{\mathbf{x}} = \begin{pmatrix} 0 & 1 \\ -\omega_0^2 & 0 \end{pmatrix} \mathbf{x}, \quad \mathbf{x}(0) = \begin{pmatrix} x_{10} & x_{20} \end{pmatrix}^T .$$

Given numerical values $\omega_0 = 1$ and $\mathbf{x}(0) = \begin{pmatrix} 1 & 0 \end{pmatrix}^T$, the true solution is $x(t) = \cos(t)$.

In Fig. 3.8, this true solution is shown along with *approximated solutions* $\hat{x}_k$ using a fourth-order RUNGE-KUTTA method (RK4) and the state transition matrix (LIN) for different (fixed) step sizes h . It is easy to see that RK4 only reproduces the steady oscillation with sufficiently small step sizes (and thus sufficiently small GDE), whereas LIN very accurately simulates the actual dynamics independent of step size. For high-frequency eigenmodes in a mechanical structure, the step size for an RK4 implementation would thus not only have to lie within the region of absolute stability, but would have to be sufficiently small to avoid undesirable damping effects (leading to even longer computation times!).

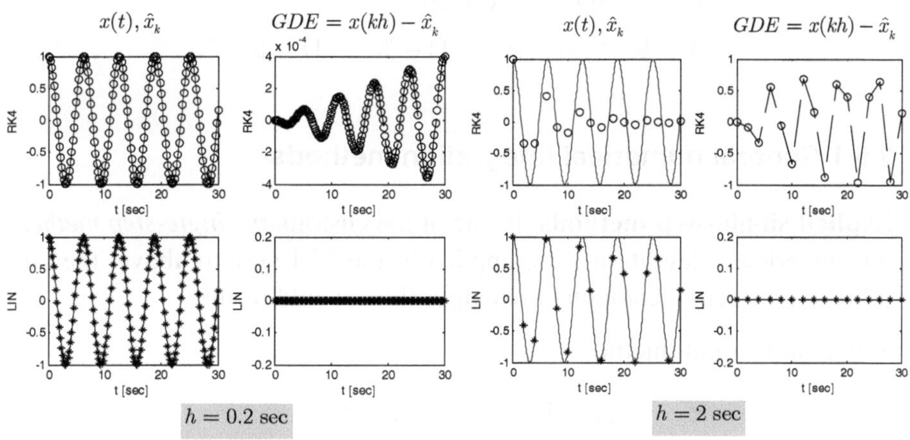

Fig. 3.8. Approximated solutions for a harmonic oscillator $(\omega_0 = 1)$

■

3.5 High-Order Linear Systems

Multibody systems The dynamic models of physical phenomena relevant to the design of mechatronic systems presented in this book are primarily linear. In particular, the equations of motion of multibody systems are generally available in a linear form (FEM), though for sufficiently accurate modeling, they will have high system order. As has been discussed several times already, in a simulation, all eigenmodes (i.e. also high-frequency, often weakly-damped ones) should be incorporated. This generally results in linear, weakly-damped, stiff high-order systems. The general limitations of and difficulties in numerical integration methods for such system models have already been discussed in detail in the previous sections.

Solution approach: state transition matrix This section presents an alternative approach for the numerical solution of dynamic system equations. This approach employs the analytical solution of a linear system of ordinary differential equations via the *state transition matrix*. This method is marked by a significantly reduced computational load (computation time) with arbitrarily high computational accuracy.

For the following discussion, consider the *linear*, *time-invariant system*

$$\begin{aligned} \dot{\mathbf{x}} &= \mathbf{A}\mathbf{x} + \mathbf{B}\mathbf{u} \qquad \mathbf{x}(t_0) = \mathbf{x}_0 \in \mathbb{R}^n \\ \mathbf{y} &= \mathbf{C}\mathbf{x} + \mathbf{D}\mathbf{u} \end{aligned} \tag{3.12}$$

where $\begin{array}{l} \mathbf{x} \in \mathbb{R}^n, \mathbf{u} \in \mathbb{R}^m, \mathbf{y} \in \mathbb{R}^p \,, \\ \mathbf{A} \in \mathbb{R}^{n\times n}, \mathbf{B} \in \mathbb{R}^{n\times m}, \mathbf{C} \in \mathbb{R}^{p\times n}, \mathbf{D} \in \mathbb{R}^{p\times m} \,. \end{array}$

3.5.1 General numerical integration methods

Explicit single-step methods If one of the customary *single-step methods* for numerical integration were applied to the LTI system above, then the following effective computational algorithms would result:

- Forward EULER method:

$$\hat{\mathbf{x}}_{k+1} = \left\{\mathbf{I} + h\cdot\mathbf{A}\right\}\cdot\hat{\mathbf{x}}_k + h\cdot\mathbf{B}\cdot\mathbf{u}_k \,,$$

$$\hat{\mathbf{x}}_{k+1} = \tilde{\mathbf{\Phi}}^{(1)}\cdot\hat{\mathbf{x}}_k + \tilde{\mathbf{H}}^{(1)}\cdot\mathbf{u}_k \,, \tag{3.13}$$

- Fourth-order RUNGE-KUTTA method:

$$\hat{\mathbf{x}}_{k+1} = \left\{ \mathbf{I} + h\mathbf{A} + \frac{h^2}{2}\mathbf{A}^2 + \frac{h^3}{6}\mathbf{A}^3 + \frac{h^4}{24}\mathbf{A}^4 \right\} \hat{\mathbf{x}}_k + $$
$$+ \left\{ h\mathbf{B} + \frac{h^2}{2}\mathbf{A}\,\mathbf{B} + \frac{h^3}{6}\mathbf{A}^2\,\mathbf{B} + \frac{h^4}{24}\mathbf{A}^3\,\mathbf{B} \right\} \mathbf{u}_k ,$$

$$\hat{\mathbf{x}}_{k+1} = \tilde{\mathbf{\Phi}}^{(4)} \cdot \hat{\mathbf{x}}_k + \tilde{\mathbf{H}}^{(4)} \cdot \mathbf{u}_k . \tag{3.14}$$

Time-invariant recursion Applying these common numerical integration methods to the linear system, the ordinary differential equations in Eq. (3.12) can thus be reduced to the simple recursion formulas (3.13) or (3.14). It is worth noting that, due to the time-invariance of Eq. (3.12), the *coefficient matrices* of these recursions are *constant at all time steps*, which in turn means that they only need to be calculated *once before* the actual simulation run.

3.5.2 Solution via the state transition matrix

Analytical solution A deeper understanding of this problem is obtained by comparing with the *exact analytical solution* of the system of ordinary differential equations (3.12) using the *state transition matrix* (or *fundamental matrix* or *matrix exponential*) $\mathbf{\Phi}(t)$ (see (Ogata 2010)):

$$\mathbf{x}(t) = \mathbf{\Phi}(t - t_0) \cdot \mathbf{x}(t_0) + \int_{t_0}^{t} \mathbf{\Phi}(t - \tau) \cdot \mathbf{B} \cdot \mathbf{u}(\tau)\, d\tau . \tag{3.15}$$

If Eq. (3.15) is applied over an integration interval $[t_k, t_k + h]$, it follows that

$$\mathbf{x}(t_k + h) = \mathbf{\Phi}(h) \cdot \mathbf{x}(t_k) + \int_{t_k}^{t_k+h} \mathbf{\Phi}\left(t_k + h - \tau\right) \cdot \mathbf{B} \cdot \mathbf{u}(\tau)\, d\tau . \tag{3.16}$$

Under the assumption

$$\mathbf{u}(t) = \mathbf{u}(t_k) = const. \quad \text{for} \quad t \in \left[t_k, t_k + h\right) \tag{3.17}$$

in Eq. (3.16), the following *recursion formula* for the *exact solution* with *arbitrary step size* h results:

$$\mathbf{x}(t_k + h) = \mathbf{\Phi}(h) \cdot \mathbf{x}(t_k) + \left[\int_0^h \mathbf{\Phi}(\tau) \cdot \mathbf{B} \cdot d\tau \right] \cdot \mathbf{u}(t_k),$$

$$\mathbf{x}(t_k + h) = \mathbf{\Phi}(h) \cdot \mathbf{x}(t_k) + \mathbf{H}(h) \cdot \mathbf{u}(t_k), \tag{3.18}$$

$$\text{where} \quad \mathbf{H}(h) := \int_0^h \mathbf{\Phi}(\tau) \cdot \mathbf{B} \cdot d\tau . \tag{3.19}$$

The following well-known relations hold for the *state transition matrix* $\mathbf{\Phi}(h)$ and the *discrete input matrix* $\mathbf{H}(h)$:

$$\mathbf{\Phi}(h) := e^{\mathbf{A} \cdot h} = \mathbf{I} + h \cdot \mathbf{A} + \frac{h^2}{2!} \cdot \mathbf{A}^2 + \frac{h^3}{3!} \cdot \mathbf{A}^3 + \dots , \tag{3.20}$$

$$\mathbf{H}(h) = \left\{ h \cdot \mathbf{I} + \frac{h^2}{2!} \mathbf{A} + \frac{h^3}{3!} \mathbf{A}^2 + \frac{h^4}{4!} \mathbf{A}^3 + \dots \right\} \cdot \mathbf{B}. \tag{3.21}$$

Consequences for general numerical integration methods

Approximating the state transition matrix A comparison of Eqs. (3.20) and (3.21) with Eqs. (3.13) and (3.14) shows that the numerical integration methods presented here employ *approximations* of the true solution. That is, in these methods, the *state transition matrix* $\mathbf{\Phi}(h)$ and the *discrete input matrix* $\mathbf{H}(h)$ are approximated by a corresponding (small) number of summation terms.

This makes the connection to the method order p directly apparent: e.g. RK4 has $p = 4$ with a local discretization error $LDE = \mathcal{O}\left(h^5\right)$, meaning that $\tilde{\mathbf{\Phi}}^{(4)}$ approximates the exact state transition matrix $\mathbf{\Phi}(h)$ to fourth order h^4. Implicit methods (e.g. the trapezoid method) approximate $\mathbf{\Phi}(h)$ via PADÉ approximations (cf. Table 3.1).

General recursion for single-step methods The *general recursion formula* for single-step methods with *linear time-invariant systems* can then be simply formulated as follows:

$$\begin{aligned} \hat{\mathbf{x}}(t_k + h) &= \tilde{\mathbf{\Phi}}(h) \cdot \hat{\mathbf{x}}(t_k) + \tilde{\mathbf{H}}(h) \cdot \mathbf{u}(t_k), \\ \hat{\mathbf{y}}(t_k) &= \mathbf{C} \cdot \hat{\mathbf{x}}(t_k) + \mathbf{D} \cdot \mathbf{u}(t_k). \end{aligned} \tag{3.22}$$

3.5.3 Accuracy of the simulation solutions

Factors influencing simulation accuracy The *accuracy* of the state transition matrix simulation solution $\hat{\mathbf{x}}(t_k)$ thus fundamentally depends only on the following factors:

- the *accuracy of the approximations* $\tilde{\mathbf{\Phi}}(h)$, $\tilde{\mathbf{H}}(h)$, and
- the *validity* of assumption (3.17), i.e. that inputs $\mathbf{u}(t)$ are *constant over the integration interval* $[t_k, t_k + h]$.

Thus, the choice of step size h no longer has the same elementary significance for simulation accuracy as it does for general numerical integration methods.

Offline computation of $\tilde{\mathbf{\Phi}}, \tilde{\mathbf{H}}$ Determination of sufficiently accurate approximations $\tilde{\mathbf{\Phi}}, \tilde{\mathbf{H}}$ from Eqs. (3.20), (3.21) is completely decoupled from the actual simulation algorithm (3.22). Thus, even for *high integration accuracy*, the *computational burden* for simulation is determined solely by the *system order*, that is, the dimensions of the system matrices $\mathbf{A}, \mathbf{B}, \mathbf{C}, \mathbf{D}$ or $\tilde{\mathbf{\Phi}}, \tilde{\mathbf{H}}$.

The computation of $\tilde{\mathbf{\Phi}}$ and $\tilde{\mathbf{H}}$ can take place offline using appropriate, numerically stable algorithms.

However, in the case of *time-varying systems* (e.g. systems with variable structure), $\tilde{\mathbf{\Phi}}$ and $\tilde{\mathbf{H}}$ must be recomputed whenever the system parameters $\mathbf{A}$ and $\mathbf{B}$ change.

Constant inputs The second fundamental factor in the accuracy of the state transition matrix method concerns *constancy of the inputs* over the integration interval. In general, this condition will of course not be met exactly. In such cases, the integration step size h should be chosen to be sufficiently short given the rate of change of the inputs and the overall required accuracy.

In the following configurations, however, the requirement in Eq. (3.17) is met *exactly* so that an *arbitrarily accurate* solution is achievable:

1. $u_i(t)$ are step inputs (constant after $t = 0^+$) (Fig. 3.9), i.e. $u_i(t) = c_i \cdot \sigma(t)$, $i = 1, \ldots, m$. There are no restrictions on the step size h.

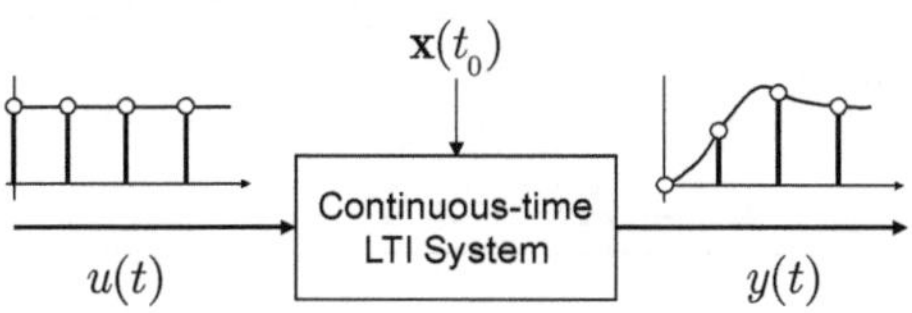

Fig. 3.9. Constant inputs

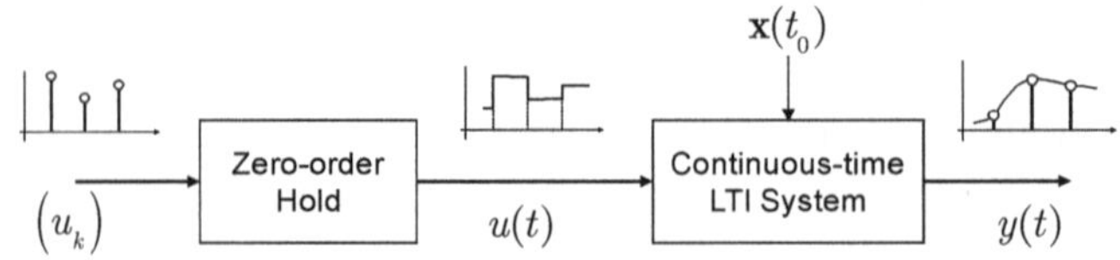

Fig. 3.10. Staircase inputs resulting from sampled controllers

2. $u_i(t)$ are (piecewise-constant) *staircase functions*, i.e. $u_i(t) = c_i$ in $t \in [t_k, t_k + h]$, $i = 1, \ldots, m$.

   For example, this condition is met in, for example *sampled data systems* where $u_i(t)$ are the outputs of digital controllers (Fig. 3.10 with a zero-order hold element as interface between the discrete-time and the continuous-time subsystems). In such cases, the corresponding system sampling time can be used as the maximum integration step size. For several *differing* sampling times $T_{a,j}$ (e.g. in cascaded controllers), the step size h should be chosen such that

$$h_{\max} = \text{greatest common factor}\left(T_{a,j}\right),\ j = 1, 2, \ldots$$

   Using smaller step sizes, the system behavior *between* samples can also be simulated *exactly*, from the point of view of computational accuracy.

3. $u_i(t)$ can be modeled as *solutions of linear differential equations* (*signal generators, signal models*). In this case, the system of ordinary differential equations is extended with these signal generators, signal evolution is parameterized with an appropriate choice of initial values, and this *extended* system is simulated according to Eq. (3.22) (Fig. 3.11).

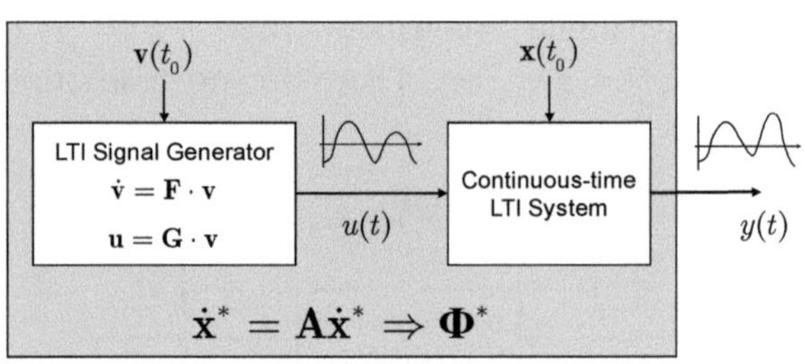

Fig. 3.11. Extended system with signal generator

General LTI signal generator:

$$\dot{\mathbf{v}} = \mathbf{F} \cdot \mathbf{v},$$
$$\mathbf{u} = \mathbf{G} \cdot \mathbf{v}.$$

Extended system:

$$\mathbf{x}^* := \begin{pmatrix} \mathbf{x} & \mathbf{v} \end{pmatrix}^T,$$

$$\dot{\mathbf{x}}^* = \underbrace{\begin{pmatrix} \mathbf{A} & \mathbf{B} \cdot \mathbf{G} \\ \mathbf{0} & \mathbf{F} \end{pmatrix}}_{\mathbf{A}^*} \mathbf{x}^*, \qquad \mathbf{x}^*(0) = \begin{pmatrix} \mathbf{x}_0 \\ \mathbf{v}_0 \end{pmatrix}.$$

In this case, the state transition matrix $\mathbf{\Phi}^*(t) = e^{\mathbf{A}^* t}$ belonging to $\mathbf{A}^*$ must be used for simulation. Note that in this case, only the homogeneous solution of the system need be computed. Here too, there are *no* restrictions on the step size *h*. A structural change in the inputs $\mathbf{u}$, however, categorically requires recomputation of $\mathbf{\Phi}^*(h)$.

Example 3.4 *Signal generator for harmonic excitation.*

$$u(t) = U_0 \cdot \sin\left(\omega_0 t + \phi\right), \quad t \geq 0.$$

LTI signal model:

$$v_1(t) := u(t) = U_0 \cdot \sin\left(\omega_0 t + \phi\right),$$
$$v_2(t) := \dot{v}_1(t) = U_0 \cdot \omega_0 \cdot \cos\left(\omega_0 t + \phi\right),$$
$$\dot{v}_2(t) = -U_0 \cdot \omega_0^2 \sin\left(\omega_0 t + \phi\right) = -\omega_0^2 \cdot v_1(t).$$

$$\Rightarrow \qquad \mathbf{v} := \begin{pmatrix} v_1 & v_2 \end{pmatrix}^T,$$

$$\dot{\mathbf{v}} = \begin{pmatrix} 0 & 1 \\ -\omega_0^2 & 0 \end{pmatrix} \mathbf{v}, \qquad \mathbf{v}(0) = \begin{pmatrix} U_0 \cdot \sin\phi \\ U_0 \cdot \omega_0 \cdot \cos\phi \end{pmatrix},$$

$$u = \begin{pmatrix} 1 & 0 \end{pmatrix} \mathbf{v}.$$

■

3.6 Numerical Integration of DAE Systems

Reference model In this section, for ease of presentation and understanding, a scalar system with *one* state variable and *one* algebraic variable is considered:

$$\dot{x} = f(x, z), \tag{3.23}$$

$$0 = g_1(x, z), \tag{3.24}$$

$$\text{or} \qquad 0 = g_2(x). \tag{3.25}$$

Vector-valued systems can be handled analogously.

In each numerical integration step, the difference equation corresponding to the differential equation (3.23) *and* the algebraic Eqs. (3.24), (3.25) must be simultaneously satisfied.

Direct DAE solution This section investigates several fundamental approaches to the *direct* numerical solution of DAE systems. Direct solution means that, without additionally reworking the model, the original DAE system—having a certain *differential index* (see Sec. 2.4.1)—is integrated.

Indeed, one alternative (presented in Sec. 2.4.3) is to use index reduction to convert a DAE system into a system of ordinary differential equations. The problems with this methodology were already discussed in detail in previous sections.

For extensions of this approach, the reader is referred to the monographs (Brenan et al. 1996) and (Cellier and Kofman 2006), as well as the problem-specific papers (Otter 1999), (Otter and Bachmann 1999), and (Otter and Bachmann 1999).

3.6.1 Explicit integration methods

Explicit solution The defining characteristic of explicit integration methods is that an approximate solution to the differential equation can be computed using a recursive system of equations, i.e. the equations can be solved *sequentially* in the appropriate form (implying a small computational load).

The fundamental problems with explicit integration methods for DAE systems can already be clearly demonstrated with the simple *explicit* EULER *method.*

Index-1 systems

By definition, $\partial g / \partial z \neq 0$, i.e. $g := g_1(x, z)$.
The EULER recursion for a *given* (*arbitrary*) initial value x_0 is:

$$k = 0: \qquad 0 = g_1(x_0, \hat{z}_0) \qquad \Rightarrow \hat{z}_0 \tag{3.26}$$

$\Rightarrow$ Suitable $\hat{z}_0$ from the algebraic Eq. (3.24) with given x_0.

$$k = 1: \begin{cases} \hat{x}_1 = x_0 + f(x_0, \hat{z}_0) \quad \Rightarrow \hat{x}_1 & (3.27) \\ \\ 0 = g_1(\hat{x}_1, \hat{z}_1) \quad \Rightarrow \hat{z}_1 & (3.28) \end{cases}$$

$\Rightarrow$ $\hat{x}_1$ from differential equation (3.27) with known x_0, $\hat{z}_0$,
$\Rightarrow$ $\hat{z}_1$ from algebraic Eq. (3.28) with known $\hat{x}_1$.

> *Index-1 systems* can be solved using an *explicit* integration method via sequential solution of the difference equation ($\rightarrow \hat{x}_{k+1}$) and the algebraic equation ($\rightarrow \hat{z}_{k+1}$).
> Additional task: determination of *consistent* initial values x_0, $\hat{z}_0$ from Eq. (3.26).
>
> An explicit integration method can be used as-is. However, at every integration step, it must be complemented with a *supplemental* algebraic equation solver.

Index-2 systems

By definition, $\partial g / \partial z = 0$, i.e. $g := g_2(x)$.
The EULER recursion given a *suitable* initial value x_0 is

$$k = 0: \qquad 0 = g_2(x_0) \qquad \Rightarrow x_0 \tag{3.29}$$

$\Rightarrow$ x_0 must satisfy the algebraic Eq. (3.25) and *cannot* be freely chosen!

$$k = 1: \left\{ \begin{aligned} \hat{x}_1 &= x_0 + f(x_0, z_0) \quad (3.30) \\ 0 &= g_2(\hat{x}_1) \quad (3.31) \end{aligned} \right\} \Rightarrow \hat{x}_1, z_0$$

$\Rightarrow$ *Not sequentially* solvable for $\hat{x}_1, z_0$. Eqs. (3.30) and (3.31) represent an *implicit*, nonlinear, algebraic system of equations; their solution requires an *implicit* solver, e.g. the NEWTON-RAPHSON method (Faires and Burden 2002).

> *DAE systems* with *index* ≥ 2 can, in principle, *not* be solved via an *explicit* integration method.
>
> An *implicit equation solver* is required to simultaneously solve the difference equation and the algebraic equation.

3.6.2 Implicit integration methods

Implicit solution As previously discussed, implicit numerical integration methods must incorporate an implicit equation solver. The corresponding increased computational load is gladly accepted for numerical integration as it guarantees absolute numerical stability even for large step sizes (which is important for stiff systems). Thus, it seems convenient to employ *implicit* integration methods for the numerical solution of DAE systems to simultaneously solve the difference equation and algebraic constraints. However, it will be seen that this is only possible for certain classes of DAE systems.

Trapezoid method for DAE systems

Implicit recursion For the DAE system in Eq. (3.23) and the algebraic equation $g = 0$ in Eqs. (3.24) and (3.25), it follows for numerical integration via the *trapezoid algorithm*[5] *TRA* (Faires and Burden 2002) with integration step size h that

$$\begin{aligned} \hat{x}_{k+1} &= \hat{x}_k + \frac{h}{2}\left[f(\hat{x}_k, \hat{z}_k) + f(\hat{x}_{k+1}, \hat{z}_{k+1})\right], \\ 0 &= g(\hat{x}_{k+1}, \hat{z}_{k+1}). \end{aligned} \tag{3.32}$$

[5] This method was chosen as an example due to its simplicity; other implicit single-step methods exhibit similar traits.

NEWTON-RAPHSON iteration Rearranging Eq. (3.32) results in a nonlinear system of algebraic equations for determining the unknown new approximate values $\hat{x}_{k+1}, \hat{z}_{k+1}$:

$$\varphi_{k+1}\left(\mathbf{p}_{k+1}\right) := \begin{pmatrix} \varphi_{1,k+1} \\ \varphi_{2,k+1} \end{pmatrix} = \begin{pmatrix} \hat{x}_{k+1} - \hat{x}_k - \frac{h}{2}\left[f(\hat{x}_k, \hat{z}_k) + f(\hat{x}_{k+1}, \hat{z}_{k+1})\right] \\ g(\hat{x}_{k+1}, \hat{z}_{k+1}) \end{pmatrix}, \tag{3.33}$$

where $\mathbf{p}_{k+1} := (\hat{x}_{k+1} \quad \hat{z}_{k+1})^T$

$$\Rightarrow \quad \varphi_{k+1}\left(\mathbf{p}_{k+1}\right) = \mathbf{0}\,. \tag{3.34}$$

The solution of the nonlinear system of equations (3.34) can be found using the well-known NEWTON-RAPHSON method, leading to the nonlinear recursion rule

$$\begin{gathered} \mathbf{p}_{k+1,i+1} = \mathbf{p}_{k+1,i} - \mathbf{J}\left(\mathbf{p}_{k+1,i}\right)^{-1} \cdot \varphi_{k+1}\left(\mathbf{p}_{k+1,i}\right). \\ \text{Iterate over } i, \text{ until } \left\|\mathbf{p}_{k+1,i+1} - \mathbf{p}_{k+1,i}\right\| \le \varepsilon\,. \end{gathered} \tag{3.35}$$

Numerical convergence A central role in the convergence of the NEWTON-RAPHSON iteration in Eq. (3.35) is played by the *JACOBIan*[6]

$$\mathbf{J}\left(\mathbf{p}\right) = \left(\begin{array}{c|c} \left.\frac{\partial \varphi_1}{\partial x}\right|_{i+1} & \left.\frac{\partial \varphi_1}{\partial z}\right|_{i+1} \\ \hline \left.\frac{\partial \varphi_2}{\partial x}\right|_{i+1} & \left.\frac{\partial \varphi_2}{\partial z}\right|_{i+1} \end{array}\right) = \left(\begin{array}{c|c} 1 - \frac{h}{2}\frac{\partial f}{\partial x} & -\frac{h}{2}\frac{\partial f}{\partial z} \\ \hline \frac{\partial g}{\partial x} & \frac{\partial g}{\partial z} \end{array}\right). \tag{3.36}$$

From Eq. (3.35), it follows that for *numerical convergence* of the NEWTON-RAPHSON iterations at integration step $(k+1)$, the *inverse* of the JACOBIan (3.36) must exist, i.e. it must hold that

$$\boxed{\det \mathbf{J}\left(\mathbf{p}\right) = \left(1 - \frac{h}{2}\frac{\partial f}{\partial x}\right)\frac{\partial g}{\partial z} + \frac{h}{2}\frac{\partial f}{\partial z}\frac{\partial g}{\partial x} \neq 0}\,. \tag{3.37}$$

The remainder of this section examines the elementary condition (3.37) for different DAE indices and arbitrarily small step sizes $h \to 0$ (required for high numerical accuracy).

---

[6] In simulation tools, the JACOBIan can be either computed numerically, or an analytical function can be made available for evaluation.

Index-1 systems

Since, by definition, the *index-1 condition* is *fulfilled*, then $\partial g_1/\partial z \neq 0$ (see Eq. (2.46)) and

$$\lim_{h\to 0}\left(\det \mathbf{J}\right) = \frac{\partial g_1}{\partial z} \neq 0\,. \tag{3.38}$$

> *Index-1 systems* are generally *well-solved* using *implicit methods (TRA)* independent of the step size chosen (even for very small step sizes).

Index-2 systems

Since, by definition, the *index-1 condition* is *violated*, then $\partial g_2/\partial z = 0$ (see Eq. (2.48)) and

$$\det \mathbf{J} = \frac{h}{2}\frac{\partial f}{\partial z}\frac{\partial g_2}{\partial x} \neq 0 \quad \text{for } h \neq 0\,. \tag{3.39}$$

Note, however, that for *very small* step sizes h

$$\det \mathbf{J} \approx 0\,, \text{ i.e. } \lim_{h\to 0}\left(\det \mathbf{J}\right) = 0\,. \tag{3.40}$$

> *Index-2 systems* can only be well-solved using *implicit methods (TRA)* given *a sufficiently large step size.*
>
> For very small step sizes, the JACOBIan becomes ill-conditioned ($\det \mathbf{J} \approx 0$) and thus can no longer be accurately inverted inside the NEWTON-RAPHSON iteration.

Index-3 systems

Since, by definition, the index-2 condition Eq. (2.48) is *violated*, it follows that, independent of the step size,

$$\det \mathbf{J} = \frac{h}{2}\frac{\partial f}{\partial z}\frac{\partial g_2}{\partial x} = 0\,.$$

> ***Index-3 systems*** are fundamentally <u>*not*</u> *solvable* using *implicit methods (TRA).*

Summary: integration algorithms for DAE systems

Index ≤ 2: *implicit* methods are always useable; for small step sizes, scaling of algebraic variables may be necessary (see Sec. 3.6.3).

Index = 3: *index reduction* by 1, i.e. differential constraints (PFAFFian form) ➔ index-2 system; proceeding further as above.

3.6.3 Scaling for index-2 systems

Avoiding singularities in the JACOBIan Ill-conditioning of the JACOBIan for index-2 systems in the case of small step sizes can be mitigated via suitable scaling of the algebraic variables according to the following scheme.

Solution approach The NEWTON-RAPHSON iteration of Eq. (3.35) at a fixed time index k (ignored here in the notation), can also be written as follows:

$$\mathbf{J}(\mathbf{p}) \cdot \left(\mathbf{p}_{i+1} - \mathbf{p}_i\right) = \varphi\left(\mathbf{p}_i\right),$$

so that for an index-2 system as in Eqs. (3.23), (3.25) it follows that

$$\tilde{\mathbf{J}} \cdot \tilde{\mathbf{p}} = \begin{pmatrix} 1 - \frac{h}{2}\frac{\partial f}{\partial x} & -\frac{\cancel{h}}{2}\frac{\partial f}{\partial z} \\ \frac{\partial g_2}{\partial x} & 0 \end{pmatrix} \begin{pmatrix} p_{1,i+1} - p_{1,i} \\ h \cdot \left(p_{2,i+1} - p_{2,i}\right) \end{pmatrix}. \tag{3.41}$$

Here, multiplication of the upper-right element of the JACOBIan by the step size h in Eq. (3.41) has been replaced by an equivalent multiplication of the second row of the parameter vector. The algebraic variable z is thus scaled by h. This trick renders the *determinant* of the modified JACOBIan $\tilde{\mathbf{J}}$ *independent of the step size* h:

$$\Rightarrow \quad \det \tilde{\mathbf{J}} = \frac{\partial f}{\partial z}\frac{\partial g}{\partial x},$$

and, as a result, the NEWTON-RAPHSON iteration becomes well-conditioned even for small step sizes.

3.6.4 Consistent initial values

Conditions In order to satisfy the algebraic constraints (3.24), *consistent initial values* $x(0)$, $z(0)$ must be chosen. In this context, the following two cases should be distinguished:

- *DAE systems with index = 1*
  $\partial g/\partial z$ is not singular (i.e. $g = g_1(x,z)$) so that $x(0)$ can be freely chosen and a consistent value for $z(0)$ can be determined by solving the algebraic constraint $g_1\left(x(0), z(0)\right) = 0$.
- *DAE systems with index ≥ 2*
  $\partial g/\partial z$ is singular (i.e. $g = g_2(x)$) so that $x(0)$ can *no longer* be *freely* chosen as $z(0)$ is no longer determined only by the algebraic constraint, but rather additional constraints (algebraic equations in x, z) must be derived from the differential equation (see Sec. 3.6.4).

Problems with inconsistency *Inconsistent* initial values lead, at a minimum, to errors in the initial simulation steps, or to a completely incorrect solution. In some tools, automatic verification of consistent initial values is carried out. However, in general, for complex problems, determining consistent initial values is a difficult and non-trivial (!) task. Thus, computer-aided solutions should always be carefully monitored as relates to this aspect.

Drift-off phenomenon Even having chosen consistent initial values, algebraic constraints are, as a rule, not satisfied exactly due to numerical rounding errors during the course of numerical integration (due to problems arising in the increased solution diversity).

Workarounds exist in the form of particular *stabilization methods* for the numerical integration of DAE systems. One elementary method employs a *linear combination* of the original and differentiated (according to the index reduction) algebraic constraints: BAUMGARTE stabilization (Baumgarte 1972) (for the practical implementation of which, see (Ascher et al. 1994)). Alternative methods employ a projection of the approximated answer into the manifold of the algebraic constraints at every integration step; these are termed *projection methods*, e.g. (Eich 1993). A comparison of these methods can be found, for example, in (Burgermeister et al. 2006).

3.7 Implementation Approaches for Simulation of Hybrid Phenomena

Discontinuities The recognition and handling of discontinuities (switching, state discontinuities) represents one of the greatest challenges for the implementation of simulations. Simple discontinuities can be handled using adaptive step size selection employing an estimation of the local discretization error (e.g. the RUNGE-KUTTA-FEHLBERG method (Faires and Burden 2002)). However, without exact knowledge of the step size algorithm actually implemented, results from this type of simulation should always be evaluated with great skepticism. For the most exact simulations of hybrid phenomena, it commends itself to employ specialized precautions and methods, e.g. (Otter et al. 1999), (Otter et al. 1999).

3.7.1 Handling discontinuities

Fundamental problem The fundamental problem is shown in Fig. 3.12. At the beginning of integration step $k+1$ (i.e. at time t_k) the vector field $\mathbf{f}^1(\mathbf{x},\mathbf{u},t)$ (the right side of the ODE) is active. If the integration step were to be carried out with a step size $h > h_s$, the result would be the approximated value $\hat{\mathbf{x}}^1(t_k + h)$ computed using only $\mathbf{f}^1(\mathbf{x},\mathbf{u},t)$. In fact, though, for $t \geq t_s$, the vector field $\mathbf{f}^2(\mathbf{x},\mathbf{u},t)$ should be taken into account in the integration, leading to the correct answer $\hat{\mathbf{x}}^2\left(t_k + h\right)$. In general, $\hat{\mathbf{x}}^1 \neq \hat{\mathbf{x}}^2$ (cf. also Fig. 2.62).

Principle-based solution approach In order to avoid this type of incorrect simulation result, the principle-based method for *event-dependent step size selection* shown in the following algorithm is recommended[7].

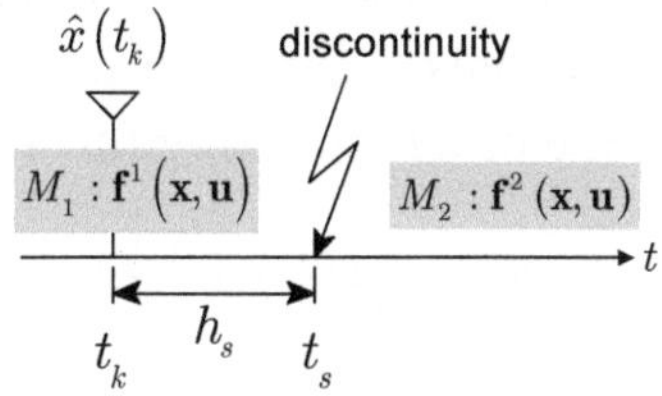

Fig. 3.12. Event-dependent step size selection

7 In "good" simulation tools, such an algorithm is generally implemented; however, in any particular case, the quality of the implementation should be verified!

- *Algorithm for event-dependent step size selection*

(1)	Identify discontinuities in following integration interval.
(2)	Determine t_s .
(3)	Match the step size: $h_s = t_s - t_k$.
(4)	Integrate step $k+1$: $\hat{x}(t_k) \xrightarrow{h_s} \hat{x}(t_s - 0)$.
(5)	Evaluate discontinuity conditions at the discontinuity time t_S : (a) discontinuous inputs: $u\left(t_s + 0\right) = new\_value$, (b) continuous state: $\hat{x}\left(t_s + 0\right) = \hat{x}\left(t_s - 0\right)$ (e.g. jump in input), (c) discontinuous state: $\hat{x}\left(t_s + 0\right) = new\_value$.
(6)	Integrate next step: $\hat{x}\left(t_s + 0\right) \xrightarrow{h_{\mathrm{NEW}}} \hat{x}\left(t_s + h_{\mathrm{NEW}}\right)$ with h_{NEW} as per given specifications and constraints (local discretization error, numerical stability, etc.)

3.7.2 Event detection

Event time In the algorithm for event-dependent step size selection given above, steps (1) and (2) present the actual challenges: *identifying* an event and determining the *event time* t_s .

Hybrid phenomena such as switching and state discontinuities can generally be modeled with the help of *discontinuity hypersurfaces* of the form

$$m(\mathbf{x}, \mathbf{u}, t) = 0 \tag{3.42}$$

and a *threshold test.*

For a net-state model (see Sec. 2.5.3), Eq. (3.42) can be used to concisely describe the C/D interface (the sets of states and their bounding surfaces). The discrete-event signals $\mathbf{v}_D(t)$ can then be described as *binary* signals using the sign change of a monitoring function

$$z\left(t_s\right) = m\left(\mathbf{x}\left(t_s\right), \mathbf{u}\left(t_s\right), t_s\right) \overset{!}{=} 0, \tag{3.43}$$

e.g.

$$v_D(t) = \begin{cases} 0, & \left|z(t)\right| \le \varepsilon \\ 1, & \left|z(t)\right| > \varepsilon \end{cases} \tag{3.44}$$

where ε represents a suitably small threshold.

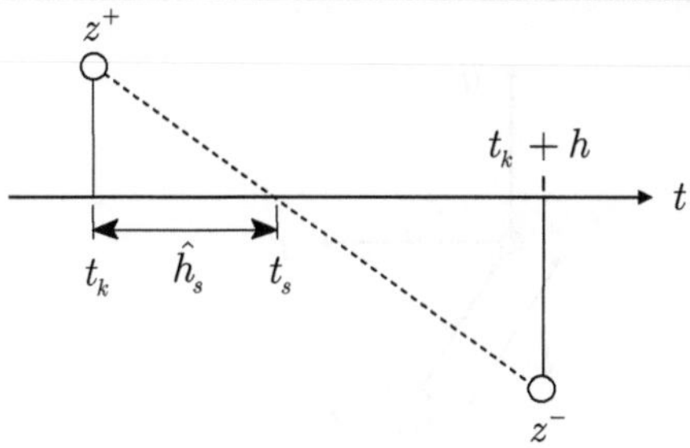

Fig. 3.13. Event detection with a monitoring function

The event time t_s can be iteratively determined via *bisection* methods according to Fig. 3.13 and the relation

$$\hat{h}_s = h \frac{|z^+|}{|z^+| + |z^-|}, \quad \text{sign } z^+ \neq \text{sign } z \,.$$

Implementation in simulation tools This type of event detection is available in good simulation tools as a stand-alone function block (e.g. SIMULINK: *hit crossing*) or as a sub-function of discrete-event blocks (e.g. SIMULINK/STATEFLOW: *edge-triggered signal inputs*). In all cases, it must be possible to access the integration algorithm or its step size selection[8].

3.8 Simulation Example: Ideal Pendulum

System configuration, problem statement As a demonstrative example of a mechanical DAE system—which at first glance has very simple dynamics—consider the *ideal pendulum* shown in Fig. 3.14. This section demonstrates the most important steps in modeling and simulating the DAE models of such systems, presenting both a *signal-* and *equation-based* approach.

---

[8] It is once more pointed out that it always commends itself to carefully test out the proper functioning of any particular implementation. The capabilities of a simulation tool promised in the user manual are not always actually delivered, see for example the behavior of SIMULINK/STATEFLOW blocks for event detection in (Buss 2002), Sec. 6.3.

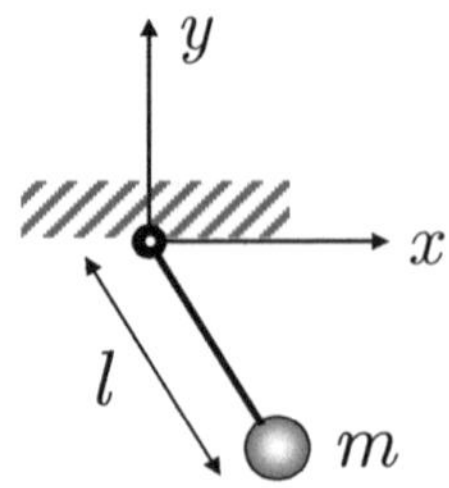

Fig. 3.14. Ideal pendulum

For the ideal pendulum shown in Fig. 3.14, find:

- the equations of motion including the constraint force in the rod (a DAE system),
- a signal-based simulation model (block-oriented, e.g. MATLAB/SIMULINK),
- an equation-based simulation model (object-oriented, e.g. MODELICA).

Solution 1: equations of motion as a DAE system

Using the generalized coordinates $\mathbf{q} = (x \quad y)^T$ and the generalized velocities $\dot{\mathbf{q}} = \mathbf{v} = (v_x \quad v_y)^T$, the resulting *kinetic co-energy* of the system is $T^* = m/2 \cdot (v_x^2 + v_y^2)$. There is a *holonomic* constraint of the form $f(\mathbf{q}) = x^2 + y^2 - l^2 = 0$. The only external force acting upon this system is gravity: $\mathbf{F} = (0 \quad -mg)^T$.

The general form of the EULER-LAGRANGE equations of the first kind is (see Sec. 2.3.2, simplified here for the present application)

$$\begin{aligned} &\frac{d}{dt}\frac{\partial T^*}{\partial \dot{\mathbf{q}}} - \frac{\partial T^*}{\partial \mathbf{q}} = \mathbf{F} + \lambda \cdot \begin{pmatrix} \dfrac{\partial f}{\partial x} \\ \dfrac{\partial f}{\partial y} \end{pmatrix}, \\ &f(\mathbf{q}) = 0, \end{aligned} \tag{3.45}$$

where λ is the LAGRANGE multiplier modeling the constraint force in the pendulum rod (maintaining the holonomic constraint).

Evaluating Eq. (3.45) gives

$$\begin{aligned} &\begin{pmatrix} m & 0 \\ 0 & m \end{pmatrix} \cdot \dot{\mathbf{v}} = \begin{pmatrix} 0 \\ -mg \end{pmatrix} + \lambda \begin{pmatrix} 2x \\ 2y \end{pmatrix}, \\ &x^2 + y^2 - l^2 = 0 \,. \end{aligned}$$

Substituting $\lambda = -\tilde{\lambda} \cdot m/2$ results in the DAE system in semi-explicit form ($u(t) = g = const.$):

$$\dot{\mathbf{x}} = \mathbf{f}(\mathbf{x}, \tilde{\lambda}, u) \quad \text{or} \quad \boxed{\begin{aligned} \dot{x} &= v_x \\ \dot{v}_x &= -\tilde{\lambda} x \\ \dot{y} &= v_y \\ \dot{v}_y &= -\tilde{\lambda} y - g \end{aligned}}, \tag{3.46}$$

$$0 = \mathbf{g}(\mathbf{x}) \qquad \text{or} \quad \boxed{0 = x^2 + y^2 - l^2}\,. \tag{3.47}$$

Solution 2: index reduction, consistent initial values

The mechanical configuration consisting of the rigid coupling of the mass m to the constraining element suggests that this is an index-3 system. This suspicion can be easily confirmed using index tests (see Sec. 2.4.2).

To determine consistent initial values for a block-oriented simulation, a systematic index reduction following Sec. 2.4.3 is carried out below.

The algebraic condition from the original index-3 system provides the *first initial value equation*

$$\boxed{0 = x^2 + y^2 - l^2}\,. \tag{3.48}$$

The first index reduction via differentiation of Eq. (3.48) gives

$$0 = \frac{d}{dt}\left(x^2 + y^2 - l^2\right) = 2x\dot{x} + 2y\dot{y}\,.$$

Substituting from Eq. (3.46) provides the algebraic condition for the index-2 system and the *second initial value equation*

$$\boxed{0 = xv_x + yv_y}\,. \tag{3.49}$$

The second index reduction via differentiation of Eq. (3.49) gives

$$0 = \frac{d}{dt}\left(xv_x + yv_y\right) = x\dot{v}_x + \dot{x}v_x + \dot{y}v_y + y\dot{v}_y\,.$$

Substituting again from Eq. (3.46) provides the algebraic condition on the index-1 system and the *third initial value equation*

$$\boxed{0 = v_x^2 + v_y^2 - gy - \tilde{\lambda} l^2}\,. \tag{3.50}$$

The third index reduction via differentiation of Eq. (3.50) gives

$$0 = \frac{d}{dt}\left(v_x^2 + v_y^2 - gy - \tilde{\lambda} l^2\right) = 2v_x \dot{v}_x + 2v_y \dot{v}_y - g\dot{y} - \dot{\tilde{\lambda}} l^2 .$$

Substituting again from Eq. (3.46) finally gives the desired *differential equation* in the *algebraic variable* $\tilde{\lambda}$ (this is equivalent to introducing a new state variable $\tilde{\lambda}$)

$$\boxed{\dot{\tilde{\lambda}} = -\frac{2\tilde{\lambda} x v_x + 2\tilde{\lambda} y v_y + 3 g v_y}{l^2}} . \tag{3.51}$$

Examining Eqs. (3.46) and (3.51), the result is thus a DAE system with *index = 0*, i.e. a *system of first-order ordinary differential equations.*

The nonlinear *system of equations* (3.48) through (3.50) is available for the determination of *consistent initial values* $x(0), v_x(0), y(0), v_y(0)$, and $\tilde{\lambda}(0)$.

Solution 3: signal-based simulation model

Differential equations (3.46) and (3.51) can be directly implemented in a signal-based simulation model. Fig. 3.15 shows an example of such an implementation using MATLAB/SIMULINK.

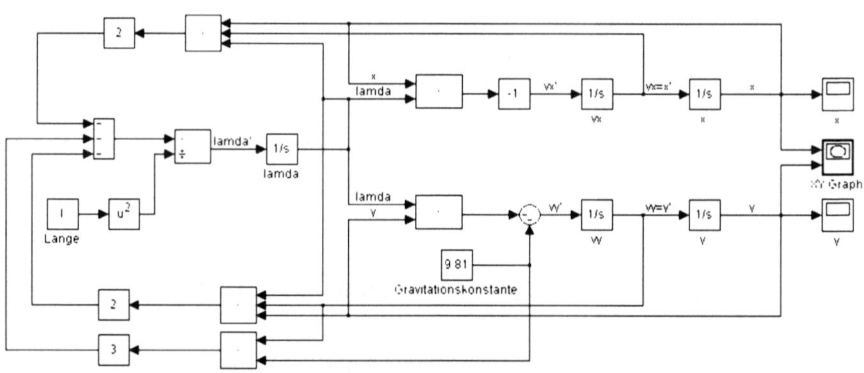

Fig. 3.15. SIMULINK model for the ideal pendulum

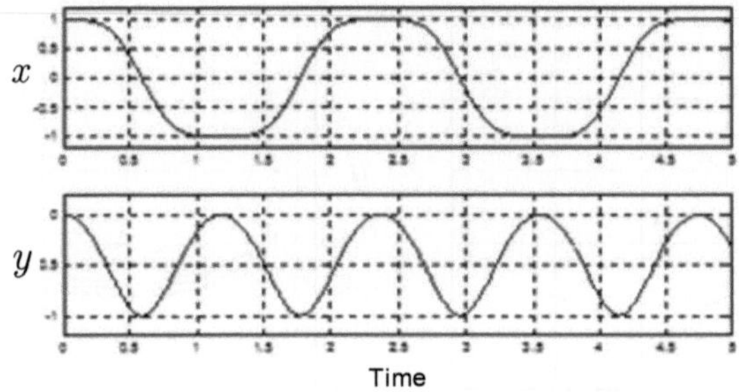

Fig. 3.16. Ideal pendulum: signal-based simulation, horizontal initial position

Solution 4: simulation experiments, signal-based model

Horizontal initial position For a rod length $l = 1$ and a horizontal initial rod position ($y_0 = 0$), giving the consistent initial values (see Eqs. (3.48) through (3.50))

$$\mathbf{x}_0 = \begin{pmatrix} x_0 & v_{x0} & y_0 & v_{y0} & \tilde{\lambda}_0 \end{pmatrix}^T = \begin{pmatrix} 1 & 0 & 0 & 0 & 0 \end{pmatrix}^T,$$

the time histories shown in Fig. 3.16 result.

45° initial position For an initial position of the pendulum at $\varphi_0 = 45°$, the resulting *consistent initial values* are

$$\mathbf{x}_0 = \begin{pmatrix} x_0 & v_{x0} & y_0 & v_{y0} & \tilde{\lambda}_0 \end{pmatrix}^T = \begin{pmatrix} \frac{1}{\sqrt{2}} & 0 & -\frac{1}{\sqrt{2}} & 0 & 6.9367 \end{pmatrix}^T. \quad (3.52)$$

As can be easily verified, the set of *inconsistent initial values*

$$\mathbf{x}_0 = \begin{pmatrix} x_0 & v_{x0} & y_0 & v_{y0} & \tilde{\lambda}_0 \end{pmatrix}^T = \begin{pmatrix} \frac{1}{\sqrt{2}} & 0 & -\frac{1}{\sqrt{2}} & 0 & 10 \end{pmatrix}^T \quad (3.53)$$

does not satisfy Eqs. (3.48) through (3.50).

The time evolution for these two different sets of initial values is depicted in Fig. 3.17. The erroneous nature of the solution trajectory with inconsistent initial values is easily observed in Fig. 3.17b[9], compared to the correct trajectory in Fig. 3.17a.

[9] Caution: this is not always so obvious, e.g. with complex models.

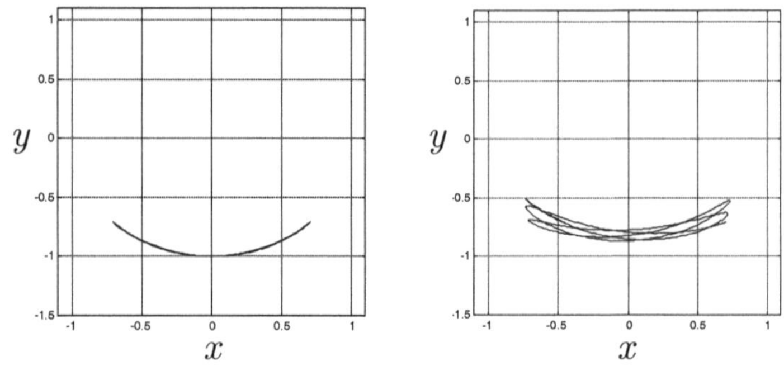

a) b)

Fig. 3.17. Ideal pendulum: signal-based simulation, initial position $\varphi_0 = 45°$: a) consistent initial values, b) inconsistent initial values

Solution 5: equation-based simulation model

The DAE system (3.46), (3.47) can be *directly* (i.e. without further transformations) implemented in an equation-based simulation tool. Fig. 3.18 shows an example of such an implementation in MODELICA. In most MODELICA-based tools, consistent initial values are also automatically generated. Given an accurate implementation, these simulation results coincide with those in Fig. 3.16 and Fig. 3.17a.

```
model pendulum
 parameter Real l=1;           /* pendulum length */
  constant Real g=9.81;        /* acceleration of gravity */

  Real x;                      /* x coordinate */
  Real y;                      /* y coordinate */
  Real lambda;                 /* Lagrange multiplier */
  Real vx;               /* x component of velocity */
  Real vy;               /* y component of velocity */
equation
  der(x)  = vx;
  der(vx) = -x*lambda;
  der(y)  = vy;
  der(vy) = -g - y*lambda;
  0 = x^2 + y^2 - l^2;
end pendulum;
```

Fig. 3.18. Ideal pendulum: equation-based implementation in MODELICA

Alternative model: system of ordinary differential equations

Polar coordinates vs. Cartesian coordinates The model of the pendulum used so far is expressed in *Cartesian* coordinates x, y (Fig. 3.14). The worthy reader may convince herself that an alternative model of the pendulum in *polar coordinates* takes the form of a system of ordinary differential equations without secondary algebraic conditions (a DAE system with index = 0) as follows:

$$\begin{aligned} \dot{\varphi} &= \omega , \\ \dot{\omega} &= -\frac{g}{l} \sin \varphi . \end{aligned} \tag{3.54}$$

The model in Eq. (3.54) can be simulated without difficulty using ordinary explicit integration algorithms.

Modeling vs. simulation aspects Along with the above concluding model treatment, let it be remarked that in the context of systems design, there is always a tradeoff between modeling and simulation. Many simulation problems can even be prevented from appearing in the first place by making an appropriate change to the model. In the above case (change of coordinates), this succeeded without any loss of modeling accuracy in the motion variables (angle, angular velocity). In the modified model (Eq. (3.54)), however, the constraint force is missing. If this latter is not of interest, then, from an implementational point of view, the second model is preferable.

Bibliography for Chapter 3

Ascher, U. M., H. Chin and S. Reich (1994). "Stabilization of DAEs and invariant manifolds." *Numerische Mathematik* **67**: 131–149.

Baumgarte, J. (1972). "Stabilization of constraints and integrals of motion in dynamical systems." *Computer Methods in Applied Mechanics and Engineering* **1**: 1-16.

Brenan, K. E., S. L. Campbell and L. R. Petzold (1996). *Numerical Solution of Initial-Value Problems in Differential-Algebraic Equations*. SIAM.

Burgermeister, B., M. Arnold and B. Esterl (2006). "DAE time integration for real-time applications in multi-body dynamics." *ZAMM - Z. Angew. Math. Mech.* **86**(10): 759–771.

Cellier, F. E. and E. Kofman (2006). *Continuous System Simulation*. Berlin. Springer.

Eich, E. (1993). "Convergence Results for a Coordinate Projection Method Applied to Mechanical Systems with Algebraic Constraints." *SIAM Journal on Numerical Analysis* **30**(5): 1467-1482.

Faires, J. D. and R. L. Burden (2002). *Numerical Methods.* Brooks/Cole.

Franklin, G. F., J. D. Powell and M. L. Workman (1998). *Digital Control of Dynamic Systems.* Addison-Wesley.

Ogata, K. (2010). *Modern Control Engineering.* Prentice Hall.

Otter, M. (1999). "Objektorientierte Modellierung Physikalischer Systeme, Teil 4." *at-Automatisierungstechnik* **47**(4): A13-A16.

Otter, M. and B. Bachmann (1999). "Objektorientierte Modellierung Physikalischer Systeme, Teil 5." *at-Automatisierungstechnik* **47**(5): A17-A20.

Otter, M. and B. Bachmann (1999). "Objektorientierte Modellierung Physikalischer Systeme, Teil 6." *at-Automatisierungstechnik* **47**(6): A21-A24.

Otter, M., H. Elmqvist and S. E. Mattson (1999). "Objektorientierte Modellierung Physikalischer Systeme, Teil 7." *at-Automatisierungstechnik* **47**(7): A25-A28.

Otter, M., H. Elmqvist and S. E. Mattson (1999). "Objektorientierte Modellierung Physikalischer Systeme, Teil 8." *at-Automatisierungstechnik* **47**(8): A29-A32.

Potthoff, U. (2003). Zur Stabilität von numerischen Integrationsverfahren. Interner Bericht, Institut für Regelungs- und Steuerungstheorie, Technische Universität Dresden.

4 Functional Realization: Multibody Dynamics

Background Mechatronic products are necessarily characterized by mechanically-oriented tasks. This means bodies *with finite mass* are to be moved in a purposeful manner subject to the effects of forces and torques. This movable structure represents the governing component of the *controlled plant* in a mechatronic system design. Thus, important aspects of mechatronic systems design include both a sufficient understanding of and a suitable abstract representational basis for the physical phenomena of dynamic mechanical structures.

Contents of Chapter 4 This chapter discusses fundamental physical phenomena in the *multibody dynamics* of *rigid-body systems* which are helpful in describing system behaviors for systems design. Following an introduction to the terminology of *multibody systems* (MBSs) including a review of important basic physical principles, the terms *degree of freedom* and *constraint*—which fundamentally determine the model order of a multibody system—are investigated in depth. *MBS equations of motion* in minimal coordinates and in linear state space form are introduced as a fundamental time domain model type. The inherent temporal dynamics of conservative multibody systems are explained using the eigenvalue problem, *eigenfrequencies*, *eigenvalues*, and *eigenmodes*. This discussion is then extended to *dissipative* multibody systems. The remainder of the chapter delves deeper into *response characteristics in the frequency domain*, and typical attributes of *MBS transfer functions* are discussed. Particular attention is paid to the physical interpretation of poles and zeros, and their relationship with MBS configuration parameters. A determining factor for system response characteristics—and thus for the achievable controller performance—is the choice of *measurement* and *actuation locations*. These concepts are used to explain *collocation* and the control theoretical ramifications of the *spatial separation* of sensors and actuators. The key role played by *zeros* (*antiresonances*) of MBS transfer functions is treated in detail. A discussion of the *migration* of MBS zeros as a function of system parameters rounds out the presentation. ■

4.1 Systems Engineering Context

The product-level purpose of a mechatronic system is centrally defined by the function *"generate motion"*. In this function, bodies are subjected to forces and torques to purposefully carry out targeted motion profiles (time histories of position/orientation, velocities, accelerations) (see Fig. 4.1). The objects to be set in motion are generally complex, composite structures of varying material properties (the mechanical structure). For handling such structures in a practical way for systems design, so-called *multibody systems* (*MBS*) offer a practical abstraction. Of concrete interest in such systems are the interactions between force/torque excitation and the motion variables; these interactions are represented by *multibody dynamics*.

Systems engineering significance From a systems engineering point of view, the function *"generate motion"* represents the core of the controlled plant in a mechatronic system. Thus, its particular significance to the design process is self-explanatory. To achieve an optimized layout of the complete system, it is important to have sufficiently precise knowledge of the relationship between the technical and physical parameters of a system and its dynamic response characteristics. In many cases, there is an opportunity to compose the mechanical structure in such a way as to enable simpler solutions for control, and thus a simpler overall system solution.

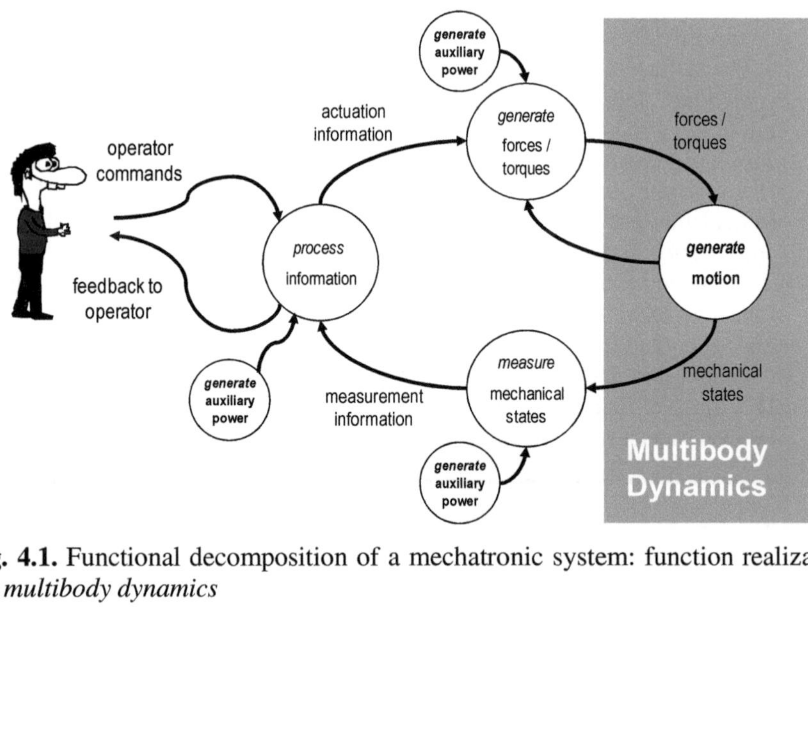

Fig. 4.1. Functional decomposition of a mechatronic system: function realization via *multibody dynamics*

Power back-effect As can be easily verified, a mutual coupling of the functions *"generate forces/torques"* and *"generate motion"* exists due to a flow of power and a physical back-effect (see Sec. 2.3.7). This back-effect is indicated in Fig. 4.1 by the feedback arrow. Thus, the function *"generate motion"* and its physical realization also play an important role in the design of the actuation system implementing the function *"generate forces/torques"*. In contrast, the connection to the function *"measure motions"* is generally feedback-free and can be modeled as a pure signal. This can be justified not least by the stipulation that the physical realization of a measurement task should influence and distort the measured physical quantity as little as possible.

4.2 Multibody Systems

Conceptually, a multibody system (MBS) is imagined as an assembly of component bodies connected to each other with a variety of connecting elements, and upon which external and internal forces act. With respect to the material properties of the bodies and connecting elements, a distinction is made between *rigid* (inflexible) and *elastic* (deformable, flexible) elements. For the mathematical description of the dynamics of the former, the concepts of *rigid-body mechanics* ($\rightarrow$ systems of differential equations, DAE systems) are used; for the latter, the concepts of *continuum mechanics* ($\rightarrow$ systems of partial differential equations) (Schwertassek and Wallrap 1999).

Model type: rigid-body system For the systems design tasks considered in this book, a sufficiently detailed yet manageable abstraction proves to be the *rigid-body system.* This term refers to a multibody system with rigid component bodies and lumped connecting elements (Fig. 4.2).

In a rigid-body system, only the bodies possess mass properties (mass, moments of inertia). All other elements are considered massless and serve to model geometric constraints and internal forces.

In the context of modeling, then, concrete physical components may need to be separated into functional and mass properties, i.e. the mass of a specific component should be suitably allocated to a rigid body.

Kinematic links Geometric relationships between bodies are modeled using kinematic linking elements. Concrete examples include *connecting rods C*, *joints* and *gearboxes J*, and *rolling wheels R*, all with potential inherent motion profiles $\gamma(t)$ (Fig. 4.2). The internal forces required to maintain these geometric constraints are called *constraint forces.*

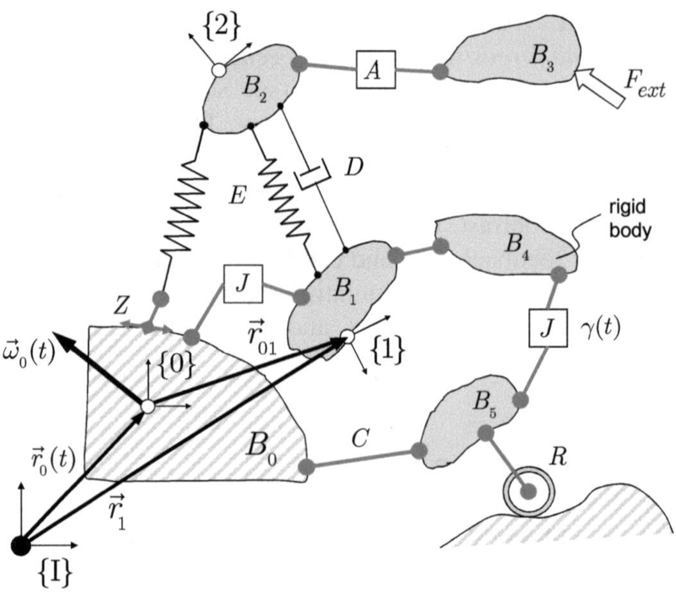

Fig. 4.2. Schematic model of a *rigid-body system*: B_i rigid bodies (mass, inertia), J joint/gearbox (massless), C connecting rod (massless), E elastic connecting element (spring, massless), D damping element (dashpot, massless), Z friction, R rolling wheel, A force/torque actuator (massless), F_{ext} external force, {I} inertial frame, {i} body-fixed coordinate system

Force elements The generation of motion-dependent internal forces is modeled using *elastic connections (spring elements) E*, velocity-dependent *damping elements D*, and motion-dependent *friction forces Z*. Assuming small motions, elastic and damping elements can often be described with sufficient accuracy by *linear relations*. General friction forces (static and kinetic friction), however, always result in nonlinear relations. In mechatronic systems, active force sources in the form of *force* and *torque actuators A* are also found (Fig. 4.2).

External forces Internal exciting forces are complemented by external forces F_{ext}, which are applied to the rigid bodies. These are understood to be forces, which—in contrast to the internal forces—lead to an increase in the total system momentum in the inertial frame, e.g. potential forces, and reaction forces (only if neglecting the loss of propellant mass).

Coordinate systems The motions of rigid bodies are described by the relative positions and orientations of reference frames with corresponding coordinate systems. The base reference frame is a suitably chosen inertial frame $\{I\}$; each body i is assigned a body-fixed coordinate system $\{i\}$.

Moving reference frame The mechanical structure of a mechatronic system can often perform large motions relative the chosen inertial frame, possibly following nonlinear laws of motion (e.g. vehicles, manipulators, telescoping arms). Often, however, it is only the motion of a few system bodies relative to a moving reference frame that is of interest (e.g. body B_3 relative to body B_0 in Fig. 4.2 or the suspension of a wheel relative to a moving vehicle frame). This also means that often the relative motions under consideration remain small and thus permit the use of linear models (springs, dampers). In such a case, the moving frame serves as the reference, e.g. coordinate system $\{0\}$ in Fig. 4.2 with translation and rotation $\vec{r}_0(t)$ and $\vec{\omega}_0(t)$, respectively. However, care should be taken in deriving the equations of motion in such cases[1].

Open and closed kinematic chains One important property of multibody systems is their *topology*. This is understood to be the spatial and action-flow arrangement of the individual MBS elements. From a systems engineering point of view, what is important in this respect is the distinction between open and closed kinematic chains (Schwertassek and Wallrap 1999).

- *Open kinematic chain, tree structure*: in this case, the MBS divides into two separate components if all kinematic connecting elements between any single arbitrary pair of rigid bodies are removed; e.g. the chain $B_0 - B_1 - B_2 - B_3$ in Fig. 4.2.
- *Closed kinematic chain, kinematic loop*: the condition for a tree structure is not satisfied; e.g. the chain $B_0 - B_1 - B_4 - B_5$ in Fig. 4.2.

When setting up equations of motion for kinematic loops, the compatibility of the motions of component bodies in the loop must be correctly guaranteed (closure condition).

1 Note: NEWTON's second law of motion holds only in an inertial space, see also *Floating Frame of Reference Formulation* in Schwertassek, R. and O. Wallrap (1999). *Dynamik flexibler Mehrkörpersysteme.* Vieweg.

4.3 Physical Fundamentals

The relevant physical foundations for rigid-body systems can be found in many good mechanics text books—e.g. (Goldstein et al. 2001), (Gregory 2006), (Pfeiffer 2008)—and belong to the standard knowledge base of an engineer. For this reason, this section only presents the most important concepts and natural laws, in order to have the most important relationships at hand, and to establish a unified nomenclature.

4.3.1 Kinematics vs. dynamics

Kinematics *Kinematics* is the study of the motion of points and bodies in space, described by the motion quantities of translational and rotational displacement, velocity, and acceleration (orientation, angular velocity, angular acceleration), without considering the causes of motion in the form of forces or torques.

Dynamics The study of *dynamics* (also named *kinetics*) describes the changes in translational and rotational motion quantities under the influence of forces and torques in space.

For the motion of a point mass in an inertial space, Fig. 4.3 illustrates the terms kinematics and dynamics (in a coordinate-free representation).

Relative kinematics A foundation for describing motion in a moving reference frame is provided by the relative kinematic relations between different coordinate systems. For example, consider the two bodies B_0 and B_1 in Fig. 4.2. Let body B_0 experience an imposed translational motion $\vec{r}_0(t)$ and rotational motion $\vec{\omega}_0(t)$. Bodies B_0 and B_1 are kinematically connected by a joint. Let the relative positions of the two body-fixed coordinate systems be $\vec{r}_{01}$.

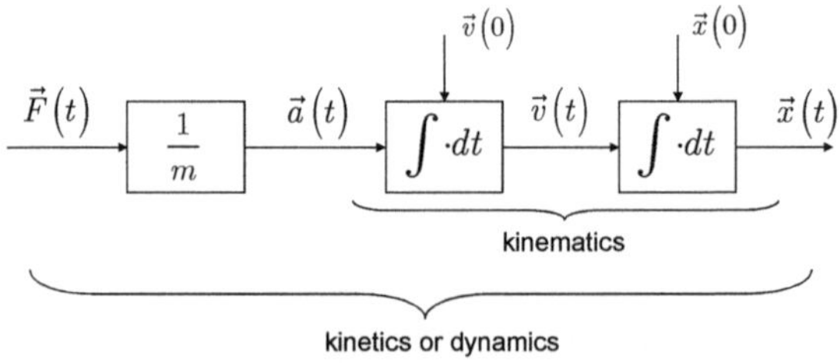

Fig. 4.3. Kinematics and dynamics of a point mass in an inertial space

The *absolute velocity* of the origin of $\{1\}$ is then[2]

$$\vec{v}_1 = \dot{\vec{r}}_1^{[I]} = \underbrace{\dot{\vec{r}}_0^{[I]} + \vec{\omega}_0 \times \vec{r}_{01}}_{\text{imposed velocity}} + \underbrace{\dot{\vec{r}}_{01}^{[0]}}_{\textit{relative velocity}} \tag{4.1}$$

and the *absolute acceleration* (also in an coordinate-free representation)

$$\vec{a}_1 = \ddot{\vec{r}}_1^{[I]} = \underbrace{\ddot{\vec{r}}_0^{[I]} + \dot{\vec{\omega}}_0^{[I]} \times \vec{r}_{01} + \vec{\omega}_0 \times \left(\vec{\omega}_0 \times \vec{r}_{01}\right)}_{\text{imposed acceleration}} + \underbrace{2\vec{\omega}_0 \times \dot{\vec{r}}_{01}^{[0]}}_{\text{Coriolis}} + \underbrace{\ddot{\vec{r}}_{01}^{[0]}}_{\textit{relative}} \tag{4.2}$$

(Coriolis and *relative* are labelled together as "acceleration".)

In the case of a *rotating* reference system, $\vec{\omega}_0 \neq \vec{0}$, so that when determining *relative translational* motions in a rotating frame, great care must be taken, as these do not arise solely from the derivatives of the corresponding inertial translation variables.

4.3.2 Rigid bodies

Rigid body A rigid body consists of a spatially-distributed quantity of point masses m_i, whose mutual separation is constant over time. Thus, no deformations of the body are possible.

Center of mass vs. center of gravity The *center of mass* of a rigid body is the point at which (conceptually and computationally) the entire mass $m_\Sigma = \Sigma m_i$ of the body can be concentrated, and to which inertial forces of the body can be reduced. The *center of gravity*, on the other hand, describes the point at which the force of gravity appears to be applied on the body. The center of mass and center of gravity are only equal when the gravitational gradient is zero over the extent of the body (constant gravitational potential). To a good approximation, this is always the case given a sufficiently small body, and only in this case is the synonymous use of *center of mass* and *center of gravity* justified. In this book, it is generally assumed that bodies are small relative to any gravity gradient, so that the term *center of mass (CM)* is used throughout.

---

2 $\dot{\vec{x}}^{[A]}$ is the time derivative of $\vec{x}$ relative to reference frame $\{A\}$, i.e. the total formal time derivatives of the components of the vector $\vec{x}$ expressed in frame $\{A\}$.

Inertia tensor The moments of inertia of a rigid body describe its inertia relative to rotational motions about fixed body axes[3]. The set of all possible moments of inertia can be concisely represented in the well-known form of a second-order tensor (here in matrix notation as a symmetric matrix $\in \mathbb{R}^{3\times3}$):

$$\mathbf{I} = \begin{pmatrix} I_{11} & I_{12} & I_{13} \\ I_{12} & I_{22} & I_{23} \\ I_{13} & I_{23} & I_{33} \end{pmatrix}, \tag{4.3}$$

with *moments of inertia* I_{11}, I_{22}, I_{33} and *products of inertia* I_{12}, I_{13}, I_{23}.

With a suitable choice of reference point (center of mass) and of axes of rotation (*principal axes of inertia*, e.g. orthogonal axes of symmetry in axis-symmetric bodies) the tensor (4.3) simplifies to

$$\mathbf{I}_{CM} = \begin{pmatrix} I_1 & 0 & 0 \\ 0 & I_2 & 0 \\ 0 & 0 & I_3 \end{pmatrix}, \tag{4.4}$$

with *principal moments of inertia* I_1, I_2, I_3.

Angular momentum The *angular momentum* $\vec{h}$ describes the moment of the linear momentum vector with respect to an arbitrary reference point[4]. In the case of a rigid body, it is advantageous to use the center of mass as the reference for the angular momentum, which is then identified as $\vec{h}_{CM}$. In this case, the angular momentum $\vec{h}_{CM}$ can be represented in a body-fixed coordinate system $\{i\}$ as an affine transform of the instantaneous angular velocity vector $\vec{\omega}$[5]:

$$^{i}\left(\vec{h}_{CM}\right) = {}^{i}\mathbf{h}_{CM} = {}^{i}\mathbf{I}_{CM} \cdot {}^{i}\left(\vec{\omega}\right) = {}^{i}\mathbf{I}_{CM} \cdot {}^{i}\omega. \tag{4.5}$$

It can be seen from Eqs. (4.3) through (4.5) that the angular momentum vector and angular velocity vector are only parallel in the special case of a spherically symmetric body.

---

3 For an exact mathematical, physical definition see e.g. (Goldstein et al. 2001). Here, only the assignment and description of definitions are of interest.

4 See Footnote 2.

5 $^{i}(\vec{x})$ is the representation of vector $\vec{x}$ in coordinate system $\{i\}$.

Newton's second law The change in the total linear momentum $\vec{p}(t)$ of a (point) mass m in an inertial frame is proportional to the sum of externally acting forces $\vec{F}_{\Sigma}(t)$ and occurs in the direction of the summed force:

$$\dot{\vec{p}}^{[\mathrm{I}]}(t) = \vec{F}_{\Sigma}(t)\,. \tag{4.6}$$

Substituting for momentum $\vec{p} = m\vec{v}$, it follows from Eq. (4.6) that

$$\dot{\vec{p}}^{[\mathrm{I}]}(t) = \dot{m}^{[\mathrm{I}]} \cdot \vec{v} + m \cdot \dot{\vec{v}}^{[\mathrm{I}]} = \vec{F}_{\Sigma}(t)\,.$$

For a constant mass, the well-known simplified relation follows:

$$m \cdot \dot{\vec{v}}^{[\mathrm{I}]} = m \cdot \vec{a}(t) = \vec{F}_{\Sigma}(t)\,. \tag{4.7}$$

Center of mass theorem For a rigid body, the momentum law (4.7) for the motion of the center of mass $\vec{r}_{CM}$ with concentrated mass m_{Σ} holds in the form

$$m_{\Sigma} \cdot \ddot{\vec{r}}_{CM}^{\;[I]} = \vec{F}_{\Sigma}(t)\,. \tag{4.8}$$

Angular momentum theorem Considering the moment of the rate of change of the linear momentum of a rigid body with respect to a fixed point in space, e.g. the center of mass, it follows from the center of mass theorem (4.8) that the time derivative of angular momentum (moment of linear momentum) under the influence of external *torques*[6] $\vec{\tau}_{CM,\Sigma}$ (i.e. moment of external forces, each also with respect to the center of mass) is

$$\dot{\vec{h}}_{CM}^{[\mathrm{I}]} = \vec{\tau}_{CM,\Sigma}\,. \tag{4.9}$$

Choosing the coordinate representation of Eq. (4.5) for the angular momentum, it follows from Eq. (4.9) that

$$\frac{d}{dt}^{[\mathrm{I}]}\left({}^{i}\mathbf{I}_{CM} \cdot {}^{i}\omega\right) = {}^{i}\tau_{CM,\Sigma}\,. \tag{4.10}$$

Note that Eq. (4.10) also holds in the case where the center of mass is in motion or accelerating.

---

6 The terms *torque* and *moment* are synonymously used in different scientific communities (physics, mechanical engineering, etc.). In this book the term *torque* is preferred.

EULER's equations For practical reasons, the reference frame for a particular equation of motion is often not chosen to be an inertial frame, but rather a body-fixed frame $\{i\}$ [7], ideally with origin at its center of mass and parallel to its principal axes. As this body-fixed frame rotates with an instantaneous angular velocity $\vec{\omega}$, Eq. (4.10) can be used to determine the time derivative of the angular momentum with respect to body-fixed coordinates[8] assuming a constant inertia tensor, giving *EULER's equations of motion*:

$$ {}^{i}\mathbf{I}_{CM} \cdot {}^{i}\dot{\omega}^{[i]} + {}^{i}\omega \times \left({}^{i}\mathbf{I}_{CM} \cdot {}^{i}\omega \right) = {}^{i}\tau_{CM,\Sigma}, \tag{4.11} $$

or in scalar form:

$$ \begin{aligned} I_1 \cdot \dot{\omega}_1 + \omega_2\omega_3 \left(I_3 - I_2 \right) &= \tau_{1,\Sigma}, \\ I_2 \cdot \dot{\omega}_2 + \omega_1\omega_3 \left(I_1 - I_3 \right) &= \tau_{2,\Sigma}, \\ I_3 \cdot \dot{\omega}_3 + \omega_1\omega_2 \left(I_2 - I_1 \right) &= \tau_{3,\Sigma}, \end{aligned} \tag{4.12} $$

where ω_i and $\tau_{i,\Sigma}$ are the projections of the vectors onto the body-fixed coordinate axes (in this case the principal axes), and the time derivatives are taken with respect to the rotating body-fixed frame.

Gyroscopic torques The cross-product term in Eq. (4.11) describes the so-called *gyroscopic torques* of a rotating body. For simplified treatments, the rotation about a fixed body axis is often considered in isolation. It can be seen from Eq. (4.12) that decoupled expression of spatial rotations is possible only if all gyroscopic terms vanish. This condition is satisfied for spherically symmetric bodies ($I_1 = I_2 = I_3$), or when the angular velocity vector $\vec{\omega}$ is exactly parallel to one of the principal axes (in this case, the mutually orthogonal components cancel). If this is not the case, bearing torques appear in the case of an artificial constraint; otherwise the body exhibits coupled motion (nutation).

Rigid body vs. point mass If, when considering the motion of a body, its orientation plays no role, then this body can be modeled as a *point mass*.

---

7 The inertia tensor is constant only in a body-fixed coordinate system.

8 In general, for a vector $\vec{x}$, the relation $\dot{\vec{x}}^{[I]} = \dot{\vec{x}}^{[i]} + \vec{\omega} \times \vec{x}$ holds, where $\vec{\omega}$ is the angular velocity of the body-fixed reference frame $\{i\}$ relative to an inertial frame $\{I\}$, and $\dot{\vec{x}}^{[i]}$ is the time derivative with respect to the body-fixed frame $\{i\}$.

This is always the case when only the motion of the center of mass is considered (e.g. the orbit of a satellite) or when the rotational degrees of freedom are artificially constrained (e.g. with guide rails).

Example 4.1 *Moving frame of reference, virtual inertial frame.*

To illustrate modeling with a moving reference frame, consider the configuration shown in Fig. 4.4 (cf. bodies B_0 and B_1 in Fig. 4.2, though connected here via force elements).

The two rigid bodies 0 (mass m_0) and 1 (mass m_1), elastically connected with a linear spring *k*, and excited by force actuator A (actuation force F_A) and external force F_1 applied to Body 1, are to undergo *purely horizontal motion* in an inertial space $\{I\}$. Let the spring be relaxed when $x_1 - x_0 = l_{01}$.

Model creation Considering free-body equations for the bodies and applying the center of mass theorem Eq. (4.8), this two-body system gives the equations of motion in inertial coordinates[9]:

$$\begin{aligned} m_0\ddot{x}_0 + kx_0 - kx_1 &= -F_A - kl_{01}\,, \\ m_1\ddot{x}_1 + kx_1 - kx_0 &= F_A + kl_{01} + F_1\,. \end{aligned} \tag{4.13}$$

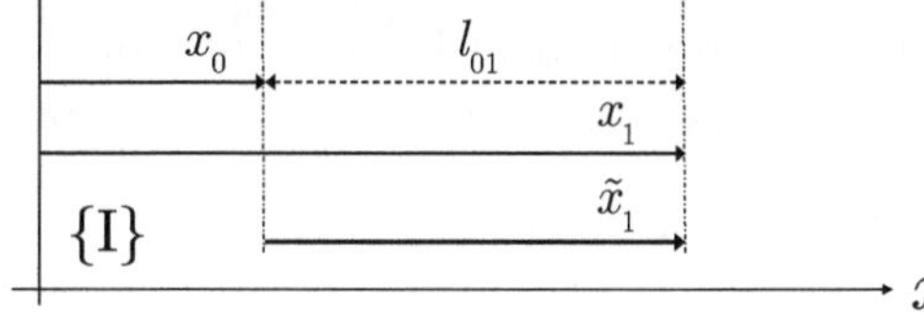

Fig. 4.4. Unconstrained two-body system with force coupling

---

9 In the two single-mass free-body systems, the actuation force F_A and the spring force are considered *external* forces; for the coupled two-mass system, however, they are *internal* forces.

Describing the motion of Body 1 relative to a body-fixed reference frame {B} attached to the center of mass of Body 0, and accounting for the coordinate transformation (relative coordinates $\tilde{x}_1$, small rotation, see Eq. (4.2)) gives

$$\tilde{x}_1 = x_1 - x_0 ,$$

with which the equations of motion can be transformed into *hybrid* coordinates $(x_0, \tilde{x}_1)$:

$$(m_0 + m_1)\ddot{x}_0 + m_1\ddot{\tilde{x}}_1 = F_1 , \tag{4.14}$$

$$m_1\ddot{\tilde{x}}_1 + k\tilde{x}_1 + m_1\ddot{x}_0 = F_A + kl_{01} + F_1 . \tag{4.15}$$

Model analysis From the *hybrid equations of motion* (4.14), (4.15) the following conclusions can now be drawn:

1. For $m_0 \gg m_1$, i.e. a moving base with *large mass* m_0, the feedback from Body 1 to the base (Body 0) remains sufficiently small, i.e. $\ddot{x}_0 \approx 0$ from Eq. (4.14). In this case, the frame of reference {B} represents a *virtual inertial frame* for the *relative motion* $\tilde{x}_1$.
2. With a controlled, *non-accelerating* motion of Body 0 (e.g. motion under position control $x_0(t) = const.$ or $\dot{x}_0(t) = const.$), the frame of reference {B} again represents a *virtual inertial frame* for the *relative motion* $\tilde{x}_1$ (since $\ddot{x}_0 = 0$ in Eq. (4.15)).

In these cases, then, the simplified *single-body system* shown in Fig. 4.5 can be considered a *substitute model.* Here, the actuator force F_A works as a quasi-external force, the spring is grounded on the virtual inertial frame, and the motion of Body 1 is considered relative to the virtual inertial frame {B}.

A further simplification results when the motion of Body 1 is defined relative to the rest length of the spring, $\tilde{\tilde{x}}_1 := \tilde{x}_1 - l_{01}$. In this case, the potential force term in Eq. (4.15) also drops out, and the equation of motion of Body 1 takes the simple (quasi-inertial) form[10]

$$m_1\ddot{\tilde{\tilde{x}}}_1 + k\tilde{\tilde{x}}_1 = F_A + F_1 . \tag{4.16}$$

Whenever possible, discussions below make use of this simplified model representation.

10 Note, however, that when employing an acceleration pick-up on Body 1, it is always the *inertial* acceleration $\ddot{x}_1$ which is measured. The same holds for measurements of angular velocities with gyroscopic sensors.

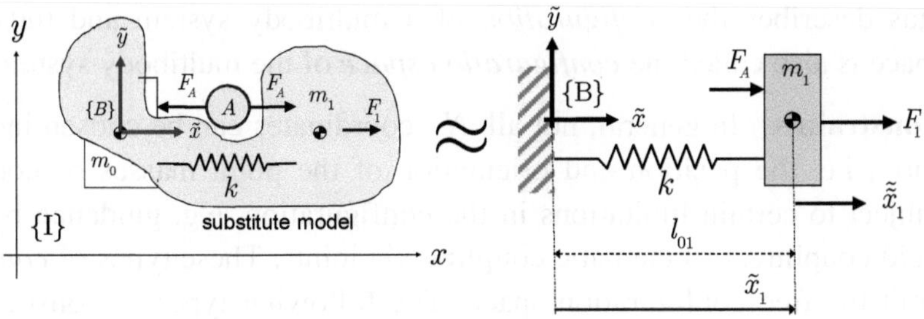

Fig. 4.5. Substitute model with virtual inertial frame given a translating base (non-accelerating)

■

4.3.3 Degrees of freedom and constraints

Generalized coordinates: configuration space The evolution of motion in space is determined for *point masses* by their position (three parameters for each mass), and for *rigid bodies* by their position (three parameters each) and orientation (three parameters each). These parameters are termed the *generalized coordinates* q_i. A secondary (though for practical reasons quite important) consideration is in which coordinate system these parameters are defined. In sum, a multibody system requires at least[11] N_q such generalized coordinates for the description of motion, where the following implications hold:

- N_P point masses $\Rightarrow N_q \geq 3N_P$,
- N_B rigid bodies $\Rightarrow N_q \geq 6N_B$.

The vector of generalized coordinates

$$\mathbf{q} = \begin{pmatrix} q_1 & q_2 & \dots & q_{Nq} \end{pmatrix}^T \tag{4.17}$$

[11] For convenience, redundant descriptions of the three orientations are often employed, e.g. the *direction cosine matrix* $\in \mathbb{R}^{3\times3}$.

thus describes the *configuration* of a multibody system and thus the **q**-space is also called the *configuration space* of the multibody system[12].

Constraints In general, not all N_q coordinates can be chosen independently, i.e. the position and orientation of the point masses or bodies are subject to certain limitations in the configuration, e.g. guidance by a rail, rigid coupling, or kinematic coupling via joints. These types of *constraints* limit the free configuration space. The following types of constraints are distinguished:

- *holonomic*: constrain the position of the system; integrable kinematic relations (see below),
- *nonholonomic*: constrain velocity and possibly position; non-integrable kinematic relations (see below),
- *rheonomic*: explicitly dependent on time,
- *scleronomic*: independent of time.

Mechanical degrees of freedom The number of mechanical degrees of freedom of a multibody system is the number generalized coordinates it possesses. As will be shown, only the $N_{C,hol}$ *holonomic* constraints reduce the degrees of freedom, so that for the number of degrees of freedoms N_{DOF}, it follows that

$$\begin{aligned} &N_P \text{ point masses} && \Rightarrow N_{DOF} = 3N_P - N_{C,hol}, \\ &N_K \text{ rigid bodies} && \Rightarrow N_{DOF} = 6N_B - N_{C,hol}. \end{aligned} \tag{4.18}$$

The number of degrees of freedom N_{DOF} plays a central role in the following discussions, and is an important model parameter.

Redundant coordinates: minimal coordinates If $N_q > N_{DOF}$, the generalized coordinates contain certain redundancies. Such a set of coordinates is thus referred to as *redundant coordinates*. For $N_q = N_{DOF}$, on the other hand, a set of *independent* generalized coordinates with minimal di-

---

[12] Generalized coordinates were also introduced in the context of energy-based modeling with the LAGRANGE formalism (Sec. 2.3.2). For didactic reasons, in that case *each* energy storage element was initially assigned a coordinate, both every mass (= kinetic energy storage) and every spring (= potential energy storage). Due to the circuit constraints on the spring elements, their coordinates can be eliminated, so that in the end, only the coordinates of the mass elements remain. This behavior was taken advantage of here, which is why the configuration space comprises only the generalized coordinates of the mass elements.

mension is obtained; these are then called *minimal coordinates*. A particularly advantageous choice is given by the minimal coordinates whose components can additionally vary throughout a given configuration space in an *unrestrained* manner. This is an important prerequisite for the use of LAGRANGE equations of the second kind (see Sec. 2.3.2).

Holonomic constraints Algebraic equations in the generalized coordinates (not velocities!) generally describe *holonomic* constraints:

$$h_j\left(q_1, q_2, \ldots, q_{N_q}, t\right) = 0, \quad j = 1, \ldots, N_{C,hol} \quad \text{or} \quad \mathbf{h}(\mathbf{q}, t) = \mathbf{0}. \qquad (4.19)$$

For each equation in (4.19), it is possible to eliminate one of the coordinates q_i using the remaining coordinates, i.e.

$$q_i = \tilde{h}_j\left(q_1, \ldots, q_{i-1}, \; q_{i+1}, \ldots, q_{N_q}, t\right).$$

Each holonomic constraint thus reduces the number of mechanical degrees of freedom by one (for examples, see Table 4.1).

Nonholonomic constraints If the constraints *cannot* be expressed as algebraic equations or integrable differential equations, e.g. *inequality constraints* in generalized coordinates

$$h\left(q_1, q_2, \ldots, q_{Nq}\right) \leq 0,$$

then these are termed *nonholonomic* constraints. Though these represent limitations on the motion of generalized coordinates, no reduction of the degrees of freedom results. All available coordinates are required for the description of the motion (for examples, see Table 4.1).

Differential constraints: PFAFFian form Distinguishing and managing holonomic and nonholonomic constraints is significantly easier when considering their *differential form*, i.e. relations between differentials (infinitesimal changes or velocities) of the generalized coordinates.

Taking the total time derivative of Eq. (4.19), it follows that

$$\frac{\partial h_j\left(\mathbf{q}, t\right)}{\partial q_1}\dot{q}_1 + \frac{\partial h_j\left(\mathbf{q}, t\right)}{\partial q_2}\dot{q}_2 + \ldots + \frac{\partial h_j\left(\mathbf{q}, t\right)}{\partial q_{Nq}}\dot{q}_{Nq} + \frac{\partial h_j\left(\mathbf{q}, t\right)}{\partial t} = 0, \qquad (4.20)$$

$$j = 1, \ldots, N_C.$$

Given the JACOBIan of $\mathbf{h}(\mathbf{q},t)$,

$$\mathbf{H}_q\left(\mathbf{q},t\right)=\left[H_{ji}\right]=\left[\frac{\partial h_j}{\partial q_i}\right]=\begin{pmatrix}\frac{\partial h_1}{\partial q_1} & \cdots & \frac{\partial h_1}{\partial q_{Nq}}\\ \vdots & \ddots & \vdots \\ \frac{\partial h_{N_{ZB}}}{\partial q_1} & \cdots & \frac{\partial h_{N_{ZB}}}{\partial q_{Nq}}\end{pmatrix},$$

$$\mathbf{h}_t\left(\mathbf{q},t\right)=\left[h_{t,j}\right]=\begin{pmatrix}\frac{\partial h_1}{\partial t}\\ \vdots \\ \frac{\partial h_{N_{ZB}}}{\partial t}\end{pmatrix}, \tag{4.21}$$

the terms in Eq. (4.20) result in the differential form of the holonomic constraints in matrix notation (also called the *PFAFFian form*):

$$\boxed{\mathbf{H}_q\left(\mathbf{q},t\right)\cdot\dot{\mathbf{q}}+\mathbf{h}_t\left(\mathbf{q},t\right)=\mathbf{0}}\,. \tag{4.22}$$

Generalized representation of constraints Eq. (4.22) formally specifies an algebraic relationship between generalized coordinates $\mathbf{q}$ and generalized velocities $\dot{\mathbf{q}}$—in other words, a system of differential equations in the generalized coordinates with a special structure. Such a formal relationship can be similarly formulated for *nonholonomic* constraints, so that Eq. (4.22) can be interpreted as the generalized representation of both holonomic *and* nonholonomic constraints.

Integrability condition for holonomic constraints For holonomic constraints, the representations as an algebraic system of equations (4.19) and as a system of differential equations (4.22) are completely equivalent. Thus, for such constraints, it is always possible to obtain the algebraic representation (4.19) *via integration* of Eq. (4.22). Eq. (4.22) is thus termed an *integrable differential equation.* To meet this condition, the following *integrability condition* for *holonomic* constraints over elements of the matrices in Eq. (4.21) must be satisfied:

$$\boxed{\frac{\partial H_{ki}}{\partial q_j}=\frac{\partial H_{kj}}{\partial q_i}}\ \underline{\text{and}}\ \boxed{\frac{\partial H_{ki}}{\partial t}=\frac{\partial h_{t,k}}{\partial q_i}}\quad \forall i,j,k\,. \tag{4.23}$$

Test for nonholonomic constraints If the integrability condition (4.23) is *not satisfied* for differential constraints of the form (4.22), then these represent *nonholonomic* constraints.

Examples In Table 4.1, a series of typical examples of holonomic and nonholonomic constraints and options for the choice of generalized coordinates are represented. Verifying the various parameters (degrees of freedom, integrability condition, etc.) is left as an exercise to the interested reader.

Table 4.1. Examples of constraints

	MBS configuration	**Generalized coordinates**	**Constraints, Degrees of freedom**
A	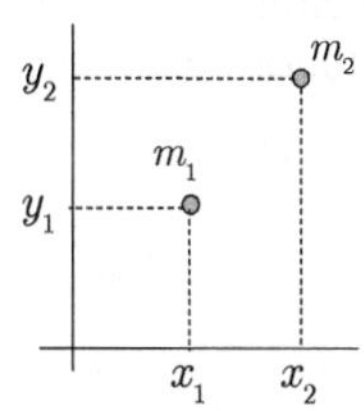	two point masses planar motion $\mathbf{q} = \begin{pmatrix} x_1 & y_1 & x_2 & y_2 \end{pmatrix}^T$ *minimal coordinates unrestrained* $N_q = 4$	no constraints $N_{C,hol} = 0$, $N_{DOF} = 4$
B	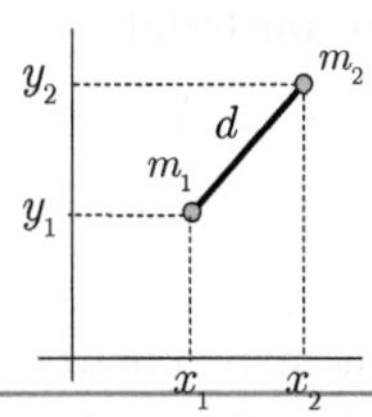	two point masses planar motion $\mathbf{q} = \begin{pmatrix} x_1 & y_1 & x_2 & y_2 \end{pmatrix}^T$ redundant coordinates $N_q = 4$	rigid link holonomic constraint $(x_2 - x_1)^2 + (y_2 - y_1)^2 = d^2$ $N_{C,hol} = 1$, $N_{DOF} = 3$
C	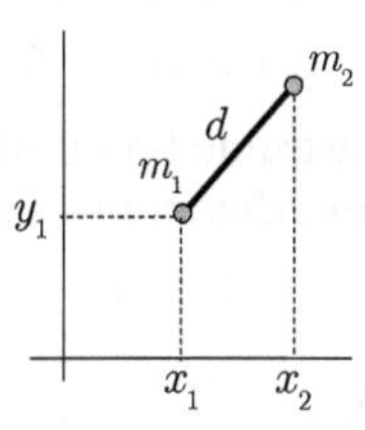	two point masses planar motion $\mathbf{q} = \begin{pmatrix} x_1 & y_1 & x_2 \end{pmatrix}^T$ minimal coordinates restrained $N_q = 3$	*nonholonomic* constraint $x_1 - d \le x_2 \le x_1 + d$ $N_{C,hol} = 0$, $N_{DOF} = 3$
D	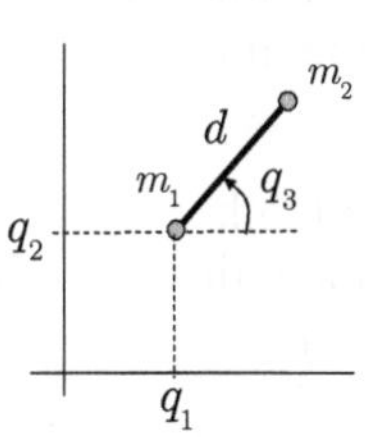	two point masses planar motion $\mathbf{q} = \begin{pmatrix} q_1 & q_2 & q_3 \end{pmatrix}^T$ *minimal coordinates unrestrained* $N_q = 3$	no constraints $N_C = 0$, $N_{DOF} = 3$

Table 4.1 (cont.). Examples of constraints

	MBS configuration	Generalized coordinates	Constraints, Degrees of freedom
E	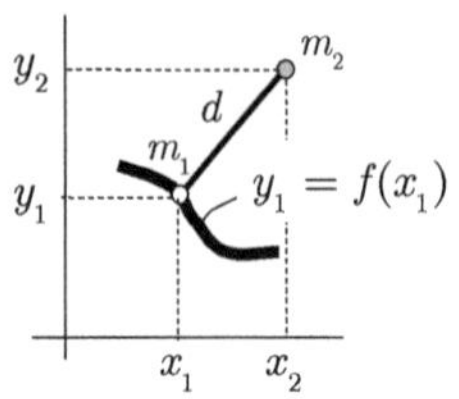	two point masses planar motion $\mathbf{q} = \begin{pmatrix} x_1 & y_1 & x_2 & y_2 \end{pmatrix}^T$ redundant coordinates $N_q = 4$	(1) rigid link holonomic constraint $\left(x_2 - x_1\right)^2 + \left(y_2 - y_1\right)^2 = d^2$ (2) kinematic control holonomic constraint $y_1 = f(x_1)$ $N_{C,hol} = 2\,,\ N_{DOF} = 2$
F	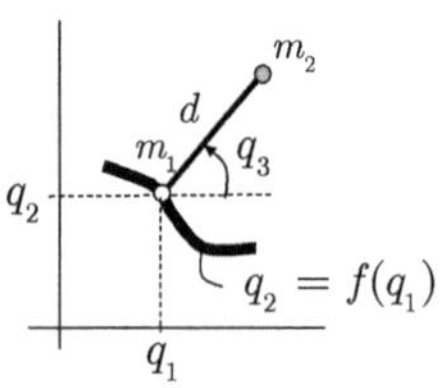	two point masses planar motion $\mathbf{q} = \begin{pmatrix} q_1 & q_2 & q_3 \end{pmatrix}^T$ redundant coordinates $N_q = 3$	kinematic control holonomic constraint $q_2 = f(q_1)$ $N_{C,hol} = 1\,,\ N_{DOF} = 2$
G	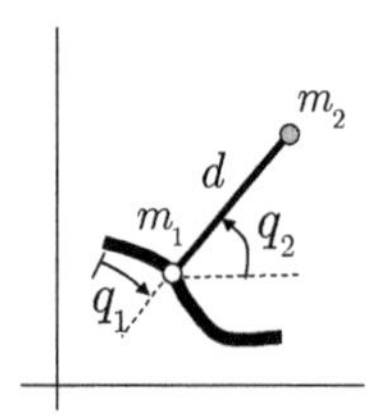	two point masses planar motion $\mathbf{q} = (q_1 \quad q_2)^T$ *minimal coordinates* *unrestrained* $N_q = 2$	no constraints $N_C = 0\,,\ N_{DOF} = 2$
H	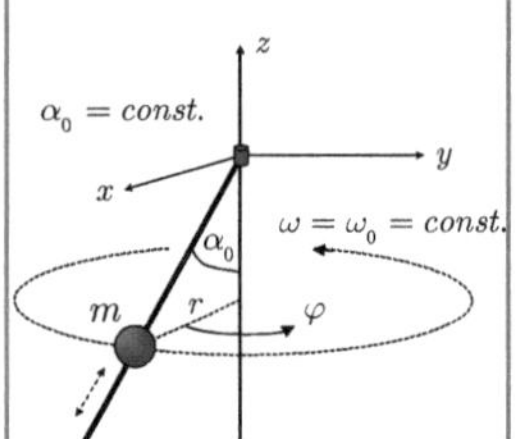	point mass 3D motion $\mathbf{q} = \begin{pmatrix} z & r & \varphi \end{pmatrix}^T$ redundant coordinates $N_q = 3$	(1) kinematic control holonomic constraint $r + z \tan \alpha_0 = 0$ (2) constrained motion holon./rheonomic constr. $\varphi - \omega_0 t = 0$ $N_{C,hol} = 2\,,\ N_{DOF} = 1$
I	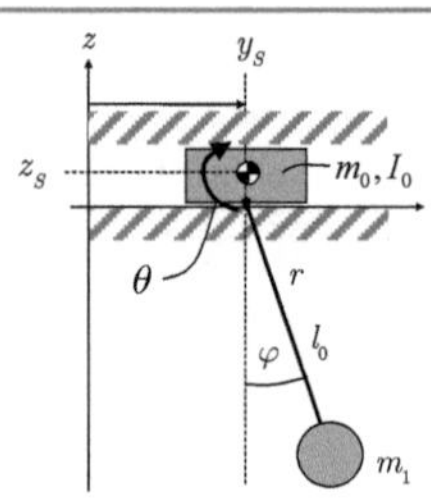	rigid body + point mass planar motion $\mathbf{q} = \begin{pmatrix} y_S & z_S & \theta & r & \varphi \end{pmatrix}^T$ redundant coordinates $N_q = 5$	(1) constrained motion holonomic constr., 2x $z_S = const, \quad \theta = const$ (2) rigid link holonomic constraint $r = l_0$ $N_{C,hol} = 3\,,\ N_{DOF} = 2$

Table 4.1. (cont.). Examples of constraints

	MBS configuration	Generalized coordinates	Constraints, Degrees of freedom
J	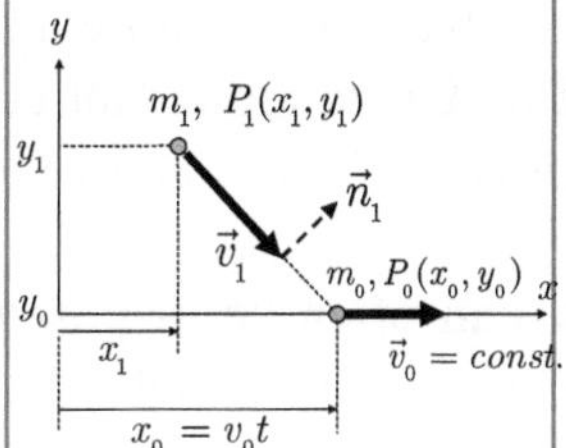	two point masses planar motion target tracking $m_1 \to m_0$ $\mathbf{q} = \begin{pmatrix} x_0 & y_0 & x_1 & y_1 \end{pmatrix}^T$ minimal coordinates constrained $N_q = 4$	cross-track velocity=0 *nonholonomic*/rheonomic constr. $\mathbf{v}_1^T \cdot \mathbf{n}_1 = 0$ $\dot{x}_1(y_1 - y_0) +$ $+\dot{y}_1(v_0 t - x_1) = 0$ $N_{C,hol} = 0\,,\ N_{DOF} = 4$
K	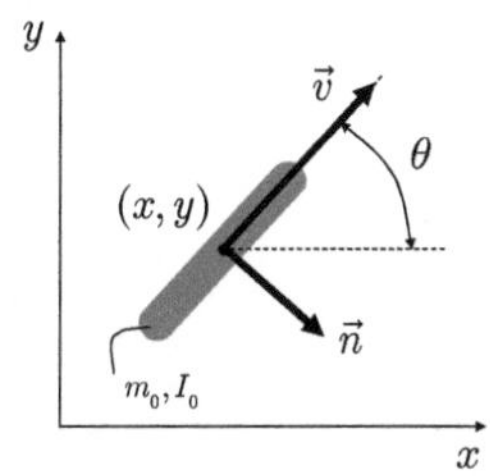	(wide) rolling wheel on plane $\mathbf{q} = \begin{pmatrix} x & y & \theta \end{pmatrix}^T$ minimal coordinates constrained $N_q = 3$	(1) normal velocity = 0 *nonholonomic* constraint $\dot{\mathbf{q}}^T \cdot \mathbf{n} = 0$ $\dot{x}\sin\theta - \dot{y}\cos\theta = 0$ $N_{C,hol} = 0\,,\ N_C = 3$
L	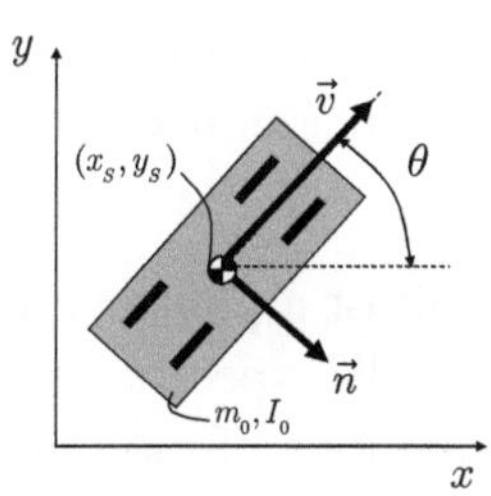	non-steerable car on plane $\mathbf{q} = \begin{pmatrix} x & y & \theta \end{pmatrix}^T$ redundant coordinates $N_q = 3$	(1) const. heading holonomic constraint $\dot{\theta} = 0$ or $\theta = \theta_0$ (2a) normal vel.=0 *nonholonomic* constraint $\dot{\mathbf{q}}^T \cdot \mathbf{n} = 0$ $\dot{x}\sin\theta - \dot{y}\cos\theta = 0$ <u>Caution:</u> this is a hidden holonomic constr.! Combined with constr. (1) results in $\theta = \theta_0 \quad \Rightarrow$ constr. (2b) (2b) holonomic constraint $\dot{x}\sin\theta_0 - \dot{y}\cos\theta_0 = 0$ $N_{C,hol} = 2\,,\ N_{DOF} = 1$

Eliminating constraints during systems design During the design phase, the goal is to work with the simplest, yet representative, design models possible. To create compact models, the designer should thus strive to make use of all available *holonomic* constraints so as to arrive at a set of *minimal coordinates*. Doing so, it is *always* possible—in the absence of *nonholonomic* constraints—to transform a DAE model in redundant coordinates into a pure *state space model* in *minimal coordinates*, and to analyze it using customary methods.

If *nonholonomic* constraints are present, they can often be disposed of using particular model assumptions (caution: this limits the model!). If, for instance, in Example C, Table 4.1, x_2 is kept within the given bounds when working with the model, the constraint never applies. In Example L, Table 4.1, holonomic constraint (1) was used to render nonholonomic constraint (2a) holonomic.

4.4 MBS Models in the Time Domain

4.4.1 Model hierarchy for systems design

LTI state space models For the tasks and methods of systems design presented in this book, easy-to-use dynamic models are desired, particularly those which enable analysis in the frequency domain (see Chapter 10: Control Theoretical Aspects, Chapter 11: Stochastic Dynamic Analysis). Eminently suited to this task are *LTI state space models* in the time domain, from which transfer functions and frequency responses of interest can be easily derived (see Sec. 2.6.3).

Physical model creation: DAE systems The result of a physical modeling process (see modeling paradigms in Sec. 2.3 or MBS-specific methods, e.g. NEWTON-EULER) will generally be a set of equations of motion in the form of a nonlinear DAE system with N_q redundant generalized coordinates, where the algebraic equations represent holonomic and nonholonomic constraints (Fig. 4.6, top). Such models generally possess a high degree of detail, and are particularly suited to verification of systems designs by simulation.

Model simplifications During the design and analysis of LTI state space models (Fig. 4.6, bottom), the model simplifications shown in Fig. 4.6 should be undertaken. Naturally, this always entails diminished representa-

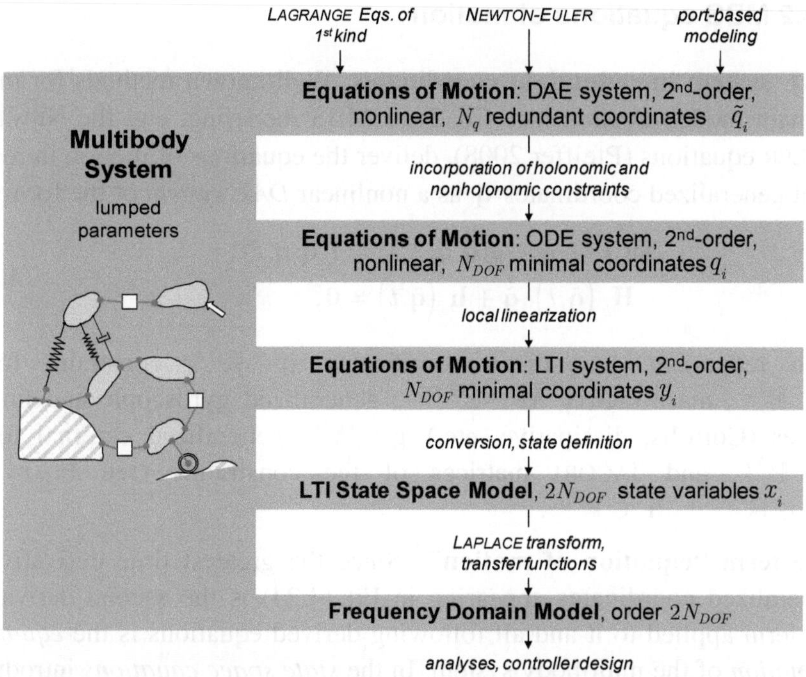

Fig. 4.6. Model hierarchy of multibody systems for systems design

tional accuracy of modeled physical system behaviors. Such circumstances should be carefully monitored at every step of the modeling process.

Equations of motion in minimal coordinates To evaluate the *spatial-temporal* dynamics of a system, linearized equations of motion in minimal coordinates are of particular significance. These take the form of a system of second-order differential equations in N_{DOF} *independent generalized coordinates*. The generalized coordinates themselves describe the spatial discretization of the multibody system. Mathematically, the spatial-temporal dynamics are clearly indicated *by the natural oscillations* (*eigenmodes*) of the system, described by the *eigenvectors* and *eigenvalues* (*eigenfrequencies*) of the LTI equations of motion (see Sec. 4.5). The number of such eigenmodes (equal to N_{DOF}), the locations of the eigenfrequencies, and the spatial configurations (eigenvectors) predictably determine the response characteristics of a multibody system and thus of the mechatronic controlled plant. For this reason, the predictions obtained from this eigenvalue problem are of central significance to systems design.

4.4.2 MBS equations of motion

DAE system in redundant coordinates Well-known methods for multi-domain model creation (see Ch. 2) or MBS modeling, e.g. the NEWTON-EULER equations (Pfeiffer 2008), deliver the equations of motion in redundant generalized coordinates $\tilde{\mathbf{q}}$ as a nonlinear *DAE system* of the form

$$\begin{aligned} \tilde{\mathbf{M}}(\tilde{\mathbf{q}},t)\cdot\ddot{\tilde{\mathbf{q}}}+\tilde{\mathbf{g}}(\tilde{\mathbf{q}},\dot{\tilde{\mathbf{q}}},\lambda,t)&=\tilde{\mathbf{f}}(\tilde{\mathbf{q}},\dot{\tilde{\mathbf{q}}},t)\,,\\ \mathbf{H}_{\tilde{q}}\left(\tilde{\mathbf{q}},t\right)\cdot\dot{\tilde{\mathbf{q}}}+\mathbf{h}_{t}\left(\tilde{\mathbf{q}},t\right)&=\mathbf{0}, \end{aligned} \tag{4.24}$$

with redundant generalized coordinates $\tilde{\mathbf{q}}\in\mathbb{R}^{Nq}$, constraint forces $\lambda\in\mathbb{R}^{N_C}$, mass matrix $\tilde{\mathbf{M}}\in\mathbb{R}^{Nq\times Nq}$, generalized gyroscopic and linking forces (Coriolis, dissipative, etc.) $\tilde{\mathbf{g}}\in\mathbb{R}^{Nq}$, generalized external forces $\tilde{\mathbf{f}}\in\mathbb{R}^{Nq}$, and JACOBI matrices of the constraints (see Eq. (4.22)) $\mathbf{H}_{\tilde{q}}\in\mathbb{R}^{N_C\times Nq}$, $\mathbf{h}_t\in\mathbb{R}^{N_C}$.

The term "equation of motion" Since the greatest time derivative of generalized coordinates appearing in Eq. (4.24) is the second derivative, the term applied to it and all following derived equations is the *equations of motion* of the multibody system. In the *state space equations* introduced further below, only first derivatives appear.

ODE system in minimal coordinates Eliminating excess coordinates and the constraint forces by solving the constraints in Eq. (4.24)[13], there remain $N_{DOF}=N_q-N_C$ independent coordinates $\mathbf{q}$, the *minimal coordinates.* Thus, the DAE system (4.24) can be represented as the system of *ordinary differential equations* (*an ODE system*)

$$\widehat{\mathbf{M}}(\mathbf{q},t)\cdot\ddot{\mathbf{q}}+\mathbf{g}(\mathbf{q},\dot{\mathbf{q}},t)=\mathbf{f}(\mathbf{q},\dot{\mathbf{q}},t)\,, \tag{4.25}$$

with *minimal coordinates* $\mathbf{q}\in\mathbb{R}^{N_{DOF}}$, *mass matrix* (symmetric positive-definite) $\widehat{\mathbf{M}}\in\mathbb{R}^{N_{DOF}\times N_{DOF}}$, generalized gyroscopic and linking forces (Coriolis, dissipative, etc.) $\mathbf{g}\in\mathbb{R}^{N_{DOF}}$, and generalized external forces $\tilde{\mathbf{f}}\in\mathbb{R}^{N_{DOF}}$.

LTV system in minimal coordinates A local linearization (see Sec. 2.6.1) about a reference trajectory $\mathbf{q}_*(t)$ with small deviations $\mathbf{y}_*(t)$ such that

$$\mathbf{q}(t)=\mathbf{q}_*(t)+\mathbf{y}_*(t)\,,$$

[13] Trivial for holonomic constraints; for *nonholonomic* constraints see the procedure of Example L, Table 4.2 and the general remarks at the end of Sec. 4.3.3.

yields a *linear time-varying* (*LTV*) system of differential equations of the form

$$\mathbf{M}_*(t) \cdot \ddot{\mathbf{y}}_* + \mathbf{P}_*(t) \cdot \dot{\mathbf{y}}_* + \mathbf{Q}_*(t) \cdot \mathbf{y}_* = \mathbf{f}_*(t), \tag{4.26}$$

with symmetric positive-definite mass matrix $\mathbf{M}_*$. The matrices $\mathbf{P}_*$ and $\mathbf{Q}_*$ describe velocity- and displacement-dependent forces, the vector $\mathbf{f}_*$ represents external forces.

LTI system in minimal coordinates If the reference for the linearization is chose to be a constant value (e.g. a resting position) so that $\mathbf{q}_*(t) = \mathbf{q}_{*0} = const.$, then with

$$\mathbf{q}(t) = \mathbf{q}_{*0} + \mathbf{y}(t), \tag{4.27}$$

the result is a *linear time-invariant* (*LTI*) system of differential equations of the form

$$\boxed{\mathbf{M} \cdot \ddot{\mathbf{y}} + \mathbf{P} \cdot \dot{\mathbf{y}} + \mathbf{Q} \cdot \mathbf{y} = \mathbf{f}(t)}. \tag{4.28}$$

Splitting each of the matrices $\mathbf{P}$ and $\mathbf{Q}$ into a symmetric and skew-symmetric component, the equations of motion can then be expressed as

$$\boxed{\mathbf{M} \cdot \ddot{\mathbf{y}} + (\mathbf{B} + \mathbf{G}) \cdot \dot{\mathbf{y}} + (\mathbf{K} + \mathbf{N}) \cdot \mathbf{y} = \mathbf{f}(t)}, \tag{4.29}$$

where the $N_{DOF} \times N_{DOF}$ matrices have the special properties

$$\mathbf{M} = \mathbf{M}^{\mathrm{T}} > 0,\ \mathbf{B} = \mathbf{B}^{\mathrm{T}},\ \mathbf{G} = \text{-}\mathbf{G}^{\mathrm{T}},\ \mathbf{K} = \mathbf{K}^{\mathrm{T}},\ \mathbf{N} = \text{-}\mathbf{N}^{\mathrm{T}}. \tag{4.30}$$

The individual terms of Eq. (4.29) represent the following physical quantities: $\mathbf{M}\ddot{\mathbf{y}}$, the inertial forces; $\mathbf{B}\dot{\mathbf{y}}$, the velocity-proportional dissipative forces; $\mathbf{G}\dot{\mathbf{y}}$, the gyroscopic forces; $\mathbf{K}\mathbf{y}$, the conservative support and linking forces; $\mathbf{N}\mathbf{y}$, the non-conservative forces (e.g. circulatory forces (Pfeiffer 2008)). The special *symmetry properties* in Eq. (4.30) can be advantageously drawn upon for model verification.

Conservative multibody system For the case $\mathbf{B} = \mathbf{0},\ \mathbf{N} = \mathbf{0},\ \mathbf{f} = \mathbf{0}$, the system is *conservative*, i.e. the energy conservation theorem $T^* + U = const.$ holds. If additionally $\mathbf{G} = \mathbf{0}$, the system is termed a *non-gyroscopic* conservative multibody system or a multibody system with *simple multibody structure* (Pfeiffer 2008):

$$\boxed{\mathbf{M} \cdot \ddot{\mathbf{y}} + \mathbf{K} \cdot \mathbf{y} = \mathbf{0}}. \tag{4.31}$$

The form of model in Eq. (4.31) is of great significance as the MBS eigenvalue problem can be directly formulated using it. The inherent energy conservation represented in the model can additionally be advantageously drawn upon for the verification of linear and nonlinear MBS models.

Example 4.2 *Planar elbow manipulator with elastic joints.*

Fig. 4.7 shows a rigid elbow manipulator (m_1, I_{10}, m_2, I_{20}) with massless, frictionless, yet *elastic* joints (k_1, k_2). In the joints, massless actuators apply torques (τ_1, τ_2) relative to the inertial frame {I} (e.g. a chain drive with shared base). The joint angles are available in the form of absolute angles (q_1, q_2) relative to {I} [14].

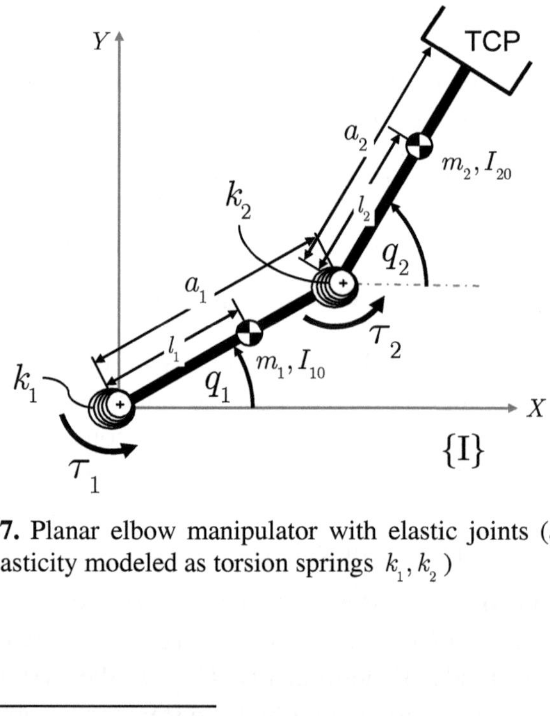

Fig. 4.7. Planar elbow manipulator with elastic joints (absolute angles, joint elasticity modeled as torsion springs k_1, k_2)

---

[14] Note the difference from a manipulator with *integrated motors* in the joints. In that case, the torques act as inner torques (*action/reaction*), *relative angles* are measured at the joints, and motor masses in the joints must be taken into account, so that a different mathematical model results.

Using any of the modeling methods previously presented (certainly, the LAGRANGE formalism can be elegantly applied here) and the *minimal coordinates* $\mathbf{q} = \begin{pmatrix} q_1 & q_2 \end{pmatrix}^T$, the following *nonlinear equations of motion* are obtained:

$$\widehat{\mathbf{M}}(\mathbf{q}) \cdot \ddot{\mathbf{q}} + \widehat{\mathbf{C}}(\mathbf{q}, \dot{\mathbf{q}}) \cdot \dot{\mathbf{q}} + \widehat{\mathbf{K}} \cdot \mathbf{q} + \widehat{\mathbf{g}}(\mathbf{q}) = \tau \,, \tag{4.32}$$

where

$$\begin{aligned}
\widehat{\mathbf{M}}(\mathbf{q}) &= \begin{pmatrix} m_1 l_1^2 + m_2 a_1^2 + I_{10} & m_2 a_1 l_2 \cos\left(q_2 - q_1\right) \\ m_2 a_1 l_2 \cos\left(q_2 - q_1\right) & m_2 l_2^2 + I_{20} \end{pmatrix}, \\
\widehat{\mathbf{C}}(\mathbf{q}, \dot{\mathbf{q}}) &= -m_2 a_1 l_2 \sin\left(q_2 - q_1\right) \begin{pmatrix} 0 & \dot{q}_2 \\ -\dot{q}_1 & 0 \end{pmatrix}, \\
\widehat{\mathbf{K}} &= \begin{pmatrix} k_1 + k_2 & -k_2 \\ -k_2 & k_2 \end{pmatrix}, \\
\widehat{\mathbf{g}}(\mathbf{q}) &= g \begin{pmatrix} \left(m_1 l_1 + m_2 a_1\right) \cos q_1 \\ m_2 l_2 \cos q_2 \end{pmatrix},
\end{aligned} \tag{4.33}$$

and $\widehat{\mathbf{C}}(\mathbf{q}, \dot{\mathbf{q}}) \cdot \dot{\mathbf{q}}$ describes the centrifugal terms, $\widehat{\mathbf{K}} \cdot \mathbf{q}$ the elastics linking forces, and $\widehat{\mathbf{g}}(\mathbf{q})$ the configuration-dependent gravitational forces. The elastic linking forces $\widehat{\mathbf{K}} \cdot \mathbf{q}$ represent the joint restoring forces given small local deviations from the current configuration q_1, q_2.

Linearizing about the *vertical equilibrium configuration* $q_{10} = q_{20} = -90°$, Eq. (4.32) gives *LTI equations of motion*

$$\mathbf{M} \cdot \ddot{\mathbf{q}} + \mathbf{K} \cdot \mathbf{q} + \mathbf{K}_g \cdot \mathbf{q} = \tau \,, \tag{4.34}$$

where

$$\begin{aligned}
\mathbf{M} &= \begin{pmatrix} m_1 l_1^2 + m_2 a_1^2 + I_{10} & m_2 a_1 l_2 \\ m_2 a_1 l_2 & m_2 l_2^2 + I_{20} \end{pmatrix}, \\
\mathbf{K} &= \begin{pmatrix} k_1 + k_2 & -k_2 \\ -k_2 & k_2 \end{pmatrix}, \mathbf{K}_g = \begin{pmatrix} \left(m_1 l_1 + m_2 a_1\right) g & 0 \\ 0 & m_2 l_2 g \end{pmatrix}.
\end{aligned} \tag{4.35}$$

■

Example 4.3 *Multiple-mass oscillator chain.*

One alternate multibody model for serial elastic structures is the mechanical *oscillator chain*. These exist as both translational and rotational chains (the latter is also termed a *torsion* oscillator), see Fig. 4.8. The system matrices for a general oscillator chain with elastic and dissipative connecting elements are presented below. Note the regular construction, and the sparse structure matrices. For a *free* (unsuspended or ungrounded) structure, k_1 or k_{N+1} are zero.

$$\mathbf{M} = diag\begin{pmatrix} m_1 & m_2 & m_3 & \dots & m_N \end{pmatrix} \text{ or } diag\begin{pmatrix} I_1 & I_2 & I_3 & \dots & I_N \end{pmatrix}$$

$$\mathbf{K} = \begin{pmatrix} k_1 + k_2 & -k_2 & & & \\ -k_2 & k_2 + k_3 & -k_3 & & \\ & -k_3 & k_3 + k_4 & -k_4 & \\ & & & \ddots & -k_N \\ & & & -k_N & k_N + k_{N+1} \end{pmatrix} \tag{4.36}$$

$$\mathbf{B} = \begin{pmatrix} b_1 + b_2 & -b_2 & & & \\ -b_2 & b_2 + b_3 & -b_3 & & \\ & -b_3 & b_3 + b_4 & & \\ & & & \ddots & -b_N \\ & & & -b_N & b_N + b_{N+1} \end{pmatrix}$$

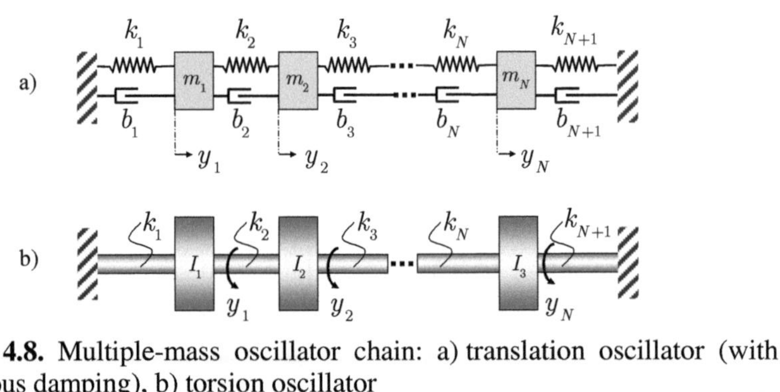

Fig. 4.8. Multiple-mass oscillator chain: a) translation oscillator (with viscous damping), b) torsion oscillator

■

4.4.3 MBS state space model

States It is well-known that the choice of states for a dynamic system is not unique. In the case of motion, and in particular for multibody systems, choosing *generalized displacement coordinates* (positions, orientations) and *velocity coordinates* (linear and angular velocities) as the states proves to be especially advantageous.

LTI state space model Starting with the LTI multibody system in minimal coordinates in Eq. (4.28), and the *state definition*

$$\mathbf{x} := \begin{pmatrix} \mathbf{y} & \dot{\mathbf{y}} \end{pmatrix}^T, \quad \mathbf{x} \in \mathbb{R}^{2N_{DOF}}, \tag{4.37}$$

the equivalent LTI state space model can be defined:

$$\dot{\mathbf{x}} = \begin{pmatrix} \mathbf{0} & \mathbf{E} \\ -\mathbf{M}^{-1}\mathbf{Q} & -\mathbf{M}^{-1}\mathbf{P} \end{pmatrix} \mathbf{x} + \begin{pmatrix} \mathbf{0} \\ \mathbf{M}^{-1} \end{pmatrix} \mathbf{f} = \mathbf{A} \cdot \mathbf{x} + \mathbf{B} \cdot \mathbf{f}, \tag{4.38}$$

where $\mathbf{E}$ is the identity matrix, $\mathbf{A} \in \mathbb{R}^{2N_{DOF} \times 2N_{DOF}}$ represents the *system matrix*, and $\mathbf{B} \in \mathbb{R}^{2N_{DOF} \times N_{DOF}}$ the *input matrix.*

Conservative LTI state space model In the case of a conservative multibody system, the state space model simplifies to

$$\dot{\mathbf{x}} = \begin{pmatrix} \mathbf{0} & \mathbf{E} \\ -\mathbf{M}^{-1}\mathbf{K} & \mathbf{0} \end{pmatrix} \mathbf{x} = \mathbf{A}_0 \cdot \mathbf{x}. \tag{4.39}$$

4.5 Natural Oscillations

4.5.1 Eigenvalue problem for conservative multibody systems

Conservative multibody system The inherent dynamics of a multibody system can be clearly explained using the conservative MBS model

$$\mathbf{M} \cdot \ddot{\mathbf{y}} + \mathbf{K} \cdot \mathbf{y} = \mathbf{0}, \tag{4.40}$$

i.e. the inputs are zero, damping ≈ 0, there are no gyroscopic effects, and excitation is solely via displacements $\mathbf{y}(0) = \mathbf{y}_0$, $\dot{\mathbf{y}}(0) = \dot{\mathbf{y}}_0$ from the resting state $\mathbf{y}_R = \dot{\mathbf{y}}_R = \mathbf{0}$.

Algebraic eigenvalue problem Due to the special structure of the differential equation (4.40), it is convenient to presume an undamped oscillation as the time evolution of the generalized coordinates:

$$\mathbf{y} = \mathbf{v} \cdot e^{j\omega t}, \tag{4.41}$$

where $\mathbf{v} = (v_1 \quad v_2 \quad \dots \quad v_{N_{DOF}})^T$ represents the vector of oscillation amplitudes of the generalized coordinates, and ω, the oscillation frequency. Substituting (4.41) into the equations of motion (4.40), gives the *algebraic* (explicit) *eigenvalue problem*:

$$\left(-\omega^2 \cdot \mathbf{M} + \mathbf{K}\right) \cdot \mathbf{v} = \mathbf{0},$$

or, equivalently[15]:

$$(\lambda \mathbf{E} - \mathbf{R}) \cdot \mathbf{v} = \mathbf{0},$$
$$\text{where} \quad \mathbf{R} = \mathbf{M}^{-1} \cdot \mathbf{K}, \quad \lambda = \omega^2. \tag{4.42}$$

As is well-known, the homogeneous algebraic system of equations (4.42) has a nontrivial solution $\mathbf{v} \neq \mathbf{0}$ for

$$\det(\lambda \mathbf{E} - \mathbf{R}) = 0. \tag{4.43}$$

Eq. (4.43) determines the N_{DOF} *eigenvalues*

$$0 \le \lambda_1 \le \lambda_2 \le \dots \le \lambda_{N_{DOF}}, \tag{4.44}$$

which, due to the special properties (4.30) of $\mathbf{M}$ and $\mathbf{K}$, are all non-negative and real, and if $\mathbf{K} > \mathbf{0}$ (suspended masses) even all positive real (repeated eigenvalues are possible).

Eigenfrequencies The N_{DOF} *eigenfrequencies* (harmonic frequencies of oscillation) are

$$\omega_j = \pm\sqrt{\lambda_j}, \quad j = 1, \dots, N_{DOF}. \tag{4.45}$$

---

15 $\mathbf{M}^{-1}$ always exists due to the requirement $\mathbf{M} > 0$, see Eq. (4.30).

Eigenvector Each eigenvalue λ_j (and thus eigenfrequency ω_j) is assigned an *eigenvector* $\mathbf{v}_j$ [16] which can be computed from the known λ_j via

$$(\lambda_j \mathbf{E} - \mathbf{R}) \cdot \mathbf{v}_j = \mathbf{0}. \tag{4.46}$$

The eigenvectors are determined up to a scaling factor, and can thus be normalized to accord with different viewpoints, e.g. the element with largest magnitude is scaled to length 1.

LTI state space model The eigenfrequencies can also be derived from the LTI state space model (4.39), where they appear as the (purely imaginary) *eigenvalues* σ_k of the system matrix $\mathbf{A}_0$:

$$\sigma_k = \pm j\omega_j, \; k = 1,\ldots,2N_{DOF}, \quad j = 1,\ldots,N_{DOF}.$$

4.5.2 Eigenmodes

Eigenmode Each eigenvalue λ_j gives a particular solution[17]

$$\mathbf{y}_j(t) = \mathbf{v}_j \left(a_j \cos \omega_j\, t + b_j \sin \omega_j\, t\right), \quad \omega_j = \sqrt{\lambda_j} \tag{4.47}$$

to the equations of motion (4.40) with a certain eigenfrequency ω_j and eigenvector $\mathbf{v}_j$.

A tuple $\{\omega_j, \mathbf{v}_j\}$ giving a solution (4.47) is termed an *eigenmode* (or *natural oscillation*) of the multibody system (4.40) as the free scale of the eigenvector $\mathbf{v}_j$ only determines the shape of the natural oscillation (the *mode shape*) and not its actual amplitude. An MBS model in the form shown in Eq. (4.40) thus possesses a total of N_{DOF} eigenmodes.

General solution The general solution of (4.40) is formed by a linear superposition of the N_{DOF} eigenmodes. In the case of differing eigenvalues λ_j, the result is

$$\mathbf{y}(t) = \sum_{j=1}^{N_{FG}} \mathbf{v}_j \left(a_j \cos \omega_j\, t + b_j \sin \omega_j\, t\right), \quad \omega_j = \sqrt{\lambda_j}. \tag{4.48}$$

---

16 Repeated eigenvector $\mathbf{v}_j$ for the eigenfrequencies $\omega_j = \pm\sqrt{\lambda_j}$.

17 The equivalent real representation of the complex solution in Eq. (4.41) with eigenfrequencies $\omega_j = \pm\sqrt{\lambda_j}$.

The $2N_{DOF}$ free parameters a_j, b_j determine the phases of the natural oscillations and are uniquely defined by the $2N_{DOF}$ given parameters $\mathbf{y}_0, \dot{\mathbf{y}}_0$. Each generalized coordinate $y_i(t)$, $i = 1,\ldots,N_{DOF}$ thus contains superposed oscillation components from all eigenfrequencies with corresponding (relative) oscillation amplitudes $v_{j,i}$.

Excitation of eigenmodes With proper choice of initial values $\mathbf{y}_0, \dot{\mathbf{y}}_0$, "pure" natural oscillations can be produced. In this case, each body moves with the same eigenfrequency ω_j given by Eq. (4.47); all other natural oscillations are suppressed. The choice of initial values producing this single-mode oscillation is the set of displacements corresponding to the components of the eigenvector, and the velocities set to zero, i.e. to excite eigenmode j with ω_j, the choice of initial values is[18]

$$\mathbf{y}_0 = \alpha \mathbf{v}_j,\ \dot{\mathbf{y}}_0 = \mathbf{0}, \quad \alpha \in \mathbb{R}.$$

Geometric interpretation of eigenvectors The components of the eigenvectors have a helpful physical, geometric interpretation, which is of great importance, particularly as relates to control (controllability and observability, see Sec. 10.6). As previously explained, the components of the eigenvector represent the relative amplitudes of oscillation of the corresponding bodies (absolute amplitudes are determined by the initial values). Identical values of eigenvector components v_{ji} and v_{jk}, i.e.

$$\mathbf{v}_j = \begin{pmatrix} \ldots & \underset{i}{\alpha} & \ldots & \underset{k}{\alpha} & \ldots \end{pmatrix}^T,$$

indicate that in the eigenmode with eigenfrequency ω_j, the generalized coordinates y_i and y_k oscillate *in phase* with *identical amplitude* (see Fig. 4.9, eigenmode 1; Fig. 4.10, eigenmodes 1 and 3). Relative measurement of these coordinates would thus not be able to detect the eigenmode ω_j (observability breakdown, see Sec. 10.6). In the same way, equal-magnitude components with opposite sign represent an *antiphase* eigenmotion (eigenmode 2 in both Fig. 4.9 and Fig. 4.10). The relative oscillation amplitudes of different bodies can also be easily extracted from the eigenvector.

---

[18] It is easy to verify that for the free coefficients in Eq. (4.48), this requires: $a_j = 1, b_j = 0,\ a_k = b_k = 0$ for $k \neq j$,

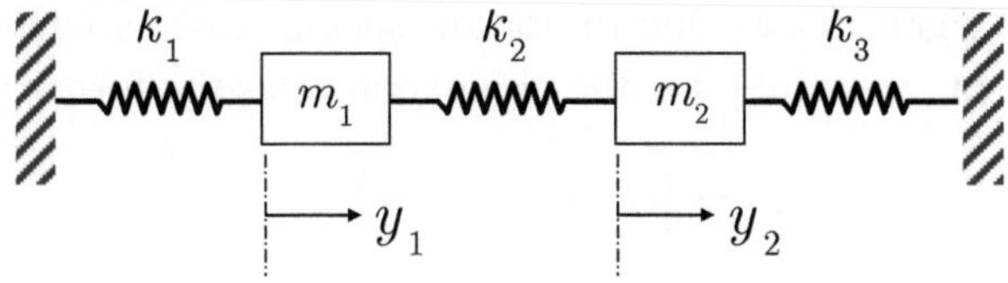

$$\mathbf{M} = \begin{pmatrix} m_1 & 0 \\ 0 & m_2 \end{pmatrix} \qquad \mathbf{K} = \begin{pmatrix} k_1 + k_2 & -k_2 \\ -k_2 & k_2 + k_3 \end{pmatrix}$$

$$\boxed{m_1 = m_2 = m, \qquad k_1 = k_2 = k_3 = k}$$

$$\omega_1 = \sqrt{\frac{k}{m}} \qquad \mathbf{v}_1 = \begin{pmatrix} 1 & 1 \end{pmatrix}^T$$

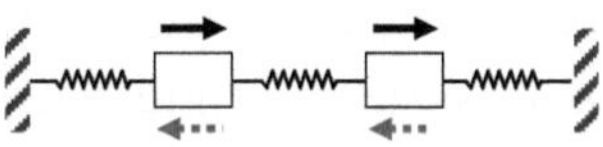

$$\omega_2 = \sqrt{3\frac{k}{m}} \qquad \mathbf{v}_2 = \begin{pmatrix} 1 & -1 \end{pmatrix}^T$$

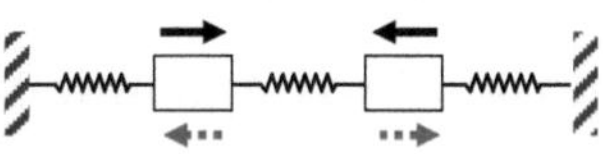

Fig. 4.9. Fully suspended two-mass oscillator: natural oscillations

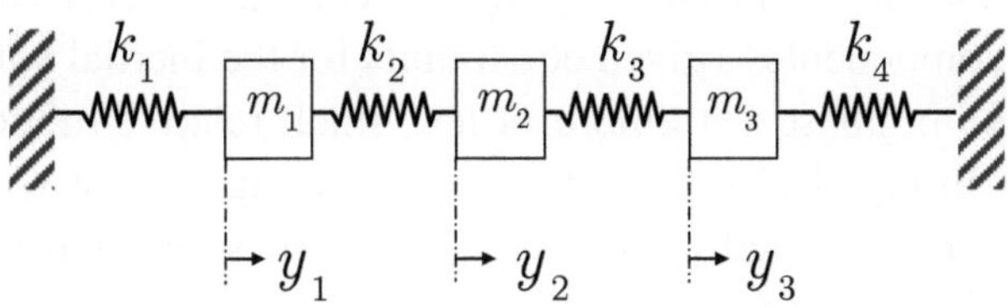

$$\mathbf{M} = \begin{pmatrix} m_1 & 0 & 0 \\ 0 & m_2 & 0 \\ 0 & 0 & m_3 \end{pmatrix} \qquad \mathbf{K} = \begin{pmatrix} k_1 + k_2 & -k_2 & 0 \\ -k_2 & k_2 + k_3 & -k_3 \\ 0 & -k_3 & k_3 + k_4 \end{pmatrix}$$

$$\boxed{m_1 = m_2 = m_3 = m\,, \qquad k_1 = k_2 = k_3 = k_4 = k}$$

$$\omega_1 = 0.765\sqrt{\frac{k}{m}} \qquad \mathbf{v}_1 = \begin{pmatrix} 0.5 & 0.707 & 0.5 \end{pmatrix}^T$$

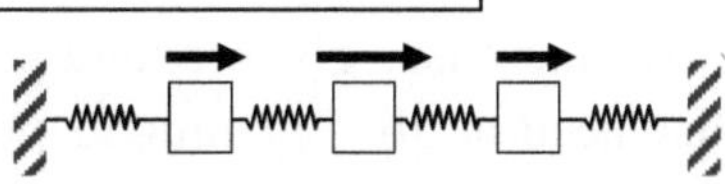

$$\omega_2 = 1.414\sqrt{\frac{k}{m}} \qquad \mathbf{v}_2 = \begin{pmatrix} -0.707 & 0 & 0.707 \end{pmatrix}^T$$

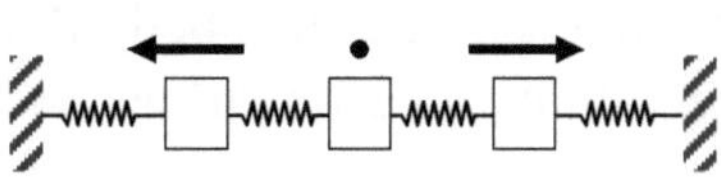

$$\omega_3 = 1.848\sqrt{\frac{k}{m}} \qquad \mathbf{v}_3 = \begin{pmatrix} -0.5 & 0.707 & -0.5 \end{pmatrix}^T$$

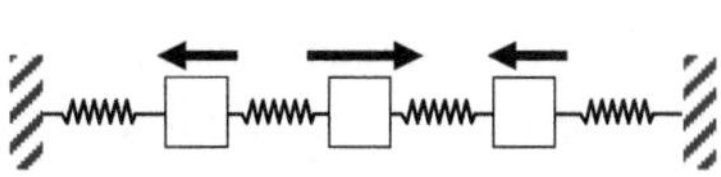

Fig. 4.10. Fully suspended three-mass oscillator: natural oscillations

Rigid body eigenmode One important special case is when all eigenvector components are equal, i.e. the eigenvector has the form

$$\mathbf{v}_j = \begin{pmatrix} \alpha & \alpha & \dots & \alpha \end{pmatrix}^T . \tag{4.49}$$

In this case, all bodies move in phase with the same amplitude, *all relative motions* of the bodies are *zero*, and the multibody system thus behaves like a rigid body (see Fig. 4.9, eigenmode 1). In this sense, a tuple $\{\omega_j, \mathbf{v}_j\}$ following Eq. (4.49) is called the *rigid-body eigenmode* (rigid-body mode) of the multibody system. Rigid-body modes always appear in *unsuspended* multibody systems at the eigenfrequency $\omega_{\text{rigid-body}} = 0$ (free motion). In *suspended* systems, rigid-body modes appear only under very specific MBS parameter configurations (see Sec. 10.6).

Pseudo-rigid-body dynamics: common modes Taking into account parameter uncertainties, for *suspended* systems under real conditions only *pseudo-rigid-body* eigenmodes exist. However, in general, the eigenmode with the *smallest* eigenfrequency always contains exclusively in-phase eigenvector components[19] (given equal units for the inertial variables), i.e. all bodies move in phase with more or less small relative motion (see e.g. eigenmode 1 in Fig. 4.10). For this reason, the eigenmode with the smallest eigenfrequency is called the *common mode* or, somewhat imprecisely, often the *rigid-body mode*.

Oscillation nodes An additional important special case occurs for zero-valued eigenvector components, i.e.

$$\mathbf{v}_j = \begin{pmatrix} \dots & \underset{i}{0} & \dots \end{pmatrix}^T . \tag{4.50}$$

In this case, given excitation of the eigenfrequency ω_j, body i is not displaced at all. Eigenmode j exhibits an *oscillation node* at this location, also referred to as a virtual support (see Fig. 4.10, eigenmode 2, mass 2). A motion sensor at this body thus cannot detect this eigenmode (another observability breakdown, see Sec. 10.6).

[19] This can be seen directly: if all bodies oscillate in phase and with approximately the same amplitude, the combined mass oscillates against the supporting springs at a frequency $\omega = \sqrt{k_{\min} / m_\Sigma}$. All other modes contain oscillating substructures having smaller composite masses.

Orthonormal eigenbasis Due to its special symmetries, the eigenvectors of the system in Eq. (4.40) are *orthogonal*[20], i.e.

$$\mathbf{v}_j^T \cdot \mathbf{v}_k = \alpha_{jk}\delta_{jk}$$
$$\delta_{jk} = 1 \text{ for } j = k\,,\ \delta_{jk} = 0 \text{ for } j \neq k\,;\quad \alpha_{jk} \neq 0\,. \tag{4.51}$$

Normalizing to $\left|\mathbf{v}_i\right| = 1$, they thus form an *orthonormal eigenbasis*. This property is useful for a variety of considerations concerning diagonizability of multibody systems (the *modal form*) and as a check on computed eigenvectors (see e.g. the eigenvectors in the examples of Fig. 4.9 and Fig. 4.10).

Modal form Using the orthonormal eigenbasis (4.51) in the form of the so-called *modal matrix*

$$\mathbf{V} = \begin{pmatrix} \mathbf{v}_1 & \mathbf{v}_2 & \dots & \mathbf{v}_{N_{DOF}} \end{pmatrix} \tag{4.52}$$

and the regular coordinate transform

$$\mathbf{y} = \mathbf{V}\mathbf{z}\,, \tag{4.53}$$

a conservative multibody system (Eq. (4.40)) can be transformed into a decoupled system of differential equations (a diagonalized form)[21] in *modal coordinates* **z**:

$$\breve{\mathbf{M}} \cdot \ddot{\mathbf{z}} + \breve{\mathbf{K}} \cdot \mathbf{z} = \mathbf{0}\,,$$
$$\breve{\mathbf{M}} = \mathbf{V}^T\mathbf{M}\mathbf{V} = diag\left(\mu_j\right),\quad \breve{\mathbf{K}} = \mathbf{V}^T\mathbf{K}\mathbf{V} = diag(\kappa_j),\quad j = 1,\dots,N_{DOF}\,. \tag{4.54}$$

---

20 When normalizing to $\left|\mathbf{v}_i\right| = 1$, then $\alpha_{jk} = 1$, and the eigenvectors are *orthonormal*. Eq. (4.51) holds only for *distinct* eigenvalues λ_j; for *repeated* eigenvalues, an orthonormal eigenbasis can be obtained in a more roundabout fashion (GRAM-SCHMIDT orthonormalization).

21 Due to the particular symmetries of $\mathbf{M}$ and $\mathbf{K}$, the modal matrix $\mathbf{V}$ diagonalizes both the mass matrix *and* the stiffness matrix.

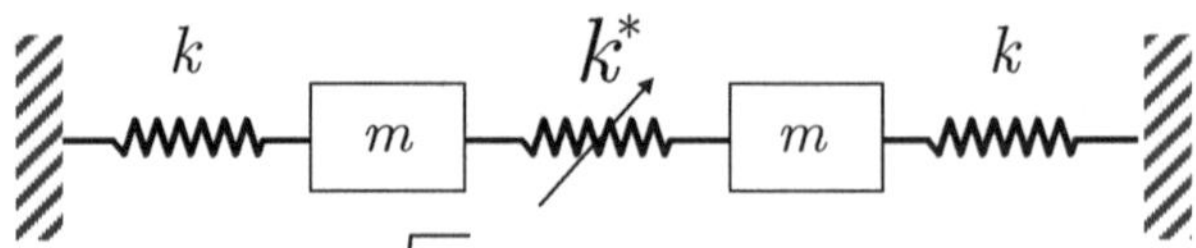

$$\omega_1 = \sqrt{\frac{k}{m}}, \quad \omega_2 = \omega_1 \sqrt{1 + 2k^* / k}$$

Fig. 4.11. Symmetric two-mass oscillator with variable stiffness, eigenfrequencies ω_1, ω_2

Here, the μ_j represent the *modal masses*, and the κ_j, the *modal stiffnesses*. It further holds that

$$\omega_j = \sqrt{\frac{\kappa_j}{\mu_j}}.$$

Representing the model in the *modal space* thus results in N_{DOF} decoupled single-mass oscillators, which represent the N_{DOF} eigenmodes. The resulting MBS parameters are the effective masses and stiffnesses for each eigenmode[22].

Migration of eigenfrequencies The eigenfrequencies of a multibody system are inherent to its configuration, and fundamentally depend on the topological arrangement, geometric parameters, and in particular on the *stiffnesses* of the elastic connecting elements and the *masses* of the bodies. In general the following relationships hold (see modal representation):

- increased masses $\rightarrow$ eigenfrequencies decrease,
- increased stiffnesses $\rightarrow$ eigenfrequencies increase.

An example of this behavior is illustrated in Fig. 4.11 using a symmetric two-mass oscillator with a variable stiffness k^*. In this case, increasing k^* has no effect on the rigid-body mode ω_1, while the second eigenfrequency successively increases. For $k^* \rightarrow \infty$, rigid coupling results with only one finite eigenfrequency.

[22] Due to the free scalability of the eigenvectors, however, the modal masses and stiffnesses are also only determined up to constant scale factors. Often, the modal masses are scaled to one, i.e. $\mu_j = 1$.

MBS natural oscillations A multibody system with N_{DOF} *degrees of freedom* (independent generalized coordinates) also has N_{DOF} *eigenfrequencies* ω_j and corresponding *eigenvectors* $\mathbf{v}_j$. The *relative oscillation amplitudes* of the generalized coordinates are represented by the components of the eigenvectors (including in-phase, anti-phase, rigid-body mode, and oscillation node components).

4.5.3 Dissipative multibody systems

Energy dissipation via viscous friction: structural damping For real mechanical structures, there is always energy dissipation due to various friction phenomena. In the context of systems design, in addition to nonlinear effects such as static and kinetic friction, viscous friction phenomena which can be described as linear, *velocity-proportional* friction forces (*structural damping*) are of particular importance. The equations of motion of a conservative multibody system are then extended to

$$\mathbf{M} \cdot \ddot{\mathbf{y}} + \mathbf{B} \cdot \dot{\mathbf{y}} + \mathbf{K} \cdot \mathbf{y} = \mathbf{0}, \tag{4.55}$$

with the (structural) damping matrix $\mathbf{B} \in \mathbb{R}^{N_{DOF} \times N_{DOF}}$.

Relative velocity vs. absolute velocity When creating a model, it must be carefully determined in what manner viscous friction forces oppose the motion of bodies. Fig. 4.12 depicts two typical cases. For damping elements between two bodies, the *relative velocity* controls the forces, i.e. the friction force on Body 2 is

$$F_{2r} = -b(\dot{y}_2 - \dot{y}_1).$$

On the other hand, when the friction interaction takes place with inertial space, the absolute velocity controls the forces, giving

$$F_{2r} = -b\dot{y}_2$$

(the *skyhook* principle, see Sec. 10.4.7). In the two cases, then, there are different damping matrices $\mathbf{B}$.

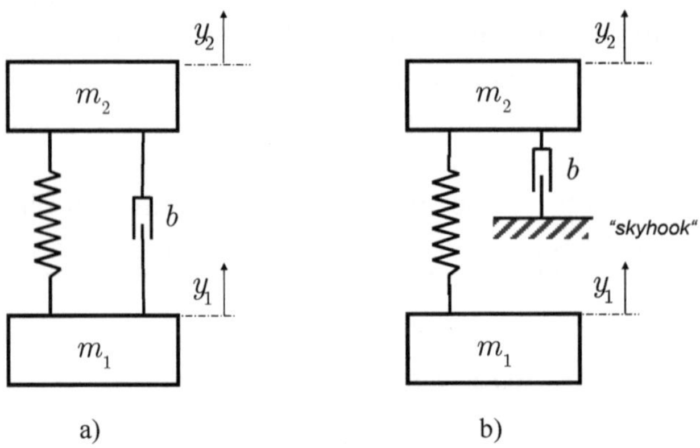

Fig. 4.12. Viscous damping in multibody systems: a) proportional to relative velocity, b) proportional to absolute velocity (*skyhook* principle)

Dynamics The principal dynamics of a dissipative multibody system can be demonstrated using the damped *single-mass oscillator model*

$$m\ddot{y} + b\dot{y} + ky = 0\,,$$

or, in normalized form,

$$\ddot{y} + 2d_0\omega_0\dot{y} + \omega_0^{\,2}y = 0\,,$$
$$\omega_0 = \sqrt{\frac{k}{m}},\quad d_0 = \frac{1}{2}\frac{b}{k}\omega_0 = \frac{1}{2}\frac{b}{m}\frac{1}{\omega_0}\,. \tag{4.56}$$

For $d_0 < 1$ (i.e. sufficiently *small* viscous damping), the differential equation (4.56) has complex conjugate eigenvalues[23]

$$\sigma_{1,2} = -\delta \pm j\Omega = -d_0\omega_0 \pm j\omega_0\sqrt{1-d_0^{\,2}}\,.$$

Given non-zero damping, though the magnitude of the eigenvalues remains equal to ω_0, the eigenvalues move into the complex left half-plane (Fig. 4.13).

---

[23] Note the slightly different definition of eigenvalues as compared to the model in Eq. (4.40), where a purely *imaginary* exponential was assumed. In the case here, to solve the differential equation, a *complex* exponential approach (non-zero real parts) has been chosen.

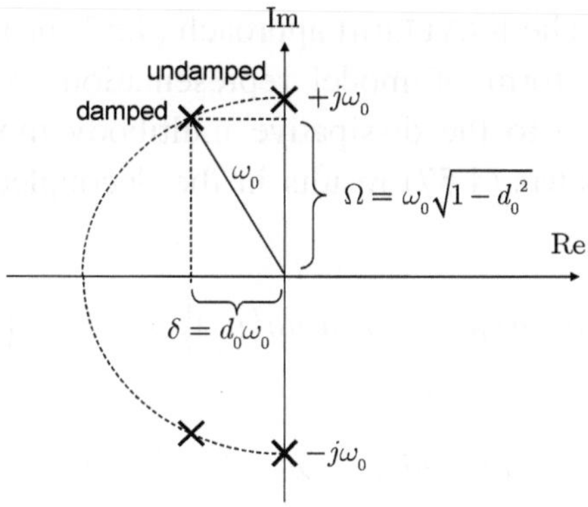

Fig. 4.13. Eigenvalues for dissipative single-mass oscillator

Due to the energy dissipation, a *decaying* oscillation of the form

$$y(t) = e^{-\delta t}(A_1 \sin \Omega t + A_2 \cos \Omega t)$$

results, with oscillation frequency $\Omega = \omega_0\sqrt{1-d_o^{\,2}}$ and a damping factor $\delta = d_0\omega_0$. For small damping, the oscillation frequency is thus approximately equal to the natural frequency (eigenfrequency) ω_0 of the undamped case $b = 0$ (i.e. $d_0 = 0$).

Passive vs. active damping The presence of structural damping induces asymptotic decay of the oscillations, though potentially with only a very small decay constant δ (*passive damping*). For this reason, *active damping* is introduced via a control loop in order to avoid undesirable structural oscillations (see Ch. 10).

Parameter uncertainty in the damping constants The damping constants b_i of structural damping are generally only approximately known, are difficult to ascertain experimentally, and often depend on environmental conditions. They should thus be considered *uncertain parameters*. This presents particular challenges, especially for controller design. Ch. 10 thus presents control theoretical approaches which particularly account for unknown—in the extreme case even vanishingly small—damping.

Proportional damping: RAYLEIGH damping Due to the presence of uncertainties, certain practical *assumptions* concerning structural damping have proven useful for analysis and design. The oft-chosen approach

$$\mathbf{B} = \alpha\mathbf{M} + \beta\mathbf{K}\,, \quad \alpha, \beta \in \mathbb{R} \tag{4.57}$$

is termed *proportional damping* or *RAYLEIGH damping* (Preumont 2002).

Modal damping The RAYLEIGH approach (4.57) proves to be a particularly advantageous form of model representation. Applying the modal transformation (4.53) to the dissipative multibody model (4.55) with the special damping matrix (4.57) results in the decoupled dissipative modal representation[24]

$$diag\left(\mu_j\right)\cdot\ddot{\mathbf{z}}+\left[\alpha\cdot diag\left(\mu_j\right)+\beta\cdot diag\left(\kappa_j\right)\right]\cdot\dot{\mathbf{z}}+diag\left(\kappa_j\right)\cdot\mathbf{z}=\mathbf{0}$$

$$diag\left(1\right)\cdot\ddot{\mathbf{z}}+diag\left(2d_j\omega_j\right)\cdot\dot{\mathbf{z}}+diag\left(\omega_j^2\right)\cdot\mathbf{z}=\mathbf{0} \tag{4.58}$$

with *modal damping ratios*

$$d_j=\frac{1}{2}\left(\frac{\alpha}{\omega_j}+\beta\omega_j\right). \tag{4.59}$$

Choice of RAYLEIGH damping constants From Eq. (4.59), it is possible to see that given *mass-proportional damping* α, the model damping decreases with increasing eigenfrequency, while given *stiffness-proportional damping* β, the model damping increases with increasing eigenfrequency. In particular, for homogeneous mechanical structures, the relative damping properties of eigenmodes near to one another are often quite well-described by a suitable weighting α,β. If the modal damping ratios d_j, d_k of two eigenmodes ω_j, ω_k ($\omega_j<\omega_k$) are known (e.g. from a frequency response measurement), then the corresponding RAYLEIGH damping constants are

$$\begin{pmatrix}\alpha\\ \beta\end{pmatrix}=2\frac{\omega_j\omega_k}{\omega_k^2-\omega_j^2}\begin{pmatrix}\omega_k & -\omega_j\\ -1/\omega_k & 1/\omega_j\end{pmatrix}\begin{pmatrix}d_j\\ d_k\end{pmatrix}. \tag{4.60}$$

Structural damping in systems design Note that during the design of a mechatronic system—for reasons of robustness if nothing else—attempting to account for the "exact" structural damping does not make

[24] Note that an arbitrary damping matrix $\mathbf{B}$ is generally *not* diagonalized by the modal matrix $\mathbf{V}$. For this reason, the special form in Eq. (4.57) is employed, making computation more convenient, so that this matrix form is also termed the *convenience hypothesis* by some authors.

sense. Rather, controller design should accommodate a wide spectrum of possible structural damping configurations.

Model analysis of dissipative multibody systems

(1) eigenvalue analysis of the conservative portion $\rightarrow \{\omega_j, \mathbf{v}_j\}$

(2) modal representation $\rightarrow \{\mu_j, \kappa_j\}$

(3) modal damping from RAYLEIGH approach $d_j = \frac{1}{2}\left(\frac{\alpha}{\omega_j} + \beta\omega_j\right)$

4.6 Response Characteristics in the Frequency Domain

Input/output behavior of multibody systems For the dynamic analysis of a mechatronic system, it is the response characteristics of the multibody plant—going from the inputs to the motions of the bodies as outputs—which are of interest. In addition to the obvious physical variables of *force* and *torque*, input excitations in the form of kinematic *imposed motions* are possible, e.g. displacement excitation of elastic elements or inertial forces. Possible mechanical state variables—in addition to the *generalized coordinates* (absolute/relative position, orientation)—include the kinematic variables of *velocity* and *acceleration.* For a concise mathematical description of the response characteristics of the system, frequency domain models in the form of *transfer functions* $G_{MBS}(s)$ or *frequency responses* $G_{MBS}(j\omega)$—which can be directly derived from the LTI equations of motion—can be used (Fig. 4.14).

Force excitation The obvious form of excitation is via external or internal forces or torques. Attention should be paid to the type of excitation: absolute with respect to inertial space or relative as an internal force between bodies (Table 4.2, left column).

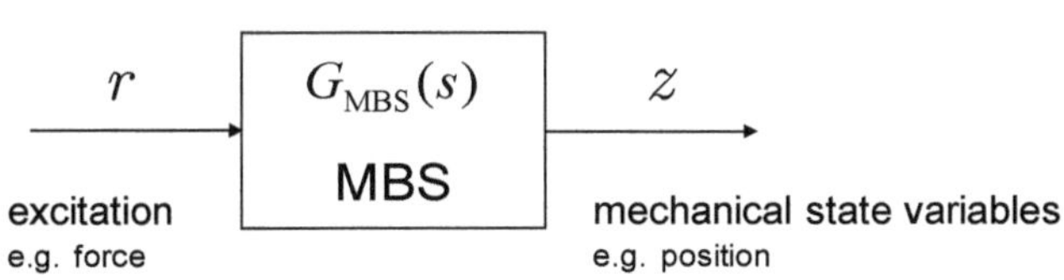

Fig. 4.14. Multibody system as a transfer function system

Table 4.2. Types of excitation for a single-mass oscillator

Type	Force excitation	Displacement excitation	Inertial excitation
Schematic	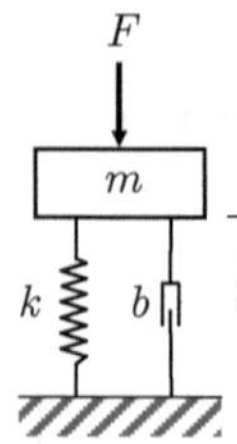	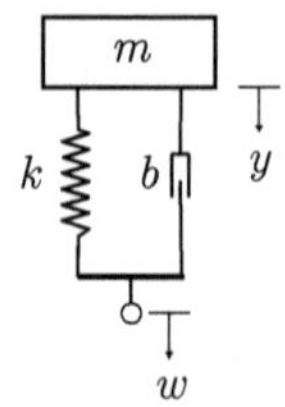	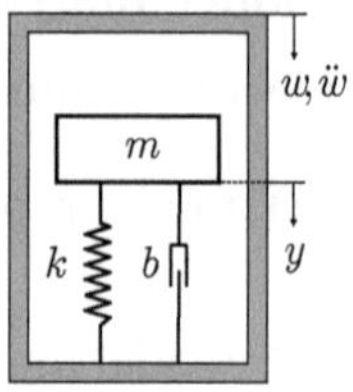
Equation of motion	$m\ddot{y} + b\dot{y} + ky = F$	$m\ddot{y} + b\left(\dot{y} - \dot{w}\right) +$ $+ k\left(y - w\right) = 0$	$m\ddot{y} + b\left(\dot{y} - \dot{w}\right) +$ $+ k\left(y - w\right) = 0$ $m\ddot{p} + b\dot{p} + kp = m\ddot{w}$
Input $r(t)$ Output $z(t)$	force F position y	displacement w position y	acceleration $\ddot{w}$ displacement $p{=}w{-}y$
Transfer function $G(s) = \frac{Z(s)}{R(s)}$	$\dfrac{K}{1 + 2d_0 \frac{s}{\omega_0} + \frac{s^2}{\omega_0^2}}$ $K = \frac{1}{k}$	$\dfrac{K(1 + \frac{b}{k}s)}{1 + 2d_0 \frac{s}{\omega_0} + \frac{s^2}{\omega_0^2}}$ $K = 1$	$\dfrac{K}{1 + 2d_0 \frac{s}{\omega_0} + \frac{s^2}{\omega_0^2}}$ $K = \frac{1}{\omega_0^2}$
	$\omega_0 = \sqrt{\frac{k}{m}}\ ,\ d_0 = \frac{1}{2}\frac{b}{k}\omega_0 = \frac{1}{2}\frac{b}{m}\frac{1}{\omega_0}$		
Frequency response $G(j\omega)$	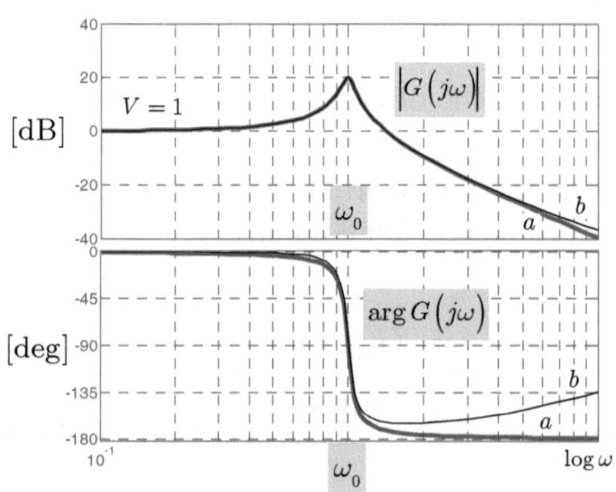 a: force, inertial excitation, b: displacement excitation		

Displacement excitation If the multibody system is subject to a guiding motion via an elastic connection, it is termed an *elastic displacement excitation.* In this type of excitation, spring forces are applied directly to system bodies. Velocity-dependent friction forces can be coupled into the system in a similar way using viscous damping elements (Table 4.2, middle column). In this case, the excitation represents a *displacement change* of the spring or damper connection point.

Inertial excitation Guiding motions including accelerations induce inertial forces in the bodies. In such cases, these accelerations represent the actual input excitation (Table 4.2, right column).

Single-mass oscillator transfer function For all the excitation types described, a single-mass oscillator produces the same type of transfer function

$$G(s) = V \frac{1}{1 + 2d_0 \dfrac{s}{\omega_0} + \dfrac{s^2}{\omega_0^2}} = \frac{V}{\{d_0;\omega_0\}},$$

where (as usual)

$$\omega_0 = \sqrt{\frac{k}{m}}$$

represents the *natural frequency* of the single-body system. In the case of *displacement excitation* via a *spring-damper element*, an additional differentiation term appears in the numerator, though for small damping ratios, it does not have a significant impact on the frequency response (Table 4.2, last row).

The three typical dynamic regimes can be recognized:

- proportional input/output behavior below the natural frequency;
- *resonance* for harmonic excitation at the natural frequency, gain inversely proportional to structural damping;
- large damping above the natural frequency.

Multibody system transfer function To specify the response characteristics of a system, the type and location of input excitations and output observations must be specified. In Fig. 4.15a, as an example, an inertial force F_μ is applied to Body μ as the excitation, and the absolute displacement y_ν of Body ν is taken as the output, whereas in Fig. 4.15b, relative input and output quantities are considered.

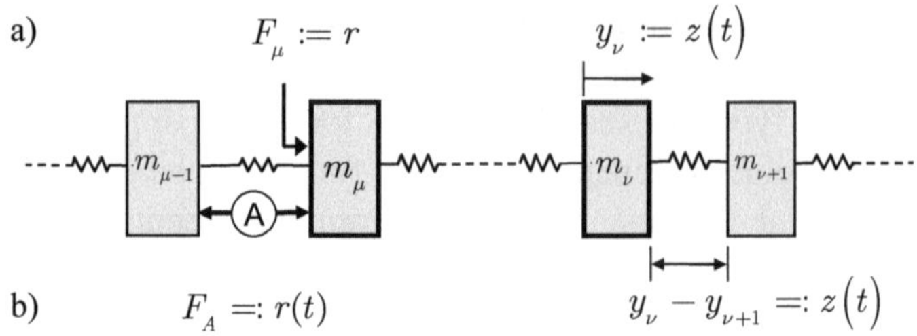

Fig. 4.15. Multibody system with force excitation and position sensing: a) absolute, b) relative

The corresponding equations of motion[25] with r as *input* and z as *output* are

$$\begin{aligned} \mathbf{M}\,\ddot{\mathbf{y}} + \mathbf{K}\,\mathbf{y} &= \mathbf{p}_f\, r \\ z &= \mathbf{p}_y^{\,T}\,\mathbf{y} \end{aligned} \quad , \qquad \begin{aligned} &\mathbf{M}, \mathbf{K} \in \mathbb{R}^{N_{DOF}\times N_{DOF}} \\ &\mathbf{y}, \mathbf{p}_f, \mathbf{p}_y \in \mathbb{R}^{N_{DOF}} \;. \end{aligned} \tag{4.61}$$

The vectors $\mathbf{p}_f$ and $\mathbf{p}_y$ represent the input and output weighting. For absolute (inertial) inputs and outputs, these are chosen following Fig. 4.15a as

$$\mathbf{p}_f = \begin{pmatrix} 0 & \ldots & \underset{\mu}{\underset{\uparrow}{1}} & \ldots & 0 \end{pmatrix}^T , \quad \mathbf{p}_y = \begin{pmatrix} 0 & \ldots & \underset{\nu}{\underset{\uparrow}{1}} & \ldots & 0 \end{pmatrix}^T .$$

For *relative* inputs and outputs, following Fig. 4.15b, they are

$$\mathbf{p}_f = \begin{pmatrix} 0 & \underset{\mu\text{-}1}{\underset{\uparrow}{-1}} & \underset{\mu}{\underset{\uparrow}{1}} & \ldots & 0 \end{pmatrix}^T , \quad \mathbf{p}_y = \begin{pmatrix} 0 & 0 & \underset{\nu}{\underset{\uparrow}{1}} & \underset{\nu+1}{\underset{\uparrow}{-1}} & 0 \end{pmatrix}^T .$$

MBS transfer function Taking the LAPLACE transform of the state space representation (4.38), the *MBS transfer function* (simplified here for the undamped case) is[26]

[25] An alternative representation uses modal coordinates as in Eq. (4.54); in this case, the eigenvectors represent the input and output weighting (actuator/sensor positions), see e.g. (Preumont 2002).

[26] In most applications, e.g. oscillator chains, the imaginary or (with finite damping) complex conjugate zeros shown here appear. However, in special cases, real zeros can also appear, see Sec. 4.7.6.

$$G_{MBS}(s) = \frac{Z(s)}{R(s)} = \begin{pmatrix} \mathbf{p}_y \\ \mathbf{0} \end{pmatrix}^T \cdot \left(s\mathbf{E} - \begin{pmatrix} \mathbf{0} & \mathbf{E} \\ -\mathbf{M}^{-1}\mathbf{K} & \mathbf{0} \end{pmatrix} \right)^{-1} \cdot \begin{pmatrix} \mathbf{0} \\ \mathbf{M}^{-1}\mathbf{p}_f \end{pmatrix},$$

$$G_{MBS}(s) = V \frac{\prod_{k=1}^{m} \left(1 + \frac{s^2}{\omega_{zk}^2} \right)}{\prod_{j=1}^{N_{DOF}} \left(1 + \frac{s^2}{\omega_{pj}^2} \right)} = V \frac{\prod_{k=1}^{m} \{\omega_{zk}\}}{\prod_{j=1}^{N_{DOF}} \{\omega_{pj}\}}, \quad m < N_{DOF}. \tag{4.62}$$

Poles of MBS transfer functions The *poles* s_j, s_j^* of the *transfer function* $G_{MBS}(s)$ can be derived from the *eigenfrequencies* ω_j of the *multibody system* (4.40), i.e.

$$\begin{aligned} s_j &= +j\omega_{pj} = +j\omega_j \\ s_j^* &= -j\omega_{pj} = -j\omega_j \end{aligned} \qquad j = 1,\ldots,N_{DOF}. \tag{4.63}$$

This can easily be seen as—due to the matrix inversion in Eq. (4.62)—the characteristic polynomial of the MBS transfer function is

$$\Delta(s) = \det\left[s\mathbf{E} - \begin{pmatrix} \mathbf{0} & \mathbf{E} \\ -\mathbf{M}^{-1}\mathbf{K} & \mathbf{0} \end{pmatrix} \right],$$

which in turn is equivalent to the algebraic eigenvalue problem (4.42)[27]. The *poles* of the transfer function are thus uniquely determined by the *eigenfrequencies* of the multibody system. In particular, in the undamped case, there are exactly N_{DOF} imaginary pairs of poles (Fig. 4.16). The poles are additionally *independent* of the locations chosen for the excitation and observation[28].

---

[27] This can also be seen via the modal representation (4.54).

[28] The last two assertions are, however, only valid under conditions of *complete observability* and *controllability* of the *multibody system* (see Sec. 10.6). In particular configurations, certain poles (eigenfrequencies) can be canceled by co-located zeros of the transfer function. Physically, this means that certain eigenfrequencies can not be observed or affected via the corresponding actuation and measurement pair.

Zeros of the MBS transfer function Unfortunately, no such concrete assertions can be made for the zeros of an MBS transfer function. However, a few general properties can be inferred given *vanishingly small damping.* In all cases, the order of the numerator polynomial is even, and can be at most $(2N_{DOF} - 2)$. Due to the lack of odd powers in the numerator, the zeros must lie symmetrically about the imaginary axis, i.e. there are either purely *imaginary pairs of zeros* (the common case), or axially symmetric real pairs of zeros (negatively/positively real $\rightarrow$ non-minimum phase system). In the case of oscillator chains, it can be shown that the location of the pairs of zeros is additionally restricted (only one zero can appear between two poles, a detailed discussion of zeros follows in Sec. 4.7).

The locations of the zeros, contrary to those of the poles, are completely dependent on the chosen excitation and observation points. A typical pole/zero configuration for a non-dissipative multibody system is shown in Fig. 4.16.

Dissipative multibody systems Applying the RAYLEIGH damping assumption, the *modal damping ratios* d_j (4.59) are found to be precisely the normalized damping ratios of the quadratic terms of the characteristic polynomial of the transfer function, i.e.

$$G_{MBS}(s) = V \frac{\prod_{k=1}^{m} \left(1 + 2d_{zk} \frac{s}{\omega_{zk}} + \frac{s^2}{\omega_{zk}^2}\right)}{\prod_{j=1}^{N_{DOF}} \left(1 + 2d_j \frac{s}{\omega_{pj}} + \frac{s^2}{\omega_{pj}^2}\right)} = V \frac{\prod_{k=1}^{m} \left\{d_{zk}; \omega_{zk}\right\}}{\prod_{j=1}^{N_{DOF}} \left\{d_j; \omega_{pj}\right\}}, \quad m < N_{DOF}. \tag{4.64}$$

With increasing structural damping, the complex conjugate pairs of poles and zeros move into the left half-plane, corresponding to the behavior in Fig. 4.13.

MBS frequency response The frequency response (BODE diagram) of a typical dissipative multibody system (4.64) is depicted in Fig. 4.17. The *eigenfrequencies* can be clearly recognized by the *resonant peaks* of the amplitude trace and the *negative-180-degree phase jumps*. The complex conjugate *zeros* can be recognized as *amplitude troughs* and *positive-180-degree phase jumps.*

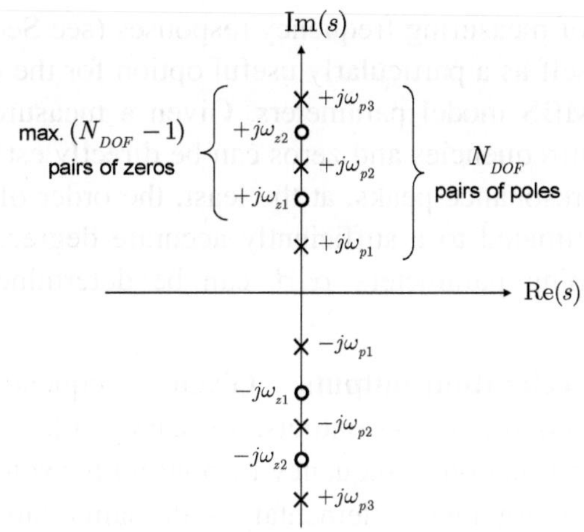

Fig. 4.16. Typical pole/zero distribution for the transfer function of a non-dissipative multibody system with three eigenmodes

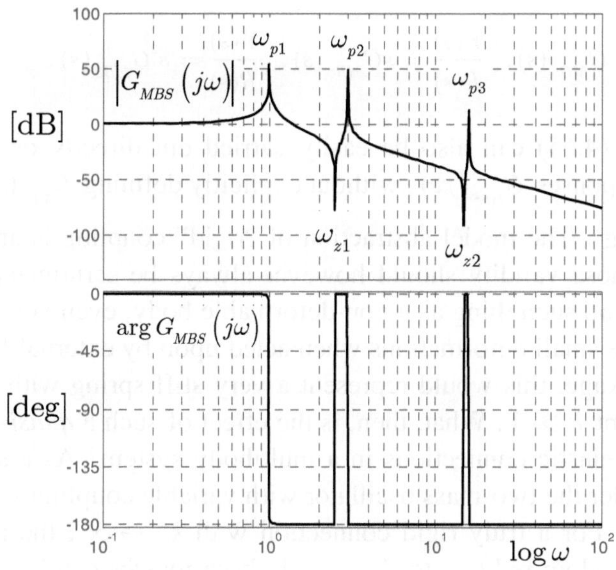

Fig. 4.17. Typical frequency response of a dissipative multibody system with three eigenmodes (cf. Fig. 4.16)

Due to the ease of measuring frequency responses (see Sec. 2.7), this model form offers itself as a particularly useful option for the experimental determination of MBS model parameters. Given a measured frequency response, the eigenfrequencies and zeros can be directly estimated with high accuracy. From resonance peaks, at the least, the order of modal damping ratios can be estimated to a sufficiently accurate degree, and in turn the RAYLEIGH damping parameters α, β can be determined directly from Eq. (4.60).

Velocity and acceleration outputs Given a frequency domain model with generalized coordinates as outputs, it is easy to obtain the corresponding MBS transfer functions (frequency responses) for velocity or acceleration outputs by employing the elementary s-domain relations

$$\mathcal{L}\{\dot{y}\} = s\mathcal{L}\{y\} \text{ and } \mathcal{L}\{\ddot{y}\} = s^2\mathcal{L}\{y\}.$$

It thus holds that

$$\frac{Y(s)}{R(s)} := G_{MBS}(s), \quad \frac{\dot{Y}(s)}{R(s)} = sG_{MBS}(s), \quad \frac{\ddot{Y}(s)}{R(s)} = s^2 G_{MBS}(s). \tag{4.65}$$

Operations (4.65) can also be easily carried out directly on measured frequency responses $G_{MBS}(j\omega)$, without explicitly defining $G_{MBS}(s)$.

Stiff coupling The model abstraction of “rigid” coupling is an ideal assumption, whose validity should however always be scrutinized. Strictly seen, there is no such thing as a non-deformable body; even very stiff bodies experience small deformations when acted upon by external forces. For an elastic linkage, this would represent a very stiff spring with very high spring constant $k \gg 1$. What, then, is the effect of such a *quasi-rigid* coupling among elastic connections in a multibody system? As a simple example, consider the two-mass oscillator with variable coupling stiffness k^* in Fig. 4.18a. For a truly rigid connection with $k^* \to \infty$, the number of degrees of freedom reduces to $N_{DOF} = 1$, both masses can be considered one combined mass, and there is only one single eigenfrequency ω_{p1}. For *finite* stiffness k^*, on the other hand, due to $N_{DOF} = 2$, there are two eigenfrequencies and one zero ω_z, whose magnitude is always located between the two poles.

Frequency separation in stiff systems These relationships can be clearly shown in a BODE diagram (Fig. 4.18b). For increasing k^*, the pair of zeros

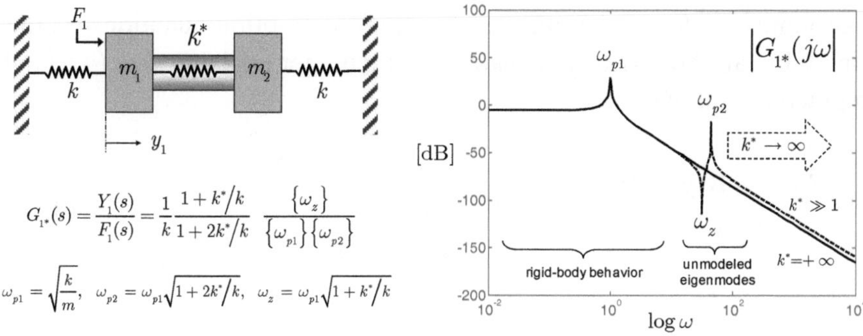

a) b)

Fig. 4.18. Effect of stiff coupling $k^* \gg 1$, fully suspended two-mass oscillator: a) mechanical configuration, b) frequency response: amplitude curve

and poles (ω_z, ω_{p2}) moves to higher frequencies without significantly affecting the response characteristic in the lower frequency domain (the rigid-body mode). *Quasi-rigid* (i.e. *very stiff*) links lead to *high-frequency* eigenfrequencies and potentially pairs of poles and zeros[29]. This behavior can also be generalized to higher-order multibody systems.

Unmodeled eigenmodes For very detailed and realistic modeling, each rigid-body system would thus need to be modeled by a large number of quasi-rigid connections. This should however be avoided for reasons of model simplicity and, in the case of simulation tasks, for reasons of numerical stability in the integration (cf. *Integrating stiff systems* in Sec. 3.3). For practical reasons, in controller design, reduced order *simplified models* in which such "stiff" eigenfrequencies are neglected are thus generally used; this is also referred to as a system with *unmodeled eigenmodes* (Fig. 4.18b). While these have no effect on the low-frequency operational domain, they must still be considered during controller design in order to avoid stability problems in the real system (see the *spillover* problem in Ch. 10).

MIMO systems: MBS transfer matrix Often, several excitation variables must be considered simultaneously—e.g. disturbances and actuator quantities—and/or the motions of several bodies in the system are of interest. In this case, the multibody system must be handled as a *multi-input multi-output* (*MIMO*) system. In the frequency domain, the relation be-

[29] Whether and at which frequencies zeros appear depends on the measurement and actuation locations (see Sec. 4.7).

tween inputs $\mathbf{r}$ (forces, torques) and outputs $\mathbf{z}$ (motion quantities) is then defined by an *MBS transfer matrix*, which, generalizing Eq. (4.61), is computed as follows:

$$\begin{matrix} \mathbf{M}\,\ddot{\mathbf{y}} + \mathbf{K}\,\mathbf{y} = \mathbf{P}_f\,\mathbf{r} \\ \mathbf{z} = \mathbf{P}_y\,\mathbf{y} \end{matrix}, \quad \begin{matrix} \mathbf{M}, \mathbf{K} \in \mathbb{R}^{N_{DOF} \times N_{DOF}}, \mathbf{y} \in \mathbb{R}^{N_{DOF}} \\ \mathbf{r} \in \mathbb{R}^{nr}, \mathbf{z} \in \mathbb{R}^{nz} \\ \mathbf{P}_f \in \mathbb{R}^{N_{DOF} \times nr}, \mathbf{P}_y \in \mathbb{R}^{nz \times N_{DOF}} \end{matrix} \tag{4.66}$$

$$\mathbf{G}_{MBS}(s) = \begin{pmatrix} \mathbf{P}_y & \mathbf{0} \end{pmatrix} \cdot \left(s\mathbf{E} - \begin{pmatrix} \mathbf{0} & \mathbf{E} \\ -\mathbf{M}^{-1}\mathbf{K} & \mathbf{0} \end{pmatrix} \right)^{-1} \cdot \begin{pmatrix} \mathbf{0} \\ \mathbf{M}^{-1}\mathbf{P}_f \end{pmatrix},$$

$$\mathbf{G}_{MBS}(s) = \begin{pmatrix} G_{1,1}(s) & \cdots & G_{1,nr}(s) \\ \vdots & G_{\nu,\mu}(s) & \vdots \\ G_{nz,1}(s) & \cdots & G_{nz,nr}(s) \end{pmatrix}. \tag{4.67}$$

Element $G_{\mu,\nu}(s)$ of the transfer matrix (4.67) describes the transfer function between the μth input and the νth output of the multibody system (4.66). Note that for a multibody system with full controllability and observability, all transfer functions $G_{\mu,\nu}(s)$ possess the same characteristic polynomial and thus the same poles. They differ only in the numerators.

Two-mass oscillator as a MIMO system In Sec. 4.7.2 (Table 4.3), below, all applicable transfer functions for a doubly-suspended two-mass oscillator are listed in a general form. This simple system can often be applied as a substitute model for baseline system considerations of more complex configurations.

4.7 Measurement and Actuation Locations

4.7.1 General multiple-mass oscillator

Measurement and actuation From the points of view of devices and construction, certain limitations for a mechanical structure must be observed as to the choice of actuators and sensors. In the general case, actuation and measurement will occur on different bodies. Fig. 4.19 depicts one such case for a general multiple-mass oscillator.

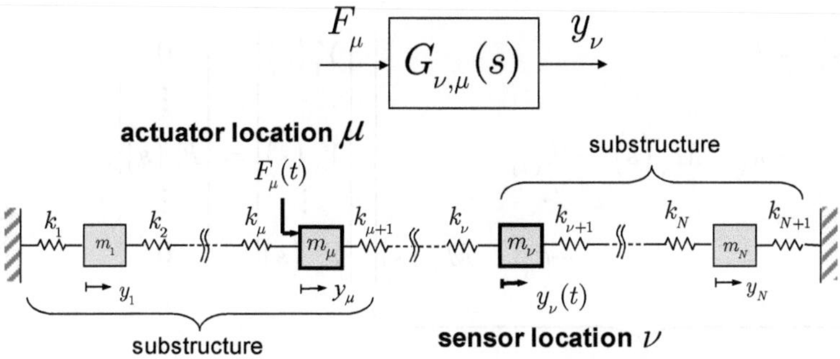

Fig. 4.19. Multiple-mass oscillator with force actuation and position measurement (here absolute, or inertial, quantities).

Transfer function Here, the transfer function between the force application at Body μ and the displacement from the rest position of Body ν is of interest:

$$G_{\nu,\mu}(s) = \frac{\mathcal{L}\{y_\nu(t)\}}{\mathcal{L}\{F_\mu(t)\}} = K_{\nu,\mu} \frac{\prod_{k=1}^{m}\left(1+\frac{s^2}{\omega_{zk}^2}\right)}{\prod_{j=1}^{N}\left(1+\frac{s^2}{\omega_{pj}^2}\right)}. \tag{4.68}$$

measurement, actuation (pointing to $G_{\nu,\mu}$); zeros (pointing to numerator); poles (pointing to denominator)

As previously mentioned in Sec. 4.6, the transfer functions for different measurement and actuation locations only differ in their static gain and the *placement* of the *zeros*. The location and number of zeros—and thus the choice of measurement and actuation locations—have a fundamental effect on response characteristics, and they equally fundamentally determine achievable controller performance. For this reason, it is worth examining this behavior more closely, using the example of the multiple-mass oscillator in Fig. 4.19.

4.7.2 Zeros of a multiple-mass oscillator

Equations of motion in the complex domain LAPLACE transformation of the equations of motion accounting for the special forms of the mass and stiffness matrices (4.36) results in

$$\underbrace{\begin{bmatrix} \tilde{M}_1(s) & -k_2 & & & \\ & \ddots & & & \\ & -k_\lambda & \tilde{M}_\lambda(s) & -k_{\lambda+1} & \\ & & & \ddots & \\ & & & -k_N & \tilde{M}_N(s) \end{bmatrix}}_{\tilde{\mathbf{M}}(s)} \cdot \begin{bmatrix} Y_1(s) \\ \vdots \\ Y_\lambda(s) \\ \vdots \\ Y_N(s) \end{bmatrix} = \begin{bmatrix} 0 \\ \vdots \\ F_\lambda(s) \\ \vdots \\ 0 \end{bmatrix}, \tag{4.69}$$

where

$$\tilde{M}_i(s) = m_i s^2 + k_i + k_{i+1} \quad \text{for} \quad i = 1, 2, \ldots, N\,. \tag{4.70}$$

The transfer function between actuation location μ and measurement location ν can be derived from (4.69), giving

$$G_{\nu,\mu}(s) = \frac{Y_\nu(s)}{F_\mu(t)} = \left[\tilde{\mathbf{M}}^{-1}(s)\right]_{\nu\mu} = \frac{(-1)^{\nu+\mu}\,\Lambda_{\nu\mu}(s)}{\det \tilde{\mathbf{M}}(s)}, \tag{4.71}$$

where $\Lambda_{\nu\mu}(s)$ is the minor (subdeterminant) of the submatrix obtained by removing the ν th row and μ th column from $\tilde{\mathbf{M}}(s)$. Due to the special structure of $\tilde{\mathbf{M}}(s)$, it holds for $\Lambda_{\nu\mu}(s)$ that

$$\Lambda_{\nu\mu}(s) = \det \mathbf{\Gamma}_1(s) \cdot \det \mathbf{\Gamma}_2(s) \cdot \det \mathbf{\Gamma}_3(s),$$

$$\mathbf{\Gamma}_1(s) = \begin{bmatrix} \tilde{M}_1(s) & -k_2 & \cdots \\ & \ddots & \\ \cdots & -k_{\mu-1} & \tilde{M}_{\mu-1}(s) \end{bmatrix}, \quad \mathbf{\Gamma}_2(s) = \begin{bmatrix} -k_{\mu+1} & & \\ & \ddots & \\ & & -k_\nu \end{bmatrix}, \tag{4.72}$$

$$\mathbf{\Gamma}_3(s) = \begin{bmatrix} \tilde{M}_{\nu+1}(s) & -k_{\nu+2} & \cdots \\ & \ddots & \\ \cdots & -k_N & \tilde{M}_N(s) \end{bmatrix}.$$

Physical interpretation of zeros The *zeros* of $G_{\nu,\mu}(s)$ and $\Lambda_{\nu\mu}(s)$ are the *eigenfrequencies* of two spring-mass subsystems: the first between masses m_1 and $m_{\mu-1}$ given *stationary* mass m_μ (substructure $\mathbf{\Gamma}_1(s)$ in Fig. 4.20), and the second between masses $m_{\nu+1}$ and m_N given *stationary* mass m_ν (substructure $\mathbf{\Gamma}_3(s)$ in Fig. 4.20) (see also (Miu 1993)).

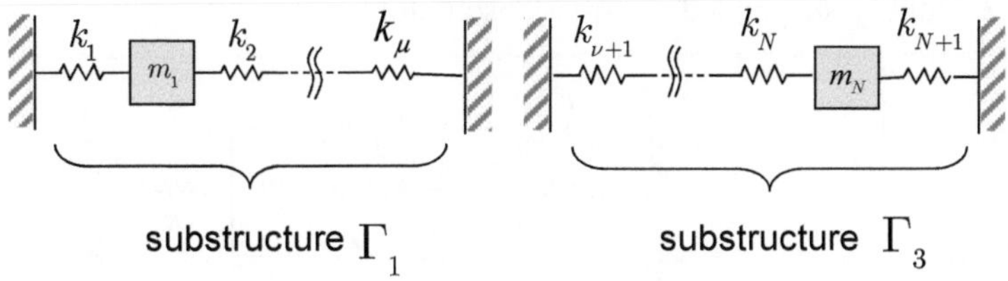

Fig. 4.20. Spring-mass subsystems for interpretation of zeros given actuation location μ and measurement location ν

Two-mass oscillator As a demonstrative example, consider the general two-mass oscillator depicted in Table 4.3. The transfer functions $G_{11}, G_{12}, G_{21}, G_{22}$ represent precisely all possible cases of inertial force inputs and absolute position measurement. For examples, in the case of G_{11} (force input *and* measurement at mass m_1), the zero ω_{z11} represents only the eigenfrequency of the single-mass oscillator m_2 given stationary mass m_1 and with the two springs k_2, k_3 connected in parallel. Since m_1 is stationary, one degree of freedom is lost, so that the order of the numerator of G_{11} is then $N_{DOF} - 1 = 2 - 1 = 1$.

Of particular interest are the cases of G_{12} and G_{21}. These have no finite zeros, i.e. the order of the numerator is zero. This confirms the physical interpretation given above, as for two stationary masses m_1, m_2 both degrees of freedom are lost and thus no possible oscillating substructure remains.

Symmetric three-mass oscillator The properties of the zeros discussed above can be verified equally well via the symmetric *three-mass oscillator* presented in Table 4.4. Thus, as has been demonstrated, the structure of the transfer function of a multiple-mass oscillator can be rather easily deduced.

Note, however, the cases (which are not necessarily apparent) of *canceling* poles (eigenfrequencies) and zeros for particular parameter configurations (here, a cancellation due to $\omega_{z2} = \omega_{p2}$, e.g. in G_{21}). Such cases represent problematic *observability* and *controllability breakdowns* and are discussed in more detail in Ch. 10.

Table 4.3. Doubly-suspended two-mass oscillator

Schematic	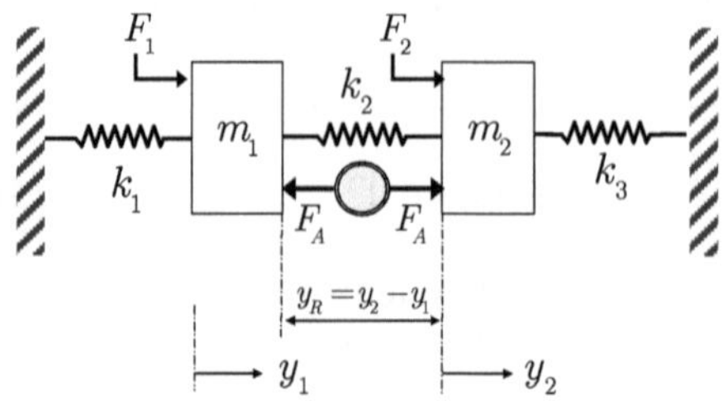
Transfer matrix	$\begin{pmatrix} Y_1(s)\\ Y_2(s)\\ Y_R(s)\end{pmatrix} = \mathbf{G}(s)\begin{pmatrix}F_1(s)\\ F_2(s)\\ F_A(s)\end{pmatrix} = \begin{pmatrix} G_{11}(s) & G_{12}(s) & G_{1A}(s)\\ G_{21}(s) & G_{22}(s) & G_{2A}(s)\\ G_{R1}(s) & G_{R2}(s) & G_{RA}(s)\end{pmatrix}\begin{pmatrix}F_1(s)\\ F_2(s)\\ F_A(s)\end{pmatrix}$ $\mathbf{G}(s) = \begin{pmatrix} K_{11}\dfrac{\{\omega_{z11}\}}{\Delta(s)} & K_{21}\dfrac{1}{\Delta(s)} & K_{1A}\dfrac{\{\omega_{z1A}\}}{\Delta(s)} \\ K_{12}\dfrac{1}{\Delta(s)} & K_{22}\dfrac{\{\omega_{z22}\}}{\Delta(s)} & K_{2A}\dfrac{\{\omega_{z2A}\}}{\Delta(s)} \\ K_{R1}\dfrac{\{\omega_{zR1}\}}{\Delta(s)} & K_{R2}\dfrac{\{\omega_{zR2}\}}{\Delta(s)} & K_{RA}\dfrac{\{\omega_{zRA}\}}{\Delta(s)} \end{pmatrix}$
Gains	$K_{11} = \dfrac{k_2 + k_3}{k^*}, \quad K_{12} = K_{21} = \dfrac{k_2}{k^*}, \quad K_{22} = \dfrac{k_1 + k_2}{k^*}$ $K_{R1} = K_{1A} = -\dfrac{k_3}{k^*}, \quad K_{R2} = K_{1A} = \dfrac{k_1}{k^*}, \quad K_{RA} = \dfrac{k_1 + k_3}{k^*}$ $k^* = k_1 k_2 + k_1 k_3 + k_2 k_3$
Poles	$\Delta(s) = \{\omega_{p1}\}\{\omega_{p2}\}, \quad \omega_{p1}, \omega_{p2}$ from $m_1 m_2 s^4 + \left[m_1\left(k_2 + k_3\right) + m_2\left(k_1 + k_2\right)\right] s^2 + k^* = 0$ $\Rightarrow s_{1,2} = \pm j\omega_{p1}, \quad s_{3,4} = \pm j\omega_{p2}$
Zeros	$\omega_{z11} = \sqrt{\dfrac{k_2 + k_3}{m_2}}, \quad \omega_{z22} = \sqrt{\dfrac{k_1 + k_2}{m_1}}, \quad \omega_{zRA} = \sqrt{\dfrac{k_1 + k_3}{m_1 + m_2}}$ $\omega_{zR1} = \omega_{z1A} = \sqrt{\dfrac{k_3}{m_2}}, \quad \omega_{zR2} = \omega_{z2A} = \sqrt{\dfrac{k_1}{m_2}}$

Table 4.4. Transfer matrix for a symmetric three-mass oscillator

Schematic	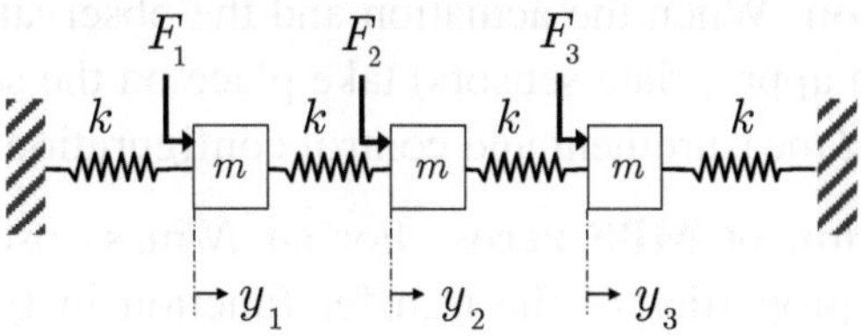
Transfer matrix[30]	$\begin{pmatrix} Y_1(s) \\ Y_2(s) \\ Y_3(s) \end{pmatrix} = \mathbf{G}(s) \begin{pmatrix} F_1(s) \\ F_2(s) \\ F_3(s) \end{pmatrix} = \begin{pmatrix} G_{11}(s) & G_{12}(s) & G_{13}(s) \\ G_{21}(s) & G_{22}(s) & G_{23}(s) \\ G_{31}(s) & G_{32}(s) & G_{33}(s) \end{pmatrix} \begin{pmatrix} F_1(s) \\ F_2(s) \\ F_3(s) \end{pmatrix}$ $\mathbf{G}(s) = \begin{pmatrix} K_{11}\dfrac{\{\omega_{z1}\}\{\omega_{z3}\}}{\Delta(s)} & K_{12}\dfrac{\cancel{\{\omega_{z2}\}}}{\cancel{\Delta}(s)} & K_{13}\dfrac{1}{\Delta(s)} \\ K_{21}\dfrac{\cancel{\{\omega_{z2}\}}}{\cancel{\Delta}(s)} & K_{22}\dfrac{\{\omega_{z2}\}^{\cancel{2}}}{\cancel{\Delta}(s)} & K_{23}\dfrac{\cancel{\{\omega_{z2}\}}}{\cancel{\Delta}(s)} \\ K_{31}\dfrac{1}{\Delta(s)} & K_{32}\dfrac{\cancel{\{\omega_{z2}\}}}{\cancel{\Delta}(s)} & K_{33}\dfrac{\{\omega_{z1}\}\{\omega_{z3}\}}{\Delta(s)} \end{pmatrix}$
Gains	$K_{11} = K_{33} = {}^{3}\!/_{4k}$, $K_{22} = {}^{1}\!/_{k}$, $K_{13} = K_{31} = {}^{1}\!/_{4k}$, $K_{12} = K_{21} = K_{23} = K_{32} = {}^{1}\!/_{2k}$
Poles	$\Delta(s) = \{\omega_{p1}\}\{\omega_{p2}\}\{\omega_{p3}\}$ $\omega_{p1} = \sqrt{\left(2-\sqrt{2}\right)\dfrac{k}{m}}$ $\omega_{p2} = \sqrt{2\dfrac{k}{m}}$, $\omega_{p3} = \sqrt{\left(2+\sqrt{2}\right)\dfrac{k}{m}}$
Zeros	$\omega_{z1} = \sqrt{\dfrac{k}{m}}$, $\omega_{z2} = \omega_{p2} = \sqrt{2\dfrac{k}{m}}$, $\omega_{z3} = \sqrt{3\dfrac{k}{m}}$

30 Note the canceling poles and zeros (marked with canceling slashes).

4.7.3 Collocated measurement and actuation

Collocation When the actuation and the observation of motion (measurement with appropriate sensors) take place on the same body[31] it is termed a *collocated* measurement and control configuration.

Distribution of MBS zeros For an N-mass oscillator, it is easy to see from the properties of the transfer function in Eq. (4.71) that in a collocated transfer function, $(N-1)$ imaginary pairs of zeros will appear—corresponding to the $(N-1)$ eigenfrequencies of the remaining substructures having $(N-1)$ bodies / degrees of freedom. Typically, this results in transfer functions of the form (see also Fig. 4.21)

$$G_C(s) = V\frac{1}{\{\omega_{p,1}\}}\cdots\frac{\{\omega_{z,i-1}\}}{\{\omega_{p,i-1}\}}\frac{\{\omega_{z,i}\}}{\{\omega_{p,i}\}}\frac{\{\omega_{z,i+1}\}}{\{\omega_{p,i+1}\}}\cdots, \tag{4.73}$$

$$\omega_{z,i-1} < \omega_{p,i-1} < \omega_{z,i} < \omega_{p,i} < \omega_{z,i+1} < \omega_{p,i+1} \,.$$

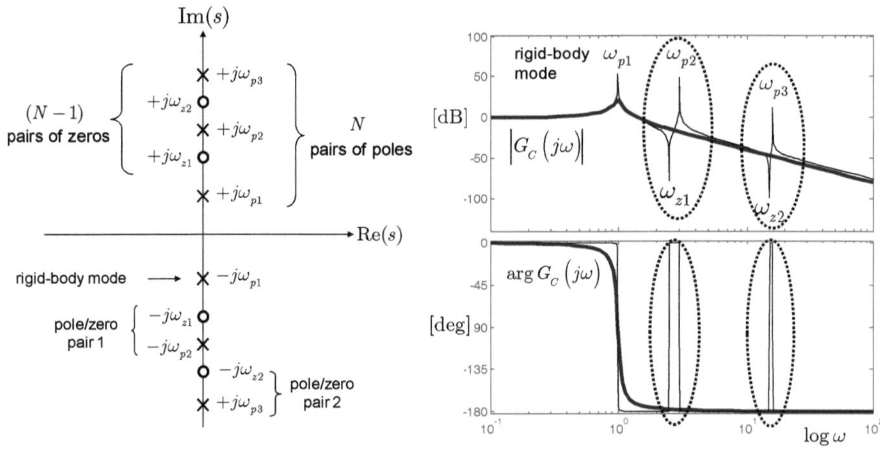

Fig. 4.21. Collocated measurement/actuation configuration for multibody system with three eigenmodes: a) pole/zero distribution, b) frequency response (BODE diagram)

[31] The assumption can be of either a rigid body or different bodies having kinematic coupling, e.g. a rigid bar, a rigid gear train.

Thus, there is always exactly one zero *between* two neighboring poles (it is "*sandwiched*"). Considering the series of eigenmodes increasing in magnitude, every case starts with the rigid-body mode (common mode) followed by $(N-1)$ *pole/zero pairs* (Fig. 4.21a). Compare also G_{11}, G_{22} in Table 4.3 and G_{11}, G_{33} in Table 4.4.

Frequency response The properties of the transfer function in Eq. (4.73) can be very easily recognized in its BODE diagram (Fig. 4.21b). Especially characteristic is the *phase evolution*. The rigid-body mode results in a $-180°$ phase jump, while the higher-frequency modes bring no further net change due to the pole/zero pairs (the zero occurring at a lower frequency than the pole). The zeros act as a type of phase lead and compensates for the negative phase jump of the eigenfrequencies. As a comparison, the amplitude and phase plots for a single-mass oscillator with the same rigid-body eigenfrequency are shown. The higher-frequency eigenmodes can considered as superimposed. With increasing stiffness, they also move to higher frequencies, while in the low-frequency band, the single-mass rigid body behavior is all that remains.

Control theoretical interpretation For the stability of a control loop, it is well-known that the phase evolution near $-180°$ is of particular importance (the NYQUIST criterion). At first glance, a collocated multibody system behaves like a single-mass oscillator, and is thus quite controllable as the only negative phase displacement is that of the rigid-body mode (the $-180°$ phase mark is never exceeded). A second looks shows, however, that ever-present parasitic phase delays (sensor/actuator dynamics, computational delays, etc.) will deflect the phase beyond $-180°$ for high-frequency eigenmodes, and at the same time, large amplitudes will appear due to the resonant peaks. Thus even in this case, fundamental stability problems always appear[32]. Suitable control approaches are discussed in detail in Ch. 10.

---

32 In the literature, collocated configurations are often imbued with "self-stable" control characteristics in that proportional feedback is always sufficient to produce a stable control loop. However, this holds only for the academic case of ideal feedback without parasitic dynamics, and is inapplicable to all practical cases. It is however correct that collocated configurations are easier to control than non-collocated ones.

4.7.4 Non-collocated measurement and actuation

Separated measurement and actuation locations In some cases, it is not possible to choose the measurement and actuation locations to be the same, i.e. the sensor and actuator are located on different bodies[33] of a multibody system. This is then termed a *non-collocated* (separated) measurement and control configuration.

Distribution of MBS zeros For an N-mass oscillator as in Eq. (4.71), determination of substructures capable of oscillation requires making two bodies stationary. This leaves at most $(N-2)$ oscillating bodies (degrees of freedom) and thus at most $(N-2)$ imaginary pairs of zeros. Typically, this results in transfer functions of the form (see also Fig. 4.22)

$$G_{NC}(s) = V\frac{1}{\{\omega_{p,1}\}}\cdots\frac{\{\omega_{z,i-1}\}}{\{\omega_{p,i-1}\}}\frac{1}{\{\omega^*_{p,i}\}}\frac{\{\omega_{z,i+1}\}}{\{\omega_{p,i+1}\}}\cdots,$$
$$\omega_{z,i-1} < \omega_{p,i-1} < \omega^*_{p,i} < \omega_{z,i+1} < \omega_{p,i+1}\,. \tag{4.74}$$

Due to the now missing pair of zeros, at some point, two poles must follow each other (in Fig. 4.22, $s = \pm j\omega_{p1}, s = \pm j\omega_{p2}$). Compare also G_{12}, G_{21} in Table 4.3.

Frequency response The difference from a collocated configuration is quite apparent in the phase evolution (Fig. 4.22b). Due to the missing zero, the now "zero-free" second eigenfrequency ω_{p2} produces an additional $-180°$ phase jump, resulting in a total phase offset of $-360°$. The remaining pole/zero pairs again do not produce a net contribution. Note that for large structural damping, the collocated and non-collocated cases are not immediately distinguishable in the amplitude plot (in this case, the zero ω_{z1} does not induce a deep trough), though certainly in the phase plot.

Separation of measurement and actuation locations The multiple-mass oscillator in Eq. (4.71) affords an additional interesting interpretation. The further the measurement and actuation locations are separated, the fewer oscillating bodies—and hence degrees of freedom—remain for the substructures which determine the zeros. Choosing, for example, the left and right outermost bodies as measurement and actuation locations, there exist no zeros at all, e.g. G_{13}, G_{31} in Table 4.4.

[33] Assuming elastic coupling of the bodies.

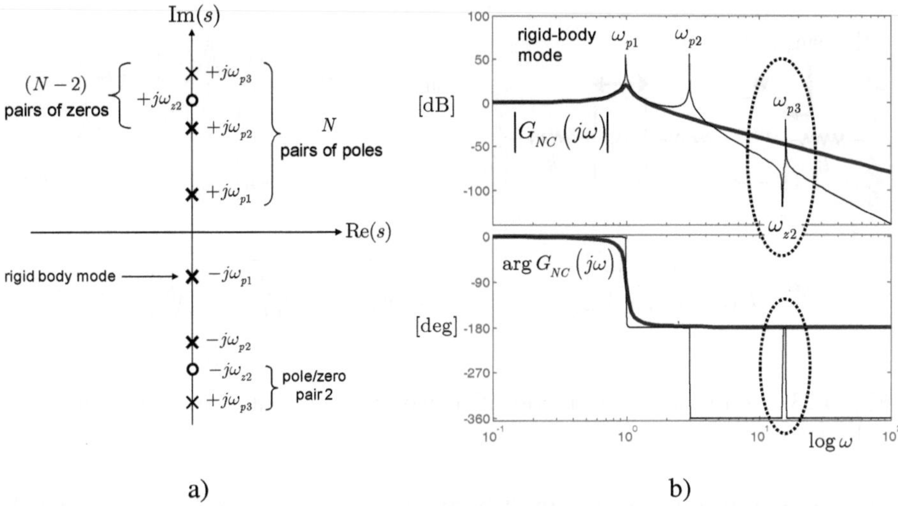

Fig. 4.22. Non-collocated measurement/actuation arrangement for multibody system with three eigenmodes: a) pole/zero distribution, b) frequency response (BODE diagram)

Control theoretical interpretation Thus, for non-collocated configurations, even without any parasitic dynamics, the $-180°$ phase mark is exceeded in all cases, with all the accompanying negative consequences for stability. In this case, it is of no consequence whether the structural damping is small or large, as even for large damping, the phase jump always occurs (though not so steeply). For these reasons, controller design for non-collocated configurations is always a particular challenge.

4.7.5 Antiresonance

One interesting interpretation of MBS zeros permits the following considerations. As a demonstrative example, consider the two-mass oscillator shown in Fig. 4.23a, with collocated measurement and actuation at Body 1. From the frequency response in Fig. 4.23b, it can be seen that, given a harmonic excitation $F_1(t) = \sin(\omega_z t)$ at the zero frequency, the amplitude of the output $y_1(t)$ in the oscillating system is zero, as $G_1(j\omega_{z1}) = 0$. Where does the input energy go?

A simple consideration shows that, though Body 1 does stand still, Body 2 oscillates precisely in resonant *antiphase* to the excitation $F_1(t)$ (*antiresonance*). The forces applied to Body 1 thus cancel and it is dynamically held stationary. In the non-dissipative case presented here, there is

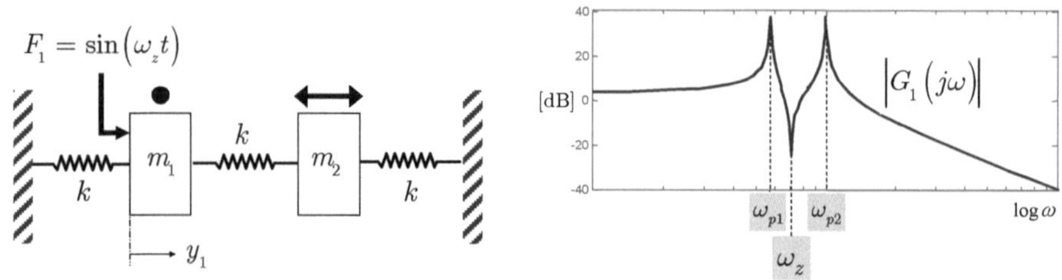

$$G_1(s) = \frac{Y_1(s)}{F_1(s)} = \frac{2}{3k}\frac{\{\omega_z\}}{\{\omega_{p1}\}\{\omega_{p2}\}}$$

$$\omega_{p1} = \sqrt{\frac{k}{m}},\quad \omega_{p2} = \sqrt{\frac{3k}{m}},\quad \omega_z = \sqrt{\frac{2k}{m}},\quad m_1 = m_2 = m$$

a) b)

Fig. 4.23. Antiresonant excitation of a two-mass oscillator: a) MBS schematic, b) frequency response

thus no energy input to the multibody system; the force F_1 does no work. With real dissipation (viscous friction), F_1 provides only the friction work.

In general, this observation confirms the dynamics for *collocated and non-collocated* configurations seen in Eq. (4.71), namely, that the bodies at the measurement and actuation locations are dynamically held stationary by the remaining, resonantly oscillating substructures (zeros $\triangleq$ *resonances* of the substructures $\triangleq$ *antiresonance frequencies*).

4.7.6 Migration of MBS zeros

Zero positions For a *non-dissipative* multibody system, the numerator of any arbitrary transfer function

$$G(s) = \frac{b_0 + b_2 s^2 + \ldots + b_{2(M-1)} s^{2(M-1)} + b_{2M} s^{2M}}{\mathcal{N}_G(s)},\quad M \le N_{DOF} - 1$$

in the MBS transfer matrix contains only *even* powers of s. This implies that the zeros of the transfer function must lie symmetrically about the imaginary axis, i.e. they lie either on the imaginary axis (*imaginary* pairs of zeros, see Fig. 4.24a, type A) or they are axially symmetric *real* pairs of zeros (see Fig. 4.24a, type B). In the case of *dissipative* systems with *small* damping, a similar situation results, the imaginary zeros simply move slightly into the left half-plane.

Parameters affecting MBS zeros The locations of zeros of an MBS transfer function fundamentally depend on both configuration parameters (geometry, masses, stiffnesses), and the *excitation* and *observation points* (cf. Eq. (4.61)). In contrast, the locations of *poles* of the transfer function are *independent* of these points and depend only on the configuration parameters. For this reason, it is to be expected that both changes in the configuration parameters (desired or undesired) and variations in the measurement and actuation locations will induce a *migration* of the zeros.

Migration possibilities for MBS zeros The principal possible forms of migration of the zeros are indicated in Fig. 4.24a: symmetrically along the imaginary axis and symmetrically along the real axis. In the process, different, varying configurations of the zeros with respect to the locations of poles can appear. These configurations have fundamental effects on the response characteristics and control characteristics of the system.

Typical configurations The following typical configurations can occur:

1. double zero at $s = 0$, Fig. 4.24b,
2. pair of zeros canceling pair of poles, Fig. 4.24c,
3. pole/zero swap given small parameter changes, Fig. 4.24d,
4. pair of zeros at $s = \pm j\infty$, Fig. 4.24e,
5. pair of real zeros $s = \pm b$, Fig. 4.24e.

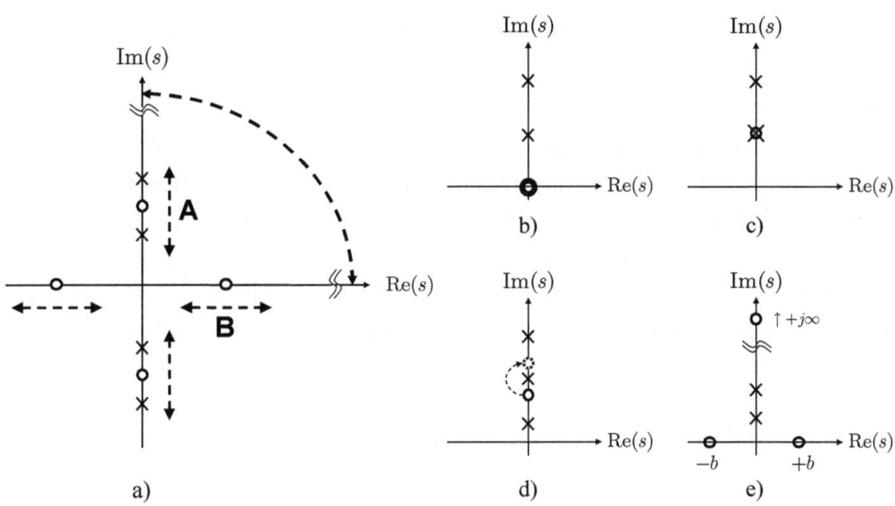

Fig. 4.24. Zeros of non-dissipative multibody systems: a) migration loci, b) double zero at s=*0*, c) pair of zeros cancels pair of poles, d) pole/zero swap, e) pair of zeros at $s = \pm j\infty$ and pair of real zeros $s = \pm b$

Double zero at $s = 0$ This case occurs when an acceleration sensor is used for the measurement, see Eq. (4.65). If there is also a double pole at $s = 0$ (i.e. the rigid-body mode of a free multibody system), there is exact canceling. Thus, in this case, the rigid-body mode is *unobservable* and *uncontrollable*.

Pole/zero cancellation If the pair of imaginary zeros exactly cancels a pair of imaginary poles $s = \pm j\omega_{pi}$, then the eigenfrequency ω_{pi} no longer appears in the transfer function. Thus, this eigenmode is also *unobservable* and *uncontrollable*. Natural oscillations induced by disturbances in this mode can not be affected by a controller.

Pole/zero swap A particularly serious case occurs when, for small parameter changes in the MBS configuration, the relative locations of poles and zeros change (Fig. 4.24d). The dramatic effects of this situation can be much better recognized in the frequency response. This situation produces a *phase uncertainty* of $360°$, as, for the migration shown in Fig. 4.24d, instead of the original $+180°$ phase jump due to the zero, the $-180°$ phase jump comes first in the alternate configuration. This completely reverses the stability conditions and can only be managed using gain stabilization (with small gain) in this frequency range. Fortunately, this behavior only appears in non-collocated configurations.

Pair of infinite zeros For certain parameter configurations, the zeros can go to infinity, resulting in a *reduction in order* of the numerator. This creates an (additional) set of neighboring pairs of poles without a separating zero, and the disadvantageous property of having a double $-180°$ phase jump.

Pair of real zeros: non-minimum phase behavior As a sort of continuous perpetuation of the variation in behavior induced by the presence of imaginary infinite zeros, pairs of zeros can migrate in from infinity along the real axis (Fig. 4.24a, dashed arc). This results in a symmetric pair of real zeros at $s = \pm b$. Due to the existence of a zero in the *right* half-plan, this results in so-called *non-minimum phase behavior* with challenging properties for control (Ogata 2010), (Horowitz 1963).

Generalized free two-mass oscillator

In order to gain some engineering intuition for the various zero configurations, consider the generalized unsuspended two-mass oscillator shown in Fig. 4.25. This example is sufficiently simple to permit the use of analytical relations, while demonstrating all important configurations.

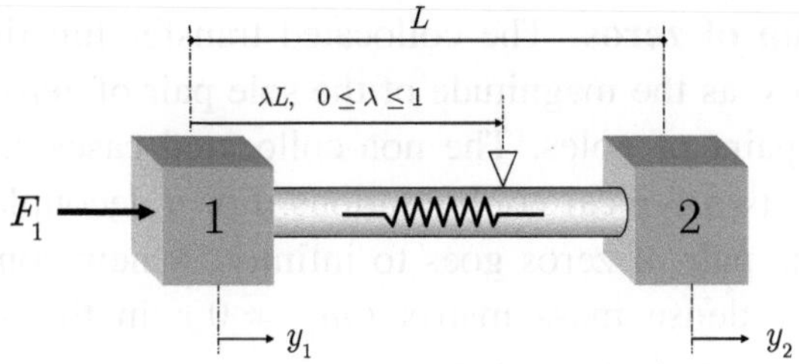

Fig. 4.25. Generalized free two-mass oscillator

Equations of motion Let the two-mass oscillator under consideration be characterized by two bodies, a massless elastic connection and generalized force application to Bodies 1 and 2. Using the generalized coordinates y_1, y_2 the resulting equations of motion are

$$\begin{pmatrix} m_{11} & m_{12} \\ m_{12} & m_{22} \end{pmatrix}\begin{pmatrix} \ddot{y}_1 \\ \ddot{y}_2 \end{pmatrix} + \begin{pmatrix} k & -k \\ -k & k \end{pmatrix}\begin{pmatrix} y_1 \\ y_2 \end{pmatrix} = \begin{pmatrix} F_1 \\ F_2 \end{pmatrix}. \tag{4.75}$$

Note the fully dense mass matrix[34], which also describes linked bodies.

Transfer matrix The transfer matrix between the input forces and generalized coordinates is[35]

$$\begin{pmatrix} Y_1(s) \\ Y_2(s) \end{pmatrix} = \mathbf{G}(s)\begin{pmatrix} F_1(s) \\ F_2(s) \end{pmatrix} = \left(\begin{array}{c|c} K\dfrac{\{\omega_{z11}\}}{s^2\{\omega_0\}} & K\dfrac{[\omega_{z12}][-\omega_{z12}]}{s^2\{\omega_0\}} \\ \hline K\dfrac{[\omega_{z12}][-\omega_{z12}]}{s^2\{\omega_0\}} & K\dfrac{\{\omega_{z22}\}}{s^2\{\omega_0\}} \end{array}\right)\begin{pmatrix} F_1(s) \\ F_2(s) \end{pmatrix}, \tag{4.76}$$

$$K = \frac{1}{m_{11} + m_{22} - 2m_{12}} \qquad \omega_{z11} = \sqrt{\frac{k}{m_{22}}}$$

$$\omega_{z22} = \sqrt{\frac{k}{m_{11}}}. \tag{4.77}$$

$$\omega_0 = \sqrt{k\frac{m_{11} + m_{22} - 2m_{12}}{m_{11}m_{22} - m_{12}^2}} \qquad \omega_{z12} = \sqrt{\frac{k}{m_{12}}}$$

---

34 One should not be disturbed by the fully dense mass matrix, which describes a general case. From the presentation which follows, it will become clear when such cases are possible.

35 As previously mentioned, $[a] := 1 + s/a$.

General discussion of zeros The collocated transfer functions G_{11}, G_{22} present no surprises, as the magnitude of the sole pair of zeros always lies between the two pairs of poles. The non-collocated cases $G_{12} = G_{21}$ already demonstrate two typical configurations. For a decoupled mass matrix ($m_{12} = 0$), the pair of zeros goes to infinity. A new configuration is present for a fully dense mass matrix ($m_{12} \neq 0$): in this case, a *non-minimal phase* pair of real zeros appears.

Generalized measurement variable Deeper insight into the migration pattern of the zeros is achieved by considering a *linear combination* of the generalized coordinates of the two bodies as a generalized measurement variable:

$$z = \lambda_1 y_1 + \lambda_2 y_2 . \qquad (4.78)$$

With a suitable choice of λ_1, λ_2, a variety of *measurement principles* can be modeled:

- *variable measurement location*: $z = \lambda L = (1-\lambda) y_1 + \lambda y_2$, $0 \leq \lambda \leq 1$, $\lambda_1 = 1-\lambda$, $\lambda_2 = \lambda$, see Fig. 4.25,
- *relative measurement*: $z = -y_1 + y_2$, $\lambda_1 = -1, \lambda_2 = 1$.

Implementation example of variable measurement location By varying the parameter λ, the measurement location can be continuously varied between Body 1 and Body 2. Physically, such a case is present in linearly elastic *longitudinally vibrating rods* (Fig. 4.26a), *torsion rods* (Fig. 4.26b) or elastic gearboxes (Fig. 4.26c). Over the length of the rods, there is a linearly increasing deformation of the form

$$z = y_1 - \frac{y_1 - y_2}{L} \lambda L = (1-\lambda) y_1 + \lambda y_2 ,$$

which can be measured with a position sensor attached at the location $y = \lambda L$ (Miu 1993).

Implementation example of coupled mass matrix In translational and rotational oscillator chains, there is always a decoupled (diagonal) mass matrix. However, considering, for example, a *two-arm manipulator* (see Example 4.2 and Fig. 4.26d), coupling terms $m_{12} \neq 0$ also come into play, cf. Eq. (4.35).

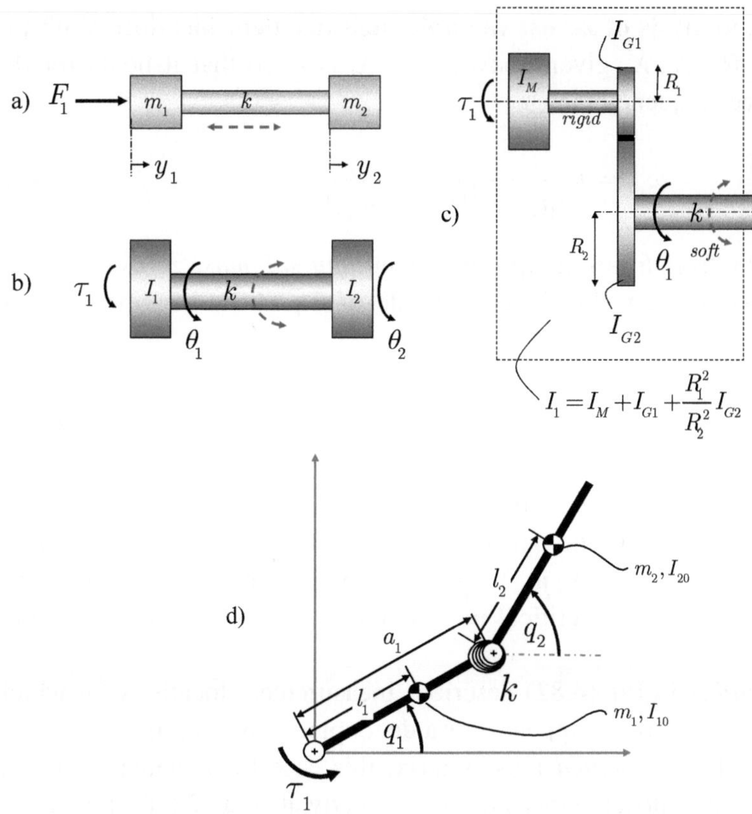

Fig. 4.26. Concrete implementational variants of the generalized free two-mass oscillator: a) longitudinally vibrating rod, b) torsion rod, c) drive train with elastic gearbox and load, d) two-arm manipulator with elastic Joint 2

Parametric response characteristics Using Eqs. (4.76) and (4.78), the generalized response given excitation by F_1 can be calculated:

$$G_{z/F1}(s) = \frac{Z(s)}{F_1(s)} = \frac{(\lambda_1 + \lambda_2)}{(m_{11} + m_{22} - 2m_{12})} \frac{\{\omega_z\}}{s^2 \{\omega_0\}}. \tag{4.79}$$

The *antiresonance frequency* which results is

$$\omega_z^{\ 2} = k \frac{\lambda_1 + \lambda_2}{\lambda_1 m_{22} - \lambda_2 m_{12}}. \tag{4.80}$$

Parametric analysis of zeros: variable measurement location With the substitution for λ_1, λ_2 given above, $\lambda_1 + \lambda_2 = 1$, so that it holds for the antiresonance frequency that

$$\omega_z^{\ 2} = k \frac{1}{m_{22} - \lambda(m_{22} + m_{12})}. \tag{4.81}$$

The transition from a *zero at infinity* to *non-minimal phase zeros* obviously occurs precisely when the right-hand side of Eq. (4.81) becomes zero or negative, i.e.

$$\lambda^* \geq \frac{m_{22}}{m_{22} + m_{12}}. \tag{4.82}$$

The *equality* in Eq. (4.82) indicates that for the measurement location $y^* = \lambda^* L$, exactly one pair of zeros lies at infinity. For a decoupled mass matrix, this is the case at $\lambda^* = 1$ (i.e. measurement at Body 2). For a coupled mass matrix, the critical measurement location slides left towards Body 1.

The *inequality* in Eq. (4.82) describes measurement locations for which *non-minimal phase zeros* appear. For a decoupled mass matrix, this case is not possible. For a *coupled* mass matrix, this case fundamentally always appears for non-collocated measurement directly at Body 2 (cf. (4.76)).

Pole/zero cancellation can, following Eq. (4.81), only appear for the flexible eigenmode ω_0. The corresponding measurement location is

$$\lambda_0 = \frac{\left(m_{22} - m_{12}\right)^2}{\left(m_{11} + m_{22} - 2m_{12}\right)\left(m_{22} + m_{12}\right)}.$$

In the symmetric case $m_{11} = m_{22}$, the critical measurement location is

$$\lambda_0 = \frac{1}{2} \frac{\left(m_{22} - m_{12}\right)}{\left(m_{22} + m_{12}\right)}. \tag{4.83}$$

In the decoupled case, this is then exactly half-way between the two bodies ($\lambda_0 = 0.5$). The eigenmode ω_0 is *unobservable* at this location. However, this is unsurprising, as the corresponding eigenmode possesses an *oscillation node* at this location (deflection is zero). For a coupled mass matrix, this point again slides further to the left.

If the measurement position is chosen to be near the oscillation node, variations in the masses can cause the oscillation node to move in such a manner that a *pole/zero swap* takes place.

Parametric analysis of zeros: relative measurements For relative measurements, $\lambda_1 = -1$ and $\lambda_2 = 1$, so that—independently of configuration parameters—the pair of zeros in Eq. (4.80) results in a *double zero* at $s = 0$. As a result, in the transfer function the rigid-body mode is canceled and is thus neither observable nor controllable.

MBS zeros (antiresonances) The *zeros* of an MBS plant play an equally important role in the control properties of a mechatronic system as the eigenmodes (eigenfrequencies, eigenvectors). While the eigenmodes represent inherent configuration properties, the zeros depend on both the MBS configuration *and* the considered measurement and actuation locations. Certain pole/zero configurations result for *non-collocated* measurement and actuation arrangements, and the difficulty of achieving targeted control of a multibody systems increases with increasing separation of measurement and actuation locations. Thus, with a suitable choice of the measurement and actuation locations, important dimensions of the design space are made available.

Bibliography for Chapter 4

Goldstein, H., C. P. Poole and J. L. Safko (2001). *Classical Mechanics*. Addison Wesley.

Gregory, R. D. (2006). *Classical Mechanics*. Cambridge University Press.

Horowitz, I. M. (1963). *Synthesis of Feedback Systems*. New York. Academic Press.

Miu, D. K. (1993). *Mechatronics: electromechanics and contromechanics*. Springer.

Ogata, K. (2010). *Modern Control Engineering*. Prentice Hall.

Pfeiffer, F. (2008). *Mechanical System Dynamics*. Springer.

Preumont, A. (2002). *Vibration Control of Active Structures - An Introduction*. Kluwer Academic Publishers.

Schwertassek, R. and O. Wallrap (1999). *Dynamik flexibler Mehrkörpersysteme*. Vieweg.

5 Functional Realization: The Generic Mechatronic Transducer

Background The functional interface between information processing—in the form of electrical signals—and mechanical structure—in the form of forces/torques and motion variables—is of key importance in a mechatronic product. The bi-directional conversion of energy from electrical to mechanical is a key component of the primary product task "*perform purposeful motions*". The multiple, diverse conversion principles available today can also be integrated into a mechatronic product in a functionally and physically compact manner in the form of *mechatronic transducers*. Along with a basic grasp of these conversion principles, understanding the influences which parameters describing the transducer have on its transfer characteristics is of particular interest in systems design.

Content of Chapter 5 In this chapter, using the concept of a *generic mechatronic transducer*, general commonalities in the power coupling and transfer characteristics of a variety of transducers—including their *force generation*, *electrical properties*, *causal structures*, and *dynamic models*—are discussed independent of the physical transducer phenomena. The generic transducer thus forms the methodological and modeling framework for the detailed presentations of physical principles in subsequent chapters. Taking an energy-based modeling approach based on the *EULER-LAGRANGE equations* as the starting point, nonlinear and linearized *constitutive transducer equations* are derived for a lossless, unloaded transducer. For the linearized transducer, a specialized *two-port parameterization* is introduced as the central basis for subsequent *model extensions* of the electrical and mechanical circuits (e.g. *voltage vs. current sources*, *lossy transducers*, and *rigid-body vs. multibody loads*). The models presented in this chapter enable a general discussion of *generic dynamic properties* and *behaviors*—such as eigenfrequencies, transducer stiffnesses, transfer functions, and electromechanical coupling—while only requiring that a distinction be made between capacitive or inductive transducer dynamics. The general model considerations are rounded out with the most technically significant implementational issues: *oscillation damping* (the mechatronic *resonator*), *energy generation* (the mechatronic *oscillating generator*), and *self-sensing*.

■

5.1 Systems Engineering Context

Mechatronic transducers From a systems engineering point of view, the functions *"generate forces/torques"* and *"measure mechanical states"* are represented by the *actuators* and *sensors* of a mechatronic system (Fig. 5.1). These causally relate system variables from differing physical domains: *electrical* variables as the interface to information processing and *mechanical* variables as the interface to mechanical structures. An instance of such devices in an implementation is termed a *transducer.* To carry out the energy conversion, a variety of physical phenomena are employed, e.g. electrostatics, piezoelectric effects, and electromagnetism (due to their wide application, these particular phenomena are also discussed in more detail in following chapters). Due to the generally high spatial and functional density of their integration (see Sec. 1.1), such transducers are referred to below as *mechatronic transducers.*

Causal structure For both energy conversion tasks, the transfer responses in the causal directions shown in Fig. 5.1 are of interest. Often, to enable compact construction, transducer elements are integrated directly into the mechanical structure. Thus, in addition to common functional properties such as *linearity* and *dynamics*, structural *parameter dependencies* also play an important role in the analysis of closed-loop dynamics.

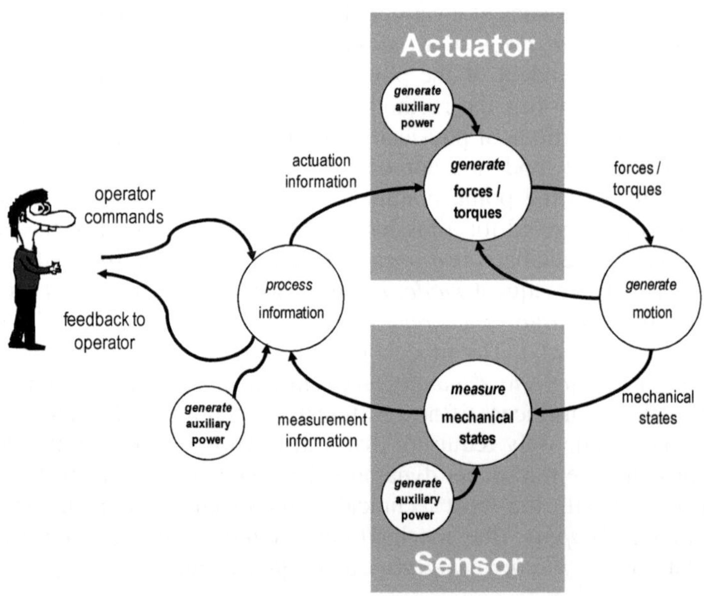

Fig. 5.1. Functional decomposition of a mechatronic system: *mechatronic transducers* realizing sensor and actuator functionalities

Power back-effect Since, for the *generation of forces*, a transformation from *electrical to mechanical* power—and thus coupling of power into the mechanical structure—takes place, any *power back-effect* that occurs must be incorporated into the transducer dynamic model (note the feedback loop in Fig. 5.1). The predictive dynamic model of an actuator must thus always incorporate any accelerated mechanical load. Initial exploratory investigations may, however, limit themselves to modeling only a single mass, though these should be exchanged for multibody models in more detailed considerations.

For functionality as a *sensor*, by definition, the actual value of the sensed variable should be affected as little as possible by the measurement system. Motion measurement generally proceeds using internal test masses which are (if possible) rigidly or (sometimes inevitably) elastically suspended in the mechanical structure. Given a negligible sensor test mass, there will be practically no power back-effect on the mechanical structure. The motion variables to be measured can then be modeled as a *back-effect-free displacement excitation* on the test masses inside the sensor (note the unidirectional, causal structure of the sensor function in Fig. 5.1). At the local transducer level, however, the power back-effect of the *sensor test mass* must well be accounted for. For this latter task, simple single-mass models generally suffice.

Generic mechatronic transducer Independent of the concrete physical transducer phenomenon under consideration, general properties of an abstract *generic mechatronic transducer* model can be formulated for its power coupling and transfer behavior, including:

- *force generation* as a function of the geometry, mechanical effects, and electric drive;
- *electrical properties* as a function of the geometry and mechanical effects;
- *causality structures* and *dynamic models*.

Thus, the generic transducer creates a methodological and modeling framework for the detailed representations of physical transducer principles to be discussed in later chapters. In order to avoid redundancy in the presentation, this chapter forgoes any explanatory physically-oriented examples; these can be found in subsequent chapters. The center of attention here is a unified formal description and notation, forming the reference point for subsequent chapters.

5.2 General Generic Transducer Model

5.2.1 System configuration

For systems design, it is the dynamic interaction between the excitation and system variables of a transducer which is important. Fig. 5.2 shows the generalized system configuration for a loaded generic mechatronic transducer with *one* mechanical degree of freedom, which can model both *actuator* and *sensor functionality*.

Mechanically, the generic transducer consists of a stationary element—the *stator*—and a mobile element—the *armature*—moving parallel to the transducer force direction. The stator is rigidly attached to a base structure supplying the reaction force. The armature is connected to a mobile structure having mass—the *load*. In Fig. 5.2, a *rigid body* with mass m is depicted as the load; for more complex arrangements, a *multibody system (MBS)* as described in Ch. 4 can be imagined in its place.

To generate the transducer force F_T, electrical *auxiliary energy* in the form of a controlled *voltage source* u_S or *current source* i_S is required.

Often, due to the particular physics of an implementation, only unidirectional force generation is possible, so that the armature must be provided with a restoring force. In Fig. 5.2, this is achieved with an *elastic suspension* with stiffness k; for a *free* mass, $k = 0$.

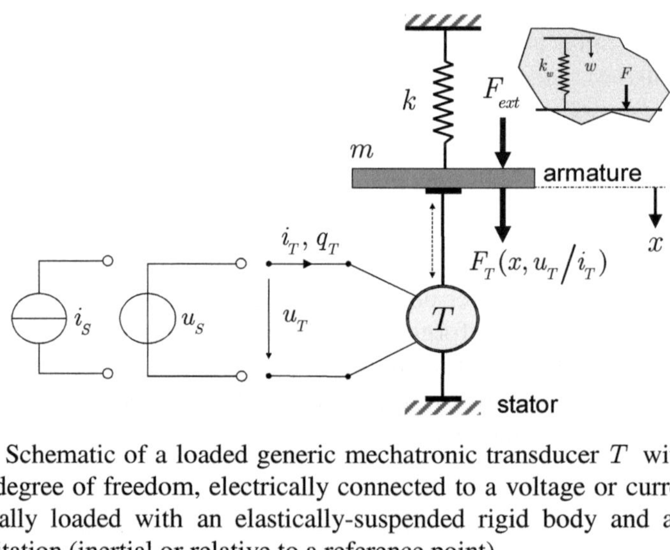

Fig. 5.2. Schematic of a loaded generic mechatronic transducer T with one mechanical degree of freedom, electrically connected to a voltage or current source, mechanically loaded with an elastically-suspended rigid body and an external force excitation (inertial or relative to a reference point)

The mechanical excitation is provided by a generalized external *force excitation* $F_{ext}(t)$. An imposed (back-effect-free) displacement excitation $w(t)$ can be modeled *elastically* with $F_{ext} = k_w \cdot (w - x)$ or *rigidly* with $x(t) = w(t)$ corresponding to $k_w \to \infty$.

System-oriented causal structures The configuration in Fig. 5.2 can be used to derive the system-oriented causal structure of the generic mechatronic transducer shown in Fig. 5.3. Excitations are the independent variable $u_S(t)$ or $i_S(t)$ of the auxiliary energy source, and the mechanical force $F_{ext}(t)$. Affected variables can include the armature displacement $x(t)$ and electrical power variables dependent on the varying transducer geometry, i.e. the transducer current $i_T(t)$ or voltage $u_T(t)$.

Transducer as an actuator If the transducer is driven as a *force actuator*, then either $u_S(t)$ or $i_S(t)$ is the time-varying control input. The connected mechanical structure to be moved is modeled—with back-effect—as a rigid body m or, potentially, a multibody system (see Ch. 4). The external excitation force describes disturbances applied to the structure. The affected variables include the motion variable $x(t)$ of the mechanical load and the dependent (uncontrolled) variable of the electrical auxiliary energy source.

Transducer as a sensor As a sensor, the transducer in Fig. 5.2 is capable of representing forces $F_{ext}(t)$ or displacements $x(t)$ affecting the transducer geometry. This can occur directly as a measurement of voltage $u_T(t)$ or current $i_T(t)$, or indirectly as a displacement measurement $x(t)$. In this latter operating mode, the transducer is operated at a steady-state electrical set point $u_S = U_0 = const.$ or $i_S = I_0 = const.$. The only transducer mass m which need be considered is then the armature mass (e.g. the electrode for capacitive transducers). The excitation variables $F_{ext}(t)$ can often be modeled using a *back-effect-free* connection as in Fig. 5.2.

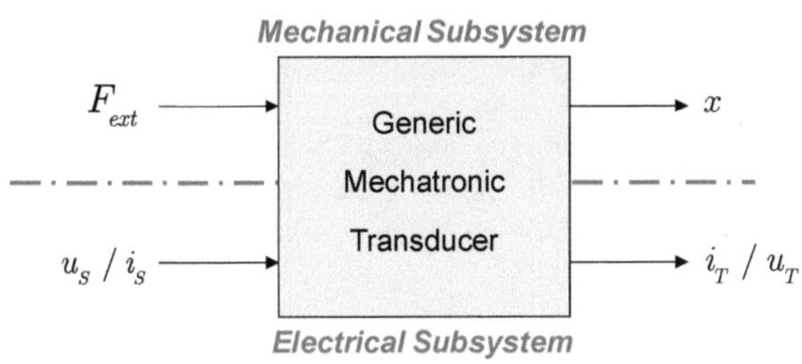

Fig. 5.3. System-oriented causal structure for a loaded generic mechatronic transducer

Controlled auxiliary energy sources On the electrical side, transducers generally require an auxiliary energy source. For operation as an actuator, this provides the necessary (low to very high) power for mechanical motion. For operation as a sensor, the transducer is driven to a steady-state operating point with, as a rule, low to very low power use. Table 5.1 presents the symbol, two-port circuit, and an example realization for a *controlled voltage source* and a *controlled current source*. For actuator operation, the control input (control voltage) u_{in} is a *zero-power* signal (control signal). For sensor operation, it can be used to set an operating point. In most cases, these sources are considered ideal, i.e. lossless. If this is not the case, it is specially indicated.

Table 5.1. Controlled electrical auxiliary energy sources (ideal, lossless)

<table>
<tr><th></th><th>Controlled voltage source</th><th>Controlled current source</th></tr>
<tr><td>Symbol for ideal source</td><td>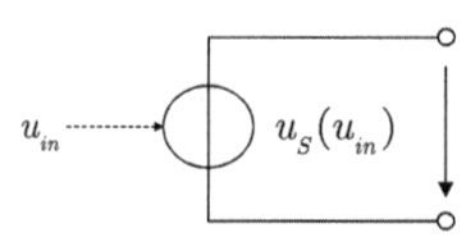</td><td>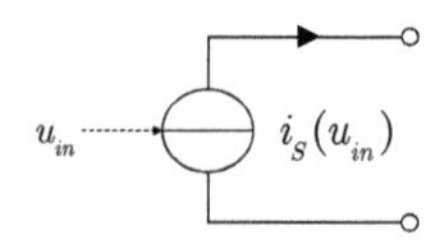</td></tr>
<tr><td>Two-port equivalent circuit</td><td>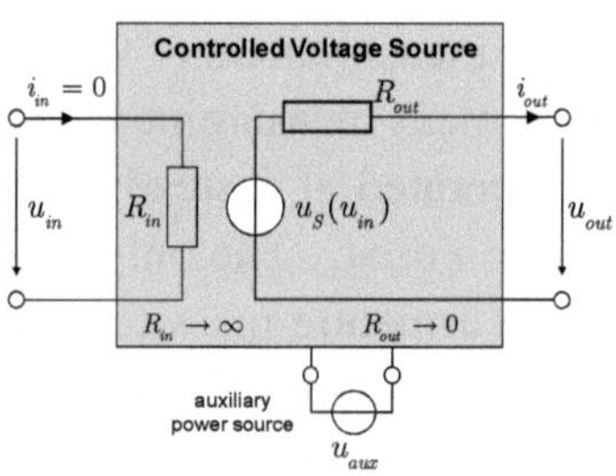
</td><td>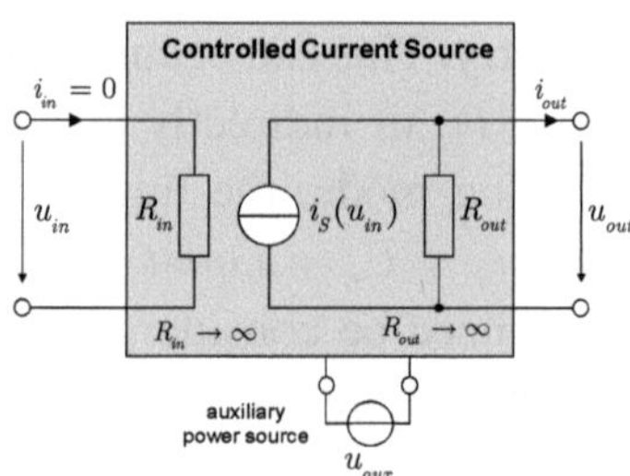
</td></tr>
<tr><td>Example realization</td><td>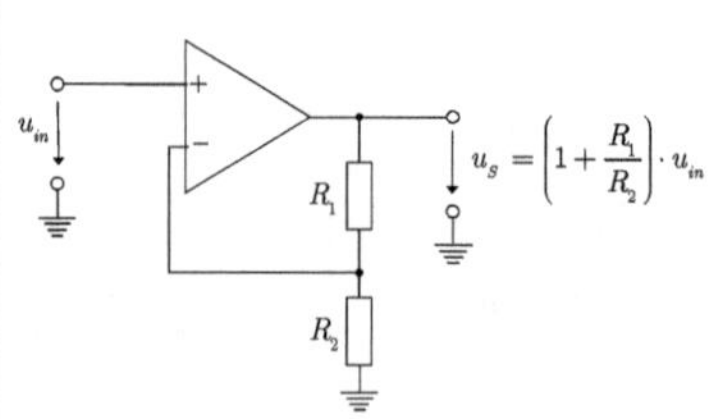</td><td>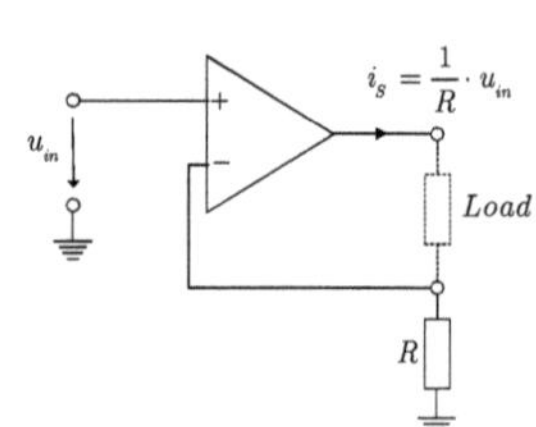
</td></tr>
</table>

In this chapter, the control input u_{in} is omitted for reasons of clarity, and only (controllable) auxiliary energy sources $u_S(t)$, $i_S(t)$ are considered.

5.2.2 Modeling approach

Model requirements In the context of systems design, the following aspects are important for modeling:

- *nonlinear large-signal responses* for determining possible operational limitations (e.g. unstable operating regimes),
- *stable equilibria* for determining steady-state operating points,
- *small-signal responses* in the time and frequency domains for characterizing the response and for controller design and dynamic analysis,
- *clarity in the structure* of the multi-domain transduction phenomena,
- *clear assignments* of physical and implementation parameters to model parameters.

From a methodological point of view, a *general modeling framework* which can be used to discuss diverse physical transducer principles in a unified form is also desirable.

Model hierarchy The discussions in later chapters of various energy conversion principles using particular physical phenomena will refer to the *model hierarchy* shown in Fig. 5.4. The present chapter discusses the *generalizable* relationships therein, forming a unified methodological scaffold via the generic mechatronic transducer of Fig. 5.2.

Reference configuration Fundamental investigations are carried out using a *lossless* transducer configuration with a simple, elastically-suspended rigid-body load. This model forms a conservative system, encompasses all significant multi-domain properties, and enables clear understanding with well-defined analytical relationships for the system dynamics. Model extensions which lead towards realization of the abstract model—in particular concerning dissipative phenomena and multibody loads—are presented at the end of this chapter.

Modeling methods The tasks of systems design call for the ability to represent various aspects of the behavior of a physical system implementation. The modeling paradigms and model formulations presented in Ch. 2 can be seen as a set of building blocks for the design of systems. The particular modeling process presented here can serve as *one* example of how the individual strengths of different methods can be adeptly put to use in the tasks of systems design[1]. Specifically, this section combines three different modeling approaches.

At the physical level, *energy-based modeling* using *energy functions* is applied and, using the *EULER-LAGRANGE equations*, a fundamental nonlinear dynamic model is generated.

For presentational rigor, a *multi-port model* in the form of a linear *two-port* (four-terminal element) with *constitutive two-port parameters* is introduced to represent the core functionality of the transducer.

Finally, considering concrete external electrical and mechanical loads on the transducer and using the constitutive two-port parameters, *signal-based linear time- and frequency-domain models* are derived, offering an optimal compromise between model clarity and ease-of-use for the controller design and dynamic analysis approaches preferred in this book.

Model Branch A: the unloaded generic transducer

Fundamental transducer functionalities are discussed using an *unloaded* (unconnected) *transducer model* (Model Branch A in Fig. 5.4, left).

Constitutive ELM basic equations: A1 The starting point is the *constitutive electromechanical (ELM) basic equations (A1)* relating energy and power variables of the physical domains under consideration (both electrical and mechanical, see Fig. 5.2). It can be shown that there exists a *domain-independent* general approach for setting up these equations, so that all further considerations can largely proceed *generically*.

ELM energy functions: A2 Formal integration of the constitutive ELM basic equations (A1) gives the *ELM energy functions (A2)* in the form of energies and co-energies, which are then combined in the LAGRANGian.

---

1 From a methodological point of view, a "continuous" workflow within a single modeling paradigm may by all means reach the highest level of modeling aesthetics. However, such a practice is usually accompanied with compromises and contortions. From the pragmatic, systems-level point of view, and in the practical experience of the author, an artful playing upon the "method keyboard" gives access to decidedly more comprehensible solutions for multifaceted systems design tasks. Thus, the obliging reader may take the particular modeling approach presented here as a well-intended attempt at a melodious "method harmony", without precluding the possibility that other modeling approaches may be equally "melodious".

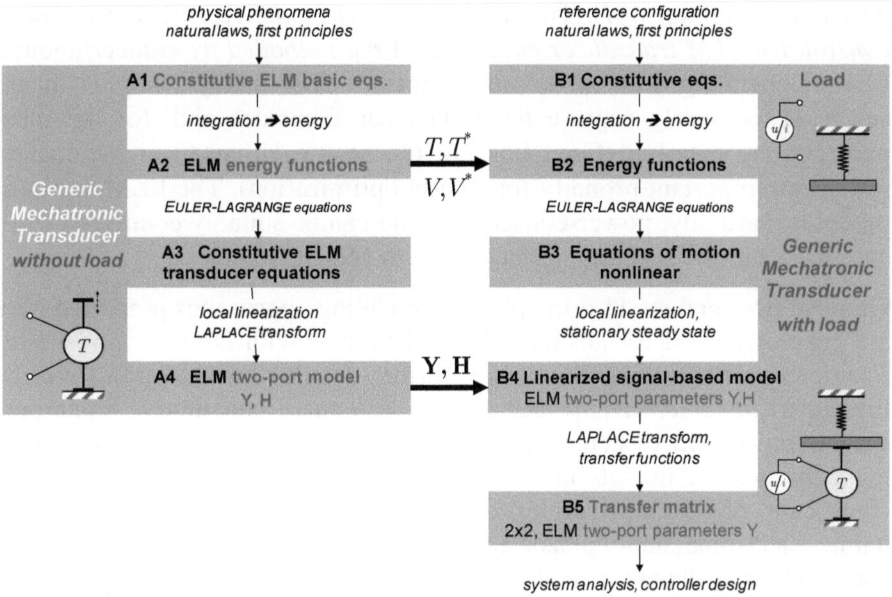

Fig. 5.4. Model hierarchy of a generic *mechatronic transducer* for systems design (ELM = electromechanical)

These functions can be easily re-used in scalar form for a loaded transducer (forming an interface point to Model Branch B).

Constitutive ELM transducer equations: A3 For the unloaded transducer, the ELM energy functions provide the basis for formal calculation of *transducer forces* via the *EULER-LAGRANGE equations*. As a side effect, the constitutive ELM basic equations (A1) defined at the starting point are also reconstructed[2]. The resulting models are the *nonlinear constitutive ELM transducer equations (A3)* for the *unloaded transducer*[3]. These describe the complete, bidirectional, electromechanical coupling in the transducer. They are power-conserving and possess an as-yet indeterminate causal structure.

ELM two-port model: A4 For concise representation of the small-signal response, local linearizations of equations (A3) are transformed into a *two-*

---

[2] This property can also be profitably employed as a check on calculations.

[3] If there is no additional dependence of the energy and power variables on variables of the *external* load—e.g. nonlinear electrical behaviors (hysteresis), or a non-collocated multibody system as the load—equations (A3) could also be directly transferred to (B3).

port representation (A4). This *ELM two-port model* represents the *linear constitutive ELM transducer equations* of the *unloaded transducer* about a general operating point in a general form. The usually complex matrices $\mathbf{Y}, \mathbf{H}$ concisely capture the transducer dynamics and, for the most part, contain clearly-defined links to the physical transducer parameters (this is an important property for design optimization). The ELM two-port model is, naturally, power-conserving, and can be suitably connected to an external load (the second interface point to Model Branch B).

Generic properties In principle, the modeling approaches presented here are generally valid for abstract physical models with *lumped parameters*. For pragmatic and didactic reasons, further detailed considerations will be limited to *linear electrical and electromechanical* constitutive equations, i.e. saturation and hysteresis are excluded. Given these limitations, however, the generic models in Model Branch A—when applied to the most significant physical principles of transduction (electrostatic, electrodynamic, electromagnetic, piezoelectric, or hydraulic)—deliver quite useable and predictive dynamic models. These are obtained simply by inserting concrete instances of constitutive ELM basic equations (A1) into the general model parameters in the entire Model Branch A.

Model Branch B: the loaded generic transducer

The actual transducer functionality of interest for a *mechatronic product*—namely the interaction between moving mechanical structures and electrical interfaces to information processing functions—is discussed using a *loaded* (i.e. connected to electrical and mechanical loads) *transducer model* (Model Branch B, Fig. 5.4, right). The procedure mirrors the methodology of Model Branch A.

Reference configuration For pragmatic, didactic reasons, the chosen reference configuration is limited to the external loading indicated in Fig. 5.2 with *lossless auxiliary energy sources* (voltage or current) and a non-dissipative *elastic connection* to a *rigid-body load.*

Constitutive equations: B1 The *physical constitutive relations (B1)* must also be defined for the external loading. For the reference loads (lossless electrical source, elastically-suspended rigid body), this is trivial. For external components deviating from these assumptions, corresponding relations must be defined.

Energy functions: B2 Formal integration of the constitutive equations (B1) again gives *energy functions (B2)* in the form of energy and co-energy. These are supplemented with the ELM energy functions (A2) of the unloaded transducer.

Equations of motion: B3 Using the total energy function (B1 + A2) in the *EULER-LAGRANGE equations*, gives the *nonlinear equations of motion (B3)* of the loaded transducer. These are of fundamental importance, as it is only at this point that the complete transducer dynamics as indicated in Fig. 5.3 are defined. In particular, these equations can be used to analyze *steady-state behavior* and often to ascertain any *unstable* operating regimes. If present, stable *equilibrium positions* defined by (B3) represent sensible operating points for the transducer in small-signal applications.

Linear signal-based model with ELM two-port parameters: B4 With the equilibrium positions obtained from the nonlinear equations of motion (B3), local linearization can be used to create a *linear signal-based model (B4)*. Simple manipulation of this equation allows the linear constitutive ELM equations of the unloaded transducer to be separated out, so that the unloaded ELM two-port parameters (A4) can be used. When employing the linear signal-based models (B4) for time-domain analysis, minor accommodations must be made for any two-port parameters that are complex.

Transfer matrix: B5 Applying the LAPLACE transform to the linear signal-based model (B4) gives the *transfer matrix (B5)* for the *(2×2) MIMO* causal structure shown in Fig. 5.3. Separating out the unloaded ELM two-port parameters (A4) reveals a clear correspondence between *transfer properties* (gains, poles, and zeros) and *electromechanical*, physically-oriented parameters (e.g. stiffnesses, active capacitances/inductances, coupling, or electrical and mechanical configuration parameters). The transfer matrix (B5) thus serves as the primary working basis for subsequent steps of dynamic analysis, controller design, and system optimization.

Generic properties All steps in Model Branch B deliver *generic models* using the generalized transducer parameters from Model Branch A and the equally general model parameters of the reference configuration. Thus, the general results from Model Branches A and B can be used immediately in specific applications by substituting *concrete physical* parameters (the constitutive equations (A1), (B1)).

The model hierarchy presented here and the modeling methods employed also present a general framework for specific *model extensions*. Extensions to the transducer and its loading configuration can be easily incorporated at the level of the constitutive equations (A1) and (B1). The formal computational steps can then be adopted and applied in the same manner.

5.3 The Unloaded Generic Transducer

5.3.1 Energy-based model

Model hierarchy The modeling approach taken here is based on the transducer arrangement shown in Fig. 5.5 having no external sources and sinks, i.e. it is electrically and mechanically unloaded. The modeling steps to make Model Branch A from Fig. 5.4 concrete are shown in Fig. 5.6, and are discussed in detail in this section.

Conservative system The configuration for the electrically and mechanically unloaded generic transducer shown in Fig. 5.5 contains no dissipative elements, so that this is a *conservative* system.

Mechanical subsystem Let the mechanical subsystem consist of a massless stator, and contain no internal kinetic energy storage elements (inertial masses). If needed, let a transducer internal potential energy storage element (a spring, e.g. for solid-state transducers) be present.

Mechanical coordinates Energy-based modeling requires suitable generalized coordinates satisfying the properties defined in Sec. 2.3.1. Following the specifications laid out there, the *armature position* x suggests itself as the mechanical generalized coordinate.

Electrical subsystem Let the electrical subsystem for electromechanical force generation be characterized by the presence of only *one electrical energy storage element*[4], i.e. there is storage in an electric field (capacitive dynamics) *or* a magnetic field (inductive dynamics).

---

4 This does not imply a limitation within the set of transducer implementations considered here. Assuming *quasi-static electromagnetic* fields (with temporal changes sufficiently small relative to spatial propagation effects in the observed time period), the transducing effects of electromagnetic fields can be considered as spatially concentrated (a model with lumped parameters): (a) *electric* fields in which—compared to elsewhere—particularly high electrical energy densities are achieved (e.g. between two electrodes separated by a dielectric), or (b) *magnetic* fields in which—compared to elsewhere—particularly high magnetic energy densities are achieved (e.g. in a coil with many windings). This distinction enables a very clear representation of fundamental transducer properties. Additional parasitic energy storage and dissipative elements, which are always present in real transducers, are considered separately in subsequent sections.

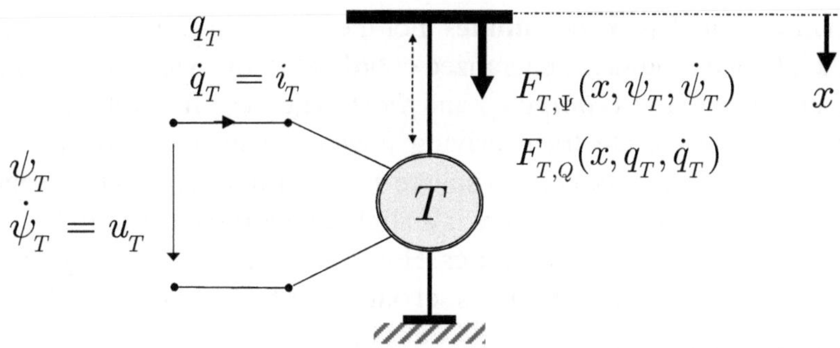

Fig. 5.5. Unloaded generic mechatronic transducer: principle of operation

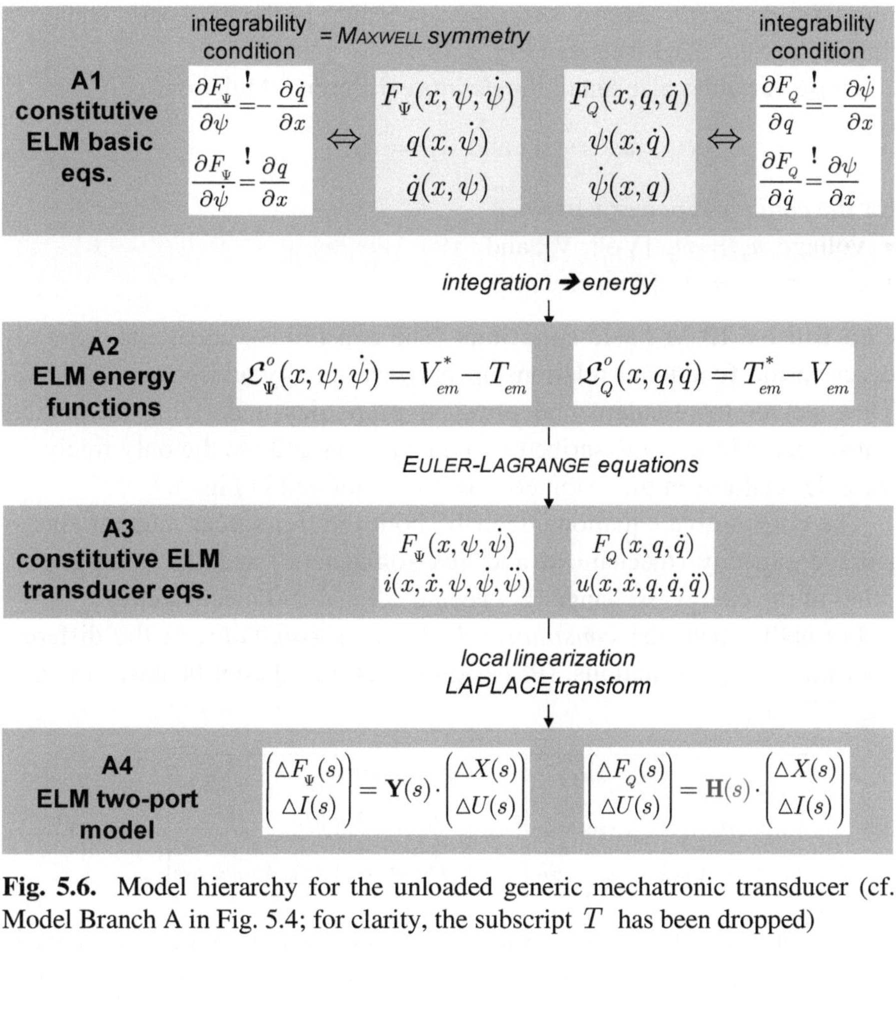

Fig. 5.6. Model hierarchy for the unloaded generic mechatronic transducer (cf. Model Branch A in Fig. 5.4; for clarity, the subscript T has been dropped)

Generalized electrical coordinates For the *electrical* subsystem, there are in principle two candidate generalized coordinates: the two conjugate generalized energy variables *charge* q_T and *flux linkage* ψ_T. In addition, there are two options for the electrical driving power: either a *controlled voltage* source u_S or a *controlled current* source i_S (see Table 5.1). As these options can be combined arbitrarily, there is a slight embarrassment of riches in selecting between the four possibilities. How to most advantageously combine these attributes in terms of the goals set out in this book is described below.

Electrical coordinates The two electrical coordinates

- *charge* q_T [Coulomb, C = As], and
- *flux linkage* ψ_T [Weber, Wb = Vs]

represent *conjugate energy variables* according to the axiomatic construct defined in Sec. 2.3.1 (see also Fig. 5.7b).

These variables also entail the *defining* differential relations

$$u_T := \dot{\psi}_T \quad \text{and} \quad i_T := \dot{q}_T \tag{5.1}$$

for the *conjugate power variables*

- voltage $u_T = \dot{\psi}_T$ [Volt, V], and
- current $i_T = \dot{q}_T$ [Ampere, A].

Constitutive ELM basic equations The constitutive equations generally describe the functional relationship between energy and power variables. They depict the fundamental physical properties underlying a concrete transducer. These are described by *natural laws* and are the only freely selectable relations in the axiomatic construct defined in Fig. 5.7.

The constitutive equations describe both the transducer internal energy storage capacity (mechanical and electromagnetic) and the internal coupling of the energy exchange (between electrical and mechanical).

Formally, then, the *constitutive ELM basic equations*, in the different coordinate representations (PSI-coordinates or Q-coordinates, see also Fig. 5.7), are

$$\boxed{\begin{aligned} F_{T,\Psi} &= F_{T,\Psi}(x, \psi_T, \dot{\psi}_T) \\ q_T &= q_T(x, \dot{\psi}_T) \ \text{capacitive} \\ \dot{q}_T &= \dot{q}_T(x, \psi_T) \ \text{inductive} \end{aligned}} \quad \text{or} \quad \boxed{\begin{aligned} F_{T,Q} &= F_{T,Q}(x, q_T, \dot{q}_T) \\ \psi_T &= \psi_T(x, \dot{q}_T) \ \text{inductive} \\ \dot{\psi}_T &= \dot{\psi}_T(x, q_T) \ \text{capacitive} \end{aligned}}\,. \tag{5.2}$$

The sets of equations with the two coordinate representations given in Eqs. (5.2) are equivalent, and, depending on the preferred representation for a particular physical phenomenon, can be used in either one form or the other.

Note also that not all equations must be given for every case. Due to the assumption of a *conservative system*, for example, *one* electrical constitutive basic equation suffices for a complete functional description if there is no internal mechanical energy storage element present. In this case, the corresponding constitutive relation for transducer force can be systematically reconstructed as shown below.

Thus, in this book, the constitutive equations (5.2) are termed the *basic* equations, as they describe the underlying physical phenomena. In contrast, the fully reconstructed constitutive equations are termed constitutive *transducer* equations, as they describe the complete transducer dynamics.

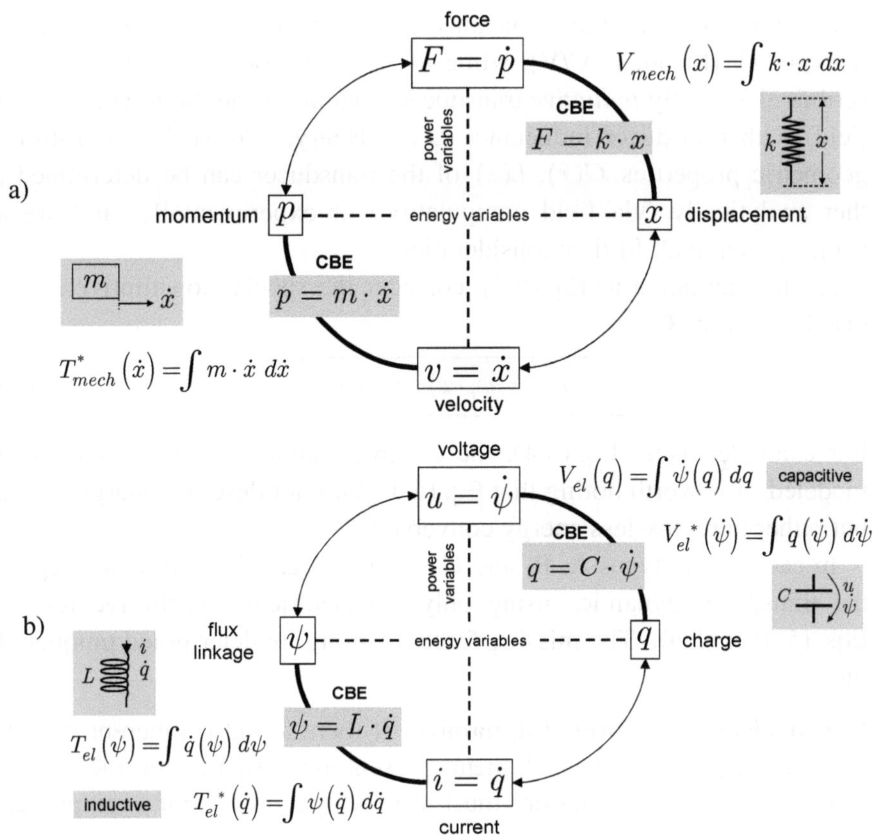

Fig. 5.7. Generalized coordinates: a) *mechanical* energy and power variables with linear mechanical constitutive equations, b) *electrical* energy and power variables with linear electrical constitutive equations (CBE = constitutive basic equations)

Table 5.2 lists examples of constitutive *ELM basic equations* for several cases of practical importance. As can be seen, for many cases, natural laws or empirically derived relations describe only the electrical constitutive dynamics.

Linear electrical dynamics As a concretization of Eq. (5.2), let *linear electrical dynamics* be assumed for the electrical subsystem, i.e. the *electrical constitutive* relations are

$$\boxed{q_T(x, \dot{\psi}_T) := C(x) \cdot \dot{\psi}_T}, \text{ or } \boxed{\psi_T(x, \dot{q}_T) := L(x) \cdot \dot{q}_T}, \tag{5.3}$$

where $C(x)$ and $L(x)$ are generally *nonlinear* functions of the armature position x (Table 5.2, types A and B, see also Fig. 5.7b).

The equation on the left of (5.3) describes capacitive electrical transducer dynamics (coupling via an electric field) with transducer capacitance $C(x)$ [Farad, F=C/V], while the equation on the right of (5.3) describes electrically *inductive* transducer dynamics (coupling via a magnetic field) with transducer inductance $L(x)$ [Henry, Wb/A]. The constitutive geometric properties $C(x)$, $L(x)$ of the transducer can be determined either analytically (via field computation) or experimentally, and are assumed given in all further considerations.

As an alternative to Eq. (5.3), consider the special constitutive relation (Table 5.2, type C)

$$\boxed{\psi_T(x, i_T) = \psi_T(x) := g_\psi(x)}. \tag{5.4}$$

For example, using Eq. (5.4), the electrodynamic LORENTZ *force* can be modeled. It is worth noting that Eq. (5.4) does not describe energy storage, but rather pure, lossless energy conversion.

In each of the types A, B, and C of Table 5.2, it is sufficient to specify the transducer dynamics using only the electrical constitutive relations Eqs. (5.3) or (5.4). The missing force relations are determined uniquely by them.

Linear electromechanical dynamics If there are also inherent potential mechanical energy storage elements present in the transducer, then the mechanical constitutive relation[5] must also be specified. For example, con-

[5] At least the portion describing mechanical energy storage, as this is not reflected in the electrical coupling relation.

sider the important practical case of both *mechanically* and *electrically linear* relations, e.g. for capacitive electrical dynamics in electrical PSI-coordinates:

$$\begin{aligned} F_{T,\Psi}(x, \dot{\psi}_T) &= a \cdot x + b \cdot \dot{\psi}_T \,, \\ q_T(x, \dot{\psi}_T) &= b \cdot x + c \cdot \dot{\psi}_T \,. \end{aligned} \tag{5.5}$$

Solid-state transducers (e.g. *piezoelectric transducers*) in their linear operating regime are typically described using Eq. (5.5). The control-independent coefficients a, c describe the mechanical and electrical storage capacity, while the coefficient b describes the electromechanical energy exchange. Additionally, the symmetry of the two equations relative to the coupling b is required for conservative systems, as will be shown below.

The Q-coordinate representation equivalent to Eq. (5.5) can be found in the right-hand column of Table 5.2, type D.

Polynomial electromechanical dynamics One practical extension of linear ELM dynamics is a *polynomial* description of the constitutive equation, indicated in Table 5.2 by the type E. Such a model can be imagined to have been created, for example, via experimentation with the large-signal dynamics of a solid-state transducer, i.e. a nonlinear model extension of Eq. (5.5).

For example, with second-order polynomials in the variables $x, \dot{\psi}$, the model equations

$$\begin{aligned} F_\Psi &= a_0 + a_1 x + \mathbf{c}_1 \dot{\psi} + \mathbf{c}_2 x \dot{\psi} + a_2 \frac{x^2}{2} + \mathbf{c}_3 \frac{\dot{\psi}^2}{2}, \\ q &= b_0 + \mathbf{c}_1 x + b_1 \dot{\psi} + \mathbf{c}_3 x \dot{\psi} + \mathbf{c}_2 \frac{x^2}{2} + b_2 \frac{\dot{\psi}^2}{2} \end{aligned} \tag{5.6}$$

offer a set of *consistent constitutive* equations for $F_{T,\Psi}(x, \dot{\psi})$ and $q_T(x, \dot{\psi})$ exhibiting *capacitive* transducer dynamics. Here too, the symmetry of the coupling factors $\mathbf{c}_1, \mathbf{c}_2, \mathbf{c}_3$ in Eq. (5.6) is required to maintain the assumption of a conservative system, and must be accounted for when approximating from series of measurements[6].

[6] Given sufficiently accurate parameters for the electrical constitutive equation, it would thus be sufficient to determine the mechanical parameters a_0, a_1, a_2 via measurement.

Table 5.2. Examples of constitutive ELM basic equations

Type	PSI-coordinates (x, ψ)	Q-coordinates (x, q)
A. Electrical, linear, capacitive	$q = C(x) \cdot \dot{\psi}$	$\dot{\psi} = \frac{1}{C(x)} \cdot q$
B. Electrical, linear, inductive	$\dot{q} = \frac{1}{L(x)} \cdot \psi$	$\psi = L(x) \cdot \dot{q}$
C. Electrical, linear, inductive, no storage	---	$\psi = g(x)$
D. Electro-mechanical, linear, capacitive	$F_\Psi = a \cdot x + b \cdot \dot{\psi}$ $q = b \cdot x + c \cdot \dot{\psi}$	$F_Q = \left(a - \frac{b^2}{c}\right) \cdot x + \frac{b}{c} q$ $\dot{\psi} = -\frac{b}{c} x + \frac{1}{c} q$
E. Electro-mechanical, polynomial[7]	- capacitive - $F_\Psi = \text{polynomial}_F(x, \dot{\psi}; N)$ $q = \text{polynomial}_q(x, \dot{\psi}; N)$	- inductive - $F_Q = \text{polynomial}_F(x, \dot{q}; N)$ $\psi = \text{polynomial}_\psi(x, \dot{q}; N)$

Flux linkage vs. electrical voltage At first glance, the use of flux linkage as a generalized coordinate may seem somewhat alien. Normally, this quantity is only used in connection with magnetic fields. What does it signify in connection with a capacitive storage element? Why is electrical voltage not used as the generalized coordinate?

It can be seen from Fig. 5.7b that, using the defining relation for a capacitive energy storage element C, there is a direct connection from voltage to the flux linkage ψ via its time differential $\dot{\psi}$. Thus, capacitive electrical dynamics can be equally well represented using the time derivative of flux linkage.

If, in contrast, the electrical voltage were to be used as the generalized coordinate, then, to maintain the physical relationships shown in Fig. 5.7b,

---

[7] $\text{polynomial}(x, y; N)$ is the polynomial of order N in the variables x, y.

the constitutive equation for inductive dynamics would have to be written in an unwieldy form using the integral of the voltage.

Energy functions The EULER-LAGRANGE modeling approach requires determining the scalar energy and co-energy functions. While in mechanical systems (given the customary choice of position as generalized coordinate), the tuple consisting of kinetic *co-energy* T^* and potential *energy* V is sufficient, in electrical systems, for representational reasons (as further discussed below), *all* energy functions T, T^*, V, V^* are required.

Electromechanical energy As indicated in Fig. 5.7, computation of the energy functions occurs via integration of the constitutive equations (5.2). Due to the dependence of the path integrals on two independent variables, great care must be taken when performing this integration.

Consider, for example, the *linear electrical constitutive* relation for *capacitive* dynamics with the substitution $u_T = \dot{\psi}_T$

$$q_T(x, u_T) = C(x) \cdot u_T \,. \tag{5.7}$$

Using the defining relations shown in Fig. 5.7, the potential energy is determined from the path integral of the constitutive equations, the integral being taken over the generalized coordinates, here x and q_T.

Formally, for this integral, the mechanical constitutive relation $F_T = F_T(x, u_T)$ is also required. Practically speaking, however, it is sufficient to know the general property (proven below)

$$F_T\left(x, u_T = 0\right) = 0\,, \tag{5.8}$$

i.e. the transducer produces no force as electrical excitation vanishes.

Due to the requirement of a conservative system, the value of the path integral is *independent* of the chosen integration path. This general property can be taken advantage of to freely choose an integration path convenient for the calculation. Fig. 5.8b shows one possible convenient path. For the potential energy function $V^*_{elm}(x, u_T)$ (here the potential co-energy), the result is thus (see Fig. 5.8a, arrow from left to right)

$$V^*_{elm}(x, u_T) = V^*_{elm,1} + V^*_{elm,2} = \underbrace{\int_0^x F_T(\tilde{x}, u_T{=}0)\, d\tilde{x}}_{0} + \underbrace{\int_0^{u_W} q_T(x, \tilde{u})\, d\tilde{u}}_{\frac{1}{2} C(x) \cdot u_T^{\,2}} \,. \tag{5.9}$$

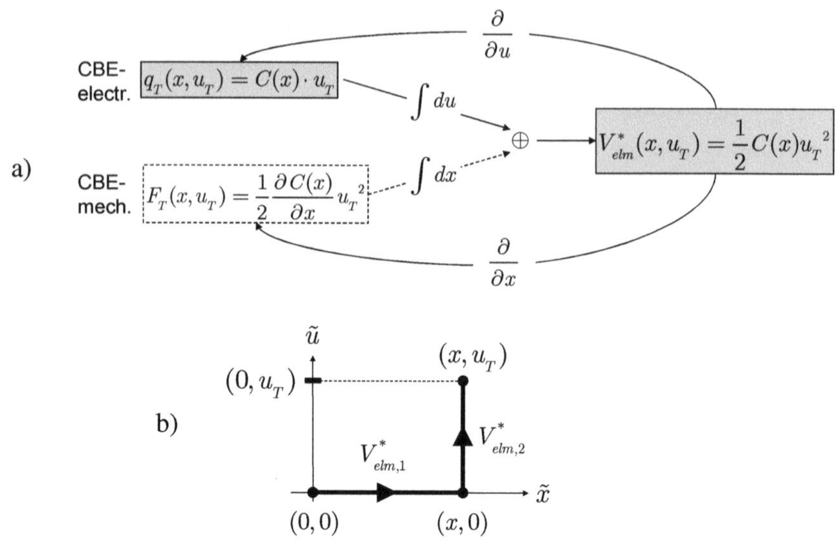

Fig. 5.8. Integration of the constitutive ELM basic equations for the example of a linear electrical capacitive transducer: a) relationships between constitutive basic equations (CBEs) and the potential co-energy, b) example of a convenient integration path for the constitutive equation (5.7)

The first path integral (the mechanical work) follows $u = 0$, and is thus—due to the property in Eq. (5.8)—equal to zero. For the second path integral (the electrical work), the coordinate x is kept constant—so that $C(x)$ is also constant and can be decoupled—and the integral is taken over the electrical voltage u.

For $C(x) = C_0 = const.$, the well-known relation for the stored energy of a capacitor is obtained. Due to the particular, linear relation in Eq. (5.7), a formally equivalent result is also obtained in the coupled case: the change in energy is proportional to the change in the path-dependent capacitance function $C(x)$. However, this is no longer the case when the capacitance function also depends on electrical quantities, i.e. when $C(x,u)$ holds. In this case, the path integral can no longer be decoupled in the manner shown.

Integrability condition The path independence of the energy integral (5.9) is fundamental and bears a tight relation to special properties of the constitutive equations.

The following representation as a *total differential* is equivalent to the energy integral (5.9):

$$dV^*_{elm}(x,u_T) = F_T(x,u_T)\cdot dx + q_T(x,u_T)\cdot du_T\,, \tag{5.10}$$

$$\boxed{F_T(x,u_T) = \frac{\partial V_{elm}^*(x,u_T)}{\partial x}, \quad q_T(x,u_T) = \frac{\partial V_{elm}^*(x,u_T)}{\partial u_T}}. \tag{5.11}$$

In the case of a *conservative potential field*—a precondition for $V_{elm}^*(x,u_T)$—it must hold that (Bronshtein et al. 2005)

$$\frac{\partial^2 V_{elm}^*(x,u_T)}{\partial x\,\partial u_T} \stackrel{!}{=} \frac{\partial^2 V_{elm}^*(x,u_T)}{\partial u_T\,\partial x}. \tag{5.12}$$

This relation, known as the *integrability condition* of the total differential, can also be written using the *constitutive equations* (5.11) as

$$\boxed{\frac{\partial F_T(x,u_T)}{\partial u_T} \stackrel{!}{=} \frac{\partial q_T(x,u_T)}{\partial x}}. \tag{5.13}$$

General integrability conditions The *integrability condition* (5.13) on the *constitutive transducer equations* represents their energy-conserving electromechanical coupling[8] (see also (Karnopp et al. 2006)).

The procedure presented above can also be directly transferred to the Q-coordinate representation. In summary, for the two possible coordinate representations, the following *general integrability conditions* are obtained for the constitutive equations with capacitive and inductive dynamics, respectively:

- *PSI-coordinates*: $\begin{matrix} F_\Psi(x,\psi,\dot\psi) \\ \text{cap: } q(x,\dot\psi) \\ \text{ind: } \dot q(x,\psi) \end{matrix} \Leftrightarrow \boxed{\begin{matrix} \text{cap: } \dfrac{\partial F_\Psi}{\partial \dot\psi} \stackrel{!}{=} \dfrac{\partial q}{\partial x} \\ \text{ind: } \dfrac{\partial F_\Psi}{\partial \psi} \stackrel{!}{=} -\dfrac{\partial \dot q}{\partial x} \end{matrix}}$, (5.14)

---

[8] This is sometimes also referred to as *MAXWELL symmetry* or the *MAXWELL reciprocity condition*. An equally common equivalent statement to Eq. (5.13) is that the *JACOBIan* of the constitutive equations must be *symmetric*. For this latter formulation, however, note that when using q and ψ as coordinates (i.e. not $\dot q, \dot\psi$), a *skew-symmetric* JACOBIan results. This is due to the relation (established below) $F_T = -\partial V_{elm}(q) / \partial x$ or $F_T = -\partial T_{elm}(\psi) / \partial x$ (negative sign!) using the energy functions $V(q), T(\psi)$ instead of the co-energy functions $V^*(\dot\psi), T^*(\dot q)$.

- *Q-coordinates*:

$$\begin{array}{l} F_Q(x,q,\dot{q}) \\ \text{cap: } \dot{\psi}(x,q) \\ \text{ind: } \psi(x,\dot{q}) \end{array} \Leftrightarrow \boxed{\begin{array}{l} \text{cap: } \dfrac{\partial F_Q}{\partial q} \overset{!}{=} -\dfrac{\partial \dot{\psi}}{\partial x} \\ \text{ind: } \dfrac{\partial F_Q}{\partial \dot{q}} \overset{!}{=} \dfrac{\partial \psi}{\partial x} \end{array}}. \tag{5.15}$$

Reciprocity: reciprocal transducers Using conditions (5.14) and (5.15), the previously-mentioned special symmetry properties of the models in Eq. (5.5) and Eq. (5.6) can now also be easily explained and equivalent symmetries predicted for the transducer types A through C. Concretely, this means that the electromechanical coupling is symmetric. Similar symmetries are known in two-port theory. In that theory, there is *reciprocity* of two ports ν and μ if, given an excitation at the ν th port, the same response is measured at the μ th port as would be measured at the ν th port given excitation at the μ th port (Reinschke and Schwarz 1976). In this sense, the properties in Eq. (5.14) and Eq. (5.15) are also termed *reciprocity conditions*, and a transducer for which they hold is called a *reciprocal transducer*.

Reconstruction of the constitutive equations Eq. (5.11) describes a fundamental property of conservative systems. Knowing the energy function $V^*_{elm}(x,u_T)$, the complete constitutive equations can be reconstructed using Eq. (5.11). In the case being considered here, Eq. (5.7) only defined the constitutive electrical dynamics. Due to the special property in Eq. (5.8) and the path independence of the path integral, it was possible to calculate the energy function $V^*_{elm}(x,u_T)$ relying exclusively on the constitutive electrical equation. Eq. (5.11), however, allows the corresponding *constitutive mechanical* dynamics $F_T(x,u_T)$ of the electromechanical coupling to be determined, so to speak for free (see Fig. 5.8a CBE-2, lower arrow from right to left).

Thus, in *the reference case considered here*, the constitutive electrical equation (5.7) is, on its own, sufficient to describe the complete electromechanical dynamics of the transducer. This is certainly not always the case. It can be easily seen that this would *no longer* be possible if the condition in Eq. (5.8) were not satisfied, or if an internal mechanical energy storage element (spring) were present in the transducer. In these cases, the mechanical constitutive properties must also be explicitly defined.

ELM energy functions Due to the conditions specified above, the result in Eq. (5.9) holds analogously for all energy terms resulting from the constitutive ELM equations (5.2), so that it follows in general that:

- the *ELM energy functions* $T^*_{elm}(x,\dot{q}_T)$, $V_{elm}(x,q_T)$ are expressed in *Q-(charge-) coordinates* with generalized coordinates x, q_T, while
- the *ELM energy functions* $T_{elm}(x,\psi_T)$, $V^*_{elm}(x,\dot{\psi}_T)$ are in *PSI- (flux linkage-) coordinates* with generalized coordinates x, ψ_T.

For the special case of *linear electrical* constitutive ELM basic equations, the following relations are obtained:

$$T^*_{elm}(x,\dot{q}_T) = \frac{1}{2}L(x)\cdot\dot{q}_T^2, \quad V_{elm}(x,q_T) = \frac{1}{2}\frac{1}{C(x)}\cdot q_T^2, \tag{5.16}$$

$$T_{elm}(x,\psi_T) = \frac{1}{2}\frac{1}{L(x)}\cdot\psi_T^2, \quad V^*_{elm}(x,\dot{\psi}_T) = \frac{1}{2}C(x)\cdot\dot{\psi}_T^2. \tag{5.17}$$

Note the combinations of energy and co-energy terms in Eqs. (5.16) and (5.17), resulting from the choice of coordinates. In addition, the LEGENDRE transformation between energy and co-energy terms is easy to verify (see Sec. 2.3.1).

LAGRANGian The energy terms (5.16), (5.17) now completely describe the unloaded generic mechatronic transducer of Fig. 5.5. For further calculations, they will be combined as previously described in the LAGRANGian $\mathcal{L}$ (Goldstein et al. 2001), (Wellstead 1979). In the notation below, the initial assumption that only one either capacitive or inductive energy storage element is present is taken into account ("$\vee$" should be interpreted as *OR*).

- *LAGRANGian* in *Q-coordinates* (generalized coordinates x, q_T):

$$\mathcal{L}^o_Q(x,q_T,\dot{q}_T) = \left[T^*_{elm}(x,\dot{q}_T)\right]_{ind} \vee \left[-V_{elm}(x,q_T)\right]_{cap}. \tag{5.18}$$

- *LAGRANGian* in *PSI-coordinates* (generalized coordinates x, ψ_T):

$$\mathcal{L}^o_\Psi(x,\psi_T,\dot{\psi}_T) = \left[V^*_{elm}(x,\dot{\psi}_T)\right]_{cap} \vee \left[-T_{elm}(x,\psi_T)\right]_{ind}. \tag{5.19}$$

The two LAGRANGians $\mathcal{L}^o_Q$ and $\mathcal{L}^o_\Psi$ have exactly the same value, they differ only in their coordinate representation.

Remark on co-energy The combination of energy terms in Eq. (5.19) is particularly worth noting. Due to the chosen definitions of energy variables, the *LAGRANGian* must always be set up as the difference *"co-energy minus energy"*.

Given the choice of generalized momentum coordinates (here the flux linkage ψ_T) as generalized coordinates, potential energy takes the form of a co-energy V^*_{elm}.

In contrast, making the choice customary in mechanical systems of generalized displacement as a generalized coordinate (in the electrical case here, the charge q_T) results instead in the kinetic energy taking the form of a co-energy T^*_{elm} (see Eq. (5.18)).

It is this latter formulation that results in the commonly-used (though in the general sense, incorrect) formulation of the LAGRANGian as the "difference between kinetic and potential energy" (this holds only for one particular choice of generalized coordinates).

This subtle, but immensely important, distinction vanishes if the constitutive equations are used from the start to define *electrical* and *magnetic* energy and co-energy functions corresponding to the physical setup, and the co-energy is then used consistently for subsequent calculations, e.g. (Preumont 2006), (Senturia 2001).

5.3.2 Constitutive ELM transducer equations

EULER-LAGRANGE equations As previously described, the equations of motion of the unloaded generic transducer are obtained using the *EULER-LAGRANGE equations*, which, in their general form for the *Q-coordinates* representation, appear as follows:

$$
\begin{aligned}
x: &\quad \cancel{\frac{d}{dt}\left(\frac{\partial \mathcal{L}^o_Q(x,q_T,\dot{q}_T)}{\partial \dot{x}}\right)} - \frac{\partial \mathcal{L}^o_Q(x,q_T,\dot{q}_T)}{\partial x} = F_{gen}, \\
q_T: &\quad \frac{d}{dt}\left(\frac{\partial \mathcal{L}^o_Q(x,q_T,\dot{q}_T)}{\partial \dot{q}_T}\right) - \frac{\partial \mathcal{L}^o_Q(x,q_T,\dot{q}_T)}{\partial q_T} = u_{gen}.
\end{aligned}
\tag{5.20}
$$

The generalized excitation variables F_{gen}, u_{gen} depend on the external loading of the transducer. As $\mathcal{L}^o_Q(x,q_T,\dot{q}_T)$ does not depend on $\dot{x}$, the left term in the x-equation of Eq. (5.20) is zero.

Constitutive ELM transducer equations The left-hand sides of Eq. (5.20) now represent precisely the mechanical force F_T and electrical voltage u_T generated by the transducer as a function of the generalized coordinates x, q_T. Substituting Eq. (5.18) and taking the derivative gives the *constitutive ELM transducer equations* in *Q-coordinates* in the general form[9]

$$\boxed{\begin{aligned} F_{T,Q}(x, q_T, \dot{q}_T) &= \left[\frac{\partial T^*_{elm}(x, \dot{q}_T)}{\partial x}\right]_{ind} \vee \left[-\frac{\partial V_{elm}(x, q_T)}{\partial x}\right]_{cap}, \\ u_T(x, \dot{x}, q_T, \dot{q}_T, \ddot{q}_T) &= \left[\frac{d}{dt}\left(\frac{\partial T^*_{elm}(x, \dot{q}_T)}{\partial \dot{q}_T}\right)\right]_{ind} \vee \left[\frac{\partial V_{elm}(x, q_T)}{\partial q_T}\right]_{cap}. \end{aligned}} \tag{5.21}$$

Eqs. (5.21) now completely describe the *bidirectional electromechanical* coupling in the transducer; they conserve power and have an as-yet indeterminate causal structure. Thus they can be arbitrarily interpreted as a sensor or actuator with either a voltage or current source.

A transducer description equivalent to Eqs. (5.21) in *PSI-coordinates* can be derived with the same basic procedure. In this case, however, in the EULER-LAGRANGE equation, the electrical coordinate is defined to be the induced *transducer current* i_T as a function of the generalized coordinates x, ψ_T.

In Table 5.3, the constitutive ELM transducer equations for both representations (Q-coordinates and PSI-coordinates) are once again contrasted. Table 5.3 also presents the derived equations for *linear electrical* transducer dynamics (cf. Eq. (5.3)).

---

[9] Note that in Eq. (5.20), the first equation describes the balance of forces $\Sigma F_i = F_T + F_{gen} = 0$ or $-F_T = F_{gen}$, i.e. the left-hand side of Eq. (5.20) is equal to $-F_T$, which accounts for the sign placement in Eq. (5.21). It can be easily seen that as a result—in accordance with the coordinate definition in Fig. 5.2—the positive transducer force also works in the positive x-direction. The second equation describes the transducer voltage $\dot{\psi}_T$, in other words, the time-differential of the underlying constitutive equation. In the case of coupling via the magnetic field (inductive dynamics), this implies $d/dt\left[\psi_T(x, \dot{q}_T)\right]$.

Table 5.3. Constitutive ELM transducer equations in various coordinate representations (ELM = electromechanical, "$\vee$" should be interpreted as *OR*), see also Fig. 5.5

	PSI-coordinates
General	$F_{T,\Psi}(x,\psi_T,\dot{\psi}_T)=\left[\dfrac{\partial V^*_{elm}(x,\dot{\psi}_T)}{\partial x}\right]_{cap}\vee\left[-\dfrac{\partial T_{elm}(x,\psi_T)}{\partial x}\right]_{ind}$ $i_T(x,\dot{x},\psi_T,\dot{\psi}_T,\ddot{\psi}_T)=\left[\dfrac{d}{dt}\left(\dfrac{\partial V^*_{elm}(x,\dot{\psi}_T)}{\partial\dot{\psi}_T}\right)\right]_{cap}\vee\left[\dfrac{\partial T_{elm}(x,\psi_T)}{\partial\psi_T}\right]_{ind}$
Linear, electrical dynamics see Eq. (5.3)	$F_{T,\Psi}(x,\psi_T,\dot{\psi}_T)=\left[\dfrac{1}{2}\dfrac{\partial C(x)}{\partial x}\dot{\psi}_T^2\right]_{cap}\vee\left[\dfrac{1}{2}\dfrac{1}{L(x)^2}\dfrac{\partial L(x)}{\partial x}\psi_T^2\right]_{ind}$ $i_T(x,\dot{x},\psi_T,\dot{\psi}_T,\ddot{\psi}_T)=\left[C(x)\ddot{\psi}_T+\dfrac{\partial C(x)}{\partial x}\dot{x}\dot{\psi}_T\right]_{cap}\vee\left[\dfrac{1}{L(x)}\psi_T\right]_{ind}$
	Q-coordinates
General	$F_{T,Q}(x,q_T,\dot{q}_T)=\left[\dfrac{\partial T^*_{elm}(x,\dot{q}_T)}{\partial x}\right]_{ind}\vee\left[-\dfrac{\partial V_{elm}(x,q_T)}{\partial x}\right]_{cap}$ $u_T(x,\dot{x},q_T,\dot{q}_T,\ddot{q}_T)=\left[\dfrac{d}{dt}\left(\dfrac{\partial T^*_{elm}(x,\dot{q}_T)}{\partial\dot{q}_T}\right)\right]_{ind}\vee\left[\dfrac{\partial V_{elm}(x,q_T)}{\partial q_T}\right]_{cap}$
Linear, electrical dynamics see Eq. (5.3)	$F_{T,Q}(x,q_T,\dot{q}_T)=\left[\dfrac{1}{2}\dfrac{\partial L(x)}{\partial x}\dot{q}_T^2\right]_{ind}\vee\left[\dfrac{1}{2}\dfrac{1}{C(x)^2}\dfrac{\partial C(x)}{\partial x}q_T^2\right]_{cap}$ $u_T(x,\dot{x},q_T,\dot{q}_T,\ddot{q}_T)=\left[L(x)\ddot{q}_T+\dfrac{\partial L(x)}{\partial x}\dot{x}\dot{q}_T\right]_{ind}\vee\left[\dfrac{1}{C(x)}q_T\right]_{cap}$

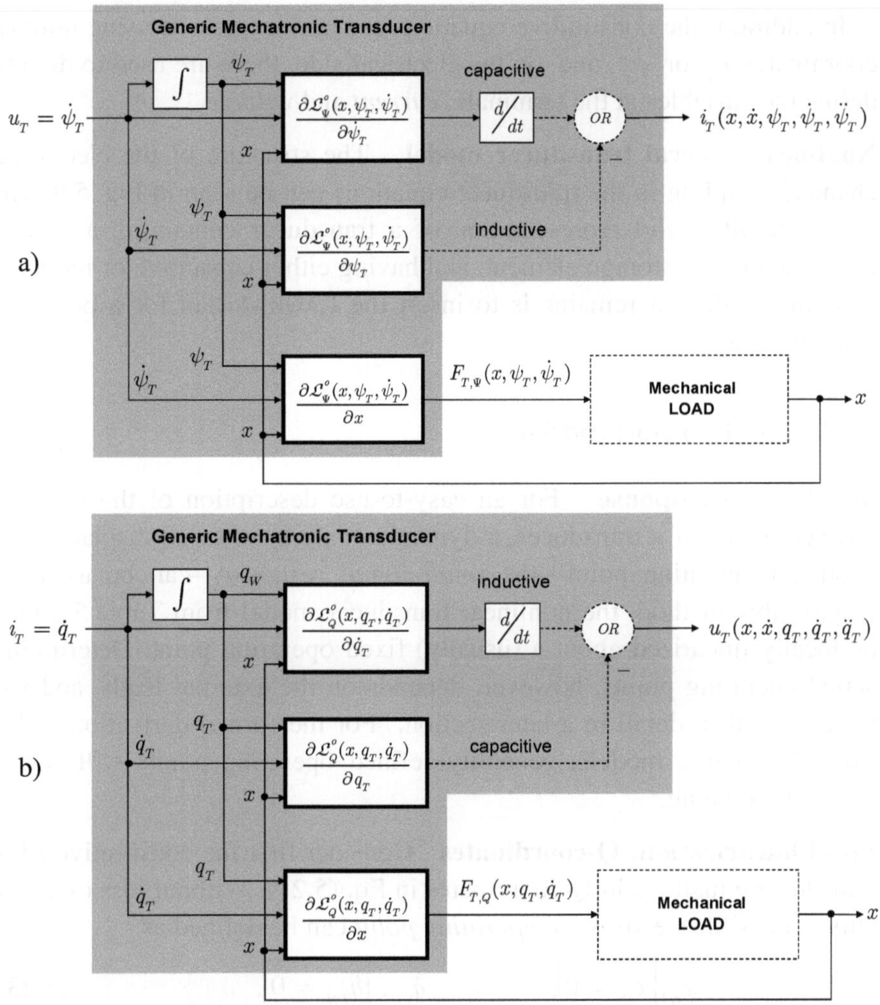

Fig. 5.9. General nonlinear model of the unloaded generic transducer: a) PSI-coordinates, b) Q-coordinates

Implications From Table 5.3, it is possible to recognize the following important general property of the unloaded generic transducer:

$$F_{T,Q}(x, 0, 0) = F_{T,\Psi}(x, 0, 0) = 0\,, \tag{5.22}$$

i.e. the transducer becomes force-free as the electrical excitation vanishes (this property was already exploited in Eq. (5.8) in the discussion of path independence of the energy integral).

In addition, the constitutive equations are now expressed using uniform coordinates q_T or ψ_T, and, on the electrical side, these are used to directly define the variables at the terminals: *current* and *voltage*.

Nonlinear general transducer model The structure of the electromechanical coupling in the transducer equations can be seen in Fig. 5.9. This is a generally valid representation of a transducer containing a lumped electrical energy storage element, and having either capacitive or inductive dynamics—all that remains is to insert the LAGRANGian for a particular concrete case.

5.3.3 ELM two-port model

Small-signal response For an easy-to-use description of the response characteristics of a transducer, a dynamic analysis of small displacements about an operating point—the *small-signal response*—can be used. To employ this method, the nonlinear transducer model from Table 5.3 must be locally linearized about a (usually) fixed operating point. Determining actual operating points, however, depends on the external loads, and will be discussed in detail in a later section. For the formal derivation of linearized dynamic models, generally-defined operating points will suffice for the time being.

Local linearization: Q-coordinates Consider first the constitutive ELM transducer equations in Q-coordinates in Eq. (5.21). Without loss of generality, a possible steady-state *operating point* can be defined as

$$x_R, \boxed{\dot{x}_R = 0}, \qquad q_{TR}, \; \dot{q}_{TR}, \boxed{\ddot{q}_{TR} = 0}. \tag{5.23}$$

This gives the constitutive ELM transducer equations *linearized* about the operating point in Eq. (5.23) in the form

$$\begin{aligned} \Delta F_{T,Q}(x, q_T, \dot{q}_T) &= K_{Fx}\Delta x + \left[K_{Fq}\Delta q_T\right]_{cap} \vee \left[K_{F\dot{q}}\Delta\dot{q}_T\right]_{ind}, \\ \Delta u_T(x, \dot{x}, q_T, \dot{q}_T, \ddot{q}_T) &= \left[K_{Ux}\Delta x + K_{Uq}\Delta q_T\right]_{cap} \vee \\ &\quad \left[K_{U\dot{x}}\Delta\dot{x} + K_{U\dot{q}}\Delta\dot{q}_T + K_{U\ddot{q}}\Delta\ddot{q}_T\right]_{ind}, \end{aligned} \tag{5.24}$$

with constant (generally operating-point-dependent) *transducer coefficients*

$$K_{F\lambda} := \frac{\partial}{\partial \lambda} F_{T,Q}(x, q_T, \dot{q}_T) \Big|_{x=x_R,\, q_T=q_{TR},\, \dot{q}_T=\dot{q}_{TR}},$$

$$K_{U\mu} := \frac{\partial}{\partial \mu} u_T(x, \dot{x}, q_T, \dot{q}_T, \ddot{q}_T) \Big|_{x=x_R,\, \dot{x}=0,\, q_T=q_{TR},\, \dot{q}_T=\dot{q}_{TR},\, \ddot{q}_T=0}, \tag{5.25}$$

$$\lambda = x, q_T, \dot{q}_T; \quad \mu = x, \dot{x}, q_T, \dot{q}_T, \ddot{q}_T; \qquad K_{U\dot{q}}=0\ .$$

Integrability conditions: reciprocity relations It is easy to verify that—due to the integrability conditions (5.15)—the following reciprocity relations must hold for the transducer coefficients:

$$\boxed{K_{Fq} \overset{!}{=} -K_{Ux}} \quad \text{and} \quad \boxed{K_{F\dot{q}} \overset{!}{=} K_{U\dot{x}}}\ . \tag{5.26}$$

Algebraic constitutive equations: Q-coordinates The somewhat unwieldy time derivatives of x and q_T in Eq. (5.24) can be easily eliminated by applying the LAPLACE transform, which—noting that $K_{U\dot{q}}=0$—results in the *algebraic constitutive equations* with *complex* coefficients

$$\Delta F_{T,Q}(s) = K_{Fx}\Delta X(s) + \left[K_{Fq}\frac{1}{s}\Delta\dot{Q}_T(s)\right]_{cap} \vee \left[K_{F\dot{q}}\Delta\dot{Q}_T(s)\right]_{ind},$$

$$\Delta U_T(s) = \left[K_{Ux}\Delta X(s) + K_{Uq}\frac{1}{s}\Delta\dot{Q}_T(s)\right]_{cap} \vee \tag{5.27}$$

$$\left[K_{U\dot{x}}s\Delta X(s) + K_{U\ddot{q}}s\Delta\dot{Q}_T(s)\right]_{ind},$$

or in a generalized form, substituting $\Delta i_T = \Delta\dot{q}_T$ and $\Delta I_T(s) = \Delta\dot{Q}_T(s)$, and using matrix notation:

$$\begin{pmatrix}\Delta F_{T,Q}(s)\\ \Delta U_T(s)\end{pmatrix} = \left(\begin{array}{c|c} K_{Fx} & \left[\dfrac{K_{Fq}}{s}\right]_{cap} \vee \left[K_{F\dot{q}}\right]_{ind} \\ \hline \left[K_{Ux}\right]_{cap} \vee \left[K_{U\dot{x}}s\right]_{ind} & \left[\dfrac{K_{Uq}}{s}\right]_{cap} \vee \left[K_{U\ddot{q}}s\right]_{ind} \end{array}\right)\begin{pmatrix}\Delta X(s)\\ \Delta I_T(s)\end{pmatrix}. \tag{5.28}$$

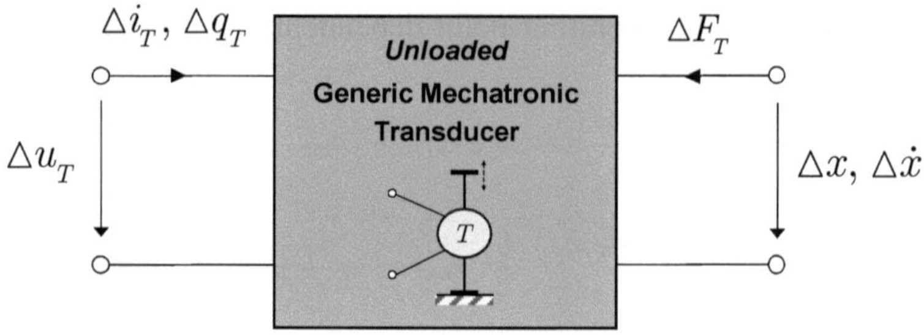

Fig. 5.10. Two-port representation of the constitutive transducer equations of a generic mechatronic transducer

Two-port representation The form of the constitutive equations (5.28) suggests their interpretation as an electromechanical *two-port* network (see Fig. 5.10, cf. Sec. 2.3.4). In the case considered here, on the mechanical side, the *transducer force* $F_{T,Q}$ is interpreted as the *flow variable*, and the *armature position* x as the *displacement variable*. Equivalently, however, the armature velocity $\dot{x}$ can be interpreted as the mechanical displacement variable and the charge q_T as the electrical flow variable[10].

Two-port hybrid formulation The form of Eq. (5.28) is known in two-port theory as the *first hybrid formulation*, with coefficient matrix $\mathbf{H}$ (Thomas et al. 2009):

$$\boxed{\begin{pmatrix} \Delta F_{T,Q}(s) \\ \Delta U_T(s) \end{pmatrix} = \mathbf{H}(s) \cdot \begin{pmatrix} \Delta X(s) \\ \Delta I_T(s) \end{pmatrix} = \begin{pmatrix} H_{11}(s) & H_{12}(s) \\ H_{21}(s) & H_{22}(s) \end{pmatrix} \begin{pmatrix} \Delta X(s) \\ \Delta I_T(s) \end{pmatrix}}. \tag{5.29}$$

Local linearization in PSI-coordinates Formally, the local linearization of the transducer model in PSI-coordinates from Table 5.3 is performed completely analogously to the linearization in Q-coordinates presented above.

10 It is worth noting at this point that a fundamentally different assignment of mechanical flow and displacement variables is also possible, namely that the *flow variable = velocity (or position)* and *displacement variable = force*. Which variant is chosen is fundamentally arbitrary. The choice used in the text is for reasons of convenience and a certain compatibility of the resulting model to modular, network-based modeling concepts (cf. Sec. 2.3.9).

Without loss of generality, the following possible steady-state *operating points* can be defined:

$$x_R,\ \boxed{\dot{x}_R = 0}, \qquad \psi_{TR},\ \dot{\psi}_{TR},\ \boxed{\ddot{\psi}_{TR} = 0}. \tag{5.30}$$

This gives the constitutive ELM transducer equations *linearized* about the operating point in Eq. (5.30) in the form

$$\begin{aligned}\Delta F_{T,\Psi}(x,\psi_T,\dot{\psi}_T) &= K^*_{Fx}\Delta x + \left[K^*_{F\psi}\Delta\psi_T\right]_{ind} \vee \left[K^*_{F\dot{\psi}}\Delta\dot{\psi}\right]_{cap},\\ \Delta i_T(x,\dot{x},\psi_T,\dot{\psi}_T,\ddot{\psi}_T) &= \left[K_{Ix}\Delta x + K_{I\psi}\Delta\psi_T + K_{I\dot{\psi}}\Delta\dot{\psi}\right]_{ind} \vee\\ &\quad \left[K_{I\dot{x}}\Delta\dot{x} + K_{I\ddot{\psi}}\Delta\ddot{\psi}_T\right]_{cap},\end{aligned} \tag{5.31}$$

with constant (generally operating-point-dependent) *transducer coefficients*

$$\begin{aligned}K^*_{F\lambda} &:= \frac{\partial}{\partial\lambda}F_{T,\Psi}(x,\psi_T,\dot{\psi}_T)\ \Big|_{x=x_R,\ \psi_T=\psi_{TR},\ \dot{\psi}_T=\dot{\psi}_{TR}},\\ K_{I\mu} &:= \frac{\partial}{\partial\mu}i_T(x,\dot{x},\psi_T,\dot{\psi}_T,\ddot{\psi}_T)\ \Big|_{x=x_R,\ \dot{x}=0,\ \psi_T=\psi_{TR},\ \dot{\psi}_T=\dot{\psi}_{TR},\ \ddot{\psi}_T=0},\\ \lambda &= x,\psi_T,\dot{\psi}_T;\quad \mu = x,\dot{x},\psi_T,\dot{\psi}_T,\ddot{\psi}_T;\quad K_{I\dot{\psi}} = 0\,.\end{aligned} \tag{5.32}$$

Integrability conditions: reciprocity relations It is easy to verify that due to the integrability conditions (5.14), the following reciprocity relations must hold for the transducer coefficients:

$$\boxed{K_{F\psi} \overset{!}{=} -K_{Ix}} \quad \text{and} \quad \boxed{K_{F\dot{\psi}} \overset{!}{=} K_{I\dot{x}}} \tag{5.33}$$

Algebraic constitutive equations: PSI-coordinates The same considerations as before—using LAPLACE transforms, noting $K_{I\dot{\psi}} = 0$, and the substitution $u_T = \dot{\psi}_T$—lead to the *algebraic constitutive equations* with *complex* coefficients in matrix notation:

$$\begin{pmatrix}\Delta F_{T,\Psi}(s)\\ \Delta I_T(s)\end{pmatrix} = \left(\begin{array}{c|c} K^*_{Fx} & \left[\dfrac{K^*_{F\psi}}{s}\right]_{ind} \vee \left[K^*_{F\dot{\psi}}\right]_{cap}\\ \hline \left[K_{Ix}\right]_{ind} \vee \left[K_{I\dot{x}}s\right]_{cap} & \left[\dfrac{K_{I\psi}}{s}\right]_{ind} \vee \left[K_{I\ddot{\psi}}s\right]_{cap}\end{array}\right)\begin{pmatrix}\Delta X(s)\\ \Delta U_T(s)\end{pmatrix}. \tag{5.34}$$

Two-port admittance formulation The form of Eq. (5.34) is known in two-port theory as the *admittance or conductance formulation*, with coefficient matrix $\mathbf{Y}$ (Thomas et al. 2009):

$$\begin{pmatrix} \Delta F_{T,\Psi}(s) \\ \Delta I_T(s) \end{pmatrix} = \mathbf{Y}(s) \cdot \begin{pmatrix} \Delta X(s) \\ \Delta U_T(s) \end{pmatrix}. \tag{5.35}$$

Duality in the two-port equations The two-port equations (5.28), (5.34) exhibit *duality* in two respects.

Since the two descriptions represent the same physical system, and both formulations describe the same dynamics at the terminals, the admittance and hybrid formulation must be completely equivalent. In consequence of this equivalence, there is also a *representational equivalence* between admittance and hybrid formulations. Naturally, this is in complete accord with two-port theory; these well-known correspondences between admittance and hybrid formulations (Thomas et al. 2009) are listed in Table 5.4. The consequent correspondences between the physical transducer parameters follow from Eqs. (5.25), (5.32).

Behavior at the terminals The two-port equations comprehensively describe the electromechanical coupling in the transducer at the small-signal level. By using complex coefficients, it is even possible to describe differential or integral dependencies between the terminal variables in simple manner. To this extent, it is also secondary whether the position or the velocity is chosen as the mechanical displacement variable, or the current or charge as the electrical flow variable.

Table 5.4. Correspondences between admittance and hybrid formulations of the constitutive ELM transducer equations (ELM = electromechanical)

Admittance formulation	Hybrid formulation
$Y_{11} = \dfrac{\det \mathbf{H}}{H_{22}} = H_{11} - \dfrac{H_{12} \cdot H_{21}}{H_{22}}$	$H_{11} = \dfrac{\det \mathbf{Y}}{Y_{22}} = Y_{11} - \dfrac{Y_{12} \cdot Y_{21}}{Y_{22}}$
$Y_{12} = H_{12} / H_{22}$	$H_{12} = Y_{12} / Y_{22}$
$Y_{21} = -H_{21} / H_{22}$	$H_{21} = -Y_{21} / Y_{22}$
$Y_{22} = 1 / H_{22}$	$H_{22} = 1 / Y_{22}$

Table 5.5. ELM two-port matrices for transducer implementations

	$\begin{pmatrix} F_{T,\Psi} \\ I_T \end{pmatrix} = \mathbf{Y}(s) \cdot \begin{pmatrix} X \\ U_T \end{pmatrix}$ $\mathbf{Y}(s)$	$\begin{pmatrix} F_{T,Q} \\ U_T \end{pmatrix} = \mathbf{H}(s) \cdot \begin{pmatrix} X \\ I_T \end{pmatrix}$ $\mathbf{H}(s)$
Electrostatic transducer Variable plate separation see Ch. 6	$\begin{pmatrix} k_{el,U} & K_{el,U} \\ sK_{el,U} & sC_R \end{pmatrix}$	$\begin{pmatrix} k_{el,U} - \frac{K_{el,U}^2}{C_R} & \frac{K_{el,U}}{sC_R} \\ -\frac{K_{el,U}}{C_R} & \frac{1}{sC_R} \end{pmatrix}$
Piezoelectric transducer see Ch. 7	$\begin{pmatrix} -k_{pz,U} & -K_{pz,U} \\ -sK_{pz,U} & sC_{pz} \end{pmatrix}$	$\begin{pmatrix} -k_{pz,U} - \frac{K_{pz,U}^2}{C_{pz}} & -\frac{K_{pz,U}}{sC_{pz}} \\ \frac{K_{pz,U}}{C_{pz}} & \frac{1}{sC_{pz}} \end{pmatrix}$
Electromagnetic transducer (Reluctance force) see Ch. 8	$\begin{pmatrix} k_{em,U} & \frac{K_{em,U}}{s} \\ -K_{em,U} & \frac{1}{sL_R} \end{pmatrix}$	$\begin{pmatrix} k_{em,U} + L_R K_{em,U}^2 & L_R K_{em,U} \\ sL_R K_{em,U} & sL_R \end{pmatrix}$
Electrodynamic transducer (LORENTZ force) see Ch. 8	$\begin{pmatrix} -\frac{K_{ED}^2}{L_{sp}} & \frac{K_{ED}}{sL_{sp}} \\ -\frac{K_{ED}}{L_{sp}} & \frac{1}{sL_{sp}} \end{pmatrix}$	$\begin{pmatrix} 0 & K_{ED} \\ sK_{ED} & sL_{sp} \end{pmatrix}$
Back-effect-free actuator e.g. servo-hydraulic	$\begin{pmatrix} k & \frac{K}{K_{U/I}} \\ \boxed{0} & \frac{1}{K_{U/I}} \end{pmatrix}$	$\begin{pmatrix} k & K \\ \boxed{0} & K_{U/I} \end{pmatrix}$

Transducer families Based on the dynamics at their electrical terminals, the following *transducer families* can be defined (see Table 5.5):

- *capacitive* transducers: *electrostatic*, *piezoelectric*,
- *inductive* transducers: *electrodynamic*, *electromagnetic*,
- *back-effect-free* transducers: the electrical drive is decoupled from mechanical motion.

Interpreting the two-port parameters The two-port parameters have very clear physical meanings, so that, in any concrete case, significant insight into the system dynamics can already be gleaned from them.

- *ELM stiffness*, *differential transducer stiffness*, or *mechanical impedance*[11]: $H_{11} = F_x$ or $Y_{11} = F_x^*$ describe the inherent, operating-point-dependent stiffness of the *unloaded* transducer, i.e. the force generated proportional to the armature motion. These coefficients are always real. Positive stiffness produces behavior corresponding to that of an ordinary linear spring. Negative stiffness, by reducing stiffness in the external elastic connection, causes so-called *electromechanical softening*. As stiffness goes to zero, the generated force becomes independent of the armature displacement.

- *Electrical impedance* $H_{22}(s)$ or *electrical admittance* $Y_{22}(s)$ describe the electrical dynamics at the terminals of the *unloaded* transducer (the *terminal dynamics*). Depending on the underlying constitutive ELM basic equations, *capacitive* or *inductive* electrical behaviors result.

- *ELM coupling elements* $H_{12}(s), H_{21}(s)$, and $Y_{12}(s), Y_{21}(s)$ describe the electromechanical coupling in the unloaded transducer. At the mechanical level, $H_{12}(s), Y_{12}(s)$ describe the ELM force generation mechanisms, while at the electrical level, $H_{21}(s), Y_{21}(s)$ describe the electrical quantities induced by motion (polarization currents, induced voltages).

---

11 In the literature, *mechanical impedance* does not refer uniquely to either the relationship between *force and displacement* or that between *force and velocity*. The *differential stiffness* is the linear approximation (tangent) to the force-displacement curve of the transducer at the current operating point x_R.

5.4 The Loaded Generic Transducer

5.4.1 Energy-based model

Mechanical energy Building upon the unloaded transducer in Fig. 5.4, the elastically-suspended, rigid-body load shown in Fig. 5.2 introduces a kinetic energy storage element (mass m) and a potential energy storage element (spring k). These have the linear constitutive relations, independent of the electrical subsystem (see Table 2.2),

$$F = kx, \quad p = m\dot{x}\,.$$

Integrating these relations is unproblematic, and, following the method of Fig. 5.7a, results in the well-known energy functions

$$V_{mech}(x) = \frac{1}{2}kx^2\,, \quad T^*_{mech}(\dot{x}) = \frac{1}{2}m\dot{x}^2\,. \tag{5.36}$$

LAGRANGian For the complete, loaded transducer system, the LAGRANGian of the unloaded transducer in Eqs. (5.18), (5.19) need only be supplemented with the mechanical energy terms, giving for the *loaded transducer*:

- the *LAGRANGian* in *Q-coordinates* with generalized coordinates x, q_T:

$$\mathcal{L}_Q(x,\dot{x},q_T,\dot{q}_T) = \mathcal{L}^o_Q(x,q_T,\dot{q}_T) + T^*_{mech}(\dot{x}) - V_{mech}(x)\,, \tag{5.37}$$

- the *LAGRANGian* in *PSI-coordinates* with generalized coordinates x, ψ_T:

$$\mathcal{L}_\Psi(x,\dot{x},\psi_T,\dot{\psi}_T) = \mathcal{L}^o_\Psi(x,\psi_T,\dot{\psi}_T) + T^*_{mech}(\dot{x}) - V_{mech}(x)\,. \tag{5.38}$$

5.4.2 Nonlinear equations of motion

EULER-LAGRANGE equations The equations of motion of the loaded generic transducer are obtained analogously to the unloaded case via the *EULER-LAGRANGE equations*, all that is required is to substitute the LAGRANGians (5.37) or (5.38) in place of $\mathcal{L}^o_Q$ or $\mathcal{L}^o_\Psi$. Some additional thought must, however, be given to the excitation functions.

Causal structure: voltage/current source In contrast to the unloaded transducer, the definition of external excitation variables (right-hand side

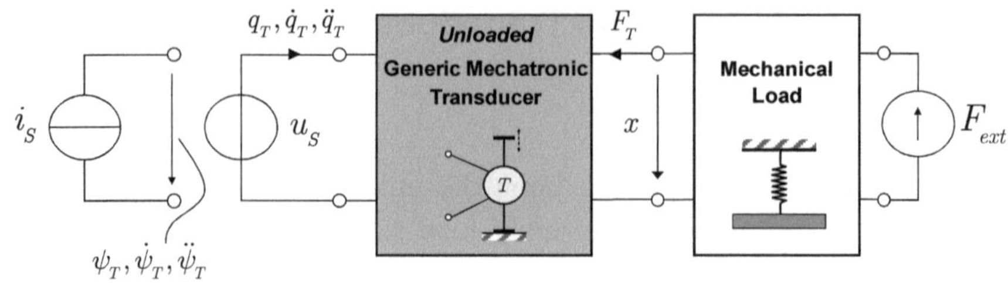

Fig. 5.11. Loaded generic transducer in two-port representation

of the EULER-LAGRANGE equations) requires establishing a causal structure. On the *mechanical* side, this is rather simple: the *external force* $F_{ext}(t)$ acts as the excitation (Fig. 5.2) and represents the generalized force F_{gen}.

However, on the *electrical* side, as shown in Fig. 5.2, there is a choice of two possible excitations: a controlled *voltage source* $u_S(t)$ or a controlled *current source* $i_S(t)$. Fig. 5.11 shows a two-port representation equivalent to Fig. 5.2. This results in a total of four model families, as each excitation type $u_S(t)$ or $i_S(t)$ can be represented in either Q- or PSI-coordinates. These four variants are sketched out in Table 5.6, and will now be discussed further.

Voltage source With a controlled voltage source, the voltage $u_S(t)$—and thus the flux linkage $\psi_S(t)$ and its time derivatives $\dot{\psi}_S(t)$, $\ddot{\psi}_S(t)$—can be forced *independently* of the loading at the terminals.

The ensuing flow of charge is represented by $q_T(t)$, $\dot{q}_T(t) = i_T(t)$, and $\ddot{q}_T(t)$, and is *dependent* on the loading at the terminals. This has important implications for the model representation.

In *Q-coordinate representation*, the charge coordinate q_T is (along with the armature position x) also a dependent variable, and thus a generalized coordinate. When evaluating the EULER-LAGRANGE equations, a derivative must thus also be taken with respect to q_T, giving an additional differential equation (left-hand side: the constitutive ELM transducer equation for u_T; right-hand side: $u_{gen} = u_S$) defining q_T given the excitation $u_S(t)$. To determine $x(t)$ and $q_T(t)$, then, the *coupled* system of differential equations in Table 5.6, first row, right column, must be solved.

In *PSI-coordinate representation*, the situation is different, and simpler. The electrical coordinate $\psi_T(t) = \psi_S(t)$ is uniquely determined by the source voltage $u_S(t)$, so that it acts as an independent input, and thus is *not* a

Table 5.6. Nonlinear equations of motion of the loaded generic transducer: electrical and mechanical coordinates vs. auxiliary energy sources

Coordinates	**PSI-coordinates** (x, ψ_T)	**Q-coordinates** (x, q_T)
Ideal voltage source $u_S(t)$	Gen. coordinate: x $x: \ \frac{d}{dt}\left(\frac{\partial \mathcal{L}_\Psi}{\partial \dot{x}}\right) - \frac{\partial \mathcal{L}_\Psi}{\partial x} = F_{ext}$ $\boxed{i_T = i_T\left(x, \dot{x}, \psi_S(t), \dot{\psi}_S(t), \ddot{\psi}_S(t)\right)}$	Gen. coordinates: x, q_T $x: \ \frac{d}{dt}\left(\frac{\partial \mathcal{L}_Q}{\partial \dot{x}}\right) - \frac{\partial \mathcal{L}_Q}{\partial x} = F_{ext}$ $u_T(x, \dot{x}, q_T, \dot{q}_T, \ddot{q}_T) = u_S(t)$
Ideal current source $i_S(t)$	Gen. coordinates: x, ψ_T $x: \ \frac{d}{dt}\left(\frac{\partial \mathcal{L}_\Psi}{\partial \dot{x}}\right) - \frac{\partial \mathcal{L}_\Psi}{\partial x} = F_{ext}$ $i_T(x, \dot{x}, \psi_T, \dot{\psi}_T, \ddot{\psi}_T) = i_S(t)$	Gen. coordinate: x $x: \ \frac{d}{dt}\left(\frac{\partial \mathcal{L}_Q}{\partial \dot{x}}\right) - \frac{\partial \mathcal{L}_Q}{\partial x} = F_{ext}$ $\boxed{u_T = u_T\left(x, \dot{x}, q_S(t), \dot{q}_S(t), \ddot{q}_S(t)\right)}$

generalized coordinate. The dependent electrical terminal variable $i_T(t)$ can be easily computed via an algebraic relation (the constitutive ELM transducer equation for i_T, Table 5.3) from the source flux $\psi_S(t)$ and its time derivatives, and the mechanical coordinate $x(t)$ and its time derivative. Thus, all that remains as a generalized coordinate is the armature position x and, consequently, a single differential equation in x, with the nonlinear transducer forces as the excitation from the electrical side and the external force as excitation from the mechanical side (Table 5.6, first row, left column).

Voltage source: PSI-representation The two representations in Q- and PSI-coordinates are naturally completely equivalent with respect to their input/output behavior. The *PSI-representation* is, however, representationally clearer, as the electrical excitation only acts over a forward path, and the electromechanical feedback acts only within the mechanical system. For this reason, when considering *voltage-drive* transducers, the discussion here prefers the *PSI-coordinate* representation. Fig. 5.12a sketches the gen-

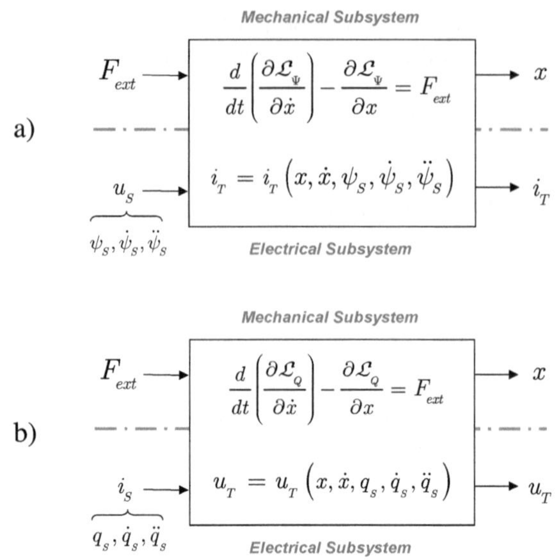

Fig. 5.12. Causal structures for the loaded generic transducer: a) ideal voltage source, PSI-coordinates; b) ideal current source, Q-coordinates

eral causal structure for a voltage-drive transducer along with the applicable model variables.

Current source When using the controlled current source option $i_S(t)$, as shown in Fig. 5.2 and Fig. 5.11, the dual considerations to those for voltage drive can be directly applied.

In *PSI-coordinate representation*, there are two generalized coordinates x, ψ_T, and thus a coupled system of differential equations (Table 5.6, second row, left column).

In this case, the *Q-coordinate representation* leads to the simpler variant, as the dependent transducer voltage u_T can be directly computed from an algebraic relation in the source charge and its time derivatives, and the armature position and its time derivative (Table 5.6, second row, right column).

Current source: Q-representation Due to its superior representational clarity, when considering current-drive transducers, the discussion here prefers the Q-coordinate representation. Fig. 5.12b sketches the general causal structure of a current-drive transducer along with the applicable model variables.

Table 5.7. Nonlinear equations of motion of the loaded generic transducer for voltage and current sources in different coordinate representations for *linear electrical* dynamics, see Eq. (5.3)

	Voltage source $u_S(t)$
PSI-coordinates	$m\ddot{x} + kx + \left[-\frac{1}{2}\frac{\partial C(x)}{\partial x}\dot{\psi}_S^2\right]_{cap} \vee \left[-\frac{1}{2}\frac{1}{L(x)^2}\frac{\partial L(x)}{\partial x}\psi_S^2\right]_{ind} = F_{ext}$ $\boxed{i_T = \left[C(x)\cdot\ddot{\psi}_S + \frac{\partial C(x)}{\partial x}\cdot\dot{x}\cdot\dot{\psi}_S\right]_{cap} \vee \left[\frac{1}{L(x)}\cdot\psi_S\right]_{ind}}$
Q-coordinates	$m\ddot{x} + kx + \left[-\frac{1}{2}\frac{\partial L(x)}{\partial x}\dot{q}_T^2\right]_{ind} \vee \left[-\frac{1}{2}\frac{1}{C(x)^2}\frac{\partial C(x)}{\partial x}q_T^2\right]_{cap} = F_{ext}$ $\left[L(x)\cdot\ddot{q}_T + \frac{\partial L(x)}{\partial x}\cdot\dot{x}\cdot\dot{q}_T\right]_{ind} \vee \left[\frac{1}{C(x)}\cdot q_T\right]_{cap} = u_S$
	Current source $i_S(t)$
PSI-coordinates	$m\ddot{x} + kx + \left[-\frac{1}{2}\frac{\partial C(x)}{\partial x}\dot{\psi}_T^2\right]_{cap} \vee \left[-\frac{1}{2}\frac{1}{L(x)^2}\frac{\partial L(x)}{\partial x}\psi_T^2\right]_{ind} = F_{ext}$ $\left[C(x)\cdot\ddot{\psi}_T + \frac{\partial C(x)}{\partial x}\cdot\dot{x}\cdot\dot{\psi}_T\right]_{cap} \vee \left[\frac{1}{L(x)}\cdot\psi_T\right]_{ind} = i_S$
Q-coordinates	$m\ddot{x} + kx + \left[-\frac{1}{2}\frac{\partial L(x)}{\partial x}\dot{q}_S^2\right]_{ind} \vee \left[-\frac{1}{2}\frac{1}{C(x)^2}\frac{\partial C(x)}{\partial x}q_S^2\right]_{cap} = F_{ext}$ $\boxed{u_T = \left[L(x)\cdot\ddot{q}_S + \frac{\partial L(x)}{\partial x}\cdot\dot{x}\cdot\dot{q}_S\right]_{ind} \vee \left[\frac{1}{C(x)}\cdot q_S\right]_{cap}}$

Linear electrical dynamics For the special case of *linear electrical* transducer dynamics assumed here (cf. Eq. (5.3)), the *nonlinear equations of motion* given external loading as in Fig. 5.2 and Fig. 5.11 are listed in Table 5.7. The preferred model variants (because of their simpler structure) are shaded gray.

5.4.3 Equilibrium positions: operating points

Determining equations for equilibrium positions When implementing the general transducer models in Tables 5.4 and 5.7, *steady-state* operating points for steady-state excitations are of interest. Limiting further consideration to the preferred representation forms in Tables 5.4 and 5.7, only the mechanical equation of motion is relevant in determining possible equilibrium positions for operation.

At a steady-state operating point, the armature must be at rest, i.e. $\dot{x} = \ddot{x} = 0$, which immediately leads to the (trivial) general *equilibrium position condition*

$$\boxed{\dot{x}_R = 0} \tag{5.39}$$

on the armature velocity.

This leaves a nonlinear, algebraic determining equation in one of the below forms as the sole condition requiring evaluation to find possible *steady-state armature positions (equilibrium positions)* x_R for steady-state mechanical excitation $F_{ext}(t) = F_0$:

- PSI-coordinates, steady-state source variables $\psi_{S,0}, \dot{\psi}_{S,0}$

$$\boxed{kx_R + F_{T,\Psi}(x_R, \psi_{S,0}, \dot{\psi}_{S,0}) = F_0}, \tag{5.40}$$

- Q-coordinates, steady-state source variables $q_{S,0}, \dot{q}_{S,0}$

$$\boxed{kx_R + F_{T,Q}(x_R, q_{S,0}, \dot{q}_{S,0}) = F_0}, \tag{5.41}$$

where the transducer forces F_T are determined by the constitutive transducer equations in Table 5.3.

Steady-state electrical source variables Detailed discussion of the determination of equilibrium positions only makes sense for concrete transducer configurations with explicitly defined constitutive relations, for which see subsequent chapters. In particular, meaningful definitions of steady-state source variables depend on the constitutive electrical traits of the transducer.

Force balance One general attribute of the equilibrium position conditions Eqs. (5.40), (5.41) can however already be discussed in a general form at this point. In the steady-state case, Eqs. (5.40) and (5.41) describe the *force balance* between the spring force of the elastic suspension and the difference between external force and transducer force, i.e.

$$F(x_R) = kx_R = F_0 - F_T(x_R, elec.vars.)\,.$$

In general, it is to be expected that for a nonlinear force-displacement curve in the transducer, there may exist *multiple intersections* with the linear spring curve (Fig. 5.13).

Stability of equilibrium positions vs. differential transducer stiffness The stability of the equilibrium positions can be easily determined from the behavior of the transducer *differential stiffness* (see Sec. 5.3.3 and Fig. 5.13)

$$K_{elm} := \frac{\partial}{\partial x} F_T(x, elec.vars)\Big|_{x=x^*} \tag{5.42}$$

at the equilibrium position $x^* = x_{R,i}$ under consideration. The linearized mechanical equation of motion has the general form

$$m \cdot \Delta\ddot{x} + \left(k - K_{elm}\right) \cdot \Delta x = \Delta F_{ext}\,, \tag{5.43}$$

with the *resulting stiffness* of the electrically activated transducer

$$K_T = k - K_{elm}\,. \tag{5.44}$$

This results in the following, generally-valid statements about the stability[12] of an equilibrium position:

- *stable* equilibrium position: $\boxed{K_T > 0 \Leftrightarrow k > K_{elm}}$ (x_{R1} in Fig. 5.13a),
- *unstable* equilibrium position: $K_T \leq 0 \Leftrightarrow k \leq K_{elm}$ ($x_{R2}, \ldots, x_{R5}$ in Fig. 5.13a).

---

12 The concept to apply is that of *state stability* (Ogata 2010). The dynamic system (5.43) with $K_T = 0$ is (state) *unstable* due to the double eigenvalue $\lambda_{1,2} = 0$ and the non-diagonal system matrix. With finite mechanical damping and $K_T = 0$, there is a single eigenvalue $\lambda = 0$ and thus *marginal stability*. However, this is only of academic interest; in practice, this latter case can also be considered *unstable*.

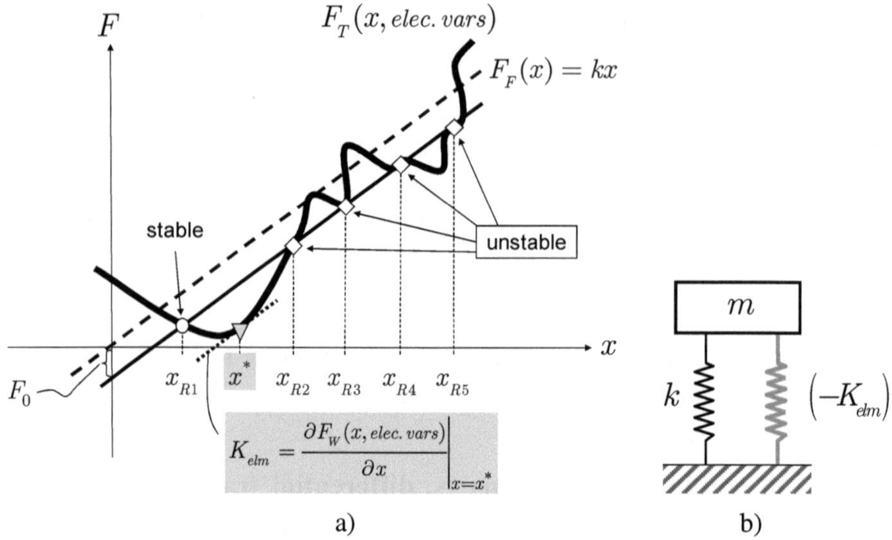

Fig. 5.13. Force relations for the loaded generic mechatronic transducer: a) force-displacement diagram given constant electrical transducer variables, b) mechanically equivalent element

Mechanically, the (differential) transducer stiffness acts like a spring of stiffness $-K_{elm}$ connected in parallel with the elastic suspension (Fig. 5.13b).

In the *stable* case, negative K_{elm} increases the total stiffness, while *positive* K_{elm} decreases it. The latter case is termed that of *electromechanical softening*.

In the *unstable* case $K_T < 0$, given a differential change in the armature displacement, the transducer force increases faster than the spring force, so that the transducer force dominates and the armature is pulled in its direction without opposition (known as the *pull-in* phenomenon in electrostatic transducers).

The case $K_T = 0$ is, so-to-speak, the entryway to the unstable domain. Here, the transducer and spring forces are just in balance and the smallest displacement of the armature in the direction of the positive transducer force gradient results in unopposed motion, as described above.

Force map As shown in Table 5.3, the transducer force depends both on the armature position x, and on the electrical drive variable (voltage or current, depending on the chosen representation). This allows its static behavior to be described using a *characteristic force map* $F_T(x,\lambda)$, where λ describes the applicable electrical working variable for a particular case.

Mathematically, the characteristic force map is a higher-dimensional surface. The transducer force-displacement curve shown in Fig. 5.13 is just a section curve $F_T(x,\lambda_0)$ of the transducer characteristic map at a constant value λ_0 of the electrical working variable.

As a demonstrative example of a transducer implementation, the characteristic force map of an *electrostatic transducer* (a plate transducer) is depicted in Fig. 5.14a. For this case, let the plate voltage be controllable.

The corresponding section for a constant transducer voltage can be seen in Fig. 5.14b. Three equilibrium positions can be seen, of which only two (x_{R1}, x_{R3}) are stable. In the end, however, only the equilibrium position x_{R1} is actually useable as an operating point, as x_{R3} lies outside the operating regime.

Geometric interpretation of equilibrium position equations The equilibrium position equations (5.40), (5.41) can be interpreted concretely as defining the intersection of the transducer map with the spring force plane (Fig. 5.14b).

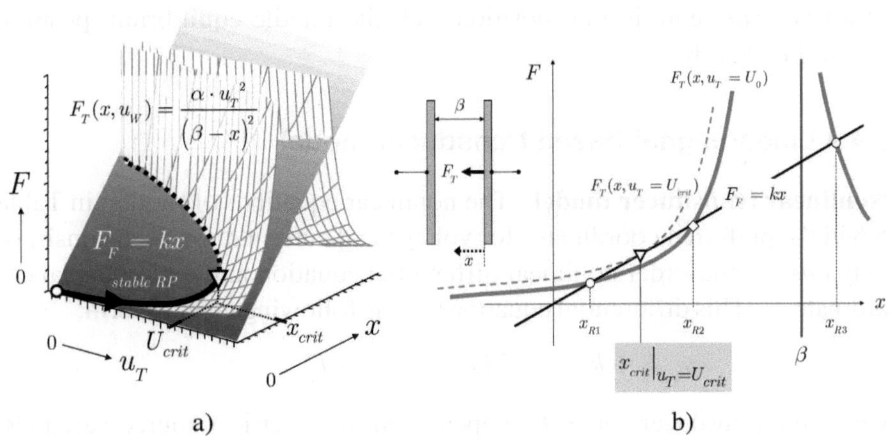

Fig. 5.14. Example of a characteristic force map of an electrostatic transducer with variable electrode separation and voltage drive (e.g. a plate transducer, see Sec. 6.4): a) force map, b) force-displacement curves for a constant transducer voltage

Initializing operating points Given a fixed stiffness k of the elastic suspension and a steady-state mechanical excitation (external force), the active operating point (equilibrium position) depends solely on the electrical drive variable (voltage or current). For example, consider the *electrostatic transducer* in Fig. 5.14. As the transducer force in the electrically non-active state is equal to zero (see Eq. (5.22)), increasing values of the electrical drive variable (here the voltage) cause the operating point to move along the voltage-force curve as shown in Fig. 5.14a (the solid intersection of the transducer force surface with the spring force plane). This voltage-force curve forms the set of all *stable* equilibrium positions (quasi-statically) reachable from the electrically inactive state (see Fig. 5.14a,b).

Critical electrical drive As can be seen in the example of the electrostatic transducer in Fig. 5.14, there is a *critical voltage* U_{crit} beyond which there are no more reachable equilibrium positions. It can be seen in Fig. 5.14b that this is due to precisely the previously-mentioned *marginally stable* case. The bounding value U_{crit} represents the maximum operating drive value; the corresponding x_{crit} characterizes the maximum controllable armature displacement. As the intersection curve changes with the angle of the spring force plane, these operational bounds also depend on the elastic suspension k.

The dashed intersection curve indicated in Fig. 5.14a represents the set of all *unstable* equilibrium positions (cf. the middle equilibrium position x_{R2} in Fig. 5.14b).

5.4.4 Linear signal-based transducer model

Nonlinear transducer model The nonlinear equations of motion in Table 5.6 in the preferred coordinates for voltage and current drive each consist of only *one* second-order nonlinear differential equation in the mechanical coordinate x. This differential equation has the following general form:

$$m \cdot \ddot{x} + k \cdot x - F_T(x, u_S \,/\, i_S) = F_{ext}, \tag{5.45}$$

where the transducer force F_T depends on the electrical source variables. In addition, there is also one *algebraic output equation* for the dependent electrical terminal variable—i_T for voltage drive, or u_T for current drive (see Table 5.6, Table 5.7).

Local linearization: small-signal response For an easy-to-use description of the transducer small-signal response characteristics, a linearized model can be obtained from Eq. (5.45) via the previously-described *local linearization* for sufficiently small deviations from *equilibrium positions* (see Sec. 2.6.1), i.e.

$$x = x_R + \Delta x, \quad \dot{x} = \dot{x}_R + \Delta\dot{x} = \Delta\dot{x}\,, \tag{5.46}$$

and for small deviations from steady-state electrical source and mechanical excitation variables,

$$u_S = U_0 + \Delta u_S, \quad i_S = I_0 + \Delta i_S, \quad F_{ext} = F_0 + \Delta F_{ext}\,. \tag{5.47}$$

It now turns out that the fundamental groundwork has already been laid in the previous sections. The linear terms in Eq. (5.45) can be directly adopted into the linearized model. The partial derivatives of the transducer forces and the dependent electrical terminal variables were already calculated while setting up the *ELM two-port model* in Sec. 5.3.3 and are defined in Eqs. (5.24), (5.25), (5.31), and (5.32).

Linear transducer model in the complex plane As the linear transducer model will not be further used in the time-domain representation, its explicit form is omitted here for reasons of space[13].

Instead, the LAPLACE transform of the linear transducer model is examined, and the ELM two-port parameterization using the *hybrid formulation* $\mathbf{H}(s)$ of Eq. (5.29) and the *admittance formulation* $\mathbf{Y}(s)$ of Eq. (5.35) are introduced. A comparison with the preferred coordinate representations for the electrical drive immediately demonstrates an advantageous assignment of ELM two-port parameters in the loaded linear transducer model:

- *voltage drive* ⇔ *admittance formulation* $\mathbf{Y}$,
- *current drive* ⇔ *hybrid formulation* $\mathbf{H}$.

Voltage-drive transducer: signal-based model Considering the system configuration in Fig. 5.2, the linearized transducer equations, and the constitutive transducer equations (5.35) results in the following model for a loaded generic transducer with *voltage drive* (lossless, ideal electrical subsystem):

$$\text{Mechanical load:} \quad ms^2 \cdot \Delta X(s) = -k \cdot \Delta X(s) + \Delta F_{T,u}(s) + \Delta F_{ext}(s)\,,$$

$$\text{Electromechanical coupling:} \quad \begin{pmatrix} \Delta F_{T,u}(s) \\ \Delta I_T(s) \end{pmatrix} = \mathbf{Y}(s) \cdot \begin{pmatrix} \Delta X(s) \\ \Delta U_S(s) \end{pmatrix}. \tag{5.48}$$

[13] The time-domain representation can be derived directly from Eqs. (5.24), (5.31), and (5.45).

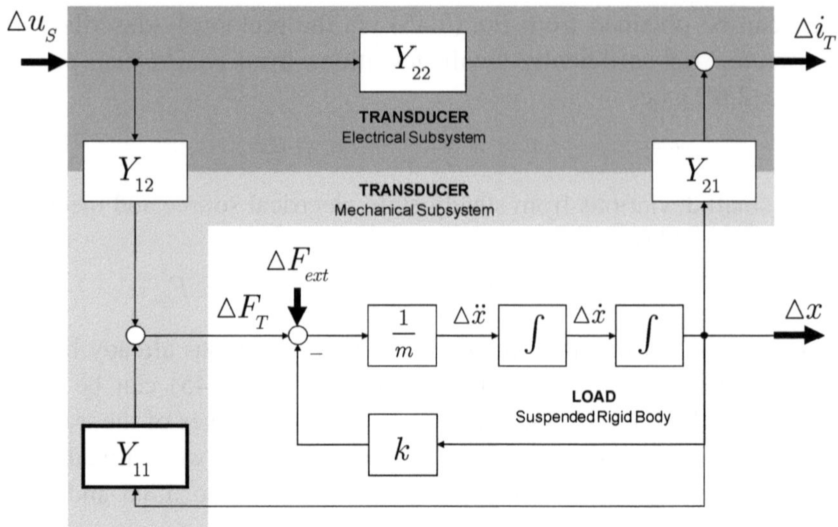

Fig. 5.15. Signal flow diagram of a loaded generic mechatronic transducer with *voltage drive* (lossless, mechanical load: elastically-suspended rigid body)

The corresponding signal flow diagram is shown in Fig. 5.15. The physical interpretation of the two-port parameters can be directly seen there:

- Y_{11} represents the *electromechanical transducer stiffness*; $Y_{11} > 0$ implies electromechanical softening; positive $Y_{11} \geq k$ leads to instability in the transducer.
- Y_{12}, Y_{21} represent the electromechanical energy exchange.
- Y_{22} represents the *electrical admittance* of the transducer.

Current-drive transducer: signal-based model As the dual model to the voltage-drive transducer, using equivalent assumptions, the constitutive transducer equations (5.29) can be used to obtain a signal-based model of a loaded generic transducer with *current drive* (ideal electrical subsystem):

$$\text{Mechanical load:} \quad ms^2 \cdot \Delta X(s) = -k \cdot \Delta X(s) + \Delta F_{T,i}(s) + \Delta F_{ext}(s)\,,$$

$$\text{Electromechanical coupling:} \quad \begin{pmatrix} \Delta F_{T,i}(s) \\ \Delta U_T(s) \end{pmatrix} = \mathbf{H}(s) \cdot \begin{pmatrix} \Delta X(s) \\ \Delta I_S(s) \end{pmatrix}. \tag{5.49}$$

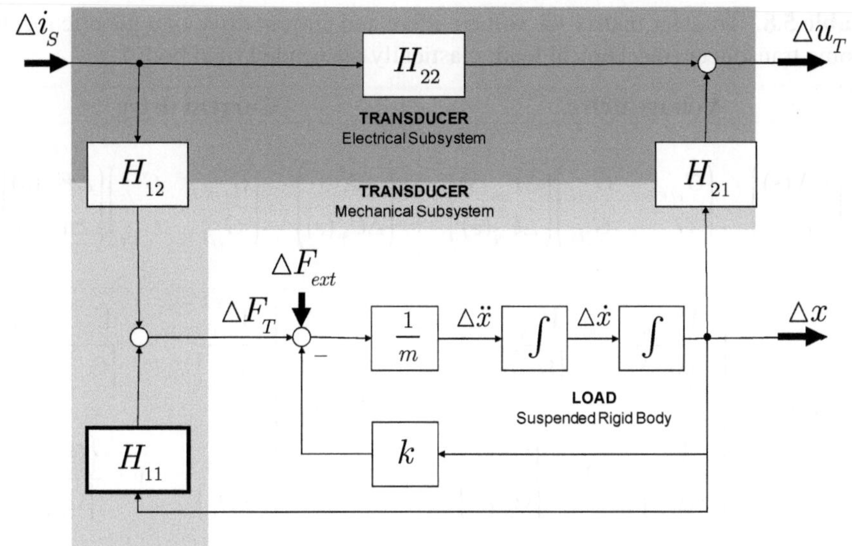

Fig. 5.16. Signal flow diagram for a loaded generic mechatronic transducer with *current drive* (lossless, mechanical load: elastically-suspended rigid body)

From the signal flow diagram in Fig. 5.16, the electrically dual structure to voltage drive, as well as the equivalent physical interpretations of the two-port parameters can be discerned:

- H_{11} represents the *electromechanical transducer stiffness;* $H_{11} > 0$ implies electromechanical softening; positive $H_{11} \geq k$ leads to instability of the loaded transducer.
- H_{12}, H_{21} represent the electromechanical energy exchange.
- H_{22} represents the *electrical impedance* of the transducer.

5.4.5 Transfer matrix

Frequency domain model For a clear evaluation of the response characteristic of the transducer, it is expedient to employ the frequency domain representation.

The *transfer matrix* $\mathbf{G}(s)$ between the transducer inputs ΔF_{ext}, Δu_S / Δi_S, and outputs Δx, Δi_T / Δu_T, as shown in Fig. 5.3, completely describes the sensor/actuator dynamics of the transducer. The transfer matrix for voltage or current drive is obtained via simple manipulation of the linear transducer models in Eq. (5.48) and (5.49), see Table 5.8.

Table 5.8. Transfer matrix for voltage drive and current drive of a generic mechatronic transducer (mechanical load: elastically-suspended rigid body)

Voltage drive	**Current drive**
$\begin{pmatrix}\Delta X(s)\\ \Delta I_T(s)\end{pmatrix}=\begin{pmatrix}G_{x/F,U} & G_{x/u}\\ G_{i/F} & G_{i/u}\end{pmatrix}\begin{pmatrix}\Delta F_{ext}(s)\\ \Delta U_S(s)\end{pmatrix}$	$\begin{pmatrix}\Delta X(s)\\ \Delta U_T(s)\end{pmatrix}=\begin{pmatrix}G_{x/F,I} & G_{x/i}\\ G_{u/F} & G_{u/i}\end{pmatrix}\begin{pmatrix}\Delta F_{ext}(s)\\ \Delta I_S(s)\end{pmatrix}$
$\mathbf{G}_U(s)=\begin{pmatrix}\dfrac{V_{x/F,U}}{\{\Omega_U\}} & \dfrac{V_{x/u}}{\{\Omega_U\}}\\ \dfrac{V_{i/F}}{\{\Omega_U\}} & V_{i/u}\dfrac{\{\Omega_I\}}{\{\Omega_U\}}\end{pmatrix}$	$\mathbf{G}_I(s)=\begin{pmatrix}\dfrac{V_{x/F,I}}{\{\Omega_I\}} & \dfrac{V_{x/i}}{\{\Omega_I\}}\\ \dfrac{V_{u/F}}{\{\Omega_I\}} & V_{u/i}\dfrac{\{\Omega_U\}}{\{\Omega_I\}}\end{pmatrix}$
$k_{T,U}:=k-Y_{11}=k_{T,I}+\dfrac{H_{12}\cdot H_{21}}{H_{22}}$ $\boxed{\Omega_U^{\;2}:=\dfrac{k_{T,U}}{m}}$ $\dfrac{Y_{12}\cdot Y_{21}}{Y_{22}}\in\mathbb{R}\begin{cases}<0: & \Omega_U>\Omega_I \text{ ind.}\\ =0: & \Omega_U=\Omega_I\\ >0: & \Omega_U<\Omega_I \text{ cap.}\end{cases}$	$k_{T,I}:=k-H_{11}=k_{T,U}+\dfrac{Y_{12}\cdot Y_{21}}{Y_{22}}$ $\boxed{\Omega_I^{\;2}:=\dfrac{k_{T,I}}{m}}$ $\dfrac{H_{12}\cdot H_{21}}{H_{22}}\in\mathbb{R}\begin{cases}<0: & \Omega_U<\Omega_I \text{ cap.}\\ =0: & \Omega_U=\Omega_I\\ >0: & \Omega_U>\Omega_I \text{ ind.}\end{cases}$
$V_{x/F,U}:=\dfrac{1}{k_{T,U}},\quad V_{x/u}:=\dfrac{Y_{12}}{k_{T,U}}$ $V_{i/F}:=\dfrac{Y_{21}}{k_{T,U}},\quad V_{i/u}:=Y_{22}\dfrac{k_{T,I}}{k_{T,U}}$	$V_{x/F,I}:=\dfrac{1}{k_{T,I}},\quad V_{x/i}:=\dfrac{H_{12}}{k_{T,I}}$ $V_{u/F}:=\dfrac{H_{21}}{k_{T,I}},\quad V_{u/i}:=H_{22}\dfrac{k_{T,U}}{k_{T,I}}$

5.4.6 Discussion of the response characteristics

The following discussion assumes real-valued two-port elements Y_{11}, H_{11}. Given the previous assumptions and the transducer types investigated in subsequent chapters, this is always the case.

Transducer eigenfrequency In all transfer channels of the transfer matrices $\mathbf{G}_U(s)$ and $\mathbf{G}_I(s)$ in Table 5.8, the drive-variable-dependent transducer eigenfrequency Ω_U or Ω_I appears. For voltage drive (current drive) and a fixed load mass m, this is determined by the resulting transducer stiffness $k_{T,U} = k - Y_{11}$ ($k_{T,I} = k - H_{11}$), see also Fig. 5.15 and Fig. 5.16. The electromechanical transducer stiffnesses Y_{11} and H_{11} thus play a central role in the mechanical behavior of the transducer.

In certain cases, it can certainly happen that, for example, $Y_{11} = 0$. This would imply for *voltage drive*, that there is no shift in the mechanical eigenfrequency $\Omega_0 = \sqrt{k/m}$, which is then solely determined by the elastic suspension. For the same transducer, however, *current drive* would—since $H_{11} = \det \mathbf{Y}/Y_{22} \neq 0$—certainly result in a shift in the electromechanical eigenfrequency.

Mechanical interpretation of transducer eigenfrequencies A physical, mechanical interpretation of transducer eigenfrequencies along with an indication for how to experimentally determine them is apparent from Fig. 5.17 for *specific* transducer types[14]. Consider a mechanical excitation F_{ext} of the transducer mass with simultaneous zero electrical excitation.

In the case of *shorted electrical terminals* ($u_S = 0$), the transducer mass oscillates due to the ELM stiffness Y_{11} with the eigenfrequency Ω_U (Fig. 5.17a). If the electrical terminals are unconnected—an *open electrical circuit*—then the transducer mass oscillates due to the ELM stiffness H_{11} with the eigenfrequency Ω_I (Fig. 5.17b).

Relative locations of resonance and antiresonance Table 5.8 presents the conditions for determining the relative locations of the resonance Ω_U or Ω_I, and antiresonance Ω_I or Ω_U. As shown, this depends on the quotient of the product of the two-port coupling terms and the two-port electrical term. For the transducer types considered here, this quotient is always real (even for complex two-port parameters), so that the relations presented in Table 5.8 represent generic transducer attributes.

Proportional regime The frequency response is affected in all transducer channels by the electromechanical resonance. Thus for both sensor and actuator operation, the basic frequency response shown in Fig. 5.18a results, where $\tilde{G}$ indicates one of $G_{x/F}$, $G^{cap}_{x/u}$, $G^{ind}_{x/i}$, $G^{cap}_{u/F}$, $G^{ind}_{i/F}$ (*cap* or *ind* respectively indicate capacitive or inductive dynamics at the electrical terminals).

[14] Condition: even without auxiliary electrical energy, there should be a current flow in the presence of an active external force, e.g. *piezoelectric* and *electrodynamic* transducers.

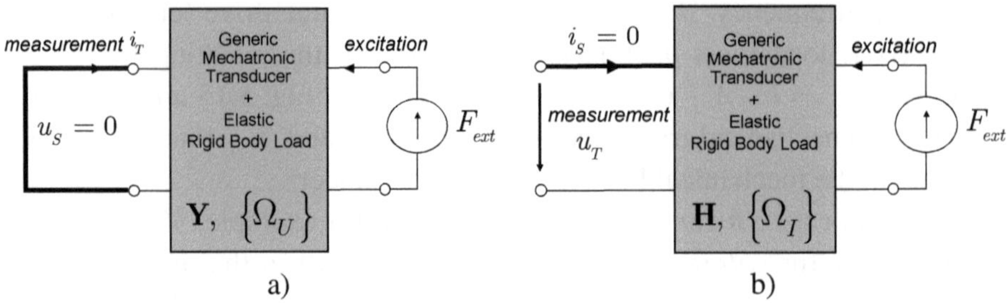

Fig. 5.17. The physical significance and measurement of transducer eigenfrequencies: a) $\{\Omega_U\}$: eigenfrequency for a shorted input; b) $\{\Omega_I\}$: eigenfrequency for an open input

Note that only in certain cases is the gain factor V in Table 5.8 real (with the resulting proportional quasi-static transducer response characteristics, see Chs. 6 through 8).

For operation both as a *sensor* ($G_{x/F}, G^{cap}_{u/F}, G^{ind}_{i/F}$, recording mechanical quantities) and as an *actuator* ($G^{cap}_{x/u}, G^{ind}_{x/i}$, generating forces), *proportional* mapping from inputs to outputs is desirable. However, this is only the case in the frequency band *below* the eigenfrequency (Fig. 5.18a).

Thus, for the maximum useable bandwidth, the transducer eigenfrequency $\Omega = \sqrt{k_T/m}$ should be chosen to be as *large* as possible. For a given load, the eigenfrequency can be selected with an suitable choice of the transducer stiffness k_T , depending on the elastic attachment k and the transducer parameters Y_{11} or H_{11}.

Dynamic stiffness: mechanical impedance Of particular importance to the dynamics at the mechanical terminals is the transfer function $G_{x/F}$. Its reciprocal

$$\frac{1}{G_{x/F}(s)} = \frac{\Delta F(s)}{\Delta X(s)} = k_T \cdot \left(1 + \frac{s^2}{\Omega^2}\right) := Z_{mech}(s) \qquad (5.50)$$

defines the so-called *dynamic stiffness* or *mechanical impedance*[15] of the transducer (Fig. 5.18b). Using this property, the mechanical relationship between changes in force and displacement can be clearly described.

---

[15] Formally, a distinction should be made between the *static* mechanical impedance H_{11} or Y_{11} (see Sec. 5.3.3) and the *dynamic* mechanical impedance $G^{-1}_{x/F}(j\omega)$ considered here.

In the case under consideration, for small frequencies sufficiently below the natural frequency Ω, the transducer behaves like a spring with stiffness k_T. Near the natural frequency, the transducer becomes infinitely soft, and above the natural frequency, the transducer acts mechanically like a very stiff spring.

Electrical terminal dynamics The electrical terminal dynamics are described by the transfer functions $G_{i/u}, G_{u/i}$. Here, $G_{u/i}(s)$ represents the *electrical impedance*, and $G_{i/u}(s)$, the *electrical admittance* of the transducer.

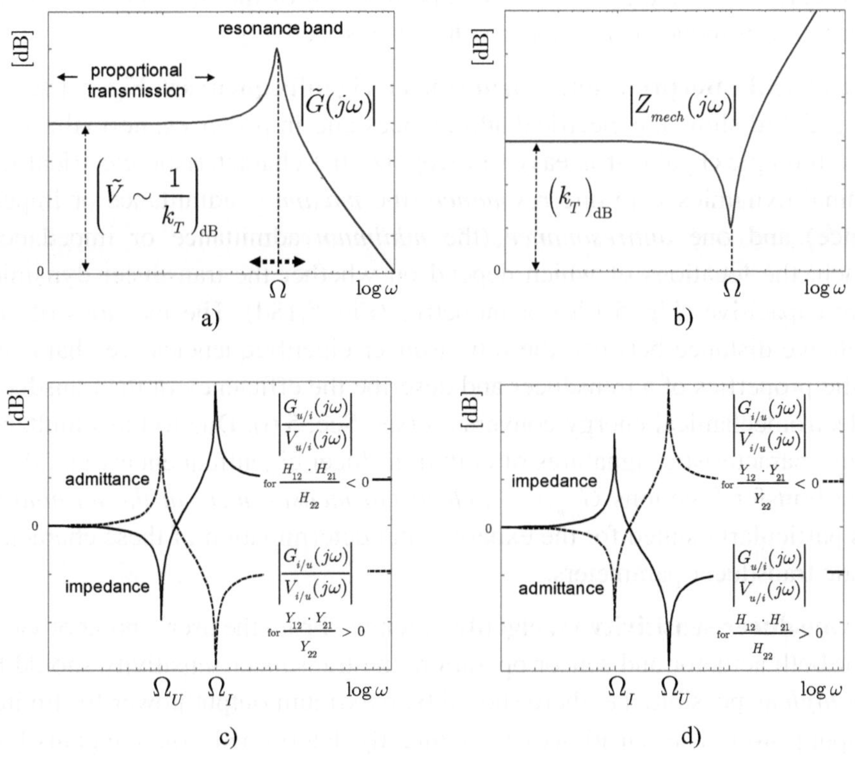

Fig. 5.18. General dynamic behavior of a *loaded generic mechatronic transducer* (load: suspended rigid body, mechanically damped): a) actuator or sensor dynamics with $\tilde{G}: G_{x/F}, G^{cap}_{x/u}, G^{ind}_{x/i}, G^{cap}_{u/F}, G^{ind}_{i/F}$, b) dynamic stiffness or mechanical impedance $Z_{mech}(j\omega)$, c) electrical admittance and impedance—normalized by the complex gain—for *capacitive* transducer dynamics $(\Omega_U < \Omega_I)$, d) electrical admittance and impedance—normalized by the complex gain—for *inductive* transducer dynamics $(\Omega_U > \Omega_I)$

As the transfer functions all apply to the same transducer, and only the view of its terminals differs, it naturally holds that

$$G_{u/i}(s) = \frac{1}{G_{i/u}(s)}.$$

The dynamics here are also dominated by the respective transducer eigenfrequency Ω_U or Ω_I. Near this resonant frequency, large input currents or voltages should be expected, putting a significant demand on the auxiliary energy source.

However, it is also worth noting the appearance of an *antiresonance* at the respective *dual* eigenfrequency Ω_I or Ω_U. Here, there is a canceling of the current or voltage due to the superposition of the motion-induced and electrically-induced currents or voltages, respectively.

Electrical interpretation of transducer eigenfrequencies Fig. 5.18c and Fig. 5.18d show the electrical admittances and impedances, normalized by the (complex) gain. It is easy to recognize the characteristic electrical terminal dynamics with one *resonance* (the *maximum* admittance or impedance) and one *antiresonance* (the *minimum* admittance or impedance) each, the locations of which depend on whether the transducer dynamics are capacitive (Fig. 5.18c) or inductive (Fig. 5.18d). The locations of and relative distance between these transducer eigenfrequencies are characteristic properties of a transducer and describe the efficiency of the transducer electromechanical energy conversion (see Sec. 5.6). Due to the simultaneous characteristic signatures of both transducer eigenfrequencies Ω_U, Ω_I in the transfer functions $G_{i/u}, G_{u/i}$, *electrical measurement at the terminals*[16] is particularly suited for the experimental determination of these characteristic transducer parameters.

Transducer sensitivity vs. eigenfrequency From the user's point of view, for both actuator and sensor operation, the transducer sensitivity should be as *high* as possible, i.e. there should be maximum output power for limited input power. The sensitivity can be directly determined from the gains V of the transfer functions in Table 5.8. In all transfer functions there is a reciprocal dependence of the gain on transducer stiffness k_T. For high sensitivity, the transducer stiffness k_T should thus be made as *small* as possible. However, this simultaneously entails a reduction of the eigenfrequency Ω

[16] No sensor of mechanical quantities is required.

(Fig. 5.18a). As a further consequence, the proportional operating regime of the transducer (where gain $\approx$ constant) is noticeably reduced.

Thus, for a given transducer, to achieve the best-possible sensitivity, a low natural frequency must be accepted, and its disadvantageous properties countered with appropriate closed-loop control measures (see Ch. 10).

Resonant operation: electrical tuning of eigenfrequencies In some applications, the transducer is to be operated as an actuator with harmonic mechanical displacements of *constant frequency*, e.g. a *scanning mirror* (Schuster et al. 2006), (Schenk et al. 2000), or *vibrating gyroscope* (Apostolyuk 2006). Such cases take advantage of extremely high sensitivities (gains) near the resonant frequency Ω. To precisely select the gain and phase of the oscillation for an application and compensate parameter variations due to the structure or operation of the transducer, the eigenfrequency can be tuned with an appropriate choice of the *resting voltage* U_0 or resting current I_0 (Fig. 5.18a).

Transducer stability: characteristic polynomial From the signal flow diagrams Fig. 5.15 and Fig. 5.16, the internal transducer feedback structure can be seen. The mechanical feedback via the spring force is in and of itself always stable. However, it is worth noting the electromechanical feedback occurring via Y_{11} or H_{11}. Formally, this is a positive feedback, and, for positive transducer stiffnesses, reduces the negative feedback of the spring force. For $Y_{11} \geq k$ (voltage drive) or $H_{11} \geq k$ (current drive), the positive electromechanical feedback dominates and the transducer becomes unstable. This behavior can also be clearly seen in the *characteristic polynomial* $\Delta(s)$ of the transfer matrices in Table 5.8:

$$
\begin{aligned}
&\text{voltage drive:}\quad \Delta_U(s) := k_{T,U} + ms^2 = k - Y_{11} + ms^2\,,\\
&\text{current drive:}\quad \Delta_I(s) := k_{T,I} + ms^2 = k - H_{11} + ms^2\,.
\end{aligned}
\tag{5.51}
$$

For the critical cases considered, it can be seen from Eq. (5.51) that for $Y_{11} = k$ (voltage drive) or $H_{11} = k$ (current drive), there is a *repeated pole* at $s = 0$ (marginal stability), and with further increasing transducer stiffnesses, one pole moves along the positive real axis (exponential instability).

In both cases, however, the transducer can certainly be employed as a completely useable actuator (e.g. in an electrostatic or magnetic bearing). However, the inherent instability of the transducer must then be actively countered with suitable control measures (see Ch. 10).

5.5 Lossy Transducer

5.5.1 General transducer behavior

Dissipative phenomena In contrast to the ideal assumptions made so far, in real transducers, dissipative phenomena on both the mechanical and electrical sides must be accounted for. Inside the *mechanical* subsystem, *viscous friction phenomena* appear. Such effects can be incorporated into the load as additional mechanical damping (Fig. 5.19). This maintains the order (number of states) of the system, and, relative to the undamped case, only the eigenfrequencies change slightly: the purely imaginary pairs of poles at the eigenfrequencies move slightly into the left half-plane (see Sec. 4.5.3). This behavior can easily be qualitatively taken into account in the complete model without requiring much additional calculation.

Resistive losses For real electrical systems, resistive losses must always be accounted for. These arise from non-negligible internal resistances in the controlled auxiliary energy sources, resistance in the conductors, or insulation losses (Fig. 5.19). Normally, resistive losses are considered unwelcome parasitic effects, and they are brought as close to idealized conditions as possible.

Resistive feedback From a system theoretical point of view, electrical resistance in certain system configurations induces *(analog) electrical feedback.* This property can be exploited in a targeted manner in a mechatronic transducer by cleverly manipulating the physical feedback properties to advantageously influence the *dynamic frequency response* at a local level.

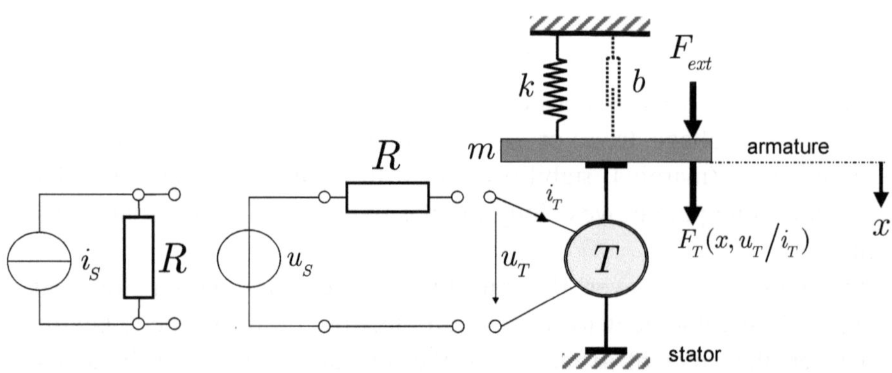

Fig. 5.19. Lossy generic mechatronic transducer

This choice of electrical resistance makes available an important *design degree of freedom* for optimizing the system behavior with *minimal* implementation effort.

Resistive configurations Resistive losses can be fundamentally modeled as a *serial resistance* on a terminal lead (lossless at $R = 0$) or as a *parallel resistance* across a terminal pair (lossless as $R \to \infty$) (Fig. 5.19). The resulting effect on the system dynamics and the model description fundamentally depends on the choice of auxiliary energy source.

Serial resistance, voltage drive For the electrical configuration shown in Fig. 5.20a, the transducer voltage u_T is no longer—as in the lossless case—equal to the independent source voltage. Rather, it now depends on the load current i_T. This behavior, which differs from the lossless transducer, must be incorporated into the dynamic model and leads to load-dependent *feedback* via the electrical subsystem of the transducer.

Serial resistance, current drive From Fig. 5.20b, it can be seen that serial resistance has no effect on the independent transducer input $i_T = i_S$. Thus, there is *no* change in the dynamics compared to the lossless case.

Parallel resistance, voltage drive This configuration, shown in Fig. 5.20c, is the dual of the previous case of serial resistance with current drive. In this case, the parallel resistance has no effect on the independent transducer input $u_T = u_S$ and induces *no* change in the dynamics compared to the lossless case.

Parallel resistance, current drive For this configuration, shown in Fig. 5.20d, there is again a dual behavior to the first case of serial resistance with voltage drive. The transducer current i_T is now no longer equal to the independent source current, but rather depends on the current transducer terminal voltage u_T. This in turn causes load-dependent *feedback* via the transducer electrical subsystem and must thus be incorporated into the dynamic model.

Resistance-insensitive configurations From the above analysis, the important, generally-applicable result follows that the configurations [*serial resistance + current source*] and [*parallel resistance + voltage source*] are insensitive to electrical resistive losses.

If there are *disturbing, parasitic* resistive phenomena present, these configurations thus represent *convenient* and *robust* system configurations, and could therefore be preferred.

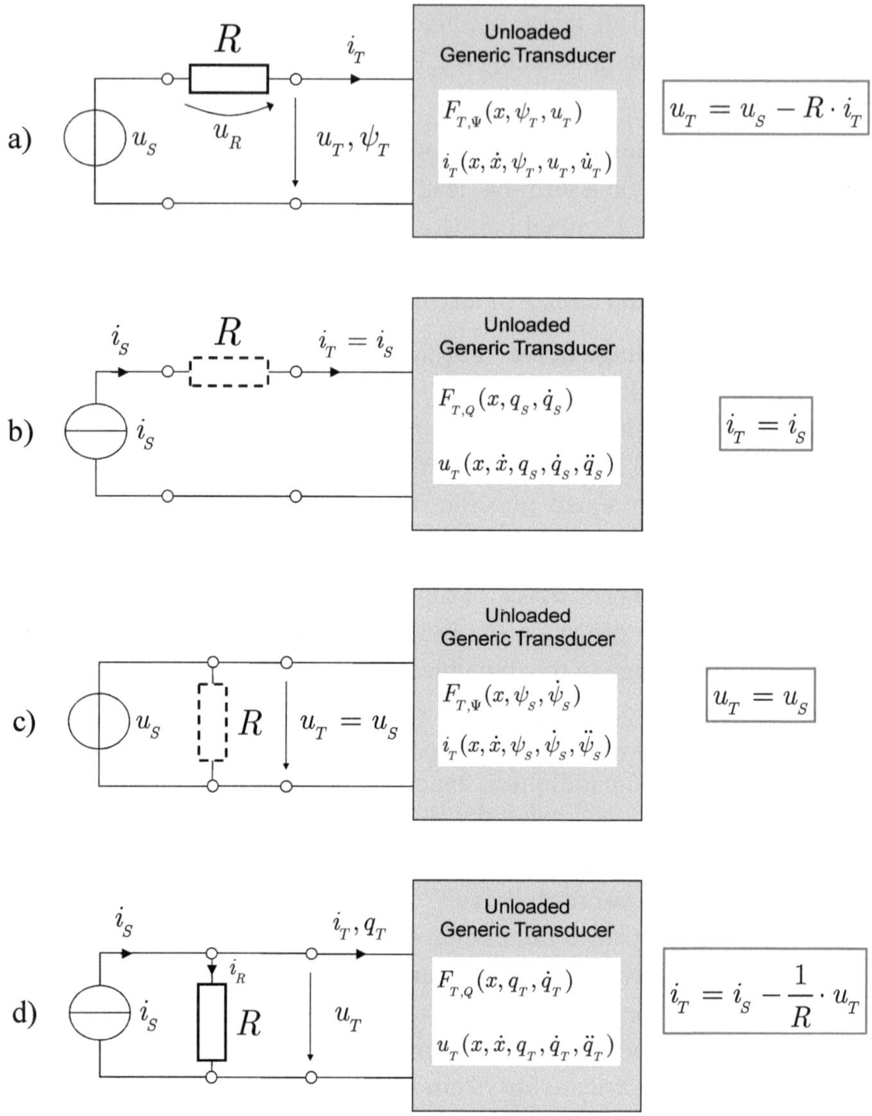

Fig. 5.20. Electrical configurations for lossy generic mechatronic transducers: a) serial resistance + voltage source, b) serial resistance + current source, c) parallel resistance + voltage source, d) parallel resistance + current source

As has yet to be shown, resistance feedback also possesses quite *positive* system properties. In this light, then, the two configurations highlighted above must be categorized as *disadvantageous*, as they block the positive properties of feedback. These configurations will thus not be further considered here.

5.5.2 Nonlinear model: equilibrium positions

Serial resistance, voltage drive

Equations of motion Combining the nonlinear equations of motion of the lossless transducer in Table 5.8 and the voltage relation from the mesh rule applied to Fig. 5.20a, the following nonlinear model for the generic *voltage-drive* transducer with *serial resistance* is obtained:

$$\boxed{\begin{aligned} & m\ddot{x} + kx + \left[-\frac{1}{2}\frac{\partial C(x)}{\partial x}\dot{\psi}_T^2\right]_{cap} \vee \left[-\frac{1}{2}\frac{1}{L(x)^2}\frac{\partial L(x)}{\partial x}\psi_T^2\right]_{ind} = F_{ext}, \\ & \dot{\psi}_T + \left[RC(x)\ddot{\psi}_T + R\frac{\partial C(x)}{\partial x}\dot{x}\cdot\dot{\psi}_T\right]_{cap} \vee \left[R\frac{1}{L(x)}\psi_T\right]_{ind} = u_S, \end{aligned}} \tag{5.52}$$

$$i_T = \left[C(x)\cdot\ddot{\psi}_T + \frac{\partial C(x)}{\partial x}\cdot\dot{x}\cdot\dot{\psi}_T\right]_{cap} \vee \left[\frac{1}{L(x)}\cdot\psi_T\right]_{ind}.$$

In contrast to the lossless transducer, there is now a *coupled* system of differential equations in the two coordinates x and ψ_T. The transducer current is obtained from an algebraic equation in the flux linkage and its derivatives.

Equilibrium positions The change in steady-state dynamics from the lossless case can be easily assessed as follows. The resistance only affects applicable system variables when a current flows. In the steady-state case, at most a constant current can flow, i.e. $i_{T,R} = const.$

In the case of *capacitive* transducer dynamics, steady-state current flow is not possible, giving the *identical* equilibrium position condition as in the lossless case.

However, there is a difference for *inductive* transducer dynamics. In this case, it is only for $u_S \neq 0$ that the serial resistance permits steady-state dynamics, which then exhibit the *steady-state rest current*

$$i_{T,R} = \frac{U_0}{R}.$$

Using

$$\psi_{T,R} = L(x_R)\cdot i_{T,R} = L(x_R)\frac{U_0}{R},$$

the equation of motion for the armature position then gives the *equilibrium position condition*

$$kx_R - \frac{1}{2}\frac{\partial L(x)}{\partial x}\bigg|_{x=x_R} \cdot \frac{U_0^2}{R^2} = F_0 . \tag{5.53}$$

Parallel resistance, current drive

Equations of motion Combining the nonlinear equations of motion of the lossless transducer in Table 5.8 and the current relation from the node rule applied to Fig. 5.20c, the following nonlinear model for the generic *current-drive* transducer with *parallel resistance* is obtained:

$$\boxed{\begin{aligned} m\ddot{x} + kx + \left[-\frac{1}{2}\frac{\partial L(x)}{\partial x}\dot{q}_T^2\right]_{ind} \vee \left[-\frac{1}{2}\frac{1}{C(x)^2}\frac{\partial C(x)}{\partial x}q_T^2\right]_{cap} &= F_{ext} , \\ \dot{q}_T + \left[\frac{1}{R}L(x)\cdot\ddot{q}_T + \frac{1}{R}\frac{\partial L(x)}{\partial x}\cdot\dot{x}\cdot\dot{q}_T\right]_{ind} \vee \left[\frac{1}{R}\frac{1}{C(x)}\cdot q_T\right]_{cap} &= i_S , \end{aligned}} \tag{5.54}$$

$$u_T = \left[L(x)\cdot\ddot{q}_T + \frac{\partial L(x)}{\partial x}\cdot\dot{x}\cdot\dot{q}_T\right]_{ind} \vee \left[\frac{1}{C(x)}\cdot q_T\right]_{cap} .$$

In contrast to the lossless transducer, there is again a *coupled* system of differential equations in the two coordinates x and q_T. The transducer terminal voltage is obtained from an algebraic equation in the transported transducer charge and its derivatives.

Equilibrium positions Here again, the above considerations hold analogously as relates to the effect of the resistance on possible equilibrium positions. It need only be noted that in the lossless case, $R \to \infty$.

In the case of *inductive* transducer dynamics, for finite R in the steady-state state, the only current flow is due to the inductance; the resistance is for all intents and purposes nonexistent, so that an *identical* equilibrium position situation results as for the lossless transducer.

For *capacitive* transducer dynamics, the parallel resistance induces a *steady-state transducer voltage*

$$u_{T,R} = R\cdot I_0 ,$$

which in turn leads to the *equilibrium position condition*

$$kx_R - \frac{1}{2}\frac{\partial C(x)}{\partial x}\bigg|_{x=x_R} \cdot R^2 I_0^{\,2} = F_0 \,. \tag{5.55}$$

5.5.3 Linear signal-based model

Calculation From the nonlinear equations of motion (5.52) and (5.54), and using applicable equilibrium positions, a linear transducer model can be set up as previously explained using a local linearization. It is easy to see that no changes in the linearization coefficients result from connecting a linear resistance R compared to the lossless case. Thus, the model of the lossless case (see Figs. 5.15, 5.16) can be directly used to derive a signal-based model for the lossy transducer.

Serial resistance, voltage drive: impedance feedback Fig. 5.21a shows the block diagram of the voltage-drive transducer with serial resistance. The two-port parameters are computed as in the lossless case. At most, an updated equilibrium position need be applied.

From Fig. 5.21a, it is quite easy to see the load-dependent feedback structure due to the serial resistance; this is *impedance feedback*.

Parallel resistance, current drive: admittance feedback The block diagram of the current-drive transducer with parallel resistance is shown in Fig. 5.21b. For the two-port parameters, the same properties as discussed above hold, and in particular, the load-dependent feedback structure due to the parallel resistance; this is *admittance feedback*.

Calculating the frequency response From the block diagram shown in Fig. 5.21, the transfer matrix can be computed for any particular concrete case via elementary manipulations, and in turn, the frequency response can be analyzed. The advantages of the two-port parameterization become apparent for general predictions using general parameter correspondences. The intermeshed feedback structure—which at first glance appears unmanageable—can, upon closer inspection, be quite straightforwardly analyzed using the two-port parameters.

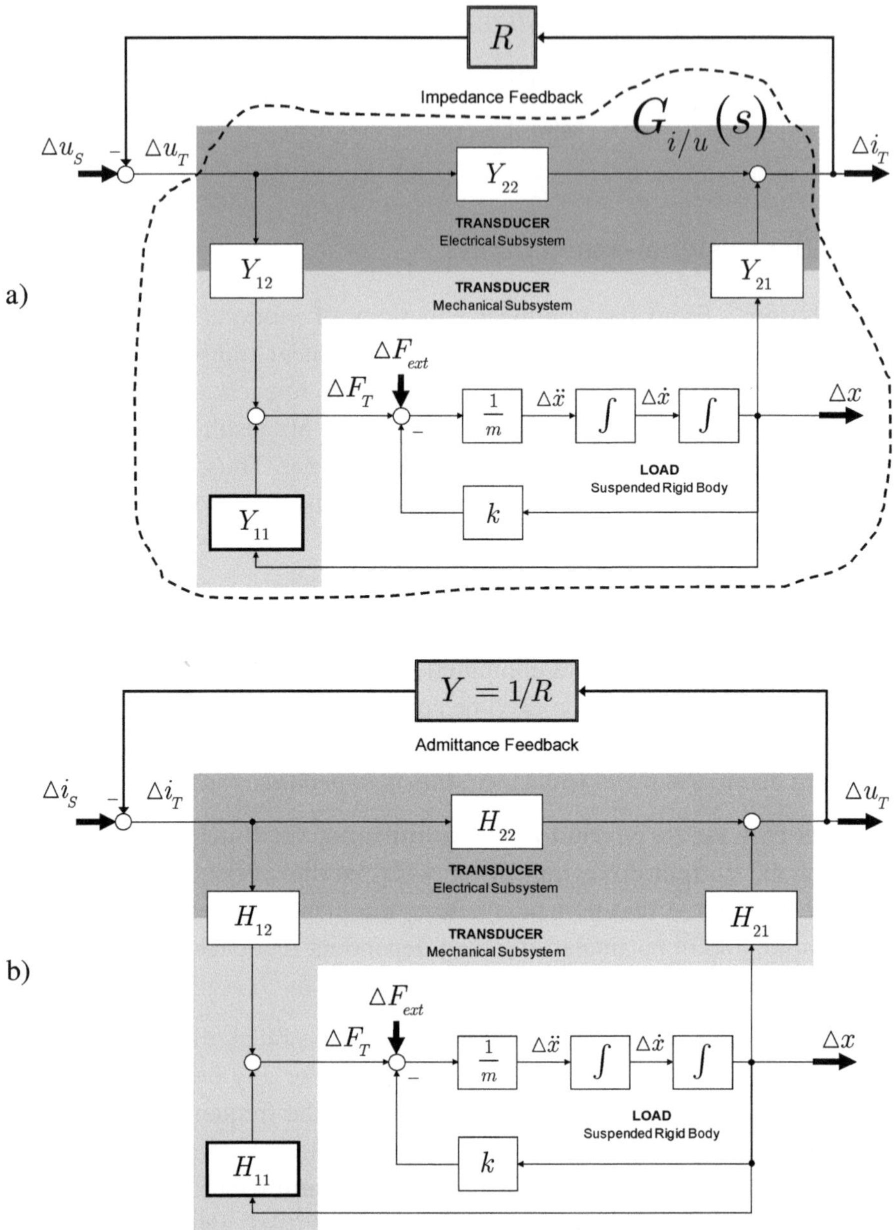

Fig. 5.21. Linear signal-based model of a lossy generic mechatronic transducer: a) serial resistance, voltage drive $\rightarrow$ *impedance feedback*; b) parallel resistance, current drive $\rightarrow$ *admittance feedback* ($\mathbf{Y}, \mathbf{H}$ are the two-port parameters of the lossless transducer)

5.5.4 Constitutive two-port equations with dissipative resistors

Lossy two-port element The two-port representation of the unloaded generic transducer (see Fig. 5.10) can be easily extended with *resistors*, as show in Fig. 5.22. The dynamics at the terminals of the resulting *lossy unloaded generic transducer*, can, after some computation, be described with extended two-port parameters $\tilde{\mathbf{Y}}, \tilde{\mathbf{H}}$ (see Table 5.9). Formally, $\tilde{\mathbf{Y}}, \tilde{\mathbf{H}}$ can then be manipulated in exactly the same way as the lossless two-port parameters $\mathbf{Y}, \mathbf{H}$.

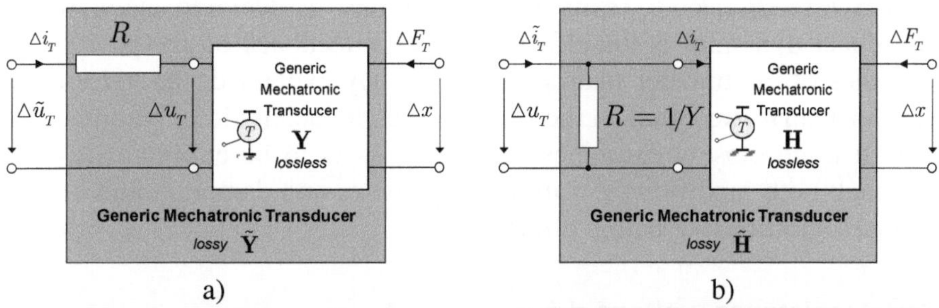

Fig. 5.22. Two-port representation of the constitutive transducer equations of a *lossy*, unloaded generic mechatronic transducer: a) serial resistance R, b) parallel resistance $R = 1 / Y$

Table 5.9. Correspondences for the two-port parameters of the constitutive transducer equations for a lossy transducer (variables without "∼" describe the lossless transducer)

Serial resistance R	**Parallel resistance** $R = 1/Y$
$\tilde{Y}_{11} = \dfrac{Y_{11} + R \cdot \det \mathbf{Y}}{1 + R \cdot Y_{22}}$	$\tilde{H}_{11} = \dfrac{H_{11} + Y \cdot \det \mathbf{H}}{1 + Y \cdot H_{22}}$
$\tilde{Y}_{12} = \dfrac{Y_{12}}{1 + R \cdot Y_{22}}$	$\tilde{H}_{12} = \dfrac{H_{12}}{1 + Y \cdot H_{22}}$
$\tilde{Y}_{21} = \dfrac{Y_{21}}{1 + R \cdot Y_{22}}$	$\tilde{H}_{21} = \dfrac{H_{21}}{1 + Y \cdot H_{22}}$
$\tilde{Y}_{22} = \dfrac{Y_{22}}{1 + R \cdot Y_{22}}$	$\tilde{H}_{22} = \dfrac{H_{22}}{1 + Y \cdot H_{22}}$

5.5.5 Linear dynamic analysis

Transfer matrix via two-port parameters Due to the formal equivalence of the *lossy* two-port parameters $\tilde{\mathbf{Y}}, \tilde{\mathbf{H}}$ and the *lossless* two-port parameters $\mathbf{Y}, \mathbf{H}$, the relations for the *transfer matrix* in Table 5.8 can be used directly. It should however be noted that, in particular, $\tilde{Y}_{11}(s)$ and $\tilde{H}_{11}(s)$ are now no longer real, so that a few additional conversions are required. These are demonstrated in this section for a few typical transducer types in, as far as possible, a general form, so that they can be applied for concrete technical implementations.

Resistive feedback vs. transducer dynamics A deciding factor for the transducer dynamics is the effect of resistive feedback on the positions of the poles of the transfer matrix. As previously explained, the poles of the transfer matrix are equal to the roots of the characteristic polynomial; this section more closely examines the latter. As is to be expected, resistive feedback both modifies the natural frequencies and damping, and also introduces additional dynamic effects.

Serial resistance, voltage drive

Characteristic polynomial Following Eq. (5.51), the *characteristic polynomial* for the voltage-drive transducer with serial resistance is formally[17]

$$\tilde{\Delta}_U(s) := \tilde{k}_{T,U} + ms^2 = k - \tilde{Y}_{11}(s) + ms^2 .$$

With the value for $\tilde{Y}_{11}(s)$ from Table 5.9 and some elementary rearranging[18],

$$\tilde{\Delta}_U(s) = \left(\frac{k - Y_{11}}{m} + s^2\right) + R \cdot Y_{22}(s) \cdot \left(\frac{k - \det \mathbf{Y}(s)/Y_{22}(s)}{m} + s^2\right). \quad (5.56)$$

Substituting the values for the lossless transducer in Table 5.8, the characteristic polynomial Eq. (5.56) can be simplified using the *transducer eigenfrequencies* Ω_U, Ω_I for *voltage drive* or *current drive*:

$$\boxed{\tilde{\Delta}_U(s) = \left(\Omega_U^{\,2} + s^2\right) + R \cdot Y_{22}(s) \cdot \left(\Omega_I^{\,2} + s^2\right)}. \quad (5.57)$$

---

[17] Note: $\tilde{Y}_{11}(s)$ is complex, see Table 5.9.

[18] Note that Y_{11} and $\det \mathbf{Y}(s) \,/\, Y_{22}(s) = H_{11}$ are always real for the transducer types considered here.

Electrical impedance feedback A physical interpretation of the characteristic equation (5.57) can be obtained from closer inspection of the feedback structure in Fig. 5.21a. The impedance feedback induces negative electrical feedback to the transducer electrical subsystem, represented by the transfer function (see Table 5.8)

$$G_{i/u}(s) = \frac{\mathcal{Z}_{i/u}(s)}{\mathcal{N}_{i/u}(s)}.$$

To improve clarity, this configuration is depicted once more in Fig. 5.23 as a control loop. It is easy to verify that the denominator of the closed-loop transfer function

$$T_{i/uq}(s) = \frac{\mathcal{Z}_{i/u}(s)}{\mathcal{N}_{i/u}(s) + R\cdot \mathcal{Z}_{i/u}(s)} = \frac{\mathcal{Z}_{i/u}(s)}{\tilde{\Delta}_U(s)} \tag{5.58}$$

is, as expected, equal to the characteristic equation (5.57). Further, all other transfer functions of the lossy transducer also have the same denominator, and thus the same poles (assuming no pole-zero cancelation from unobservable or uncontrollable states).

Additional dynamics From Eq. (5.57), it can be seen that a complex $Y_{22}(s)$ results in an increase in the system order. In most cases, $Y_{22}(s)$ represents *capacitive* or *inductive* dynamics ($\sim s$ or $\sim 1/s$), so that an additional real pole appears for the lossy transducer, i.e.

$$\tilde{\Delta}_U(s) = \left(\tilde{\Omega}_U^{\,2} + 2\tilde{d}_U\tilde{\Omega}_U \cdot s + s^2\right)\left(\tilde{\omega}_U + s\right). \tag{5.59}$$

Even though it is only a third-order polynomial, explicit analytical determination of the roots of Eq. (5.57) or the parameters $\tilde{\Omega}_U, \tilde{d}_U, \tilde{\omega}_U$ in Eq. (5.59) and their dependence on the physical transducer parameters is neither very practical nor enlightening. For improved clarity, another path, which still remains easily generalizable, is chosen here.

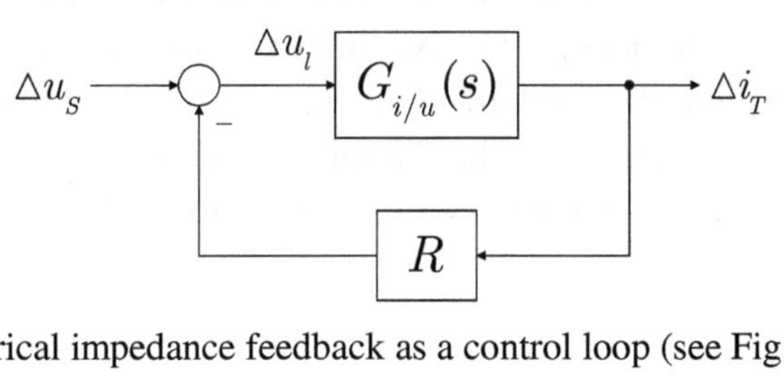

Fig. 5.23. Electrical impedance feedback as a control loop (see Fig. 5.21a)

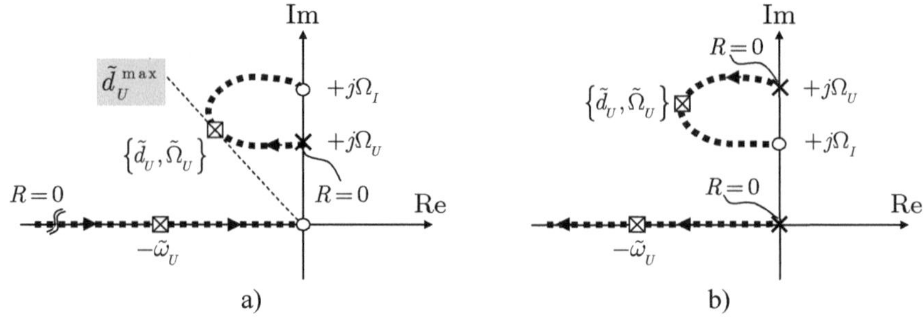

Fig. 5.24. Root loci for the poles of a *voltage-drive* generic transducer with *serial resistance* R: a) *capacitive* transducer dynamics, b) *inductive* transducer dynamics (only upper s-half-plane shown, relative locations of Ω_U, Ω_I according to the general relationships from Table 5.8)

Root locus for variable resistance For fixed transducer parameters, the roots of the characteristic equation depend solely on the (variable) resistance R. This dependence can be nicely determined qualitatively—and for concrete cases, also quantitatively—with the well-known method of *root loci* (Ogata 2010), (Horowitz 1963).

Since, for the case under consideration, the roots of $\mathcal{Z}_{i/u}(s)$ and $\mathcal{N}_{i/u}(s)$ are known explicitly, *generally-valid qualitative* statements about the distribution of roots of $\tilde{\Delta}_U(s)$ can be easily made. For concrete cases with available numerical parameters, further *quantitative* predictions are also possible (a part of system optimization).

The motion of the poles of the lossy transducer can be quite clearly described using the parameters $\Omega_U, \Omega_I, Y_{22}$ of the lossless transducer. The poles of the transfer function ($\tilde{\Omega}_U, \tilde{d}_U, \tilde{\omega}_U$, respectively) as a function of the dissipative resistance R exhibit the qualitative behavior shown in Fig. 5.24.

Electromechanical damping of eigenfrequencies One noteworthy property of impedance feedback can be seen from Fig. 5.24: the *mechanical undamped* natural frequency Ω_U becomes *damped* by the electrical feedback and moves into the left half-plane, see $\{\tilde{d}_U, \tilde{\Omega}_U\}$. However, even with an arbitrarily high resistance, the damping *cannot* be set *arbitrarily high*, as the roots of $\tilde{\Delta}_U(s)$ eventually move back towards the imaginary axis at $\pm j\Omega_I$.

Maximum damping Due to the simple geometry of the root locus, a little bit of calculation[19] gives a condition *for the maximum achievable damping* $\tilde{d}_U^{\max}$ for capacitive and inductive transducer dynamics (see also (Preumont 2006)):

$$\boxed{\tilde{d}_U^{\max} = \frac{|\Omega_U - \Omega_I|}{2\min(\Omega_U, \Omega_I)}}. \tag{5.60}$$

The corresponding *optimal resistances* are then

$$\boxed{R_{cap}^{\max} = \frac{1}{C_T}\frac{1}{\Omega_I}\sqrt{\frac{\Omega_U}{\Omega_I}}, \quad R_{ind}^{\max} = L_T\Omega_U\sqrt{\frac{\Omega_U}{\Omega_I}}}, \tag{5.61}$$

where C_T and L_T represent the effective capacitance and inductance, respectively, at the electrical terminals of the transducer with the armature *held fixed*[20].

Analog passive damping Resistive feedback offers a simple option for incorporating *passive* damping into the mechanical structure using *analog electrical* principles. This can be significant when the transducer structure makes mechanical intervention difficult (e.g. for a solid-state piezoelectric transducer).

Such feedback is an inherent system property and does not require any active control. As can be seen in Fig. 5.23, it also does not induce any sort of stability problems, as long as additional parasitic effects do not have to be taken into account (see Ch. 10).

Delay dynamics An additional effect recognizable in Fig. 5.24 is the motion of the additional real pole $s = -\tilde{\omega}_U$ as a function of the serial resistance. This motion results in different behaviors for the capacitive and inductive transducer dynamics. With *capacitive* dynamics, for increasing R, the electrical time constant $\tau_{el} = 1/\tilde{\omega}_U = RC_{eff}$ increases (Fig. 5.24a), while for *inductive* dynamics, the electrical time constant $\tau_{el} = 1/\tilde{\omega}_U = L_{eff}/R$ decreases (Fig. 5.24b).

---

[19] Applying the angle condition for a point on the root locus (Ogata 2010) and the geometric meaning of $\tilde{d}_U$, see Sec. 4.5.3.

[20] These are the two-port elements $Y_{22}(s)$, $H_{22}(s)$.

In addition, Fig. 5.24b also demonstrates the well-known fact that an inductive transducer with voltage drive is only stable with serial resistance (there is an unstable pole $s = 0$ for $R = 0$).

Design optimization An optimized system layout using a voltage-drive transducer should thus not consider the *effective serial resistance* as a parasitic phenomenon. Instead, this resistance, which cannot be avoided in reality, should be purposefully used as an additional *design degree of freedom* to satisfy applicable cost- and effort-related system requirements.

Parallel resistance, current drive

Characteristic polynomial For a current-drive transducer with parallel resistance, all the above considerations for the voltage-drive transducer can be applied in turn. For convenience, it is better here to use the conductance $Y = 1/R$ of the parallel resistance, giving, following Eq. (5.51), the *characteristic polynomial*

$$\boxed{\tilde{\Delta}_I(s) = \left(\Omega_I^{\,2} + s^2\right) + Y \cdot H_{22}(s) \cdot \left(\Omega_U^{\,2} + s^2\right)}. \tag{5.62}$$

Root locus for variable resistance The roots of the characteristic polynomial $\tilde{\Delta}_I(s)$ can once again be determined by considering the root locus. Fig. 5.25 depicts the qualitative curves for *capacitive* and *inductive* transducer dynamics. It is possible to discern (unsurprisingly) a *dual* behavior to that of the voltage-drive transducer with serial resistance.

Analog passive damping In a current-drive transducer, analog passive damping of the mechanical natural frequency can be realized using a resistance R parallel to the current source (see Eq. (5.61)).

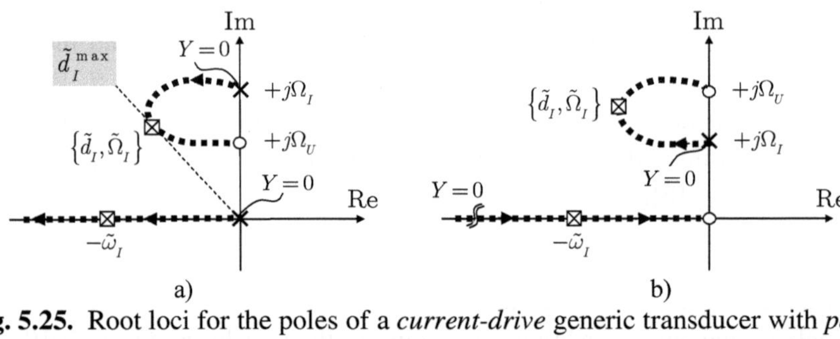

Fig. 5.25. Root loci for the poles of a *current-drive* generic transducer with *parallel resistance R*: a) *capacitive* transducer dynamics, b) *inductive* transducer dynamics (only upper s-half-plane shown, relative locations of Ω_U, Ω_I according to the general relationships from Table 5.8)

Resistive shunting One interesting variant can be derived from this principle for *solid-state piezoelectric transducers*. Removing the current source, i.e. $i_S = 0$, a resistor applied across the transducer terminals will convert the polarization current generated in the piezoelectric transducer by mechanical voltages (forces) into heat, thus creating precisely the mechanical damping described above (called *resistive shunting*) (Preumont 2006).

5.5.6 General impedance and admittance feedback

General impedances and admittances *Electrical feedback* via input serial impedances or parallel admittances as in Fig. 5.20 does not necessarily have to limit itself to purely resistive effects. The relations in Sec. 5.5.5 so far derived for resistances R can be directly applied to general electrical input circuits if, in all applicable equations, the following substitutions are made:

$$\text{voltage drive:} \quad R \rightarrow Z_{FB}(s)\,,$$

$$\text{current drive:} \quad Y = 1/R \rightarrow Y_{FB}(s)\,.$$

The *general serial impedance* $Z_{FB}(s)$ or *parallel admittance* $Y_{FB}(s)$ describe both passive and active electrical networks.

Characteristic polynomial The effects of a complex input circuit can be investigated using the example of a *voltage-drive* transducer (Fig. 5.26). Replacing the resistance in Eq. (5.58) with the general serial impedance $Z_{FB}(s)$ gives the following transfer function for the transducer electrical subsystem:

$$T_{i/uq}(s) = \frac{\mathcal{Z}_{i/u}(s)}{\mathcal{N}_{i/u}(s) + Z_{FB}(s) \cdot \mathcal{Z}_{i/u}(s)} = \frac{\mathcal{Z}_{i/u}(s)}{\breve{\Delta}_U(s)}\,,$$

where the denominator (i.e. characteristic polynomial) now fundamentally depends on $Z_{FB}(s)$.

Additional design degrees of freedom The considerations in Sec. 5.5.5 regarding the effects of the resistance R on the location of the poles of the fed-back transducer showed that motion of the poles representing undamped mechanical natural frequencies is only possible in a relatively small area.

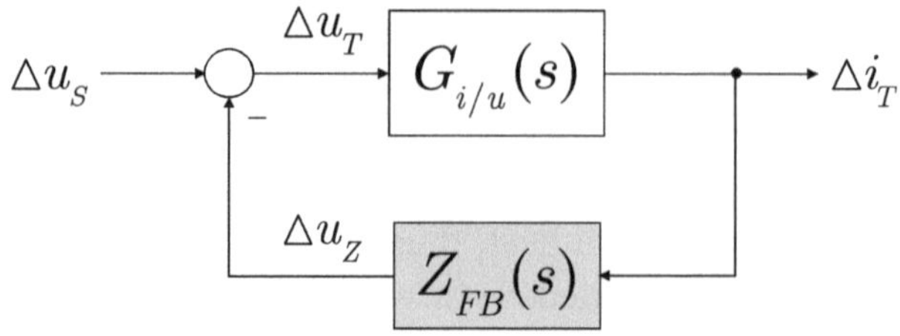

Fig. 5.26. Impedance feedback with general serial impedance for a voltage-drive transducer

The maximum reachable damping is thus structurally dependent on the transducer resonances. With a complex-valued $Z_{FB}(s)$, additional zeros and poles can now be incorporated, and, with a suitable choice, produce advantageous distortions of the root locus. Depending on the implementation complexity, this increases the number of design degrees of freedom (see Example 5.1).

Passive vs. active electrical networks The easiest networks to implement are passive RLC circuits, which can easily be placed locally on and in the transducer at both the macroscopic and microscopic level. The disadvantage to note here, however, is that only relatively limited pole-zero configurations can be realized. Using active networks employing semiconductor circuit elements, on the other hand, the realization of beneficial pole-zero configurations becomes significantly less restricted. Nowadays, this latter option is also possible at both the macroscopic and microscopic level.

Example 5.1 *Capacitive transducer with RL impedance feedback.*

Fig. 5.27 shows the schematic of the electrical port of a voltage-drive capacitive transducer. At the selected operating point, let the electrical input be represented by the effective transducer capacitance C_T (e.g. the equilibrium position capacitance).

At the input terminals, there is an RL one-port with the general serial impedance

$$Z_{FB}(s) = \frac{\Delta U_Z(s)}{\Delta I_T(s)} = R + L \cdot s \,.$$

Problem Find a qualitative description of the dynamic properties of the loaded transducer as a function of the circuit parameters R and L.

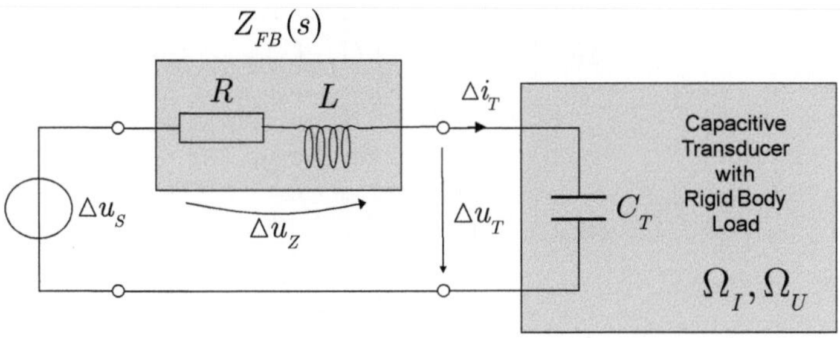

Fig. 5.27. Voltage-drive capacitive transducer with resistive-inductive serial impedance

Solution Substituting the values in Table 5.5 into Eq. (5.57), the *characteristic polynomial* of the transfer matrix becomes

$$\breve{\Delta}_U(s) = \left(\Omega_U^2 + s^2\right) + \left(R + L \cdot s\right) \cdot C_T \cdot s \cdot \left(\Omega_I^2 + s^2\right). \tag{5.63}$$

The roots of $\breve{\Delta}_U(s)$ now depend on two design parameters: R and L.

As expected, for $R = L = 0$, the undamped transducer resonance Ω_U results. The object of the RL circuit is the greatest possible damping of this eigenfrequency using an optimal choice L^{opt}, R^{opt} of the circuit impedance. The characteristic polynomial is now forth-order, so that a parametric determination of roots in order to find L^{opt}, R^{opt} is not feasible, even if computer algebra programs are employed. Thus, even in this simple case, the root locus will be used to ascertain the qualitative dependence of the roots of Eq. (5.63) on the circuit parameters R and L.

In the current case, a root locus procedure cannot be directly applied to Eq. (5.63), as the dependence is on *two parameters*. A minor rearrangement of the block diagram of the electrical feedback, however, suggests the following procedure (Fig. 5.28). By *sequentially closing* the parallel inductance and resistance feedback loops, it is possible to vary only one design parameter at each step.

For a particular order of feedback loop closure, the characteristic polynomial can be partitioned as follows:

$$\breve{\Delta}_U(s) = \underbrace{\Big[\underbrace{\left(\Omega_U^2 + s^2\right)}_{\mathcal{N}_1(s)} + L \cdot C_T s^2 \underbrace{\left(\Omega_I^2 + s^2\right)}_{\mathcal{Z}_1(s)}\Big]}_{\mathcal{N}_2(s)} + \underbrace{R \cdot C_T s\left(\Omega_I^2 + s^2\right)}_{\mathcal{Z}_2(s)}$$

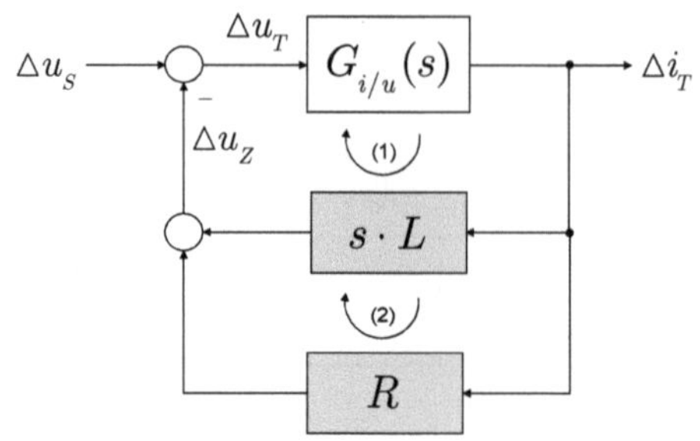

Fig. 5.28. Impedance feedback: sequential closing of parallel feedback loops

Step 1: $R = 0$, *varying L*
The root locus for $\mathcal{N}_1(s)$, $\mathcal{Z}_1(s)$ with *varying* L is sketched in Fig. 5.29a. For increasing L, an additional imaginary pair of poles moves in from infinity. The physical interpretation for this can be seen by considering the electrical circuit in Fig. 5.27. The series connection of L and C_T represents an *undamped serial oscillator* and the imaginary pair of poles represents the corresponding electrical eigenfrequency. In general, the electrical eigenfrequency $\Omega_{el} \geq \Omega_I$. For a fixed $L = L^*$, the active eigenfrequencies Ω_U^*, Ω_{el}^* result, which still remain undamped.

Step 2: fixed L^*, *varying R*
Possible root loci $\mathcal{N}_1(s)$, $\mathcal{Z}_1(s)$, with varying R and fixed L^*, are sketched in Fig. 5.29b-d. Depending on the relative magnitudes of the parameters $L^*, C_R, \Omega_U, \Omega_I$, different root loci result. For fixed $R = R^{**}$, the active eigenfrequencies Ω_U^{**}, Ω_{el}^{**} result, which now have damping d_U^{**}, d_{el}^{**} and describe the definitive dynamics of the loaded transducer.

Discussion of dynamics All variants (Fig. 5.29b-d) lead to *stable* dynamics for the *closed-loop* transducer. Since both eigenfrequencies $\Omega_U^{**}, \Omega_{el}^{**}$ appear in all transducer transfer functions, it is highly important that both be maximally damped. The circuit with $R \neq 0$ steers the root locus branches into the left half-plane, generating the desired dissipative dynamics. From the two variants in Fig. 5.29b and Fig. 5.29c, it can be seen that simultaneous maximization of the damping of both eigenfrequencies cannot be achieved offhand by independently selecting the circuit parameters R and L. Given a change in the relative magnitudes of the parameters, though the shape of the root locus changes, one of the two eigenfrequencies still remains significantly less damped. It can be shown that the *optimal* circuit L^{opt}, R^{opt} qualitatively leads to a root locus as in Fig. 5.29d (Preumont 2006). In this case, the electrical and mechanical transducer eigenfrequencies collapse ($\Omega_{el}^{opt} = \Omega_{U}^{opt}$) and possess identical damping $d_{el}^{opt} = d_{U}^{opt}$. For the *optimal parameters* the following relations hold (Preumont 2006):

$$\boxed{\begin{aligned} \Omega_U^{opt} &= \Omega_I, \quad d^{opt} = \frac{1}{2}\sqrt{\frac{\Omega_I^{\,2}}{\Omega_U^{\,2}} - 1}, \\ L^{opt} &= \frac{1}{C_T}\frac{\Omega_U^{\,2}}{\Omega_I^{\,4}}, \quad R^{opt} = \frac{2}{C_T}\frac{\Omega_U^{\,2}}{\Omega_I^{\,3}}\sqrt{\frac{\Omega_I^{\,2}}{\Omega_U^{\,2}} - 1}. \end{aligned}} \tag{5.64}$$

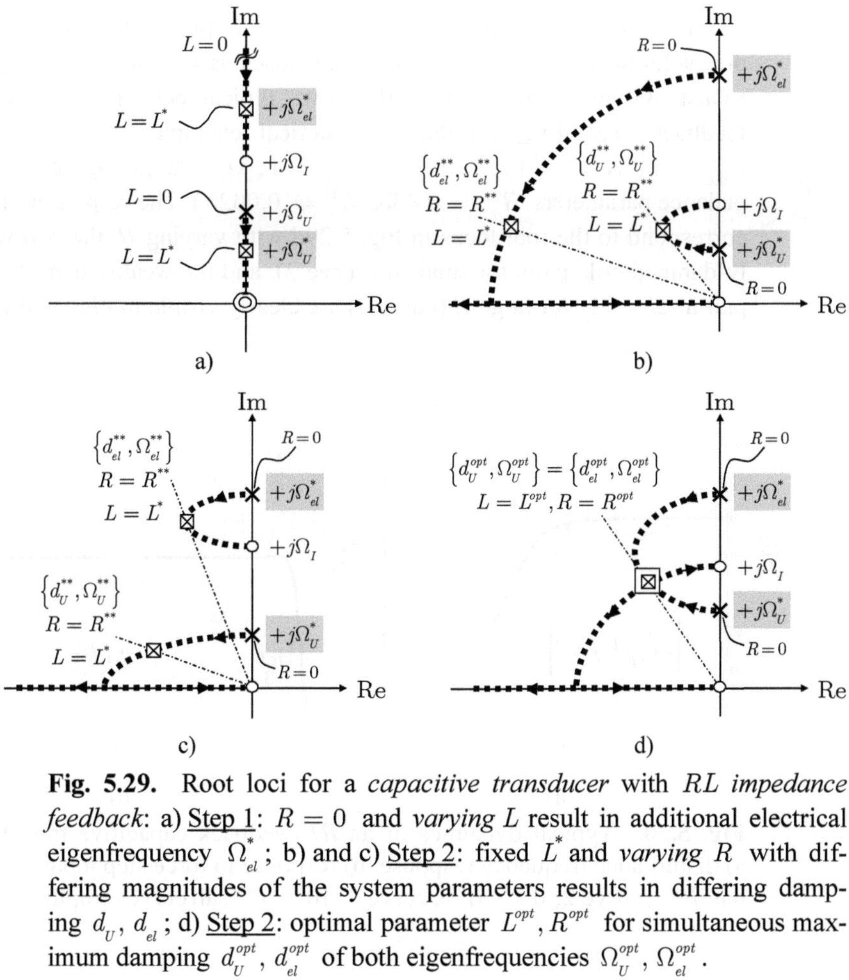

Fig. 5.29. Root loci for a *capacitive transducer* with *RL impedance feedback*: a) Step 1: $R = 0$ and *varying* L result in additional electrical eigenfrequency Ω_{el}^{*}; b) and c) Step 2: fixed L^{*} and *varying* R with differing magnitudes of the system parameters results in differing damping d_U, d_{el}; d) Step 2: optimal parameter L^{opt}, R^{opt} for simultaneous maximum damping d_U^{opt}, d_{el}^{opt} of both eigenfrequencies Ω_U^{opt}, Ω_{el}^{opt}.

A comparison of the maximum achievable damping with RL impedance feedback (see Eq. (5.64)) and purely *resistive* impedance feedback (see Eq. (5.60)) demonstrates the greater effectiveness of the complex-valued impedance, since

$$d_{R-L}^{opt} \approx \sqrt{d_R^{\max}}\ .$$

However, this *higher damping* of *mechanical* eigenfrequencies is paid for with the additional electrical eigenfrequency. Its effects must naturally also be accounted for when designing a controller for the complete mechatronic system (see Ch. 10). It can be further shown that the greater damping is only effective in a constrained tuning neighborhood of the transducer eigenfrequency (Preumont 2006), and when detuned, quickly falls below the resistive damping.

Given possible parameter uncertainty or variation, in the end, the simpler solution of resistive impedance feedback is the more effective and robust system solution. Fig. 5.30 shows typical behaviors of an RL-feedback capacitive transducer (numerical example: $k = 2,\ m = 1,\ k_{el,U} = 1,\ K_U = \sqrt{3},\ C_R = 1 \Rightarrow \Omega_U = 1,\ \Omega_I = 2$ giving optimal impedance parameters $R^{opt} = 0.433,\ L^{opt} = 0.0625$). The responses shown correspond to the root locus in Fig. 5.29d with varying R: the two weakly-damped pole pairs for small R (curve 2), and the weakly-damped pole pair at $\omega \approx \Omega_I$ for large R (curve 3) are clearly visible in Fig. 5.30a.

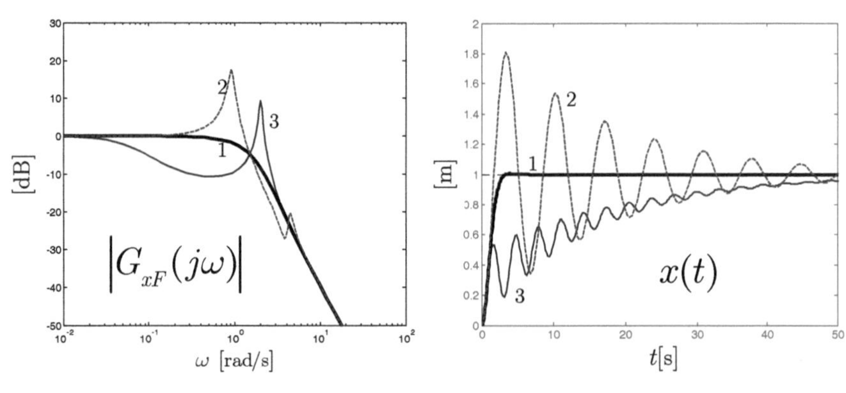

a) b)

Fig. 5.30. Typical dynamics of an RL-feedback capacitive transducer: a) disturbance frequency response, b) response to force step input; curve 1 has R^{opt}; curve 2, $0.1 \times R^{opt}$; curve 3, $10 \times R^{opt}$; all curves employ L^{opt}

Discussion of the analysis Fundamentally, when constructing the root locus, a different partitioning (swapping the order of loop closures with L and R) could have swapped the order of the parameter variation. However, it is easy to verify that in this case, first R is varied in Step 1, resulting in the root locus in Fig. 5.24a. If the poles are fixed at $R = R^{**}$, constructing the root locus in Step 2 with varying L no longer gives as clear an answer as in the case described here. Which order is more convenient depends on the particular configuration and must be verified for any single case.

An alternative demonstrative perspective is offered by the *frequency response* using BODE *diagram* or NICHOLS *diagram* representations. The application of these analysis methods is discussed in detail in Ch. 10.

■

Analog electrical feedback Modifying the electrical input to a transducer using impedance and admittance feedback influences the power coupling to the mechanical system and has a noticeable, and in the best case, beneficial impact on the mechatronic plant by increasing the mechanical damping. In fact, this feedback acts as a *local control loop*, approximately equivalent to the inner loop of a cascaded controller.

From the *design perspective*, with this measure, the design enters the border zone of *control systems design* for mechatronic systems. This again demonstrates the intermeshed nature of mechatronic systems design. Compared to other representations known in the literature, it is precisely the control theoretical perspective applied here (using feedback and a root locus) which offers clear insights into system interrelationships.

From the *realization perspective*, it is above all worth noting that this method employs *minimally-complex analog feedback*, requiring neither a sensor nor an explicit control device, and additionally avoiding any complications associated with digital control.

Here, then, *analog control*—so often already declared deceased—proves itself, as if through a back door, to be an enduring keystone solution in high-tech mechatronic products.

5.6 Electromechanical Coupling Factor

5.6.1 General significance and attributes

Electromechanical energy conversion The essential attribute of a mechatronic transducer is its capability to convert electrical energy to mechanical energy and vice versa. However, even under the idealizing assumption of *lossless* energy conversion, all of the energy flowing into the electrical or mechanical terminals is not transformed into the respective complementary form of energy. The reason for this lies in the capacity of a mechatronic transducer to store both electrical and mechanical energy. Consider, for example, the ELM two-port equations in the admittance formulation (see Eq. (5.35)):

$$\begin{pmatrix} \Delta F_{T,\Psi}(s) \\ \Delta I_T(s) \end{pmatrix} = \mathbf{Y}(s) \cdot \begin{pmatrix} \Delta X(s) \\ \Delta U_T(s) \end{pmatrix} = \begin{pmatrix} Y_{11} & Y_{12} \\ Y_{21} & Y_{22} \end{pmatrix} \cdot \begin{pmatrix} \Delta X(s) \\ \Delta U_T(s) \end{pmatrix}.$$

The element Y_{11} represents the dynamics at the mechanical terminals of the transducer in the form of a spring stiffness, and thus the capacity of the transducer to internally store potential mechanical energy. On the other hand, the element Y_{22} represents the dynamics at the electrical terminals of the transducer in the form of a capacitance or inductance, and thus its capability to internally store electrical or magnetic energy. Only the two diagonal elements Y_{12}, Y_{21} describe the internal electromechanical interaction and thus the actual energy conversion.

Electromechanical coupling factor: definition To quantitatively describe the capacity of a (lossless, mechanically loaded) mechatronic transducer to deliver the energy furnished to it on to a mechanical or electrical load, a non-negative measure—the *electromechanical (ELM) coupling factor*[21] κ^2—is defined:

$$\boxed{\kappa^2 := \frac{\text{output energy (mechanical or electrical)}}{\text{input energy (electrical or mechanical)}}}, \quad 0 \leq \kappa^2 \leq 1. \tag{5.65}$$

[21] Customarily, the symbol for the coupling factor is k^2. As this section deals with the electromechanical coupling factor for a transducer with an *elastic rigid-body* mechanical load (m, k), the discussion falls back onto the neutral symbol κ for clarity.

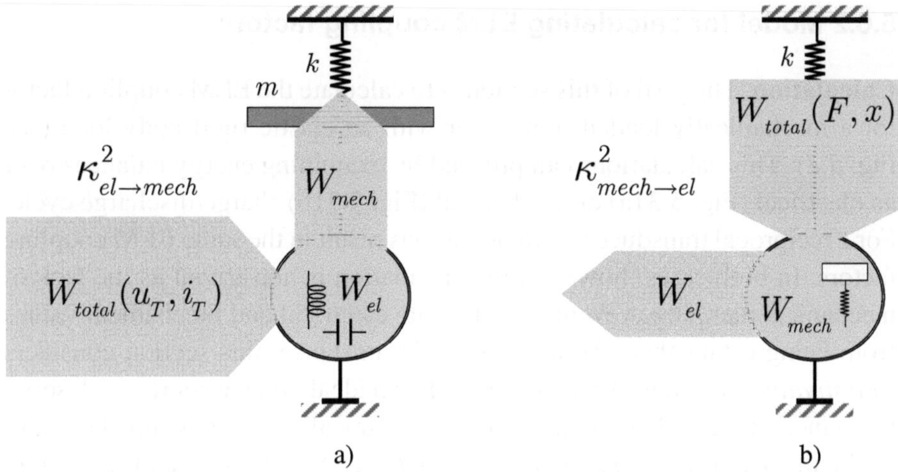

Fig. 5.31. Definition of electromechanical coupling factors: a) actuator operation, b) sensor operation; transducer is reciprocal if $\kappa^2_{el\to mech} = \kappa^2_{mech\to el} = \kappa^2$

Fig. 5.31 shows a schematic of the energy flows for a mechatronic actuator and sensor. If $\kappa^2_{el\to mech} = \kappa^2_{mech\to el} = \kappa^2$, the transducer is termed *reciprocal.*

Note that the coupling factor defined in Eq. (5.65) does *not* represent the *efficiency* of the transducer, but is only a measure of the loss of energy during the actual conversion. As the coupling factor describes the energy conversion in a lossless transducer, it is always greater than the efficiency.

ELM coupling factor: general representation In the mechatronic transducer literature, the electromechanical (ELM) coupling factor is found primarily in connection with *piezoelectric* transducers—e.g. (Mohammed 1966), (DIN 1988), (IEEE 1988), (Senturia 2001), (Preumont 2006)—and only sporadically in connection with other transducer types—e.g. (Tilmans 1996), (Senturia 2001), (Yaralioglu et al. 2003), (Preumont 2006)—for which representations are mostly tailored to the parameters and notation of the particular transducer type.

The following section demonstrates that the general transducer description chosen in this book and based on the ELM two-port parameters enables a *generalized* representation of electromechanical coupling factors, which is completely *independent* of the particular *transducer type.*

5.6.2 Model for calculating ELM coupling factors

Calculation The goal of this section is to calculate the ELM coupling factor for a mechanically loaded transducer with an elastic rigid-body load (see Fig. 5.2). This calculations can proceed by examining energy balances over an electrical (Fig. 5.31a) or mechanical (Fig. 5.31b) charge/discharge cycle. For a reciprocal transducer, both operations result in the same ELM coupling factors. In both cases, however, the calculation is non-trivial as the lack of mechanical damping prevents steady-state electrical and mechanical values from being established. To avoid this circumstance, this section considers a *dissipative electrical drive*—concretely, an ideal voltage source with serial resistance[22] (Fig. 5.32a). This ensures mechanically stable, damped dynamics in all cases due to *impedance feedback* (see Sec. 5.5). In addition, independent of the electrical terminal characteristics, finite energy variables (charge for capacitive dynamics, and flux linkage for inductive dynamics) result, so that an energy balance is possible using steady-state quantities.

Experiment description The schematic experimental configuration is shown in Fig. 5.32a, and the corresponding time evolution can be seen in Fig. 5.32b. As an example, the ELM coupling factor is to be determined via an *electrical charge/discharge cycle.*

In the process, it is assumed that no external force excitation acts on the transducer, and that initially, it is electrically and mechanically fully discharged.

If, for example, an operating point is selected using a constant voltage (e.g. in an electrostatic transducer or electromagnetic transducer), the energy balance is set up for changes relative to this operating point (differential energy balance). This also suggests the use of linear transducer models.

Terminal-relative energy balance With the transducer configuration in Fig. 5.32a, the energy balance relative to the electrical terminal variables Δu_T, Δi_T is of interest. Even though the electrical drive is itself dissipative, the transducer in Fig. 5.32a represents a *conservative* system at the electrical terminals. This is significant to the extent that the applicable *energy integrals* over the electrical coordinates $\Delta\psi_T$, Δq_T thus remain *path-independent.*

[22] All considerations can also be applied equivalently, and with equivalent results, to find the ELM coupling factor for an ideal current source and parallel resistance.

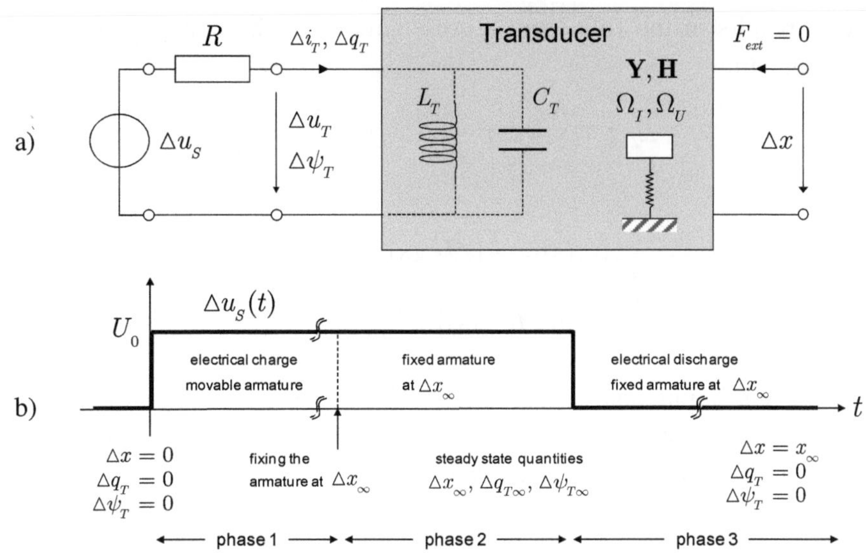

Fig. 5.32. Model for calculating an electromechanical coupling factor: a) transducer configuration, b) time evolution of electrical charge/discharge

For calculation, it is thus completely sufficient to consider *steady-state* values; the "path" to reach them via the transient trajectory has no further relevance. As impedance feedback ensures overall stable dynamics of the electrical energy variables $\Delta\psi_T, \Delta q_T$, it becomes possible to easily compute steady-state system variables via the final value theorem[23] of the LAPLACE transform using the linear transducer model.

Phase 1: Electrical charging with free-moving armature

Steady-state electrical terminal variables To calculate the electrical terminal variables as the transducer is charged, the transfer function $T_{i/uq}(s)$ for the transducer with resistive feedback in Eq. (5.58) is employed (see Fig. 5.23). Applying the ELM two-port parameters from Table 5.8 gives

$$T_{i/uq}(s) = \frac{\Delta I_T(s)}{\Delta U_S(s)} = \frac{\dfrac{k_{T,I}}{k_{T,U}} Y_{22}(s) \cdot \{\Omega_I\}}{\{\Omega_U\} + R \cdot \dfrac{k_{T,I}}{k_{T,U}} Y_{22}(s) \cdot \{\Omega_I\}}.$$

[23] The specified conditions ensure the existence of the final value in the time domain and thus enable equivalent calculation of the final value in the s-domain.

This results in the following representation for the energy variables in the s-domain:

$$\Delta Q_T(s) = \frac{1}{s}\Delta I_T(s) = \frac{1}{s} T_{i/uq}(s) \cdot \Delta U_S(s)\,,$$

$$\Delta \Psi_T(s) = \frac{1}{s}\Delta U_T(s) = \frac{1}{s}\left(\Delta U_S(s) - R \cdot \Delta I_T(s)\right) = \\ = \frac{1}{s}\left(1 - R \cdot T_{i/uq}(s)\right) \cdot \Delta U_S(s)\,.$$

For a step-input source voltage $u_S(t) = U_0 \cdot \sigma(t)$, a few intermediate calculations and application of the LAPLACE transform final value theorem (Ogata 2010) give the following *steady-state* values[24]:

$$\Delta Q_{T\infty} = U_0 \frac{k_{T,I}}{k_{T,U}} \cdot \lim_{s \to 0} \left[\frac{1}{s} \frac{Y_{22}(s)}{1 + R \cdot \frac{k_{T,I}}{k_{T,U}} Y_{22}(s)} \right], \tag{5.66}$$

$$\Delta \Psi_{T\infty} = U_0 \cdot \lim_{s \to 0} \left[\frac{1}{s} \frac{1}{1 + R \cdot \frac{k_{T,I}}{k_{T,U}} Y_{22}(s)} \right]. \tag{5.67}$$

Note that the steady-state electrical energy variables plainly depend on the ELM transducer admittance $Y_{22}(s)$, but—via the transducer parameters $k_{T,U}, k_{T,I}$—are also determined by the *mechanical coupling* (Y_{11}, Y_{12}, Y_{21}) and the *external mechanical* load (k). The electrical charge supplied by the voltage source is thus also used to "mechanically charge" the transducer.

---

24 The existence of final values is fundamentally ensured by the resistive impedance feedback; all that remains to note is that the finite energy variable corresponding to either the case of capacitive or inductive transducer impedance should be examined, i.e. q_T for capacitive transducers, ψ_T for inductive transducers.

Steady-state armature position The constant electrical excitation results in some constant armature displacement x_∞, so that mechanical potential energy is stored in the transducer spring Y_{11} and the load suspension k. For further computations, however, it is not necessary to explicitly determine x_∞.

Charging energy: capacitive dynamics For capacitive dynamics, $Y_{22}(s) = C_T \cdot s$, so that the *accumulated charge* is[25]

$$\Delta Q_{T\infty} = C_T \frac{k_{T,I}}{k_{T,U}} U_0 . \tag{5.68}$$

Due to the previously mentioned path-independence of the energy integral, it follows from Eq. (5.68) that the *total stored energy* in the transducer is

$$\boxed{W_{el} = V^*_{charge} = C_T \frac{k_{T,I}}{k_{T,U}} \frac{U_0^2}{2}}. \tag{5.69}$$

Charging energy: inductive dynamics For inductive dynamics, $Y_{22}(s) = 1/(L_T \cdot s)$, so that the *accumulated flux linkage* is[26]

$$\Delta \Psi_{T\infty} = L_T \frac{k_{T,U}}{k_{T,I}} \frac{U_0}{R} . \tag{5.70}$$

Analogously to the capacitive transducer, the *total stored energy* in the transducer is

$$\boxed{W_{el} = T^*_{charge} = L_T \frac{k_{T,U}}{k_{T,I}} \frac{1}{2} \left(\frac{U_0}{R} \right)^2}. \tag{5.71}$$

Phase 2: Fixing the armature under electrical excitation

After steady-state values are achieved, the armature is mechanically fixed at its steady-state displacement x_∞, while a constant excitation $u_S(t) = U_0$ is maintained, i.e. $x(t) = x_\infty = const.$ during this phase. As a result, there are no changes in the energy balance.

---

25 The flow variable $\psi_T(t)$, as the integral of the transducer voltage, does not have a finite value in the steady state, as $u_{T\infty} = U_0$. However, this also not further relevant, as in the capacitive case, energy is stored in the electric field.

26 U_0/R describes the steady-state current in the transducer inductance.

Phase 3: Electrical discharge with fixed armature

If, in the electrically-charged state with armature fixed at $x(t) = x_\infty$, the source voltage is reset to $\Delta u_S = 0$ (a short circuit, or quasi-short for set point operation), then—due to the mechanical stop—only the energy contained in the transducer electrical storage (capacitance or inductance) can be recovered.

Discharge energy: capacitive dynamics In the steady-state, mechanically-fixed condition, the transducer terminal voltage remains unchanged at U_0, giving the electrically-recoverable energy stored by the transducer capacitance C_T:

$$\boxed{V^*_{discharge} = C_T \cdot \frac{U_0^{\,2}}{2}}. \tag{5.72}$$

Discharge energy: inductive dynamics For inductive transducer dynamics, in the steady-state, mechanically-fixed condition, the rest current U_0 / R flows unchanged, giving the electrically-recoverable energy stored by the transducer inductance:

$$\boxed{T^*_{discharge} = L_T \cdot \frac{1}{2}\left(\frac{U_0}{R}\right)^2}. \tag{5.73}$$

Energy balance: ELM coupling factor

Mechanical energy A comparison of the input electrical energy in Eqs. (5.69), (5.71) and the recovered electrical energy in Eqs. (5.72), (5.73) shows that the difference must be stored as mechanical energy in the transducer, i.e.

$$\begin{aligned} \text{capacitive:}\quad W_{mech} &= V^*_{charge} - V^*_{discharge}, \\ \text{inductive:}\quad W_{mech} &= T^*_{charge} - T^*_{discharge}. \end{aligned} \tag{5.74}$$

ELM coupling factor Using the definition in Eq. (5.65) and the calculated energy balances for electrical charge and discharge gives the *ELM coupling factor* for:

- *capacitive* transducer dynamics, $k_{T,I} > k_{T,U}$, $\Omega_I > \Omega_U$ (see Table 5.8)

$$\kappa_{cap}^2 = \frac{W_{mech}}{W_{el}} = \frac{C_T \dfrac{k_{T,I}}{k_{T,U}} - C_T}{C_T \dfrac{k_{T,I}}{k_{T,U}}} = \frac{k_{T,I} - k_{T,U}}{k_{T,I}} = \frac{\Omega_I^2 - \Omega_U^2}{\Omega_I^2}, \tag{5.75}$$

- *inductive* transducer dynamics, $k_{T,U} > k_{T,I}$, $\Omega_U > \Omega_I$ (see Table 5.8)

$$\kappa_{ind}^2 = \frac{W_{mech}}{W_{el}} = \frac{L_T \dfrac{k_{T,U}}{k_{T,I}} - L_T}{L_T \dfrac{k_{T,U}}{k_{T,I}}} = \frac{k_{T,U} - k_{T,I}}{k_{T,U}} = \frac{\Omega_U^2 - \Omega_I^2}{\Omega_U^2}. \tag{5.76}$$

General ELM coupling factor The type-specific coupling factors (5.75) and (5.76) can be transformed into a single, *type-independent* ratio, so that for the *ELM coupling factor*, the following *general* relation results:

$$\boxed{\kappa^2 = \frac{\left|k_{T,I} - k_{T,U}\right|}{\max\left\{k_{T,I}, k_{T,U}\right\}} = \frac{\left|\Omega_I^2 - \Omega_U^2\right|}{\max\left\{\Omega_I^2, \Omega_U^2\right\}}}. \tag{5.77}$$

ELM coupling factor with two-port parameters Substituting the transducer variables in Eq. (5.77) with the *ELM two-port parameters* (see Table 5.8), the following equivalent relation for the general ELM coupling factor results:

$$\boxed{\begin{aligned} \kappa^2 &= \frac{\left|Y_{11} - H_{11}\right|}{\max\left\{k - Y_{11}, k - H_{11}\right\}} = \\ &= \frac{\left|\dfrac{Y_{12} \cdot Y_{21}}{Y_{22}}\right|}{\max\left\{k - Y_{11}, k - H_{11}\right\}} = \frac{\left|\dfrac{H_{12} \cdot H_{21}}{H_{22}}\right|}{\max\left\{k - Y_{11}, k - H_{11}\right\}}. \end{aligned}} \tag{5.78}$$

Note that the mixed product terms are all real, and carry the physical units of a spring stiffness.

Reciprocal transducer It is easy to verify that the derivation of the ELM coupling factors can be performed analogously via a mechanical excitation, i.e. a constant force excitation with open and shorted electrodes. In the case of a *reciprocal* transducer (all transducer types in Table 5.5 except for the back-effect-free actuator), the same result as in Eqs. (5.77), (5.78) is obtained.

5.6.3 Discussion of ELM coupling factors

Systems engineering significance ELM coupling factors offer a convenient and concise method to compare different transducers of the same family—and moreover even transducers of *different*—families as concerns their energy conversion capabilities. If an ELM coupling factor is not directly available from data sheets, it can be relatively easily determined from the constitutive transducer parameters (see Table 5.5 or the respective chapters on functional realizations of transducers), or even experimentally.

Experimental determination of coupling factors The description of the ELM coupling factor in Eq. (5.77) suggests its experimental determination via admittance or impedance measurements at the electrical terminals of a mechanically loaded transducer. The maximum and minimum of the measured admittance (impedance) give the characteristic transducer eigenfrequencies Ω_U, Ω_I and thus, via Eq. (5.77), the ELM coupling factor. All that remains to verify is whether non-negligible parasitic electrical impedances are present. If so, these must be determined separately and taken into account in the calculation (via transducer eigenfrequency shifts, see Sec. 5.5).

Spread of transducer eigenfrequencies The spread of the transducer eigenfrequencies Ω_U, Ω_I can be represented using Eqs. (5.75) and (5.76) as follows:

$$\text{capacitive: } \Omega_I^{\;2} = \frac{\Omega_U^{\;2}}{1-\kappa^2}, \qquad \text{inductive: } \Omega_U^{\;2} = \frac{\Omega_I^{\;2}}{1-\kappa^2}. \tag{5.79}$$

Electrical admittance with mechanical loading The electrical admittance of an unloaded transducer is given by the two-port element $Y_{22}(s)$ in Eq. (5.35) (for *capacitive dynamics*: $Y_{22}(s) = s \cdot C_T$, *inductive* dynamics: $Y_{22}(s) = 1/(s \cdot L_T)$). For the loaded transducer, on the other hand, due to the presence of electromechanical coupling, the electrical storage capacity fundamentally increases. This can be read off of the (complex) gain of the transfer function $G_{i/u}(s)$ (admittance):

$$K_{i/u} = Y_{22}(s)\frac{k_{T,I}}{k_{T,U}} = \begin{cases} s\cdot\dfrac{C_T}{1-\kappa^2} = s\cdot\tilde{C}_T\,, \\[2ex] \dfrac{1-\kappa^2}{s\cdot L_T} = \dfrac{1}{s\cdot\tilde{L}_T}\,. \end{cases} \tag{5.80}$$

The electrical parameters $\tilde{C}_T$, $\tilde{L}_T$ represent the electrical storage elements active at the electrical terminals, accounting for the electromechanical coupling.

Reciprocally, the electrical *impedances* are given by inverting the relations in Eq. (5.80).

Mechanical damping for impedance/admittance feedback In Sec. 5.5.5 and Sec. 5.5.6, the fundamental relationship between the maximum achievable mechanical damping and the ELM coupling factor was already identified. As discussed there, energy from the electromechanical energy exchange is dissipated by a resistance in the electrical loop, which damps the mechanical subsystem (itself an undamped spring-mass-oscillator).

For purely *resistive impedance feedback* (voltage drive) and *resistive admittance feedback* (current drive), Eq. (5.60) and the ELM coupling factors for the maximum achievable mechanical damping imply (equally for both *capacitive and inductive* dynamics[27]) that

$$\boxed{d_R^{\max} = \frac{|\Omega_U - \Omega_I|}{2\min(\Omega_U,\Omega_I)} \simeq \frac{1}{4}\frac{\kappa^2}{1-\kappa^2}}\,. \tag{5.81}$$

In Sec. 5.5.6, the maximum achievable damping for a capacitive transducer with *RL impedance feedback* as in Eq. (5.64) was determined. This result can be generalized for transducers with electrically resonant load

---

27 Approximation: $\dfrac{x^2 - y^2}{y^2} \simeq \dfrac{2\cdot(x-y)}{y}$.

circuits (see Sec. 5.7) giving the general maximum achievable mechanical damping for *electrically resonant loading*

$$\boxed{d_{resonant}^{\max} = \frac{1}{2}\sqrt{\frac{\kappa^2}{1-\kappa^2}}}. \tag{5.82}$$

It is worth noting that the maximum achievable mechanical damping for a mechatronic transducer with electrical feedback thus *solely* depends on the ELM coupling factor.

Effect of spring stiffness From the general two-port relation (5.78), it follows that the ELM coupling factor is inversely proportional[28] to the difference of the load stiffness and transducer stiffness, i.e.

$$\boxed{\kappa^2 \sim \frac{\alpha_1}{k - Y_{11} + \alpha_2}} \text{ or } \boxed{\kappa^2 \sim \frac{\beta_1}{k - H_{11} + \beta_2}}. \tag{5.83}$$

From Eq. (5.83), it can be seen that *stiff attachment of the load* (large k) leads to a *reduction* of the ELM coupling factors. On the other hand, though softer attachment of the load increases both the ELM coupling factor and the transducer sensitivities $K_{x/F}, K_{x/u}, K_{i/F}$, it also decreases the useable bandwidth due to the resulting smaller eigenfrequency Ω_U.

The *transducer stiffnesses* Y_{11}, H_{11} play a deciding role here. In the case of a *negative* transducer stiffness ($Y_{11} < 0,\ H_{11} < 0$, e.g. electrodynamic and piezoelectric transducers, see Table 5.5), it acts as a spring in parallel with the load stiffness and *increases* the effective stiffness, though with the disadvantageous ramifications discussed above.

For a *positive* transducer stiffness ($Y_{11} > 0,\ H_{11} > 0$, e.g. electrostatic and electromagnetic transducers, see Table 5.5), though a mechanically parallel load stiffness again results, the effective stiffness is decreased (*electromechanical softening*). This inherently destabilizing effect notably leads to an *increase* in the ELM coupling factor.

Effect of two-port coupling terms Eq. (5.78) shows that for a large ELM coupling factor, the two-port coupling terms Y_{12}, Y_{21} or H_{12}, H_{21} should be chosen to have the largest possible magnitude. This is also not unexpected, as it is precisely these terms which describe the electromechanical coupling of the unloaded transducer.

---

[28] The parameters α_i, β_i result from Eq. (5.78).

Effect of transducer electrical energy storage The electrical storage capacity of the transducer (capacitive and inductive) is described by the transducer parameters C_T, L_T, which can be found both in the two-port elements $Y_{22}(s)$ or $H_{22}(s)$ and in the coupling elements (see Table 5.5). However, the determining factors for the ELM coupling factor are the mixed products, for which in general (see Table 5.4),

$$\frac{Y_{12} \cdot Y_{21}}{Y_{22}} = -\frac{H_{12} \cdot H_{21}}{H_{22}}, \tag{5.84}$$

or, accounting for the type-specific transducer characteristics,

$$\left| \frac{Y_{12} \cdot Y_{21}}{Y_{22}} \right| = \left| \frac{H_{12} \cdot H_{21}}{H_{22}} \right| = \begin{cases} \dfrac{\gamma_C}{C_T} & \text{capacitive}, \\ \dfrac{\gamma_L}{L_T} & \text{inductive}. \end{cases} \tag{5.85}$$

From Eq. (5.78) it follows that, in general, the product terms of Eq. (5.84) must be chosen to have *the largest possible magnitude* to achieve a high ELM coupling factor. This, in turn, means for the two transducer types that—following Eq. (5.85)—the *electrical transducer storage* C_T, L_T should be chosen *as small as possible.*

This also agrees with the initial energy considerations for deriving the ELM coupling factor. The energy stored in the electrical storage element does not remain in the transducer during the electrical charge/discharge cycle and thus is not available as mechanical energy.

Semi-active augmentation of ELM coupling factors An ELM coupling factor can be increased via external circuitry which reduces the transducer electrical storage. For this purpose, suitable *negative capacitances* or *negative inductances* are connected into the electrical drive loop of the transducer (Oleskiewicz et al. 2005), (Preumont 2006), (Marneffe and Preumont 2008) and (Funato et al. 1997). For the configurations depicted in Fig. 5.33, the resulting effective transducer storage parameters (see correspondences in Table 5.9) are

$$C_T^* = C_T - C_{comp}, \quad L_T^* = L_T - L_{comp},$$

and, given the remaining transducer parameters remain unchanged, the *modified* ELM coupling factors use C_T^*, L_T^* instead of C_T, L_T. Negative

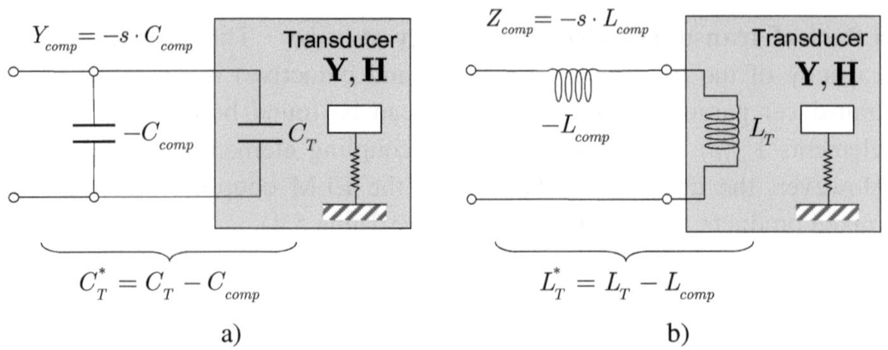

Fig. 5.33. Electrical transducer circuitry with negative impedances/admittances: a) capacitive transducer with negative compensation capacitance, b) inductive transducer with negative compensation inductance

impedances can be realized in specialized circuits incorporating operational amplifiers (semi-active networks), and in all cases require auxiliary energy.

When employing this type of circuitry, however, attention should paid to the fact that *stability problems* can appear if the capacitance is reduced too much, see (Marneffe and Preumont 2008).

5.7 Transducers with Multibody Loads

5.7.1 Frequency response

Multibody load In some applications, modeling the load connected to the transducer as an elastically-suspended rigid body is not sufficient, e.g. if the connection is to a flexible structure or if the mechanical load itself is flexible. A typical arrangement for a transducer connected to a multibody load is shown in Fig. 5.34a. This arrangement assumes that the armature is at the same time the electrode to which the internal electromechanical force $F_T(x_{armat.}, u_T \,/\, i_T)$ is applied, i.e. the electromechanical force and electrical/magnetic field quantities are *collocated.*

Linear transducer model To model the dynamics of a transducer having a multibody load, consider the example of a *voltage-drive* transducer in Fig. 5.15. Replacing the single-mass oscillator with an *MBS state-space model* gives the state-space block diagram in Fig. 5.34b. Note that the application of the *transducer force* and the *external force* can take place on different bodies $(\mathbf{p}_{armat.}, \mathbf{p}_{ext})$.

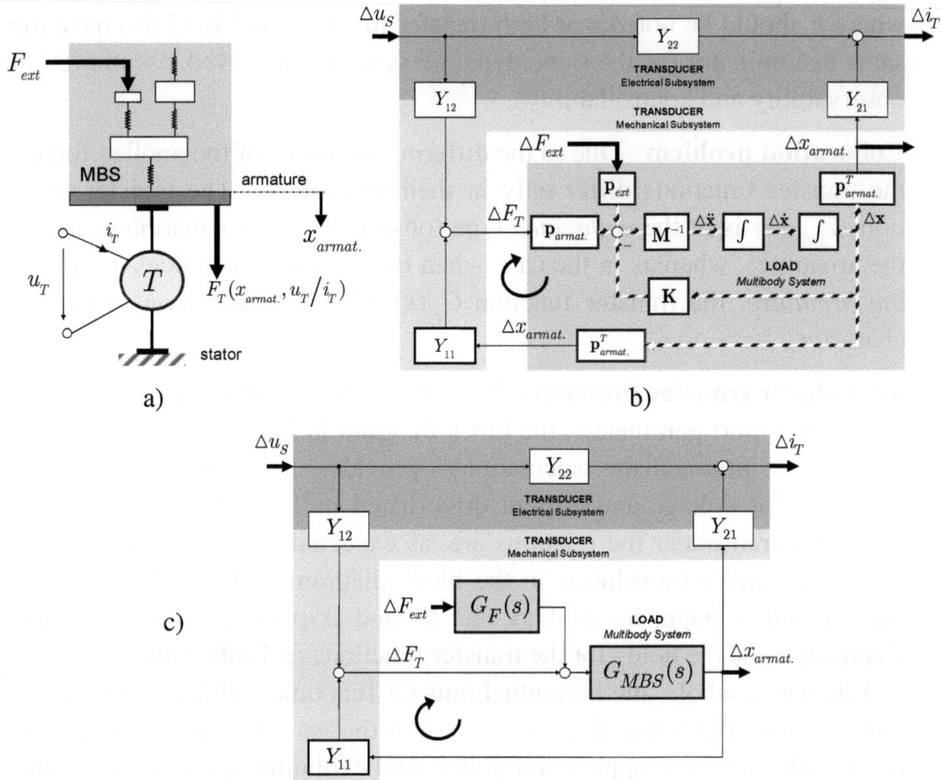

Fig. 5.34. Generic mechatronic transducer with multibody load: a) general configuration, b) block diagram of linearized loaded transducer with state-space MBS model, c) block diagram of linearized loaded transducer with MBS transfer functions

Signal-based transducer model To enable easy evaluation of the response characteristics, it is expedient to transform the MBS state-space model into an equivalent representation using transfer functions (Fig. 5.34c). From the signal flow in Fig. 5.34b, it can be seen that the only relevant transfer functions are those between transducer force and armature position, and external force and armature position.

For these two transfer functions, it holds that

$$\begin{aligned}
\frac{X_{armat.}(s)}{F_T(s)} &= G_{MBS}(s) = \frac{\mathcal{Z}_{MBS}(s)}{\mathcal{N}_{MBS}(s)}, \\
\frac{X_{armat.}(s)}{F_{ext}(s)} &= G_F(s) \cdot G_{MBS}(s) = \frac{\mathcal{Z}_{ext}(s)}{\mathcal{Z}_{MBS}(s)} G_{MBS}(s) = \frac{\mathcal{Z}_{ext}(s)}{\mathcal{N}_{MBS}(s)},
\end{aligned} \tag{5.86}$$

where it should be noted that both transfer functions in Eq. (5.86) have the same denominator (as the same dynamic system is involved, assuming full observability and controllability).

Collocation problem Due to the differing locations of the applied forces, the transfer functions differ only in their numerators. The transfer function $G_{MBS}(s)$ is of the *collocated* type (observation and actuation occurs at the armature), whereas in the case when the *external* force does *not act at the armature*, the transfer function $G_F(s) \cdot G_{MBS}(s)$ has a *non-collocated* character.

Transducer transfer functions Due to the simple transducer structure using the two-port parameters, the block diagram in Fig. 5.34c—following a few quick intermediate calculations—provides the transducer transfer functions for voltage and current drive listed in Table 5.10. For the *current-drive* transducer the relations are, as expected, the *duals* of those for the *voltage-drive* transducer: in the block diagram of Fig. 5.34c only the current and voltage, as well as the *H*- and *Y*-parameters need be exchanged; the same holds for the transfer functions in Table 5.10.

Taken as a whole, the individual transfer functions exhibit a clear structure. For one, the *zeros* $\mathcal{Z}_{MBS}(s)$, $\mathcal{Z}_{ext}(s)$ of the *multibody load* correspond to the differing force application points, and explicitly appear in the corresponding transfer channels. In addition, all transfer functions, regardless of the control type, possess *the same poles*. Finally, the *dynamics at the electrical terminals* are strictly *dual* for voltage and current drive (admittance vs. impedance). It is only in these transfer functions at the terminals that zeros different from those of the MBS transfer functions appear—though these do correspond to the respective poles of the dual structure (with characteristic polynomials $\Delta_U(s)$, $\Delta_I(s)$).

Characteristic polynomials A detailed analysis of the characteristic polynomials $\Delta_U(s)$, $\Delta_I(s)$ is interesting as it makes accessible fundamental dynamic and stability predictions.

The first property to make itself apparent is the fact that the poles of the loaded transducer (roots of the characteristic polynomials) are affected exclusively by the parameters Y_{11} and H_{11}. This property can also be quite clearly recognized in the block diagrams of Fig. 5.34b and Fig. 5.34c.

The qualitative location of the roots of the characteristic polynomial can be easily estimated in a general form by considering the *root locus* (Ogata

Table 5.10. Transfer matrix for voltage and current drive of a generic mechatronic transducer with a *multibody load* (see Fig. 5.34c, Eq. (5.86))

Voltage drive	Current drive
$\begin{pmatrix} \Delta X(s) \\ \Delta I_T(s) \end{pmatrix} = \begin{pmatrix} \tilde{G}_{x/F,U} & \tilde{G}_{x/u} \\ \tilde{G}_{i/F} & \tilde{G}_{i/u} \end{pmatrix} \begin{pmatrix} \Delta F_{ext}(s) \\ \Delta U_S(s) \end{pmatrix}$	$\begin{pmatrix} \Delta X(s) \\ \Delta U_T(s) \end{pmatrix} = \begin{pmatrix} \tilde{G}_{x/F,I} & \tilde{G}_{x/i} \\ \tilde{G}_{u/F} & G_{u/i} \end{pmatrix} \begin{pmatrix} \Delta F_{ext}(s) \\ \Delta I_S(s) \end{pmatrix}$
$\tilde{\mathbf{G}}_U(s) = \begin{pmatrix} \dfrac{\mathcal{Z}_{ext}(s)}{\Delta_U(s)} & \dfrac{\mathcal{Z}_{MBS}(s)\cdot Y_{12}}{\Delta_U(s)} \\ \dfrac{\mathcal{Z}_{ext}(s)\cdot Y_{21}}{\Delta_U(s)} & Y_{22}\dfrac{\Delta_I(s)}{\Delta_U(s)} \end{pmatrix}$	$\tilde{\mathbf{G}}_I(s) = \begin{pmatrix} \dfrac{\mathcal{Z}_{ext}(s)}{\Delta_I(s)} & \dfrac{\mathcal{Z}_{MBS}(s)\cdot H_{12}}{\Delta_I(s)} \\ \dfrac{\mathcal{Z}_{ext}(s)\cdot H_{21}}{\Delta_I(s)} & H_{22}\dfrac{\Delta_U(s)}{\Delta_I(s)} \end{pmatrix}$

$$\Delta_U(s) := \mathcal{N}_{MBS}(s) - Y_{11}\cdot \mathcal{Z}_{MBS}(s)$$

$$\Delta_I(s) := \mathcal{N}_{MBS}(s) - H_{11}\cdot \mathcal{Z}_{MBS}(s)$$

2010). In the characteristic polynomials $\Delta_U(s)$, $\Delta_I(s)$ shown in Table 5.10, the transducer stiffnesses Y_{11}, H_{11} function as feedback gains on the armature position, as evident from Fig. 5.34c.

As a *collocated* structure can be assumed for $G_{MBS}(s)$ (see introductory remarks for this section), $G_{MBS}(s)$ possesses one imaginary pole pair $\{\omega_{p0}\}$ representing the common mode, and additional, larger-magnitude, alternating imaginary zero/pole pairs $\{\omega_{zi}\},\{\omega_{pi}\}$. The resulting root locus for the transducer poles as a function of the transducer stiffnesses Y_{11}, H_{11} is depicted in Fig. 5.35.

Pole and zero locations Relative pole and zero locations can be easily read off of Fig. 5.35 according to the transducer type (see Table 5.5). Relative to the free structure, the structural eigenfrequencies are pushed toward *lower* frequencies for *positive* transducer stiffnesses, while for negative transducer stiffnesses, they move toward higher frequencies. During this movement of the poles, however, no zeros of the free structure are ex-

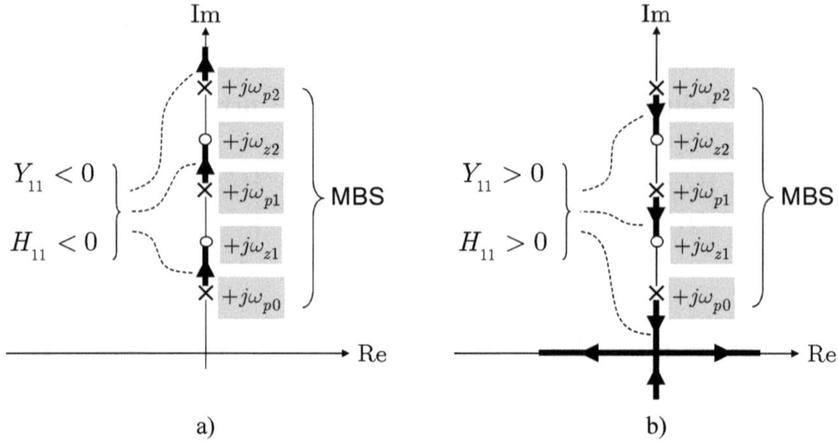

Fig. 5.35. Root loci for the poles of the transducer with a multibody load (3 DOF): a) $Y_{11} < 0$ or $H_{11} < 0$, b) $Y_{11} > 0$ or $H_{11} > 0$

ceeded, so that for a flexible load, the relative locations of poles and zeros—and thus the collocated characteristics—remain intact.

It is easy to verify the generally-valid properties for the already extensively discussed elastically-suspended rigid-body load (single-mass oscillator) with $\omega_{p0} = \Omega_0 = \sqrt{k/m}$ and the values in Table 5.8.

Stability properties From the root locus, it can be seen that, due to the collocated multibody configuration, no stability problems can appear due to the *mechanical feedback*[29].

The lowest branch of the root locus for positive transducer stiffnesses does, however, extend into the right half-plane; this represents the previously-discussed *electromechanical softening* and always appears, regardless of the mechanical configuration. The increasing transducer stiffness cancels the spring stiffness of the elastic suspension. With complete cancellation, this leads to an ungrounded mechanical load (with marginal stability and a repeated pole at the origin), and with a continued increase of the transducer stiffness, to an unstable system.

---

[29] Only this type of feedback is considered here, as a lossless transducer is assumed. In case of *electrical* feedback (impedance or admittance), other considerations apply.

5.7.2 Impedance and admittance feedback

Electrical terminal dynamics As previously discussed, the dynamics at the electrical terminals are represented by the transfer functions $G_{i/u}$, $G_{u/i}$ (see Fig. 5.20). The zeros of these transfer functions are precisely the poles of the respective dual transfer functions, or pairwise Ω_U, Ω_I for the single-mass oscillator. It is easy to verify, using the root locus in Fig. 5.35 and the two-port parameters for transducer implementations from Table 5.5, that a *collocated* pattern of alternating pole and zero pairs also results for $G_{i/u}$, $G_{u/i}$.

Impedance/admittance feedback In the end, since the multibody load is simply a generalization of the elastically-suspended rigid-body load (single-mass oscillator), no fundamental differences from the results found in Sec. 5.5 are to be expected. However, the effects of structural eigenmodes do merit some examination.

Due to the collocated structure of the feedback transfer functions $G_{i/u}, G_{u/i}$, all previously mentioned properties do in fact reappear. The transducer eigenfrequencies (here the structural eigenfrequencies) are all damped, though to a greater or lesser extent. The design formulas for rigid-body loads are also approximately applicable here.

To illuminate these concepts, Example 5.2 considers the capacitive transducer previously discussed in Example 5.1.

Example 5.2 *Capacitive transducer with multibody load and impedance feedback.*

Consider again the *voltage-drive capacitive* transducer depicted in Fig. 5.27, though now mechanically connected to an undamped *two-mass oscillator* (see Table 4.3, collocated configuration) with

$$G_{MBS}(s) = V_{MBS} \frac{\{\omega_{z1}\}}{\{\omega_{p0}\}\{\omega_{p1}\}}, \quad \omega_{p0} < \omega_{z1} < \omega_{p1}. \tag{5.87}$$

This example investigates to what extent this *MBS transducer configuration* for *(A) resistive feedback* and *(B) RL feedback* differs from the rigid-body transducer.

Electrical terminal dynamics The transfer function $G_{i/u}(s)$ of the transducer with the MBS structure of Eq. (5.87) as the mechanical load can be easily obtained from Fig. 5.35. However, this does require assuming a particular transducer type.

For example, for an *electrostatic* transducer, Table 5.5 gives the ordering of transducer stiffnesses $Y_{11} > H_{11} > 0$, and thus the pole/zero configuration for $G_{i/u}(s)$ shown in Fig. 5.36a.

As can be easily verified, for the second possible capacitive transducer type—a *piezoelectric transducer* with transducer stiffnesses $H_{11} < Y_{11} < 0$—a qualitatively equivalent pole/zero configuration would result, though the branches of the root locus in this case would point upwards, indicating higher eigenfrequencies for the loaded transducer.

Due to the MBS structure, alternating pole/zero pairs $(\Omega_{U0}, \Omega_{I0})$, $(\Omega_{U1}, \Omega_{I1})$ result, similar to the rigid-body load case.

(A) Resistive impedance feedback For resistive impedance feedback, there is only a serial resistor R inserted into the auxiliary energy voltage loop (see Fig. 5.20a or Fig. 5.23). The corresponding root locus for the poles of the transducer with feedback is sketched in Fig. 5.36b.

The expected damping of the structural eigenfrequencies (poles ⊠) is recognizable. Both poles of the MBS have moved into the left half-plane, though with differing damping. It is thus not possible to maximally damp both poles with a single resistance.

(B) *RL* impedance feedback For RL feedback, a resistor R and an inductor L are inserted in series into the auxiliary energy voltage loop as shown in Fig. 5.27. The implementational effects of this measure can be determined analogously to Example 5.1 by sequentially closing first the L-feedback and subsequently the R-feedback loops (Fig. 5.28). The corresponding root loci are sketched in Fig. 5.37. After closing the L-feedback loop, a new imaginary pole pair appears, arising from the LC_T serial oscillator loop.

After closing the R-feedback loop, the root locus variants shown in Fig. 5.29b-d are possible, depending on the parameterization. With a suitable choice of L^*, R^*, the root locus shown in Fig. 5.37b can also be selected, realizing maximum damping at the lowest structural eigenfrequency. However, in this case, the second structural eigenfrequency is significantly less damped. As for resistive impedance feedback, here too, compromises in the maximum damping of several eigenfrequencies must be made.

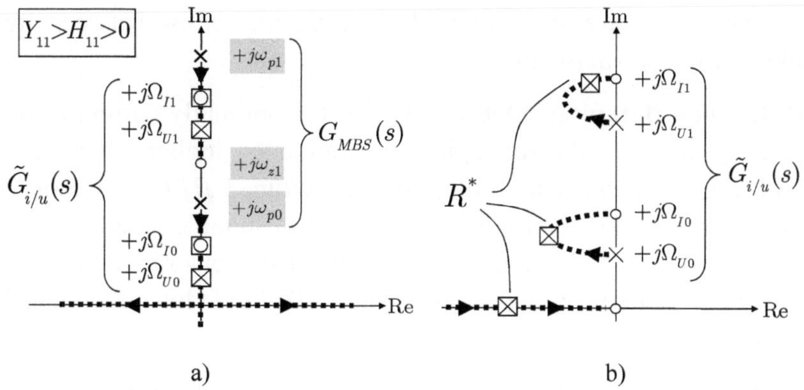

Fig. 5.36. Voltage-drive capacitive transducer with a two-mass oscillator as load: a) root locus for electrical dynamics corresponding to an electrostatic transducer, b) root locus for poles of an MBS transducer with *resistive* impedance feedback

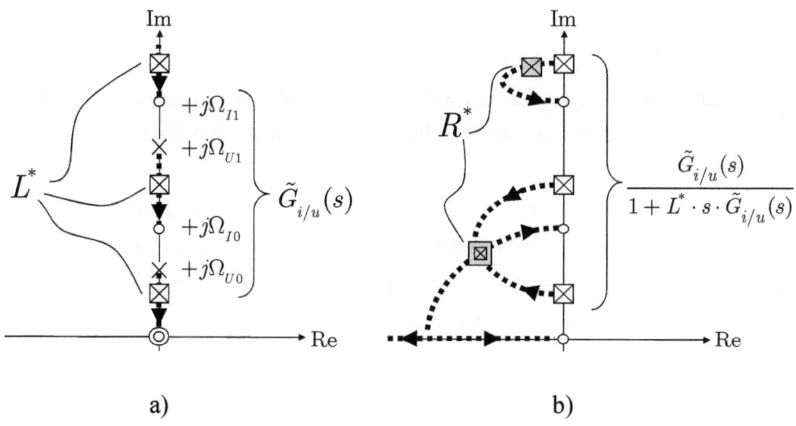

Fig. 5.37. Voltage-drive capacitive transducer with a two-mass oscillator as load and RL impedance feedback: a) root locus for poles of the MBS transducer after closing the L-feedback loop, b) root locus for poles of the MBS transducer after closing the R-feedback loop

■

General MBS damping behavior In the case of multiple eigenfrequencies, resistive or general impedance feedback can strongly damp only one of them. Due to its higher selectivity, general impedance feedback (using an RLC oscillator loop) damps the remaining eigenfrequencies more weakly than does R-feedback with its significantly wider bandwidth. For any particular case, there must thus be a compromise between frequency selectivity and bandwidth.

Multimode damping One possibility for optimally damping several eigenfrequencies at the same time is offered by higher-order passive networks. These employ multiple, differently-tuned RLC oscillator loops inserted into the electrical transducer circuit in suitable parallel architectures (Hollkamp 1994), (Fleming et al. 2002), or higher-order admittance functions directly realized with digital filters (DSPs) (Moheimani and Behrens 2004).

General stability From the above discussion and from Example 5.2, it should be quite evident that for the *internally lossless* transducer with a collocated MBS as the load, there are no stability problems induced by the multibody system itself[30]. This is in fact the case as long as no significant parasitic effects come to bear.

Stability can be imperiled in the presence of parasitic phase delays in the electrical feedback due to leakage capacitances and inductances. These can cause high-frequency eigenmodes—in particular those of the mechanical structure—to become unstable due to electrical feedback, e.g. (Marneffe and Preumont 2008).

However, the quantitative effects of parasitic phase delays are difficult to determine from the root locus plots used up to now. A much clearer approach in this respect is offered by the *frequency response*—particularly in the case of a weakly-damped multibody system—the representation of which in a *NICHOLS diagram* (*gain-phase plot*) is discussed in detail in Ch. 10.

5.8 Mechatronic Resonator

Mechanical oscillation damping One important task when dealing with multibody systems consists of the artificial damping of mechanical eigenfrequencies. Often size, weight, or other physical restrictions prevent building dissipative elements into the mechanical structure. One recourse is offered by *mechanical resonance dampers* or *harmonic absorbers*. Fig. 5.38a demonstrates the principle using an undamped single-mass os-

[30] Attention must naturally be paid to inherent transducer stability problems arising from electrostatic softening and *pull-in* effects, though these also appear in the case of an elastically-suspended rigid-body load.

cillator (m, k). The eigenmode of the mass is excited by an external disturbance force. This disturbance is then to be countered by a coupled *dissipative* single-mass oscillator, or *tuned mass damper*, (m_T, k_T, b_T). For this, the eigenfrequency of the damper is tuned in such a way that the superimposed oscillations of the two coupled oscillators (a *double resonator*) cause kinetic energy from the load mass m to be coupled into the damping loop and dissipated there in the damper b_T (VDI 2006). This principle has been well known for many years (Lehr 1930), (Hartog 1947) and has established itself in a wide range of industries, including harmonic absorbers in buildings and bridges.

Mechatronic double resonator Some of the advantages of connecting passive networks to the electrical terminals of a transducer have already been made clear in the two previous sections. In particular, recall the RL resonance circuit for the capacitive transducer of Example 5.1. This concept can be generalized and employed as a mechatronic equivalent—the *mechatronic resonator* (Fig. 5.38b)—to the passive two-mass damper of Fig. 5.34a.

Applying this concept to the case of a capacitive transducer using RL impedance feedback, a (damped) electrical resonator (R, L, C_T) is connected to the mechanical resonator (m, k), resulting in a *mechatronic double resonator* similar to the mechanical double resonator of Fig. 5.38a (Hagood and Flotow 1991), (Preumont 2002), (Moheimani 2003), (Neubauer et al. 2005).

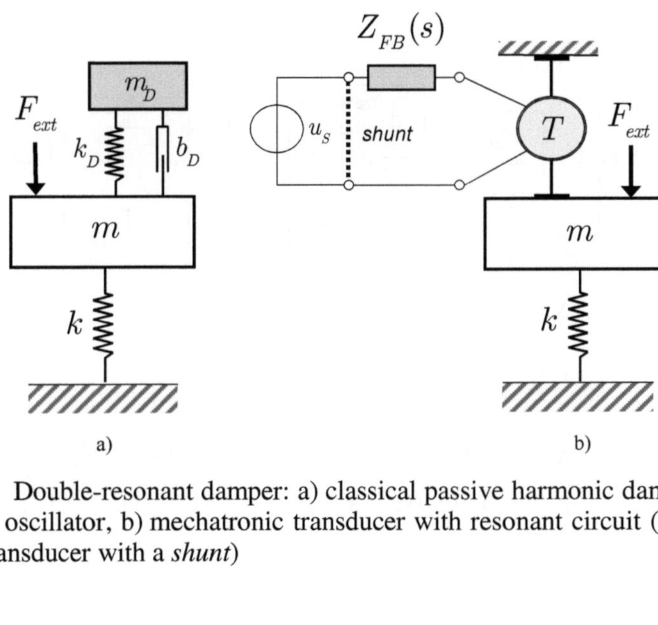

Fig. 5.38. Double-resonant damper: a) classical passive harmonic damper using a two-mass oscillator, b) mechatronic transducer with resonant circuit (optionally a passive transducer with a *shunt*)

Table 5.11. Equivalent configurations (having the same mathematical model) for a mechatronic double resonator (a coupled electromechanical oscillator for passive harmonic damping)

<table>
<tr><td></td><td>Capacitive transducer
$\Omega_I > \Omega_U$</td><td>Inductive transducer
$\Omega_U > \Omega_I$</td></tr>
<tr><td colspan="3">$\Omega_I \leftrightarrow \Omega_U, \quad R \leftrightarrow Y$
$L \leftrightarrow C, \quad C_T \leftrightarrow L_T$</td></tr>
<tr><td>Ideal voltage source
$u_S(t)$</td><td>$Z_{FB}(s) = R + Ls$
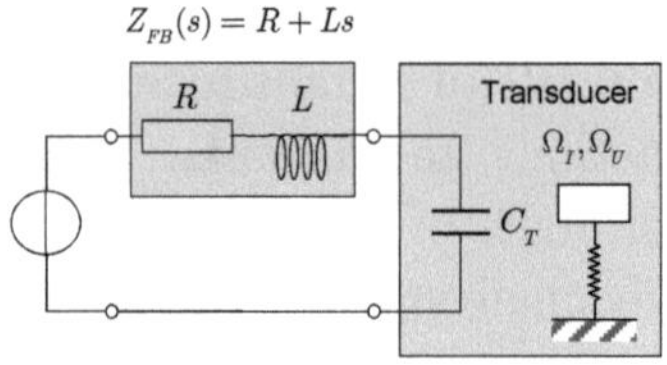
Optimal parameters see Eq. (5.64)
$\tilde{\Delta}_U(s) = (\Omega_U{}^2 + s^2) + C_T s(R+Ls)(\Omega_I{}^2 + s^2)$</td><td>$Z_{FB}(s) = \dfrac{1}{Y + Cs}$
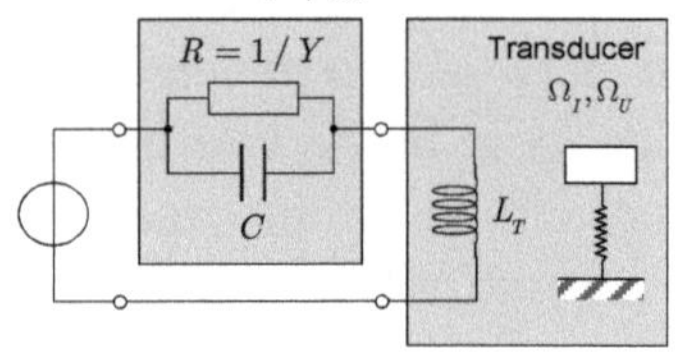
$\tilde{\Delta}_U(s) = (\Omega_U{}^2 + s^2) + \dfrac{1}{L_T s}\dfrac{1}{(Y+Cs)}(\Omega_I{}^2 + s^2)$
$\tilde{\tilde{\Delta}}_U(s) = (\Omega_I{}^2 + s^2) + L_T s(Y+Cs)(\Omega_U{}^2 + s^2)$</td></tr>
<tr><td>Ideal current source
$i_S(t)$</td><td>$Y_{FB}(s) = \dfrac{1}{R + Ls}$
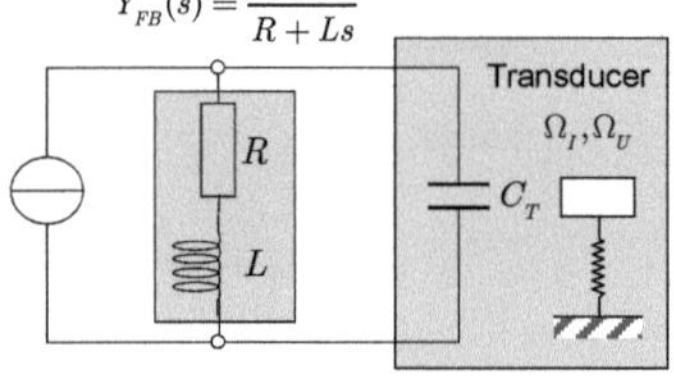
$\tilde{\Delta}_I(s) = (\Omega_I{}^2 + s^2) + \dfrac{1}{C_T s}\dfrac{1}{(R+Ls)}(\Omega_U{}^2 + s^2)$
$\tilde{\tilde{\Delta}}_I(s) = (\Omega_U{}^2 + s^2) + C_T s(R+Ls)(\Omega_I{}^2 + s^2)$</td><td>$Y_{FB}(s) = Y + Cs$
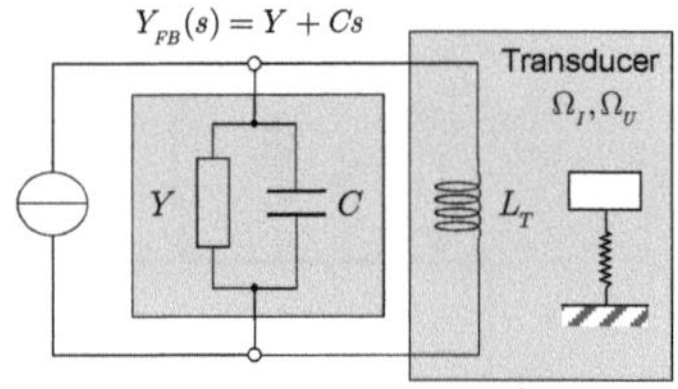
$\tilde{\Delta}_I(s) = (\Omega_I{}^2 + s^2) + L_T s(Y+Cs)(\Omega_U{}^2 + s^2)$</td></tr>
</table>

The functioning of the mechatronic double resonator—whose theory of operation was discussed in detail in Example 5.1—corresponds precisely to that of the mechanical analog. By suitably tuning the impedance parameters of $Z_{FB}(s)$, maximum damping of both eigenmodes can be achieved.

Equivalent configurations The configuration consisting of a voltage-drive capacitive transducer with impedance feedback discussed in Example 5.1 is only one of many possibilities. Table 5.11 shows four configurations for *capacitive* and *inductive* transducers with *voltage* and *current drive*, all of which have equivalent damping behaviors. With the transformations and values represented there, all four configurations lead to *the same characteristic polynomial,* and thence, the derived *design formulas* of Eq. (5.64).

In all four cases, damped RLC oscillator circuits are implemented at the transducer electrical side. Depending on the transducer type, the circuit is augmented with an external electrical storage element complementing the internal transducer electrical storage element.

Remarks on RLC tuning The goal of electrical tuning here is to damp the mechanical eigenfrequency $\Omega_0 = \sqrt{k / m}$. From the design discussion in Example 5.1, it should be clear that the eigenvalues of optimally-tuned resonators are the eigenvalues of the coupled double resonator (m, k) and (R, L, C) (see root locus in Fig. 5.29d). In particular, from Eq. (5.64), for optimal tuning, the following relation holds for the electrical oscillator circuit parameters

$$\Omega_{el}^{opt} = \frac{1}{\sqrt{LC}} = \frac{\Omega_I^2}{\Omega_U},$$

and *not*, as is often *falsely* stated, $\Omega_{el} = 1/\sqrt{LC} = \Omega_0$.

5.9 Mechatronic Oscillating Generator

Electromechanical energy conversion The bidirectional conversion capability of a mechatronic transducer between mechanical and electrical forms of energy suggests its use as an electromechanical energy generator. Of course, this principle has been long applied for the generation of electrical energy at the large scale using electromagnetic generators. In such cases, the mechanical energy is input in very particular forms, e.g. via turbines fed by water pressure or thermal processes.

Ambient mechanical energy: energy harvesting Due to the physically compact nature and the unconventional conversion mechanisms of mechatronic transducers, recently, concepts for exploiting ambient mechanical energy have become a focus of scientific and engineering research. As such processes do not require specialized sources of energy, but rather take advantage of sources of energy available in the everyday environment, they are often termed *energy harvesting* or *energy scavenging*. This energy generation principle has particular significance for the self-sufficiency of mobile electronic devices such as mobile telephones, sensors, and medical implants (Priya 2007).

Technical layout The working principles of a mechatronic oscillating generator are depicted in Fig. 5.39. A seismic mass is elastically suspended in a housing via a mechatronic transducer. The housing is made to move by an external excitation $w(t)$, applying a displacement excitation to the mass (see Fig. 5.2). The energy W_{mech} stored in the mass-spring system is converted into electrical energy W_{el} in the mechatronic transducer, and can be used at a *load impedance* $Z_L(s)$, or—following transformation in a rectifier—can be stored electrically (e.g. in a battery) for later use (Mateu and Moll 2007).

By using a seismic mass, no provision need be made to input forces; an oscillating generator can be easily placed onto any moving mechanical structure, e.g. automobiles, bicycles, shoes, prostheses, etc.

To implement energy harvesting, suitable transducer types are first and foremost those which function without an auxiliary energy supply, i.e. *electrodynamic* and *piezoelectric* transducers.

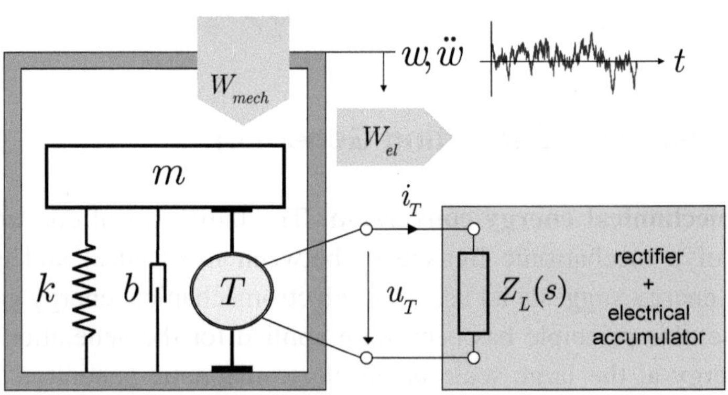

Fig. 5.39. *Energy harvesting*: overview of a mechatronic oscillating generator with seismic mass (displacement/acceleration excitation)

Equivalence to mechatronic accelerometers The arrangement in Fig. 5.39 is identical (excluding the electrical energy storage element) to a mechatronic accelerometer. In such a sensor, the electrical charge or voltage generated by the mechanical subsystem is used as the measurement information. For such an application, the question of an auxiliary energy source customarily plays no great role, so that any capacitive or inductive transducer principle can be employed.

Engineering discussion The energy conversion for this type of system will be described here using the frequency response from the acceleration input to the electrical terminal variables u_T / i_T when connected to the impedance $Z_L(s)$. All models necessary for this description are available in the previous sections.

Depending on the model type, the source must be a short circuit—i.e. $u_Q = 0$—for a voltage-drive transducer, or an open circuit—i.e. $i_Q = 0$—for a current-drive transducer.

For a given load impedance $Z_L(s)$, the transfer function $G^*_{i/u}(s)$ or $G^*_{u/i}(s)$ for the transducer with impedance or admittance feedback is then calculated, for which the simple relations for an elastic rigid-body load (see Sec. 5.5) suffice. Using $G^*_{i/u}(s)$ or $G^*_{u/i}(s)$, the transfer function $G^*_{i/\ddot{w}}(s)$ of the energy conversion can be easily determined.

From the discussion of electromechanical coupling factors (Sec. 5.6), it is clear that for the most efficient conversion of energy, the largest possible coupling factor κ^2 should be sought. Corresponding design criteria relating to the transducer type, parameters, and mechanical loading were discussed in detail in Sec. 5.6.

An important aspect of the design is the *dynamic characteristics* of the *mechanical excitation* $w(t)$ as they relate to the *transfer characteristics* of the electrically loaded *transducer*. As previously explained, for an elastic rigid-body load, the transducer exhibits a clear resonance which depends on the transducer parameters and the mechanical load (k, m). From the structure of the transducer transfer functions, it is clear that a maximum in the conversion of energy from the force or acceleration input to the current or voltage output occurs precisely at this resonant frequency.

On the other hand, mechanical excitation sources often possess pronounced maxima at particular excitation frequencies in their signal spectrum, e.g. structural vibrations in a vehicle, or motor cruise RPMs.

The goal of a systems design is thus to tune the transducer (the electromechanical circuit) to take advantage of the available excitation spectrum (Shu and Lien 2006), (Ward and Behrens 2008), (Twiefel et al. 2008).

5.10 Self-Sensing Actuators

5.10.1 Principle of operation

Problem description To solve the primary task of a mechatronic system—"*generate purposeful motions*"—generally requires a closed loop with sensor feedback of kinematic quantities (position, velocity, acceleration, see Ch. 1). In space- or cost-constrained applications, it is generally desirable to reduce device complexity as much as possible. As force generation and information processing can hardly be dispensed with, the interesting question poses itself of under what conditions an explicit *motion sensor is unnecessary* while maintaining the possibility of constructing a complete control loop.

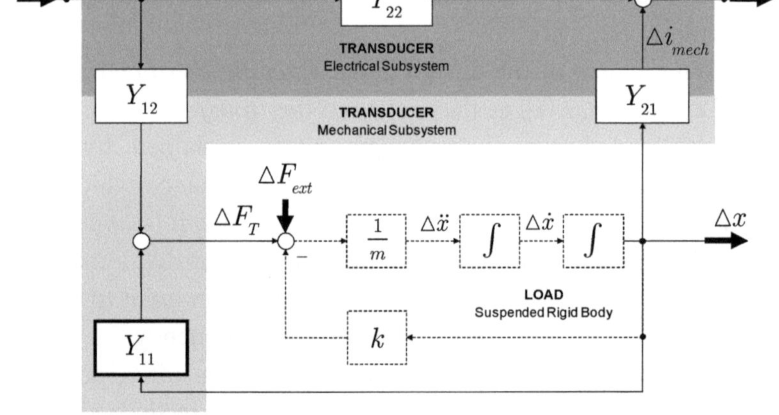

Fig. 5.40. *Self-sensing*: fundamental principles using the example of a linearized generic mechatronic transducer with voltage drive

Fundamental concept of a *self-sensing* actuator From the signal flow diagram of a linearized *voltage-drive*[31] generic mechatronic transducer depicted in Fig. 5.40 (cf. Fig. 5.15), it can be clearly seen that two component currents are superimposed within the transducer current Δi_T : the current $Y_{22}\,\Delta u_T$ provided by the voltage source, and the mechanically induced current $Y_{21}\,\Delta x$. By observing the transducer current Δi_T , it should thus fundamentally be possible to *reconstruct* the *armature displacement* Δx and *armature velocity* $\Delta \dot{x}$. Strictly speaking, this is a model-*aided measurement*; in the technical jargon, when implemented in *reciprocal transducers*, this is commonly termed *self-sensing.*

For this *reconstruction*—also called *estimation*—it is advantageous to exploit both of the available electrical terminal variables Δu_T and Δi_T , giving the fundamental structure for a *reconstruction filter* shown in Fig. 5.40. The output signal estimates of $\Delta \hat{x}$ for the armature displacement, and $\Delta \hat{\dot{x}}$ for the armature velocity can then stand in as measurements for control.

This type of component is then termed a *self-sensing actuator*, i.e. the actuation and measurement functionalities are unified in a single device. From the viewpoint of system functionalities, this principle is also known as *sensorless control*, as the control loop can be closed without an explicit sensor.

This principle has been applied for many years in *electrical machines* (DC, AC motors; reconstruction via the back-EMF (Lorenz 1999)). Since the beginning of the 1990s, self-sensing approaches have been intensively investigated, particularly for *piezoelectric* transducers (Anderson et al. 1992), (Dosch et al. 1992) and *magnetic bearings* (Vischer and Bleuler 1993). Though, as is shown below, this principle is applicable to all reciprocal transducers, comparatively few applications are known for other transducer types.

The following section demonstrates that the self-sensing solution concepts known in the literature are naturally and straightforwardly made available by the generic mechatronic transducer model presented in this chapter.

---

[31] For a current-drive transducer, completely analogous conditions result. Also in the case of a *nonlinear* transducer model, equivalent relations hold.

5.10.2 Signal-based self-sensing solution approach

Direct signal reconstruction Given known electromechanical parameters of the transducer (see ELM two-port parameters), the straightforward signal-based reconstruction shown in Fig. 5.41a results. The reconstruction filter is fed the input Δu_T of the transducer (Δu_T is already known from the control algorithm) and the *measured transducer current* $\Delta \tilde{i}_T$, which are processed as shown ($\hat{Y}_{22}$, $\hat{Y}_{21}$ represent estimates of the transducer parameters). Depending on the transducer type, the (complex) two-port parameters $\hat{Y}_{22}(s)$, $\hat{Y}_{21}(s)$ require differentiation or integration in the time domain, which must be taken into account in the implementation (see ELM two-port matrices for technical transducer principles in Table 5.5).

For capacitive transducers (electrostatic, piezoelectric), for example, the integration of $1/\hat{Y}_{21}$, which would be subject to a drift error, can be avoided by accounting for only the real coefficients of $\hat{Y}_{21}(s)$ (pure proportional behavior). This reconstructs the armature velocity, which is completely sufficient for velocity feedback (see active harmonic isolation in Sec. 10.6.2).

State observer, LUENBERGER observer From a control theoretical point of view, the obvious solution is the *observer structure* shown in Fig. 5.41b following LUENBERGER (Franklin et al. 1998). Here, the (loaded) generic transducer is interpreted as a dynamic system and recreated as a model in the observer filter. As before, the observer is fed the input Δu_T to the transducer as well as the *measured transducer current* $\Delta \tilde{i}_T$. By comparing the reconstructed transducer current $\Delta \hat{i}_T$ and measurement $\Delta \tilde{i}_T$, the observer can generate an internal correction signal to update the internal model variables. With an accurate observer layout, sufficiently accurate estimates of the transducer variables can be obtained, and, above all, drift er-

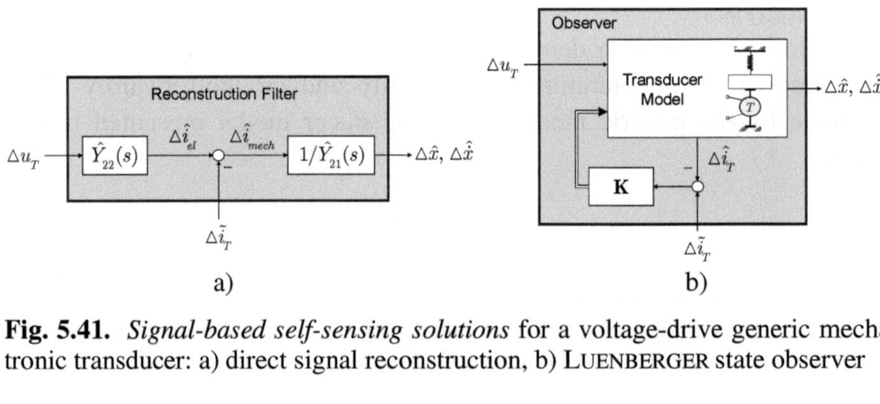

Fig. 5.41. *Signal-based self-sensing solutions* for a voltage-drive generic mechatronic transducer: a) direct signal reconstruction, b) LUENBERGER state observer

rors avoided. With noisy measurements, the observer algorithm can be extended with stochastic considerations, e.g. the KALMAN filter (Franklin et al. 1998). For example, such an approach was suggested quite early on for sensorless control of a *magnetic bearing* (Vischer and Bleuler 1993).

Realization problem: current measurement The signal-based solution approach presented above has one principal obstacle to realization in the simple form shown: current measurement. In both cases, back-effect-free measurement of the transducer current Δi_T is assumed. However, for realistic transducer conditions, this is not possible. This type of measurement requires intervention at the analog electrical level, which must fundamentally influence the transducer dynamics. Established solution variants for analog current measurement along with completely analog electrical reconstruction are discussed below.

5.10.3 Analog electrical self-sensing solutions

Current measurement via serial resistance The most practical possibility for current measurement consists of *voltage measurement* over a known *measurement resistance* (Fig. 5.42a). High-resistance—and thus practically back-effect-free—voltage measurement can be realized using simple operational amplifier circuits (Thomas et al. 2009).

In the configuration shown in Fig. 5.42a, the resistance R_m connected in series to the transducer terminal serves as the measurement resistance. Structurally, the same picture results as for the impedance feedback discussed in Sec. 5.5. Electrically, the measurement resistance implies a feedback voltage relative to the transducer voltage Δu_T. Electromechanically, this feedback entails *modified dynamics* of the loaded transducer, i.e. a shift of the transducer eigenfrequency (see transducer models in Sec. 5.5.5).

From the measurement voltage $u_m = R_m \cdot \Delta i_T$, a current measurement $\Delta \tilde{i}_T = u_m / R_m$ can be easily reconstructed and, for example, used in the signal-based self-sensing solutions presented above. In the process, however, the transducer models should be modified to include the R_m impedance feedback.

WHEATSTONE bridge One very elegant option for both current measurement—which must be undertaken with analog electronics in any case—and *analog electrical signal reconstruction* employs a *WHEATSTONE bridge* as shown in Fig. 5.42b.

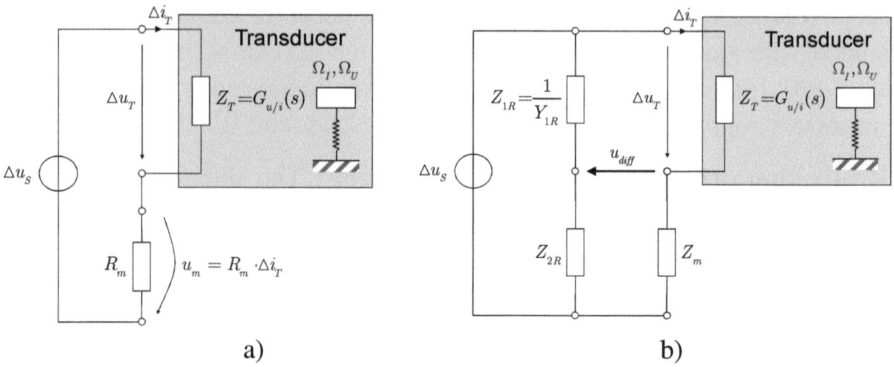

Fig. 5.42. *Analog electrical self-sensing solution* for a voltage-drive generic mechatronic transducer: a) current measurement with a serial resistance (implies impedance feedback), b) WHEATSTONE bridge (reference leg (Z_{1R}, Z_{2R}), measurement leg (Z_T, Z_m))

Well-known in electrical measurement domains, the WHEATSTONE bridge structure (Thomas et al. 2009) consists of a *measurement leg* and a *reference leg* which are each furnished with a voltage divider. In the case considered here, the transducer—represented by the impedance Z_T—is placed together with a serial impedance Z_m in the *measurement leg* (Fig. 5.42b, right). The *reference leg* (Fig. 5.42b, left) consists of a serial connection of reference impedances Z_{1R}, Z_{2R}, which mirror the impedances of the measurement legs up to a common scale factor. The source voltage Δu_S is considered as the input; for the output, the *bridge voltage* u_{diff} is of interest.

Bridge comparison Applying KIRCHHOFF's node and mesh equations (see Sec. 2.3.4) gives the fundamental *bridge equation* in the s-domain:

$$U_{diff}(s) = Z_m(s) \cdot \Delta I_T(s) - \frac{Z_{2R}(s)}{Z_{1R}(s) + Z_{2R}(s)} \Delta U_S(s). \tag{5.88}$$

The transducer current Δi_T depends on the physical transducer parameters. The (generally complex) bridge impedances Z_m, Z_{1R}, Z_{2R} are suitably chosen so that for an electrically balanced bridge, $u_{diff} = 0$. However, as Z_T also possesses an *electromechanical* component not reflected in its electrical copy Z_{1R}, electrical balance in the bridge may result in $u_{diff} \neq 0$, revealing the electromechanical component. Fundamentally, the

WHEATSTONE bridge in Fig. 5.42b realizes the signal-based computational flow in Fig. 5.41a in analog electronics.

The remainder of this section sheds more light on the balance conditions and description of the dynamics of various configurations.

WHEATSTONE self-sensing without mechanical back-effect

Dynamic model Consider first the simple case in which the transducer has no mechanical back-effect effect on the connected mechanical structure. This is approximately the case when the armature motion can be described as an *imposed motion*. Such conditions are assumed for customary operation of a transducer as a *sensor*.

For these operational conditions, the two-port model of the *unloaded* transducer can be employed (see Eq. (5.35)). The transducer current is composed of two independent components (Δx is considered an independent imposed quantity):

$$\Delta I_T(s) = Y_{21}(s)\cdot\Delta X(s) + Y_{22}(s)\cdot\Delta U_T(s).$$

From the mesh equation, it further follows that (omitting the complex variable s for clarity)

$$\Delta U_S = \Delta U_T + Z_m \Delta I_T ,$$

and, substituting and rearranging, the transducer current is found to be

$$\Delta I_T = \frac{Y_{22}}{1+Y_{22}Z_m}\Delta U_S + \frac{Y_{21}}{1+Y_{22}Z_m}\Delta X . \tag{5.89}$$

For the *bridge voltage*, it follows from Eqs. (5.88), (5.89) that

$$U_{diff} = \left[\frac{Y_{22}Z_m}{1+Y_{22}Z_m} - \frac{Y_{1R}Z_{2R}}{1+Y_{1R}Z_{2R}}\right]\Delta U_S + \frac{Y_{21}Z_m}{1+Y_{22}Z_m}\Delta X . \tag{5.90}$$

Bridge balance The bridge voltage becomes independent of the input voltage Δu_S when the bracketed expression in Eq. (5.90) is zero, giving the following *electrical balance condition*:

$$\boxed{Y_{22}\cdot Z_m \overset{!}{=} Y_{1R}\cdot Z_{2R} \quad \text{or} \quad Z_m = \frac{Y_{1R}\cdot Z_{2R}}{\hat{Y}_{22}}} . \tag{5.91}$$

For the optimal choice of Z_m in Eq. (5.91), the fact that only a (more or less accurate) estimate $\hat{Y}_{22}$ is available for the electrical transducer admittance was taken into account. The remaining network elements are assumed to be known completely.

Using the defining *admittance relation*[32] $\sigma_Y := Y_{22} / \hat{Y}_{22}$, and the value for Z_m from Eq. (5.91), it finally follows for the bridge voltage that

$$U_{diff} = \frac{Y_{1R}Z_{2R}\left(\sigma_Y - 1\right)}{\left(1+\sigma_Y \cdot Y_{1R}Z_{2R}\right)\left(1+Y_{1R}Z_{2R}\right)} \Delta U_S + \frac{Y_{21}}{\hat{Y}_{22}} \frac{Y_{1R}Z_{2R}}{1+\sigma_Y \cdot Y_{1R}Z_{2R}} \Delta X .$$

Given sufficiently accurate knowledge of the transducer admittance, i.e. $\sigma_Y \approx 1$, the bridge voltage is independent of Δu_S and is proportional to the armature displacement:

$$\boxed{U_{diff} \approx \frac{Y_{21}}{\hat{Y}_{22}} \frac{Y_{1R}Z_{2R}}{1+Y_{1R}Z_{2R}} \Delta X} . \tag{5.92}$$

Choosing the bridge elements The bridge balance condition (5.91) clearly shows that the reference leg should be constructed as a mirror image of the measurement leg. However, the network elements can be freely scaled by a common factor α, i.e.

$$Y_{1R} = \frac{1}{\alpha} Y_{22}, \quad Z_{2R} = \alpha Z_m . \tag{5.93}$$

The transducer admittance Y_{22} is set as a parameter of the concrete transducer. However, an interesting design degree of freedom is available in the measurement impedance Z_m.

For example, in a *capacitive transducer* (electrostatic or piezoelectric, see Table 5.5) different system behaviors result depending on the choice of Z_m. With $Y_{22}(s) = sC_T$, $Y_{21}(s) = sK_T$, ideal balancing gives the possibilities

$$\boxed{Z_m = R_m} \Rightarrow U_{diff}(s) = \frac{K_T}{C_T} \frac{sC_T R}{1+sC_T R} \Delta X \sim s \cdot \Delta X(s) , \tag{5.94}$$

$$\boxed{Z_m = \frac{1}{sC_m}} \Rightarrow U_{diff}(s) = \frac{K_T}{C_T} \frac{C_T/C_m}{1+C_T/C_m} \Delta X \sim \Delta X(s) . \tag{5.95}$$

---

[32] For lossless capacitive or inductive dynamics, this can be assumed real.

In a *capacitive-resistive* bridge, the bridge voltage u_{diff} according to Eq. (5.94), is—at least for sufficiently small frequencies—proportional to the *armature velocity*, whereas for a purely *capacitive* bridge, the voltage is proportional to the *armature displacement* (Eq. (5.95)).

This relationship has been exploited for many years in piezoelectric transducers (Dosch et al. 1992); for electrostatic transducers, on the other hand, it has been largely ignored.

WHEATSTONE self-sensing with mechanical back-effect

Mechanically loaded generic transducer For operation as an actuator subject to mechanical back-effect, the model equations of the loaded generic transducer must be employed; see the transfer matrix in Table 5.8.

For this configuration, the alternate circuit diagram of Fig. 5.42b also applies, as well as the bridge equation (5.88). In contrast to the case without mechanical back-effect handled above, here the only independent variable is the source voltage Δu_S. The current Δi_{mech} induced by the armature motion is implicitly contained in the transfer matrix.

Note that between the transducer impedance Z_T or admittance Y_T and the transfer matrix in Table 5.8, the following respective relationships hold:

$$Z_T(s) = \frac{\Delta U_T(s)}{\Delta I_T(s)} = G_{u/i}(s) \text{ or } Y_T(s) = \frac{\Delta I_T(s)}{\Delta U_T(s)} = G_{i/u}(s). \tag{5.96}$$

For the *bridge equation*, accounting for Eq. (5.96), a quick calculation gives

$$\boxed{U_{diff} = \frac{G_{i/u} Z_m - Y_{1R} Z_{2R}}{\left(1 + G_{i/u} Z_m\right)\left(1 + Y_{1R} Z_{2R}\right)} \Delta U_S}. \tag{5.97}$$

Bridge balance Eq. (5.97) describes the transfer dynamics from the input voltage of the transducer (the drive voltage) Δu_S to a fictitious measurement value, represented by the bridge voltage u_{diff}. The transducer admittance $G_{i/u}(s)$ represents the complete electromechanical coupling of the transducer. In this case, for practical reasons, no attempt is made to zero out the right-hand side of Eq. (5.97) with a suitable choice of Z_m, Z_{1R}, Z_{2R}. With allusion to the previously-discussed, mechanical-back-effect-free case, only the purely electrical component $Y_{22}(s)$ of the transducer admittance need be compensated in the bridge to reveal the mechanically-induced current in the bridge voltage (see also Fig. 5.40). As a result, Eq. (5.91) again follows as the *electrical balance condition.*

Discussion Eq. (5.97) offers a convenient opportunity to investigate the dynamics of a loaded self-sensing actuator.

The first noteworthy property is a separation of the poles of the loaded transducer (zeros of $(1+G_{i/u}Z_m)$, see impedance feedback in Sec. 5.5.5) and the reference leg (zeros of $(1+Y_{1R}Z_{2R})$). Assuming capacitive or inductive transducer dynamics and a resistive measurement impedance Z_{2R}, the reference leg then introduces an additional real pole. This is not unexpected considering that the reference leg represents a parallel leg to the terminals of the voltage source, and thus does not affect the transducer leg.

With the choice of bridge elements, analogously to the mechanical-back-effect-free case, either *displacement-* or *velocity-proportional* behavior can be effected in the low-frequency band. For reasons of space, a deeper discussion of this choice is not included here. The behaviors discussed in the literature—e.g. for piezoelectric transducers (Dosch et al. 1992), (Brusa et al. 1998), (Preumont 2006)—can be deduced with few calculation steps from Eq. (5.97) and the relations in Table 5.5 and Table 5.8.

Multibody load: collocation One important property of self-sensing actuators in relation to multibody loads should be noted. As the self-sensing principle presented here fundamentally implies force application and motion measurement on the same body (here the armature), there is an *inherently collocated* measurement and actuation arrangement with all accompanying advantages.

The associated special structure of the transfer function $U_{diff}/\Delta U_S$, including alternating resonant and antiresonant frequencies, can be directly derived from Eq. (5.97) if the corresponding MBS formulation from Sec. 5.7 is used for $G_{i/u}(s)$.

Implementation difficulties For all the described advantages of the self-sensing principle, the implementation difficulties which exist in real applications should also not be kept secret. The experienced engineer may justifiably surmise that a solution *based on compensation*, as presented here, reacts very sensitively to parameter variations and model uncertainties. Indeed, a lack of robustness due to model errors related to internal transducer losses or non-modeled parasitic electrical effects, and to a strong dependence of transducer parameters on the environment has been reported. In such cases, under certain conditions, *adaptive approaches* can result in definite improvements, e.g. (Chan and Liao 2009).

Self-sensing vs. impedance feedback Closely considered, the *impedance feedback* discussed in detail in Sec. 5.5 relies precisely upon the self-sensing properties of a reciprocal transducer. Though the armature motion is not explicitly reconstructed in that case, using electrical feedback, a feedback signal proportional to the armature motion is also generated. Thus the literature sometimes also refers to self-sensing in connection with an instance of impedance feedback, e.g. (Paulitsch et al. 2006).

Bibliography for Chapter 5

Anderson, E. H., N. W. Hagood and J. M. Goodliffe (1992). Self-sensing piezoelectric actuation - Analysis and application to controlled structures *Proceedings of the 33rd AIAA/ASME/ASC/AHS Structures, Structural Dynamics and Materials Conference* Dallas, TX. pp.2141-2155.

Apostolyuk, V. (2006). Theory and Design of Micromechanical Vibratory Gyroscopes. *MEMS/NEMS Handbook, Techniques and Applications*. C. T. Leondes. Springer. **1:** 173-195.

Bronshtein, I. N., K. A. Semendyayev, G. Musiol and H. Mühlig (2005). *Handbook of Mathematics*. Springer.

Brusa, E., S. Carabelli, F. Carraro and A. Tonoli (1998). "Electromechanical Tuning of Self-Sensing Piezoelectric Transducers." *Journal of Intelligent Material Systems and Structures* **9**(3): 198-209.

Chan, K. and W. Liao (2009). "Self-sensing actuators with passive damping for adaptive vibration control of hard disk drives." *Microsystem Technologies* **15**(3): 355-366.

DIN (1988). Guide to dynamic measurements of piezoelectric ceramics with high electromechanical coupling; identical with IEC 60483, edition 1976. DIN IEC 60483: 1988-04. DIN.

Dosch, J. J., D. J. Inman and E. Garcia (1992). "A Self-Sensing Piezoelectric Actuator for Collocated Control." *Journal of Intelligent Material Systems and Structures* **3**(1): 166-185.

Fleming, A. J., S. Behrens and S. O. R. Moheimani (2002). "Optimization and implementation of multimode piezoelectric shunt damping systems." *Mechatronics, IEEE/ASME Transactions on* **7**(1): 87-94.

Franklin, G. F., J. D. Powell and M. L. Workman (1998). *Digital Control of Dynamic Systems*. Addison-Wesley.

Funato, H., A. Kawamura and K. Kamiyama (1997). "Realization of negative inductance using variable active-passive reactance (VAPAR)." *Power Electronics, IEEE Transactions on* **12**(4): 589-596.

Goldstein, H., C. P. Poole and J. L. Safko (2001). *Classical Mechanics*. Addison Wesley.

Hagood, N. W. and A. v. Flotow (1991). "Damping of structural vibrations with piezoelectric materials and passive electrical networks." *Journal of Sound and Vibration* **146**(2): 243-268.

Hartog, J. P. D. (1947). *Mechanical Vibrations*. McGraw-Hill.

Hollkamp, J. J. (1994). "Multimodal Passive Vibration Suppression with Piezoelectric Materials and Resonant Shunts." *Journal of Intelligent Material Systems and Structures* **5**(1): 49-57.

Horowitz, I. M. (1963). *Synthesis of Feedback Systems*. New York. Academic Press.

IEEE (1988). "IEEE standard on piezoelectricity." *ANSI/IEEE Std 176-1987*.

Karnopp, D. C., D. L. Margolis and R. C. Rosenberg (2006). *System dynamics: modeling and simulation of mechatronic systems*. John Wiley & Sons, Inc.

Lehr, E. (1930). "Untersuchung der erzwungenen Koppelschwingungen eines elektromechanischen Systems unter Verwendung eines graphischen Verfahrens." *Archiv für Elektrotechnik* **XXIV.**: 330-348.

Lorenz, R. D. (1999)._Advances in electric drive control. *Electric Machines and Drives, 1999. International Conference IEMD '99*. pp.9-16.

Marneffe, B. d. and A. Preumont (2008). "Vibration damping with negative capacitance shunts: theory and experiment." *Smart Materials and Structures* **17**(035015): 9.

Mateu, L. and F. Moll (2007)._System-Level Simulation of a Self-Powered Sensor with Piezoelectric Energy Harvesting. *Sensor Technologies and Applications, 2007. SensorComm 2007. International Conference on*. pp. 399-404.

Mohammed, A. (1966). "Expressions for the Electromechanical Coupling Factor in Terms of Critical Frequencies." *The Journal of the Acoustical Society of America* **39**(2): 289-293.

Moheimani, S. O. R. (2003). "A survey of recent innovations in vibration damping and control using shunted piezoelectric transducers." *Control Systems Technology, IEEE Transactions on* **11**(4): 482-494.

Moheimani, S. O. R. and S. Behrens (2004). "Multimode piezoelectric shunt damping with a highly resonant impedance." *Control Systems Technology, IEEE Transactions on* **12**(3): 484-491.

Neubauer, M., R. Oleskiewicz and K. Popp (2005). "Comparison of Damping Performance of Tuned Mass Dampers and Shunted Piezo Elements." *PAMM* **5**(1): 117-118.

Ogata, K. (2010). *Modern Control Engineering*. Prentice Hall.

Oleskiewicz, R., M. Neubauer, T. Krzyzynski and K. Popp (2005). "Synthetic Impedance Circuits in Semi-Passive Vibration Control with Piezo-Ceramics - Efficiency and Limitations." *PAMM* **5**(1): 121-122.

Paulitsch, C., P. Gardonio and S. J. Elliott (2006). "Active vibration damping using self-sensing, electrodynamic actuators." *Smart Materials and Structures* **15**: 499–508.

Preumont, A. (2002). *Vibration Control of Active Structures - An Introduction*. Kluwer Academic Publishers.

Preumont, A. (2006). *Mechatronics, Dynamics of Electromechanical and Piezoelectric Systems*. Springer.

Priya, S. (2007). "Advances in energy harvesting using low profile piezoelectric transducers." *Journal of Electroceramics* **19**(1): 167-184.

Reinschke, K. and P. Schwarz (1976). *Verfahren zur rechnergestützten Analyse linearer Netzwerke*. Akademie Verlag Berlin.

Schenk, H., P. Durr, T. Haase, D. Kunze, et al. (2000). "Large deflection micromechanical scanning mirrors for linear scans and pattern generation." *Selected Topics in Quantum Electronics, IEEE Journal of* **6**(5): 715-722.

Schuster, T., T. Sandner and H. Lakner (2006)._Investigations on an Integrated Optical Position Detection of Micromachined Scanning Mirrors. *Photonics and Microsystems, 2006 International Students and Young Scientists Workshop*. pp.55-58.

Senturia, S. D. (2001). *Microsystem Design*. Kluwer Academic Publishers.

Shu, Y. C. and I. C. Lien (2006). "Analysis of power output for piezoelectric energy harvesting systems." *Smart Materials and Structures* **15**: 1499–1512.

Thomas, R. E., A. J. Rosa and G. J. Toussaint (2009). *The Analysis and Design of Linear Circuits*. John Wiley and Sons, Inc.

Tilmans, H. A. C. (1996). "Equivalent circuit representation of electromechanical transducers: I. Lumped-parameter systems." *Journal of Micromechanics and Microengineering* (6): 157–176.

Twiefel, J., B. Richter, T. Sattel and J. Wallaschek (2008). "Power output estimation and experimental validation for piezoelectric energy harvesting systems." *Journal of Electroceramics* **20**(3): 203-208.

VDI (2006). Schwingungsdämpfer und Schwingungstilger - Schwingungstilger und Schwingungstilgung (in German) - Dynamic damper and dynamic vibration absorber - Dynamic vibration absorber and dynamic vibration absorption (in English). V. D. I. VDI. **3833 Blatt 2::2006-12**.

Vischer, D. and H. Bleuler (1993). "Self-sensing active magnetic levitation." *Magnetics, IEEE Transactions on* **29**(2): 1276-1281.

Ward, J. K. and S. Behrens (2008). "Adaptive learning algorithms for vibration energy harvesting." *Smart Materials and Structures* **17**: 035025 (035029pp).

Wellstead, P. E. (1979). *Introduction to Physical System Modelling*. London. Academic Press Ltd.

Yaralioglu, G. G., A. S. Ergun, B. Bayram, E. Haeggstrom, et al. (2003). "Calculation and measurement of electromechanical coupling coefficient of capacitive micromachined ultrasonic transducers." *Ultrasonics, Ferroelectrics and Frequency Control, IEEE Transactions on* **50**(4): 449-456.

6 Functional Realization: Electrostatic Transducers

Background Electrostatics represents the most ancient known manifestation of electricity, but it has only been since the end of the 20th century that this principle has achieved true engineering significance as a key component of *microelectromechanical systems* (*MEMS*). Due to the physically-constrained, micro-scale force generation available from this principle, its many possible sensor and actuator applications can also only be realized at the micro-scale. Electrostatic transducers are particularly attractive due to the relative ease of their construction. All they require is some conductive materials for the electrodes. Thus, using only a little wiggle room and air as the dielectric, *high-precision, highly dynamic* mechatronic systems moving the *smallest of masses* can be created in minimal volume. Interestingly enough, many of the phenomena relevant to the implementation of such devices became the subject of detailed scientific investigation only in the last decade of the last century, so that this class of mechatronic systems is certain to hold many scientific and technical surprises for the future.

Contents of Chapter 6 This chapter discusses fundamental physical phenomena and technical peculiarities in the behavior of electrostatic transducers. To begin, the generalized model of the *generic mechatronic transducer* of Ch. 5 is made *concrete* for electrostatic transduction principles. Next, detailed considerations regarding different transducer configurations—*transverse* and *longitudinal directions of motion* relative to the electrode surface, and control via *voltage* or *current sources*—are presented. The occurrence of *pull-in*, which has fundamental significance for the steady-state and dynamic behaviors of electrostatic transducers, is discussed in detail. The discussion focuses on various *differential transducer configurations* (comb structures, *electrostatic bearings*). This includes detailed considerations concerning the dynamics and construction of practical *comb structures*, which enable *force multiplication*—an important technique for physical implementations—in a compact structure.

■

6.1 Systems Engineering Context

Electrostatic transducers Power transducers using *electrostatic* phenomena are among the standard components of mechatronic systems and are employed both for generating forces and moments (actuators) and for measuring motions (sensors). However, due to their relatively low energy densities, only weak forces can be generated. Electrostatic principles can, however, be easily miniaturized, and are thus almost exclusively employed in *microelectromechanical systems* (*MEMS*). From the point of view of implementation, electrostatic transducer principles are particularly attractive for micro-systems as (in contrast to electromagnetic or piezoelectric principles) no specialized materials are required. At most, some *conductive* materials are required for the electrodes (e.g. silicon)—in other words, materials which are already part of the standard palette of micro-system technology. Such transducers can thus achieve high-precision, dynamic motion of the smallest of masses in a minimal volume. Electrostatic transducers require electrical auxiliary energy in the form of DC voltage sources.

Systems engineering significance From a systems engineering point of view, the functions *"generate forces/torques"* and *"measure mechanical states"* realizable with electrostatic phenomena represent the *actuators* and *sensors* of a mechatronic system (see Fig. 6.1). For both tasks, the transfer characteristics in the causal directions shown in Fig. 6.1 are of interest. Of-

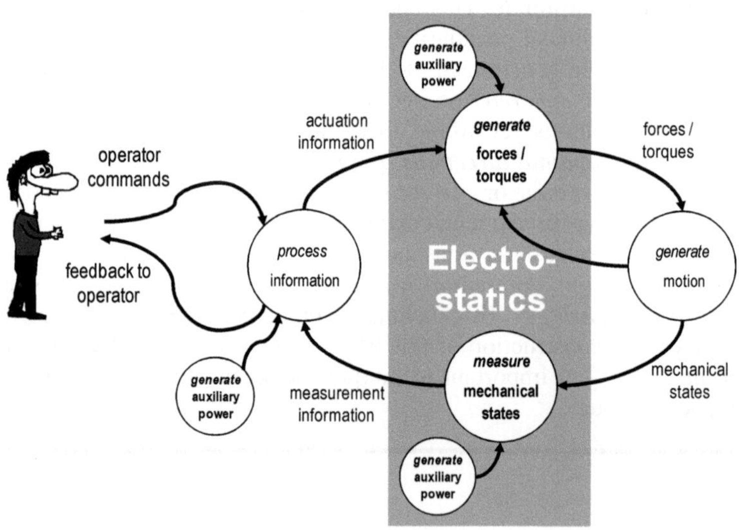

Fig. 6.1. Functional decomposition of a mechatronic system: functional realization using *electrostatics*

ten, to enable compact construction, transducer elements are integrated directly into the mechanical structure. Thus, in addition to common functional properties such as *linearity* and *dynamics*, structural *parameter dependencies* related to the transfer characteristics play an important role in controller design.

Mechatronic phenomena Electrostatic transducers can store *electric charges*. This storage capacity can be described by the *capacitance* of a configuration. Suitable configurations consist of conductive *electrodes* and an enclosed, non-conductive medium (the *dielectric*). The *electrostatic force* acting between the stored charges can be transferred to mechanical structures with movable electrodes. Both the capacitance and the direction and magnitude of the electrostatic forces depend on the geometry and material properties of the transducer configuration. For systems design, the following model relationships are thus of interest:

- *capacitance* as a function of geometry and materials;
- *electrostatic forces* as a function of geometry, materials, and electrical input;
- *transfer characteristics* including mechanical power feedback.

6.2 Physical Foundations

Electrostatic field An electrostatic field is generally understood to be the electric field generated by charges *at rest*, i.e. there is no electric current field[1]. Such static fields develop between *electrodes* separated by a *non-conducting medium* (a dielectric). For freely-moving electrodes, air is chosen as the dielectric. For the mechatronic phenomena considered here, it is the *field pattern* in the insulator between the electrodes which is important.

Generic transducer Computation of the electric field is a non-trivial task for real configurations, and far exceeds the scope of this book. This section considers simple, elementary configurations—*generic electrostatic transducers*—with a homogeneous field pattern, demonstrating the relationships and behaviors relevant for systems engineering. It will be shown that these simple configurations are also quite usable for initial design considerations on real configurations. Naturally, for real, continuously-refined designs,

[1] A more complete assumption posits the absence or near-absence of time-varying magnetic fields, i.e. the existence of locally stationary magnetic fields does not change the results presented here.

more exact field models must be employed at later stages. For these, and a generally more thorough presentation of the underlying physics, the reader is referred to corresponding textbooks, e.g. (Jackson 1999).

MAXWELL's equations for an electrostatic field The fundamental physical relations applicable here are defined by *MAXWELL's equations*[2] for *electrostatic* fields in isotropic *insulators* (no charge flow between electrodes). They are, in integral form[3] (where s is the bounding curve of area A)

field equation: $$\oint_s \vec{E} \cdot d\vec{s} = 0\,, \qquad (6.1)$$

continuity equation: $$\oint \vec{D} \cdot d\vec{A} = q_C\,, \qquad (6.2)$$

material equation: $$\vec{D} = \varepsilon \vec{E}\,, \qquad (6.3)$$

with the quantities

- *field intensity* $\vec{E}$ $[\mathrm{V/m}]$;
- flux density $\vec{D}$ $[\mathrm{As/m^2}]$;
- electrical *charge* q_C $[\mathrm{As}]$, stored within the area A;
- *dielectric constant, permittivity* $\varepsilon = \varepsilon_r \varepsilon_0$ $[\mathrm{A \cdot s/V \cdot m}]$,
  $\varepsilon_0 = 8{,}854 \cdot 10^{-12}\,\mathrm{A \cdot s/V \cdot m}$, $\varepsilon_r \geq 1$ is material-dependent.

---

2 James Clerk MAXWELL, 1831-1879, Scottish physicist. The equations bearing his name and their corrections were the first to summarize a variety of empirical laws concerning electromagnetic phenomena known in his time into a unified, axiomatic construct describing electromagnetism (in several versions around 1865).

3 In the *integral form*, MAXWELL's equations link physical quantities at different locations: quantities acting inside an area to other quantities acting at the edges of this area (*action at a distance*). In contrast to the integral form, the *differential* equations $\operatorname{curl} \vec{E} = \vec{0}$, $\operatorname{div} \vec{D} = \rho$ only link physical quantities acting at the same location and at the same time (*action at close range*). As the geometry is significant when describing transducers, it is the integral form which is applied here (and in Chapter 8).

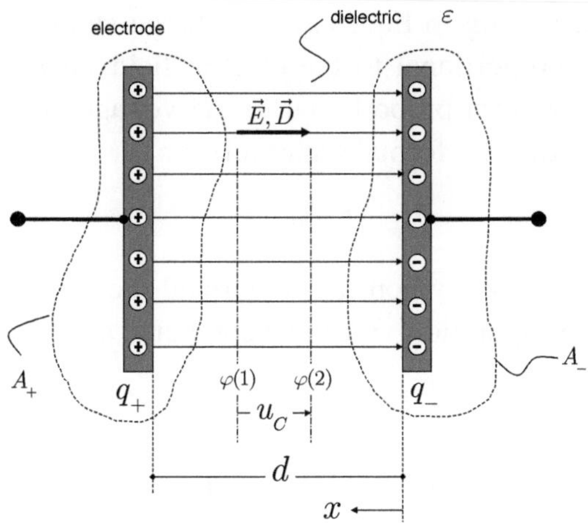

Fig. 6.2. Electrostatic field between two planar electrodes (plate capacitor, homogeneous field)

Using Eq. (6.1), the following relations can be established (Fig. 6.2):

- the scalar *potential field* $\varphi(\vec{r})$ [V] of the electrostatic field[4]

$$\vec{E} = -\nabla\varphi \ , \tag{6.4}$$

- the *potential difference*, electrical tension, or *voltage*[5] u_C [V]

$$\int_{P_1}^{P_2} \vec{E} \cdot d\vec{s} = \varphi(P_1) - \varphi(P_2) =: u_C \ . \tag{6.5}$$

Capacitor The configuration shown in Fig. 6.2 is able to store charge in the electrodes. This general arrangement is called a *capacitor*. Due to the planar shape of the electrodes, the configuration shown in Fig. 6.2 is termed a *plate capacitor*. One attribute particular to this rather regular configuration is that the field lines in the electrode gap are homogeneous and parallel for an isotropic dielectric (i.e. the material properties are independent of direction).

---

4 Note: from the curl-free property of Eq. (6.1), it follows that $\operatorname{curl} \nabla\varphi = \vec{0}$.

5 Also termed the *polarization voltage*, as the field propagates in non-conductors via dielectric polarization.

Capacitance According to Eq. (6.2), the charge stored on the electrodes of a capacitor is proportional to the electric field intensity, and—due to Eq. (6.5)—this is in turn proportional to the voltage applied between the two electrodes, so that the formal scalar relation

$$q_C = C \cdot u_C \tag{6.6}$$

results. The proportionality constant is termed the *capacitance* C $[\mathrm{As/V}]$ of the electrode configuration. It can be computed as

$$C = \frac{q_C}{u_C} = \frac{\oint\limits_A \vec{D} \cdot d\vec{A}}{\int\limits_s \vec{E} \cdot d\vec{s}} = \varepsilon \, \frac{\oint\limits_A \vec{E} \cdot d\vec{A}}{\int\limits_s \vec{E} \cdot d\vec{s}} . \tag{6.7}$$

The capacitance thus depends both on the *material properties* of the dielectric (ε) and on the *electrode geometry*.

Capacitance of a plate capacitor For a plate capacitor with a *moving electrode* (Fig. 6.2, x is the displacement from the rest position), assuming a homogeneous field in Eq. (6.7) gives the elementary capacitance relation

$$C(x) = \frac{q_C}{u_C} = \frac{\left\|\vec{D}\right\| A}{\left\|\vec{E}\right\| \left(d - x\right)} = \frac{\varepsilon \left\|\vec{E}\right\| A}{\left\|\vec{E}\right\| \left(d - x\right)} = \frac{\varepsilon A}{d - x} , \tag{6.8}$$

where A denotes the area of an electrode (plate).

6.3 Generic Electrostatic Transducer

6.3.1 System configuration

Electrode configuration Electrostatic force generation can be exploited in a variety of ways to implement transducer functions. Fundamentally, it is the movable electrode (armature) which transforms mechanical energy into electrical energy, or vice versa. This movable electrode is then suitably connected to a mechanical structure to apply or receive forces. Fig. 6.3 depicts a schematic electrode configuration, where motion occurs perpendicular to the electrodes (a variable electrode separation). Often, longitudinal motion of the electrodes maintaining constant separation is also employed.

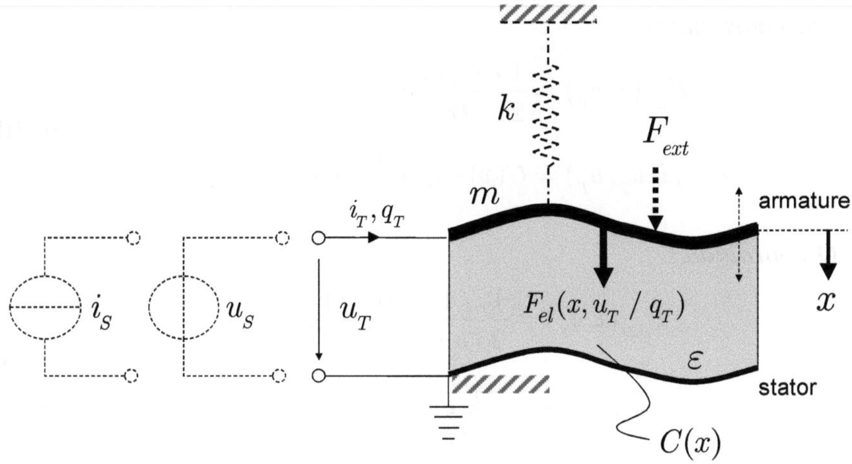

Fig. 6.3. Principle of operation of a generic electrostatic transducer with one mechanical degree of freedom (electrode moving in one dimension—here, perpendicular to the electrode surface). Dashed lines represent the external loading of a voltage or current source, and an elastic suspension

Dielectric In general, air is employed as the dielectric, and the air gap geometry determines the potential range motion of the electrode and thus the available transducer travel.

Suspension Due to the attractive nature of the electrostatic force, the moving electrode must be elastically suspended (with the exception of *differential transducers*, see Sec. 6.6.4).

6.3.2 Electrostatic constitutive transducer equations

Electrostatic constitutive basic equation The fundamental constitutive relation shown in Fig. 5.7b between the electrical energy variable q_T and the electrical power variable $\dot{\psi}_T = u_T$ is given by Eq. (6.6) (coupling via an electric field), i.e.

$$q_T = C(x) \cdot \dot{\psi}_T = C(x) \cdot u_T \,. \tag{6.9}$$

Constitutive ELM transducer equations The constitutive relation (6.9), combined with the results from Sec. 5.3.2 (see Table 5.3), directly leads to the *constitutive ELM transducer equations* for the electrostatic transducer in the different coordinate representations:

- *PSI-coordinates*

$$F_{el,\Psi}(x, u_T) = \frac{1}{2}\frac{\partial C(x)}{\partial x} u_T^{\,2}\,,$$
$$i_T(x, \dot{x}, u_T, \dot{u}_T) = C(x)\cdot \dot{u}_T + \frac{\partial C(x)}{\partial x}\cdot \dot{x}\cdot u_T\,, \tag{6.10}$$

- *Q-coordinates*

$$F_{el,Q}(x, q_T) = \frac{1}{2}\frac{1}{C(x)^2}\frac{\partial C(x)}{\partial x} q_T^{\,2}\,,$$
$$u_T(x, q_T) = \frac{1}{C(x)}\cdot q_T\,. \tag{6.11}$$

Eqs. (6.10), (6.11) describe the electrostatic force curve and the dynamics at the terminals of the unloaded transducer as a function of one of the two (assumed independent) electrical terminal variables.

Electrostatic force As can be seen in Eqs. (6.10), (6.11), the force between the charged electrodes—the *electrostatic force* F_{el}—is *unidirectional*, independent of the polarity of the electrical terminal variables.

Given the known, geometrically parameterized capacitance $C(x)$ for an electrode configuration, relations (6.10), (6.11) can thus be used to directly compute the electrostatic transducer force.

Force direction In general, the following holds for the *force direction* in an electrostatic transducer:

> **Proposition 6.1.** *Electrostatic force direction* (Jackson 1999)
>
> The electrostatic force is always directed so as to *increase* the *capacitance* of a configuration.

Electrostatic force in a plate transducer To determine an illustrative electrostatic force curve, consider the simple plate transducer in Fig. 6.4, in which the right electrode is assumed movable. According to Eqs. (6.8) and (6.10), the force acting on this electrode as a function of the variable transducer voltage u_T is given by

$$F_{el}(x, u_T) = \frac{1}{2}u_T^{\,2}\frac{\partial}{\partial x}\left(\frac{\varepsilon A}{d - x}\right) = \frac{1}{2}u_T^{\,2}\frac{\varepsilon A}{(d - x)^2}\,. \tag{6.12}$$

Substituting the transducer charge expression in Eq. (6.8) for the capacitor voltage in Eq. (6.12) (or equivalently, evaluating Eq. (6.11)) gives the COULOMB force previously seen in Example 2.1:

$$F_{el}(q_T) = \frac{1}{2}\frac{{q_T}^2}{\varepsilon A}. \tag{6.13}$$

Note that—in contrast to the voltage-dependent electrostatic force of Eq. (6.12)—the charge-dependent COULOMB force in a plate transducer is *independent* of the *electrode separation.*

Generalization: forces at boundary surfaces The electrodes in Fig. 6.4 each create a boundary surface between a conductor and a dielectric. Eqs. (6.12) and (6.13) describe the forces acting on these boundary surfaces.

This concept can be generalized for boundary surfaces between differing dielectrics in an electrostatic field. In general, the following holds:

> **Proposition 6.2.** *Force direction at boundary surfaces* (Jackson 1999)
>
> The total force acting on a boundary surface is always *perpendicular* to the surface. The force direction is independent of the direction of the field and is always directed toward the dielectric having the *smaller* dielectric constant.

Using the example of a parallel electrode configuration, Fig. 6.5 depicts two cases of practical importance: one having the boundary surface normal to, and the other, parallel to the electric field. For both (geometrically simple) cases, the resulting force can be easily computed.

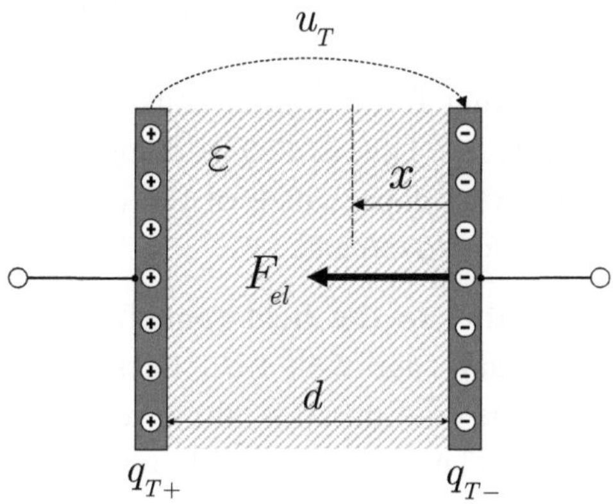

Fig. 6.4. Electrostatic force between two charged electrodes (plate transducer)

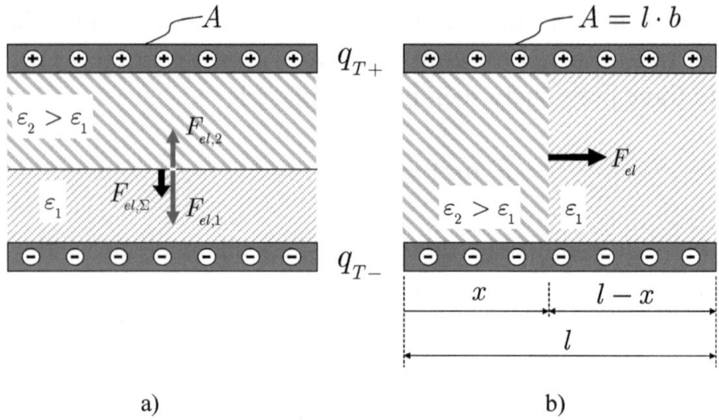

Fig. 6.5. Forces at boundary surfaces in the dielectric for the example of a plate capacitor: a) boundary surface normal to the field, b) boundary surface parallel to the field

Boundary surface normal to the field (parallel electrodes) In Fig. 6.5a, a thin metal foil can be imagined between the two dielectrics $\varepsilon_1, \varepsilon_2$. Such a foil will not affect the electric field, and charges can only be generated on it via electrostatic induction (Jackson 1999). The forces on this virtual electrode can be computed directly from Eq. (6.13):

$$F_{el,1} = \frac{1}{2}\frac{q_T^{\,2}}{\varepsilon_1 A}, \quad F_{el,2} = \frac{1}{2}\frac{q_T^{\,2}}{\varepsilon_2 A},$$
$$F_{el,\Sigma} = F_{el,1} - F_{el,2} = \frac{1}{2}\frac{q_T^{\,2}}{A}\left(\frac{1}{\varepsilon_1} - \frac{1}{\varepsilon_2}\right). \tag{6.14}$$

For $\varepsilon_2 > \varepsilon_1$ the resulting force $F_{el,\Sigma}$ is thus positive for the direction shown, pointing towards ε_1. This is also in complete agreement with the principle of increasing total capacitance of the configuration, i.e. the dielectric ε_2 has the tendency to expand.

Boundary surface parallel to the field (parallel electrodes) Fig. 6.5b depicts the complementary case with field lines running parallel to the boundary surface. For this configuration, the force is obtained by evaluating the elementary force relation in Eq. (6.10), taking into account the total capacitance resulting from the two component capacitances connected in parallel:

$$C = C_1 + C_2 = \frac{1}{d}\left(\varepsilon_1 A_1 + \varepsilon_2 A_2\right) \;\Rightarrow\; C(x) = \frac{\varepsilon_1 bl}{d} + \frac{b}{d} x\left(\varepsilon_2 - \varepsilon_1\right),$$

$$F_{el} = \frac{1}{2} u_T^{\,2} \frac{\partial C(x)}{\partial x} = \frac{1}{2} u_T^{\,2} \frac{b}{d}\left(\varepsilon_2 - \varepsilon_1\right). \qquad (6.15)$$

The force direction for $\varepsilon_2 > \varepsilon_1$ again naturally agrees with the above propositions, and reveals a noteworthy property. If dielectric 1 consists of air ($\varepsilon_1 \approx \varepsilon_0$), then a movable dielectric 2 (ε_2) will be constantly pulled into the plate capacitor. For such a configuration, only a *centered* dielectric can be at a *stable* rest position.

6.3.3 ELM two-port model

Local linearization The constitutive electrostatic force equations (6.10) and (6.11) are always quadratic and nonlinear in the independent electrical terminal variable u_T or q_T, and often also nonlinear in the armature displacement x. For this reason, for small-signal analysis, a local linearization about a steady-state operating point should be performed. Without loss of generality, possible *steady-state operating points* can be defined at

$$x_R,\ \dot{x}_R = 0, \quad u_{T,R},\ \dot{u}_{T,R} = 0, \text{ or } q_{T,R},\ \dot{q}_{T,R} = 0, \qquad (6.16)$$

where, for a concrete electrode configuration, the non-zero steady-state quantities in Eq. (6.16) should be calculated as a function of $C(x)$ (see examples in later sections).

Two-port admittance form Starting with the constitutive transducer equations (6.10) and applying the generally-valid results of Sec. 5.3.3 gives the *two-port admittance form* of the unloaded electrostatic transducer

$$\begin{pmatrix} \Delta F_{el,\Psi}(s) \\ \Delta I_T(s) \end{pmatrix} = \mathbf{Y}_{el}(s) \cdot \begin{pmatrix} \Delta X(s) \\ \Delta U_T(s) \end{pmatrix} = \begin{pmatrix} k_{el,U} & K_{el,U} \\ s \cdot K_{el,U} & s \cdot C_R \end{pmatrix} \begin{pmatrix} \Delta X(s) \\ \Delta U_t(s) \end{pmatrix}, \qquad (6.17)$$

with the *operating-point-dependent transducer parameters*

$$
\begin{aligned}
&\bullet\ \textit{electrostatic voltage stiffness} && k_{el,U} := \frac{1}{2} u^2 \left.\frac{\partial^2 C(x)}{\partial x^2}\right|_{\substack{u=u_{T,R}\\ x=x_R}}, \\
&\bullet\ \textit{voltage coefficient} && K_{el,U} := u \left.\frac{\partial C(x)}{\partial x}\right|_{u=u_{T,R},\ x=x_R}, \\
&\bullet\ \textit{rest capacitance} && C_R := C(x)\big|_{x=x_R}.
\end{aligned}
\tag{6.18}
$$

Two-port hybrid form In the same way, the constitutive transducer equations (6.11) can be used to arrive at the *two-port hybrid form* for the unloaded electrostatic transducer

$$
\begin{pmatrix} \Delta F_{el,Q}(s) \\ \Delta U_T(s) \end{pmatrix} = \mathbf{H}_{el}(s) \cdot \begin{pmatrix} \Delta X(s) \\ \Delta I_T(s) \end{pmatrix} = \begin{pmatrix} k_{el,I} & \dfrac{K_{el,I}}{s} \\ -K_{el,I} & \dfrac{1}{s \cdot C_R} \end{pmatrix} \begin{pmatrix} \Delta X(s) \\ \Delta I_T(s) \end{pmatrix}, \tag{6.19}
$$

with the *operating-point-dependent transducer parameters*

- *electrostatic current stiffness*

$$
k_{el,I} := \frac{q^2}{C^2(x)} \left[\frac{1}{2} \frac{\partial^2 C(x)}{\partial x^2} - \frac{1}{C(x)} \left(\frac{\partial C(x)}{\partial x} \right)^2 \right] \Bigg|_{q=q_{T,R},\ x=x_R}, \tag{6.20}
$$

- *current coefficient* $\quad K_{el,I} := q \dfrac{1}{C^2(x)} \left.\dfrac{\partial C(x)}{\partial x}\right|_{q=q_{T,R},\ x=x_R},$

and the *rest capacitance* C_R defined as in Eq. (6.18).

Relationship between the two-port parameters It is easy to verify that the following relations hold for the parameters of the admittance and hybrid forms (see Table 5.4):

$$
\begin{aligned}
k_{el,U} &= k_{el,I} + C_R K_{el,I}^{\ 2}, & K_{el,U} &= C_R K_{el,I}, \\
k_{el,I} &= k_{el,U} - \frac{K_{el,U}^{\ 2}}{C_R}, & K_{el,I} &= \frac{K_{el,U}}{C_R}.
\end{aligned}
\tag{6.21}
$$

6.3.4 Loaded electrostatic transducer

Mechanical suspension Due to the unidirectional action of the electrostatic force (with quadratic dependence on the electrical terminal variables, and force direction always increasing the transducer capacitance), *elastic suspension* of the armature electrode is *essential* (see Fig. 6.3). The spring force of the suspension must compensate the transducer force. This and subsequent sections calculate the rest positions and analyze the steady-state behaviors of a few common electrode configurations.

Linear dynamic model The linear dynamic model of the loaded electrostatic transducer can be easily obtained from the generic model in Sec. 5.4.4 using the two-port parameters in Eqs. (6.17), (6.19). The signal-flow diagrams for a voltage-drive and a current-drive transducer are depicted in Fig. 6.6 and Fig. 6.7, respectively (cf. Fig. 5.15, Fig. 5.16).

ELM coupling factor Applying the relations from Sec. 5.6 results in the operating-point-dependent general formula for the *ELM coupling factor* of an *electrostatic* transducer (only defined for an elastically suspended armature electrode)

$$\kappa_{el}^2 = \frac{1}{1 + \frac{C_R}{K_{el,U}^2}\left(k - k_{el,U}\right)} = \frac{C_R \cdot K_{el,I}^2}{k - k_{el,I}} \,. \tag{6.22}$$

Transducer stiffness: electrostatic softening The (differential) electrostatic transducer stiffnesses in Eqs. (6.18), (6.20) exhibit four noteworthy properties:

- The transducer stiffness $k_{el,I}$ for current drive is always *less* than the stiffness $k_{el,U}$ for voltage drive.
- As a rule, both stiffnesses are greater than zero, leading—as described in Sec. 5.4.3—to a decrease in the total stiffness $(k - k_{el})$ of the transducer (here, electrostatic softening).
- Under certain conditions, the differential transducer stiffness for *current drive* can become zero or negative.
- The stiffnesses change with the rest position, i.e. with the electrical control applied to the transducer. On one hand, this can be considered a *parameter variation* during controller design for a system including an electrostatic transducer. On the other hand, combined with electrostatic softening, this implies a danger of *instability* at the operating point (see *pull-in*).

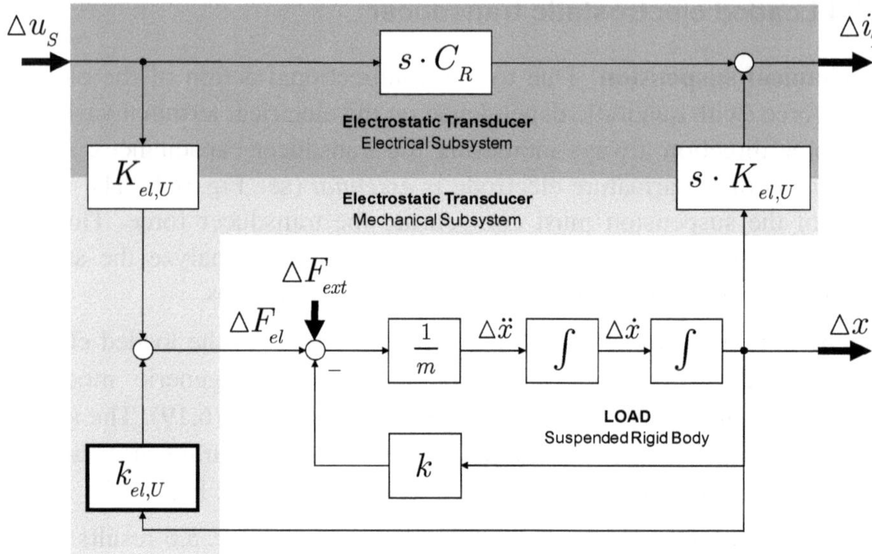

Fig. 6.6. Signal-flow diagram of a loaded *electrostatic* transducer with *voltage drive* (model linearized about a stable operating point, lossless, ideal voltage source, mechanical load: elastically suspended rigid body, cf. Fig. 6.3)

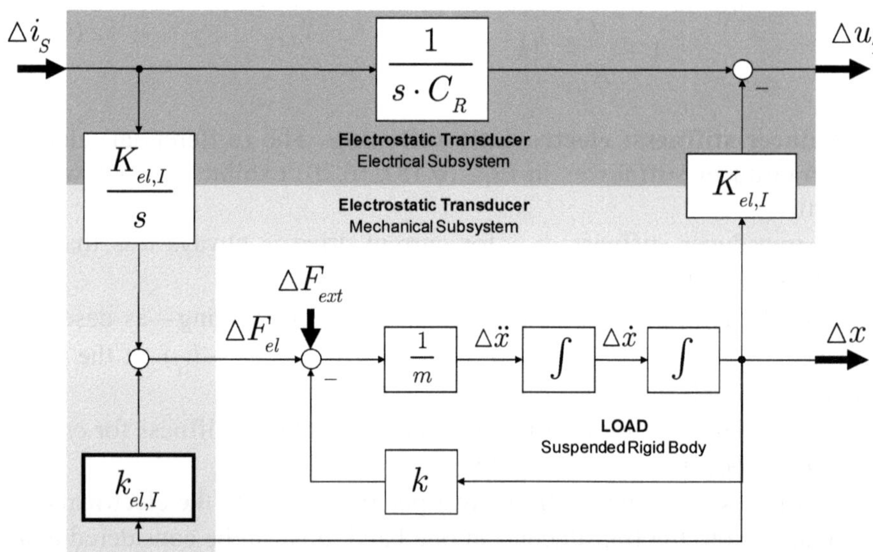

Fig. 6.7. Signal-flow diagram of a loaded *electrostatic* transducer with *current drive* (linearized model about a stable operating point, lossless, ideal current source, mechanical load: elastically suspended rigid body, cf. Fig. 6.3)

6.3.5 Structural principles

Table 6.1 presents a variety of transducer configurations having one mechanical degree of freedom. Fundamentally, transducers can be divided into types having a *variable* or a *constant* electrode separation.

Variable electrode separation The Types A and B in Table 6.1 have variable electrode separation and exploit perpendicular forces on the electrode surfaces. The motion degree of freedom is thus parallel to the electric field. For the tip-plate transducer (Type B), it should be noted that the field inside the electrode gap is no longer homogeneous, resulting in its more complicated description. The two approximations listed can be used for rough estimates. Approximation 1—corresponding to Type A—gives slightly lower forces than actually occur. The force is fundamentally *inversely proportional* to the *square* of the electrode separation.

Constant electrode separation For a constant plate separation, either the overlapping electrode area or the dielectric can be varied. In Types C and D, constant translational or rotational forces act on the capacitor plates as long as the electrodes do not completely overlap. Again, the force is directed in such a way as to maximize the capacitance of the arrangement. In Type E, forces on boundary surfaces of the dielectric are exploited. Here again, the constant electrostatic force acts until the dielectric completely fills the electrode gap.

Geometric scaling To form an idea of the order of magnitude of electrostatic forces, consider a typical application of the Type B transducer configuration for microscope mirrors. For a mirror area of $500\,\mu m \times 500\,\mu m$, an electrode separation of $20\,\mu m$, and an operating voltage of $100\ V$, an electrostatic force $F_{el} \approx 30\,\mu N$ results. This extremely small force is completely sufficient to move a microgram-scale silicon wafer. However, due to the force relation (6.12) and (6.15), geometric scaling of the electrodes by a factor λ would have no effect on the generated force, while the mass to be moved would increase by a factor of λ^3, assuming homogeneous density. Any significant increase in the force can only be achieved by an absolute increase in the operating voltage, or by a relative decrease in the electrode separation. Both solutions clearly quickly hit technical limits. Maximum operating voltages are determined both by the dielectric strength of the insulator and by practical operating constraints. Reduced electrode separation and larger lateral dimensions is subject to manufacturing limitations, and even so, would not even be usable due to the resulting restricted range of motion.

Table 6.1. Generic electrostatic transducers with one mechanical degree of freedom x or φ

	Transducer type	Transducer model
A	*plate transducer, transverse* variable electrode separation 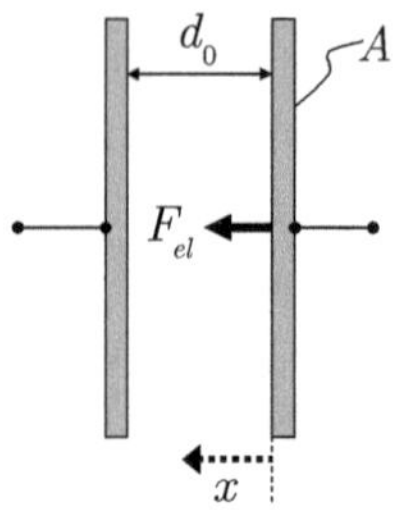	$x < d_0$ $C(x) = \dfrac{\varepsilon A}{d_0 - x}$ $F_{el}(x,U) = \dfrac{1}{2} U^2 \dfrac{\varepsilon A}{(d_0 - x)^2}$
B	*tip-plate transducer* variable electrode separation 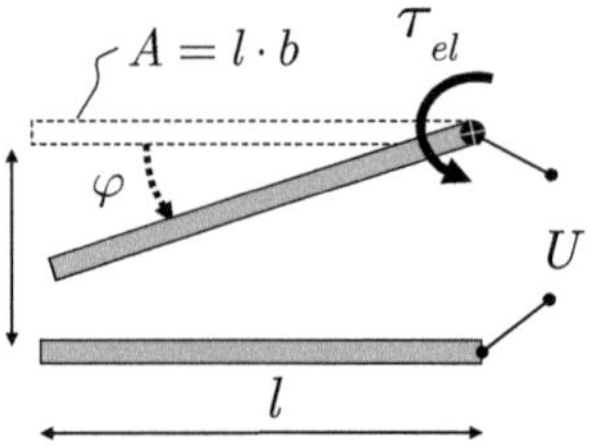	$l \gg d$ $C(\varphi) = -\dfrac{\varepsilon b}{\varphi} \ln\left(1 - \dfrac{l\varphi}{d}\right)$ $\tau_{el}(\varphi,U) = \dfrac{\varepsilon A}{2} U^2 \left[\dfrac{\ln\left(1 - \dfrac{l\varphi}{d}\right)}{l\varphi^2} + \dfrac{1}{d\varphi\left(1 - \dfrac{l\varphi}{d}\right)}\right]$ Approximation 1 for $\varphi \ll 1$ $C(\varphi) \approx \dfrac{\varepsilon A}{d - \dfrac{l}{2}\varphi}$ $\tau_{el}(\varphi,U) \approx \dfrac{\varepsilon A l}{2} U^2 \dfrac{1}{\left(d - \dfrac{l}{2}\varphi\right)^2}$ Approximation 2 for $\varphi \ll 1$ $C(\varphi) \approx \dfrac{\varepsilon b l}{d}\left[1 + \dfrac{1}{2}\dfrac{l\varphi}{d} + \dfrac{1}{3}\left(\dfrac{l\varphi}{d}\right)^2\right]$ $\tau_{el}(\varphi,U) \approx \dfrac{1}{2}\dfrac{\varepsilon b l^2}{d^2} U^2 \left[\dfrac{1}{2} + \dfrac{2}{3}\dfrac{l\varphi}{d}\right]$

Table 6.1. (cont.) Generic electrostatic transducers with one mechanical degree of freedom x or φ

	Transducer type	Transducer model
C	*plate transducer, longitudinal* variable area 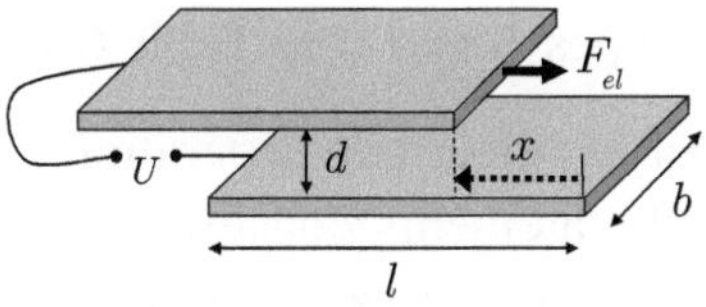	$0 \le x \le l$ $C(x) = \dfrac{\varepsilon b\left(l - x\right)}{d}$ $F_{el}(U) = -\dfrac{1}{2}\dfrac{\varepsilon b}{d}U^2$
D	*rotary capacitor* variable area 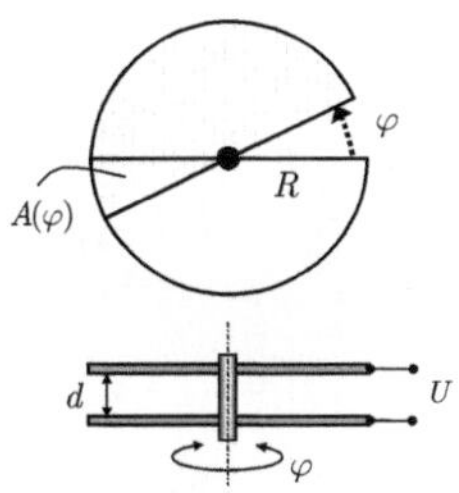	$0 \le \varphi \le \pi$ $C(\varphi) = \dfrac{\varepsilon R^2}{2d}\varphi$ $T_{el}(U) = \dfrac{1}{2}\dfrac{\varepsilon R^2}{2d}U^2$
E	*linear transducer* variable dielectric 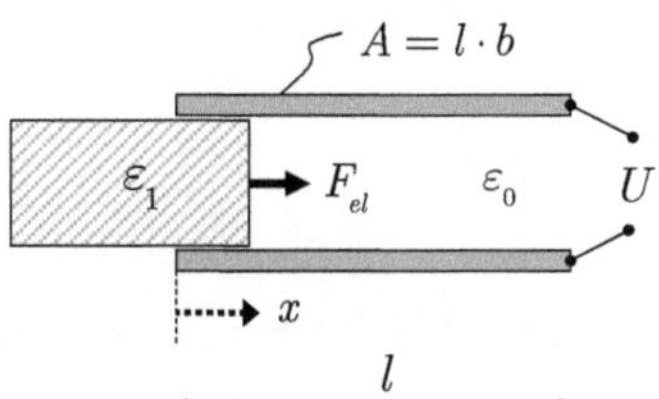	$0 \le x \le l$ $C(x) = \dfrac{\varepsilon_0 bl}{d} + \dfrac{b}{d}x\left(\varepsilon_1 - \varepsilon_0\right)$ $F_{el}(U) = \dfrac{1}{2}\dfrac{b}{d}\left(\varepsilon_1 - \varepsilon_0\right)U^2$

These attributes also explain the extremely limited application of electrostatic transducer principles at the macroscopic scale (e.g. electrical measurement instruments such as the needle electrometer (Jackson 1999)). A breakthrough in mechatronic applications has only been made possible by the advances in microsystem technology of the recent past. Only in the micro-scale structures created in this arena can the extremely small forces generated by electrostatics be meaningfully exploited (Senturia 2001).

6.4 Transducer with Variable Electrode Separation and Voltage Drive

6.4.1 General dynamic model

Transducer type Electrostatic transducers with variable electrode separation enable motion of the armature *perpendicular* to the electrode surface and have a configuration according to Types A and B in Table 6.1.

Configuration equations With a *variable* electrode separation and a homogeneous field, the following general statements concerning *transducer capacitance* and electrostatic *force generation* are valid for voltage drive:

$$C_T(x) := \frac{\alpha}{\beta - x} \quad , \quad C_0 := C_T(0) = \frac{\alpha}{\beta}, \tag{6.23}$$

$$\Rightarrow \quad \frac{\partial CT(x)}{\partial x} = \frac{\alpha}{\left(\beta - x\right)^2}, \qquad F_{el}(x, u_T) = \frac{1}{2} u_T^{\ 2} \frac{\alpha}{\left(\beta - x\right)^2}.$$

Steady-state behavior

Rest positions Applying Eq. (5.40), the rest positions x_R for static excitations $u_T(t) = U_0$, $F_{ext}(t) = F_0$ can be found from the cubic determining equation

$$\frac{2}{\alpha}\left(kx_R - F_0\right)\left(\beta - x_R\right)^2 = U_0^{\ 2}. \tag{6.24}$$

From the three possible solutions of Eq. (6.24) only one—as is shown below—results in a fundamentally stable rest position, though also only for particular configurations (see the pull-in phenomenon). For further considerations, parameterizing the rest positions with the tuple (x_R, F_0) proves to be convenient[6].

Pull-in phenomenon In steady state, for a vanishing mechanical excitation $F_{ext} = 0$, the spring force $F_F = kx_R$ and the electrostatic force $F_{el}(x_R, U_0)$

[6] The corresponding U_0 can be determined via Eq. (6.24). Manipulating and interpreting the rest displacement x_R as an operational parameter is more descriptive than using the source voltage for this purpose.

must balance. This corresponds precisely to the rest position condition Eq. (6.24). However, a stable rest position results only when the *differential stiffness* $\partial F_\Sigma / \partial x$ of the summed force acting on the electrode $F_\Sigma = F_F - F_{el}$ is positive. In the optimal case, there is a single stable rest position, as can be seen in Fig. 6.8 (cf. Fig. 5.13).

For $x_R = \beta/3$, there is a marginally stable equilibrium due to $\partial F_\Sigma / \partial x = 0$, while for $x_R > \beta/3$, the gradient dominates the electrostatic force. At this point, the mobile electrode is pulled toward the static electrode without a counteracting force—a phenomenon termed *pull-in* (Senturia 2001).

The critical rest displacement is thus called the *pull-in displacement* (see Fig. 6.8)

$$x_{pi} = \frac{\beta}{3}. \tag{6.25}$$

Using Eqs. (6.25) and (6.24), the corresponding rest voltage U_{pi}—the *pull-in voltage*—at the pull-in electrode displacement x_{pi} is then

$$U_{pi} = \sqrt{\frac{8}{27}\frac{\beta^3}{\alpha}k}. \tag{6.26}$$

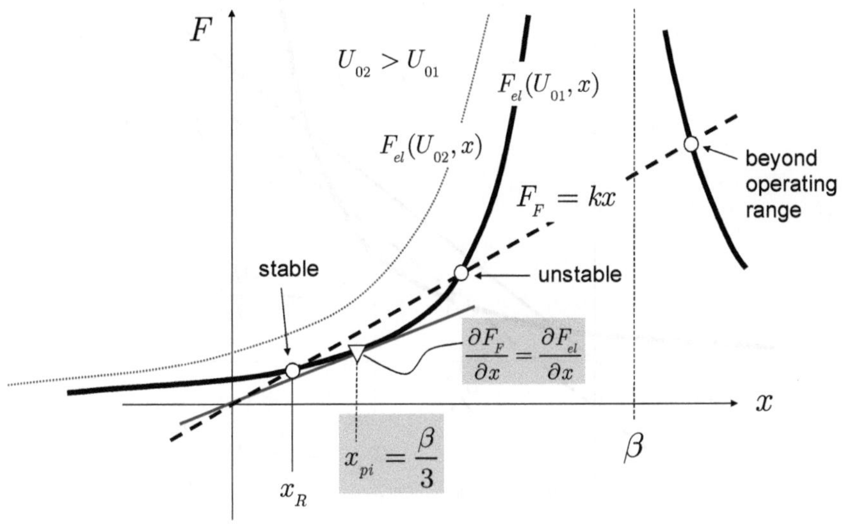

Fig. 6.8. Pull-in: steady-state force curve with *vanishing* mechanical excitation $F_{ext} = 0$

For *stable operation* avoiding pull-in, the steady-state operating point is thus subject to the requirement

$$x_R < x_{pi}\text{, or equivalently, } U_0 < U_{pi}. \tag{6.27}$$

Thus, due to the threat of pull-in, when using voltage drive, only *one third* of the electrode gap can be exploited for targeted motions. One possibility for electrically increasing the range of motion is described in Sec. 6.4.2.

Pull-in limits for static mechanical excitation Fig. 6.9 depicts the applicable force curve in the presence of an additional *static mechanical force* F_0. It can be seen that, due to the shift in the spring force curve by the amount $(-F_0)$, both the rest position and pull-in limits become

$$\tilde{x}_{pi} = \frac{1}{3}(\beta + \frac{2}{k}F_0), \quad \tilde{U}_{pi} = \sqrt{\frac{8}{27}\frac{\left(k\beta - F_0\right)^3}{\alpha k^2}}. \tag{6.28}$$

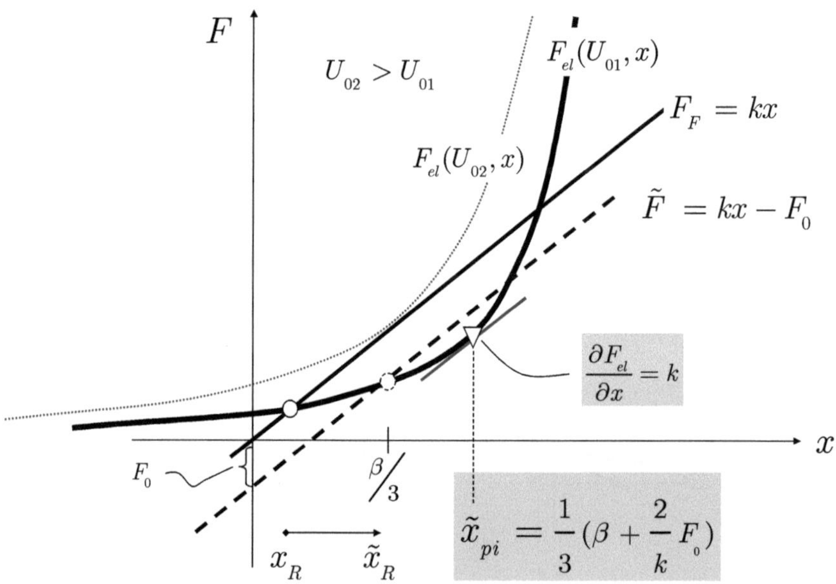

Fig. 6.9. Pull-in: steady-state force curve with *static* mechanical excitation $F_{ext} = F_0 = const. \neq 0$

Dynamic behavior

Transducer parameters Substituting $C_T(x)$ from Eq. (6.23) into Eq. (6.18), and defining the *relative rest displacement* $X_R := x_R / \beta$, gives the operating-point-dependent two-port parameters

$$\begin{aligned}
&\bullet\ \textit{electrostatic voltage stiffness} && k_{el,U} := \frac{2(k \cdot X_R - F_0/\beta)}{1 - X_R} \left[\frac{\text{N}}{\text{m}}\right], \\
&\bullet\ \textit{voltage coefficient} && K_{el,U} := \frac{\sqrt{2C_0(k \cdot X_R - F_0/\beta)}}{1 - X_R}, \\
&\bullet\ \textit{rest capacitance} && C_R := C_0 \frac{1}{1 - X_R} \quad \left[\frac{\text{As}}{\text{V}}\right].
\end{aligned} \tag{6.29}$$

Transfer matrix From Eq. (6.29) and Table 5.8, the *transfer matrix* $\mathbf{G}(s)$ is then

$$\begin{pmatrix} \Delta X(s) \\ \Delta I_T(s) \end{pmatrix} = \mathbf{G}(s) \begin{pmatrix} \Delta F_{ext}(s) \\ \Delta U_S(s) \end{pmatrix} = \begin{pmatrix} V_{x/F,U} \dfrac{1}{\{\Omega_U\}} & V_{x/u} \dfrac{1}{\{\Omega_U\}} \\ V_{i/F} \dfrac{s}{\{\Omega_U\}} & V_{i/u} \cdot s \dfrac{\{\Omega_0\}}{\{\Omega_U\}} \end{pmatrix} \begin{pmatrix} \Delta F_{ext}(s) \\ \Delta U_S(s) \end{pmatrix}, \tag{6.30}$$

with the parameters

$$\begin{gathered}
k_{T,U} = k - k_{el,U}, \quad \Omega_U{}^2 = \frac{k_{T,U}}{m}, \quad \Omega_0{}^2 = \frac{k}{m} \quad \text{where } \Omega_U < \Omega_0 \\
V_{x/F,U} = \frac{1}{k_{T,U}}, \quad V_{x/u} = V_{i/F} = \frac{K_{el,U}}{k_{T,U}}, \quad V_{i/u} = C_R \frac{k}{k_{T,U}}.
\end{gathered} \tag{6.31}$$

It is worth noting that the *antiresonance* of the electrical terminal transfer function (the electrical admittance) is given by the mechanical eigenfrequency Ω_0. As this zero is in turn equal to the transducer poles for current drive, the differential transducer stiffness is clearly zero for this drive configuration. This property is further discussed in the next section.

Electrostatic softening vs. gain variation The transducer stiffness $k_{T,U}$ in Eq. (6.31) depends on the operating point, i.e. it changes with the rest displacement x_R according to

$$k_{T,U} = \frac{k(1-3X_R)+2F_0/\beta}{\left(1-X_R\right)}.$$

As x_R increases, $k_{T,U}$ goes to zero (electrostatic softening); at pull-in, $k_{T,U} = 0$ (cf. Eqs. (6.25) and (6.28)).

This further implies that the *gains* of the transfer functions in Eq. (6.31) quickly increase near the pull-in armature displacement. For one, this means that increasing transducer sensitivity depends on the operating point. Additionally, this extreme degree of parameter variation must be carefully considered during controller design if the transducer is to be used as an actuator in a closed loop.

Characteristic polynomial: transducer stability An important component of stability analysis is the *characteristic polynomial* $\Delta_U(s)$ of the transfer matrix $\mathbf{G}(s)$

$$\Delta_U(s) = s^2 + \Omega_U{}^2 = s^2 + \frac{k_{T,U}}{m} = s^2 + \frac{k(1-3X_R)+2F_0/\beta}{m\left(1-X_R\right)}. \tag{6.32}$$

The pull-in conditions (6.25) and (6.28) can be precisely recovered as stability limits of Eq. (6.32) (double pole at $s = 0$, i.e. $k_{T,U} = 0$).

ELM coupling factor Substituting the transducer parameters (6.29) into Eq. (6.22) gives the *ELM coupling factor* for the voltage-drive plate transducer

$$\kappa_{el}{}^2 = \frac{2\left(kX_R - F_0/\beta\right)}{k\left(1-X_R\right)} = \frac{k_{el,U}}{k}. \tag{6.33}$$

According to Eq. (6.33), the ELM coupling factor thus fundamentally depends on the *relative motion geometry* of the transducer, i.e. on the rest displacement x_R relative to the zero-voltage electrode separation β. In some cases, there is an additional offset originating in the static mechanical excitation F_0. The right-hand expression in Eq. (6.33) relates the ELM coupling factor to the transducer stiffnesses.

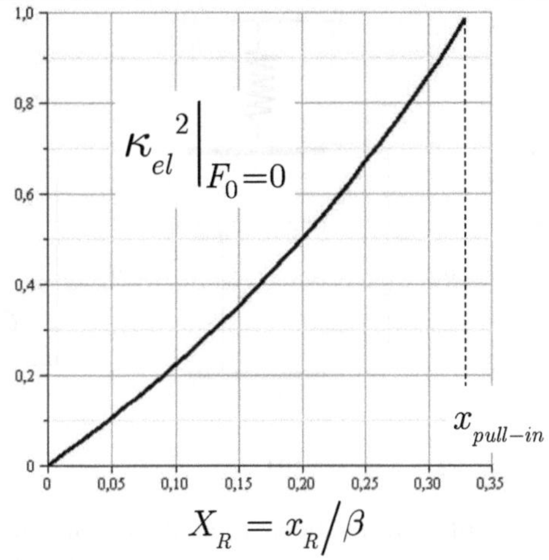

Fig. 6.10. ELM coupling factor κ_{el}^2 for $F_0 = 0$ as a function of the rest displacement x_R for a voltage-drive electrostatic plate transducer with variable electrode separation

Fig. 6.10 shows the ELM coupling factor as a function of the stable rest displacement x_R for $F_0 = 0$. For an increasing rest displacement, the applied mechanical power clearly climbs as well. However, in this case, electrostatic softening also renders the transducer increasingly unstable. Thus, in order to prevent pull-in, large coupling factors are only possible with strict limits on the range of motion.

6.4.2 Increasing the range of motion with serial capacitors

Geometric motion restrictions Pull-in appreciably reduces the range of motion of a voltage-drive transducer. The transducer in Fig. 6.11a can be stably operated only up to $x_{pi} = \beta/3$. If an armature travel of $x_{\max} = \beta$ were desired, the uncharged electrode separation would have to be physically increased to 3β (Fig. 6.11b). However, according to Eq. (6.26), the tripling of the air gap implies that this range of motion requires a $\sqrt{3^3} \approx 5$-fold increase in the control voltage relative to the original design in Fig. 6.11a. The increased gap results in a smaller total capacitance, which can be thought of as a serial connection of the original plate arrangement and the extended version in Fig. 6.11c.

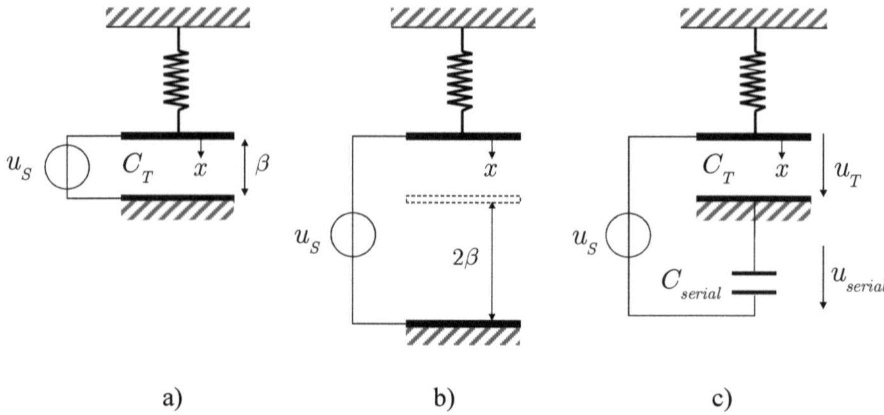

Fig. 6.11. Voltage-drive plate transducer: a) electrode separation of β: electrode travel $0 \leq x < \beta/3$, b) increased electrode separation of 3β: increased electrode travel $0 \leq x < \beta$, c) smaller electrode separation of β with serial capacitance C_{serial}: increased electrode travel $0 \leq x < \beta$

Serial capacitance One practical option for increasing the stable range of motion while maintaining *the same electrode geometry* is the use of a *serial capacitor* as shown in Fig. 6.11c (Seeger and Crary 1997), (Chan and Dutton 2000). This allows the entire available electrode travel β of the original transducer to be exploited.

The total capacitance of the arrangement in Fig. 6.11c is

$$C_\Sigma(x) = \frac{C_T \cdot C_{serial}}{C_T + C_{serial}} = \frac{\alpha}{\beta + \dfrac{\alpha}{C_{serial}} - x} = \frac{\alpha}{\tilde{\beta} - x}. \tag{6.34}$$

From Eq. (6.34), a *fictitious increase* in the electrode separation of the uncharged transducer to $\tilde{\beta}$ can be seen. Without affecting the system behavior, the pull-in limits can now be computed relative to $\tilde{\beta}$.

Thus, to achieve the electrostatic armature travel β shown in Fig. 6.11a, a serial capacitance

$$C_{serial} \geq \frac{1}{2}\frac{\alpha}{\beta} = \frac{1}{2}C_0$$

should be chosen.

Mechanism The mechanism through which the serial capacitance acts is providing a voltage divider on the drive voltage according to

$$u_T = \frac{C_{serial}}{C_T + C_{serial}} u_S = \frac{\beta - x}{\dfrac{\alpha}{C_{serial}} + \beta - x} u_S .$$

Thus, for an increasing source voltage u_S, and resulting increasing displacement x and transducer capacitance C_T, the polarization voltage u_T at the transducer is automatically reduced (through electrical feedback). The electrostatic force thus now follows the rule

$$F_{el}(x, u_S) = \frac{\alpha}{2} \frac{1}{(\beta - x)^2} \left(\frac{(\beta - x)}{\left(\dfrac{\alpha}{C_{serial}} + \beta - x \right)} u_S \right)^2 = \frac{\alpha}{2} \frac{1}{(\dfrac{\alpha}{C_{serial}} + \beta - x)^2} u_S^{\,2} .$$

Advantages, disadvantages, limitations Electrically increasing the armature travel maintains a *compact geometry*. Due to the voltage divider behavior of this method, however, a *higher control voltage* is required than in the case of direct transducer control. In addition, *parasitic capacitances* still limit stable operation (Chan and Dutton 2000).

6.4.3 Passive damping with serial resistance

Transducer with a voltage source and serial resistance Fig. 6.12 depicts an arrangement consisting of a voltage-drive plate transducer with a series resistor R. In Sec. 5.5, the dynamics of a lossy elementary transducer were discussed in detail; these can now be demonstrated concretely in this transducer type. As previously explained, mechanical energy is dissipated in the electrical loop via electromechanical coupling, resulting in passive damping of the mechanical single-mass oscillator. Thus, the resistance value should be considered an important design degree of freedom.

Steady-state behavior At steady state, the current in the transducer is zero (there is voltage balance, $u_T = u_S$). Thus, there is no change in rest positions or pull-in conditions compared to the lossless plate transducer.

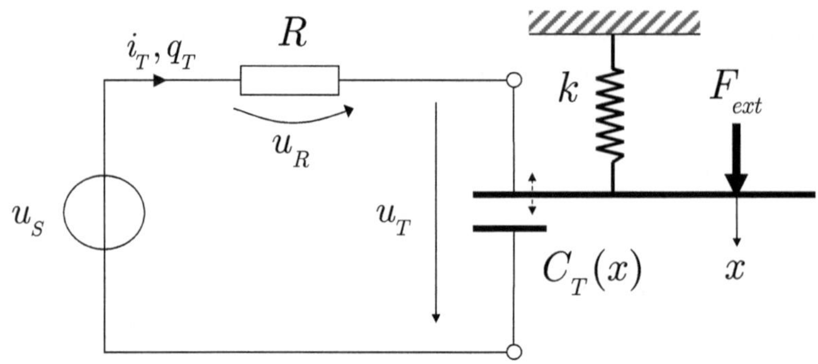

Fig. 6.12. Voltage-drive electrostatic plate transducer with variable electrode separation and series resistor

Dynamic behavior: small-signal dynamics To calculate the linearized transducer models at the stable rest position (U_0, x_R), the values from Tables 5.7 and 5.8 along with the ELM two-port admittance parameters of Eq. (6.17) or Eq. (6.29) can be applied directly, giving, after a few manipulations[7],

$$\begin{pmatrix} \Delta X(s) \\ \Delta I_T(s) \end{pmatrix} = \begin{pmatrix} V_{x/F,U} \dfrac{[\tilde{\omega}_Z]}{\tilde{\mathcal{N}}(s)} & V_{x/u} \dfrac{1}{\tilde{\mathcal{N}}(s)} \\ V_{i/F} \dfrac{s}{\tilde{\mathcal{N}}(s)} & V_{i/u} \cdot s \dfrac{\{\Omega_0\}}{\tilde{\mathcal{N}}(s)} \end{pmatrix} \begin{pmatrix} \Delta F_{ext}(s) \\ \Delta U_S(s) \end{pmatrix},$$

$$\tilde{\mathcal{N}}(s) := \left(1 + 2\tilde{d}_U \frac{s}{\tilde{\Omega}_U} + \frac{s^2}{\tilde{\Omega}_U{}^2}\right) \cdot \left(1 + \frac{s}{\tilde{\omega}_U}\right), \quad \tilde{\omega}_Z = \frac{1}{R \cdot C_R}. \tag{6.35}$$

where the gains are identical to those of the lossless transducer in Eq. (6.31)[8].

A change in the pole locations is observed. In addition to the now-damped complex pole pair $\{\tilde{d}_U, \tilde{\Omega}_U\}$ of the mechanical eigenfrequency, an

7 The equations can be easily evaluated with a computer algebra program (e.g. MAPLE, MATHEMATICA). These can also be used to easily find closed-form equations for the transfer matrix as a function of the physical parameters for other electrode configurations.

8 This is not unexpected, as the above explanations give the same steady-state values for the transducer with resistance as for the lossless transducer.

additional real pole $[\tilde{\omega}_U]$ appears due to the RC input circuit. In addition, derivative dynamics with a real zero $[\tilde{\omega}_Z]$ can be discerned in the mechanical transfer channel.

Root locus as a function of resistance The dependence of the poles $\{\tilde{d}_U, \tilde{\Omega}_U\}$, $[\tilde{\omega}_U]$ on the resistance R was discussed in depth in Sec. 5.5.5. For the current case, the root locus in Fig. 5.24a applies exactly, confirming the expected damping behavior of the electrical feedback.

Maximum damping Defining $X_R := x_R / \beta$, and from the considerations in Sec. 5.5.5, the *maximum* reachable *damping* of the transducer eigenfrequency is

$$d^{\max} = \frac{1}{2}\left(\sqrt{\frac{1 - X_R}{1 - 3X_R + \dfrac{2F_0}{k\beta}}} - 1\right). \tag{6.36}$$

The *optimal resistance* can be found according to Sec. 5.5.5 to be

$$R^{\max} = \frac{1 - X_R}{\Omega_0 \cdot C_0}\left(\frac{1 - 3X_R + \dfrac{2F_0}{k\beta}}{1 - X_R}\right)^{\frac{1}{4}}. \tag{6.37}$$

Interestingly, the maximum achievable damping of the transducer eigenfrequency depends solely on the *relative motion geometry* of the transducer, and increases with increasing rest position x_R (or X_R), as shown in Fig. 6.13a for $F_0 = 0$.

This behavior is also thoroughly plausible, when compared with the behavior of the ELM coupling factors (see Fig. 6.10). A high ELM coupling factor implies that a large portion of the mechanical energy is transformed into electrical energy and thus available to be dissipated in the resistance R.

The corresponding optimal resistance in Eq. (6.37) is plotted in Fig. 6.13b as a function of the rest displacement, again for $F_0 = 0$.

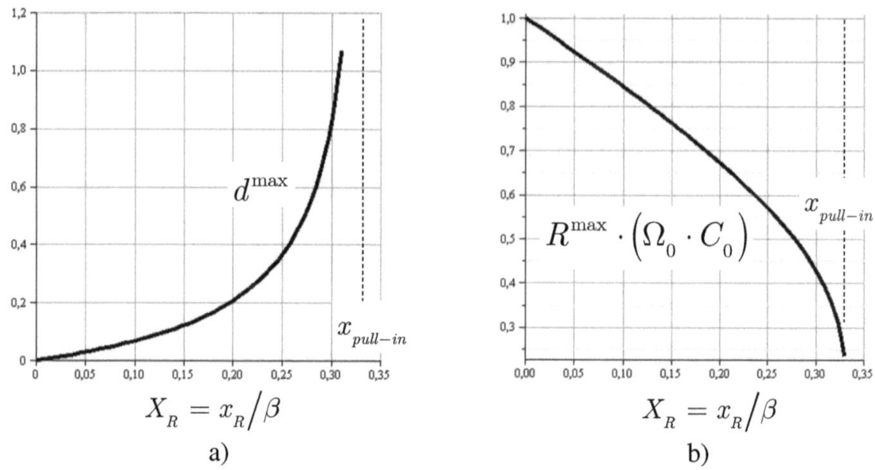

Fig. 6.13. Voltage-drive electrostatic plate transducer with variable electrode separation and series resistance: a) maximum achievable damping Eq. (6.36) for $F_0 = 0$, b) optimal resistance for maximum damping Eq. (6.37) for $F_0 = 0$

Root locus as a function of rest displacement For the lossy transducer, there is an interesting behavior in the dependence of the transducer eigenfrequency on the rest displacement x_R. To see this, the *characteristic polynomial* $\Delta(s)$ of the transfer matrix $\mathbf{G}(s)$ is rearranged as follows (to simplify the presentation, the case $F_0 = 0$ is considered):

$$\Delta(s) = \left(s^2 + \frac{k}{m}\right)\left(s + \frac{\beta}{R\alpha}\right) - x_R \frac{1}{R\alpha}\left(s^2 + \frac{3k}{m}\right). \tag{6.38}$$

Eq. (6.38) can be used to give the following representation for constructing the *root locus* of $\Delta(s)$ as a function of the normalized rest displacement $\tilde{X}_R := x_R/R\alpha$:

$$\underbrace{\left(s^2 + \Omega_0^{\,2}\right)\left(s + 1/RC_0\right)}_{\mathcal{N}(s)} - \tilde{X}_R \cdot \underbrace{\left(s^2 + 3\Omega_0^{\,2}\right)}_{\mathcal{Z}(s)} = 0. \tag{6.39}$$

Depending on the relationship between the mechanical and electrical time constants, two different dynamic behaviors result.

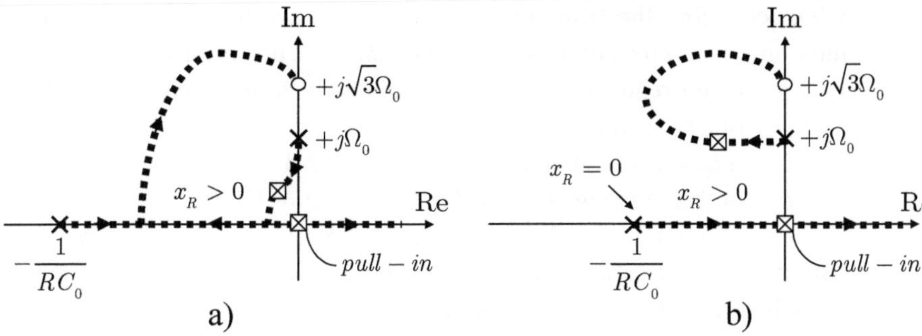

Fig. 6.14. Root loci for the eigenvalues of a voltage-drive plate transducer (variable electrode separation) as a function of the rest displacement x_R: a) configuration with $1/RC_0 \geq 3\Omega_0$, b) configuration with $1/RC_0 < 3\Omega_0$ (only upper half-plane shown)

With a *low* resistance $1/RC_0 \geq 3\Omega_0$, electrostatic softening (decrease in the eigenfrequency) and a minor increase in the damping of the complex pole pairs can be clearly seen (Fig. 6.14a).

For a *higher* resistance, i.e. $1/RC_0 < 3\Omega_0$ (Fig. 6.14b), the mechanical eigenfrequency Ω_0 remains almost constant for increasing rest displacement x_R. The impedance feedback essentially stiffens the system.

In both cases, the pull-in effect is vividly discernible, as at $x_R = x_{pi}$, a real pole moves into the right half-plane.

Example 6.1 *Electrostatic plate transducer with variable electrode separation.*

Transducer task Consider a voltage-drive plate transducer with variable electrode separation. The transducer is to be used to measure external forces (e.g. a pressure sensor).

Physical model The given *physical parameters* mimic those of a MEMS transducer in 6 μm polysilicon layer technology with integrated 0.8 μm CMOS electronics (Seeger and Boser 2003):

$$\beta = 1.5\ \mu m,\ C_0 = 0.4\ pF,\ k = 8\ N/m,\ m = 5.63\ \mu g\,.$$

Let the transducer have negligible mechanical damping.

Characterizing the transducer To evaluate operational possibilities for the transducer, *elementary specification data* are determined:

- eigenfrequency of the *short-circuited transducer*: $f_0 = 6\ kHz$ or $\Omega_0 = 37700\ rad/s$,
- theoretically available travel: $x_{pi} = \beta/3 = 0.5\ \mu m$,
- pull-in voltage at $F_0 = 0$: $U_{pi,0} = 3.65\ V$,
- maximum transducer force at x_{pi}: $F_{W,pi} = k \cdot x_{pi} = 4\ \mu N$.

An additional parameter of interest is the *maximum* permissible *external force* before pull-in. For this, solving Eq. (6.28) for F_0 gives

$$F_{0,\max} = k\beta - \sqrt[3]{\frac{27}{8}\beta \cdot C_0 \cdot k^2 \cdot U_0^{\,2}}\,. \tag{6.40}$$

Eq. (6.40) holds for the steady-state case, but also provides a reference point for the maximum force amplitude $F_0 + \Delta F_{ext}$ for dynamic excitation. For increasing rest displacement (proportional to U_0), the permissible external force thus decreases. At $U_0 = U_{pi}$, following Eq. (6.26), $F_{0,\max} = 0$.

A. No static force excitation $F_0 = 0 \;\Rightarrow\; F_{ext} = \Delta F_{ext}(t)$

As the mechanical subsystem is undamped, the overall system damping at different rest displacements is of interest for dynamic operations. The mechanical subsystem is damped using *resistive impedance feedback*. The maximum possible damping for any operating point is determined by Eq. (6.36).

Fig. 6.15 shows the *frequency responses* $G_{x/F}(j\omega)$ and the *step responses* for *small-signal dynamics* at the *rest displacements* specified in Table 6.2.

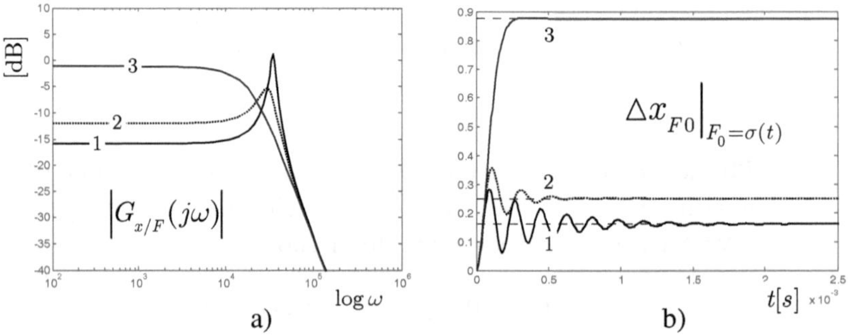

Fig. 6.15. Voltage-drive electrostatic plate transducer with $F_0 = 0$ and various rest displacements: a) frequency responses $G_{x/F}(j\omega)$, b) step responses (to units steps, note the differing gains); for legend, see Table 6.2

Table 6.2. Operating parameters of the electrostatic transducer with vanishing static force excitation $F_0 = 0$

Curve in Fig. 6.15	$X_R = x_R / \beta$	$d^{\max}$	$R^{\max}$ $[M\Omega]$	U_0 $[V]$
1	0.1	0.07	56	2.7
2	0.2	0.21	45	3.4
3	0.3	0.82	29	3.64

Cleary visible are the increasing damping and the increasing transducer sensitivity (gain $V_{x/F}$ in Fig. 6.15a or equivalently the static gain of the step responses in Fig. 6.15b for increasing rest displacement. Thus, for a sensible response, a large rest displacement should be preferred. However, for the option with the most attractive dynamics—$X_R = 0.3$— Eq. (6.40) shows that there is only an (impractically small) permissible force range of 0.02 μN.

A practical compromise between sufficient mechanical damping and sufficient force could be the option $X_R = 0.2$ with $F_{0,\max} = 0.5\ \mu N$.

B. Static force excitation $F_{ext} = F_0 + \Delta F_{ext}(t)$

Based on the above considerations, the armature rest position $X_R = 0.2$ ($x_R = 0.3\ \mu m$) is chosen as a compromise for non-zero static force excitation $F_0 \neq 0$.

To illustrate the *large-signal dynamics* of the transducer, Fig. 6.16 shows the results of a nonlinear simulation of typical phases of operation. For the entire simulation, the serial resistance of Table 6.2, row 2 is applied; the operational voltage $u_s(t)$ is set to the corresponding steady-state values in Table 6.2 in sequential steps.

As the simulation starts, the transducer armature is driven from the zero-voltage state to different rest positions—the *initialization of operating points.* To enable comparison with the small-signal dynamics shown in Fig. 6.15, the same rest displacements were chosen. The expected damping behavior with increasing rest displacement is confirmed.

After $t = 5\ ms$, at the rest displacement $X_R = 0.2$, an external force F_0 is applied to the armature, enabling a *force measurement.* After transients have decayed, the transducer response is a steady-state armature displacement Δx_{F0}, which can be evaluated to measure the applied force (left as an exercise to the reader so inclined; solution: $F_0 = 0.2\ \mu N$).

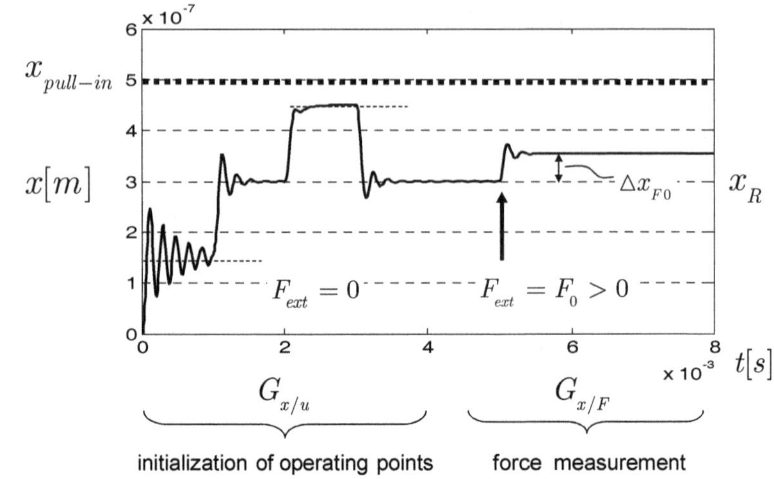

Fig. 6.16. Voltage-drive electrostatic plate transducer with $F_0 \neq 0$: large-signal dynamics, nonlinear simulation; operating parameters from Table 6.2, row 2

6.5 Transducer with Variable Electrode Separation and Current Drive

Configuration equations Driving a plate transducer with a current source results in different, *simpler* conditions as compared to voltage drive. With variable electrode separation and a homogeneous field, the *transducer capacitance* also follows the calculations in Eq. (6.23). However, electrostatic *force generation* under direct charge control is governed by the *separation-independent* relation

$$F_{el}(q_T) = \frac{1}{2}\frac{q_T^{\ 2}}{\alpha}. \tag{6.41}$$

As the active electrostatic COULOMB force (6.41) only depends on the stored charge q_T—which is in turn directly controlled by the current source—employing charge transport (current) allows the electrostatic force to be directly applied to the mass of the electrode *without feedback*.

Steady-state behavior

Rest positions For a steady-state charge $q_S(t) = Q_0$, Eq. (5.41) directly give the rest position condition

$$x_R = \frac{1}{2\alpha k} Q_0^{\ 2} + \frac{1}{k} F_0 \tag{6.42}$$

for the static mechanical excitation $F_{ext}(t) = F_0$.

As long as the rest current $I_0 = 0$ is maintained, a suitable charge Q_0 can thus be used to *stably* set *any* desired electrode separation x_R following Eq. (6.42).

Pull-in under current drive Under these ideal assumptions, a plate transducer with current drive thus exhibits *no pull-in*. This allows the entire electrode separation gap to be exploited for the electrode travel.

However, this attribute should not be falsely extended to imply that, under current/charge control, pull-in never occurs. The statement truly holds *only for parallel electrode configurations* with a *homogeneous electric field.*

For example, (Elata 2006) shows that in a *tip-plate transducer* as shown in Table 6.1 B, even with charge control, pull-in behaviors are quite possible. This is due to the nonhomogeneous electric field, so that a more accurate analysis of a tip-plate transducer should also make use of the more detailed models in Table 6.1.

With non-negligible *parasitic* capacitances, pull-in behaviors also manifest themselves in current-drive transducers with parallel electrodes (Seeger and Boser 1999), (Seeger and Boser 2003).

Pull-in limits with voltage vs. current drive The possibility of pull-in thus cannot be completely excluded, even in the case of current drive. However, it can be shown that in general the stable control range for *current drive* is generally *larger* than for voltage drive (Elata 2006), (Bochobza-Degani et al. 2003).

From the stability requirement that the differential stiffness of the complete transducer configuration always be positive (cf. Sec. 6.4.1) follows the requirement

$$\frac{\partial F_{el}}{\partial x} < \frac{\partial F_F}{\partial x},$$

i.e. the electrostatic stiffness should be a small as possible over as large a controllable range x as possible. Comparing electrostatic stiffnesses for voltage and current drive in the force relations in Eq. (6.10) and Eq. (6.11) directly results in the generally valid relation

$$\frac{\partial F_{el}(x,q_T)}{\partial x} = \frac{\partial F_{el}(x,u_T)}{\partial x} - \frac{{q_T}^2}{C_T(x)^3}\left(\frac{\partial C_T(x)}{\partial x}\right)^2. \tag{6.43}$$

Since, for implementable configurations, it can be assumed that $C_T(x) > 0$, the electrostatic stiffness of a current-drive transducer is thus always less than that of a comparable voltage-drive transducer. This in turn implies the larger stable controllable range of the current-drive transducer described above.

Dynamic behavior

Transducer parameters Substituting $C_T(x)$ from Eq. (6.23) into Eq. (6.20), and defining the *relative rest displacement* $X_R := x_R \,/\, \beta$ gives the operating-point-dependent two-port parameters

- *electrostatic current stiffness* $k_{el,I} := 0 \quad \left[\frac{\mathrm{N}}{\mathrm{m}}\right]$,
- *current coefficient* $K_{el,I} := \sqrt{\frac{2}{C_0}(k \cdot X_R - F_0/\beta)} \quad \left[\frac{\mathrm{V}}{\mathrm{m}}\right], \left[\frac{\mathrm{N}}{\mathrm{As}}\right]$, (6.44)
- *rest capacitance* $C_R := C_0 \frac{1}{1 - X_R} \quad \left[\frac{\mathrm{As}}{\mathrm{V}}\right]$.

Transfer matrix From Eq. (6.44) and Table 5.8 the *transfer matrix* $\mathbf{G}(s)$ can be calculated as

$$\begin{pmatrix} \Delta X(s) \\ \Delta U_T(s) \end{pmatrix} = \mathbf{G}(s) \begin{pmatrix} \Delta F_{ext}(s) \\ \Delta I_S(s) \end{pmatrix} = \begin{pmatrix} V_{x/F,I} \dfrac{1}{\{\Omega_0\}} & V_{x/i} \dfrac{1}{s \cdot \{\Omega_0\}} \\ V_{u/F} \dfrac{1}{\{\Omega_0\}} & V_{u/i} \cdot \dfrac{\{\Omega_U\}}{s \cdot \{\Omega_0\}} \end{pmatrix} \begin{pmatrix} \Delta F_{ext}(s) \\ \Delta I_S(s) \end{pmatrix}, \tag{6.45}$$

with the parameters

$$k_{T,U} = k - k_{el,U}, \quad \Omega_U^{\,2} = \frac{k_{T,U}}{m}, \quad \Omega_0^{\,2} = \frac{k}{m} \quad \text{where } \Omega_U < \Omega_0$$
$$V_{x/F,I} = \frac{1}{k}, \quad V_{x/i} = \frac{K_{el,I}}{k}, \quad V_{u/F} = -\frac{K_{el,I}}{k}, \quad V_{u/i} = \frac{1}{C_R}\frac{k_{T,U}}{k}. \tag{6.46}$$

Dynamic behavior The dynamics are shaped by the drive-*independent* mechanical eigenfrequency Ω_0, and the integrating nature of the current source control channel (cf. Fig. 6.7). There is *no electromechanical feedback*, so that the system dynamics are significantly clearer than for voltage drive. Controller design need only take into account the drive-independent sensitivity (gain) $V_{x/i}$ as a variable parameter (with positive gradient for increasing x_R).

6.6 Differential Transducers

6.6.1 Generic transducer configuration

Principle of operation An extension of the plate transducer with variable electrode separation is the *differential configuration* shown in Fig. 6.17. The movable electrode is located between two static electrodes, which can, in general, be controlled separately with voltages u_{TI} and u_{TII}. Having the two capacitances

$$C_I(x) = \frac{\alpha}{\beta_I - x}, \qquad C_{II}(x) = \frac{\alpha}{\beta_{II} + x}, \qquad \alpha = \varepsilon_0 A \tag{6.47}$$

on its two sides, the moving electrode (armature) is acted on by two *oppositely directed* electrostatic forces (force direction shown in Fig. 6.17)

$$F_{el,I}(x, u_{TI}) = \frac{\alpha}{2}\frac{u_{TI}^{\;2}}{\left(\beta_I - x\right)^2}, \qquad F_{el,II}(x, u_{TII}) = \frac{\alpha}{2}\frac{u_{TII}^{\;2}}{\left(\beta_{II} + x\right)^2}. \tag{6.48}$$

Depending on the control laws for u_{TI}, u_{TII} different behavioral properties can be achieved for the differential transducer. These are more closely considered below.

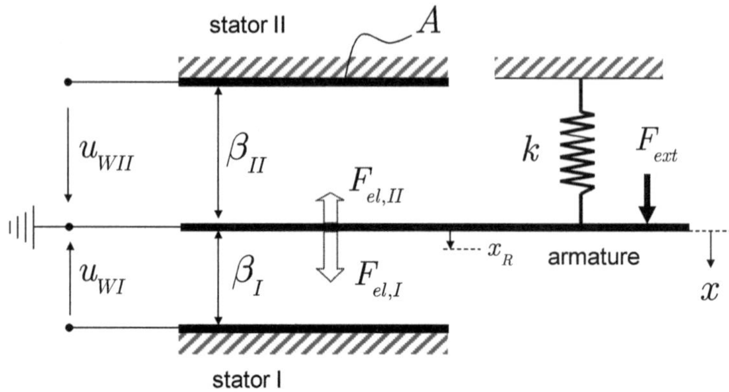

Fig. 6.17. General configuration of a differential electrostatic transducer

Table 6.3. Configurations for differential electrostatic transducers with variable electrode separation (N/A = not applicable)

Geometric/mechanical configuration		Electrical configuration	
		Symmetric	Asymmetric
$\beta_{II} = \lambda \cdot \beta_I$		*push-push* $u_{WI} = u_{WII} = U_0 + \Delta u$	*push-pull* $u_{WI} = U_{0I} + \Delta u$ $u_{WII} = U_{0II} - \Delta u$
symmetric $\lambda = 1$	free $k = 0$	-- N/A -- electrostat. bifurcation Sec. 6.6.2	*electrostatic bearing* Sec. 6.6.4
	suspended $k \neq 0$	-- N/A -- electrostat. bifurcation Sec. 6.6.2	*monotransducer* poor cost-benefit ratio
asymmetric $\lambda > 1$	free $k = 0$	-- N/A -- unidirectional force generation, no countering force	*electrostatic bearing* Sec. 6.6.4
	suspended $k \neq 0$	*N-fold transverse comb transducer* Sec. 6.6.3	*axisymmetric N-fold transverse double comb transducer* Sec. 6.6.5

Generic configurations Differential transducers are found in the literature with a variety of layouts. These can be classified according to their *geometric* and *electrical symmetries*, and by the method of *suspension* of the armature (see Table 6.3). Interesting system properties for differential configurations include the possibility of N-fold cascading in a limited volume—giving an N-fold multiplication of the electrostatic force—as well as the generation of bidirectional forces (only in asymmetric configurations). Important fundamental principles and types are discussed in more detail in subsequent sections.

6.6.2 Push-push control: mechanically symmetric configuration

Symmetric configurations As a special case, consider a *geometrically symmetric* configuration with $\beta_I = \beta_{II} = \beta$ and simultaneously *electrically symmetric* control $u_{TI} = u_{TII} = U_0 + \Delta u$. This is known as *push-push control.* For ease of presentation, no external force is assumed ($F_{ext} = 0$), i.e. for a relaxed spring and uncharged electrodes, the mobile armature electrode is located precisely half-way between the stator electrodes.

Rest positions In the steady-state case, the balance of the spring force and the summed electrostatic forces gives the two rest position conditions (cf. Eq. (6.23), second row)

$$x_R = 0, \tag{6.49}$$

$$U_0^2 = k\frac{\beta^3}{2\alpha}\left(1 - \left(\frac{x_R}{\beta}\right)^2\right)^2. \tag{6.50}$$

It can be shown that of the three rest positions defined by Eqs. (6.49) and (6.50), only $x_R = 0$ is *stable.*

Transducer stiffness To determine the stable range for control voltages at $x_R = 0$, the condition $K_\Sigma(x_R = 0, U_0) < 0$ is evaluated for the transducer stiffness $K_\Sigma(x,u) = \partial F_\Sigma / \partial x$, giving the stability condition

$$U_0^2 < k\frac{\beta^3}{2\alpha} = U_{pi}^2 \quad \text{for } x_R = 0. \tag{6.51}$$

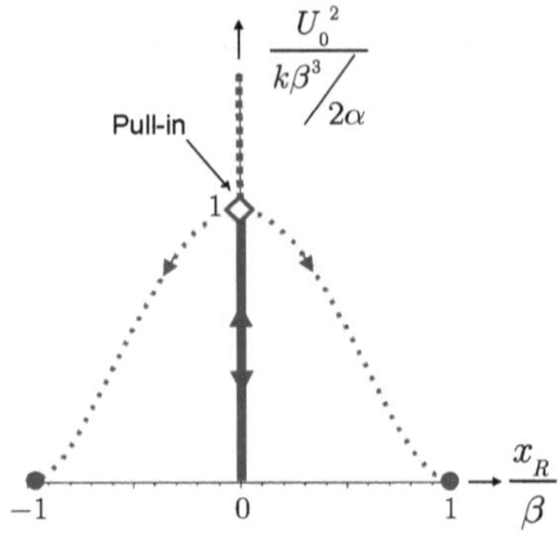

Fig. 6.18. Electrostatic bifurcation in a symmetric differential transducer: locus of the rest positions x_R, solid line = stable, dotted line = unstable, from (Elata 2006)

Electrostatic bifurcation The relationships in Eqs. (6.49) through (6.51) are illustrated in Fig. 6.18. For a rest voltage $|U_0| < U_{pi}$, the transducer has a single stable rest position $x_R = 0$ (solid line).

When U_{pi} is exceeded, any minor disturbance causes the armature electrode to be pulled toward one of the two stator electrodes (dotted lines). Thus, the pull-in in this case is *non-deterministic*; this is also referred to a *bifurcation* of the rest positions (Elata 2006). For these reasons, a differential configuration as in Fig. 6.17 which is electrically and mechanically *symmetric* is *not* useable in practice.

6.6.3 Push-push control: transverse comb transducer

Armature offset Targeted motion of a differential transducer can be achieved using the push-push control input $u_{TI} = u_{TII} = U_0 + \Delta u$ with an armature offset (see Fig. 6.17), i.e.

$$\beta_I = \beta, \quad \beta_{II} = \lambda \cdot \beta, \quad \lambda \geq 1. \tag{6.52}$$

This results in two unequal electrostatic forces acting on the armature, where the force $F_{el,I}$ acting at the smaller electrode gap is primarily used for motion generation, and the force $F_{el,II}$ acting on the other side weakens the resulting transducer force. In general, to minimize the reduction in total force, λ is thus chosen as large as possible. At first glance, considered as a single component, this differential transducer configuration with systematic force weakening is nothing other than an "inferior" plate transducer. This initial impression is, however, deceiving, as will be shown below.

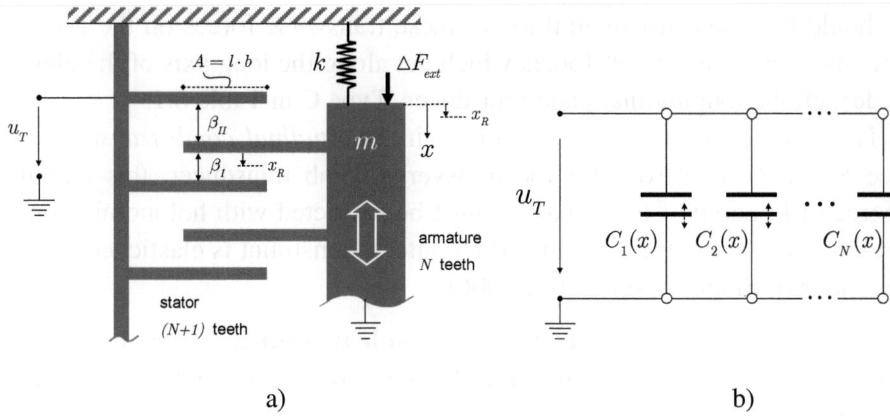

Fig. 6.19. Transverse comb transducer with N armature teeth and $(N+1)$ stator teeth (spring relaxed at $x = 0$): a) schematic, b) electrical equivalent circuit

Force multiplication: comb structure From a systems viewpoint, the asymmetric push-push configuration offers the interesting possibility of *force multiplication* in a *limited volume*. Electrostatic forces can be summed in the structure by physically connecting several movable electrodes in a row and rigidly coupling them to the load mass, creating an *armature comb structure* with N teeth (Fig. 6.19). On the opposing side, N corresponding pairs of stator electrodes are required. Creating an assembly of explicit electrode pairs would require a large structural volume, and the neighboring outer electrodes would each have to be insulated from one another. To optimally exploit the space available, each electrode is thus used multiple times, resulting in a similar *stator comb structure*, having at least $(N+1)$ electrodes (Fig. 6.19).

Push-push control The minimum-volume mechanical arrangement of comb structures as shown in Fig. 6.19, however, requires on the electrical side that all stator electrode pairs have *push-push* control, i.e. $u_{TI}(t) = u_{TII}(t) = U_0 + \Delta u(t)$.

Transverse motion profile Due to the electrostatic force directions visible in Fig. 6.17 and Fig. 6.19, the resulting direction of motion is perpendicular to the electrode orientation—resulting in a *transverse comb transducer* (Imamura et al. 1996).

It should be noted that in addition to these transverse forces on the armature, there are also lateral[9] forces which act along the long axis of the electrodes, cf. the longitudinal plate transducer, Type C in Table 6.1.

These forces are explicitly exploited in *longitudinal comb transducers* (see Sec. 6.7). However, for the transverse comb transducer, this lateral degree of freedom of the armature must be restricted with holonomic constraints imposed by the structure. If the lateral constraint is elastic, coupled equations of motion result (Elata 2006).

Configuration equations: transverse comb transducer For the transverse comb transducer shown in Fig. 6.19, the following configuration equations result ($\alpha = \varepsilon_0 A$):

- *transducer capacitance* $C_T(x) = N \cdot \alpha \cdot \left(\frac{1}{\beta - x} + \frac{1}{\lambda \cdot \beta + x} \right)$ (6.53)

- *transducer force* $F_{el}(x, u_T) = N \frac{\alpha}{2} u_T^{\,2} \cdot \left(\frac{1}{(\beta - x)^2} - \frac{1}{(\lambda \cdot \beta + x)^2} \right)$ (6.54)

as well as the *rest position condition*

$$U_0^{\,2} = k \cdot x_R \cdot \left(N \frac{\alpha}{2} \left(\frac{1}{(\beta - x_R)^2} - \frac{1}{(\lambda \cdot \beta + x_R)^2} \right) \right)^{-1}. \tag{6.55}$$

From the configuration equations (6.53), (6.54), the positive and negative attributes of this structure can be inferred. The N pairs of teeth result in an N-fold increase in the electrostatic force, compared to a simple differential transducer. However, this comes at the price of the resulting N parallel capacitances $C_i(x)$ (Fig. 6.18b), which induce correspondingly large polarization currents.

Configuration optimization Eq. (6.54) shows that in order to maximize the transducer force, the number of teeth N and the electrode separation

---

9 Note that "transverse" indicates the direction perpendicular to the long axis of the plate electrodes, i.e. it specifies the primarily exploited mechanical degree of freedom of an electrostatic plate transducer. As a consequence, "lateral" is understood in relation to this transverse axis (i.e. parallel to the electrodes' long axis).

ratio λ should both be selected to be as large as possible. In addition, the largest possible armature travel x is generally desired. If all three of these design parameters were to be maximized, unacceptably large structure sizes would result (a geometric explosion).

Thus, there is also a binding limit on the *structure size*, i.e. the *length* L of the comb structure, as well as a desired armature travel x_{max}. Free design parameters are the *tooth count* N and the *electrode separation ratio* λ. A happy medium must be found between minimizing force reduction from the back-side force component (large λ) and maximizing force multiplication (large N).

An optimal value for the electrode separation ratio λ is obtained from the following simple consideration. The size limit—the *comb length L*—gives the geometric boundary condition[10]

$$N \cdot \beta \cdot \left(1 + \lambda\right) = L\,, \tag{6.56}$$

where β is determined by the required armature travel ($\beta \approx 3x_{max}$ to avoid pull-in). Eliminating N in the force equation (6.54) using Eq. (6.56) and assuming constant comb length gives the transducer force at $x = 0$ [11]

$$F_{el}(0, u_T) = \frac{\alpha}{2} u_T^{\ 2} \frac{L}{\beta^3} \cdot \left[\frac{1}{1+\lambda}\left(1 - \frac{1}{\lambda^2}\right)\right] = \frac{\alpha}{2} u_T^{\ 2} \frac{L}{\beta^3} \cdot \Lambda_F(\lambda)\,. \tag{6.57}$$

The value of the dimensionless geometry factor $\Lambda_F(\lambda)$ as a function of the free design parameter λ is shown in Fig. 6.20a. It can be seen that the *maximum possible transducer force* is achieved at $\lambda_{opt} = 2.4$. For an optimal realization, λ_{opt} is varied slightly so as to give a *whole-numbered tooth count* $N_{opt} \in \mathbb{N}$ in Eq. (6.56) (gray-shaded area in Fig. 6.20a).

Pull-in behavior The relation $\partial F_{el} / \partial x = k$ and the rest position condition (6.55) give a determining equation for the pull-in limit $x_{pull-in}$ as a function of λ. The graphical evaluation of this somewhat unwieldy equation is shown in Fig. 6.20b. Compared to a simple plate transducer (large λ) a slight *reduction* of the pull-in limit results in the optimal λ range.

---

[10] The thickness of the teeth is negligible relative to the gap. The results for optimal values of λ are not affected by it to any significant extent.

[11] It is easy to verify that for $0 \le x/\beta < 0.33$, there are no appreciable qualitative differences from this case.

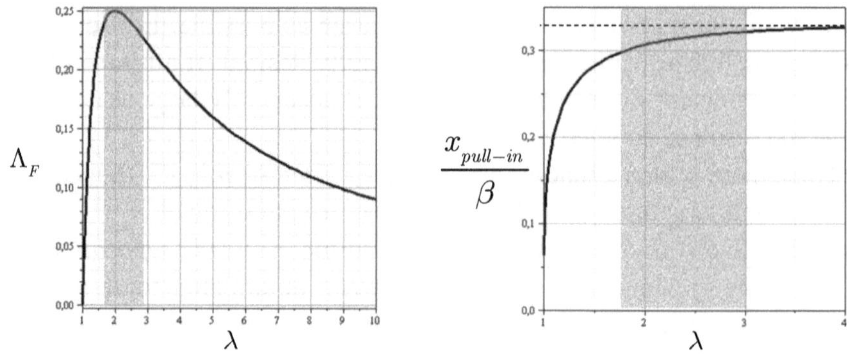

Fig. 6.20. Transverse comb transducer with push-push control: a) *geometry factor* for the *transducer force* at $x = 0$ (normalized), b) pull-in limit on travel as a function of the electrode separation ratio λ (grey shading shows favorable/optimal values of λ)

Transducer parameters The dynamics are again described by the operating-point-dependent *two-port parameters* ($X_R := x_R/\beta$):

- *electrostatic voltage stiffness*

$$k_{el,U} := k\frac{2X_R}{1-X_R}\frac{(\lambda-1)(\lambda+3X_R)+1+3X_R^{\,2}}{(\lambda+X_R)(\lambda-1+2X_R)}\quad\left[\frac{\mathrm{N}}{\mathrm{m}}\right],$$

- *voltage coefficient*

$$K_{el,U} := \frac{\sqrt{2C_0\cdot k\cdot X_R}}{1-X_R}\frac{\sqrt{N\cdot(\lambda+1)(\lambda-1+2X_R)}}{(\lambda+X_R)}\quad\left[\frac{\mathrm{N}}{\mathrm{V}}\right],\left[\frac{\mathrm{As}}{\mathrm{m}}\right],\qquad(6.58)$$

- *rest capacitance* $\quad C_R := N\cdot C_0\left(\frac{1}{1-X_R}+\frac{1}{\lambda+X_R}\right)\quad\left[\frac{\mathrm{As}}{\mathrm{V}}\right],$

- *ELM coupling factor* $\quad \kappa_{el}^{\,2} = \frac{2X_R}{1-X_R}\frac{(\lambda-1+2X_R)^2}{(\lambda-1)(\lambda+X_R)}\,.$

Compared to the simple plate transducer, the two-port parameters are corrected here with a factor which depends on λ.

For the *transfer matrix*, the previously presented relations in Eq. (6.30) hold.

Free armature For the considerations up to this point, it was of critical importance that a countering force—here the spring force $F_F = k \cdot x$—balance the unidirectional total electrostatic force F_{el}. For push-push control, only the magnitude of F_{el} can be changed, the force direction is always directed toward the smaller air gap. For this reason, this arrangement is generally *not suitable* to a free armature.

6.6.4 Push-pull control: electrostatic bearing

Free armature In some applications, there is no structural way to elastically suspend the armature. In this case, the armature essentially floats freely in space, i.e. $k = 0$ in Fig. 6.17. To control the motion of the armature in this case (e.g. for positioning, or to compensate disturbances), it must be possible to change the sign of the resultant electrostatic force. This is fundamentally possible with the differential electrode configuration shown in Fig. 6.17.

Electrostatic push-pull principle The transducer configuration shown in Fig. 6.17 generates a resulting transducer force ($\alpha = \varepsilon_0 A$)

$$F_{el}(x, u_{TI}, u_{TII}) = \frac{\alpha}{2}\left[\frac{u_{TI}^{2}}{\left(\beta_I - x\right)^2} - \frac{u_{TII}^{2}}{\left(\beta_{II} + x\right)^2}\right]. \tag{6.59}$$

With a suitable *asymmetric* choice of transducer voltages u_{TI}, u_{TII} or of electrode separations β_I, β_{II}, the magnitudes of the two terms in Eq. (6.59) can be modified, allowing both *positive* and *negative* transducer forces F_{el} to be generated.

Without loss of generality, the following approach can be chosen for control of the electrical transducer:

$$\begin{aligned} u_{TI} &= U_{0I} + \Delta u\,, \\ u_{TII} &= U_{0II} - \Delta u\,. \end{aligned} \tag{6.60}$$

Using the two static rest voltages U_{0I} and U_{0II}, the operating point (rest position x_R and static bearing force $F_{el,0} = F_{el}(x_{R,} U_{0I}, U_{0II})$) can be freely set, while the dynamic component $\Delta u(t)$ can be used to generate bidirectional dynamic bearing forces.

Steady-state behavior: rest positions For a constant external force $F_{ext}(t) = F_0$, Eqs. (6.59), (6.60) and $\beta_I = \beta$, $\beta_{II} = \lambda \cdot \beta$ give the rest position condition

$$\frac{\alpha}{2}\frac{U_{0I}^{\;2}}{\left(\beta - x_R\right)^2} - \frac{\alpha}{2}\frac{U_{0II}^{\;2}}{\left(\lambda \cdot \beta + x_R\right)^2} + F_0 = 0 \;. \tag{6.61}$$

Without loss of generality, let $x_R = 0$, as the rest position of the armature can be freely defined via the asymmetry ratio λ.

Depending on the *sign* of the static *disturbance force* F_0, Eq. (6.61) gives one of the following relations for rest voltages:

$$\begin{aligned}
F_0 > 0: &\quad U_{0II}^{\;2} = \lambda^2 \cdot \left[U_{0I}^{\;2} + F_0 \cdot \frac{2\beta^2}{\alpha}\right],\\
F_0 < 0: &\quad U_{0I}^{\;2} = \left[U_{0II}^{\;2}\frac{1}{\lambda^2} - F_0 \cdot \frac{2\beta^2}{\alpha}\right],\\
F_0 = 0: &\quad \frac{U_{0II}}{U_{0I}} = \lambda \;.
\end{aligned} \tag{6.62}$$

Transducer parameters The dynamics of sufficiently small motions about the rest position x_R are again described by the operating-point-dependent *two-port parameters*. For $x_R = 0$ and $C_0 := \alpha/\beta$, the following values result:

- *electrostatic voltage stiffness*

$$k_{el,U} := \frac{C_0}{\beta^2} \cdot \left[U_{0I}^{\;2} + \frac{1}{\lambda^3}U_{0II}^{\;2}\right],$$

- *voltage coefficient*

$$K_{el,U} := \frac{C_0}{\beta} \cdot \left[U_{0I} + \frac{1}{\lambda^2}U_{0II}\right], \tag{6.63}$$

- *rest capacitance*

$$C_R := C_0 \cdot \left[1 + \frac{1}{\lambda}\right].$$

The *transfer matrix* is again obtained from the previously presented relations in Eq. (6.30).

Unstable transfer characteristics It can be seen that the dominant attribute of the dynamics of an electrostatic bearing having a *free* armature, is *unstable* transfer characteristics, since—using the transducer parameters (6.63) in Eq. (6.32)—the *characteristic polynomial* of the transfer matrix is

$$\Delta_U(s) = s^2 - \frac{k_{el,U}}{m} = s^2 - \Omega_{U0}{}^2 = \left(s + \Omega_{U0}\right)\left(s - \Omega_{U0}\right), \tag{6.64}$$

where

$$\Omega_{U0} := \frac{1}{\beta}\sqrt{\frac{C_0}{m} \cdot \left[U_{0I}{}^2 + \frac{1}{\lambda^3} U_{0II}{}^2\right]}. \tag{6.65}$$

In all channels of the transducer transfer matrix, an *unstable pole* $s = +\Omega_{U0}$ thus appears in the right half-plane.

As a result, an electrostatic bearing can not be operated under open loop control; *stabilization* always requires (local) *closed-loop control*. Due to its exponentially unstable character and moreover nonlinear dynamics, the design of suitable control laws for this type of transducer presents a particular challenge for systems design (Han et al. 2005), (Han et al. 2006).

Example 6.2 *Electrostatically suspended gyro.*

Fig. 6.21 shows a schematic representation of the vertical suspension of an electrostatically supported gyro. In a vacuum chamber (housing), a spherical rotor with a conductive surface is set rotating and is held in frictionless suspension between the two electrodes using an electrostatic field. The angular momentum axis of the rotor remains inertially fixed when the housing rotates. By measuring the orientation of this axis relative to the housing, the rotation of the latter relative to inertial space can be measured (inertial sensor), (Damrongsak et al. 2008), (Bencze et al. 2007), (Han et al. 2005).

Let the following *technical data* be given (see Fig. 6.21, values taken from an implemented gyro (Han et al. 2005)):

rotor mass $m = 10$ g, gap width (nominal) $d = 70\ \mu$m,

electrode surface $A = 6.6\ \text{cm}^2$, voltage source $u_{S,\max} = 1500$ V.

Find the parameters for electrical control and the model data for the uncontrolled configuration.

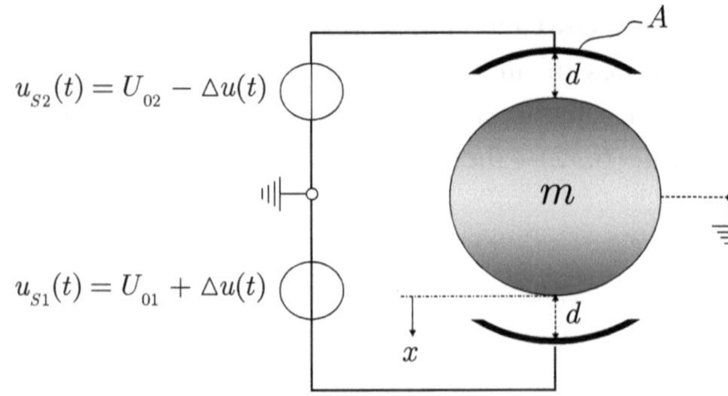

Fig. 6.21. Schematic of an electrostatically suspended gyro

Rest voltages The electrostatic bearing must always compensate the weight of the rotors $F_0 = m \cdot g = 0.0981\ \mathrm{N}$. Based on this, Eq. (6.62) gives the condition for the rest voltages ($\lambda = 1$)

$$U_{02}^{\ 2} = U_{01}^{\ 2} + mg\frac{2d^2}{\varepsilon_0 A} = U_{01}^{\ 2} + \left(405\right)^2\ \left[\mathrm{V}^2\right]. \tag{6.66}$$

The rest voltage U_{01} is a free design parameter. It can be seen from Eq. (6.63) that the rest voltage determines the bearing stiffness $k_{el,U}$. For good disturbance rejection (cf. Eqs. (6.30), (6.31)), a sufficiently *large stiffness* (i.e. large rest voltage) is required. Here, following (Han et al. 2005), $U_{01} = 750\ \mathrm{V}$ is chosen, so that Eq. (6.66) gives $U_{02} = 852\ \mathrm{V}$.

Dynamic bearing forces The maximum supply voltage $u_{S,\max} = 1500\ \mathrm{V}$ gives the maximum (one-sided) bearing force

$$F_{el,\max} = \frac{\varepsilon_0 A}{2d^2} u_{S,\max}^2 = 1.34\ \mathrm{N}.$$

Subtracting the steady-state bearing force of the upper bearing, the remaining *maximum upward acceleration* during dynamic operation is then

$$a_{\max} = \frac{1}{m}\left(F_{el,\max} - \frac{\varepsilon_0 A}{2d^2} U_{02}^{\ 2}\right) = 9.3\ \mathrm{g}.$$

Transducer parameters The *rest capacitance* of each electrode is calculated to be $C_0 = 83\ \mathrm{pF}$, so that a bearing capacitance of $C_R = 166\ \mathrm{pF}$ acts as a load on the voltage source.

The following values are found for the operating-point-dependent parameters: *bearing stiffness* $k_{el,U} = 2.2 \cdot 10^4\ \mathrm{N/m}$, *voltage coefficient* $K_U = 1.9 \cdot 10^{-3}\ \mathrm{N/V}$, *bearing bandwidth* $\Omega_{U,0} = 1500\ \mathrm{rad/s}$. ■

6.6.5 Push-pull control: axisymmetric double-comb transducer

Transducer configuration For push-pull control, the two differential transducer stator electrodes must be acted upon by different voltages. This renders direct implementation of this concept in a transverse comb arrangement for N-fold force multiplication unattractive due to physical constraints on the geometry. The structurally simple comb configuration presented in Sec. 6.6.3 implies push-push control. However, it is possible to combine the advantages of push-pull control with the advantages of a transverse comb transducer with a slight change to the structure.

For push-pull operation, two transverse *push-push* comb transducers A, B are used and their two armatures are rigidly coupled. The electrodes of the two sub-transducers A, B are arranged to be geometrically *axisymmetric* so as to result in equally directed forces to achieve push-pull operation. Electrically, the two sub-transducers A, B are asymmetrically controlled for the push-pull arrangement (Horsley et al. 1998), (Horsley et al. 1999). Fig. 6.22 depicts such a configuration in the form of an *axisymmetric double-comb transducer.*

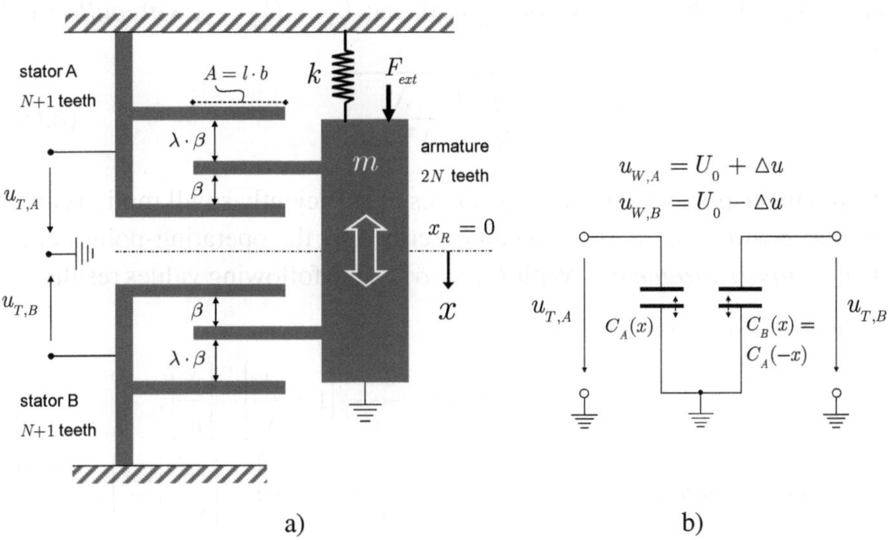

Fig. 6.22. Axisymmetric double-comb transducer with push-pull control: a) schematic diagram, b) electrically equivalent circuit

Configuration equations For the axisymmetric double-comb transducer depicted in Fig. 6.22, the following configuration equations result ($\alpha = \varepsilon_0 A$):

- *transducer capacitance*

$$C_{T,A}(x) = N\alpha\left(\frac{1}{\beta - x} + \frac{1}{\lambda\beta + x}\right),$$
$$C_{T,B}(x) = N\alpha\left(\frac{1}{\beta + x} + \frac{1}{\lambda\beta - x}\right), \quad (6.67)$$
$$C_T(x) = C_{T,A}(x) + C_{T,B}(x),$$

- *transducer force*

$$F_{el,A}(x, u_{T,A}) = N\frac{\alpha}{2}u_{T,A}^2\left(\frac{1}{(\beta - x)^2} - \frac{1}{(\lambda\beta + x)^2}\right),$$
$$F_{el,B}(x, u_{T,B}) = N\frac{\alpha}{2}u_{T,B}^2\left(\frac{1}{(\lambda\beta - x)^2} - \frac{1}{(\beta + x)^2}\right), \quad (6.68)$$
$$F_{el}(x, u_{T,A}, u_{T,B}) = F_{el,A}(x, u_{T,A}) + F_{el,B}(x, u_{T,B}).$$

Due to the axisymmetric geometric configuration, the stable *rest position* is found to be $x_R = 0$, for $U_0 \neq 0$ and $U_0 < U_{pull-in}$, with *pull-in voltage*

$$U_{pull-in} = \sqrt{\frac{k\beta^3}{2\alpha}\frac{\lambda^3}{\lambda^3 + 1}}. \quad (6.69)$$

Transducer parameters The dynamics of sufficiently small motions about the rest position $x_R = 0$ are again described by the operating-point-dependent *two-port parameters*. With $C_0 := \alpha/\beta$, the following values result:

- *electrostatic voltage stiffness*

$$k_{el,U} := NU_0^2\frac{2C_0}{\beta^2}\cdot\left[1 + \frac{1}{\lambda^3}\right] \quad \left[\frac{\text{N}}{\text{m}}\right],$$

- *voltage coefficient*

$$K_{el,U} := NU_0\frac{2C_0}{\beta}\cdot\left[1 - \frac{1}{\lambda^2}\right] \quad \left[\frac{\text{N}}{\text{V}}\right],\left[\frac{\text{As}}{\text{m}}\right], \quad (6.70)$$

- *rest capacitance*

$$C_R := N\cdot 2C_0\cdot\left[1 + \frac{1}{\lambda}\right] \quad \left[\frac{\text{As}}{\text{V}}\right].$$

The *transfer matrix* is obtained from the previously-mentioned relations in Eq. (6.30).

Setting an operating point In this transducer, setting the operating point using a rest voltage $U_0 \neq 0$ has no effect on the rest displacement, which always remains $x_R = 0$. Here, the rest voltage U_0 only determines—via the electrostatic stiffness $k_{el,U}$ and voltage coefficient K_U—the transducer eigenfrequency and the gain factors, and the corresponding aspects of the system dynamics.

Geometric layout The geometry factor λ follows the same relations as in the case of the transverse comb transducer with push-push control discussed in Sec. 6.6.3. An optimal value given a limited structure length again occurs at $\lambda \approx 2$ (cf. Fig. 6.20). It can also be seen from Eq. (6.70) that for a nearly symmetric electrode geometry $\lambda \approx 1$, the electrical controllability (gain via K_U) approaches zero, rendering the transducer unusable. For an example layout for the read head of a CD drive, the reader is referred to (Horsley et al. 1998).

6.7 Transducers with Constant Electrode Separation

6.7.1 General dynamic model

Configuration equations Transducers with constant plate separation use forces perpendicular to the electric field, as shown in the geometric configuration Types C through E in Table 6.1. In these cases, either the electrode area or the overlap with the dielectric is varied, so that the following generalized statement can be applied for the transducer capacitance and the electrostatic force (cf. Table 6.1):

$$C_T(x) := C_0 + \gamma \cdot x \quad \Rightarrow \quad \frac{\partial C_T(x)}{\partial x} = \gamma \,. \tag{6.71}$$

Transducer forces Inserting the capacitance function (6.71) into the ELM transducer equations (6.10), (6.11) the transducer force is then, in

- *PSI-coordinates* ($u_T = \dot{\psi}$):

$$F_{el,\Psi}(u_T) = \frac{1}{2}\gamma \cdot u_T^{\,2}, \tag{6.72}$$

- *Q-coordinates*:

$$F_{el,S}(x,q_T) = \frac{1}{2}\frac{\gamma}{\left(C_0 + \gamma x\right)^2} \cdot q_T^{\,2}. \tag{6.73}$$

Note that for *voltage drive*, the transducer force in Eq. (6.72) is *independent* of displacement, while for current drive, a nonlinear dependence on the electrode setting results. In a manner of speaking, this then represents the dual of the plate transducer with variable separation.

Transducer with voltage drive

Rest positions With the static excitations $u_S(t) = U_0$, $F_{ext}(t) = F_0$, Eq. (5.40) and Eq. (6.72) give the rest position condition

$$x_R = \frac{1}{2}\frac{\gamma}{k}U_0^{\,2} + \frac{1}{k}F_0. \tag{6.74}$$

Using a suitable rest voltage U_0, *any* desired electrode position x_R can be *stably* selected using Eq. (6.74).

Transducer parameters The dynamics are again described by the operating-point-dependent *two-port parameters* (in admittance form, Eqs. (6.17), (6.18)), with

- *electrostatic voltage stiffness* $k_{el,U} := 0 \quad \left[\frac{\mathrm{N}}{\mathrm{m}}\right]$,
- *voltage coefficient* $K_{el,U} := \gamma \cdot U_0 \quad \left[\frac{\mathrm{N}}{\mathrm{V}}\right], \left[\frac{\mathrm{As}}{\mathrm{m}}\right]$, (6.75)
- *rest capacitance* $C_R := C_0 + \gamma \cdot x_R \quad \left[\frac{\mathrm{As}}{\mathrm{V}}\right]$.

The *transfer matrix* is obtained from the previously presented relations in Eq. (6.30).

Dynamic behavior Due to the vanishing electrostatic stiffness, the transducer eigenfrequency depends solely on the elastic suspension and is thus operating-point-*independent.* Thus, along with the voltage coefficient, it is constant, resulting in constant transfer functions (no operating-point-dependent parameter variations, an important aspect for controller design). On the other hand, it can also not be tuned using the electrical control input, which precludes most oscillator applications.

Pull-in with voltage drive As can be seen from the rest position condition and the electrostatic stiffness, *no pull-in* occurs under the assumed conditions.

Transducer with current drive

Rest positions With the static excitations $q_S(t) = Q_0$ and $F_{ext}(t) = F_0$, Eq. (5.41) and Eq. (6.73) give the rest position condition

$$2\left(k \cdot x_R - F_0\right)\left(\frac{C_0}{\gamma} + x_R\right)^2 = Q_0^{\,2}. \tag{6.76}$$

Of the three possible solutions, only one leads to a stable rest position, which can be set using Q_0.

Transducer parameters For the dynamics, the operating-point-dependent *two-port parameters* in hybrid form are obtained from Eqs. (6.19), (6.20), giving

- *electrostatic current stiffness* $k_{el,I} := -\dfrac{\gamma^2}{\left(C_0 + \gamma x_R\right)^3} Q_0^{\,2} \quad \left[\dfrac{\mathrm{N}}{\mathrm{m}}\right]$,
- *current coefficient* $K_{el,I} := \dfrac{\gamma}{\left(C_0 + \gamma x_R\right)^2} Q_0 \quad \left[\dfrac{\mathrm{V}}{\mathrm{m}}\right], \left[\dfrac{\mathrm{N}}{\mathrm{As}}\right]$, (6.77)
- *rest capacitance* $C_R := C_0 + \gamma \cdot x_R \quad \left[\dfrac{\mathrm{As}}{\mathrm{V}}\right]$.

The *transfer matrix* is obtained analogously to the previously-mentioned relations Eq. (6.30) or Eq. (6.45).

Dynamic behavior Using current drive, all transducer parameters are now operating-point-dependent and thus vary with the selected rest position. The larger the overlap of the two electrodes, the smaller the magnitudes of the electrostatic stiffness and the current coefficient. For large armature travel, provision must be made for substantial parameter variations.

One advantageous attribute, however, is the *negative* sign on the *electrostatic stiffness*, which implies a fundamental increase in the resulting transducer stiffness $k_{T,I} = k - k_{el,I}$ (electrostatic stiffening).

Pull-in with current drive From the fundamental relation for a stable rest position $k_{el,I} < k$ (cf. Sec. 5.4.3) and Eq. (6.77), it directly follows that for all $x_R > 0$, there can fundamentally be no pull-in.

6.7.2 Longitudinal comb transducer

Transducer configuration For force multiplication, the *comb configuration* already known from Sec. 6.6.3 can be employed directly. In the case considered in this section, however, it is the degree of freedom of longitudinal motion along the comb electrodes which is exploited (the y-direction in Fig. 6.23).

Two-sided suspension To enable frictionless motion, the armature can be elastically suspended on two sides (x- and y-directions, see Fig. 6.23). Clearly, the spring constant in the *transverse (lateral)* direction of motion (here, the x-direction) should be chosen as large as possible, while in the *longitudinal* primary direction of motion (here, the y-direction), a smaller spring constant matching the transducer forces should be chosen. Two pos-

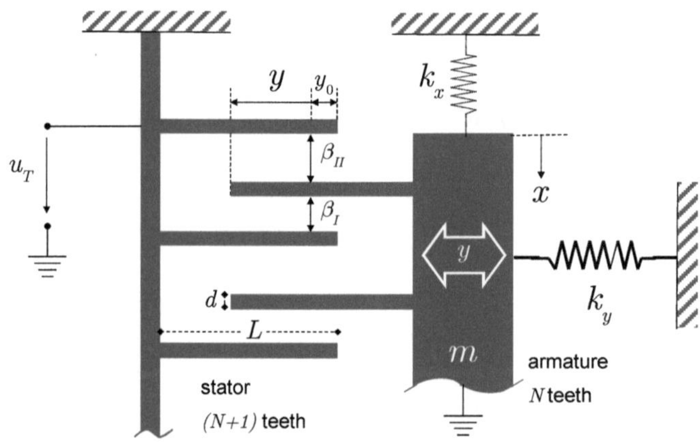

Fig. 6.23. Longitudinal comb transducer: primary motion in y -direction (longitudinal relative to comb electrodes), stiff transverse suspension in x -direction

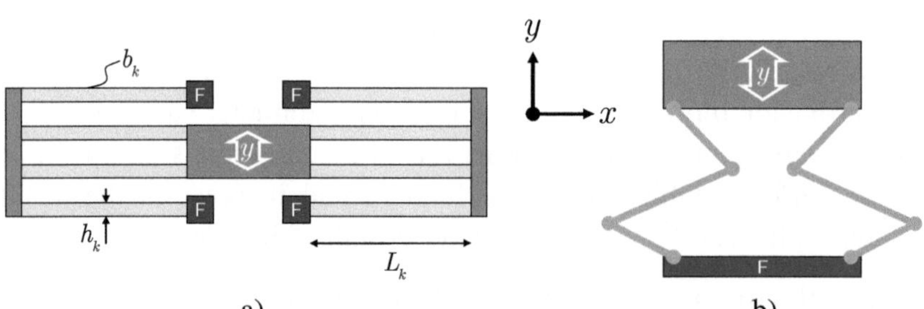

Fig. 6.24. Two-sided elastic suspension: a) folded beam springs (from (Legtenberg et al. 1996)), b) crab-leg springs; F is the attachment point of a spring

sible layouts for two-sided elastic armature suspension are shown in Fig. 6.24.

Configuration equations For a comb configuration with voltage drive as in Fig. 6.23, the relations already presented in previous sections give the transducer capacitance and transducer forces (with comb width b, comb overlap at zero voltage y_0, tooth count N, and negligible comb height)

- *transducer capacitance*

$$C_T(x,y) = N\varepsilon_0 b(y_0 + y)\cdot\left(\frac{1}{\beta_I - x} + \frac{1}{\beta_{II} + x}\right), \tag{6.78}$$

- *transducer force in x-direction*

$$F_{el,x}(x,y,u_T) = \frac{N}{2}\varepsilon_0 b(y_0 + y)\cdot\left(\frac{1}{\left(\beta_I - x\right)^2} - \frac{1}{\left(\beta_{II} + x\right)^2}\right)\cdot u_T^{\ 2}, \tag{6.79}$$

- *transducer force in y-direction*

$$F_{el,y}(x,u_T) = \frac{N}{2}\varepsilon_0 b\cdot\left(\frac{1}{\beta_I - x} + \frac{1}{\beta_{II} + x}\right)\cdot u_T^{\ 2}. \tag{6.80}$$

Lateral instability: side pull-in In the *longitudinal direction* (y-direction parallel to the electrode surface) *no pull-in* occurs under the idealizing assumptions (no field leakage, negligible comb height, cf. Sec. 6.7.1). However, in the lateral direction x, the configuration is equivalent to a push-push differential transducer, so that pull-in may well occur (cf. Sec. 6.6.2). This coupled behavior will be further examined below (for which see also (Legtenberg et al. 1996), (Chen and Lee 2004), (Huang and Lu 2004), (Borovic et al. 2006), (Elata 2006)).

For practical reasons, an electrode configuration symmetric in the x-direction (with $\beta_I = \beta_{II} = \beta$) is examined[12], so that Eq. (6.51) gives the *lateral (side) pull-in* voltage

$$U_{x,pi}^2 = k_x \frac{\beta^3}{2\varepsilon_0 b\left(y_0 + y_R\right)}, \tag{6.81}$$

[12] This does not restrict generality, as can be gathered from the explanations in Sec. 6.6.3.

where y_R indicates the longitudinal rest position for an applied voltage $U_{x,pi}$.

The rest position y_R is calculated from Eq. (6.74) for the given transducer configuration as

$$y_R = \frac{\varepsilon_0 b}{k_y \beta} \cdot U_{x,pi}^2. \tag{6.82}$$

Substituting y_R from Eq. (6.82) into Eq. (6.81) and rearranging gives the *lateral pull-in* condition

$$U_{x,pi}^2 = \frac{\beta^2 k_y}{2b\varepsilon_0}\left[\sqrt{2\frac{k_x}{k_y} + \frac{y_0^{\,2}}{\beta^2}} - \frac{y_0}{\beta}\right]. \tag{6.83}$$

With the assumption—always fulfilled in practice—that $k_x \gg k_y$, the *lateral pull-in condition* simplifies to

$$U_{x,pi}^2 \approx \frac{\beta^2 k_y}{2b\varepsilon_0}\left[\sqrt{2\frac{k_x}{k_y}} - \frac{y_0}{\beta}\right].$$

Further, from a comparison with Eq. (6.82), the *longitudinal displacement* for *lateral pull-in* is also found to be

$$\boxed{y_{x,pi} = \beta\sqrt{\frac{k_x}{2k_y}} - \frac{y_0}{2}}. \tag{6.84}$$

Design conclusions The longitudinal working range of the transducer is thus fundamentally determined by the ratio of stiffnesses in the longitudinal and lateral directions. A second important design parameter can be seen to be the electrode separation β. On the one hand, to achieve *large transducer forces*, this should be kept as *small* as possible; however, it follows from Eq. (6.84) that such a tight tooth separation in fact leads to *lateral pull-in.*

Stiffness variation in folded beam springs Closer inspection of the stiffness properties of beam spring structures shows that when stretched, the configurations depicted in Fig. 6.24 exhibit an unfavorable change in stiffness ratio k_x / k_y.

For the folded beam spring shown in Fig. 6.24a, it holds for very *small* deformations that (Legtenberg et al. 1996) (with modulus of elasticity E)

$$\begin{aligned} k_x &= 2Eb_k h_k / L_k \\ k_y &= 2Eb_k h_k^3 / L_k^3 \end{aligned} \quad \Rightarrow \frac{k_x}{k_y} = \left(\frac{L_k}{h_k}\right)^2 .$$

For larger deformations δ_y in the y-direction (precisely what is desired in this transducer type), contraction of the beam spring causes a displacement-dependent reduction of the lateral stiffness k_x to appear (Legtenberg et al. 1996), where

$$\delta k_x \sim 1/\delta_y . \tag{6.85}$$

Taking into account the displacement-dependent softening of Eq. (6.85) and the relation (6.84), the stable working range (avoiding *lateral pull-in*) is reduced even further.

Dynamic behavior As previously discussed, the dynamics are determined by the ELM two-port parameters, which can be straightforwardly derived taking into consideration the current stiffness and rest position relations from Sec. 6.6.

Transverse vs. longitudinal comb transducer The comb structures presented here can fundamentally be applied in cases of transverse and longitudinal motion. However, they exhibit fundamental dynamic differences in their armature travel limits and transducer forces. The *longitudinal transducer* allows for a significantly larger controlled displacement in the direction of the comb teeth, in the end limited only by the *lateral pull-in* (the bending of long teeth will cause deviations from the nominal electrode separation).

A comparison of longitudinal and transverse transducer forces in an equivalent geometry having $\beta_I = \beta$, $\beta_{II} = \lambda\beta$ (see Eqs. (6.79), (6.80)), shows clear advantages for the transverse transducer since

$$\frac{F_{el,trans}}{F_{el,long}} = \frac{L}{\beta}\frac{\lambda^2 - 1}{\lambda(\lambda - 1)} . \tag{6.86}$$

For the optimal neighborhood $\lambda \approx 2.4$ of a technically reasonable comb geometry with $L \gg \beta$, the transverse transducer force is at least an order of magnitude larger than longitudinal one.

Design recommendation For large force requirements (taking into account small achievable displacements), a transverse comb transducer is to be preferred, while for large displacement requirements (taking into account the moderate transducer forces achievable), a longitudinal comb transducer is recommend.

6.7.3 Comb transducer with linearly stepped teeth

Tuning problem for resonator applications Under voltage drive in the longitudinal direction, the comb structure with equal-length teeth presented in the previous sections exhibits the convenient property that no electrostatic softening occurs, i.e. *regardless* of the voltage level, $k_{el,U} = 0$. However, for use as a resonator, this does eliminate one critical design degree of freedom: the ability to tune the transducer eigenfrequency (via the transducer stiffness $k_T = k_{mech} - k_{el,U}(U_0)$) by changing the source voltage.
Linearly stepped comb teeth The reason for the lack of electrostatic softening lies in the linear dependence of capacitance on the armature displacement in Eq. (6.71), causing the second spatial derivative (responsible for the electrostatic stiffness) to be zero. For non-zero electrostatic stiffness, the capacitance must be at least quadratic in the armature displacement. This dependence can be achieved with clever electrode geometries. One easily realizable option consists of a comb configuration with *linearly stepped teeth*, as suggested by (Lee and Cho 1998) (Fig. 6.25).

Configuration equations The configuration depicted in Fig. 6.25 having symmetric electrode separation β gives the following approximate *capacitance function* for the overlapping electrode area (shaded), with comb width b, comb overlap at zero voltage y_0, armature tooth count in the overlapping area n, and negligible comb height:

$$C_T(y) = n\frac{\varepsilon_0 b}{\beta}\left(y_0 + y\right). \tag{6.87}$$

In addition, n can be approximated as

$$n \approx \frac{h(y)}{\beta} = \frac{H}{\beta D}\left(y_0 + y\right). \tag{6.88}$$

Eqs. (6.87) and (6.88) then result in the following relations for the applicable transducer parameters:

- *transducer capacitance* $\quad C_T(y) \approx \dfrac{\varepsilon_0 b}{\beta^2}\dfrac{D}{H}\left(y_0 + y\right)^2,$

- *transducer force* $\quad F_{el}(y, u_T) \approx \dfrac{\varepsilon_0 b}{\beta^2}\dfrac{D}{H}\left(y_0 + y\right)\cdot {u_T}^2,$

- *electrostatic voltage stiffness* $\quad k_{el,U}(U_0) \approx \dfrac{\varepsilon_0 b}{\beta^2}\dfrac{D}{H}{U_0}^2.$ (6.89)

The electrostatic stiffness Eq. (6.89) now exhibits the desired dependence on the rest voltage U_0. Note that though Eq. (6.89) does not describe any direct dependence on the control variable, such is in fact indirectly present due to the rest position condition (6.74).

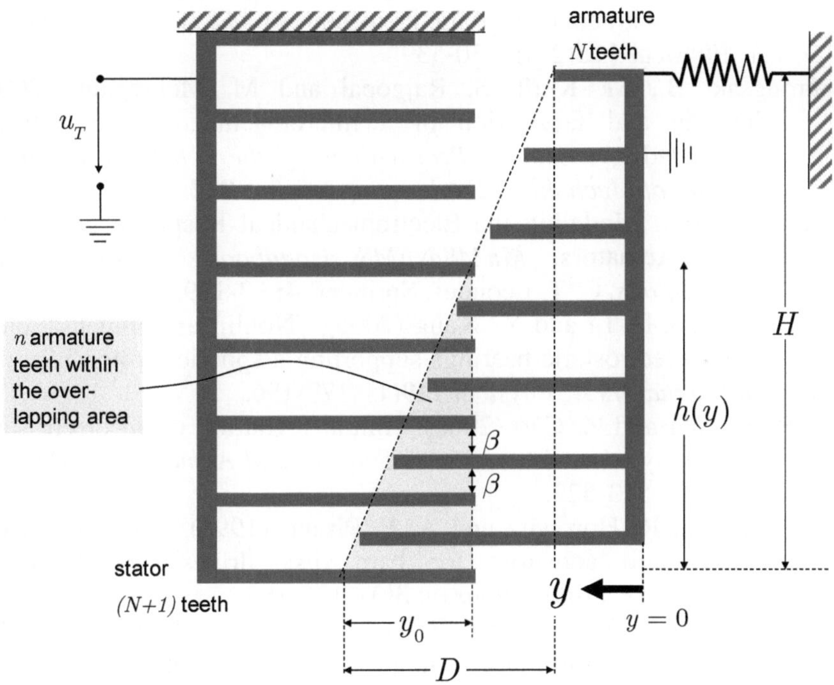

Fig. 6.25. Configuration for a longitudinal comb transducer with linearly stepped teeth for tuning the electrostatic stiffness with a source voltage, from (Lee and Cho 1998).

Bibliography for Chapter 6

Bencze, W. J., M. E. Eglington, R. W. Brumley and S. Buchman (2007). "Precision electrostatic suspension system for the Gravity Probe B relativity mission's science gyroscopes." *Advances in Space Research* **39**(2): 224-229.

Bochobza-Degani, O., D. Elata and Y. Nemirovsky (2003). "A general relation between the ranges of stability of electrostatic actuators under charge or voltage control." *Appl. Phys. Lett.* **82**: 302-304.

Borovic, B., F. L. Lewis, A. Q. Liu, E. S. Kolesar, et al. (2006). "The lateral instability problem in electrostatic comb drive actuators: modeling and feedback control." *Journal of Micromechanics and Microengineering*(7): 1233.

Chan, E. K. and R. W. Dutton (2000). "Electrostatic micromechanical actuator with extended range of travel." *Microelectromechanical Systems, Journal of* **9**(3): 321-328.

Chen, C. and C. Lee (2004). "Design and modeling for comb drive actuator with enlarged static displacement." *Sensors and Actuators A: Physical* **115**(2-3): 530-539.

Damrongsak, B., M. Kraft, S. Rajgopal and M. Mehregany (2008). "Design and fabrication of a micromachined electrostatically suspended gyroscope." *Proceedings of the I MECH E Part C Journal of Mechanical Engineering Science* **222**: 53-63.

Elata, D. (2006). Modeling the Electromechanical Response of Electrostatic Actuators. *MEMS/NEMS Handbook, Techniques and Applications*. C. T. Leondes. Springer. **4:** 93-119.

Han, F., Z. Gao, D. Li and Y. Wang (2005). "Nonlinear compensation of active electrostatic bearings supporting a spherical rotor." *Sensors and Actuators A: Physical* **119**(1): 177-186.

Han, F., Q. Wu and Z. Gao (2006). "Initial levitation of an electrostatic bearing system without bias." *Sensors and Actuators A: Physical* **130-131**: 513-522.

Horsley, D. A., R. Horowitz and A. P. Pisano (1998). "Microfabricated electrostatic actuators for hard disk drives." *Mechatronics, IEEE/ASME Transactions on* **3**(3): 175-183.

Horsley, D. A., N. Wongkomet, R. Horowitz and A. P. Pisano (1999). "Precision positioning using a microfabricated electrostatic actuator." *Magnetics, IEEE Transactions on* **35**(2): 993-999.

Huang, W. and G. Lu (2004). "Analysis of lateral instability of in-plane comb drive MEMS actuators based on a two-dimensional model." *Sensors and Actuators A: Physical* **113**(1): 78-85.

Imamura, T., T. Koshikawa and M. Katayama (1996). Transverse mode electrostatic microactuator for MEMS-based HDD slider. *Micro Electro Mechanical Systems, 1996, MEMS '96, Proceedings. 'An Investigation of Micro Structures, Sensors, Actuators, Machines and Systems'. IEEE, The Ninth Annual International Workshop on.* pp.216-221.

Jackson, J. D. (1999). *Classical Electrodynamics, Third Edition.* John Wiley and Sons, Inc.

Lee, K. B. and Y. H. Cho (1998). "A triangular electrostatic comb array for micromechanical resonant frequency tuning." *Sensors and Actuators A: Physical* **70**(1-2): 112-117.

Legtenberg, R., A. W. Groeneveld and M. Elwenspoek (1996). "Comb-drive actuators for large displacements." *Journal of Micromechanics and Microengineering* **6**: 320-329.

Seeger, J. I. and B. E. Boser (1999). Dynamics and control of parallel-plate actuators beyond the electrostatic instability. *Tech. Dig. 10th Intl. Conf. Solid-State Sensors and Actuators (Transducers '99)*, Sendai, Japan. pp.pp. 474-477.

Seeger, J. I. and B. E. Boser (2003). "Charge control of parallel-plate, electrostatic actuators and the tip-in instability." *Microelectromechanical Systems, Journal of* **12**(5): 656-671.

Seeger, J. I. and S. B. Crary (1997). Stabilization of electrostatically actuated mechanical devices. *Solid State Sensors and Actuators, 1997. TRANSDUCERS '97 Chicago., 1997 International Conference on.* pp.1133-1136 vol.1132.

Senturia, S. D. (2001). *Microsystem Design.* Kluwer Academic Publishers.

7 Functional Realization: Piezoelectric Transducer

Background Hardly any other principle of electromechanical transduction has contributed more to the advance of consumer mechatronic products than *piezoelectricity*. The most prominent instantiations of so-called *unconventional* transduction principles, piezo elements are *solid-state transducers*, and can be incorporated directly into mechanical components without complicated structural requirements (forming *smart structures*); generate high to very high forces in a minimal volume; have high dynamic response; and can even be operated without auxiliary energy in some applications (oscillation damping, *shunting*) or employed in the generation of electricity (*energy harvesting*).

Contents of Chapter 7 This chapter discusses fundamental physical phenomena of piezoelectric transducers and several attributes of their implementational variants. First, using the linear *constitutive piezoelectric material equations*, applicable solid-state properties are presented, along with *commonly used descriptive formulations*. Next, based on the *generic mechatronic transducer* introduced in Ch. 5, standardized dynamic models (the *ELM two-port models* and *transfer matrix*) are derived for *voltage* and *current drive*, and used to demonstrate relationships between structural parameters, material parameters, and dynamic properties. A further focus is the representation of various common transducer structures such as *disk transducers*, *stack transducers*, *lever transmissions*, and *strip transducers*, along with a discussion of their dynamics. An entire section is devoted to a discussion of the concept of electrical loading with *impedance feedback* (i.e. *shunting*), which is demonstrated in detail with an application example for an *accelerometer*. Finally, of particular importance to piezoelectric transducers, the concept of *mechanical resonant operation* is elucidated using the example of commercially successfully *ultrasonic piezo motor* implementations, whose dynamic and operational properties are discussed in detail. ■

7.1 Systems Engineering Context

Piezoelectric transducers Next to "classical" transducer concepts using electrostatic and electromagnetic phenomena, a series of so-called *unconventional* transducer concepts, exploiting other physical phenomena, have established themselves in the literature and in industry. Among these phenomena, *piezoelectricity* has proven to be extraordinarily useful. Demonstrated in 1880 by the brothers Pierre and Jacques CURIE, the *piezoelectric effect* was soon put to technical use, e.g. in crystal oscillators and ultrasonic transducers. A true breakthrough in broadly-applicable *piezoelectric transducers* was made possible with the 1950 patent by Walter P. Kistler for a *charge amplifier* with associated *microelectronic* drive and measurement circuits. Today, piezoelectric transducers have taken on key roles in many technological applications, fundamentally determining their achievable performance, e.g. piezo-injection technology for diesel and gasoline engines (see Deutscher Zukunftspreis 2005). As *solid-state transducers*, these devices can be directly integrated into mechanical structures without significant additional fixtures or other elements (creating *smart structures*), and in such applications act equally as sensors (for deformation sensing) and highly-dynamic actuators (for force generation).

Systems engineering significance From a systems engineering point of view, the functions *"generate forces/torques"* and *"measure mechanical states"* realizable with piezoelectric phenomena represent the *actuators* and

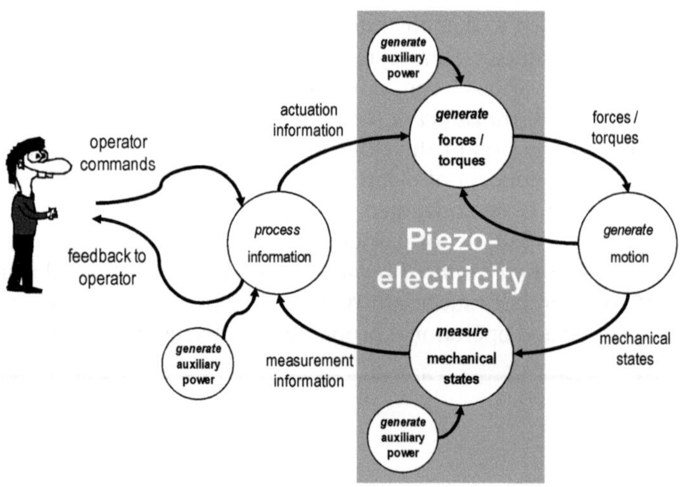

Fig. 7.1. Functional decomposition of a mechatronic system: functional realization using *piezoelectricity*

sensors of a mechatronic system (see Fig. 7.1). For both tasks, it is the transfer characteristics in the causal directions shown in Fig. 7.1 which are of interest. As is the case for other transducer principles, functional properties such as *linearity*, *dynamics*, and structural *parameter dependencies* related to the transfer characteristics play important roles in controller design.

Mechatronic phenomena Due to their *solid-state nature*, piezoelectric transducers do not require suspension of the electrodes per se, they can be directly coupled to a mechanical structure (simply stated: "piezo element is the transducer"). Thus, all of the electrical and mechanical properties (including electromechanical coupling) of such transducers are primarily described by their material properties. The dynamics at the electrical terminals of a transducer are characterized by capacitive energy storage, and at the mechanical terminals, by potential energy storage (a spring)—both independent of the material and geometry.

For many applications, a *linear* dynamic model description is thoroughly sufficient, though in some cases additional hysteresis effects must be taken into account.

It is worth noting that in certain applications, piezoelectric transducers can also be operated without auxiliary electrical energy (short-circuit operation, or *shunting*). Otherwise, a comparatively high voltage (100–1000V) is required.

For systems design, the following model relationships are thus of interest:

- *capacitance*, *spring stiffness*, and *piezoelectric coupling parameter* as a function of geometry and material;
- *piezoelectric forces* as a function of geometry, materials, and electrical drive;
- *transfer characteristics* including mechanical power feedback.

7.2 Physical Foundations

Solid-state transducers The term "solid-state" designates a material in the solid phase of matter. Solid-state elements are characterized by a high stability of the ordering of the material's building blocks, and are mechatronically interesting for their *electrical* and *mechanical* properties[1]—here,

[1] Additional technically-interesting properties are, for example, transduction via magnetic fields (*magnetostrictive* transducers) and temperature dependence (e.g. *shape memory alloys*). However, these are not further considered in this book.

in particular, their pronounced capability for electromechanical *transduction*. This allows a solid material to be employed directly as the working element for electromechanical energy conversion in the form of a *solid-state transducer* (Senturia 2001), (Janocha 2004). In the most favorable applications of this concept, transducer elements can be directly integrated into a mechanical structure with no, or at most minimal, additional structural complexity (forming a *smart structure*).

In this book, due to their great practical significance, solid-state transducers employing the *piezoelectric* effect—i.e. transduction via an electric field and continuum mechanical material properties—are singled out for discussion. The *nomenclature* here follows (IEEE 1988).

Solid-state electromechanical properties Fig. 7.2 depicts a solid-state element made of a *non-conducting* material (dielectric) and a pair of electrodes. Independent of the geometric dimensions, the *electrical* behavior is characterized by the electric field quantities introduced in Ch. 6:

- *electric field* $E = \frac{u}{L}$ [V/m],
- *electric displacement* $D = \frac{q}{A}$ [As/m²];

and the *mechanical* behavior by the two related quantities

- *strain* $S = \frac{\Delta l}{L}$ [1],
- *stress* $T = \frac{F}{A}$ [N/m²].

Direct piezoelectric effect The *direct piezoelectric* effect describes an *electrical* solid-state property, namely the generation of electrical polarization (a displacement of positive and negative charges) in an insulator by mechanical means.

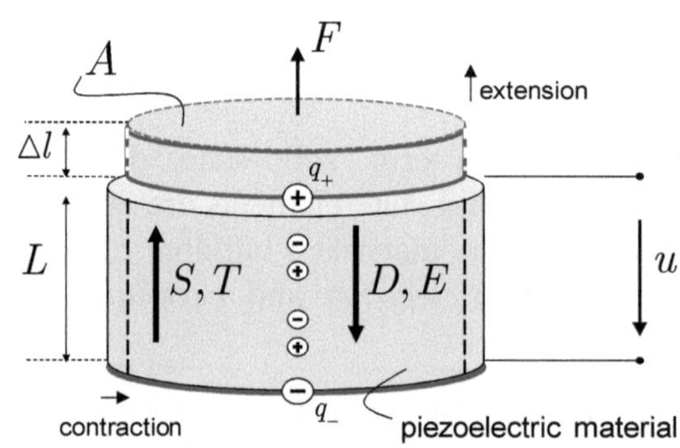

Fig. 7.2. Piezoelectric solid-state element: single-axis tension

The linear constitutive material equation for locally small strains and one-dimensional tension is

$$D = \varepsilon \cdot E + e \cdot S\,, \tag{7.1}$$

with the parameters of

- *permittivity (dielectric constant)* ε $[\mathrm{As/Vm}]$,
- *piezoelectric force constant* e $[\mathrm{As/m^2 = N/Vm}]$.

The first term in Eq. (7.1) describes the well-known separation of charges and the alignment of electrical dipoles (polarization) in a non-conducting solid resulting from an applied electric field E and inducing an electric flux density D via the material-dependent permittivity ε, (Fig. 7.2, cf. also the plate capacitor in Ch. 6).

In a class of special materials, polarization can also be induced by a *mechanical displacement* (here characterized by the *strain* S) creating the *direct piezoelectric effect*—the second term in Eq. (7.1). The corresponding coefficient e is commonly called the *piezoelectric (force) constant.* This effect can be straightforwardly applied to *sensing* tasks (electrical displacement sensing, or indirect force sensing via a displacement measurement).

Reciprocal piezoelectric effect The *reciprocal (inverse) piezoelectric* effect describes a *mechanical* solid-state property, namely the generation of mechanical stress by an electric field.

The linear constitutive material equation for locally small strains and one-dimensional tension is

$$T = c \cdot S - e \cdot E \tag{7.2}$$

with the

- *modulus of elasticity (YOUNG's modulus)* c $[\mathrm{N/m^2}]$.

The first term in Eq. (7.2) described the well-known linear elastic behavior of a solid body, according to which the strain S and mechanical stress T are proportional according to the modulus of elasticity c, (*HOOKE's Law*).

In piezoelectric solids, *mechanical stress* can also be imposed by an *electric field* E in the *reciprocal piezoelectric effect*—the second term in Eq. (7.2). The negative sign results from the definition of directions for mechanical stress and strain (with longitudinal deformation Δl).

It is immediately worth noting that the same coefficient e appears here as in the direct piezoelectric effect in Eq. (7.1). As will be shown, this is always the case assuming lossless electromechanical energy conversion (i.e. a *conservative system*, cf. the integrability condition in Sec. 5.3.1).

This effect can be straightforwardly applied to *actuation* tasks requiring electrical force generation.

Anisotropic constitutive material equations: tensor form Piezoelectric materials generally exhibit a directional dependency (*anisotropy*) in their electromechanical behavior. This implies that, in the general case, the variables and coefficients in the constitutive equations (7.1), (7.2) must be interpreted as *tensors* (Bronshtein et al. 2005).

Further, it should be noted that the two effects described by Eqs. (7.1), (7.2) are always active simultaneously, as can be seen from the coupling factor e. Thus, the coupled system of equations must naturally also always be considered as a whole.

The transducer is customarily set in a Cartesian coordinate system $(1,2,3)$ such that the polarization direction (and the electric field) points along the 3-axis (see Fig. 7.3).

The (T,D)-*tensor form* of the constitutive material equations corresponding to Eqs. (7.1), (7.2) is then

$$\begin{aligned} \boldsymbol{T}(\boldsymbol{S},\boldsymbol{E}) &= \boldsymbol{c}^{E} \cdot \boldsymbol{S} - \boldsymbol{e} \cdot \boldsymbol{E}, \\ \boldsymbol{D}(\boldsymbol{S},\boldsymbol{E}) &= \boldsymbol{e} \cdot \boldsymbol{S} + \boldsymbol{\varepsilon}^{S} \cdot \boldsymbol{E}, \end{aligned} \tag{7.3}$$

$\boldsymbol{c}^{E}$: YOUNG's modulus when $\boldsymbol{E} = \boldsymbol{0}$, i.e. electrical short circuit;

$\boldsymbol{\varepsilon}^{S}$: permittivity when $\boldsymbol{S} = \boldsymbol{0}$, i.e. mechanically braced;

$\boldsymbol{e}$: piezoelectric force constants $[\mathrm{As/m^2{=}N/Vm}]$.

Equivalently to Eq. (7.3), the (S,D)-*tensor form* of the constitutive material equations are found to be

$$\begin{aligned} \boldsymbol{S}(\boldsymbol{T},\boldsymbol{E}) &= \boldsymbol{s}^{E} \cdot \boldsymbol{T} + \boldsymbol{d} \cdot \boldsymbol{E}, \\ \boldsymbol{D}(\boldsymbol{T},\boldsymbol{E}) &= \boldsymbol{d} \cdot \boldsymbol{T} + \boldsymbol{\varepsilon}^{T} \cdot \boldsymbol{E}, \end{aligned} \tag{7.4}$$

$\boldsymbol{s}^{E}$: compliance when $\boldsymbol{E} = \boldsymbol{0}$, i.e. electrical short circuit;

$\boldsymbol{\varepsilon}^{T}$: permittivity when $\boldsymbol{T} = \boldsymbol{0}$, i.e. no mechanical load;

$\boldsymbol{d}$: piezoelectric charge constants $[\mathrm{As/N{=}m/V}]$.

In Eqs. (7.3) and (7.4), $\boldsymbol{E},\boldsymbol{D}$ represent rank-1 tensors (3 components); $\boldsymbol{T},\boldsymbol{S},\boldsymbol{\varepsilon}$ are rank-2 tensors (9 components, of which 6 are independent); $\boldsymbol{e},\boldsymbol{d}$ are rank-3 tensors (27 components, 18 independent); and $\boldsymbol{c},\boldsymbol{s}$ are rank-4 tensors (81 components, 21 independent). The number of independent tensor components results from crystal symmetries and the choice of coordinate system (Jordan and Ounaies 2001), (IEEE 1988).

Material parameters Note the special significance of the material parameters $\boldsymbol{c}^E, \boldsymbol{s}^E, \boldsymbol{\varepsilon}^S, \boldsymbol{\varepsilon}^T$ in Eqs. (7.3), (7.4). These can be experimentally determined under the conditions specified by the superscripts (the superscripted quantity is set to zero). In particular, care is required when specifying permittivity, as the two values refer to two completely dissimilar operating conditions (the superscript/operating condition should thus always be indicated).

Most piezo materials, particularly piezo ceramics, exhibit *transverse isotropy*. This means that there is anisotropy only in the polarization direction while all orthogonal directions are isotropic (cylindrical symmetry). This implies that a number of the material tensor components are zero (Jordan and Ounaies 2001).

Luckily, for cases important to implementation, only a few of the directional dependencies are significant, so that for further considerations in this book, the entire tensor description can be ignored and only scalar equations such as Eqs. (7.1), (7.2) employed.

Longitudinal effect, transverse effect The two directional dependencies most important in practice are shown in Fig. 7.3: the *longitudinal effect*, where the electric field and corresponding mechanical quantities T, S act in parallel directions (Fig. 7.3a); and the *transverse effect*, where the electrical and mechanical directions of action are orthogonal (Fig. 7.3b).

The piezoelectric behavior is described by the tensor components e_{33} and d_{33} for the *longitudinal effect*, and the tensor components e_{31} and d_{31} for the *transverse effect*[2] (the remaining components are all zero). These parameters, like the remaining material parameters, can be determined from data sheets for the piezo material (or transducer), generally also directly with the indicated superscript designation.

[2] The first index of the d-parameters and e-parameters indicates the respective "electrical" component $(E \text{ or } D)$, the second index specifies the "mechanical" component $(T \text{ or } S)$. Due to crystal symmetries and the choice of coordinate system, the components of the rank-3 tensors $\boldsymbol{d}, \boldsymbol{e}$ can have a simplified representation requiring only two indices (IEEE 1998).

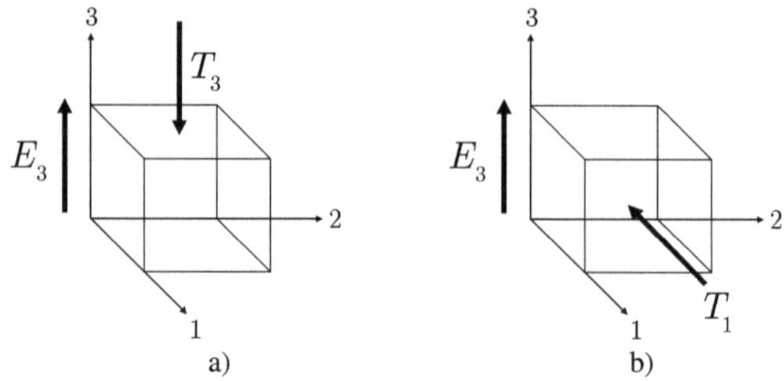

Fig. 7.3. Anisotropic piezo material: a) longitudinal effect (here $e_{33} > 0$, $d_{33} > 0$), b) transverse effect (here due to contraction: $e_{31} < 0$, $d_{31} < 0$)

Extension vs. contraction of the material As already hinted at in Fig. 7.2, during the deformation of a solid-state element, simultaneous extension and contraction effects are always observed, e.g. lengthening always entails a simultaneous reduction in the orthogonal cross section (see longitudinal effect in Fig. 7.2). Thus, for example, $e_{31} < 0$ and $d_{31} < 0$.

Scalar constitutive material equations Assuming limited directional dependencies (transverse isotropy, and longitudinal effect or transverse effect) and noting the corresponding active (i.e. non-zero) tensor components, the tensor equations (7.3), (7.4) can be represented in a simplified scalar form. The two most commonly-used representations (T, D) and (S, D) are listed in Table 7.1 along with the corresponding material parameters. These two representations are equivalent; depending on the application one or the other version will be easier to manipulate. For the less common equation combinations (T, E) and (S, E), other material pa-

Table 7.1. Scalar constitutive piezoelectric material equations

(T, D)-form	(S, D)-form
$\begin{pmatrix} T \\ D \end{pmatrix} = \begin{pmatrix} c^E & -e \\ e & \varepsilon^S \end{pmatrix} \begin{pmatrix} S \\ E \end{pmatrix}$	$\begin{pmatrix} S \\ D \end{pmatrix} = \begin{pmatrix} s^E & d \\ d & \varepsilon^T \end{pmatrix} \begin{pmatrix} T \\ E \end{pmatrix}$

material parameters: $s^E = \dfrac{1}{c^E}$, $d = \dfrac{e}{c^E} = e \cdot s^E$, $\varepsilon^T = \varepsilon^S + \dfrac{e^2}{c^E}$

ELM coupling factor: $\kappa^2 = \dfrac{e^2}{e^2 + c^E \cdot \varepsilon^S} = \dfrac{d^2}{s^E \cdot \varepsilon^T}$

rameters result, though these can be derived from the prior case (IEEE 1988). It should, however, always be carefully noted which system of equations underlies a particular set of parameters (e.g. ε^S vs. ε^T !).

The ELM coupling factor κ^2 also shown in Table 7.1 conforms to the general definition of the coupling factor of Ch. 5. This equivalence will be made plain in the next section. It is worth noting that the ELM coupling factor is independent of the physical geometry of an element, as it only contains material properties.

Piezo materials In contrast to electrostatic and electromagnetic transducers, in piezo transducers, solid-state parameters are determining factors in the system dynamics, alongside the transducer geometry. For this reason, in any concrete application, the material data must be considered (i.e. the data sheets should be consulted).

Roughly speaking, a distinction can be made between natural (crystalline) piezo materials (e.g. *quartz*) and synthetic materials (*piezoceramics, piezoelectric plastics*). Among piezoceramics, *lead zirconate titanate (PZT)* is very commonly found as a transducer material.

In the production of piezoceramics (during the sintering process), an electric field is applied at a sufficiently high temperature (above the CURIE temperature) to induce a polarization of the ceramic which is maintained after cooling. This leaves the ceramic anisotropically piezoelectric with a particular configuration. This property must be taken into consideration during operation, as the CURIE temperature cannot be exceeded again without destroying the piezoelectric nature of the ceramic.

In general, piezoceramics exhibit larger piezoelectric coupling parameters, whereas quartzes possess better temperature stability. For a more detailed discussion of piezo materials, the reader is referred e.g. to (Lenk et al. 2011).

As a brief overview, typical material parameters for quartz and PZT have been gathered together in Table 7.2 (Lenk et al. 2011).

Electromechanical directions of action Using the constitutive material equations in Table 7.1 and the particular material parameters in Table 7.2, the electromechanical directions of action can again be illustrated. The direction predetermined by the polarization process can be indicated at the corresponding electrode terminal (a positive "+" or negative "–" terminal).

As shown in Eq. (7.1), for *open electrodes* $(D = 0)$, a *positive strain* (lengthening, tensile loading) and a positive force constant $e > 0$ generate

Table 7.2. Typical material parameters for piezoelectric materials, from (Lenk et al. 2011), $\varepsilon_0 = 8.85 \cdot 10^{-12}$ As/Vm

Material parameters	Quartz	PZT
c^E $[10^{10}\ \mathrm{N/m^2}]$	7.8	6
$\varepsilon^S/\varepsilon_0$, $\varepsilon^T/\varepsilon_0$ [1]	4.7, 4.7	900, 1400
e_{33}, e_{31} $[\mathrm{As/m^2{=}N/Vm}]$	0.18, -0.18	16, -5
d_{33}, d_{31} $[10^{-12}\ \mathrm{As/N{=}m/V}]$	2.3, -2.3	270, -80

a *negative* electric field $E = -(e / \varepsilon) \cdot S$, i.e. a field directed against the inherent polarization direction (for $e < 0$, the opposite direction of action applies). Correspondingly, a contraction or compressive loading with $e > 0$ induces a positive electric field in the polarization direction.

As shown in Eq. (7.2), as the external force vanishes $(T = 0)$, a *positive electric field* in the polarization direction and $e > 0$ in turn induce a *positive strain* (extension) $S = (e / c) \cdot E$, while for $e > 0$, a negative electric field will generate a negative strain (contraction) of the piezo element.

Maximum electric field To avoid destroying the polarization structure (i.e. inducing *depolarization*), the electric field may not exceed the following typical upper limits:

- maximum field along the polarization direction: $1 - 2\ \mathrm{kV/mm}$,
- maximum field against the polarization direction: $300\ \mathrm{V/mm}$.

Piezoelectric laminated structures The linear constitutive material equations described above hold in general for piezoelectric materials. The situations illustrated in Fig. 7.2 and 7.3 show the relationships for compact piezo materials with locally homogeneous electrical and mechanical properties (forming discrete transducers describable using lumped parameter models).

If the piezo material possesses a spatially *extended structure*, e.g. a *piezo foil laminate*, then its description must take into account the spatial dependence of the properties (describable using models with distributed parameters). In the important practical case of elastic mechanical structures upon which the piezo foil is applied, the *material relations* (7.3), (7.4) remain completely valid, though a *spatial dependence* must be taken into account for the *strain* S. This leads to a *continuum mechanics* description using partial differential equations. For reasons of space, such infinite-dimensional modeling approaches are not further pursued in this book; the interested reader is referred to extending literature, e.g. (Preumont 2006), (Moheimani et al. 2003), (Kugi et al. 2006).

7.3 Generic Piezoelectric Transducer

7.3.1 System configuration

Discrete transducer This section considers transducer configurations in which—to a sufficiently good approximation—active electrical and mechanical quantities are locally homogeneous. This allows a piezo element to be represented using models with lumped, spatially-independent parameters. Fig. 7.4 depicts such an arrangement forming a generic piezoelectric transducer. The directions of the electrical and mechanical coordinates correspond to the generic mechatronic transducer in Ch. 5; the piezoelectric force F_{pz} is the transducer force F_T.

External suspension The external elastic suspension shown in Fig. 7.4 is not absolutely necessary in this case, and should be considered optional. Due to the solid-state nature of piezo materials, the coupling of the trans-

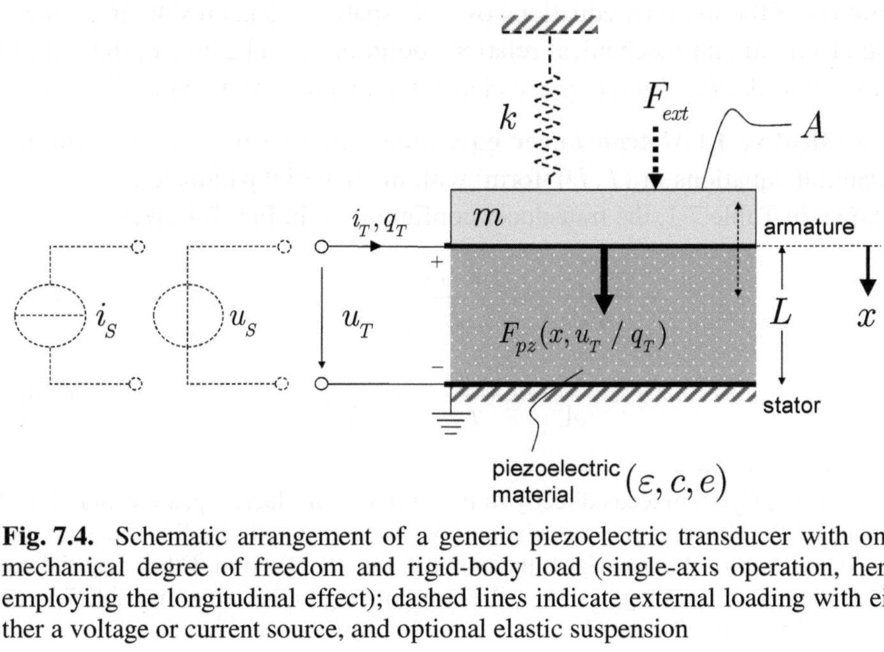

Fig. 7.4. Schematic arrangement of a generic piezoelectric transducer with one mechanical degree of freedom and rigid-body load (single-axis operation, here employing the longitudinal effect); dashed lines indicate external loading with either a voltage or current source, and optional elastic suspension

ducer to a mechanical structure can take place directly at the stator and armature electrodes (i.e. with direct integration into the structure). Thus, piezo transducers can be employed with minimal mechanical complexity.

Analogy to electrostatic plate transducer Considering the arrangement in Fig. 7.4 reveals its resemblance to an electrostatic plate transducer (see Ch. 6). The primary difference to that arrangement is in the dielectric between the electrode plates. In the *electrostatic transducer*, the dielectric is air, entailing elastic suspension (with a spring k) of the armature. In the *piezo transducer*, the solid-state dielectric functions as an elastic separator between the transducer electrodes.

A piezo transducer can certainly be described as an electrostatic transducer with an internal spring, as it stores both electrical and mechanical potential energy. Thus, as concerns their electrical dynamics, an extensive analogy between these two transducer types is to be expected.

7.3.2 Constitutive piezoelectric transducer equations

Geometry vs. material equations The constitutive material equations presented in Sec. 7.2 describe geometry-independent physical piezoelectric relationships. For a given concrete transducer configuration, formal integration of the material equations over its spatial extents results in geometric electrical and mechanical relations defining its behavior. In the current case, consider the discrete piezo element geometry presented in Fig. 7.4.

Constitutive ELM transducer equations Starting from the constitutive material equations in (T, D)-form[3] with the material parameters (c^E, ε^S, e) shown in Table 7.1, the transducer configuration in Fig. 7.4 gives

$$\frac{-(-F_{pz})}{A} = c^E \frac{(-x)}{L} - e\frac{u_T}{L}, \tag{7.5}$$

$$\frac{q_T}{A} = e\frac{(-x)}{L} + \varepsilon^S \frac{u_T}{L}. \tag{7.6}$$

3 The (T, D)-form leads directly to the form of transducer equations introduced in Ch. 5. However, the calculation could also proceed equally well with any other form of the material equations, e.g. the (S, D)-form. Other material parameters, such as, e.g., (s^E, ε^T, d) corresponding to the values in Table 7.1 can also be easily substituted.

As the strain S is positive for positive Δl, it holds in Eqs. (7.5) and (7.6) that $\Delta l = -x$. The left-hand side of Eq. (7.5) requires a bit more consideration. By definition, the mechanical stress T is positive for tension, i.e. it has the same sign as the strain S. This implies the minus sign inside the parentheses (positive T opposing the positive F_T direction). However, the mechanical stress T also describes the external force acting to create a deformation Δl (cf. Fig. 7.2) and generating an internal opposing (thus the outer minus sign) reaction force, i.e. the transducer force F_{pz} [4].

Rearranging Eqs. (7.5), (7.6) and differentiating Eq. (7.6) with respect to time gives the *constitutive piezoelectric ELM transducer equations* in

- *PSI-coordinates*

$$\begin{aligned} F_{pz,\Psi}(x, u_T) &= -k^E \cdot x - K_T \cdot u_T\,, \\ i_T(\dot{x}, \dot{u}_T) &= -K_T \cdot \dot{x} + C^S \cdot \dot{u}_T\,, \end{aligned} \tag{7.7}$$

- *Q-coordinates*

$$\begin{aligned} F_{pz,Q}(x, q_T) &= -\left(k^E + \frac{K_T^{\,2}}{C^S}\right) \cdot x - \frac{K_T}{C^S} \cdot q_T\,, \\ u_T(x, q_T) &= \frac{K_T}{C^S} \cdot x + \frac{1}{C^S} \cdot q_T\,, \end{aligned} \tag{7.8}$$

with *transducer parameters*

- *piezoelectric stiffness* (elec. short circuit) $\quad k^E := \dfrac{c^E \cdot A}{L}$,
- *piezoelectric transducer constant* $\quad K_T := \dfrac{e \cdot A}{L}$, (7.9)
- *piezo capacitance* (mech. braced) $\quad C^S := \dfrac{\varepsilon^S \cdot A}{L}$.

Integrability condition The two forms of the constitutive ELM transducer equations (7.7), (7.8) are precisely the Type D ELM basic equations shown in Table 5.2 with *electromechanically linear dynamics*. This also explains the symmetry and anti-symmetry of the coupling terms in

---

4 When comparing the transducer equations of different authors, attention should be paid to the respective choices of coordinate system (resulting in different signs for otherwise equal physical transducer parameters).

Eqs. (7.7), (7.8) and the material equations in Table 7.1, as this symmetry is in fact defined by the integrability condition Eq. (5.13).

7.3.3 ELM two-port model

Linear transducer equations Due to the assumption of *linear piezoelectricity*, the constitutive ELM transducer equations (7.7), (7.8) can be used directly as the basis for the ELM two-port model of the *unloaded* piezoelectric transducer[5].

Two-port admittance form Starting with the constitutive transducer equations (7.7) and applying the general results from Sec. 5.3.3 gives the *two-port admittance form* for the unloaded piezoelectric transducer

$$\begin{pmatrix} F_{pz,\Psi}(s) \\ I_T(s) \end{pmatrix} = \mathbf{Y}_{pz}(s) \cdot \begin{pmatrix} X(s) \\ U_T(s) \end{pmatrix} = \begin{pmatrix} -k_{pz,U} & -K_{pz,U} \\ -s \cdot K_{pz,U} & s \cdot C_{pz} \end{pmatrix} \begin{pmatrix} X(s) \\ U_T(s) \end{pmatrix}, \tag{7.10}$$

with defining relations for the *transducer parameters* from Eq. (7.9):

- *piezoelectric voltage stiffness* $k_{pz,U} := k^E = \dfrac{c^E \cdot A}{L} \quad \left[\dfrac{\mathrm{N}}{\mathrm{m}}\right]$,
- *voltage coefficient* $K_{pz,U} := K_T = \dfrac{e \cdot A}{L} \quad \left[\dfrac{\mathrm{N}}{\mathrm{V}}\right], \left[\dfrac{\mathrm{As}}{\mathrm{m}}\right]$, (7.11)
- *piezo capacitance* $C_{pz} := C^S = \dfrac{\varepsilon^S \cdot A}{L} \quad \left[\dfrac{\mathrm{As}}{\mathrm{V}}\right]$.

Two-port hybrid form Equivalently, using the constitutive transducer equations (7.8) gives the *two-port hybrid form* of the unloaded piezoelectric transducer

$$\begin{pmatrix} F_{pz,Q}(s) \\ U_T(s) \end{pmatrix} = \mathbf{H}_{pz}(s) \cdot \begin{pmatrix} X(s) \\ I_T(s) \end{pmatrix} = \begin{pmatrix} -k_{pz,I} & -\dfrac{K_{pz,I}}{s} \\ K_{pz,I} & \dfrac{1}{s \cdot C_{pz}} \end{pmatrix} \begin{pmatrix} X(s) \\ I_T(s) \end{pmatrix}, \tag{7.12}$$

---

5 Thus, a detour via the LAGRANGian does not have to be embarked upon here. As can be easily verified, this would take the following form: $\mathcal{L}_\Psi^o(x, u_T) = -\tfrac{1}{2} k^E x^2 - K_T x u_T + \tfrac{1}{2} C^S u_T^2$.

with defining relations for the *transducer parameters* from Eq. (7.11):

- *piezoelectric current stiffness*

$$k_{pz,I} := k_{pz,U} + \frac{K_{pz,U}^2}{C_{pz}} = \frac{A}{L}\left(c^E + \frac{e^2}{\varepsilon^S}\right) \quad \left[\frac{\mathrm{N}}{\mathrm{m}}\right], \tag{7.13}$$

- *current coefficient* $K_{pz,I} := \dfrac{K_{pz,U}}{C_{pz}} = \dfrac{e}{\varepsilon^S} \quad \left[\dfrac{\mathrm{V}}{\mathrm{m}}\right], \left[\dfrac{\mathrm{N}}{\mathrm{As}}\right],$

where the *piezo capacitance* C_{pz} is defined as in Eq. (7.11). It always holds for the piezoelectric current stiffness that $k_{pz,I} > k_{pz,U}$.

Relation between two-port parameters It is easy to verify that the correspondences in Table 5.4 hold for the parameters of the admittance and hybrid forms.

Unloaded ELM coupling factor In contrast to the electrostatic transducer, the inherent mechanical potential energy storage of a piezoelectric transducer already allows the *ELM coupling factor* to be defined for the *unloaded* configuration. Considering the values from Sec. 5.6 gives (cf. also Eq. (6.21) and Table 7.1) the unloaded ELM coupling factor

$$\kappa_{pz,0}^2 = \frac{1}{1 + \dfrac{C_{pz}}{K_{pz,U}^2} k_{pz,U}} = \frac{C_{pz} \cdot K_{pz,I}^2}{k_{pz,I}} = \frac{1}{1 + \dfrac{c^E \cdot \varepsilon^S}{e^2}}. \tag{7.14}$$

To maximize the coupling factor of a transducer, a piezo material having the greatest possible piezoelectric force constant e should be chosen.

7.3.4 Loaded piezoelectric transducer

Linear dynamic model The linear dynamic model of the *loaded* piezoelectric transducer can be easily derived from the generic model in Sec. 5.4.4 using the two-port parameters of Eqs. (7.10) and (7.12). The signal flow diagrams for voltage- and current-drive transducers are shown in Figs. 7.5 and 7.6, respectively (cf. Fig. 5.15, Fig. 5.16). The transducer exhibits electrically capacitive dynamics, and has great phenomenological similarity to an electrostatic plate transducer.

Piezoelectric stiffness The most substantial difference from an electrostatic plate transducer lies in the *negative* piezoelectric stiffness $(-k_{pz,U})$ or $(-k_{pz,I})$. In the signal flow diagrams Fig. 7.5 and 7.6 it is quite apparent that as a result—even with an unsuspended load—a negative restoring force F_T always results and *no* type of pull-in effect is possible.

Optional mechanical suspension Due to the solid attachment via the piezo element, external suspension is not necessarily required; it is shown as optional in Figs. 7.5 and 7.6.

Dynamic behavior

Transfer matrix for voltage drive From Eq. (7.10) and Table 5.8, the *transfer matrix* $\mathbf{G}(s)$ is obtained, where (see also Fig. 7.5)

$$\begin{pmatrix} X(s) \\ I_T(s) \end{pmatrix} = \mathbf{G}(s) \begin{pmatrix} F_{ext}(s) \\ U_S(s) \end{pmatrix} = \begin{pmatrix} V_{x/F,U} \dfrac{1}{\{\Omega_U\}} & V_{x/u} \dfrac{1}{\{\Omega_U\}} \\ V_{i/F} \dfrac{s}{\{\Omega_U\}} & V_{i/u} \cdot s \dfrac{\{\Omega_I\}}{\{\Omega_U\}} \end{pmatrix} \begin{pmatrix} F_{ext}(s) \\ U_S(s) \end{pmatrix}, \tag{7.15}$$

with the parameters

$$k_{T,U} = k + k_{pz,U}, \quad \Omega_U^2 = \frac{k_{T,U}}{m}, \quad k_{T,I} = k + k_{pz,I}, \quad \Omega_I^2 = \frac{k_{T,I}}{m},$$
$$\text{where } \Omega_U < \Omega_I,$$

$$\begin{aligned} V_{x/F,U} &= \frac{1}{k_{T,U}} = \frac{1}{k_{pz,U}} \frac{1}{1 + \dfrac{k}{k_{pz,U}}}, \\ V_{x/u} &= V_{i/F} = -\frac{K_{pz,U}}{k_{T,U}} = -\frac{e}{c^E} \frac{1}{1 + \dfrac{k}{k_{pz,U}}}, \\ V_{i/u} &= \tilde{C}_{pz} = C_{pz} \frac{k_{T,I}}{k_{T,U}} = C_{pz} \frac{1 + \dfrac{e^2}{c^E \cdot \varepsilon^S} + \dfrac{k}{k_{pz,U}}}{1 + \dfrac{k}{k_{pz,U}}} = C_{pz} \frac{1}{1 - \kappa_{pz}^2}. \end{aligned} \tag{7.16}$$

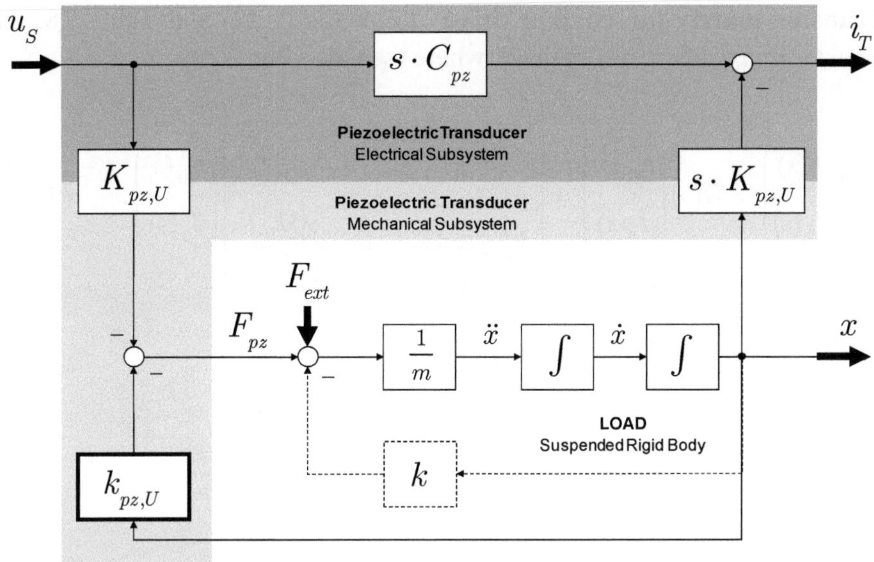

Fig. 7.5. Signal flow diagram for a loaded *piezoelectric* transducer with *voltage drive* (lossless, ideal voltage source, mechanical load: rigid body, optionally elastically suspended, cf. Fig. 7.4)

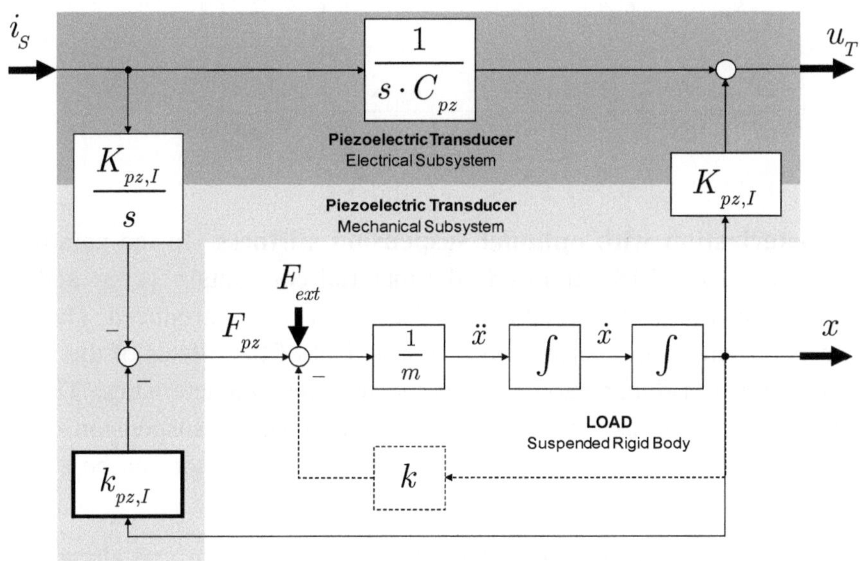

Fig. 7.6. Signal flow diagram for a loaded *piezoelectric* transducer with *current drive* (lossless, ideal current source, mechanical load: rigid body, optionally elastically attached, cf. Fig. 7.4)

Transfer matrix for current drive From Eq. (7.12) and Table 5.8, the *transfer matrix* $\mathbf{G}(s)$ is obtained, where (see also Fig. 7.6)

$$\begin{pmatrix} X(s) \\ U_T(s) \end{pmatrix} = \mathbf{G}(s) \begin{pmatrix} F_{ext}(s) \\ I_S(s) \end{pmatrix} = \begin{pmatrix} V_{x/F,I} \dfrac{1}{\{\Omega_I\}} & V_{x/i} \dfrac{1}{s \cdot \{\Omega_I\}} \\ V_{u/F} \dfrac{1}{\{\Omega_I\}} & V_{u/i} \dfrac{\{\Omega_U\}}{s \cdot \{\Omega_I\}} \end{pmatrix} \begin{pmatrix} F_{ext}(s) \\ I_S(s) \end{pmatrix}, \tag{7.17}$$

with the parameters

$$k_{T,I} = k + k_{pz,I}, \; \Omega_I^2 = \frac{k_{T,I}}{m}, \quad k_{T,U} = k + k_{pz,U}, \; \Omega_U^2 = \frac{k_{T,U}}{m},$$
$$\text{where } \Omega_U < \Omega_I,$$

$$V_{x/F,I} = \frac{1}{k_{T,I}} = \frac{1}{k_{pz,U}} \frac{1-\kappa_{pz}^2}{1+\dfrac{k}{k_{pz,U}}},$$
$$V_{x/i} = -\frac{K_{pz,I}}{k_{T,I}} = -\frac{1}{e}\frac{L}{A}\kappa_{pz}^2, \qquad V_{u/F} = \frac{K_{pz,I}}{k_{T,I}} = \frac{1}{e}\frac{L}{A}\kappa_{pz}^2 = \frac{\kappa_{pz}^2}{K_{pz,U}}, \tag{7.18}$$
$$V_{u/i} = \frac{1}{\tilde{C}_{pz}} = \frac{1}{C_{pz}}\frac{k_{T,U}}{k_{T,I}} = \frac{1}{C_{pz}} \frac{1+\dfrac{k}{k_{pz,U}}}{1+\dfrac{e^2}{c^E \cdot \varepsilon^S}+\dfrac{k}{k_{pz,U}}} = \frac{1-\kappa_{pz}^2}{C_{pz}}.$$

Parameterization with optional suspension stiffness In the parameter equations (7.16), (7.18), alongside the material constants c^E, ε^S, e and the geometric quantities L, A, the ratio $k / k_{pz,U}$ was also introduced. The first two sets of parameters permit direct assessment of the effects of the material and the transducer geometry on the transfer characteristics. The ratio $k / k_{pz,U}$ can be used to assess the effect of elastic suspension of the armature. For a transducer without this feature, this factor can be simply set to zero.

Loaded ELM coupling factor The ELM coupling factor was already extensively discussed in Sec. 5.6. The properties for a capacitive transducer found there can be nicely verified using the piezoelectric transducer considered here.

For the *loaded* transducer, the *ELM coupling factor* following Eq. (5.77) is

$$\kappa_{pz}^2 = \frac{k_{T,I} - k_{T,U}}{k_{T,I}} = \frac{1}{1 + \dfrac{c^E \cdot \varepsilon^S}{e^2}\left(1 + \dfrac{k}{k_{pz,U}}\right)} \le \kappa_{pz,0}^2 . \tag{7.19}$$

The additional suspension stiffness k thus reduces the ELM coupling factor relative to the unsuspended case—$\kappa_{pz,0}^2$ in Eq. (7.14)—as already generally demonstrated in Sec. 5.6.

Effective transducer capacitance It can be seen in Eqs. (7.16) and (7.18) that the effective transducer capacitance $\tilde{C}_{pz}$ is clearly greater than the transducer capacitance C_{pz} for the mechanically braced state in Eq. (7.9) since

$$\tilde{C}_{pz} = C^T = \frac{\varepsilon^T \cdot A}{L} = C_{pz} \frac{1}{1 - \kappa_{pz}^2} . \tag{7.20}$$

The effective transducer capacitance is thus equal to the transducer capacitance *without mechanical loading* (cf. Eq. (7.4)).

A thoroughly desirable increase in the ELM coupling factor thus also results in a decidedly higher transducer current (and greater demands on the auxiliary energy source).

Transducer eigenfrequencies In general, from Sec. 5.4, it holds for the eigenfrequencies that for (cf. Fig. 5.17)

- *voltage drive*: eigenfrequency Ω_U with an electrical *short circuit*,
- *current drive*: eigenfrequency Ω_I with an electrical *open circuit.*

Due to the capacitive transducer dynamics, in general (see Eq. (5.75))

$$\frac{\Omega_U}{\Omega_I} = \sqrt{1 - \kappa^2} \quad \text{and} \quad \Omega_U < \Omega_I .$$

Characteristic polynomial: transducer stability As can be seen from the characteristic polynomials of the transfer matrices $\mathbf{G}(s)$ in Eqs. (7.15) and (7.17), the lossy case always exhibits *marginally stable dynamics* with an imaginary pair of poles corresponding to the transducer eigenfrequency, independently of the use of voltage or current drive. As previously explained, the transducer eigenfrequencies Ω_U and Ω_I depend on the transducer stiffness $k_{pz,U}$, $k_{pz,I}$. Given additional external elastic suspension k, the transducer stiffness increases, due to the parallel springs. In contrast to an electrostatic transducer, however, a linear piezoelectric transducer can not become unstable.

Transducer gain, transducer sensitivity For *actuator operation*, the highest possible *transducer gain* $V_{x/u}, V_{x/i}$, and for *sensor operation*, the largest possible *transducer sensitivity* $V_{i/F}, V_{u/F}$ is desired.

For *voltage drive*, following Eq. (7.16), the applicable gains are directly proportional to the piezoelectric force constant e. For this case, the "best" possible piezo material having a large value of e should be chosen.

For *current drive*, following Eq. (7.18), large values of e only exhibit an approximately constant or slightly inversely proportional relationship to applicable gains.

7.3.5 Structural principles

Disk transducer

Layout Discrete transducers are usually laid out as disk transducers with a round or square cross-section exploiting the *longitudinal effect* (see Fig. 7.2).

Disk thickness vs. operating voltage The phenomenon limiting the disk thickness L, and thus the achievable transducer size, is the maximum permissible electric field intensity. With the previously defined maximum electric fields, there result two large classes of piezo transducers:

- *Low-voltage transducers*: maximum voltage typically $100\ \mathrm{V}$, disk thickness typically 20–100 $\mu\mathrm{m}$,
- *High-voltage transducers*: maximum voltage typically $1000\ \mathrm{V}$, disk thickness 0.5–1 mm.

Displacements The achievable actuator displacements for a single disk transducer are in the range of $0.1\ \%$ of disk thickness L. For this reason, in many applications, additional measures must be taken to extend the available displacement, of which two commonly-used options are discussed in more detail below.

Stack transducers: translators

Increasing displacement by cascading For fundamental reasons, piezoelectric transducers directly applying the longitudinal and transverse effects can only realize very small displacements (ca. 0.1 % of the transducer thickness). One straightforward extension for discrete transducers is to spatially cascade multiple transducer elements so that their individual displacements sum.

Stack transducers Fig. 7.7a shows the cascading of N piezo disk transducers using the longitudinal effect, creating a *stack transducer.* The transducer elements are connected *mechanically in series* and *electrically in parallel* (Fig. 7.7b). Note that in this variant, adjacent electrodes are each at the same electrical potential. This solves the problem of insulation without any further effort, as electrodes at different potentials are all separated by the dielectric piezo material.

ELM two-port model Due to the series-parallel cascading of this structure, the same electrical voltage is applied to all transducer elements, and they are all permeated by the same flux. This arrangement can be expediently described by modifying the two-port admittance form of Eq. (7.10) into the *series-parallel form* or *second hybrid form* (Thomas et al. 2009):

$$\begin{pmatrix} X(s) \\ I_t(s) \end{pmatrix} = \mathbf{D}_{pz}(s) \cdot \begin{pmatrix} F_{pz,\Psi}(s) \\ U_t(s) \end{pmatrix} = \begin{pmatrix} -\dfrac{1}{k_{pz,U}} & -\dfrac{K_{pz,U}}{k_{pz,U}} \\ -s\dfrac{K_{pz,U}}{k_{pz,U}} & s \cdot \left[\dfrac{K_{pz,U}^2}{k_{pz,U}} + C_{pz}\right] \end{pmatrix} \begin{pmatrix} F_{pz,\Psi}(s) \\ U_t(s) \end{pmatrix}. \quad (7.21)$$

When cascading N transducer elements, the result is the two-port series-parallel matrix

$$\mathbf{D}_{N,pz} = N \cdot \mathbf{D}_{pz}(s) = \begin{pmatrix} -\dfrac{N}{k_{pz,U}} & -\dfrac{N}{k_{pz,U}} K_{pz,U} \\ -s\dfrac{N}{k_{pz,U}} K_{pz,U} & s \cdot \left[\dfrac{N}{k_{pz,U}} K_{pz,U}^2 + NC_{pz}\right] \end{pmatrix}. \quad (7.22)$$

Then, equating coefficients of Eq. (7.22) and the *two-port admittance matrix* (7.10) straightforwardly gives the *admittance and hybrid two-port parameters* of the N *-stack transducer*

$$\begin{aligned} k_{N,pz,U} &= \frac{k_{pz,U}}{N}, & k_{N,pz,I} &= \frac{k_{pz,I}}{N}, \\ K_{N,pz,U} &= K_{pz,U}, & K_{N,pz,I} &= \frac{K_{pz,I}}{N}, \end{aligned} \tag{7.23}$$
$$C_{N,pz} = N \cdot C_{pz} \;.$$

The mechanical and electrical equivalent circuit diagrams are depicted in Fig. 7.7c and d.

Discussion The multiplication of displacement is achieved at the price of a *reduction* of the *transducer stiffness* according to Eq. (7.23). This in turn reduces the eigenfrequency of the transducer, and thus also limits the range of frequencies having constant transducer sensitivity or gain.

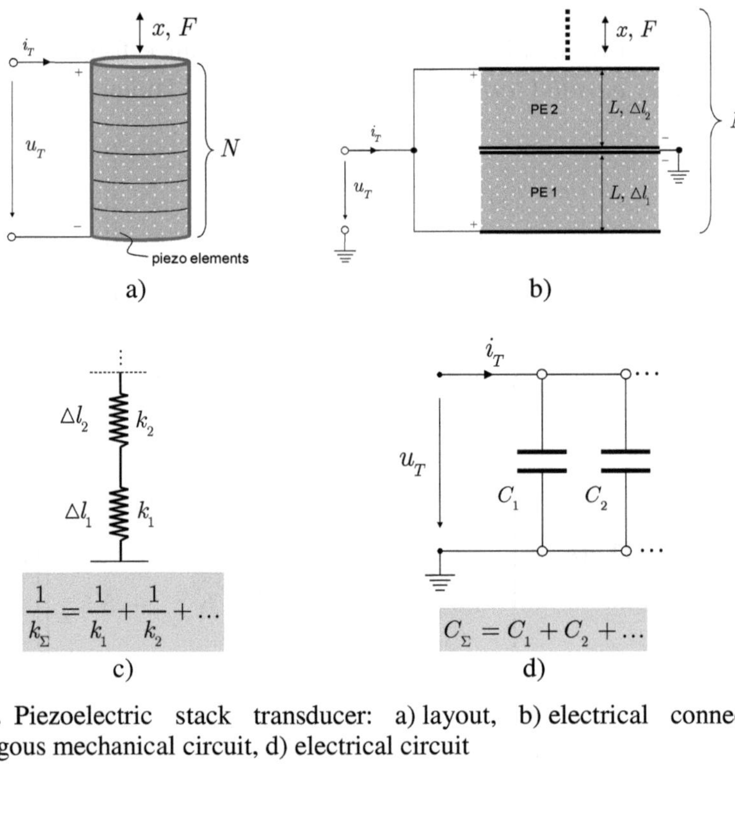

Fig. 7.7. Piezoelectric stack transducer: a) layout, b) electrical connections, c) analogous mechanical circuit, d) electrical circuit

Electrically, there is a *multiplication* of the effective *transducer capacitance*, implying high transducer currents (and demand on the auxiliary energy source).

Evaluating Eqs. (7.19) and (7.23) easily shows that the ELM coupling factor does not change due to cascading, i.e. it is independent of the number of disk elements.

Example 7.1 *Piezoelectric stack transducer.*

A. Single disk

Consider a *high-voltage disk transducer* (made of PZT) with a disk thickness $L = 1$ mm and square cross-section with edge length $D = 10$ mm (area $A = 10^{-4}\ \mathrm{m}^2$).

Transducer parameters From the transducer equations (7.15) through (7.20) and Table 7.2 the following transducer parameters result:

$$k_{pz,U} = 6\ \mathrm{kN}/\mu\mathrm{m}, \quad C_{pz} = 0.8\ \mathrm{nF}, \quad \kappa_{pz} = 0.6\,.$$

The effective transducer capacitance (usually specified in the data sheet) can be found using Eq. (7.20) to be $\tilde{C}_{pz} = 1.2\ \mathrm{nF}$, i.e. decidedly greater than C_{pz} in the braced state.

Voltage drive The *transducer gain* and *maximum displacement* are then

$$V_{x/u} = \text{-}2.7 \cdot 10^{-10}\ \mathrm{m/V} \Rightarrow U_{Q,0} = 1000\ \mathrm{V}: \quad \Delta x_{\max} = 0.27\ \mu\mathrm{m}\,.$$

As can be seen, this rather high voltage ($1\ \mathrm{kV}$) only effectuates a lengthening of 0.027 %.

Current drive The *transducer gain* for the integrating drive behavior of Eq. (7.17) is then

$$V_{x/i} = \text{-}0.22\ \mathrm{m/As}\,. \tag{7.24}$$

B. Stack configuration

In a stack assembly with $N = 10$ disks and resulting transducer thickness $L_{\Sigma} = 10\ \mathrm{mm}$, the following transducer parameters result for *voltage drive*:

$$k_{N,pz,U} = 600\ \mathrm{N}/\mu\mathrm{m}, \quad \tilde{C}_{N,pz} = 12\ \mathrm{nF}, \quad \kappa_{N,pz} = \kappa_{pz} = 0.6\,,$$

$$V_{N,x/u} = \text{-}27 \cdot 10^{-10}\ \mathrm{m/V} \Rightarrow U_{Q,0} = 1000\ \mathrm{V}: \quad \Delta x_{\max} = 2.7\ \mu\mathrm{m}\,.$$

■

Lever transmission

Displacement multiplication One obvious design option for multiplying the displacement is to employ kinematic transmissions. However, more conventional gear train configurations should be avoided so as to retain the advantages of the structural simplicity of solid-state transducers. Thus, simple kinematic structures exploiting only the lever law are often employed in the form of *flexure joints.*

Lever structures Fig. 7.8a depicts the forces and kinematic relationships in a simple lever structure. The rotating joint is realized as a flexure in which the elastic properties of a minimally lossy material are exploited without requiring an intricate structure. Using the *transmission ratio* $r = r_2/r_1$ gives the well-known relations

$$\Delta l_2 = r \cdot \Delta l_1 , \quad F_2 = \frac{1}{r} \cdot F_1 , \quad k_2 = \frac{1}{r^2} \cdot k_1 . \tag{7.25}$$

The desired multiplication of displacement occurs at the cost of a reduction in the available force and a quadratic reduction in stiffness, resulting in a drastic reduction in the transducer eigenfrequency.

A space-saving implementation of a lever transmission is shown in Fig. 7.8b.

Mechanical pre-tensioning

Compression vs. tension Due to their crystal structure, piezoelectric materials are primarily suited for compressive deformation. If shear and tensile stresses become too large, there is a risk of material failure. Thus, the maximum tensile loading of a piezo elements is at most 5-10% of the maximum compressive load (Physikinstrumente 2006).

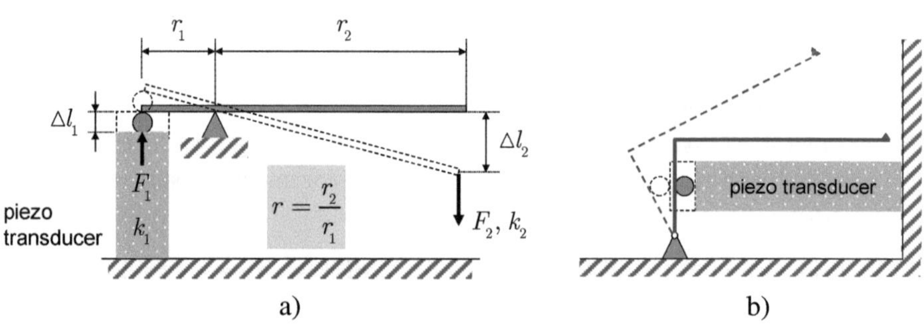

Fig. 7.8. Lever transmission for displacement multiplication: a) schematic arrangement, b) implementation example, from (Physikinstrumente 2006)

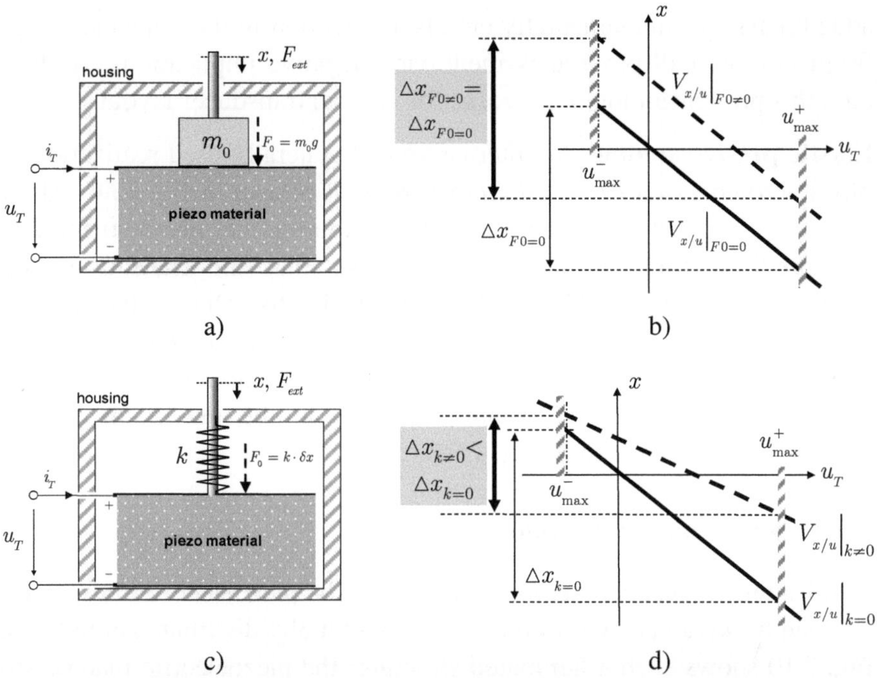

Fig. 7.9. Piezoelectric transducer with mechanical pre-tensioning: a) schematic arrangement for constant weight pre-tensioning, b) static gain $V_{x/u}$ for constant weight pre-tensioning, c) schematic arrangement for elastic pre-tensioning with spring constant k, d) static gain $V_{x/u}$ for elastic pre-tensioning with spring constant k

Pre-tensioning With a constant pre-applied compressive force $F_0 > 0$, a bipolar dynamic force ΔF can be accommodated as the summed force $F_{ext} = F_0 + \Delta F$ acting on the piezo material stays within allowable limits $[F^-_{max}, F^+_{max}]$.

Pre-tensioning with constant weight A constant pre-tensioning force can be realized using the constant weight of an *added load mass* m_0 (Fig. 7.9a). This causes the static voltage/displacement line to translate upward (Fig. 7.9b, translation is equal to the gain $V_{x/u}$). Within the permitted operating voltage range[6], however, the *same displacement* Δx is possible as without pre-tensioning. A disadvantage, however, is that the

[6] For excessively large negative operating voltages (opposing the polarization imposed during manufacture), depolarization would occur, thus $u^-_{max} < u^+_{max}$.

added mass m_0 fundamentally entails a reduction in the transducer eigenfrequency with all previously-mentioned negative implications. Additionally, this pre-tensioning only works for vertical transducer layouts.

Elastic pre-tensioning An alternative and generally-used realization employs a *compressed pre-tensioning spring*, as shown in Fig. 7.9c. The effective spring stiffness is now the sum of the internal piezo stiffness and the pre-tensioning spring stiffness k, so that all transducer equations (7.15) through (7.23) can be applied directly. Due to the increased stiffness, there is a favorable increase in the transducer eigenfrequency, but simultaneously a decrease in the transducer gain or sensitivity. Another disadvantage is the concomitant *reduction* in the maximum *displacement* Δx, as can be seen in Fig. 7.9d.

Strip transducers: contractors

Laminated structures: transverse effect The piezoelectric transverse effect can be advantageously employed in a spatially distributed architecture. Fig. 7.10 shows such a laminated structure: the piezoelectric material (*active material*) takes the form of a *strip element* tightly bound to a substrate. By applying an electrical voltage along the 3-coordinate, the transverse effect (e_{31}, d_{31}) can be used to generate a change in length $\Delta\lambda$ along the orthogonal 1-coordinate. It follows from the constitutive material equations in (S, D) form that

$$\Delta\lambda = d_{31} \frac{\lambda}{\delta} u_T . \tag{7.26}$$

As in the disk transducer (exploiting the longitudinal effect), the realizable displacement depends on the transducer geometry; in particular, it is proportional to the strip length λ.

Since, when subjected to a positive electrical voltage (i.e. an electric field along the polarization direction), the transverse effect induces a contraction ($e_{31} < 0$ and $d_{31} < 0$), strip transducers also often called *contractors*.

Unimorphs, bimorphs Strip transducers in the layout shown are commonly employed in laminated structures. The arrangement shown in Fig. 7.10 is termed a *unimorph*, as only *one* piezoelectric laminated layer is fused to the substrate. A *bimorph* consists of two active laminate layers, optionally separated by a substrate (only suggested in Fig. 7.10).

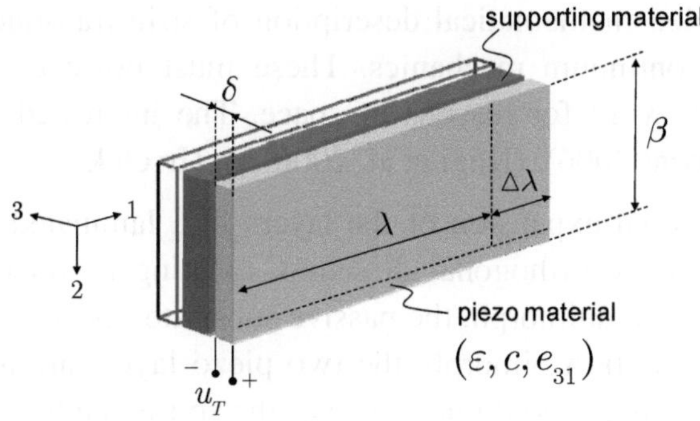

Fig. 7.10. Piezoelectric strip transducer

Electrode shape Using a suitable electrode shape, the type and manner of force generation can be precisely selected. Fig. 7.11 shows two common variants. With a *rectangular* electrode geometry, an applied voltage induces mechanical *bending moments* at the electrode edges (Fig. 7.11a). With a *triangular* electrode geometry, a voltage induces a *point force* at the triangle tip (Fig. 7.11b).

Note further that only the piezo material covered by the electrode area actively contributes to the force/moment. If the transducer is fixed at one end, the bending moments are absorbed by the support.

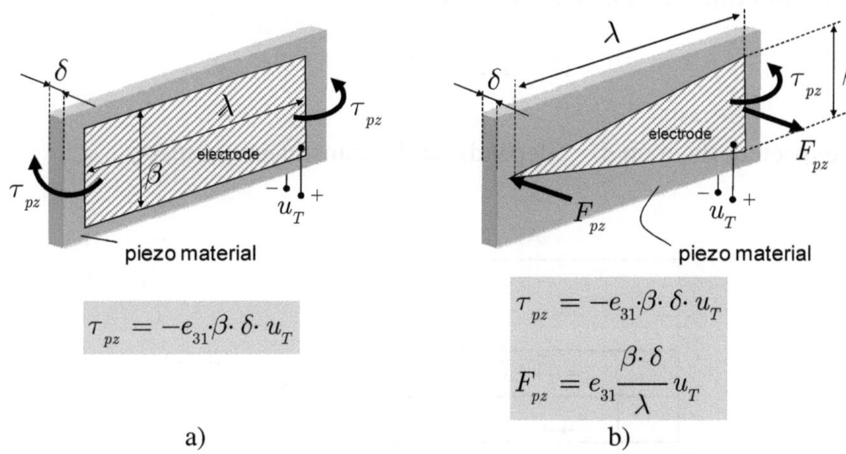

Fig. 7.11. Electrode variants for piezoelectric strip transducers: a) rectangular shape inducing bending moments, b) triangular shape inducing a point force at the electrode tip (pictures from (Preumont 2006))

A more detailed mathematical description of strip transducers requires methods from continuum mechanics. These must however be excluded from the present work for reasons of space. The interested reader is referred to (Preumont 2006), (Kugi et al. 2006), or (Irschik et al. 1997).

Benders An unequal expansion of the layers in a laminated structure induces bending in the orthogonal direction, creating a so-called *bending beam* or *bender*. In a unimorph, the passive substrate acts as an elastic base for the piezo layer. In a bimorph, the two piezo layers are arranged in a suitable complementary configuration, i.e. the top extends while the bottom contracts (see Fig. 7.12).

Cantilever bimorph One oft-used arrangement of piezoelectric strip transducers is the cantilever *bimorph* (Fig. 7.12). For the derivation of its detailed mathematical model, the reader is referred to, for example, (Lenk et al. 2011). The equivalent two-port transducer parameters for a discrete transducer are (Lenk et al. 2011)

$$k_{pz,U} = \frac{1}{4} c_{11}^E \frac{\beta\delta^3}{\lambda^3}, \quad K_{pz,U} = -\frac{3}{4} e_{31} \frac{\beta\delta}{\lambda}, \quad C_{pz} = C^S = 4\varepsilon_{33}^S \frac{\beta\lambda}{\delta} \quad . \tag{7.27}$$

Note that different active areas appear here for the electrical and mechanical field quantities.

As an example, the *static gain* for *actuator operation* under *voltage drive* from Eqs. (7.27), (7.16) is then

$$V_{x/u} = 3 \frac{e_{31}}{c_{11}^E} \frac{\lambda^2}{\delta^2} .$$

As expected, the gain $V_{x/u}$ depends on the transducer geometry.

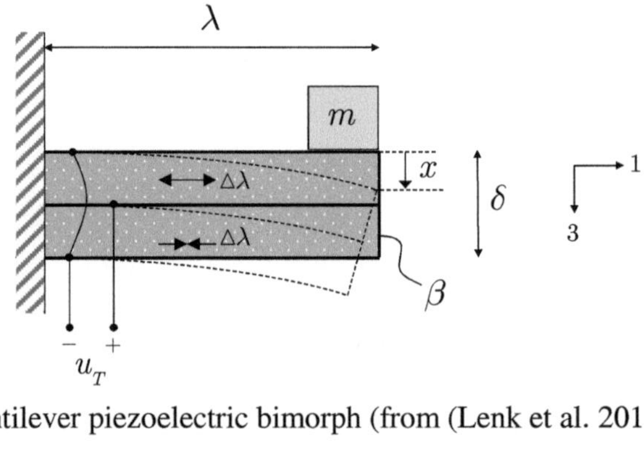

Fig. 7.12. Cantilever piezoelectric bimorph (from (Lenk et al. 2011))

7.4 Transducers with Impedance Feedback

Electrical open circuit: high-voltage generator One particularly dangerous mode of operation for a piezoelectric transducer is with an *electrical open circuit* and simultaneous mechanical excitation by external forces (e.g. shock loading). Due to the polarization, very high voltages can be generated, posing a danger both externally (to users or other devices) and internally (transducer damage).

The induced voltages can be easily evaluated using the current-controlled transducer model of Eq. (7.17), setting $i_S = 0$ (see Fig. 7.6), and examining the transducer gain

$$V_{u/F} = \frac{K_{pz,I}}{k_{T,I}} = \frac{1}{e}\frac{L}{A}\kappa_{pz}^2 \quad [\mathrm{V/N}]. \tag{7.28}$$

For example, for the *disk transducer* considered in Example 7.1[7], a value of $V_{u/F} = 0.22\ \mathrm{V/N}$ results.

Impedance shunting To avoid a state of electrical open circuit, the transducer terminals must be connected with a suitable (low-resistance) impedance $Z(s)$ creating an *impedance shunt.* Fig. 7.13 depicts suitable model configurations for voltage and current drive.

Passive impedance shunting Piezoelectric transducers are peculiar in that an impedance shunt is fully active even without an auxiliary energy source (i.e. with $u_S{=}0$ or $i_S{=}0$); a purely *passive system* results without any need for auxiliary energy, creating a *passive impedance shunt* (see Fig. 7.13). The great attraction of piezoelectric transducers is due not least to this purely passive operating mode.

Layout and application In the simplest case, $Z(s)$ is chosen to be a purely *resistive shunt* (e.g. (Preumont 2006)), adding one parameter to the design space. Concrete circuits, possible extensions using complex impedances $Z(s)$, and potential applications (electromechanical damping, mechatronic resonators, or energy harvesting) for such shunting were already discussed in depth in Ch. 5. All of the considerations pursued there regarding *capacitive* transducer dynamics (e.g. inductive impedances for resonators) can be directly applied to piezoelectric transducers.

---

7 It is left as an exercise to the reader so disposed to show that in a *stack transducer*, the mechanically induced voltage is independent of the number of layers.

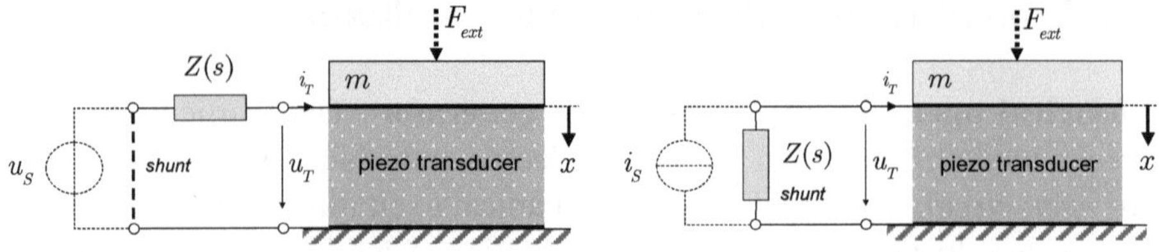

Fig. 7.13. Impedance feedback for a piezoelectric transducer: a) voltage drive, passive impedance shunt with $u_S = 0$; b) current drive, passive impedance shunt with $i_S = 0$

Example 7.2 *Piezoelectric accelerometer.*

Sensor configuration Consider the piezoelectric accelerometer shown in Fig. 7.14a. A disk transducer is fixed to a housing, and mechanically loaded with an inertial mass m. The electrical terminals are connected with a resistance R (*resistive shunting*).

Operation Due to imposed motions $z(t)$ of the housing, the inertial mass undergoes a relative motion $\Delta x = x - z$, resulting in a force on the disk transducer. The induced polarization current i_T (or, equivalently, the voltage drop u_T over the resistance R) is a measure of the imposed housing acceleration $a_z := \ddot{z}(t)$.

Model Compared to the generic piezoelectric transducer in Fig. 7.4, there are only two particularities to account for in applying the previously introduced dynamic models.

Displacement excitation The generic transducer model describes the armature motion *relative* to a *resting* (stator) reference point in a (virtual) inertial system (cf. Example 4.1). In the present case, these models can be easily reused if the motion of the housing (moving reference) is accounted for as an inertial force $F_{ext} = -m\ddot{z}$ [8].

Load resistance The incorporation of electrical load impedances was dealt with in detail in Sec. 5.5. It is easiest to work with the lossy constitutive two-port equations in Table 5.9.

---

[8] This corresponds to the term $m_1\ddot{x}_0$ in Eq. (4.15), where x_0 represents the imposed displacement z.

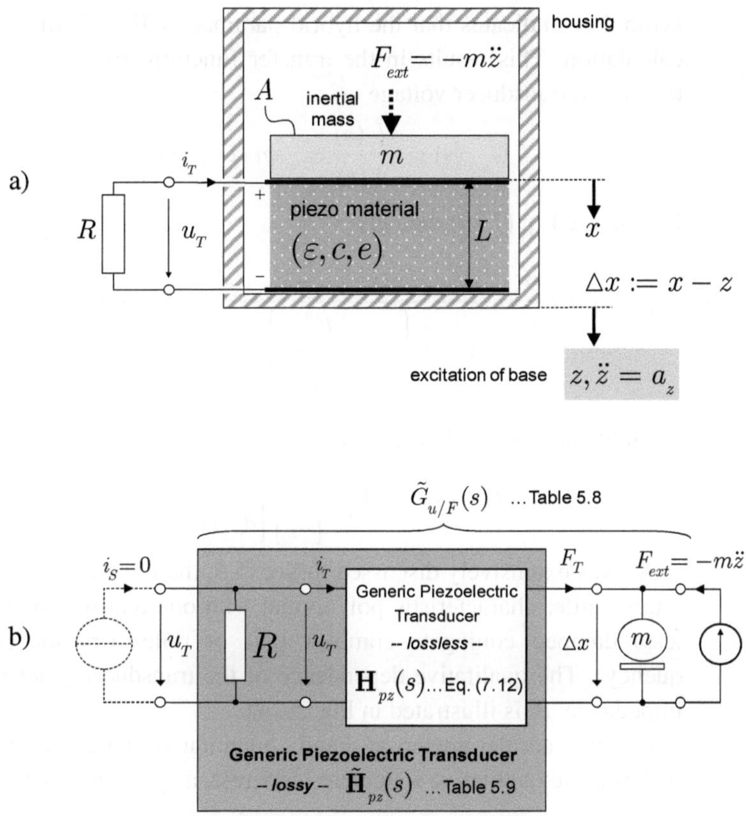

Fig. 7.14. Piezoelectric accelerometer: a) schematic arrangement, b) circuit diagram in two-port hybrid form ($\mathbf{H}_{pz}$ from Eq. (7.12))

In the present case, the hybrid parameters $\mathbf{H}_{pz}$ can be determined using the current-controlled transducer model with source current $i_S = 0$ (Fig. 7.14b). With the hybrid matrix $\tilde{\mathbf{H}}_{pz}(s)$ composed using the two-port parameters $\mathbf{H}_{pz}(s)$ of the lossless transducer in Eq. (7.12) and the load impedance R from Table 5.9, the desired transducer transfer functions can then be directly calculated from the formulas in Table 5.8[9].

Transfer characteristics In the generic transducer notation, it is the transfer function $\tilde{G}_{u/F}(s)$ which is of interest here (Fig. 7.14b), where the

[9] In the present case of a resistive load, this can still be carried out by hand. However, particularly for complex impedances (accounting for parasitic effects using inductances and capacitances), it is advisable to employ a computer algebra tool (e.g. MAPLE, MATHEMATICA).

symbol $\sim$ indicates that the hybrid parameters $\tilde{\mathbf{H}}_{pz}(s)$ were used in the calculation. This results in the transfer function from housing acceleration a_z to transducer voltage u_T

$$G_{u/a}(s) = \frac{U_T(s)}{A_z(s)} = -m \cdot \tilde{G}_{u/F}(s)\,. \tag{7.29}$$

Evaluating Eq. (7.29) then gives

$$G_{u/a}(s) = -mR\frac{K_{pz,U}}{k_{pz,U}} \frac{s}{1 + R\left(C_{pz} + \frac{K_{pz,U}^2}{k_{pz,U}}\right)s + \frac{m}{k_{pz,U}}s^2 + RC_{pz}\frac{m}{k_{pz,U}}s^3}, \tag{7.30}$$

or, in the notation of Sec. 5.5.5,

$$G_{u/a}(s) = -V_{u/a}\frac{s}{\left[\tilde{\omega}_U\right]\left\{\tilde{d}_U, \tilde{\Omega}_U\right\}}\,. \tag{7.31}$$

As was extensively discussed in Sec. 5.5, the transfer function now has a third-order characteristic polynomial with one real pole and a (more or less) damped conjugate complex pair of poles (transducer eigenfrequency). The qualitative dependence of the transducer poles on the load impedance R is illustrated in Fig. 5.24a.

For the accelerometer at hand, quantitative analysis of Eq. (7.30) makes these qualitative statements concrete, as presented below.

(a) *Electrical open circuit* ($R \to \infty$):

$$\lim_{R\to\infty} G_{u/a}(s) = -m \cdot G_{u/F}(s) = -m\frac{\kappa_{pz}^2}{K_{pz,U}}\frac{1}{\left\{\Omega_I\right\}}\,. \tag{7.32}$$

As expected, Eq. (7.32) coincides with the transfer function of the lossy transducer in Eq. (7.17).

(b) *Real transducer pole* $\tilde{\omega}_U$: for frequencies $\omega \ll \sqrt{k_{pz,U}/m} = \Omega_U$, the two right-most terms of the characteristic polynomial in Eq. (7.30) are negligibly small relative to the first two terms, so that in the lower frequency band, the approximation

$$G_{u/a}(s) \approx -mR\frac{K_{pz,U}}{k_{pz,U}} \frac{s}{1 + R\left(C_{pz} + \frac{K_{pz,U}^2}{k_{pz,U}}\right)s} = -mR\frac{K_{pz,U}}{k_{pz,U}}\frac{s}{\left[\tilde{\omega}_U\right]} \tag{7.33}$$

holds, where

$$\boxed{\tilde{\omega}_U :\approx \frac{1}{R\tilde{C}_{pz}}}, \quad \tilde{C}_{pz} = C_{pz} + \frac{K^2_{pz,U}}{k_{pz,U}}. \tag{7.34}$$

The effective capacitance is thus the capacitance of the transducer with a mechanical open circuit (cf. Eq. (7.20)).

(c) *Transducer eigenfrequency*: for sufficiently *large* R, it holds approximately for frequencies $\omega \gg \tilde{\omega}_U$ that

$$G_{u/a}(s) \approx -m \frac{K_{pz,U}}{k_{pz,U}\tilde{C}_{pz}} \frac{1}{[\Omega_I]} = -m \frac{\kappa^2_{pz}}{K_{pz,U}} \frac{1}{[\Omega_I]}, \tag{7.35}$$

so that the transducer eigenfrequency obeys the approximate condition

$$\boxed{\tilde{\Omega}_U \approx \Omega_I}. \tag{7.36}$$

Operational bandwidth of the sensor Eqs. (7.31) through (7.36) result in the typical frequency response depicted in Fig. 7.15. Due to the differentiating dynamics, static accelerations can not be detected. For high frequencies, the transducer eigenfrequency limits operation as a sensor.

In the *frequency band* $\tilde{\omega}_U < \omega < \tilde{\Omega}_U$, there is a frequency-independent *transducer sensitivity* of (g is Earth's gravitational acceleration).

$$\boxed{K_{u/a} = -9.81 \cdot m \cdot \frac{\kappa^2_{pz}}{K_{pz,U}} \quad [\mathrm{V}/g]}. \tag{7.37}$$

Transducer layouts Rearranging Eq. (7.37) according to the physical and structural transducer parameters offers an alternative representation of transducer sensitivity

$$K_{u/a} = -\underbrace{m}_{\text{inertial mass}} \times \underbrace{\frac{L}{A}}_{\text{geometry}} \times \underbrace{\frac{e}{e^2 + c^E \cdot \varepsilon^S}}_{\text{material}}. \tag{7.38}$$

A variety of configuration options for maximizing transducer sensitivity can be discerned from Eq. (7.38). Though *increasing* the inertial mass with a small disk area and large disk thickness increases the transducer sensitivity, all of these measures simultaneously *reduce* the transducer eigenfrequency, and thus the operating bandwidth. The right-most term contains purely *material parameters*. Here, options are limited, as combinations of these parameters can only appear within small regimes of variability. If possible, softer materials with lower permittivity should be preferred (rendering the product $c^E \cdot \varepsilon^S$ small), though a small c^E (soft material) does also reduce the transducer eigenfrequency.

The load resistance R regulates only the *lower bandwidth limit* $\tilde{\omega}_U$, though it does so in a straightforward manner. Thus, large values can be selected for the resistance. However, this is only able to induce very small damping at the transducer eigenfrequency (cf. Fig. 5.24a).

In summary, during design, a *compromise* must be met between *transducer sensitivity* and *working bandwidth.*

Numerical example To illustrate one implementation, consider the disk transducer $(N = 1)$ from Example 7.1 along with an inertial mass $m = 10$ g. The remaining transducer parameters can be calculated as $K_{pz,U} = 1.6$ N/V, $\tilde{C}_{pz} = 1.2$ nF. Evaluating Eqs. (7.34) and (7.36) gives the bandwidth bounds $\tilde{\omega}_U = 6.17$ rad/s (1 Hz) and $\tilde{\Omega}_U = 9.5 \cdot 10^5$ rad/s (150 kHz), so that Eq. (7.37) gives a transducer sensitivity $K_{u/a} = -21.6$ mV/g. Again, the frequency response is depicted in Fig. 7.15.

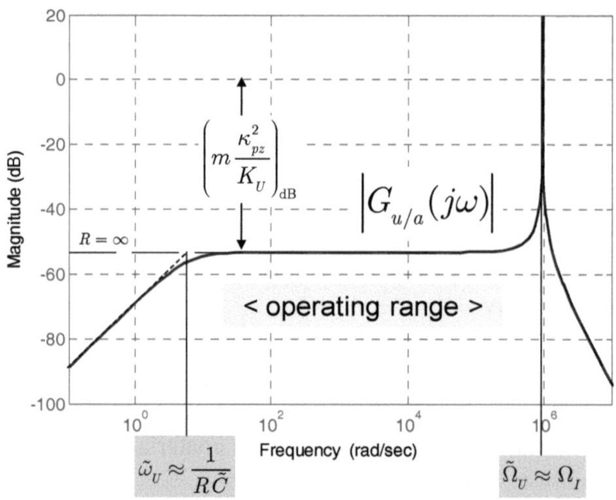

Fig. 7.15. Piezoelectric accelerometer: frequency response from housing acceleration to transducer voltage; numerical values for disk transducer of Example 7.1 with $m = 10$ g.

■

7.5 Mechanical Resonators

7.5.1 Proportional operation vs. resonant operation

Proportional operating regime The frequency response of a piezoelectric transducer is characterized by a generally very weakly damped resonance, which does, however, as a rule lie at rather high frequencies (10–200 kHz). At frequencies below the transducer eigenfrequency, there is *proportional* gain with low phase lag to the armature position (Fig. 7.16). In this frequency range, the transducer can be profitably applied to measurement and actuation tasks.

Static transducer gain The *static* transducer gains of a piezoelectric transducer are generally very small (e.g. $V_{x/u}$ for actuator operation is on the order of 0.1–1 nm/V). Since the operating voltage is also limited to at most a few kilovolts for physical and practical reasons, only very small displacements are generally possible.

Resonant operation: ultrasonic transducer It can be seen from Fig. 7.16, that in the resonant neighborhood of the transducer eigenfrequency substantially higher gain—the *dynamic transducer gain* $V^*_{x/u}$—is available.

Thus, for a constant amplitude of the drive voltage, significantly greater displacements are achieved at this frequency. This feature can be taken advantage of in specialized applications where the transducer is driven in *re-*

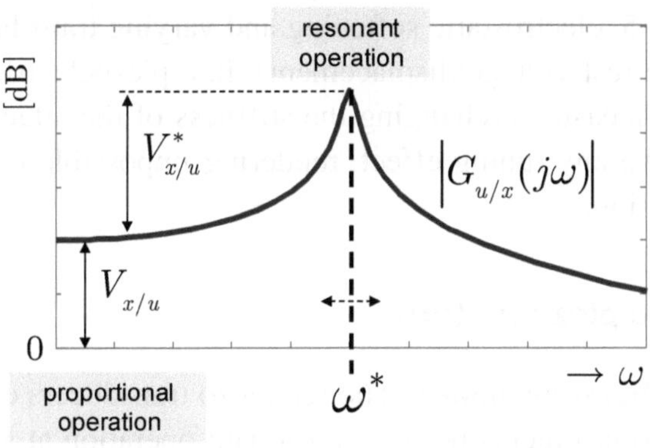

Fig. 7.16. Frequency response of a generic piezoelectric transducer: resonant operation (here as an actuator)

sonant operation within a restricted neighborhood of the transducer eigenfrequency ω^*.

Typical transducer eigenfrequencies lie in the range of 20–200 kHz, so that for resonant operation, acoustic emissions of the transducer lie in the human-inaudible ultrasonic range. Thus, these are also referred to as *ultrasonic transducers*.

Resonance tuning Naturally, for resonant operation, high mechanical oscillator loop efficiency (i.e. the smallest possible mechanical damping) is desired. Thus, the electrical input impedance should be kept as lossless as possible (i.e. only a small internal resistance R is allowed in the voltage source).

However, high oscillator loop efficiency also implies a very narrow resonant peak, so that precise tuning is required for the excitation frequency coming from the voltage source. Piezoelectric solid-state properties are also particularly dependent on temperature (Luck and Agba 1998) so that the resonant frequency ω^* can change during operation and online adaptation of the drive frequency is generally required.

An additional difficulty for resonant operation is posed by imprecise prediction and environment-dependent variation of the dynamic transducer gain $V^*_{x/u}$. For this reason, the transducer cannot meet high-precision demands inherently; if required, positioning accuracy must be ensured with a local control loop. In resonant operation, it is predominantly the high efficiency of the power transfer which is of interest.

There is one additional operational limitation on targeted adjustment (tuning) of the transducer eigenfrequency. In contrast to an electrostatic transducer, here, fine-tuning by selecting the transducer rest voltage is not possible (see Ch. 5, electrostatic softening and varying transducer stiffness depending on the rest voltage/displacement). In a piezoelectric transducer, only mechanical measures (changing the stiffness of the attachment or armature mass) have any tuning effect, rendering impossible online adaptation during operation.

7.5.2 Ultrasonic piezo motors

Concept of an ultrasonic motor Though piezo transducers can only realize very small displacements per se, in resonant operation at a correspondingly high frequency, considerably high velocities can be generated. As a typical example, for an amplitude of $1\,\mu\text{m}$ and resonant frequency of 50 kHz, a velocity of 0.3 m/s is achieved. If this driving velocity is suitably transferred to a linkage for sufficient time, arbitrarily large displacements with very fine granularity can be achieved. Implementation of this principle is particularly straightforward using a *rotor*, leading to the term—borrowing from the action of an electric motor—*ultrasonic motor*.

The first technical implementation of this idea (Barth 1973) is sketched in Fig. 7.17. The longitudinal oscillation of the left and right piezo transducers induce clockwise and counter-clockwise rotation in the rotor, respectively.

Though this idea appears conceptually quite simple, physically implementing and operating such a device is extremely demanding.

This section discusses the important concepts for a basic understanding of *elliptical rotation* and *friction drives*. For a detailed presentation of such drives, the reader is referred to the literature, e.g. (Uchino 1998), (Uchino 1997), (Ueha and Tomikawa 1993).

Traveling wave vs. standing wave To drive the rotor or slider of a motor, a spatially and temporally advancing *traveling wave* (Uchino 1998).

$$p_t(x,t) = A \cdot \cos(kx - \omega t) \tag{7.39}$$

is required, with phase velocity $v_p = \omega/k$.

A *standing wave*, having localized periodic amplitudes, is described by

$$p_s(x,t) = A \cdot \cos kx \cdot \cos \omega t\,. \tag{7.40}$$

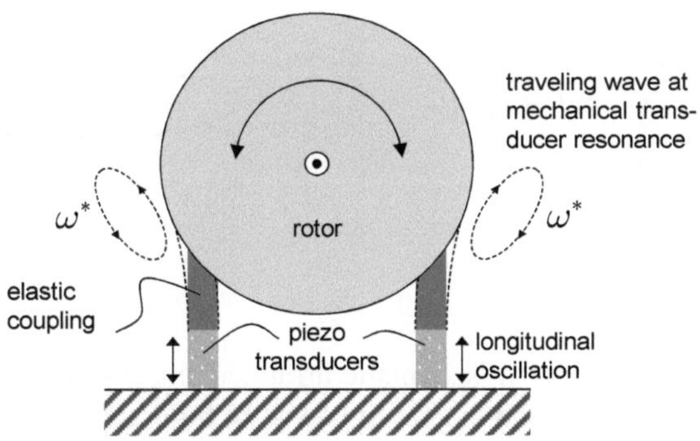

Fig. 7.17. Concept for a piezoelectric ultrasonic motor, from (Barth 1973)

Elementary rearranging of Eq. (7.39) shows that a *traveling wave* can be generated by the *superposition* of two standing waves from Eq. (7.40) with spatial and temporal phase offsets of 90°:

$$p_t(x,t) = A \cdot \cos kx \cdot \cos \omega t + A \cdot \cos\left(kx - \pi/2\right) \cdot \cos\left(\omega t - \pi/2\right). \quad (7.41)$$

As standing waves can be quite easily realized with stationary actuators, Eq. (7.41) represents the key to generating mechanical traveling waves using stationary transducer elements in the smallest of volumes.

Elliptical trajectories Kinematically speaking, for a limited volume, Eq. (7.41) is interpreted as the *elliptical* motion trajectory of a selected actuator point (usually the actuator tip or head), as is elucidated below.

Fig. 7.18a shows the schematic arrangement of a spatially orthogonal actuator configuration, in which temporally orthogonal driving inputs ($\phi = 90°$) precisely fulfill Eq. (7.41). In this case, an *elliptical orbit* results—also termed an *elliptical rotation* (Fig. 7.18b).

Formally, when excited using the same frequency ω^*, this trajectory can be described in a coordinate system (x, y) with a LISSAJOUS figure defined by (for variable definitions, see Fig. 7.18)

$$\frac{x^2}{A_x^2} - \frac{2\cos\phi}{A_x A_y} xy + \frac{y^2}{A_y^2} = \sin^2 \phi .$$

For $\phi = 90°$, the axisymmetric ellipse illustrated in Fig. 7.18b results.

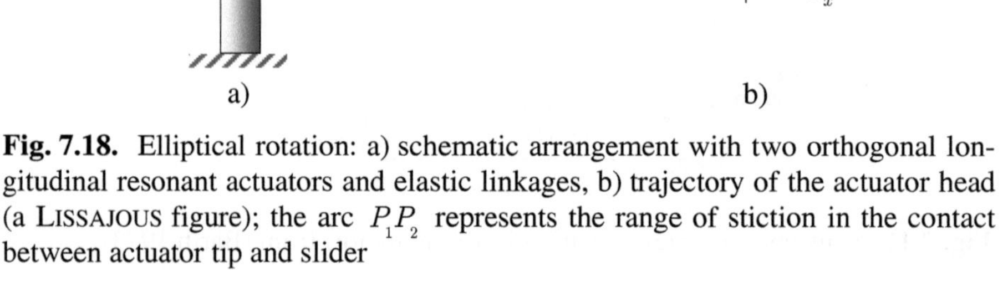

Fig. 7.18. Elliptical rotation: a) schematic arrangement with two orthogonal longitudinal resonant actuators and elastic linkages, b) trajectory of the actuator head (a LISSAJOUS figure); the arc P_1P_2 represents the range of stiction in the contact between actuator tip and slider

One fundamental kinematic obstacle which must be overcome in any implementation can already be recognized in Fig. 7.18a. With a fixed actuator stator, the linkages (also termed *horns* in the literature) must be elastically deformable to permit an elliptical orbit (see also Fig. 7.17).

Resonant elliptical rotation The elliptical orbit shown in Fig. 7.18 can be stimulated at the transducer eigenfrequency ω^*, taking advantage of the resulting large motion amplitudes. For practical implementations, however, two principal difficulties arise in the process.

Spatial and temporal orthogonality can only be maintained with difficultly due to relatively loose *transducer tolerances* (Hemsel et al. 2006), (Bauer 2001). One further critical aspect is the elasticity of the linkages. Elasticity is thoroughly desired, and indeed absolutely necessary for operation. However, at ultrasonic operating frequencies, undesirable eigenmodes of the linkages can become active and induce potential cancellation of the phases of the motions (Bauer 2001).

Longitudinal/transverse coupler One excellent, mechatronically inspired design variant is the combination of a single *longitudinal piezo transducer* (e.g. a stack transducer) with a passive *longitudinal/transverse (L/T) coupler* (Fig. 7.19). The L/T coupler assumes the task of creating a suitable spatial phase relationship between longitudinal and transverse motion. With suitable tuning of the structural parameters and solid-state properties of the transducer, a good accord between all eigenmodes and eigenfrequencies can be achieved (Uchino 1997), (Ueha and Tomikawa 1993).

Modulated friction forces Along with the elliptical armature motion, the functioning of an ultrasonic motor is fundamentally determined by the force transfer via friction forces. Fig. 7.19 (and the simple configuration in Fig. 7.17) illustrates the tight physical contact between actuator tip (armature) and a movable slider (rotor). This contact is absolutely necessary, as ultimately, the normal force F_N generates a stiction force $F_S = \mu_S \cdot F_N$ at the contact point between the actuator tip and rotor/slider surface, with a static coefficient of friction μ_S. The details of the contact mechanics are quite complicated—see e.g. (Wallaschek 1998). To provide a basic understanding of the functioning of an ultrasonic transducer, the most significant effects are sketched out below.

The motion consists of two basic phases: *advancing* the rotor/slider in a *stiction phase* (arc P_1P_2 in Fig. 7.18b) and *retracting* the actuator head in a *sliding friction phase* (arc P_2P_1 in Fig. 7.18b). During the stiction phase, the deformation δx_K increases the normal force—and thus the stiction force F_S—as a function of the stiffness of the L/T coupler and rotor/slider.

By bending the coupler (the desired direction of bending must be ensured using appropriate measures), further longitudinal extension of the actuator applies a pushing force F_L on the slider and thus starts an advancing motion. With a contraction of the actuators, the inertia of the slider prevents it from following the backward motion of the actuator head, so that in the sliding phase, the latter can move back along the slider to point P_1 without loss of displacement in the motor. To prevent slippage of the slider, a pretension force F_0 must be consistently present. With the motion process just presented, the friction forces are thus modulated to achieve a periodic advancing of the slider.

Operation A schematic arrangement for the electrical drive of an ultrasonic motor and the time histories of applicable system variables are shown in Fig. 7.20.

The slider motion is realized by packets of excitations which are generated in a *pulse width modulator* and an *oscillator*. As the transducer requires finite time for starting and stopping oscillations, a *minimal activation time* $\tau_{\min}$ is required, resulting in a *minimum step size* $\Delta y_{\min}$ for the *slider motion* (typically, $\Delta y_{\min} = 5\text{–}50\ \text{nm}$). Achievable *slider speeds* (slope of $y_L(t)$ in Fig. 7.20b) typically lie in the range of $0.1\text{–}1$ m/s. The achievable slider displacement is limited only by the geometry of the structure.

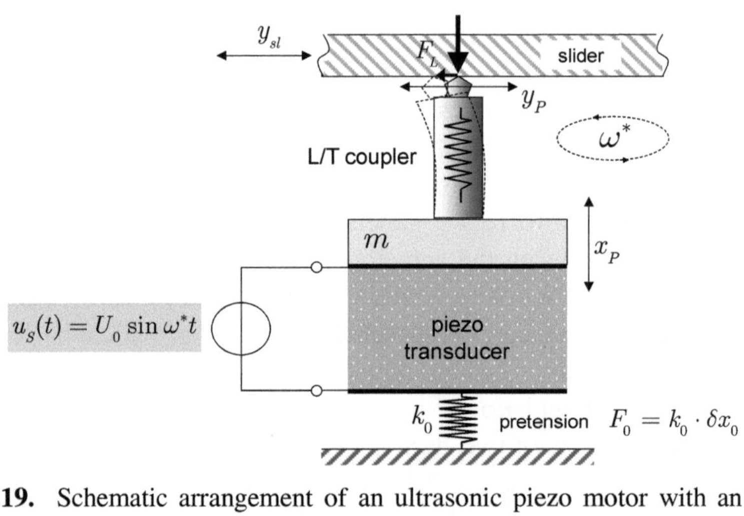

Fig. 7.19. Schematic arrangement of an ultrasonic piezo motor with an elastic longitudinal/transverse (L/T) coupler

As the electrical drive input must be generated at the mechanical transducer eigenfrequency, one unfavorable load dependency must be considered. The eigenfrequency can be approximated by

$$\omega^* = \sqrt{\frac{k_\Sigma}{m_0 + m_L}},$$

where k_Σ represents the effective stiffness of the piezo transducer, coupler, and pre-tensioning; m_0 is the effective mass of the unloaded motor; and m_L is the mass of the slider.

With a varying slider mass m_L, first, the *electric oscillator* must be *adapted* to the new resonant frequency (Fig. 7.20a). Further, for a loaded motor, the slider speed v_L is *reduced* due to the lower resonant frequency by the ratio

$$\frac{v_L}{v_0} = \sqrt{\frac{m_0}{m_0 + m_L}}.$$

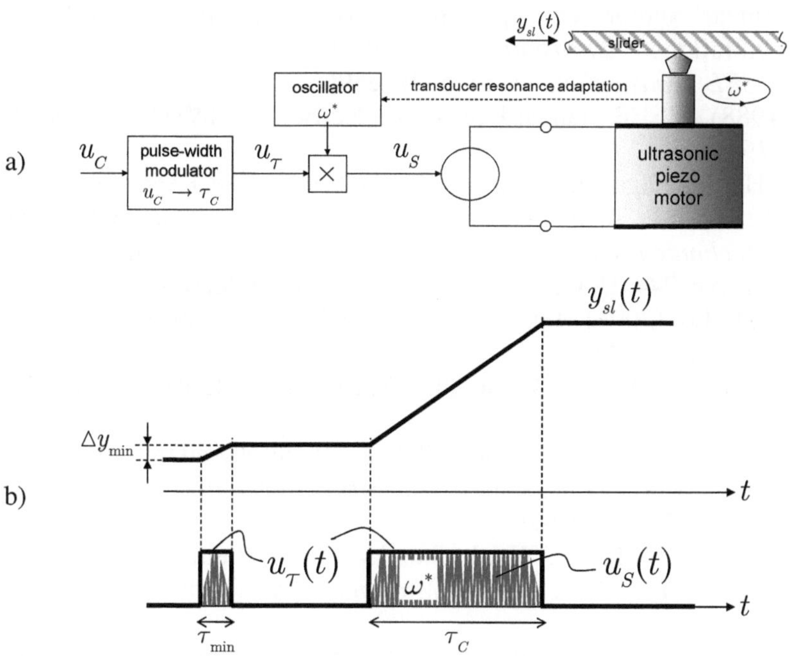

Fig. 7.20. Electrical drive of an ultrasonic piezo motor: a) schematic arrangement, b) time histories of system variables

Structural implementations The structural implementation and systems engineering layout is marked by a plethora of degrees of freedom, so that since the conceptualization of the first prototypes (Barth 1973), a myriad of differing design variants for rotating and linear drives have been imagined. A good overview of these is given for example by (Uchino 1997) or (Hemsel et al. 2006); commercially successful implementations can be found in (Physikinstrumente 2006).

Bibliography for Chapter 7

Barth, H. V. (1973). Ultrasonic driven motor. IBM Tech. Disclosure Bull. **16:** 2263.

Bauer, M. G. (2001). *Design of a Linear High Precision Ultrasonic Piezoelectric Motor.* Mechanical Engineering Department. Raleigh, North Carolina State University. **PhD**.

Bronshtein, I. N., K. A. Semendyayev, G. Musiol and H. Mühlig (2005). *Handbook of Mathematics.* Springer.

Hemsel, T., M. Mracek, J. Twiefel and P. Vasiljev (2006). "Piezoelectric linear motor concepts based on coupling of longitudinal vibrations." *Ultrasonics, Ferroelectrics and Frequency Control, IEEE Transactions on* **44**: e591–e596.

IEEE (1988). "IEEE standard on piezoelectricity." *ANSI/IEEE Std 176-1987.*

Irschik, H., K. Hagenauer and F. Ziegler (1997). An Exact Solution for Structural Shape Control by Piezoelectric Actuation. *Smart Mechanical Systems-Adaptronics, Fortschrittberichte VDI, Reihe 11, Nr. 244*. U. Gabbert. VDIVerlag, Düsseldorf**:** 93-98.

Janocha, H., Ed. (2004). *Actuators*, Springer-Verlag Berlin Heidelberg.

Jordan, T. L. and Z. Ounaies (2001). Piezoelectric Ceramics Characterization, NASA. **NASA/CR-2001-211225 ICASE Report No. 2001-28**.

Kugi, A., D. Thull and T. Meurer (2006). "Regelung adaptronischer Systeme, Teil I: Piezoelektrische Strukturen." *at-Automatisierungstechnik* **54**(6): 259-269.

Lenk, A., R. G. Ballas, R. Werthschützky and G. Pfeifer (2011). *Electromechanical Systems in Microtechnology and Mechatronics.* Springer.

Luck, R. and E. I. Agba (1998). "On the design of piezoelectric sensors and actuators." *ISA Transactions* **37**: 65-72.

Moheimani, S. O. R., D. Halim and A. J. Fleming (2003). *Spatial control of vibration : theory and experiments*. World Scientific, New Jersey

Physikinstrumente. (2006). "Die ganze Welt der Nano- und Mikropositionierung." *Gesamtkatalog*, from www.physikinstrumente.de.

Preumont, A. (2006). *Mechatronics, Dynamics of Electromechanical and Piezoelectric Systems*. Springer.

Senturia, S. D. (2001). *Microsystem Design*. Kluwer Academic Publishers.

Thomas, R. E., A. J. Rosa and G. J. Toussaint (2009). *The Analysis and Design of Linear Circuits*. John Wiley and Sons, Inc.

Uchino, K. (1997). *Piezoelectric Actuators and Ultrasonic Motors*. Boston. Kluwer Academic Publishers.

Uchino, K. (1998). "Piezoelectric Ultrasonic Motors: Overview." *Journal of Smart Materials and Structures* **7**: 273-285.

Ueha, S. and Y. Tomikawa (1993). *Ultrasonic Motors: Theory and Applications*. New York. Oxford University Press Inc.

Wallaschek, J. (1998). "Contact mechanics of piezoelectric ultrasonic motors." *Smart Materials and Structures* **7**: 369-381.

8 Functional Realization: Electromagnetically-Acting Transducers

Background The area of electromechanical energy conversion having the longest history of technical implementation is that exploiting *electromagnetic transduction*. Today, with numerous variations and sophisticated solution approaches, such transducers are present in everyday life, particularly in the macro-mechatronic domain in the form of *high-power* switching, drive, and generator devices, as well as in well-established sensor types. Current areas of research deal with questions of functional integration, e.g. *self-sensing* solutions. Due to the specialized materials required to realize electromagnetic principles, application in the micro-mechatronic world is limited to specialized areas, though showing great potential where applicable. For all such applications and extensions of current functionality, a thorough understanding of the inherent electromagnetic *transduction mechanisms* is essential.

Contents of Chapter 8 In this chapter, the fundamental physical phenomena in and behavioral peculiarities of transducers employing *electromagnetic transduction* are discussed. Based on their underlying power transformation principles, two transducer families are fundamentally distinguished: *electromagnetic (EM) transducers* using *reluctance forces*, and *electrodynamic (ED) transducers* employing the *LORENTZ force*. Starting with MAXWELL's equations, the constitutive transducer equations are derived for both families, permitting direct concretization of the general *generic mechatronic transducer* models of Ch. 5.

For *electromagnetic (EM) transducers* with *variable reluctance*, fundamental dynamic phenomena related to the *working air gap* (variable vs. constant air gaps and nonlinear dynamics) are derived and considered in detail. The class of *differential EM transducers*—which is of great practical importance—is discussed using the example of the actuator functionality of a *magnetic bearing*. The principles and properties of EM transducers with constant air gaps are thoroughly elucidated with a discussion of the nonlinear dynamics of *reluctance stepper motors*.

The dynamics of *electrodynamic (ED) transducers* are discussed in detail for both translational and rotational variants, and further extended for one important implementation type: the *voice coil transducer*.

■

8.1 Systems Engineering Context

Transducers using electromagnetic transduction Everyday products which exploit electromagnetic transduction principles extend far beyond the mechatronic products which are the focus of this chapter (e.g. electromagnetic wave propagation, energy conversion using coupled fields, etc.). In mechatronic applications, it is the conversion between electromagnetic fields and mechanical forces which is of particular importance. As a rule, the mention of electromechanical transducers first brings to mind those using *electromagnetic principles* for electromechanical power conversion. Indeed, this principle of transduction has the longest technological history, and in terms of prevalence in the commercial marketplace, is the most significant and successful energy transformation principle. Its great success lies in the high *energy densities* which can be achieved, opening up manifold applications from the macroscopic to microscopic domains (e.g. electric motors, generators, drives, switches, etc.). However, the impact has been greatest at the *large* to *medium scale*. This is likely due to the fact that effective use of electromagnetic phenomena requires special *ferromagnetic materials* (with high magnetic permeability) and electrical con-

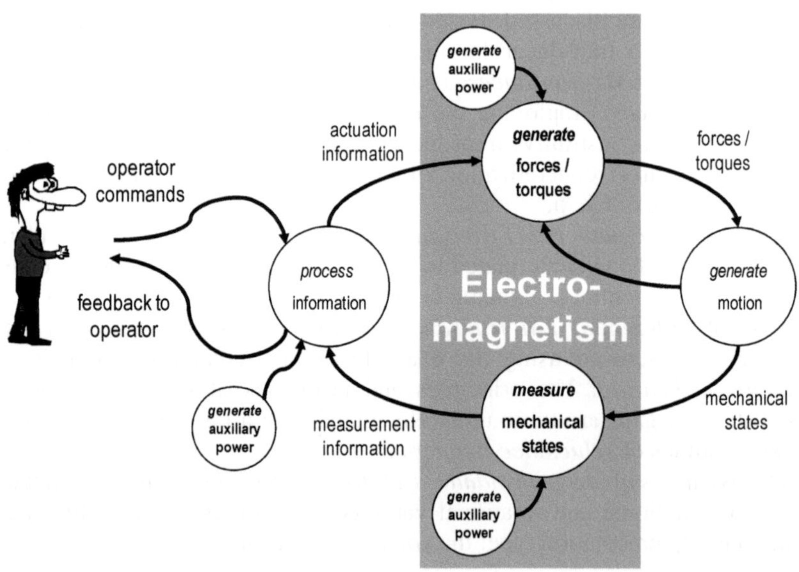

Fig. 8.1. Functional decomposition of a mechatronic system: functional realization using *electromagnetic transduction*

ductor configurations with large physical volumes (for high flux linkage). This favors larger-scale transducer structures. In contrast, in microsystems, such specialized materials must be handled as foreign objects in the semiconductor matrix, and the required volumes fundamentally hamper extreme miniaturization.

Electromagnetic vs. electrodynamic transducers Before embarking on a detailed discussion of the functional and structural principles and dynamic properties of this type of transducer, one clarification of the terminology is absolutely necessary. Electromagnetically-acting transducers are based on two completely different physical phenomena (fundamental laws): (a) *electric field forces at bounding surfaces (variable reluctance)*, and (b) the *LORENTZ force*. As a result, two different classes of transducers must also be distinguished. It is customary to designate transducers with variable reluctance as *electromagnetic (EM) transducers* and transducers based on the LORENTZ force as *electrodynamic (ED) transducers*.

Unfortunately, it has become common in many places to employ "electromagnetic transducer" as a generic term for transducers based on either principle. The ensuing confusion of terms is unmistakable. It is true that both transducer principles employ the physical phenomenon of *electromagnetic transduction*, as described by the MAXWELL's equations and the LORENTZ force. To establish a clear and unambiguous terminology, this book thus employs the *superordinate term "electromagnetically-acting transducer"* and designates the subordinate transducer types as *electromagnetic (EM) transducers* and *electrodynamic (ED) transducers* (Fig. 8.2).

In fact, it would be more consistent and unambiguous to use terms such as *reluctance transducer* or *LORENTZ transducer*, directly describing the fundamental physical principle being exploited[1].

Systems engineering significance From a systems engineering point of view, the functions *"generate forces/torques"* and *"measure mechanical states"* realizable with electromagnetic phenomena represent the *actuators* and *sensors* of a mechatronic system (see Fig. 8.1).

For both tasks, it is the transfer characteristics in the causal directions shown in Fig. 8.1 which are of interest. For optimal energy conversion,

---

[1] In the case of a few particular transducer types, this terminology has indeed established itself, e.g. the reluctance motor or LORENTZ actuator.

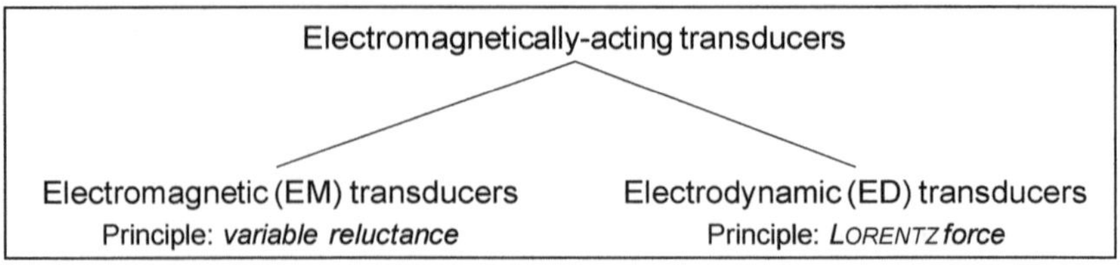

Fig. 8.2. Electromagnetically-acting transducers: transducer classes

specialized structures are required to guide magnetic flux and to create the requisite electrical conductor geometry; at least one movable transducer element (the armature) must be connected to the mechanical structure which is to be moved (for an actuator) or which is moving (for a sensor). In addition to functional attributes such as linearity and dynamic properties, it is structural parameter dependencies affecting the transfer characteristics which play a particularly important role in controller design.

Mechatronic phenomena As described above, electromagnetically-acting transducers exploit either *reluctance forces*—i.e. forces at bounding surfaces between differing magnetic reluctances (which results in a geometry-dependent *transducer inductance*)—or the *LORENTZ force* on a current-carrying conductor in a magnetic field. The dynamic properties of force generation are thoroughly different for these phenomena: they are inherently *nonlinear* for *EM transducers*, and generally *linear* for *ED transducers*. In both cases, an *air gap* (or working air gap) permeated with magnetic flux is required, either directly for force generation (EM transducers) or to allow for coil motion (ED transducers). Thus, the transducer geometry plays a determining role in transducer functionality and performance properties. Thus, for systems design, the following model relationships are of interest:

- *inductance* of the transducer configuration as a function of geometry and materials;
- *electromagnetic forces* (reluctance force, LORENTZ force) as a function of geometry, materials, and electrical drive;
- *transfer characteristics* including mechanical power feedback.

It will be shown that for the electromagnetic *reluctance transducer*, there exists a notable phenomenological and modeling equivalence to the *electrostatic* transducer. However, the physical equations are significantly more complex in the electromagnetic case.

8.2 Physical Foundations

Electromagnetic fields As is well known, moving electric charges in the form of an *electric current* $i(t)$ (including the technically important case of a *line current* $\vec{i}$ within a spatially-distributed electric conductor) generate a magnetic excitation in their neighborhood, characterized by a spatially-varying *magnetic field* $\vec{H}(\vec{r},t)$. This field exists—with a varying, material-dependent effect—in the entire neighborhood of the electric conductor. In turn, time variation of the magnetic field induces a spatially-varying electric field $\vec{E}(\vec{r},t)$ in the neighborhood of the electric conductor. The spatially-varying electric and magnetic fields are thus coupled and can be exploited for engineering purposes in a variety of ways with suitable spatial arrangement of the exciting line currents.

MAXWELL's equations for quasi-stationary fields The fundamental physical relationships of these phenomena are defined by *MAXWELL's equations* for *quasi-stationary* fields[2] (Jackson 1999). The equations applying to the magnetic field[3] are, in integral form[4], (see Fig. 8.2)

AMPÈRE's law:
$$\oint_{s_H} \vec{H} \cdot d\vec{s} = \int_{A_H} \vec{G}_\kappa \cdot d\vec{A}\,, \tag{8.1}$$

FARADAY-MAXWELL equation:
$$\oint_{s} \vec{E} \cdot d\vec{s} = -\frac{d}{dt}\int_{A} \vec{B} \cdot d\vec{A}\,, \tag{8.2}$$

Continuity equation:
$$\oint \vec{B} \cdot d\vec{A} = 0\,, \tag{8.3}$$

Material equation:
$$\vec{B} = \mu \vec{H}\,, \tag{8.4}$$

---

[2] Meaning slowly-varying fields. Given such, the free current density $\vec{G}_\kappa$ will be significantly greater than the displacement current density $\partial\vec{D}/\partial t$, so that the latter can be neglected in the formulation given in Eq. (8.1).

[3] The FARADAY-MAXWELL equation in Eq. (8.2) is not explicitly required to derive the elementary transducer equations. The corresponding induction relation results automatically from the LAGRANGE equations.

[4] See Footnote 2 in Ch. 6.

with the quantities

- *conductor current density* (free current density) $\vec{G}_\kappa$ [A/m²],
- *electrical field intensity* $\vec{E}$ [V/m],
- *magnetic field intensity* $\vec{H}$ [A/m],
- *magnetic flux density, induction* $\vec{B}$ [Vs/m²=:Tesla=T],
- *permeability* $\mu = \mu_r \mu_0 \quad [V \cdot s/A \cdot m]$,

  $\mu_0 = 4\pi \cdot 10^{-7} \; V \cdot s/A \cdot m , \qquad \mu_r \geq 1$ is material-dependent.

From AMPÈRE's law (8.1), it follows that a magnetic field is a rotational field, and always arises from moving charges (in permanent magnets this happens via molecular currents and the spatial arrangement of WEISS domains (Jackson 1999)). The action of the magnetic field in the surrounding medium is described by the material equation (8.4). The continuity equation (8.3) expresses the non-existence of magnetic field sources (monopoles), i.e. magnetic field lines are always closed (as opposed to electrostatic fields, where field lines begin and end at electric charges).

To ease the use of Eq. (8.1), the *magnetic potential difference* u_m [A] is defined as

$$u_{m,12} := \int_{P_1}^{P_2} \vec{H} \cdot d\vec{s} \; . \tag{8.5}$$

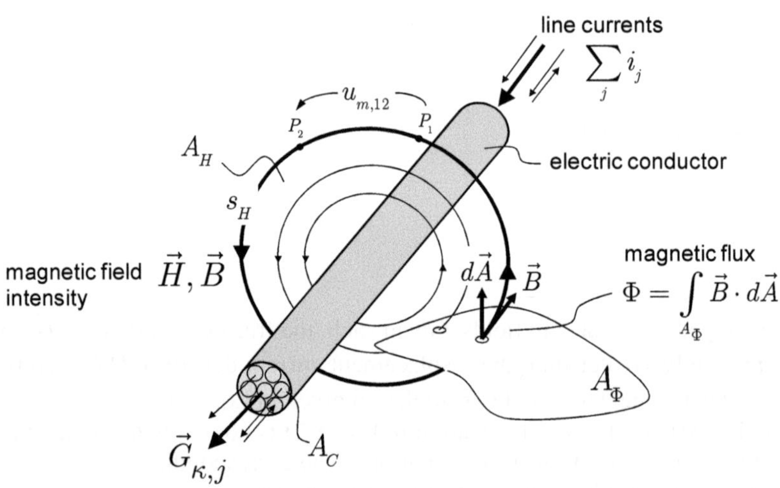

Fig. 8.2. Magnetic field around parallel linear electric conductors

Line currents AMPÈRE's law (8.1) describes how a magnetic field with a spatially-distributed field intensity $\vec{H}(\vec{r})$ arises from the flow of an electric current. In this book, it is currents carried in essentially one-dimensional conductors—so-called *line currents*—

$$i = \int_{A_L} \vec{G}_\kappa \cdot d\vec{A}_C$$

which are of interest.

Magnetomotive force (MMF) From AMPÈRE's law (8.1), it further follows that the path integral over a closed magnetic field line of length s_H equals the sum of line currents encircled by the field line (see Fig. 8.2), i.e.

$$\oint_{s_H} \vec{H} \cdot d\vec{s} = \int_{A_H} \vec{G}_\kappa \cdot d\vec{A} = \sum_j i_j =: \mathfrak{F} . \tag{8.6}$$

The quantity $\mathfrak{F}$ [A] or [AT, ampere-turns] is termed the *magnetomotive force (MMF)*

Magnetic flux Using the magnetic flux density $\vec{B}$, the material equation (8.4) describes the action of a magnetic field in a medium (e.g. air or iron). The total flux density penetrating an area A_Φ is termed the *magnetic flux* Φ $[\mathrm{V \cdot s = Weber = Wb}]$ (see Fig. 8.2), where

$$\Phi := \int_{A_\Phi} \vec{B} \cdot d\vec{A} . \tag{8.7}$$

Homogeneous magnetic field AMPÈRE's law (8.1) describes the relationship between the exciting current and the *contour integral* of the magnetic field intensity, not the field intensity itself. However, Eq. (8.1) still forms the basis magnetic field calculations (Jackson 1999), (Hughes 2006). For simple, symmetric configurations, it can be straightforwardly employed to derive predictive design models. One often easily met assumption is that of a *homogeneous* magnetic field, i.e. the field variables $\vec{H}$ and $\vec{B}$, or Φ, are independent of location (constant) within a bounded domain. For example, this is the case for the interior of a cylindrical coil (Fig. 8.3a), or within a homogeneous ferromagnetic material ($\mu_r \gg 1$) (Fig. 8.3b). In the latter case, the field lines primarily flow inside the ferromagnetic material, so that magnetic flux can be easily guided through space using physical structures.

Magnetization curves For real flux-conducting media ($\mu_r \gg 1$, ferromagnetic materials), the material equation (8.4) does not describe a linear relationship: it is well known that both *saturation* and *hysteresis effects* must be taken into account (Fig. 8.4a). The permeability μ is thus not constant, but depends on magnetic saturation (Fig. 8.4b) (Hughes 2006).

The saturation effect—characterized by the saturation flux density B_S—materially limits the maximum achievable flux density, and thus the maximum possible transducer force (typical values are $B_S = 1.2 - 1.7\ \mathrm{T}$, (Kallenbach et al. 2008)).

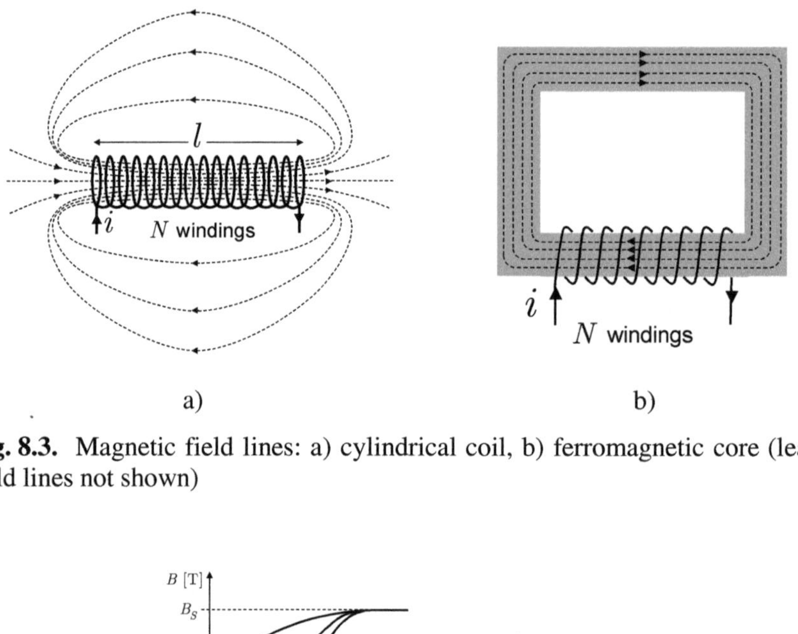

Fig. 8.3. Magnetic field lines: a) cylindrical coil, b) ferromagnetic core (leakage field lines not shown)

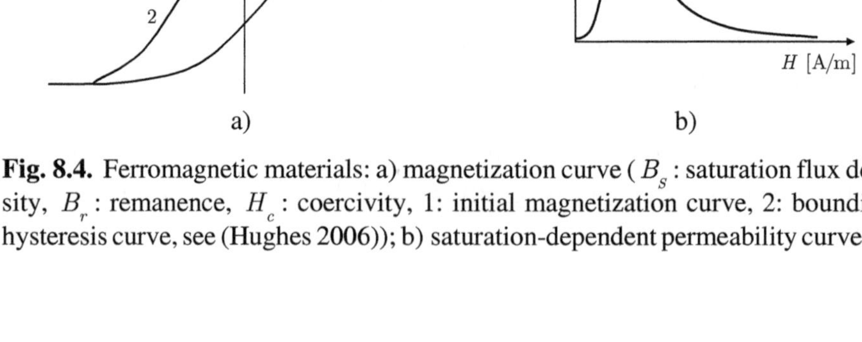

Fig. 8.4. Ferromagnetic materials: a) magnetization curve (B_S: saturation flux density, B_r: remanence, H_c: coercivity, 1: initial magnetization curve, 2: bounding hysteresis curve, see (Hughes 2006)); b) saturation-dependent permeability curve

Traversing the hysteresis curve during dynamic operations (with varying flux) results in a dissipation of energy (a transformation into heat). If present, such effects must be accounted for as losses in a transducer's dynamic model (Schweitzer and Maslen 2009).

Magnetic resistance: reluctance In their role as active quantities in the generation and spatial propagation of a magnetic field, the *magnetic potential difference* $u_{m,12}$ from Eq. (8.5) and the *magnetic flux* Φ from Eq. (8.7) play important roles as the system *effort* and *flow variables*, respectively. In analogy to electric circuits, this immediately suggests introducing the ratio between these two quantities as the *magnetic resistance* or *reluctance* R_m :

$$R_m := \frac{u_{m,12}}{\Phi} \quad [\mathrm{A/V \cdot s}]. \tag{8.8}$$

In a *homogeneous magnetic field* (with field line length l and cross-section A), R_m is then a constant, where

$$R_m = \frac{u_{m,12}}{\Phi} = \frac{Hl}{\mu HA} = \frac{l}{\mu A}. \tag{8.9}$$

It can be seen from Eq. (8.9) that the reluctance depends only on geometric parameters and the magnetic properties of the material in which the magnetic flow occurs.

Magnetic circuit, magnetic network For spatially distributed configurations with a *piecewise-homogeneous* field, using the reluctance allows a *magnetic network model* or *magnetic circuit* to be set up analogously to an electrical network (Fig. 8.5). For each spatially homogeneous section, a reluctance can be defined as a lumped network element.

The following two identical relations result from AMPÈRE's law (8.1) for the configuration in Fig. 8.5a:

$$\int_{P2}^{P1} \vec{H}_{Fe} \cdot d\vec{s} + \int_{P1}^{P2} \vec{H}_{\delta} \cdot d\vec{s} = u_{m,Fe} + u_{m,\delta} = \mathfrak{F} = Ni\,, \tag{8.10}$$

$$H_{Fe} \cdot l_{Fe} + H_{\delta} \cdot \delta = \Phi_{Fe} \frac{l_{Fe}}{\mu_{Fe} A_{Fe}} + \Phi_{\delta} \frac{\delta}{\mu_0 A_{\delta}} = \Phi_{Fe} R_{Fe} + \Phi_{\delta} R_{\delta} = \mathfrak{F}\,. \tag{8.11}$$

Analogously to electrical networks, the formulation in Eq. (8.10) can be interpreted as *KIRCHHOFF's mesh rule* for a magnetic circuit (with magnetic potential differences u_m, and the MMF $\mathfrak{F}$ as a magnetic potential source, see also Fig. 8.6a).

From the formulation in Eq. (8.11), the equivalence of magnetic flux to electric current in an electrical network can be seen.

KIRCHHOFF's *node rule* for a magnetic circuit is depicted in Fig. 8.6b.

Relative to the excitation terminals $\mathfrak{F}_S, \Phi_S$, the magnetic behavior of the entire system can be compactly described by the reluctance $R_{m\Sigma}$ at the terminals. This property ultimately offers the basis for a concise formulation of the generic transducer equations.

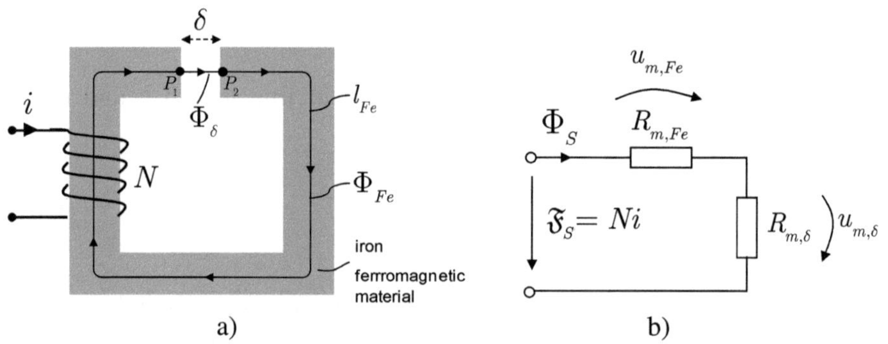

Fig. 8.5. Magnetic circuit, representative configuration: a) physical configuration, b) magnetic network with lumped parameters

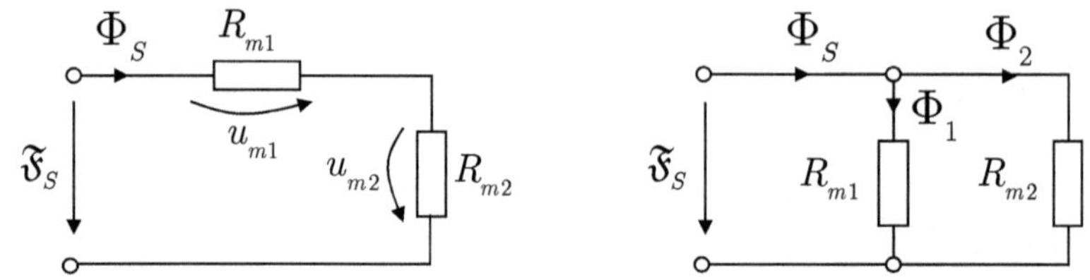

$$R_{m\Sigma} = \frac{\mathfrak{F}_S}{\Phi_S} = R_{m1} + R_{m2}$$

a)

$$\frac{1}{R_{m\Sigma}} = \frac{\Phi_S}{\mathfrak{F}_S} = \frac{1}{R_{m1}} + \frac{1}{R_{m2}}$$

b)

Fig. 8.6. Magnetic circuit, magnetic reluctances: a) series circuit (KIRCHHOFF's mesh rule), b) parallel circuit (KIRCHHOFF's node rule)

Magnetic circuit with air gap For the example configuration in Fig. 8.5a, neglecting leakage flux (i.e. $\Phi_{Fe} \approx \Phi_\delta = \Phi_S$),

$$\mathfrak{F}_S = \Phi_{Fe} R_{m,Fe} + \Phi_\delta R_{m,\delta} \approx \Phi_S \left(R_{m,Fe} + R_{m,\delta} \right), \tag{8.12}$$

and, considering customary material properties $\mu_{Fe} \gg \mu_0$,

$$R_{m\Sigma} = R_{m,Fe} + R_{m,\delta} = \frac{l_{Fe}}{\mu_{Fe} A_{Fe}} + \frac{\delta}{\mu_0 A_\delta} \approx \frac{\delta}{\mu_0 A_\delta}. \tag{8.13}$$

In a magnetic circuit with an air gap, the reluctance is thus *primarily* determined by the *air gap*. The highly permeable iron core offers negligible resistance to the magnetic flux, and simply guides the magnetic field lines along desired spatial paths.

If several air gaps are present, their reluctances should be added corresponding to the flow topology (in series or parallel).

Flux linkage The electric feedback of a time-varying magnetic field is described by the FARADAY-MAXWELL equation (flux density law) (8.2). Given a concrete spatial configuration (here a coil conductor with $k = 1, \ldots, N$ turns and a closed path s along the electric field lines with enclosed area A), this results in

$$\begin{aligned} u_{ind} &= \oint_s \vec{E} \cdot d\vec{s} = -\frac{d}{dt} \int_{A=\sum_k A_k} \vec{B} \cdot d\vec{A} = -\frac{d}{dt} \sum_k \int_{A_k} \vec{B} \cdot d\vec{A} = \\ &= -\frac{d}{dt} \sum_k \Phi_k \triangleq -\frac{d\Psi}{dt}. \end{aligned} \tag{8.14}$$

Eq. (8.14) indicates that, given temporal variation of the magnetic flux in a conductor loop, an electric voltage E_{ind} (the electromotive force or EMF) is generated (induced) proportional to the magnetic flux $\sum \Phi_k$ linked to the k current loops (s, A). The magnetic flux Φ_k describes the flux encircled by the kth winding with area A_k.

In this sense, the (magnetic) *flux linkage* Ψ is defined as

$$\Psi := \sum_{(s,A),k} \Phi_k \quad [\mathrm{V} \cdot \mathrm{s} = \mathrm{Weber} = \mathrm{Wb}]. \tag{8.15}$$

For example, for a helically-wound current loop with N *windings* (a coil with $A_k = A_W,\ k = 1,\dots,N$) and a magnetic flux of Φ_W per winding, the flux linkage according to Eq. (8.15) is

$$\Psi_N = N \cdot \Phi_W \,. \tag{8.16}$$

Flux linkage is thus *the* determining quantity for the voltage E_{ind} (the sum of the electric winding voltages) induced by the magnetic field and for the description of the countering electromagnetic action on an electric conductor in a magnetic field.

The total magnetic flux includes both the magnetic flux $\Psi_\Theta(i)$ induced electrically via a *magnetomotive force (MMF)* $\mathfrak{F}(i)$, and magnetic flux Ψ_0 coming from *permanent magnets*. Formally, then, the *flux linkage* can be written as

$$\boxed{\Psi = \Psi_0 + \Psi_\Theta(i)} \,. \tag{8.17}$$

Eq. (8.17) now represents the fundamental *electrical constitutive* relation for the energy-based transducer model introduced in Ch. 5.

Inductance From AMPÈRE's law (8.1) it follows that the magnetic flux Φ_Θ generated in a current loop—and thus also the magnetic flux linkage Ψ_Θ—is proportional the instantaneous current i at any point in time. This permits the following general statement to be formulated for the relationship between exciting current i and flux linkage Ψ_Θ:

$$\Psi_{\mathfrak{F}}(i) = L(\text{geometry} + \text{material},\ i) \cdot i \,. \tag{8.18}$$

The proportionality constant L $[\mathrm{V \cdot s/A} = \mathrm{Henry} = \mathrm{H}]$ is termed the *inductance* of the configuration.

The inductance concisely describes the electromagnetic properties of an electromechanical system. The mechanical component is reflected in the dependence of the inductance on the configuration geometry. Given a variable geometry, the inductance can thus be made to depend on motion variables (defining the electromechanical energy transfer). The electrical back effects of the magnetic field are expressed via the appropriate material properties (the permeability μ).

As a rule, nonlinear properties are accounted for via current-dependent saturation effects, i.e. $L = L(i)$, see Fig. 8.7 (cf. Fig. 8.4). There is then a distinction between the operating-point-dependent static inductance L_{st}^*

and the differential inductance L_d^* (see Fig. 8.7). The dynamics in the linear portion of the $\Psi - i$ curve are termed *electrically linear*, i.e. $\Psi = L_0 \cdot i$ (in this regime, energy = co-energy).

For the electromechanical transducers considered here, inductance generally depends both on an electrical coordinate (the current i) and a mechanical coordinate (the displacement x). Formally expressed,

$$L = L(x, i)\,. \tag{8.19}$$

Inductance vs. reluctance To calculate the inductance of a physical electromagnetic system (magnetic circuit), its relationship with reluctance can be used to great advantage (Thomas et al. 2009), (Hughes 2006).

Often, the following *assumptions* can be easily made to hold:

1. piecewise-homogeneous magnetic fields;
2. helical coil conductor;
3. all N coil windings encircle the same flux, i.e. $\Psi_\Theta = N \cdot \Phi_W$;
4. reluctance is easily computed (to a sufficient approximation).

Then, following Eq. (8.8),

$$R_m = \frac{\mathfrak{F}}{\Phi} = \frac{Ni}{\Phi} \quad \Rightarrow \quad \Phi = \Phi_W = \frac{Ni}{R_m},$$

and, substituting into Eq. (8.16),

$$\Psi_N = N \cdot \Phi_W = \frac{N^2}{R_m} i = L \cdot i\,.$$

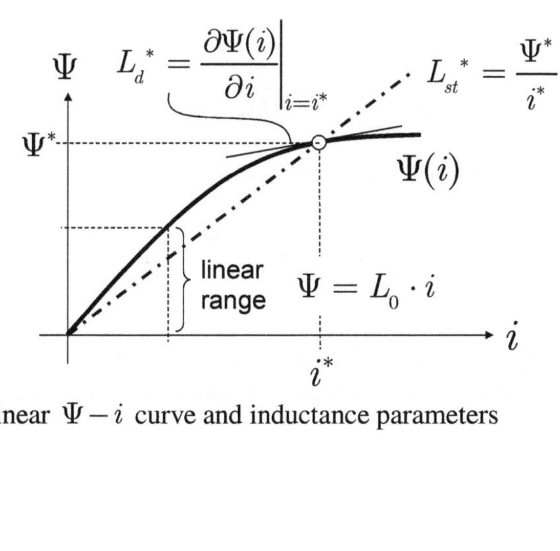

Fig. 8.7. Nonlinear $\Psi - i$ curve and inductance parameters

Thus, given a known reluctance $R_{m\Sigma}$ for the *total configuration*, it follows that the inductance is

$$\boxed{L = \frac{N^2}{R_{m\Sigma}}}. \tag{8.20}$$

If the assumptions listed above are valid, the inductance of any configuration can thus be determined as a function of the reluctance in a relatively simple manner.

Inductance for magnetic circuits with air gaps: reluctance transducers For a magnetic circuit with an air gap (corresponding to the configuration in Fig. 8.5a), the inductance at the electric terminals is

$$L = N^2 \frac{\mu_0 A_\delta}{\delta}. \tag{8.21}$$

Eq. (8.21) illustrates the typical dependence of inductance on structural geometric parameters in electromagnetic (EM) transducers. By varying the geometry of the air gap (varying δ or A_δ using a movable armature), the reluctance of the magnetic circuit—and thus the inductance of the system—can be varied, resulting in a dependence $L(\delta, A_\delta)$ and variable electromagnetic properties. This also gives rise to the alternate term for electromagnetic (EM) transducers: *reluctance transducers*.

Electrodynamic force law: the LORENTZ force In addition to the natural laws expressed by MAXWELL's equations, the *LORENTZ force law* (or *electrodynamic force law*) for a *moving charge* in a *magnetic field* is of great importance to electromagnetically-acting transducers (Thomas et al. 2009), (Hughes 2006).

Using the LORENTZ force law[5]

$$\vec{F}_L = q\left(\vec{v} \times \vec{B}\right), \tag{8.22}$$

it follows that the differential force on a current element (see Fig. 8.8) is

$$d\vec{F}_{ed} = dq\left(\vec{v} \times \vec{B}\right)\Big|_{dq\cdot\vec{v}=i\cdot d\vec{s}} = i\left(d\vec{s} \times \vec{B}\right).$$

---

[5] The second component of the LORENTZ force—the force $\vec{F} = q\vec{E}$ applied by an electric field on an electric charge—is not important here, and is thus not further considered.

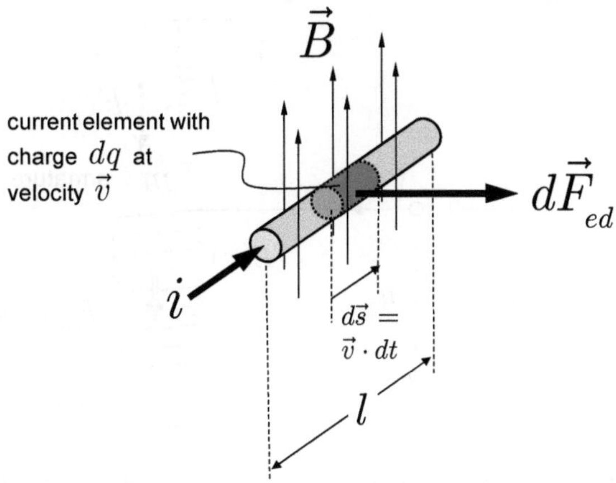

Fig. 8.8. LORENTZ force (electrodynamic force) $\vec{F}_{ed}$ on a moving charge in a magnetic field

The total force on a current-carrying one-dimensional conductor of length l (with arbitrary curvature and spatially varying $\vec{B}$) then obeys the general *electrodynamic force law*

$$\vec{F}_{ed} = i \int_l \left(d\vec{s} \times \vec{B} \right) . \tag{8.23}$$

For the special case of a *rectilinear conductor* of length l in a homogeneous magnetic field,

$$\vec{F}_{ed} = l \left(\vec{i} \times \vec{B} \right), \tag{8.24}$$

where i is directed along $\vec{l}$.

8.3 Generic EM Transducer: Variable Reluctance

8.3.1 System configuration

Magnetic circuit The principle of electromagnetic force generation (or the reluctance force) is based on *variable reluctance* in an electrically excited magnetic circuit. The variable reluctance is generated by a moving *ferromagnetic armature* located in an air gap of a generally fixed component—the stator—as part of the magnetic circuit.

The structures which form the bounding surfaces between the ferromagnetic material and the air gap and which determine the magnetic flux are

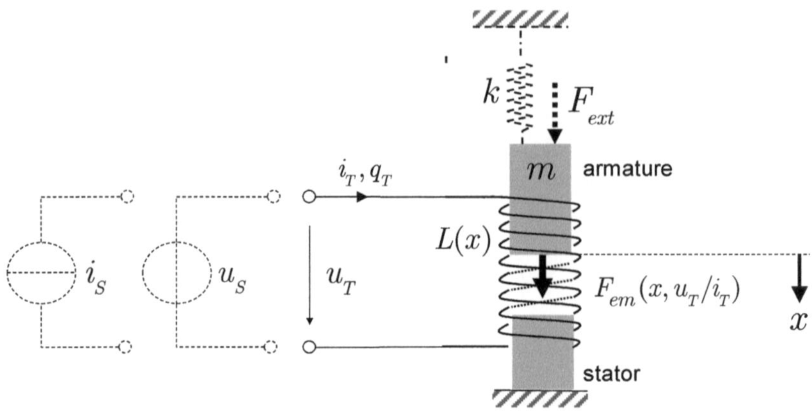

Fig. 8.9. Principle of operation for a generic electromagnetic (EM) transducer with one mechanical degree of freedom (one-dimensional moving armature in a magnetic circuit, electromagnetic force generation through variable reluctance); dashed lines indicate external loading either with a voltage or current source, and with an elastic suspension

called *pole shoes* or simply (magnetic) *poles*, and are constructed in such a manner as to minimize leakage flux.

The moving armature is suitably connected to the mechanical structure and can transmit forces or receive motions. Fig. 8.9 depicts a schematic arrangement with motion possible orthogonal (*transverse*) to the pole shoe surface (having a *variable air gap*; transducer only shown schematically—physical transducers employ a closed magnetic circuit, see Sec. 8.3.5).

Often the armature will be made to move longitudinally relative to the pole surfaces, i.e. there is a constant air gap and variable pole area. In both cases, due to the generic relationship in Eq. (8.20), the total inductance determining the magnetic flux depends on the geometry-dependent air gap reluctance.

Suspension Due to the unipolar nature of the reluctance force, elastic suspension of the moving armature is always required (except in the case of *differential EM transducers* or *electromagnetic bearings*, see Sec. 8.6).

8.3.2 Constitutive electromagnetic transducer equations

Basic electromagnetic constitutive equation The fundamental constitutive relation between the electric energy variable ψ_T and the electric power variable $\dot{q}_T = i_T$ following Fig. 5.7, assuming *lossless, electrically linear* dynamics, is given by Eq. (8.18)[6] (coupling via the magnetic field, inductance independent of the magnetomotive force or exciting current), i.e.

$$\psi_T = L(x) \cdot \dot{q}_T = L(x) \cdot i_T \,. \tag{8.25}$$

According to relation (8.13), the transducer inductance $L(x)$ primarily depends on the air gap geometry.

Constitutive ELM transducer equations The basic constitutive equation (8.25), combined with the results of Sec. 5.3.2 (see Table 5.3), directly leads to the *constitutive ELM transducer equations* for the *electromagnetic (EM) transducer* in the different coordinate representations:

- *Q-coordinates*

$$\begin{aligned} F_{em,Q}(x, \dot{q}_T) &= \frac{1}{2} \frac{\partial L(x)}{\partial x} \cdot \dot{q}_T^2 \,, \\ u_T(x, \dot{x}, \dot{q}_T, \ddot{q}_T) &= L(x) \cdot \ddot{q}_T + \frac{\partial L(x)}{\partial x} \cdot \dot{x} \cdot \dot{q}_T \,, \end{aligned} \tag{8.26}$$

- *PSI-coordinates*

$$\begin{aligned} F_{em,\Psi}(x, \psi_T) &= \frac{1}{2} \frac{1}{L(x)^2} \frac{\partial L(x)}{\partial x} \psi_T^2 \,, \\ i_T(x, \psi_T) &= \frac{1}{L(x)} \psi_T \,. \end{aligned} \tag{8.27}$$

Eqs. (8.26) and (8.27) describe the electromagnetic (EM) force laws and the dynamics at the terminals of the lossless, unloaded transducer as a function of one of the two assumed independent electric terminal variables.

Electromagnetic force: reluctance force As can be seen from Eqs. (8.26) and (8.27), the *reluctance force* F_{em} arising from the displacement-dependent inductance $L(x)$ always acts *unidirectionally*, independent of the polarity of the electric terminal variables.

6 A constant magnetic flux ψ_0 due to a *permanent magnet* may also be superimposed on the magnetic flux arising from the current flow.

Given a known, geometrically parameterized inductance $L(x)$ of the magnetic circuit, the electromagnetic transducer force (reluctance force) can thus be directly calculated from relations (8.26), (8.27).

Force direction The following holds in general for the *direction of the force* in the electromagnetic (EM) transducer:

Proposition 8.1. *Electromagnetic force direction*[7] (Jackson 1999)

The electromagnetic force is always directed so as to *increase* (*reduce*) the *inductance* (*reluctance*) of a magnetic circuit.

Electromagnetic forces between magnetic pole shoes To illustrate the electromagnetic force laws (8.26), (8.27), consider the simplified configuration of magnetic pole shoes shown in Fig. 8.10 (the electric excitation is not shown), in which the left-hand pole shoe is assumed free to move.

Neglecting leakage flux through lateral surfaces of the pole shoes, the *reluctance* of the air gap and *inductance* of the magnetic circuit can be approximated as

$$R_m(y,z) \approx \frac{y}{\mu_0 bz} \quad \text{and} \quad L(y,z) \approx N^2 \frac{\mu_0 bz}{y}.$$

The *reluctance force* $\vec{F}_{em}$ acting on this pole shoe can be decomposed into two orthogonal components (here in electrical *Q-coordinates*, i.e. with an imposed transducer current i_T, see Eq. (8.26))

$$F_{em,y}(y,z_0,i_T) = \frac{1}{2} i_T^2 \frac{\partial}{\partial y}\left(N^2 \frac{\mu_0 bz_0}{y}\right) = -\frac{1}{2} i_T^2 \frac{N^2 \mu_0 bz_0}{y^2}, \tag{8.28}$$

$$F_{em,z}(y_0,i_T) = \frac{1}{2} i_T^2 \frac{\partial}{\partial z}\left(N^2 \frac{\mu_0 bz}{y_0}\right) = \frac{1}{2} i_T^2 \frac{N^2 \mu_0 b}{y_0}. \tag{8.29}$$

The force directions for the two force components $F_{em,y}$ (the *normal force*—negative, i.e. attractive!) and $F_{em,z}$ (the *tangential force*) agree with Proposition 8.1 as expected. The moving left-hand pole shoe in Fig. 8.10

[7] Cf. the equivalent Proposition 6.1 for an *electrostatic* transducer (Sec. 6.3.2).

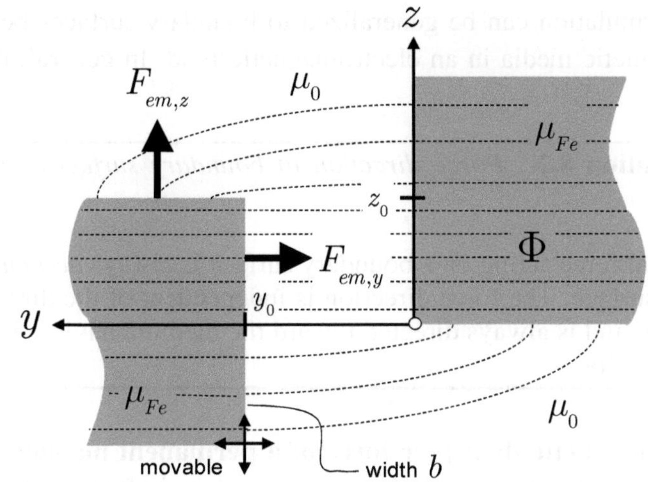

Fig. 8.10. Electromagnetic forces (reluctance forces) between two magnetic pole shoes (field lines shown schematically)

executes motions corresponding to the force directions $F_{em,y}$ and/or $F_{em,z}$, depending on the kinematic suspension. In most cases, one of the motion degrees of freedom y or z is holonomically constrained (e.g. by a low-friction guide rail).

From Eq. (8.27), the *PSI-coordinate* force components equivalent to Eqs. (8.28), (8.29) are

$$F_{em,y}(z_0, \psi_T) = -\frac{1}{2}\psi_T^{\ 2}\frac{1}{N^2\mu_0 b z_0}, \tag{8.30}$$

$$F_{em,z}(y_0, z, \psi_T) = \frac{1}{2}\psi_T^{\ 2}\frac{y_0}{N^2\mu_0 b}\frac{1}{z^2}. \tag{8.31}$$

As was already recognized in the electrostatic transducer, under certain conditions the reluctance force becomes *independent* of displacement (depending on its direction relative to the magnetic flux and chosen coordinate representation, see Eqs. (8.29) and (8.30)).

Generalization: forces on boundary surfaces The pole shoes in Fig. 8.10 form boundary surfaces between a magnetic conductor (iron, $\mu_{Fe} \gg \mu_0$) and air (μ_0). Eqs. (8.28) through (8.31) describe the forces acting on such boundary surfaces.

This formulation can be generalized to boundary surfaces between differing magnetic media in an electromagnetic field. In general, the following holds:

> **Proposition 8.2.** *Force direction at boundary surfaces*[8] (Jackson 1999)
>
> The total force acting on a boundary surface is always *perpendicular* to the surface. The force direction is independent of the direction of the field and is always directed toward the medium having the *lower* permeability.

Constant magnetic flux: pole force of a permanent magnet In a magnetic circuit, a constant and homogeneous magnetic flux can be generated either using a coil with an iron core and a constant excitation current, or, requiring significantly less energy, using a *permanent magnet.* There is a reluctance force normal to the surfaces of the permanent magnet, which can be easily computed using Eq. (8.27).

Employing Eq. (8.20) to replace inductance with reluctance, and accounting for boundary surface conditions following Eq. (8.13), gives the relation—equivalent to the force equation (8.27)—for the *pole force* of a *permanent magnet*

$$\vec{F}_{em,PM} = \frac{{B_{PM}}^2 A}{2\mu_0}\vec{n} = \frac{{\Phi_{PM}}^2}{2\mu_0 A}\vec{n}, \tag{8.32}$$

where B_{PM} is the magnetic flux density and Φ_{PM} the magnetic flux of the permanent magnet, A is the pole surface area, and $\vec{n}$ is the unit vector of the boundary surface directed into the magnet (i.e. the force is attractive).

8.3.3 ELM two-port model

Local linearization The constitutive electromagnetic force equations (8.26), (8.27) are always quadratically nonlinear with respect to the independent (imposed) electric terminal variable ψ_T or q_T, and often also nonlinear with respect to the armature displacement x. For this reason,

[8] Cf. the equivalent Proposition 6.2 for an *electrostatic* transducer (Sec. 6.3.2).

when considering small-signal dynamics, a local linearization about a steady-state operating point should be made. Without loss of generality, possible *steady-state operating points* can be defined as

$$x_R = const.,\ \dot{x}_R = 0,\quad \dot{\psi}_{T,R} = 0,\ \text{or}\ \ddot{q}_{T,R} = 0,$$
$$\dot{q}_{T,R} = i_{T,R} = const.\ \text{or}\ \psi_{T,R} = const.,\tag{8.33}$$

i.e. there is a constant magnetomotive current $i_{T,R}$ and constant rest flux $\psi_{T,R}$.

Setting operating points with permanent magnets As shown in the preceding section, the operating point of an electromagnetic transducer can be set in an energy-efficient manner using a *permanent magnet* in the magnetic circuit. Following Eq. (8.32), this allows a *static transducer force* to be generated even without an input current (an important operational aspect, e.g. for safety).

Two-port admittance form Starting with the constitutive transducer equations (8.27) and applying the generally-valid results of Sec. 5.3.3 gives the *two-port admittance form* of the unloaded electromagnetic (EM) transducer

$$\begin{pmatrix} \Delta F_{em,\Psi}(s) \\ \Delta I_T(s) \end{pmatrix} = \mathbf{Y}_{em}(s)\cdot\begin{pmatrix} \Delta X(s) \\ \Delta U_T(s) \end{pmatrix} = \begin{pmatrix} k_{em,U} & \dfrac{K_{em,U}}{s} \\ -K_{em,U} & \dfrac{1}{s}\dfrac{1}{L_R} \end{pmatrix}\begin{pmatrix} \Delta X(s) \\ \Delta U_T(s) \end{pmatrix},\tag{8.34}$$

with the *operating-point-dependent transducer parameters*

- *electromagnetic voltage stiffness*

$$k_{em,U} := \frac{\psi^2}{L(x)^2}\left[\frac{1}{2}\frac{\partial^2 L(x)}{\partial x^2} - \frac{1}{L(x)}\left(\frac{\partial L(x)}{\partial x}\right)^2\right]\Bigg|_{\psi=\psi_{T,R},\ x=x_R},$$

- *voltage coefficient* $$K_{em,U} := \psi\frac{1}{L(x)^2}\frac{\partial L(x)}{\partial x}\bigg|_{\psi=\psi_{T,R},\ x=x_R},\tag{8.35}$$

- *rest inductance* $$L_R := L(x)\big|_{x=x_R}.$$

Two-port hybrid form In the same way, the constitutive transducer equations (8.26) can be used to arrive at the *two-port hybrid form* for the unloaded electromagnetic (EM) transducer

$$\begin{pmatrix} \Delta F_{em,Q}(s) \\ \Delta U_T(s) \end{pmatrix} = \mathbf{H}_{em}(s) \cdot \begin{pmatrix} \Delta X(s) \\ \Delta I_T(s) \end{pmatrix} = \begin{pmatrix} k_{em,I} & K_{em,I} \\ sK_{em,I} & sL_R \end{pmatrix} \begin{pmatrix} \Delta X(s) \\ \Delta I_T(s) \end{pmatrix}, \tag{8.36}$$

with the *operating-point-dependent transducer parameters*

- *electromagnetic current stiffness*

$$k_{em,I} := \frac{1}{2} \dot{q}^2 \left. \frac{\partial^2 L(x)}{\partial x^2} \right|_{\substack{\dot{q}=\dot{q}_{T,R} \\ x=x_R}}, \tag{8.37}$$

- *current coefficient*

$$K_{em,I} := \dot{q} \left. \frac{\partial L(x)}{\partial x} \right|_{\dot{q}=\dot{q}_{T,R},\ x=x_R},$$

and the *rest inductance* L_R defined as in Eq. (8.35).

Relationship between the two-port parameters It is easy to verify that the following relations hold for the parameters of the admittance and hybrid forms (see Table 5.4):

$$\begin{aligned} k_{em,U} &= k_{em,I} - \frac{K_{em,I}^2}{L_R}, & K_{em,U} &= \frac{K_{em,I}}{L_R}, \\ k_{em,I} &= k_{em,U} + L_R K_{em,U}^2, & K_{em,I} &= L_R K_{em,U}. \end{aligned} \tag{8.38}$$

8.3.4 Loaded electromagnetic (EM) transducer

Mechanical suspension Due to the unidirectional action of the reluctance force (quadratic in the electrical terminal variables and always directed so as to increase the transducer inductance), *elastic suspension* of the armature electrode is *essential* (see Fig. 8.9). The suspension spring force must compensate the transducer force. In the next sections, rest positions are calculated and the steady-state behavior is analyzed for a few common electrode configurations.

Linear dynamic model The linear dynamic model of the loaded electromagnetic (EM) transducer can be easily obtained from the generic model in Sec. 5.4.4 using the two-port parameters in Eqs. (8.34), (8.36). The signal-flow diagrams for a voltage-drive and a current-drive transducer are depicted in Fig. 8.11 and Fig. 8.12, respectively (cf. Fig. 5.15, Fig. 5.16).

ELM coupling factor Applying the relations from Sec. 5.6 results in the operating-point-dependent general formula for the *ELM coupling factor* of an *electromagnetic* (EM) transducer (only defined for an elastically suspended armature electrode):

$$\kappa_{em}^{\ 2} = \frac{L_R \cdot K_{em,U}^{\ 2}}{k - k_{em,U}} = \frac{1}{1 + \dfrac{L_R}{K_{em,I}^{\ 2}}\left(k - k_{em,I}\right)}. \tag{8.39}$$

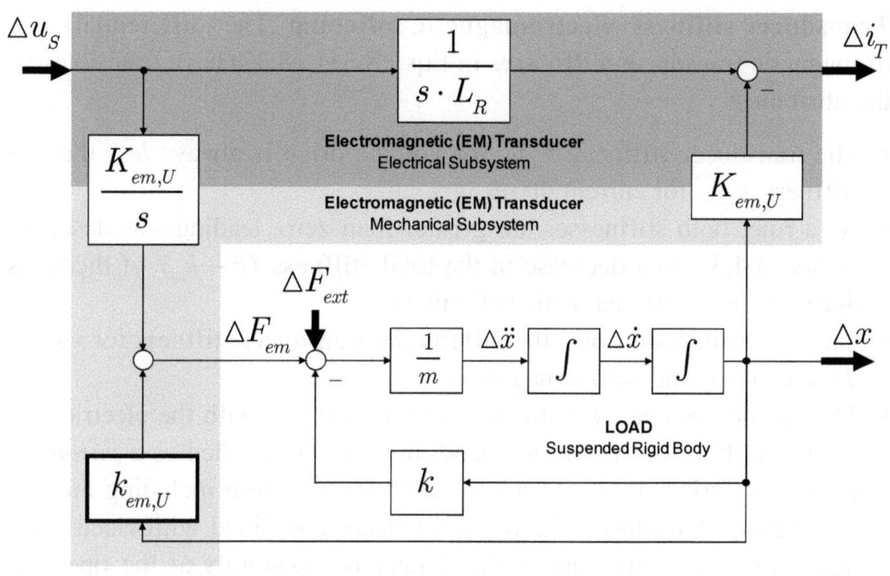

Fig. 8.11. Signal-flow diagram for a loaded *electromagnetic* (EM) transducer with *voltage drive* (model linearized about a stable operating point, lossless, ideal voltage source, mechanical load: elastically suspended rigid body, cf. Fig. 8.9)

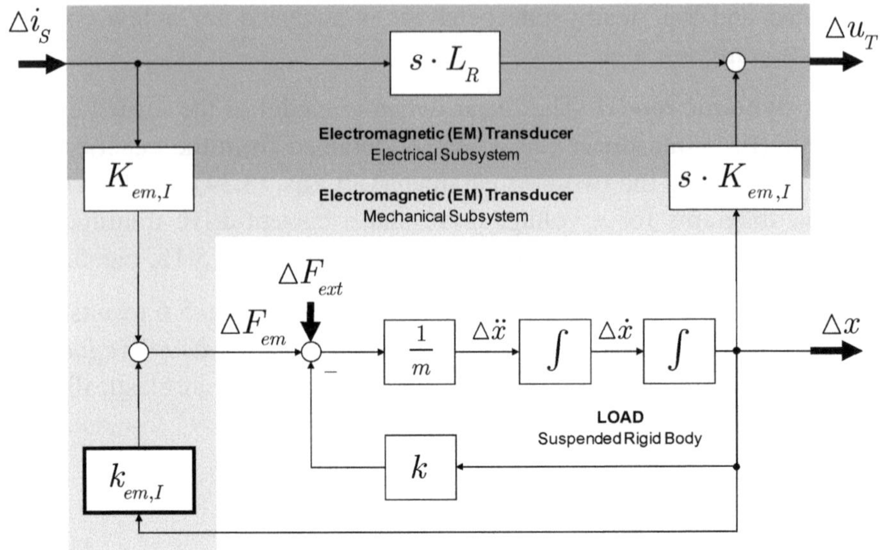

Fig. 8.12. Signal-flow diagram for a loaded *electromagnetic* (EM) transducer with *current drive* (model linearized about a stable operating point, lossless, ideal current source, mechanical load: elastically suspended rigid body, cf. Fig. 8.9)

Transducer stiffness: electromagnetic softening The (differential) electromagnetic transducer stiffnesses in Eqs. (8.35), (8.37) have four noteworthy attributes:

- The transducer stiffness $k_{em,U}$ for voltage drive is always *less* than the stiffness $k_{em,I}$ for current drive.
- As a rule, both stiffnesses are greater than zero, leading—as described in Sec. 5.4.3—to a decrease in the total stiffness $(k - k_{el})$ of the transducer (here, electromagnetic softening).
- Under certain conditions, the differential transducer stiffness for *voltage drive* can become zero or negative.
- The stiffnesses change with the rest position, i.e. with the electrical input to the transducer. On one hand, this can be handled as a *parameter variation* during design of a controller for a system including an electromagnetic transducer. On the other hand, combined with electromagnetic softening, this implies the danger of *instability* at the operating point (see *pull-in*).

8.3.5 Structural principles

Working air gap vs. secondary air gap To make use of reluctance forces, electromagnetic transducers fundamentally require an air gap to generate controlled forces. The volume employed for this purpose is thus termed the *working* (or *main*) *air gap* (Kallenbach et al. 2008). In addition, to guide the flux and particularly to allow motion of the armature in the flux field, certain transducer configurations require additional, *secondary air gaps*. Since such secondary air gaps create additional reluctance—thus decreasing the transducer inductance (see Eq. (8.20))—they should be kept to a minimum to reduce magnetic losses.

Basic types Table 8.1 presents various structural configurations with a single mechanical degree of freedom. Fundamentally, EM transducers can be divided into types having a *variable* or a *constant* working air gap. Force generation occurs according to the principles depicted in Fig. 8.10: with a force direction parallel to or orthogonal to the magnetic flux in the working air gap.

Insight into appropriate design degrees of freedom in the structural arrangement is given by the generic inductance relation Eq. (8.21) and the reluctance force relation Eq. (8.26), which can be generalized using a varying geometric configuration parameter par as follows:

$$L(\mathrm{par}) \sim \frac{A_{\mathrm{gap}}(\mathrm{par})}{\delta(\mathrm{par})} \Rightarrow F_{em} \sim \frac{\partial L(\mathrm{par})}{\partial \mathrm{par}}, \tag{8.40}$$

where A_{gap} is the common pole area overlapping in the stator and armature, and δ is the effective thickness of the gap in the magnetic flux. Thus, both *variation of the air gap pole area* A_{gap} and *variation of the air gap thickness* δ lend themselves to the *generation of forces*. In physical implementations, both of these possibilities are exploited, either exclusively or simultaneously, to achieve desired force-displacement curves.

For illustration, Table 8.1 presents applicable reluctance forces as a function of the exciting current (cf. Eq. (8.26)). It is easy to verify in all examples that the direction of the reluctance force conforms to Propositions 8.1 and 8.2. The (approximate) determination of inductances has been carried out using applicable magnetic air gap resistances according to Eq. (8.20). These simplified models thoroughly suffice as an initial approximation for system investigations, and, at a minimum, permit estimation of the order of magnitude of applicable dynamic parameters. More precise prediction requires *detailed field modeling* (see for example (Kallenbach et al. 2008)).

Variable working air gaps Types A through D shown in Table 8.1 employ a reluctance force parallel to the gap magnetic flux (the normal force). This force always pulls the armature toward the pole surface of the stator. The *tubular solenoid* (Type D) has a secondary air gap required to kinematically guide the armature (the gap is filled with a low-friction, non-magnetic material). The reluctance force exhibits the typical inverse quadratic dependence on working air gap length[9]. A *tilting armature* (Type C) has, with the same armature dimensions, a significantly smaller effective mass than the translating armature Types A and B (approximately one third), and thus enables faster movements (i.e. a lower armature inertia).

Constant working air gaps In Types E through G shown in Table 8.1, it is the reluctance force perpendicular to the air gap magnetic flux (the tangential force) which is exploited by varying the overlapping area of the pole forming the air gap. This force is always directed so as to increase the overlap. In the structurally simple case of a *plunger* as in Type E, a constant reluctance force *independent of armature displacement* results. If, however, the armature is guided through a second field, as in the plunger Type F, a displacement-dependent force-displacement curve again results.

This reluctance force generation with constant air gaps is particularly suited to *rotational transducers* such as Type G and will be briefly expanded upon here. The magnetic air gap reluctance and the inductance are respectively at a minimum and maximum when the armature and stator poles overlap exactly (i.e. the armature angle $\varphi = 0$). At the armature angle $\varphi = \pi/2$, the situation is precisely reversed. Thus, the armature-angle-dependent inductance can be represented as the sum of a constant component L_0 and an approximately harmonic component $L_V \cos 2\varphi$, giving the torque characteristic shown in Table 8.1, Type G. This principle is, for example, used in *reluctance stepper motors*.

9 Compare to an electrostatic plate transducer with variable electrode separation.

Table 8.1. Generic electromagnetic (EM) transducers with one mechanical degree of freedom x or φ (force *positive along arrow*)

	Transducer Type	Transducer Model
A	*Translating armature, U-stator* variable air gap 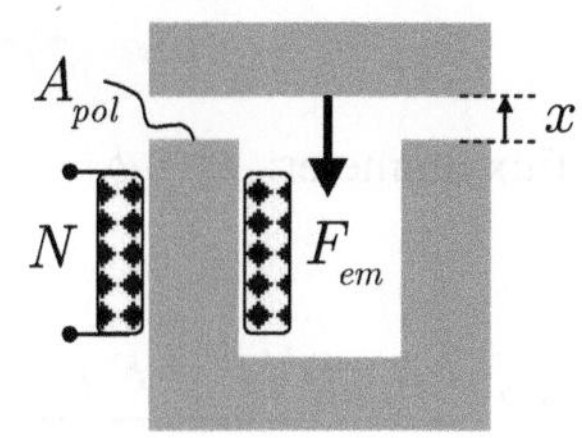	$L(x) = N^2 \frac{\mu_0 A_{pol}}{2} \frac{1}{x}$ $F_{em}(x, I) = \frac{1}{2} I^2 N^2 \frac{\mu_0 A_{pol}}{2} \frac{1}{x^2}$
B	*Translating armature, E-stator* variable air gap 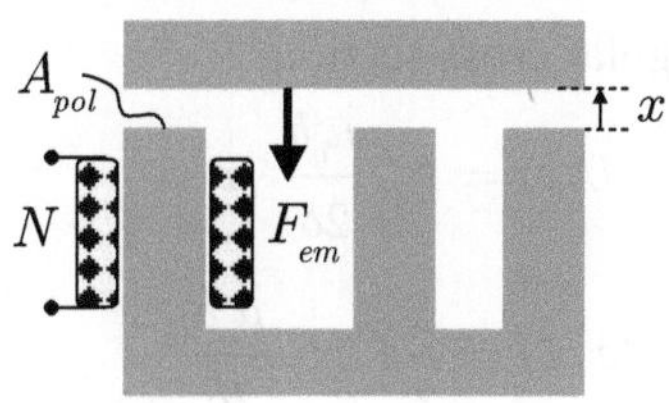	$L(x) = N^2 \frac{2\mu_0 A_{pol}}{3} \frac{1}{x}$ $F_{em}(x, I) = \frac{1}{2} I^2 N^2 \frac{2\mu_0 A_{pol}}{3} \frac{1}{x^2}$
C	*Tilting (hinged) armature, U-stator* variable air gap 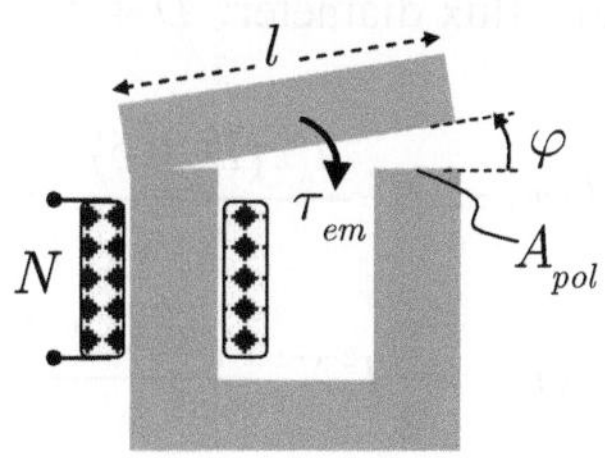	$\varphi \ll 1$ $L(\varphi) \approx N^2 \frac{\mu_0 A_{pol}}{l} \frac{1}{\varphi}$ $\tau_{em}(\varphi, I) \approx \frac{1}{2} I^2 N^2 \frac{\mu_0 A_{pol}}{l} \frac{1}{\varphi^2}$

Table 8.1. (cont.) Generic electromagnetic (EM) transducers with one mechanical degree of freedom x or φ (force *positive along arrow*)

	Transducer Type	Transducer Model
D	*Tubular solenoid* variable working air gap 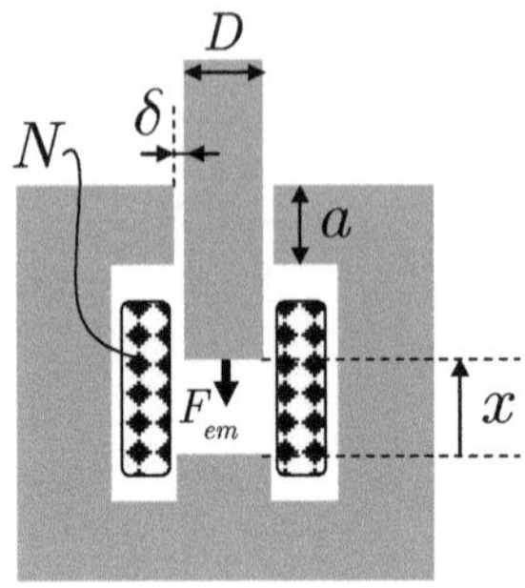 	Cylindrical cross section 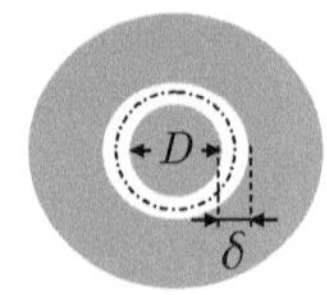 Average flux diameter: $D+\delta$ $L(x) = N^2 \dfrac{\mu_0 \pi D^2 (D+\delta) a}{\delta D^2 + 4(D+\delta) a \cdot x}$ $F_{em}(x, I) = \dfrac{1}{2} I^2 N^2 \dfrac{4\mu_0 \pi D^2 (D+\delta)^2 a^2}{\left(\delta D^2 + 4(D+\delta) a \cdot x\right)^2}$
E	*Plunger, simple stator* variable area 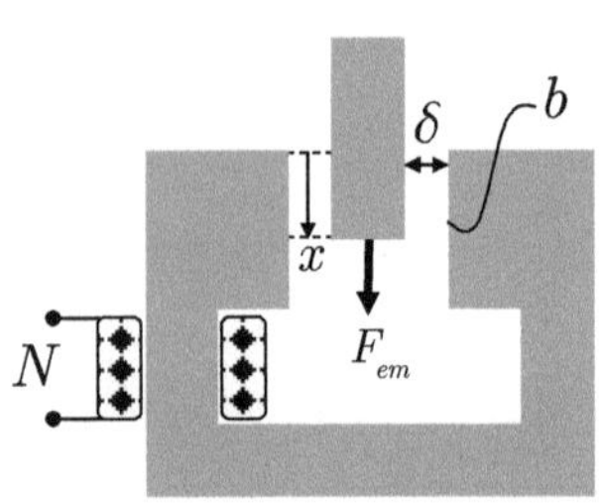 	Rectangular cross section $L(x) = N^2 \dfrac{\mu_0 b}{2\delta} \cdot x$ $F_{em}(x, I) = \dfrac{1}{2} I^2 N^2 \dfrac{\mu_0 b}{2\delta}$ Cylindrical cross section 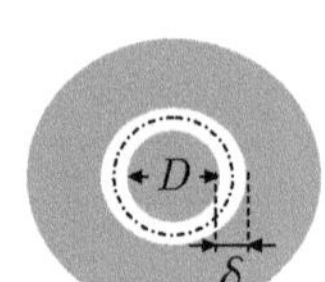 Average flux diameter: $D+\delta$ $L(x) = N^2 \dfrac{\mu_0 \pi (D+\delta)}{\delta} \cdot x$ $F_{em}(I) = \dfrac{1}{2} I^2 N^2 \dfrac{\mu_0 \pi (D+\delta)}{\delta}$

Table 8.1. (cont.) Generic electromagnetic (EM) transducers with one mechanical degree of freedom x or φ (force *positive along arrow*)

	Transducer Type	Transducer Model
F	*Plunger, U-stator* variable area 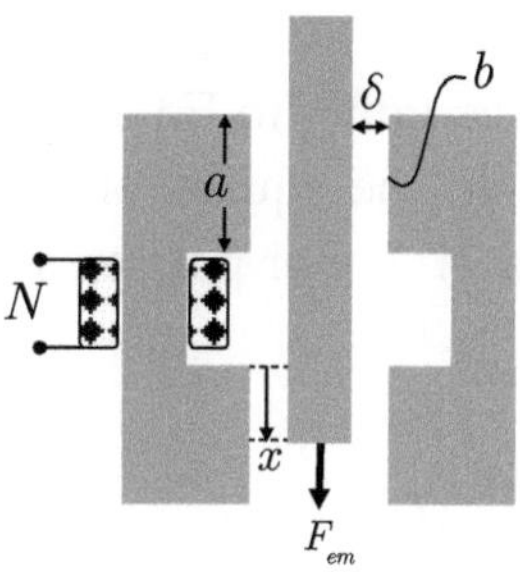	Rectangular cross section $L_0 = N^2 \frac{\mu_0 ab}{2\delta}$ Cylindrical cross section $L_0 = N^2 \frac{\mu_0 \pi (D+\delta) a}{\delta}$ Average flux diameter: $D + \delta$ $L(x) = L_0 \frac{x}{a+x}$ $F_{em}(x, I) = \frac{1}{2} I^2 L_0 \frac{a}{(a+x)^2}$
G	*Rotating armature,* *two-pole* variable area 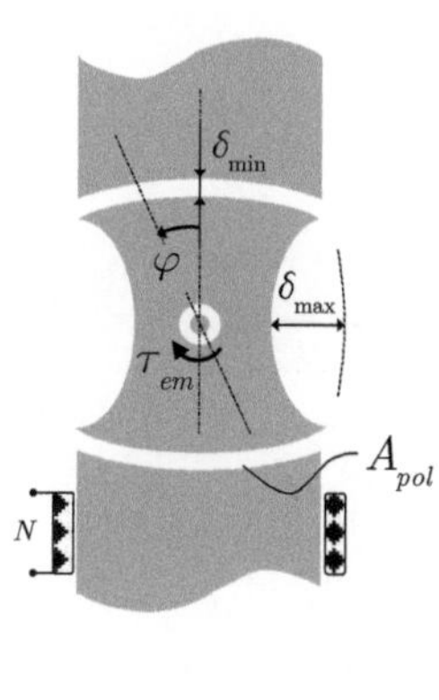	$R_{m,\min} \approx \frac{2\delta_{\min}}{\mu_0 A_{pol}} \Rightarrow L_{\max} \approx N^2 \frac{\mu_0 A_{pol}}{2\delta_{\min}}$ $R_{m,\max} \approx \frac{2\delta_{\max}}{\mu_0 A_{pol}} \Rightarrow L_{\min} \approx N^2 \frac{\mu_0 A_{pol}}{2\delta_{\max}}$ $L_0 := \frac{1}{2}\left(L_{\min} + L_{\max}\right) = \frac{N^2}{4} \mu_0 A_{pol} \left(\frac{1}{\delta_{\max}} + \frac{1}{\delta_{\min}}\right)$ $L_V := \frac{1}{2}\left(L_{\max} - L_{\min}\right) = \frac{N^2}{4} \mu_0 A_{pol} \left(\frac{1}{\delta_{\min}} - \frac{1}{\delta_{\max}}\right)$ $L(\varphi) \approx L_0 + L_V \cos 2\varphi$ $\tau_{em}(\varphi, I) \approx I^2 L_V \sin 2\varphi$

Pole shoe geometry vs. force-displacement curve The magnetic air gap reluctance, and thus both the inductance and the reluctance force, fundamentally depend on the geometry of the working air gap.

Using an appropriately designed pole shoe geometry, the two principal force-displacement characteristics of the reluctance force—inverse quadratic or displacement-independent (constant)—can be combined. Fig. 8.13b qualitatively sketches one such possibility (Fig. 8.13a corresponds to the characteristic defined in Eq. (8.28)).

Specific pole force: the pole area condition According to Eq. (8.32), the pole force of an electromagnet is proportional to the square of the flux density and directly proportional to the pole area A_{pol}. For purposes of design, the *specific pole force*

$$f_{em,pol} = \frac{F_{em}}{A_{pol}} = \frac{B^2}{2\mu_0} \quad \text{or} \quad f_{em,pol}{}^{[\mathrm{N/cm^2}]} \approx 40 \cdot \left(B^{[\mathrm{T}]}\right)^2 \tag{8.41}$$

gives a concrete measure for the magnetically achievable pole force given elementary geometric boundary conditions.

Given an assumed flux density B (less than the saturation flux density B_S, where typically $B_S \approx 1.2 - 1.7\ \mathrm{T}$, cf. Fig. 8.4), Eq. (8.41) defines the required pole surface A_{pol}. For $B = 1\ \mathrm{T}$, a specific load capacity of $f_{em,pol} = 40\ \mathrm{N/cm^2}$ is thus typical.

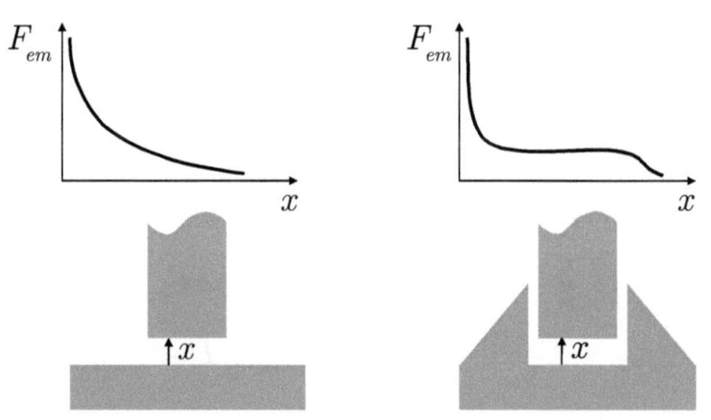

Fig. 8.13. Reluctance force-displacement curve for different pole shoe geometries: a) inversely quadratic $F_{em}(x)$ curve, b) piecewise displacement-independent $F_{em}(x)$ curve (constant force), from (Kallenbach et al. 2008)

Magnetomotive force condition, field coil The magnetomotive force required to produce a certain reluctance force can easily be estimated from the elementary magnetic circuit equation (8.12). Neglecting the reluctance $R_{m,Fe}$ in the iron core relative to the air gap reluctance R_δ, it follows from Eq. (8.12) that

$$\mathfrak{F}^{[\mathrm{AT}]} = \frac{\delta}{\mu_0} B \approx 800 \cdot \delta^{[\mathrm{mm}]} \cdot B^{[\mathrm{T}]} \ . \tag{8.42}$$

Using Eq. (8.42), it can be seen that given a flux density $B = 1\ \mathrm{T}$ and a working air gap $\delta = 1\ \mathrm{mm}$, a magnetomotive force $\mathfrak{F}$ of approximately $800\ \mathrm{ampere\text{-}turns}$ is required. Along with the necessary pole area from Eq. (8.41), the magnetomotive force condition Eq. (8.42) determines the geometric dimensions and electric properties (resistive losses) of the field coil (for more detailed aspects of coil layout, see (Kallenbach et al. 2008), (Schweitzer and Maslen 2009)).

Design guidelines for magnetic circuits

- Flux density of $B = 1\ \mathrm{T}$ results in a specific load capacity of $f_{em,pol} \approx 40\ \mathrm{N/cm^2}$.
- Magnetomotive force of $\mathfrak{F} = 800\ \mathrm{AT}$ with a working air gap of $\delta = 1\ \mathrm{mm}$ results in a flux density of $B \approx 1\ \mathrm{T}$.

Resistive coil losses One factor which can generally not be neglected as part of the electric drive design is the resistive losses in the windings of the field coil (including heating and voltage drops for voltage drive, see Fig. 8.14). As a function of the coil wire length l_{wire}, conductor cross-sectional area A_{wire}, and the conductor resistivity ρ_{Cu}, the coil resistance is given by (Hughes 2006)

$$R_C = \rho_{Cu} \frac{l_{wire}}{A_{wire}}, \quad \rho_{Cu} = 0.018 \ \frac{\Omega \cdot \mathrm{mm}^2}{\mathrm{m}} . \tag{8.43}$$

Eddy current losses In the presence of a time-varying magnetic flux, a voltage is induced in the flux-carrying, electrically conductive core, driving a short-circuit current or *eddy current*. Such eddy currents oppose the generating magnetic field and result in a higher required input power

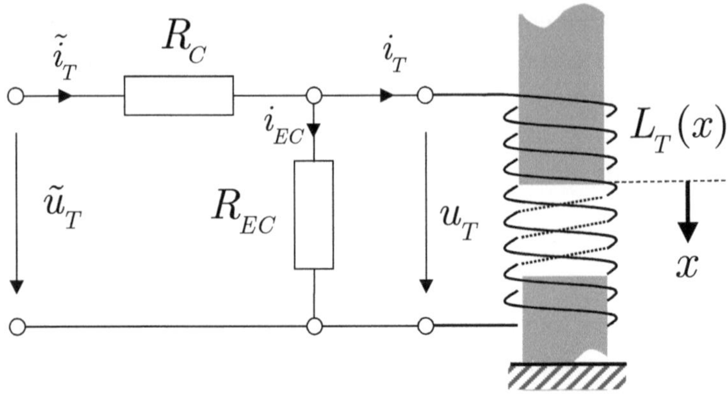

Fig. 8.14. Electrical schematic accounting for resistive coil losses R_C and eddy current losses R_{EC}

which can be approximately modeled as a *parallel resistance*[10] to the coil inductance (see Fig. 8.14). The *eddy current resistance* can be approximated as (Kallenbach et al. 2008)

$$R_{EC} \approx \rho_{Fe} \frac{l_{EC}}{A_{EC}} N^2, \quad \rho_{Fe} \approx 0.1 \frac{\Omega \cdot \text{mm}^2}{\text{m}}, \tag{8.44}$$

where l_{EC} is the length of an elementary eddy current path, and A_{EC} is the area enclosed by the eddy current path (i.e. the flux cross-section of a homogeneous iron core element). For the greatest possible eddy current resistance, small flux cross-sections are required; these can be achieved by laminating the iron core (Kallenbach et al. 2008).

8.3.6 EM transducers with variable working air gaps

Transducer type Electromagnetic (EM) transducers with variable working air gaps permit armature motion *transverse* (orthogonal) to the pole shoe surface and have a configuration according to Types A through D in Table 8.1.

[10] With current drive, this simplified model reproduces the dynamics of switching step inputs to a limited extent. For more precise considerations, see (Kallenbach, Eick et al. 2008).

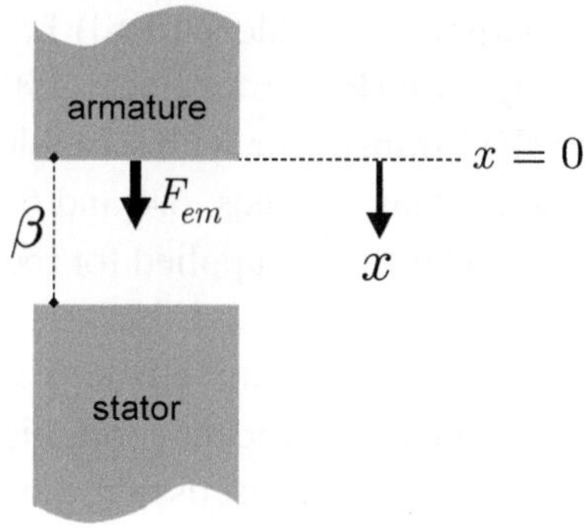

Fig. 8.15. Equivalent single-pole air gap configuration for an electromagnetic (EM) transducer with variable working air gap: β represents the *effective resting air gap thickness* of the transducer

Configuration equations With a *variable* working air gap, the effective transducer inductance can be accounted for using the single-pole equivalent configuration with an *equivalent resting air gap* β shown in Fig. 8.15. Concrete values for the generic transducer parameters α, β can be easily obtained for a given configuration, see e.g. Table 8.1, Types A through D.

The assumption of a homogeneous field distribution allows general statements to be made concerning the *transducer inductance* and the *reluctance force* with *voltage drive (PSI-coordinates)* and *current drive (Q-coordinates)*:

$$L_T(x) := \frac{\alpha}{\beta - x} \quad , \quad L_0 := L_T(0) = \frac{\alpha}{\beta} \tag{8.45}$$

$$\Rightarrow \quad \frac{\partial L_T(x)}{\partial x} = \frac{\alpha}{\left(\beta - x\right)^2},$$

$$F_{em}(x, \psi_T) = \frac{1}{2}\frac{1}{\alpha}\psi_T{}^2, \qquad F_{em}(x, i_T) = \frac{1}{2} i_T{}^2 \frac{\alpha}{\left(\beta - x\right)^2} . \tag{8.46}$$

Equivalence/duality to electrostatic transducers The configuration equations (8.45), (8.46) present a structural equivalence and duality to an electrostatic (ES) transducer with *variable electrode separation* (see Secs. 6.4, 6.5). This leads to the following correspondences (EM vs. ES):

$$L_T \triangleq C_T, \qquad q_T \triangleq \psi_T, \qquad i_T = \dot{q}_T \triangleq u_T = \dot{\psi}_T \ . \tag{8.47}$$

Phenomenologically, largely equivalent (dual) behaviors are thus to be expected for an electromagnetic (EM) transducer with variable working air gap and an electrostatic (ES) transducer with variable electrode separation. The latter was discussed in detail in Secs. 6.4 and 6.5, to which the correspondences in Eq. (8.47) can be simply applied for the case considered here.

Nevertheless, to permit this chapter and thus the dynamic properties of electromagnetic (EM) transducers to stand alone, the remainder of this section explicitly presents the most significant properties and models for EM transducers; elaborate derivation and discussion, however, is avoided. For a deeper understanding, perusal of the applicable sections in Ch. 6 (electrostatic transducers) is recommended.

EM transducer with current drive

Rest positions From Eq. (5.41) and the *saturation-dependent transducer force* F_{em} in Eq. (8.46), the rest positions x_R for static excitations $i_Q(t) = I_0$ and $F_{ext}(t) = F_0$ can be obtained in Q-coordinates from

$$\frac{2}{\alpha}\left(kx_R - F_0\right)\left(\beta - x_R\right)^2 = I_0^{\,2}\,. \tag{8.48}$$

Of the three possible solutions to Eq. (8.48)—as explained in Ch. 6—at most one leads to a fundamentally stable rest position. For further considerations, parameterization of rest positions using the tuple (x_R, F_0) again proves convenient.

Electromagnetic pull-in At a stable rest position, the *differential stiffness* $\partial F_\Sigma / \partial x$ of the total force $F_\Sigma = F_F - F_{em}$ acting on the armature must be positive. In the best case, a single stable rest position results, as can be seen in Fig. 6.8.

As in the case of the electrostatic transducer, for $F_{ext} = 0$, there is an unstable equilibrium at $x_R = \beta/3$ (due to $\partial F_\Sigma / \partial x = 0$), and for $x_R > \beta/3$, the reluctance force gradient dominates. The armature is then pulled without opposition toward the fixed stator pole shoe—a behavior termed *electromagnetic pull-in*.

The critical *pull-in values* (the marginally stable rest displacement and rest current) are

$$x_{pi} = \frac{\beta}{3}, \quad I_{pi} = \sqrt{\frac{8}{27}\frac{\beta^3}{\alpha}k}\,. \tag{8.49}$$

Pull-in limits for static mechanical excitation Given an additional external *static mechanical force* F_0, the pull-in limits become

$$\tilde{x}_{pi} = \frac{1}{3}(\beta + \frac{2}{k}F_0), \quad \tilde{I}_{pi} = \sqrt{\frac{8}{27}\frac{\left(k\beta - F_0\right)^3}{\alpha k^2}}. \tag{8.50}$$

Transducer parameters Substituting $L_T(x)$ from Eq. (8.45) into Eq. (8.35) and defining the *relative rest displacement* $X_R := x_R / \beta$ gives the operating-point-dependent *two-port parameters*

- *electromagnetic current stiffness* $k_{em,I} := \dfrac{2(k \cdot X_R - F_0/\beta)}{1 - X_R} \left[\dfrac{\text{N}}{\text{m}}\right]$,
- *current coefficient* $K_{em,I} := \dfrac{\sqrt{2L_0(k \cdot X_R - F_0/\beta)}}{1 - X_R} \left[\dfrac{\text{Vs}}{\text{m}}\right], \left[\dfrac{\text{N}}{\text{A}}\right]$, (8.51)
- *rest inductance* $L_R := L_0 \dfrac{1}{1 - X_R}$ $[\text{H}]$.

Transfer matrix From Eq. (8.51) and Table 5.8, the *transfer matrix* $\mathbf{G}(s)$ is then

$$\begin{pmatrix} \Delta X(s) \\ \Delta U_T(s) \end{pmatrix} = \mathbf{G}(s) \begin{pmatrix} \Delta F_{ext}(s) \\ \Delta I_S(s) \end{pmatrix} = \begin{pmatrix} V_{x/F,I} \dfrac{1}{\{\Omega_I\}} & V_{x/i} \dfrac{1}{\{\Omega_I\}} \\ V_{u/F} \dfrac{s}{\{\Omega_I\}} & V_{u/i} \cdot s \dfrac{\{\Omega_0\}}{\{\Omega_I\}} \end{pmatrix} \begin{pmatrix} \Delta F_{ext}(s) \\ \Delta I_S(s) \end{pmatrix}, \tag{8.52}$$

with the parameters

$$k_{T,I} = k - k_{em,I}, \quad \Omega_I^{\,2} = \frac{k_{T,I}}{m}, \quad \Omega_0^{\,2} = \frac{k}{m} \quad \text{where } \Omega_I < \Omega_0,$$
$$V_{x/F,I} = \frac{1}{k_{T,I}}, \quad V_{x/i} = V_{u/F} = \frac{K_{em,I}}{k_{T,I}}, \quad V_{u/i} = L_R \frac{k}{k_{T,I}}. \tag{8.53}$$

The *antiresonance* of the electrical terminal transfer function (the electrical impedance) is given by the mechanical eigenfrequency Ω_0 independent of the type of electrical drive, and further, corresponds to the transducer eigenfrequency under voltage drive.

Electromagnetic softening vs. gain variation The transducer stiffness $k_{T,I}$ in Eq. (8.53) depends on the operating point and is given by

$$k_{T,I} = \frac{k(1-3X_R)+2F_0/\beta}{\left(1-X_R\right)}. \tag{8.54}$$

As x_R increases, $k_{T,I}$ goes to zero (*electromagnetic softening*); at pull-in, $k_{T,I}=0$ (cf. Eqs. (8.49) and (8.50)).

The *gains* of the transfer functions in Eq. (8.53) thus quickly increase near the pull-in armature displacement. This extreme degree of parameter variation must be carefully considered during controller design if the transducer is to be used as an actuator in a closed loop.

Characteristic polynomial: transducer stability An important component of stability analysis is the *characteristic polynomial* $\Delta_I(s)$ of the transfer matrix $\mathbf{G}(s)$

$$\Delta_I(s) = s^2 + \Omega_I^{\,2} = s^2 + \frac{k_{T,I}}{m} = s^2 + \frac{k(1-3X_R)+2F_0/\beta}{m\left(1-X_R\right)}. \tag{8.55}$$

The pull-in conditions (8.49) and (8.50) can be precisely recovered as stability limits of Eq. (8.55) (double pole at $s=0$ or $k_{T,I}=0$).

ELM coupling factor Substituting the transducer parameters (8.51) into Eq. (8.39) gives the *ELM coupling factor* for the *current drive EM transducer*

$$\kappa_{em}^{\,2} = \frac{2\left(kX_R - F_0/\beta\right)}{k\left(1-X_R\right)} = \frac{k_{em,I}}{k}. \tag{8.56}$$

According to Eq. (8.56), the ELM coupling factor fundamentally depends on the *relative motion geometry* of the transducer, i.e. on the rest displacement x_R relative to the zero-voltage effective working air gap β. In some cases, there is an additional offset arising from the static mechanical excitation F_0.

The graphic representation of Eq. (8.56) is equivalent to the electrostatic transducer ELM coupling factor in Fig. 6.10.

Once again, for increasing rest displacement, the applied mechanical power climbs as well. However, electromagnetic softening also renders the transducer increasingly unstable. Thus, in order to prevent pull-in, large coupling factors are only possible within a strictly limited range of motion.

Passive damping via parallel resistance As previously explained, using a resistance in parallel with the current source in a current drive EM transducer (a *shunt*, see Fig. 8.16a), passive damping of mechanical subsystem oscillations can be achieved (see also the general presentation in Sec. 5.5 or the analogous electrostatic transducer in Sec. 6.4.3). Thus, a *parallel resistance* can again be considered among the important *design degrees of freedom.*

Steady-state dynamics without coil losses In the steady state with a finite R, there is only a current flowing in the inductance—its resistance is nearly nonexistent. Thus, rest position conditions *identical* to those of the lossless transducer result.

Dynamic behavior: small-signal dynamics neglecting coil losses To calculate a linearized transducer model at the stable rest position (I_0, x_R), the values from Tables 5.7 and 5.8 and the ELM two-port admittance parameters of Eq. (8.36) or Eq. (8.51) can be employed directly. To maintain formulae analogous to the electrostatic transducer, in place of the resistance R, the conductivity $Y = 1/R$ is employed where appropriate below. After a few manipulations, this results in[11]

$$\begin{pmatrix} \Delta X(s) \\ \Delta U_T(s) \end{pmatrix} = \begin{pmatrix} V_{x/F,I} \dfrac{[\tilde{\omega}_Z]}{\tilde{\mathcal{N}}(s)} & V_{x/i} \dfrac{1}{\tilde{\mathcal{N}}(s)} \\ V_{u/F} \dfrac{s}{\tilde{\mathcal{N}}(s)} & V_{u/i} \cdot s \dfrac{\{\Omega_0\}}{\tilde{\mathcal{N}}(s)} \end{pmatrix} \begin{pmatrix} \Delta F_{ext}(s) \\ \Delta I_S(s) \end{pmatrix}, \tag{8.57}$$

$$\tilde{\mathcal{N}}(s) := \left(1 + 2\tilde{d}_I \frac{s}{\tilde{\Omega}_I} + \frac{s^2}{\tilde{\Omega}_I^2}\right) \cdot \left(1 + \frac{s}{\tilde{\omega}_I}\right), \quad \tilde{\omega}_Z = \frac{1}{Y \cdot L_R} = \frac{R}{L_R},$$

where the gains are identical to those of the lossless transducer in Eq. (8.53).

In addition to the now-damped complex pole pair $\{\tilde{d}_I, \tilde{\Omega}_I\}$ of the mechanical eigenfrequency, an additional real pole $[\tilde{\omega}_I]$ appears due to the RL input circuit. In addition, derivative dynamics with a real zero $[\tilde{\omega}_Z]$ can be discerned in the mechanical transfer channel.

---

11 For definitions of the parameters $\tilde{d}_I, \tilde{\Omega}_I, \tilde{\omega}_I$ for the transducer with feedback, see Sec. 5.5.5.

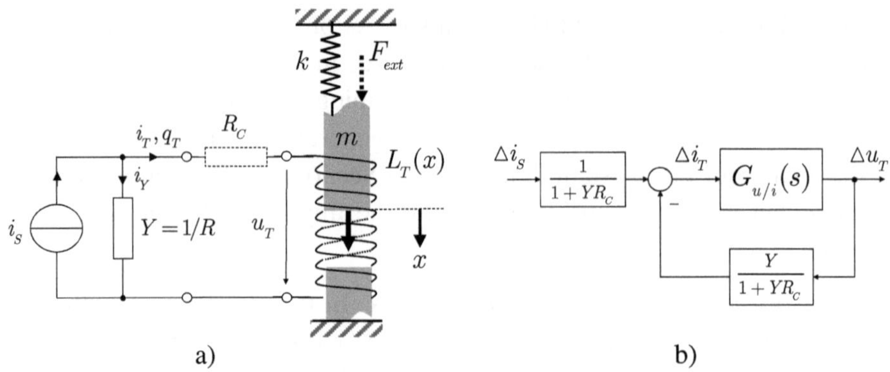

Fig. 8.16. Lossy *current drive* electromagnetic (EM) transducer with variable working air gap (parallel resistance, coil losses): a) schematic configuration (working air gap and magnetic flux guide not shown), b) signal-flow diagram for electrical gate ($G_{u/i}$ in Eq. (8.52))

Root locus as a function of resistance The dependence of the poles $\{\tilde{d}_I, \tilde{\Omega}_I\}$, $[\tilde{\omega}_I]$ on the resistance R was discussed in depth in Sec. 5.5.5. For the case considered here, the root locus in Fig. 5.24b applies exactly, confirming the expected damping produced by the electrical feedback.

Maximum damping Defining $X_R := x_R / \beta$, the considerations in Sec. 5.5.5 give the *maximum* achievable *damping* of the transducer eigenfrequency

$$d^{\max} = \frac{1}{2}\left(\sqrt{\frac{1-X_R}{1-3X_R+\dfrac{2F_0}{k\beta}}}-1\right). \tag{8.58}$$

The *optimal resistance (conductivity)* can be found according to Sec. 5.5.5 to be

$$\frac{1}{R^{\max}} = Y^{\max} = \frac{1-X_R}{\Omega_0 \cdot L_0}\left(\frac{1-3X_R+\dfrac{2F_0}{k\beta}}{1-X_R}\right)^{1/4}. \tag{8.59}$$

As in the case of the electrostatic (ES) transducer, the maximum achievable *damping* of the transducer eigenfrequency depends solely on the *relative motion geometry* of the transducer, and increases with increasing rest position x_R (or X_R), as shown in Fig. 8.17a for $F_0 = 0$.

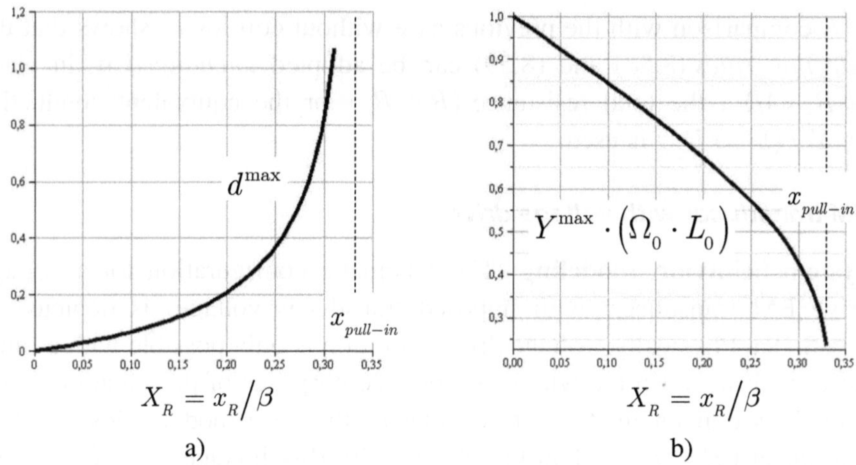

Fig. 8.17. Current drive electromagnetic (EM) transducer with variable working air gap and parallel resistance: a) maximum achievable damping Eq. (8.58) for $F_0 = 0$, b) optimal conductivity for maximum damping Eq. (8.59) for $F_0 = 0$

The corresponding optimal conductivity $Y^{max} = 1/R^{max}$ in Eq. (8.59) is plotted in Fig. 8.17b as a function of the rest displacement, again for $F_0 = 0$ (cf. Fig. 6.13).

Coil losses Non-negligible resistive losses in the windings—the *coil losses* R_C—have only a limited effect with current drive.

In normal operation without shunting ($Y=0$ in Fig. 8.16a), there is no effect on the imposed transducer current $i_T = i_S$, and the transfer characteristics remain unchanged (see Sec. 5.5). In particular, the series resistance R_T cannot effect passive electromechanical damping. If this latter is required, an additional finite *shunting resistance* $Y \neq 0$ must be introduced. This modifies the *rest current* to be $I_{T0} = I_0 / (1 + YR_T)$, and this latter should be substituted into the rest position condition (8.48) in place of I_0.

For dynamic *small-signal operation*, the block diagram shown in Fig. 8.16b describes the electric feedback. Some intermediate calculations give the *characteristic polynomial* of the transducer with feedback:

$$\begin{aligned}\Delta(s) &= (R + R_C)(k - k_{em,I}) + kL_R \cdot s + (R + R_C)m \cdot s^2 + mL_R \cdot s^3 \\ &= \frac{1 + YR_C}{Y}(k - k_{em,I}) + kL_R \cdot s + \frac{1 + YR_C}{Y} m \cdot s^2 + mL_R \cdot s^3 .\end{aligned} \tag{8.60}$$

A comparison with the previous case without coil losses shows that the *selection rules* (8.58) and (8.59) can be adopted *unchanged* if, in place of $R = 1/Y$, the total resistance $(R + R_C)$ or the equivalent conductivity $Y / (1 + YR_C)$ is used.

EM transducer with voltage drive

System behavior, modeling The schematic configuration for a voltage drive EM transducer (with imposed transducer voltage) is depicted in Fig. 8.18a. Physically, voltage drive operation is only possible with a finite series resistance (in the windings). For the purposes of presentation, however, it is convenient to first consider as the base model a lossless EM transducer ($R = R_C = 0$ in Fig. 8.18a) with flux linkage $\psi = \int \dot{\psi}(\tau) \cdot d\tau$ as the electrical coordinate, and separately introduce resistive losses in a second step.

Rest positions Given a *steady-state flux linkage* Ψ_0, Eq. (5.40) and the *drive-independent transducer force* F_{em} from Eq. (8.46) in *PSI-coordinates* give the *rest position condition* with an additional static mechanical excitation $F_{ext}(t) = F_0$

$$x_R = \frac{1}{2\alpha k} \Psi_0^{\ 2} + \frac{1}{k} F_0 . \tag{8.61}$$

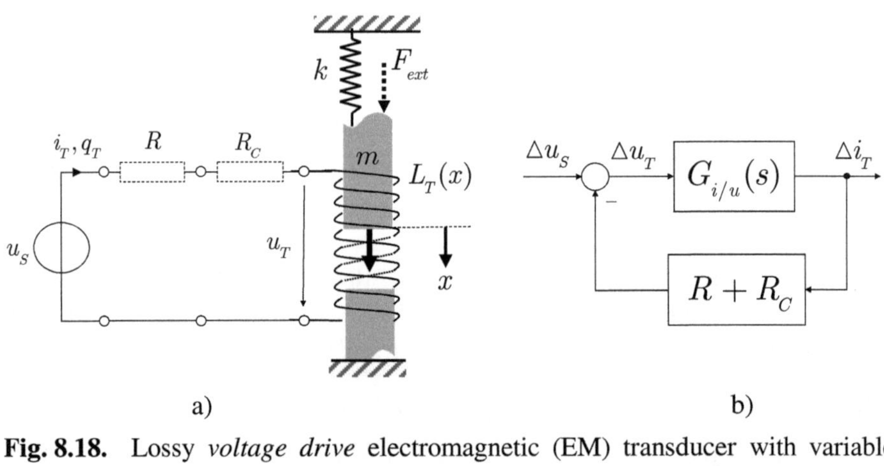

Fig. 8.18. Lossy *voltage drive* electromagnetic (EM) transducer with variable working air gap (serial resistance, coil losses): a) schematic configuration (working air gap and magnetic flux guide not shown), b) signal-flow diagram for electrical gate ($G_{i/u}$ in Eq. (8.63))

Thus, in the lossless case, by imposing a suitable flux linkage Ψ_0, *any* desired working air gap x_R can theoretically be *stably* selected using Eq. (8.61).

Pull-in with lossless voltage drive Given the ideal assumption of a homogeneous magnetic field between planar parallel pole shoes, *no* pull-in thus occurs in a *lossless* EM transducer with voltage drive. It is shown below that this idealized, purely theoretical case in no way applies to practical situations. Unfortunately, due to the resistive impedance feedback required for a voltage source, the risk of pull-in, with the previously-mentioned pull-in limit $x_R < x_{pi} = \beta/3$, is always present.

Transducer parameters for the lossless transducer Substituting $L_T(x)$ from Eq. (8.45) into Eq. (8.35) and defining the *relative rest displacement* $X_R := x_R \,/\, \beta$ gives the operating-point-dependent *two-port parameters*

- *electromagnetic voltage stiffness* $k_{em,U} := 0 \quad \left[\frac{\mathrm{N}}{\mathrm{m}}\right]$,

- *voltage coefficient* $$K_{em,U} := \sqrt{\frac{2}{L_0}(k \cdot X_R - F_0/\beta)} \quad \left[\frac{\mathrm{N}}{\mathrm{Vs}}\right], \left[\frac{\mathrm{A}}{\mathrm{m}}\right], \tag{8.62}$$

- *rest inductance* $L_R := L_0 \dfrac{1}{1 - X_R} \quad [\mathrm{H}]$.

Transfer matrix From Eq. (8.62) and Table 5.8, the *transfer matrix* $\mathbf{G}(s)$ is then

$$\begin{pmatrix} \Delta X(s) \\ \Delta I_T(s) \end{pmatrix} = \mathbf{G}(s) \begin{pmatrix} \Delta F_{ext}(s) \\ \Delta U_S(s) \end{pmatrix} = \begin{pmatrix} V_{x/F,U} \dfrac{1}{\{\Omega_0\}} & V_{x/u} \dfrac{1}{s \cdot \{\Omega_0\}} \\ V_{i/F} \dfrac{1}{\{\Omega_0\}} & V_{i/u} \cdot \dfrac{\{\Omega_I\}}{s \cdot \{\Omega_0\}} \end{pmatrix} \begin{pmatrix} \Delta F_{ext}(s) \\ \Delta U_S(s) \end{pmatrix}, \tag{8.63}$$

with the parameters

$$\begin{gathered} k_{T,I} = k - k_{em,I}, \quad \Omega_I^{\,2} = \frac{k_{T,I}}{m}, \quad \Omega_0^{\,2} = \frac{k}{m} \quad \text{where } \Omega_I < \Omega_0, \\ V_{x/F,U} = \frac{1}{k}, \quad V_{x/u} = \frac{K_{em,U}}{k}, \quad V_{i/F} = -\frac{K_{em,U}}{k}, \quad V_{i/u} = \frac{1}{L_R}\frac{k_{T,I}}{k}. \end{gathered} \tag{8.64}$$

Dynamic behavior The dynamics are shaped by the drive-*independent* mechanical eigenfrequency Ω_0, and the integrating nature of the voltage source control channel (requiring a series resistance to reach steady state). There is *no electromechanical feedback*, so that the system dynamics for the lossless case are significantly clearer than for current drive.

Electrically dissipative loading: coil losses and damping resistance As described above, a series resistance should always be considered for a voltage drive EM transducer. This is always inherently present in the (rather small) *coil resistance* R_C (Fig. 8.18a). However, to achieve a desired level of electromechanical damping, an additional series resistance R (Fig. 8.18a) can be introduced. Thus, designing the damping involves the total resistance $R_\Sigma = R + R_C$, and the coil resistance is to be regarded rather as a benefit than a parasitic property. The derivation of the corresponding design formulae is carried out analogously to the case of the current drive EM transducer.

Maximum damping Evaluating the general relations previously derived in Sec. 5.5.5, and employing the concrete transducer parameters (8.62), the relation previously described in Eq. (8.58) is obtained for the *maximum* achievable *damping*.

The *optimal resistance* is then found to be

$$R^{\max} = \frac{\Omega_0 \cdot L_0}{1 - X_R} \left(\frac{1 - X_R}{1 - 3X_R + \dfrac{2F_0}{k\beta}} \right)^{1/4}, \tag{8.65}$$

corresponding precisely to the determining equation (8.59) for the conductivity $Y^{\max} = 1 / R^{\max}$ of the shunting resistance for the current source. Thus, the graphic representations of the design formulae depicted in Fig. 8.17 also apply in this case.

Equivalence of impedance feedbacks The equivalence of the design formulae for impedance feedback under both current and voltage drive is not surprising if the nature of the impedance feedback is considered. In an *ideal voltage source*, the internal resistance is naturally zero, while in an *ideal current source*, the internal resistance is infinite (see Fig. 8.19a,b). For the system dynamics, only the electrical feedback is determining (see the feedback loops in Fig. 8.19c,d). Considering the reciprocal structure of

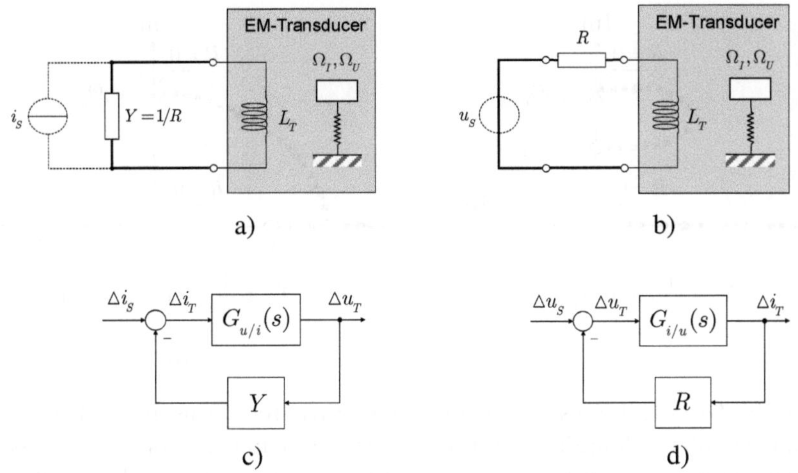

Fig. 8.19. Equivalence of impedance feedback for current and voltage drives: a) schematic, current drive; b) schematic, voltage drive; c) signal-flow diagram, current drive; d) signal-flow diagram, voltage drive

the transfer functions $G_{i/u}$, $G_{u/i}$, the equivalence of the design formulae then follows directly.

Pull-in with lossy voltage drive The fact that the lossy voltage-drive EM transducer demonstrates the same type of pull-in behavior as the current-drive EM transducer can be easily illustrated as follows. Following Eq. (8.63), the electrical frequency response is determined by

$$G_{i/u}(s) = \frac{1}{L_R} \frac{(s + \Omega_I^2)}{s(s + \Omega_0^2)}, \quad \Omega_0^2 = \frac{k}{m}, \quad \Omega_I^2 = \frac{k_{T,I}}{m}. \tag{8.66}$$

Without external mechanical excitation ($F_0 = 0$), it follows from Eq. (8.54) that

$$\Omega_I^2 = \Omega_0^2 \frac{1 - 3X_R}{1 - X_R}, \quad X_R = \frac{x_R}{\beta}. \tag{8.67}$$

For a rest displacement $x_R > x_{pi} = \beta/3$, the transfer function $G_{i/u}$ has two real zeros

$$s_{1,2} = \pm\Omega_I = \pm\Omega_0 \sqrt{(3X_R - 1) / (1 - X_R)},$$

so that—due to the impedance feedback shown in Fig. 8.19d—the *root locus* for a variable resistance R depicted in Fig. 8.20 results for the poles of

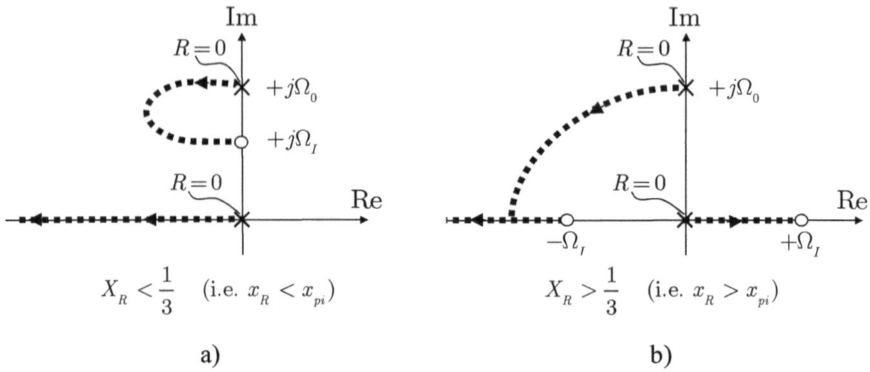

Fig. 8.20. *Lossy* EM transducer with *voltage drive*: root locus of the characteristic polynomial of the transducer transfer matrix as a function of series resistance (showing only the upper half-plane): a) armature travel below pull-in limit (stable transducer dynamics), b) armature travel above pull-in limit (unstable transducer dynamics)

the feedback system (equal to the roots of the characteristic polynomial of the transfer matrix $\tilde{\mathrm{G}}(s)$). The root locus always has one branch along the positive real axis, explaining the *unstable transducer dynamics*.

Under realistic conditions, a lossy EM transducer should thus generally also only be operated *below* the previously mentioned pull-in armature displacement $x_R < x_{pi} = \beta/3$ when operated under voltage drive.

Current drive vs. voltage drive

Controlled auxiliary voltage source, power amplifiers As explained at the beginning of Ch. 5, for an actuator, the controlled auxiliary voltage source considered here represents the power amplifier between the controller output and the transducer. In electromagnetically-acting transducers, it is customary to employ power amplifiers with current outputs, i.e. voltage-driven current sources (voltage-driven, as controller outputs are generally high-impedance, low-power voltage sources). The reasons for the wide distribution of this type of source in practice are briefly illuminated below.

Delay dynamics of field build-up A schematic view of a lossy electromagnetic (EM) transducer as seen from its electric terminals is shown in Fig. 8.21. From Eqs. (8.26) and (8.27), two deciding factors for force generation are the transducer current $i_T = \dot{q}_T$ and the flux linkage ψ_T. Both quantities can be directly imposed using auxiliary energy sources $i_S(t)$

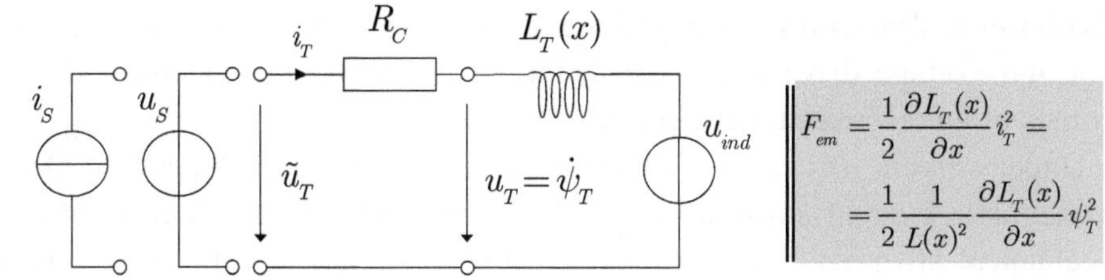

Fig. 8.21. Electrical schematic for the dynamics at the terminals of a lossy electromagnetic (EM) transducer with *current drive* (imposed drive current $i_S(t)$) or *voltage drive* (imposed drive voltage $u_S(t)$).

or $u_S(t)$, respectively. In the process, delays in the build-up of the transducer force should naturally be kept at a minimum. The electrical dynamics are described by the following equations (Fig. 8.21):

$$u_T = u_L + u_{ind} = L_T(x)\frac{di_T}{dt} + \frac{\partial L_T(x)}{\partial x} i_T \frac{dx}{dt}, \tag{8.68}$$

$$\tilde{u}_T = R_C\, i_T + u_T\,. \tag{8.69}$$

Depending on the auxiliary energy source selected, different dynamics result.

Current drive: delay-free field build-up With a *current source*, $i_T = i_S$ so that there is a *delay-free* effect on the force-generating electrical quantities (the *transducer current* $i_T = i_S$ or *transducer flux* $\psi_S = L(x) \cdot i_S$). Neither the self-induction voltage u_L nor the motion-induced voltage u_{ind} affect force generation, and the conductor resistance R_C plays no role in the field build-up. These attributes can also be nicely recognized in the linearized transducer model Fig. 8.12. This instantaneous behavior can be taken advantage of with current drive.

Voltage drive: delayed field build-up With a *voltage source*, $\tilde{u}_T = u_S$, so that Eqs. (8.68) and (8.69) determine the time evolution of the *transducer current* i_T and *transducer flux* ψ_T. For $R_T{=}0$, both the self-induction u_L and motion-induced voltage u_{ind} oppose the drive voltage. The flux can only build up with a *delay* via an integration over time $\psi_T = \int u_S \cdot d\tau$. In the lossy case ($R_T{>}0$), instead of the integral dynamics, a first-order delay with time constant $\tau_{em}{=}L_{eff}/R$ results. These attributes are also easy to recognize from the linearized transducer model in Fig. 8.11. Due to this delayed control, from an electrical point of view, voltage drive is generally avoided for electromagnetic transducers.

Mechanical dynamics As explained at the beginning of this section, current and voltage drive also have differing effects on the mechanical dynamics of an electromagnetic transducer.

Under *current drive*, as previously explained, the drive-dependent electromagnetic *current stiffness* $k_{em,I}$ (Eq. (8.51)) effects a reduction of the mechanical stiffness of the suspension (electromagnetic softening), while under *voltage drive*, the electromagnetic voltage stiffness $k_{em,U}$ in Eq. (8.62) goes to zero. This results in different eigenvalues in the coupled electromechanical model. In certain configurations, this can result in advantages for voltage drive operation, e.g. for magnetic bearings (Schweitzer and Maslen 2009).

Example 8.1 *Electromagnetic actuator.*

To develop a feel for realistic magnitudes of system parameters, consider the following generic *single-pole electromagnetic actuator* (horizontal motion, thus no effect from gravitation, i.e. $F_{ext} = F_0 = 0$).

Armature mass $m = 100\ \mathrm{g}$, stiffness of the armature suspension $k = 3 \cdot 10^4\ \mathrm{N/m}$, pole area $A_{pol} = 1\ \mathrm{cm}^2$, air gap without excitation $\delta_0 = 1\ \mathrm{mm}$, winding count of the field coil $N = 1000$, wire diameter in coil $D = 0.2\ \mathrm{mm}$.

Electrical transducer parameters The *electrical transducer parameters* are then ($l_{wire} \approx 40\ \mathrm{m}$)

$$L_0 = N^2 \frac{\mu_0 A_{pol}}{\delta_0} = 126\ \mathrm{mH}, \quad R_C = \rho_{Cu} \frac{l_{wire}}{A_{wire}} \approx 18\ \Omega .$$

Steady-state For a stable *rest displacement*, $x_R < x_{pi} = 0.33\ \mathrm{mm}$. The value $x_R = 0.15\ \mathrm{mm}$ (i.e. $X_R = 0.15$) is chosen, giving a rest current $I_0 = 0.23\ \mathrm{A}$ (for comparison, the pull-in current is $I_{pi} = 0.265\ \mathrm{A}$).

Small-signal dynamics The *transducer parameters* for linearized *small-signal dynamics* are then $\Omega_0 = 547\ \mathrm{rad/s}$ ($f_0 = 87\ \mathrm{Hz}$), $\Omega_I = 440\ \mathrm{rad/s}$ ($f_I = 70\ \mathrm{Hz}$), $V_{x/i} = 8.7\ \mathrm{mm/A}$.

Resistive impedance feedback The greatest possible *passive damping* using *resistive impedance feedback* is $d^{\max} = 0.12$, achieved with $R^{\max} = 90\ \Omega$. If using a *current source* this requires a *shunting resistance* of $R = R^{\max} - R_W = 72\ \Omega$, and if using a *voltage source*, an equally-large *series resistance*.

Vertical operation: gravity, constant rest displacement It is easy to verify that, for *vertical operation* and the same rest displacement $X_R = 0.15$, the action of gravity on the armature (i.e. $F_0 = \pm m \cdot g = \pm 0.981\ \mathrm{N}$) requires modified rest currents depending on the orientation, i.e. $I_0 = 0.2\ \mathrm{A}$ for $F_0 = +0.981\ \mathrm{N}$ and $I_0 = 0.25\ \mathrm{A}$ for $F_0 = -0.981\ \mathrm{N}$. With stable *position control* of the armature, these new rest currents would be automatically set by the control loop. The optimal damping resistance changes insignificantly to $R^{\max} = 88\ \Omega$ or $R^{\max} = 93\ \Omega$, respectively, and the pull-in-current changes to $I_{pi} = 0.253\ \mathrm{A}$ or $I_{pi} = 0.28\ \mathrm{A}$, respectively. If the optimal damping resistance $R^{\max} = 90\ \Omega$ for horizontal motion were to be retained, parameter variations from the different mounting orientations of the actuator would not be meaningful for the system dynamics, given a *position-controlled* armature: the system is robust with respect to mounting orientation.

Vertical operation: constant rest current In contrast, if the actuator were to be operated at the same rest current $I_0 = 0.23\ \mathrm{A}$ (i.e. no armature position control) then change in direction of the gravitational force when reorienting the actuator from the horizontal to the vertical would result change of the armature position to $x_R = 0.21\ \mathrm{mm}$ or $x_R = 0.11\ \mathrm{mm}$.

Linear operating regime: flux density saturation level Examining the degree of saturation of the core with Eq. (8.42) gives a rest flux density of $B_0 = NI_0\mu_0/\delta_0 = 0.29\ \mathrm{T}$. This represents a sufficiently large separation from the typical saturation flux density of $B_S \approx 1.2 \ldots 1.7\ \mathrm{T}$ that the transducer will operate in the magnetically linear regime. ■

8.3.7 Differential EM transducers: magnetic bearings

Non-suspended armature: differential transducers If structural constraints prevent elastic suspension of the armature, then the *differential principle* already presented for the electrostatic transducer may also be applied for two-sided electromagnetic suspension. Fig. 8.22 shows a two-pole configuration for vertical suspension. The armature floats freely in space; the mechanical stiffness is zero ($k = 0$). Purposeful motion of the armature (positioning or compensation of disturbance force) is accomplished via push-pull drive of the two stator coil currents.

This *magnetic bearing* principle is excellently suited to rotating shafts: it is frictionless, and thus allows for extremely high rotation speeds with minimal losses. Due to their significantly higher energy density compared to electrostatic bearings, magnetic bearings have a much greater distribution and are available in a number of implementations as established (high-tech) products (Schweitzer and Maslen 2009).

The schematic surrogate configuration shown in Fig. 8.22a serves here as a calculation aid. Physical implementations of four-pole radial bearings for electromagnetic suspension in one degree of freedom (here the vertical direction) are shown in Fig. 8.22b,c along with the resulting flow of mag-

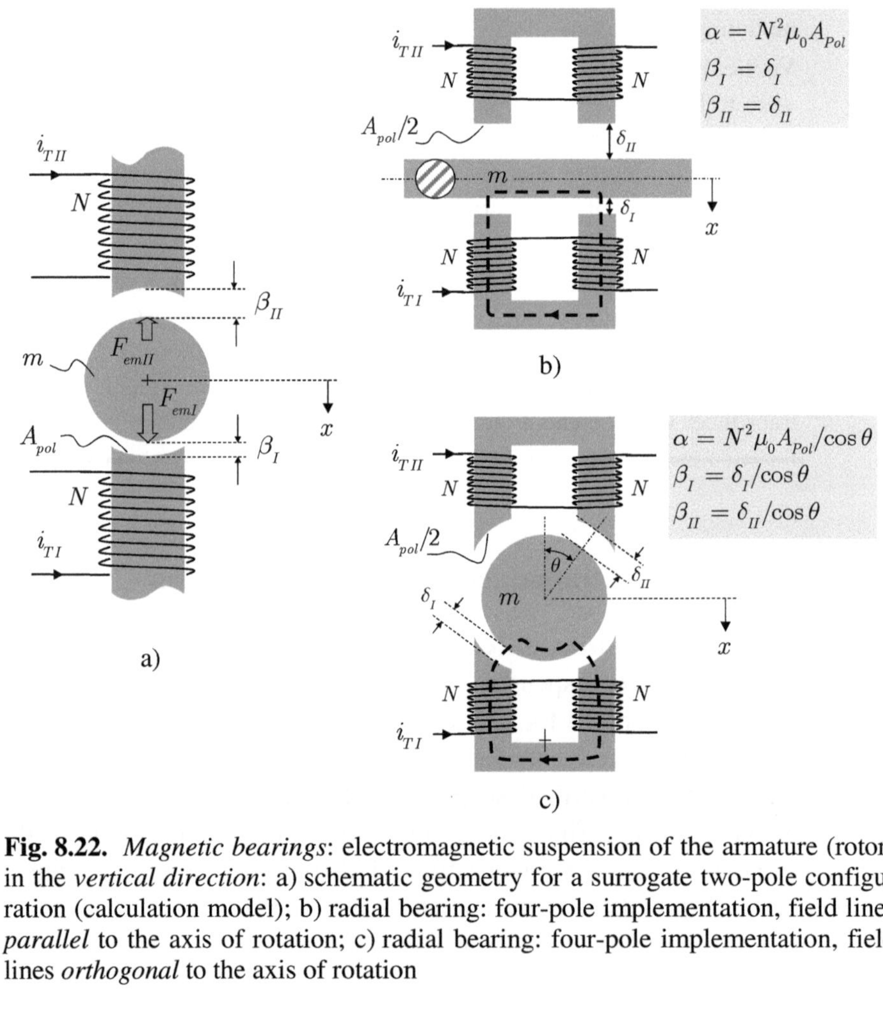

Fig. 8.22. *Magnetic bearings*: electromagnetic suspension of the armature (rotor) in the *vertical direction*: a) schematic geometry for a surrogate two-pole configuration (calculation model); b) radial bearing: four-pole implementation, field lines *parallel* to the axis of rotation; c) radial bearing: four-pole implementation, field lines *orthogonal* to the axis of rotation

netic flux. The parameters equivalent to the single-pole model (shown to the right in Fig. 8.22b,c) result from a comparison with the generic translating transducer with U-stator (Type A in Table 8.1). The depicted pole configurations can be suitably combined to suspend the armature (rotor) in multiple spatial directions (Schweitzer and Maslen 2009).

Equivalence/duality to electrostatic bearings There are clear equivalences and dualities in both the model structure and the system behavior between electrostatic bearings (Sec. 6.6.4) and electromagnetic bearings. Using the correspondences (EM vs. ES) already presented in Eq. (8.47), all results for the electrostatic bearing in Sec. 6.6.4 can in principle be directly adopted here.

Nevertheless, to permit this chapter and thus the dynamic properties of electromagnetic (EM) transducers to stand alone, the remainder of this section concisely presents the most significant properties and models for electromagnetic bearings.

Transducer configuration: current drive The moving armature is located between two fixed pole shoes, which in the general case can be independently driven using *coil currents* i_{TI}, i_{TII}. Corresponding to the two inductances

$$L_I(x) = \frac{\alpha}{\beta_I - x}, \qquad L_{II}(x) = \frac{\alpha}{\beta_{II} + x}, \qquad \alpha = N^2 \mu_0 A_{pol}, \tag{8.70}$$

of the two magnetic circuits I and II, there are two *opposing* electromagnetic forces (for the force direction, see Fig. 8.22a)

$$F_{em,I}(x, i_{TI}) = \frac{\alpha}{2} \frac{i_{TI}^{\,2}}{\left(\beta_I - x\right)^2}, \qquad F_{em,II}(x, i_{TII}) = \frac{\alpha}{2} \frac{i_{TII}^{\,2}}{\left(\beta_{II} + x\right)^2} \tag{8.71}$$

acting on the armature.

Electromagnetic push-pull principle The resulting transducer force for the transducer configuration in Fig. 8.22a with $\alpha = N^2 \mu_0 A_{pol}$ is then

$$F_{em}(x, i_{TI}, i_{TII}) = \frac{\alpha}{2} \left[\frac{i_{TI}^{\,2}}{\left(\beta_I - x\right)^2} - \frac{i_{TII}^{\,2}}{\left(\beta_{II} + x\right)^2} \right]. \tag{8.72}$$

With a suitable choice of *asymmetric* transducer currents i_{TI}, i_{TII} or working air gaps β_I, β_{II}, the magnitudes of the two terms in Eq. (8.72) can be varied, so that both *positive* and *negative* transducer forces F_{em} can be generated.

Without loss of generality, the following approach for electrical drive of the transducer, the *push-pull principle*, can be applied:

$$\begin{aligned} i_{TI} &= I_{0I} + \Delta i, \\ i_{TII} &= I_{0II} - \Delta i. \end{aligned} \tag{8.73}$$

Using the two static rest currents I_{0I}, I_{0II}, the operating point (rest position x_R and static bearing force $F_{em,0} = F_{em}(x_R, I_{0I}, I_{0II})$) can be freely selected, while the dynamic component $\Delta i(t)$ can be used to generate bidirectional dynamic bearing forces.

Steady-state behavior: rest positions For a constant external force $F_{ext}(t) = F_0$, Eqs. (8.72), (8.73), and the definitions $\beta_I = \beta$ and $\beta_{II} = \lambda \cdot \beta$ give the rest position condition

$$\frac{\alpha}{2} \frac{I_{0I}^2}{\left(\beta - x_R\right)^2} - \frac{\alpha}{2} \frac{I_{0II}^2}{\left(\lambda \cdot \beta + x_R\right)^2} + F_0 = 0. \tag{8.74}$$

Without loss of generality, the rest position can be set to $x_R = 0$ using the asymmetry ratio λ.

Depending on the *sign* of the static *disturbance force* F_0, Eq. (8.74) gives following relations for the rest currents:

$$\begin{aligned} F_0 > 0: &\quad I_{0II}^2 = \lambda^2 \cdot \left(I_{0I}^2 + F_0 \cdot \frac{2\beta^2}{\alpha} \right), \\ F_0 < 0: &\quad I_{0I}^2 = I_{0II}^2 \frac{1}{\lambda^2} - F_0 \cdot \frac{2\beta^2}{\alpha}, \\ F_0 = 0: &\quad \frac{I_{0II}}{I_{0I}} = \lambda. \end{aligned} \tag{8.75}$$

Transducer parameters The dynamics for sufficiently small motions about the rest position x_R are again described using the operating-point-dependent *two-port parameters*. For $x_R = 0$ and $L_0 := \alpha/\beta$, the following parameters result:

- *electromagnetic current stiffness*

$$k_{em,I} := \frac{L_0}{\beta^2} \cdot \left[{I_{0I}}^2 + \frac{1}{\lambda^3} {I_{0II}}^2 \right] \quad \left[\frac{\mathrm{N}}{\mathrm{m}} \right],$$

- *current coefficient*

$$K_{em,I} := \frac{L_0}{\beta} \cdot \left[I_{0I} + \frac{1}{\lambda^2} I_{0II} \right], \quad \left[\frac{\mathrm{Vs}}{\mathrm{m}} \right], \left[\frac{\mathrm{N}}{\mathrm{A}} \right], \tag{8.76}$$

- *rest inductance*

$$L_R := L_0 \cdot \left[1 + \frac{1}{\lambda} \right] \quad \left[\mathrm{H} \right].$$

The *transfer matrix* can again be obtained from the relations in Eq. (8.52).

Unstable transfer characteristics The dominant and typical dynamic property of an electromagnetic bearing with a *free* armature is *unstable* transfer characteristics, as can be seen by substituting the transducer parameters (8.76) into the *characteristic polynomial* of the transfer matrix in Eq. (8.55), giving

$$\Delta_I(s) = s^2 - \frac{k_{em,I}}{m} = s^2 - {\Omega_{I0}}^2 = \left(s + \Omega_{I0} \right)\left(s - \Omega_{I0} \right), \tag{8.77}$$

where

$$\Omega_{I0} := \frac{1}{\beta} \sqrt{\frac{L_0}{m} \cdot \left[{I_{0I}}^2 + \frac{1}{\lambda^3} {I_{0II}}^2 \right]}. \tag{8.78}$$

All channels of the transducer transfer matrix thus exhibit an *unstable pole* $s = +\Omega_{I0}$ located in the right half-plane.

Thus, an electromagnetic bearing can not be operated with open loop control—*stable operation* always requires a (local) *controller*. Despite its exponentially unstable character, the bearing can in principle be stabilized employing only a PID loop. Given high dynamic requirements, however, more complex control approaches are required (Schweitzer and Maslen 2009).

Example 8.2 *Electromagnetic bearing.*

Fig. 8.23 depicts the pole configuration for an eight-pole magnetic radial bearing. Every two pairs of poles form a bearing such as is shown in Fig. 8.22c (vertical bearings Va1-Va2, Vb1-Vb2 and horizontal bearings Ha1-Ha2, Hb1-Hb2) giving the pole spreading angle $\theta = 22.5°$.

The following *physical data* are given:

rotor mass $m = 1$ kg, symmetric air gap width $\delta = 1$ mm, $\lambda = 1$,
maximum bearing force (vertical, horizontal) $F_{em,\max} = 1$ kN,
maximum flux density $B_{\max} = 1$ T.

Find the parameters defining the magnetic and electrical layout and the model parameters for the uncontrolled configuration. For the vertical bearings, determine the rest currents compensating the weight of the rotors.

Design procedure Assignment of values proceeds using the surrogate two-pole model in Fig. 8.22a and the implementation shown in Fig. 8.22c. For this purpose, the physical air gap width δ must be converted into the surrogate air gap width $\beta = \delta/\cos\theta$. The computed parameters for the surrogate model can then be easily assigned to the physical implementation shown in Fig. 8.22c. The surrogate configuration thus completely describes both the vertical and horizontal bearings.

Pole area The required total pole area is obtained from Eq. (8.41) as $A_{pol} = 17.4$ cm^2.

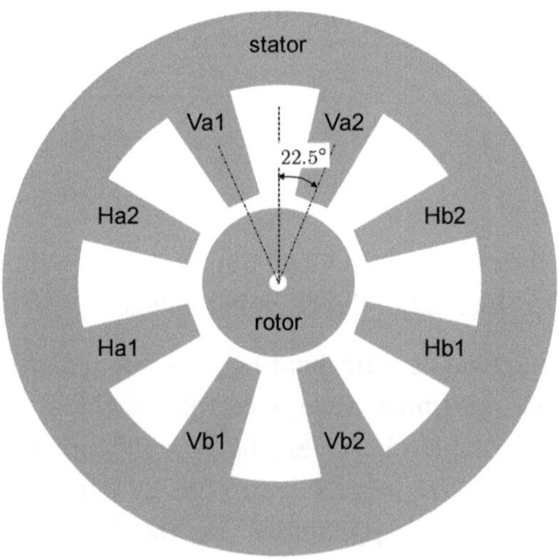

Fig. 8.23. Eight-pole magnetic radial bearing (vertical bearings Va1-Va2, Vb1-Vb2; horizontal bearings Ha1-Ha2, Hb1-Hb2; cf. Fig. 8.22c)

Magnetomotive force Allowing for the maximum allowable flux density $B_{\max}$, the surrogate air gap width β and Eq. (8.42) give a maximum magnetomotive force $\mathfrak{F}_{\max} = N \cdot I_{\max} = 1040\ \mathrm{AT}$. There is now a measure of freedom in the design to generate the requisite magnetomotive force by suitably combining the winding count N and maximum coil current $I_{\max}$. Using a design *assumption* of $I_{\max} = 4\ \mathrm{A}$ (constrained, e.g., by the power electronics), each surrogate pole requires $N = 260$ turns.

Operating-point-independent transducer parameters Using the henceforth fixed pole and coil configuration, the operating-point-independent transducer parameters can be computed. From the surrogate parameters $\alpha = 1.6 \cdot 10^{-4}\ \mathrm{Vsm/A}$ and $\beta = 1.1 \cdot 10^{-3}\ \mathrm{m}$, there follow both the rest inductance per surrogate pole / pair of poles $L_0 = 147\ \mathrm{mH}$, and—from Eq. (8.76)—the *bearing inductance* $L_R = 294\ \mathrm{mH}$ (for both vertical and horizontal bearings).

To compute the *field coil resistances*, consider the physical implementation in Fig. 8.22c. For each pole, an area of $A_{pol}/2$ must be wound with N turns. Assuming an average pole shoe circumference of $l_{\mathrm{pol}} = 16\ \mathrm{cm}$, it follows that each coil requires $l_{coil} = N \cdot l_{pol} \approx 42\ \mathrm{m}$ of wire, and assuming a coil wire diameter of $\varnothing_{wire} = 0.5\ \mathrm{mm}$, Eq. (8.43) gives a resistance $R_{coil} = 3.8\ \Omega$ per coil (i.e. pole) in Fig. 8.22c. For any pair of poles (vertical or horizontal), there is thus a *surrogate winding resistance* $R_W = 2R_{coil} = 7.6\ \Omega$.

Rest currents In the steady state, in addition to the tensioning force, the vertical bearing must counter the weight $F_0 = mg$ at the symmetric rest position $x_R = 0$ (see first equation (8.75)). Thus only one of the two rest currents I_{0I}, I_{0II} should be considered a *free parameter*. In general, it is desirable to choose the rest currents to be as large as possible, as this increases the tensioning force. However, with high rest currents, the available dynamic force is limited by $I_0 \pm \Delta i \leq I_{\max}$. As no further requirements regarding the available dynamic force are given in the present case, as an example $I_{0I} = I_{\max}/2 = 2\ \mathrm{A}$ can be chosen, so that Eq. (8.75) gives $I_{0II} = 2.036\ \mathrm{A}$.

Operating-point-dependent transducer parameters With the henceforth known rest currents, Eqs. (8.76) and (8.78) can be used to obtain the operating-point-dependent transducer parameters:
bearing stiffness $k_{em,I} = 1.02\ \mathrm{kN/mm}$, *current coefficient* $K_{em,I} = 549\ \mathrm{N/A}$, *bearing bandwidth* $\Omega_{I,0} = 1012\ \mathrm{rad/s}$ or $f_{I,0} = 161\ \mathrm{Hz}$. ■

8.3.8 EM transducers with constant working air gaps

Transducer type Electromagnetic (EM) transducers with constant working air gaps allow for armature motion *parallel* to the pole shoe surface, and have configurations according to Types E through G in Table 8.1.

Properties of the reluctance force Though all three transducer types presented here employ the same force principle, at first glance, quite different relationships between the transducer inductance, reluctance force, and armature position result

One common attribute of all three types, however, is that with *current drive* as presented here, the force acting on the armature (the tangential force) always attempts to bring the armature into a position fully overlapping the pole shoe conveying the flux, and disappears when complete overlap is achieved. In this latter case, the transducer is in a stable *rest position.* For Type E, the reluctance force is constant and independent of the armature position; for Type F it is approximately constant for $a \gg x$; and for Type G, it acts as an approximately linear restoring force for small displacements. In all three cases, however, similar dynamic properties result.

Configuration equations To serve as an example, consider a transducer configuration corresponding to *Type E* in Table 8.1 (see also Fig. 8.24). Assuming a homogeneous field, the *transducer inductance* and the general statement Eq. (8.79) give the *reluctance force* in *PSI-coordinates* and *Q-coordinates*, respectively, as

$$L_T(x) := \gamma \cdot x \quad , \quad \gamma = N^2 \frac{\mu_0 b}{2\delta} \tag{8.79}$$

$$\Rightarrow \quad \frac{\partial L_T(x)}{\partial x} = \gamma ,$$

$$F_{em}(x, \psi_T) = \frac{1}{2} \frac{1}{\gamma x^2} \psi_T^{\,2} , \qquad F_{em}(i_T) = \frac{1}{2} \gamma \, i_T^{\,2} . \tag{8.80}$$

EM transducers with current drive

Rest positions For static excitations $i_S(t) = I_0$, $F_{ext}(t) = F_0$, and the *transducer force* F_{em} from Eq. (8.80), Eq. (5.41) gives the rest position x_R in Q-coordinates (current drive):

$$x_R = \frac{1}{2} \frac{\gamma}{k} I_0^{\,2} + \frac{1}{k} F_0 . \tag{8.81}$$

Using a suitable rest current I_0, *any* desired armature position x_R can be *stably* selected following Eq. (8.81).

Transducer parameters The dynamics are described by the operating-point-dependent *two-port parameters* (hybrid form, Eqs. (8.36), (8.37))

$$\bullet\ \textit{electromagnetic current stiffness}\quad k_{em,I} := 0 \quad \left[\frac{\mathrm{N}}{\mathrm{m}}\right],$$

$$\bullet\ \textit{current coefficient}\quad K_{em,I} := \gamma \cdot I_0 \quad \left[\frac{\mathrm{Vs}}{\mathrm{m}}\right], \left[\frac{\mathrm{N}}{\mathrm{A}}\right], \tag{8.82}$$

$$\bullet\ \textit{rest inductance}\quad L_R := \gamma \cdot x_R \quad \left[\mathrm{H}\right].$$

The *transfer matrix* can be obtained from the previously mentioned relations in Eq. (8.52).

Dynamic behavior Due to the lack of electromagnetic current stiffness, the transducer eigenfrequency depends solely on the elastic suspension and is thus operating-point-*independent*. As the current coefficient is similarly constant, the resulting transfer functions are all constant as well (having no operating-point-dependent parameter variations, which is important for controller design).

Pull-in under current drive As can be seen from the rest position condition and the electromagnetic stiffness, given the assumed field conditions, *no* pull-in occurs.

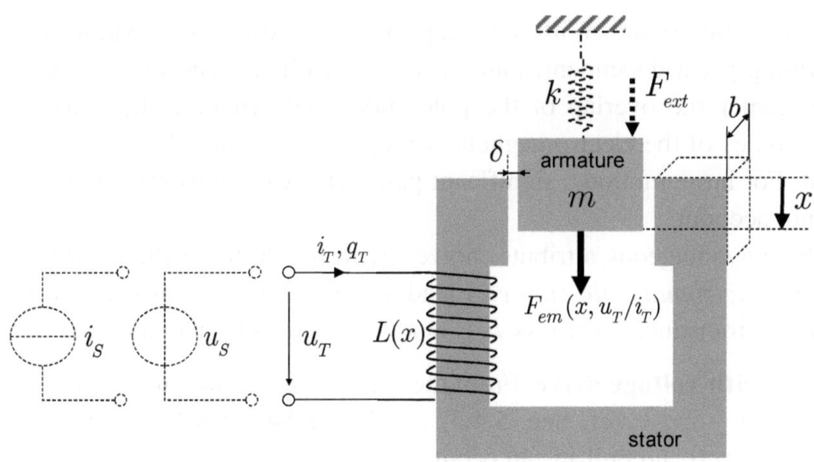

Fig. 8.24. Schematic configuration of an electromagnetic (EM) transducer with constant air gap and armature motion parallel to the pole shoes

EM transducers with voltage drive

Rest positions Taking into account the explanations in Sec. 8.3.6 regarding *steady-state flux linkage*, Eq. (5.40) and the drive-independent *transducer force* F_{em} from Eq. (8.80) result in the *PSI-coordinate rest position condition* for a static mechanical excitation $F_{ext}(t) = F_0$

$$2\gamma\,(kx_R - F_0)x_R^{\,2} = \Psi_0^{\,2}. \tag{8.83}$$

Of the three possible solutions, only one results in a stable rest position, selectable via Ψ_0.

Transducer parameters for lossless transducers The dynamics are defined by the operating-point-dependent *two-port parameters* in admittance form from Eqs. (8.34), (8.35), giving

$$\begin{aligned} &\bullet\ \textit{electromagnetic voltage stiffness} \quad k_{em,U} := -\frac{\Psi_0^{\,2}}{2\gamma\, x_R^{\,3}} \left[\frac{\mathrm{N}}{\mathrm{m}}\right], \\ &\bullet\ \textit{voltage coefficient} \quad K_{em,U} := \frac{\Psi_0}{\gamma\, x_R^{\,2}} \left[\frac{\mathrm{N}}{\mathrm{Vs}}\right], \left[\frac{\mathrm{A}}{\mathrm{m}}\right], \end{aligned} \tag{8.84}$$

and the rest inductance L_R as in Eq. (8.82).

The *transfer matrix* can be derived analogously to the previously mentioned relations in Eq. (8.63).

Dynamic behavior With a voltage drive, all transducer parameters are operating-point-dependent, and thus vary with the chosen rest position. The greater the overlap of the pole shoes and armature, the smaller the magnitudes of the electromagnetic voltage stiffness and the voltage coefficient. For large motions, significant parameter variations must thus be taken into account.

One advantageous attribute, however, turns out to be the *negative* sign on the *electromagnetic stiffness*, leading to a fundamental increase in the resulting transducer stiffness $k_{T,U} = k - k_{em,U}$ (electromagnetic stiffening).

Pull-in with voltage drive From the fundamental relation for a stable rest position $k_{em,U} < k$ (cf. Sec. 5.4.3) and Eq. (8.84), it follows directly that for all $x_R > 0$, *no* pull-in can occur.

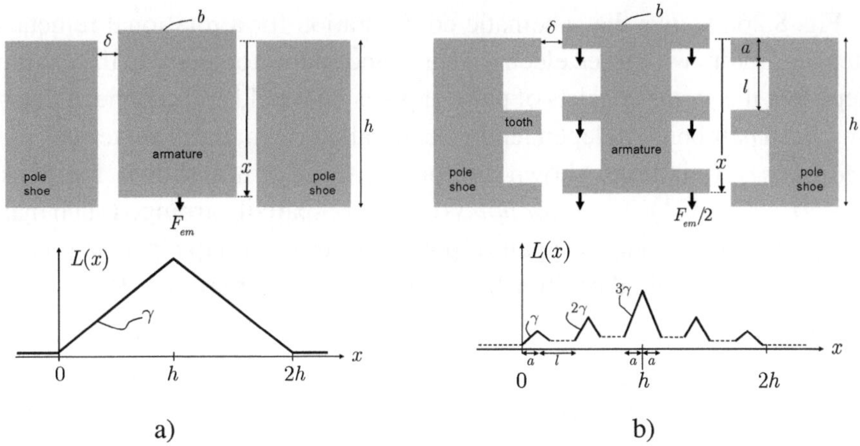

Fig. 8.25. Teeth on pole shoes and armature, configuration and inductance curve: a) straight armature, b) toothed armature

Teeth on armature and pole shoes

Force multiplication The fact that the tangential force is independent of the overlapping pole surface area $b \cdot x$ (see Fig. 8.24) can be taken advantage of for force multiplication. By creating *tooth* structures on the armature and pole shoe, as shown in Fig. 8.25b, each overlapping pair of teeth on the armature and pole shoe gives the same tangential reluctance force F_{em} as for the geometrically equivalent straight armature (Fig. 8.25a). In the armature position shown in Fig. 8.25b, three times the reluctance force is generated as for the comparable straight armature in Fig. 8.25a. However, this force is only useful over a significantly smaller displacement $x = h \pm a$.

8.3.9 Reluctance stepper motor

Layout and principle of operation The toothed armature and pole shoe structure presented in the previous section can be advantageously combined with a temporally and spatially phase-shifted electrical drive. Such configurations with an non-suspended, kinematically guided, toothed reluctance rotor (armature) are known as *reluctance stepper motors* and have been very successfully adopted with a large variety of implementations in many application areas (Hughes 2006). The electrical excitation can occur in either the stator or the rotor.

Fig. 8.26a shows the schematic configuration for a rotational reluctance stepper motors with three electric phases and four rotor teeth. In the example depicted, the *stator* consists of pairs of poles with field coils connected in series, each pair being independently electrically driven; these are termed *electric phases* (windings shown for phase $A - A'$, not shown for phases $B - B'$, $C - C'$). The *rotor poles* (teeth) are spatially arranged such that at any point in time, only one pair of poles can fully overlap a pair of poles on the stator (in the initial position here: poles $2 - 4$ overlap $B - B'$).

Single-phase drive If, given no current in coils $B - B'$ and $C - C'$, a current i_A is imposed on coils $A - A'$, this current induces the previously described *tangential reluctance forces* at rotor poles 1 and 3, respectively directed towards stator poles A and A'. This then causes the rotor to move in the direction shown by an angle $\Delta\varphi_S$—the *step angle*—until poles $1 - 3$ fully overlap $A - A'$ (final position in Fig. 8.26b). Given a sustained current flow i_A, this pole position represents a stable rest position, as further displacement causes the rotor poles to be pulled back into the original pole position. In this rest position, there is a braking moment which depends on the current i_A as described by Eq. (8.80). Thus, in this mode of operation, only one phase at a time is supplied with current.

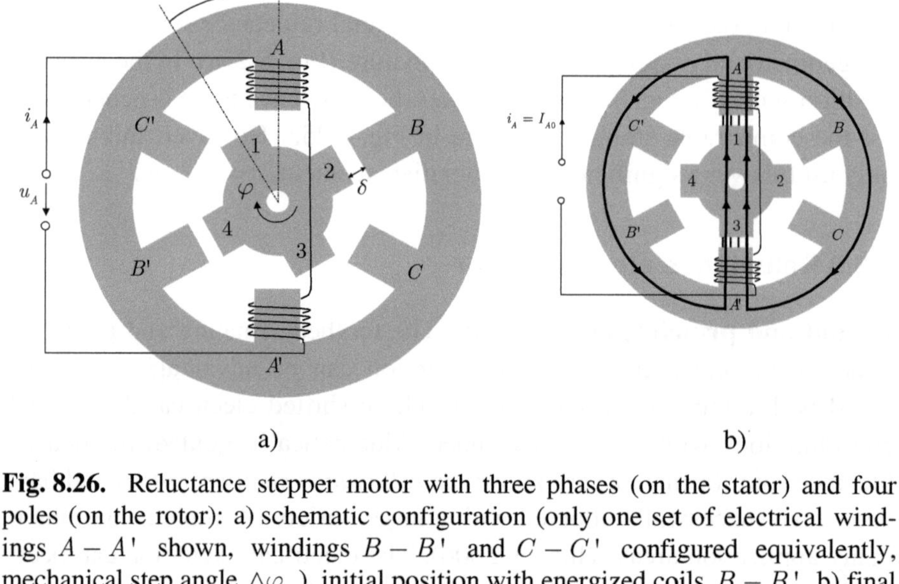

Fig. 8.26. Reluctance stepper motor with three phases (on the stator) and four poles (on the rotor): a) schematic configuration (only one set of electrical windings $A - A'$ shown, windings $B - B'$ and $C - C'$ configured equivalently, mechanical step angle $\Delta\varphi_S$), initial position with energized coils $B - B'$, b) final position for energized coils $A - A'$ (magnetic flux flow indicated)

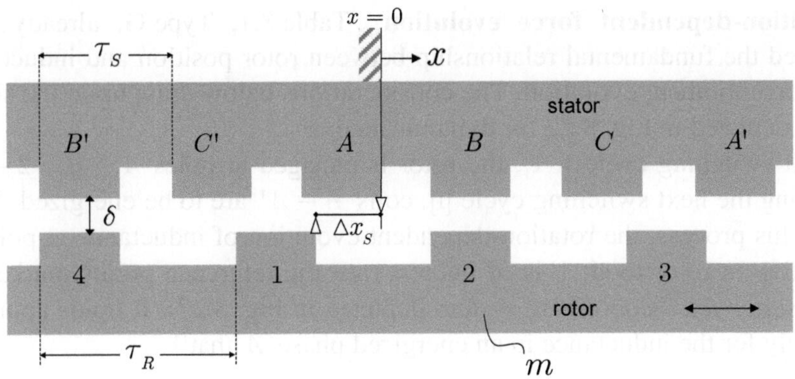

Fig. 8.27. Stator- and rotor tooth structure for a reluctance stepper motor (represents an unrolled version of the stator and rotor in Fig. 8.26, or a linear stepper motor for planar motion)

The current $i_A(t)$—upon whose temporal profile, remarkably, no particular demands need be placed—thus induces a *self-controlled* step-wise motion of the rotor by a clearly-defined step length $\Delta\varphi_S$, leading to the self-explanatory term *stepper motor.*

Reference configuration For the mathematical description of the dynamics, consider the configuration shown in Fig. 8.27. This configuration can be interpreted either as the *spatially unrolled* stator and rotor of the rotational stepper motor in Fig. 8.26 or as the schematic configuration for a *linear stepper motor*. For a generalized description, thus let the generalized motion coordinate x be introduces, the corresponding assignment for a rotating motor is obvious.

Step angle The step angle of a stepper motor depends on the spatial electromechanical layout of the stator and rotor. The mechanical *step angle* (step length) is given by the following relation (Kallenbach et al. 2008), (Kreuth 1988), (Ogata 1992)

$$\Delta x_S = \frac{\lambda}{N_{phase} \cdot N_{RT}}, \tag{8.85}$$

where λ is the rotor perimeter (for rotating motors, $\lambda = 2\pi$ or $360°$), N_{phase} is the number of electrical phases, and N_{RT} is the number of rotor teeth[12].

[12] For the stepper motor in Fig. 8.26, given $N_{phase} = 3$, $N_{RT} = 4$, $\lambda = 360°$, a mechanical step angle $\Delta\varphi_S = 30°$ results.

Position-dependent force evolution Table 8.1, Type G, already presented the fundamental relationship between rotor position and inductance or force/moment evolution. The considerations below draw upon the situation depicted in Fig. 8.27 for dynamic analysis.

At switching cycle [i-1], the rotor is engaged at poles $4-B'$, $2-B$. During the next switching cycle [i], coils $A-A'$ are to be energized. During this process, the rotation-dependent evolution of inductance at pole A relative to rotor tooth 1 is of interest (see the reference points marked in the *stator-fixed* coordinate system depicted in Fig. 8.27). It holds approximately for the inductance in an energized phase A that[13]

$$L_A(x) = L_0 + L_V \cos\left(\nu x\right), \quad \nu = \frac{2\pi}{\tau_R}. \tag{8.86}$$

The reluctance force of the pole pair $A-A'$ on the rotor can then be obtained from Eq. (8.26) as

$$F_{em,A} = -\frac{1}{2} i_A{}^2 \nu L_V \sin\left(\nu x\right). \tag{8.87}$$

The rotor position at the beginning of step [i] shown in Fig. 8.27 corresponds to an initial displacement of the rotor $x = -\Delta x_S$.

For a stator current $i_A(t) \neq 0$, a displacement $x < 0$ result in a positive reluctance force—the rotor is pulled toward pole shoe A—while for $x = 0$, *no* reluctance force is generated; and for $x > 0$, a restoring reluctance force results.

Configuration equations, single-phase drive To better understand the dynamic properties of a reluctance stepper motor, an investigation of the nonlinear *equations of motion* is helpful. Under current drive with *impedance feedback* using a shunting resistance $R = 1/Y$ (cf. Sec. 5.5.2, Eq. (5.54), and Eq. (8.86)) and an *initial value* $x(0) = -\Delta x_S$, these take the following form:

$$\begin{aligned} m\ddot{x} + \frac{1}{2} i_T^2 \,\nu\, L_V \sin\left(\nu x\right) &= F_{ext}, \\ Y\left(L_0 + L_V \cos\left(\nu x\right)\right) \frac{di_T}{dt} + \left(1 - Y \,\nu\, L_V \sin\left(\nu x\right) \dot{x}\right) i_T &= i_S. \end{aligned} \tag{8.88}$$

---

[13] By suitably shaping the pole shoes, this form can in fact be approximated. Even with differing geometries, the first spatial harmonic of the inductance $L(x)$ always has the properties defined by Eq. (8.86): maximum inductance at complete pole overlap, and the form of an even function of rotor position with an added constant component.

Qualitative model: virtual displacement excitation A simple qualitative model of the dynamic behavior during the stimulation of a step motion is provided via the following observation. With $Y=0$ and an imposed current (current drive), it can be assumed that current build-up in the in the field coil—and thus the generation of force on the armature—occurs without delay. With a constant exciting phase current $i_A(t) = I_{A0}$ and the coordinate transformation $\tilde{x} = x + \Delta x_S$, the equation of motion (8.88) without an external force ($F_{ext} = 0$) is then

$$m\ddot{\tilde{x}} = -\frac{1}{2} I_{A0}^2 \nu L_V \sin\left(\nu(\tilde{x} - \Delta x_S)\right), \quad \tilde{x}(0) = \dot{\tilde{x}}(0) = 0\,. \tag{8.89}$$

The autonomous nonlinear differential equation (8.89) has an equilibrium at $\tilde{x} = \Delta x_S$ ($x = 0$)[14].

A simple interpretation of Eq. (8.89) is given by way of the following consideration. Using the definition $F_0 := 1/2 \cdot I_{A0}^2 \nu L_V$, it follows from Eq. (8.89) that, for sufficiently small displacements[15],

$$m\ddot{\tilde{x}} \approx -F_0 \nu \tilde{x} + F_0 \nu \Delta x_S\,. \tag{8.90}$$

Eq. (8.90) takes the form of a mass-spring system with spring constant $\tilde{k}_{em} = F_0 \nu$ and a *displacement excitation* Δx_S. The displacement excitation can also be interpreted as an *external force* $\tilde{F}_{ext} = F_0 \nu \Delta x_S = \tilde{k}_{em} \Delta x_S$, so that Eq. (8.90) can also be written as

$$m\ddot{\tilde{x}} + \tilde{k}_{em} \tilde{x} = \tilde{F}_{ext}\,. \tag{8.91}$$

The transition from the initial position $\tilde{x}(0) = 0$ to the final state during a switching cycle can be approximated as an undamped single-mass oscillator with exciting force $\tilde{F}_{ext} = k_{em} \Delta x_S$. However, due to the lack of damping, the equilibrium $\tilde{x}_R = \Delta x_S$ is *not* asymptotically stable, and the rotor will oscillate about this rest position.

Transducer parameters, single-phase drive An equivalent and even more predictive description of the dynamics is given by the *operating-*

---

14 The multiple periodic equilibria are not of interest here.

15 For purely qualitative considerations, this assumption is entirely permissible. For quantitative predictions, the nonlinear model Eq. (8.89) must be consulted.

point-dependent two-port parameters in hybrid form from Eqs. (8.36), (8.37):

- *electromagnetic current stiffness*

$$k_{em,I} := -\frac{1}{2} i_{A,R}^2 \, \nu^2 \, L_V \cos\left(\nu x_R\right) \quad \left[\frac{\mathrm{N}}{\mathrm{m}}\right], \tag{8.92}$$

- *current coefficient* $$K_{em,I} := -i_{A,R} \, \nu \, L_V \sin\left(\nu x_R\right) \quad \left[\frac{\mathrm{Vs}}{\mathrm{m}}\right], \left[\frac{\mathrm{N}}{\mathrm{A}}\right], \tag{8.93}$$

- *rest inductance* $$L_R(x) := L_0 + L_V \cos\left(\nu x_R\right) \; \left[\mathrm{H}\right]. \tag{8.94}$$

In the above relations, the linearization operating points $i_{A,R}$, x_R are still kept general to facilitate the remaining discussion.

The *transfer matrix* is given by the relations in Eq. (8.52), only $k_{em,U}$ remains to be determined from Eq. (8.38).

Discussion of dynamics: single-phase drive An obvious choice of operating point for linearization is the steady-state coil current $i_{A,R} = I_{A0}$ and the rest position $x_R = 0$.

Though the case considered here is an actuator application, due to the fictitious displacement excitation $\tilde{F}_{ext} = F_0 \nu \Delta x_S$, the transfer channel $F_{ext} \rightarrow x$ is of primary interest. As expected, the electromagnetic stiffness $k_{em,I}$ here coincides with the stiffness $\tilde{k}_{em}$ found in the above approximating derivation (the differing sign results from the defining assignment $k_{em,I} := \partial F_{em,I} / \partial x$).

The transfer function

$$G_{x/F,I} = \frac{1}{k_{em,I}} \frac{1}{1 + s^2 / \Omega_I^{\,2}}, \quad \Omega_I^{\,2} = \frac{k_{em,I}}{m}$$

and the transfer function from the displacement excitation

$$G_{x/\Delta x_S} = \frac{X(s)}{\Delta X_S(s)} = \frac{1}{1 + s^2 / \Omega_I^{\,2}}$$

demonstrate the expected marginally stable dynamics, i.e. an undamped oscillation about the rest position (the final position under pole shoe A). For practical operations, this is naturally unacceptable—damping is required.

Passive damping: weak effectiveness of impedance feedback One simple solution might be to introduce passive damping via *impedance feedback*, as was done for previous transducer types. The fact that in this case this is fundamentally only *weakly effective* can be shown by examining the current coefficient $K_{em,I}$ in the neighborhood of the rest position $x_R = 0$. From Eq. (8.93), it follows that $K_{em,I} \approx 0$ in this operating regime. Thus, electromechanical coupling is practically nonexistent, as can be easily seen from the signal-flow diagram Fig. 8.12 (see also the nonlinear equation of motion (8.88)).

For $K_{em,I} \neq 0$, the motion-induced voltage $u_{ind} = K_{em,I} \cdot \dot{x}$ acts via the electrical feedback in the shunting resistor R as a velocity-proportional damping factor. From the general structure of $K_{em,I}$ shown in Eq. (8.93), it follows that at the beginning of a switching cycle [i]—i.e. when the rotor is still sufficiently far from the rest position $x_R = 0$— there is electromechanical damping, as $K_{em,I} \neq 0$. However, as the final position $x_R = 0$ is approached, this feedback is lost.

Mechanical damping Damping of the transition in a switching cycle [i] can be achieved by adjusting the *viscous mechanical bearing friction* of the rotor. However, this option obviously has certain limits: too much friction on top of a large external load can completely inhibit the motion of the rotor (Kreuth 1988).

Multi-phase drive One common option for improving the operating dynamics of reluctance stepper motors employs synchronous, coordinated drive of multiple phases, a *multi-phase drive*. For one, this makes available additional rotor rest positions. Additionally, there is the possibility of passive or active damping of the rotor resonance. This principle has been investigated and implemented in countless varieties (Krishnan 2001), (Miller 2002), (Ogata 1992).

Fig. 8.28 shows an example of *two-phase drive* of the reluctance motor in Fig. 8.26. With a simultaneous energizing of two phases, the magnetic flux forms loops as shown, making available the depicted rest positions. Since, in such cases, there are two reluctance force components acting on the rotor, suitable variation of the phase currents i_A, i_B at the rest position can be used to generate damping corrective forces, leading to *active damping* (Krishnan 2001), (Middleton and Cantoni 1986).

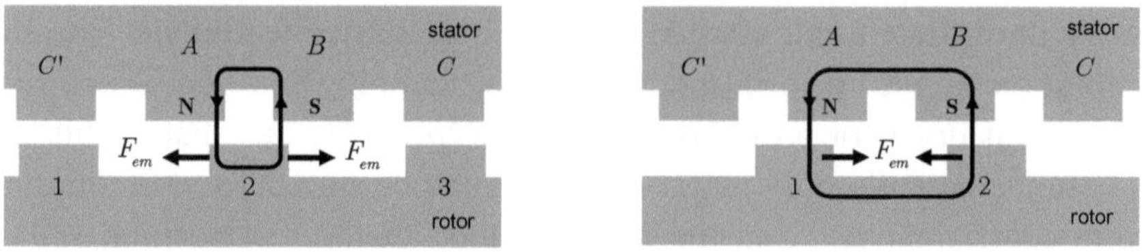

Fig. 8.28. Two-phase drive (excitation of A, B) of the reluctance transducer in Fig. 8.26: possible rest positions

Another method producing coordinated switching of the field coils consists of a single-phase drive and simultaneous shunting with impedance feedback in other phase windings, known as *passive electromagnetic damping* (Hughes and Lawrenson 1975), (Russell and Pickup 1996).

Stable set for rest positions For the reluctance stepper motor, questions of stable sets and the stability of the rest position at the end of a step deserve particular attention. This requires an analysis of the stability of the nonlinear equation of motion (8.88). For current drive with a constant coil current, this reduces to an analysis of the mechanical equation of motion (first Eq. (8.88)) as a nonlinear oscillating differential equation. It is convenient to perform this analysis in *phase space* (displacement-velocity). This enables the determination of the set of all initial states $x(0),\ \dot{x}(0)$ — the *stable set*—for which, given various external forces F_{ext} (e.g. COULOMB friction, a load), a stable transition to the rest position $x_R = 0$, i.e. the final position of step [i], results. Such analyses are well developed; for reasons of space, the reader is referred to the specialized literature (Kreuth 1988).

Example 8.3 *Reluctance stepper motor with single-phase drive.*

The step dynamics are to be investigated for a current-drive reluctance stepper motor with the following parameters:

$N_{phase} = 3,\ N_{RT} = 4 \ \Rightarrow\ \Delta\varphi_S = 30°$

$J = 100\ \mathrm{g\,cm^2},\ I_0 = 0.5\ \mathrm{A},\ L_0 = 18\ \mathrm{mH},\ L_V = 15\ \mathrm{mH}$.

Transducer parameters at the rest position For the rest position (final position at end of a step), a stiffness of $k_{em,I} = 30\ \mathrm{N/mm}$ is obtained, and thus a resonant frequency $\Omega_I = 55\ \mathrm{rad/s}$ or $f_I = 8.7\ \mathrm{Hz}$.

Passive damping: impedance feedback A suitable shunting resistance can be easily estimated using the linearized model. Impedance feedback is fundamentally effective only away from the final position at $\varphi \neq 0$, a suitable intermediate position between the final position $\varphi = 0$ and initial value $\varphi(0) = -\Delta\varphi_S = -30°$ can be chosen for linearization, e.g. $\tilde{\varphi}_R = 0.5 \cdot \varphi(0)$. Then Eqs. (8.92) through (8.94) along with (8.38) give the transducer parameters $\tilde{\Omega}_I = 39\ \mathrm{rad/s}$, $\tilde{\Omega}_U = 64\ \mathrm{rad/s}$, and $\tilde{L}_R = 25.5\ \mathrm{mH}$. The corresponding optimal resistance $R^{\max}$ for maximum damping is obtained from the general determining equation (5.61) as

$$\tilde{R}^{\max} = \tilde{L}_R \tilde{\Omega}_U \sqrt{\tilde{\Omega}_U / \tilde{\Omega}_I} \approx 2\ \Omega\,. \tag{8.95}$$

Fig. 8.29 depicts the transient behavior based on the nonlinear equations of motion (8.88) with $\tilde{R}^{\max} = 2\ \Omega$ in Curve 1. Both the effect of impedance feedback during the initial phase and the practically reduced-amplitude, undamped oscillation about the final position $\varphi = 0$ can be seen. By varying the parameter $\tilde{R}^{\max}$, it can be easily verified that the approximate value from Eq. (8.95) defines the maximum achievable damping quite well. Thus the chosen procedure and the applicability of the design formula in Eq. (5.61) are also validated for the nonlinear case.

Mechanical damping The damping achievable with simple electrical means is naturally insufficient for practical operation of the stepper motor. As a comparison, in Fig. 8.29, Curve 2 traces the transient behavior with an additional *viscous damping* of $b = 10^{-3}\ \mathrm{Nms}$. Only here is a useful dynamic stepping behavior achieved.

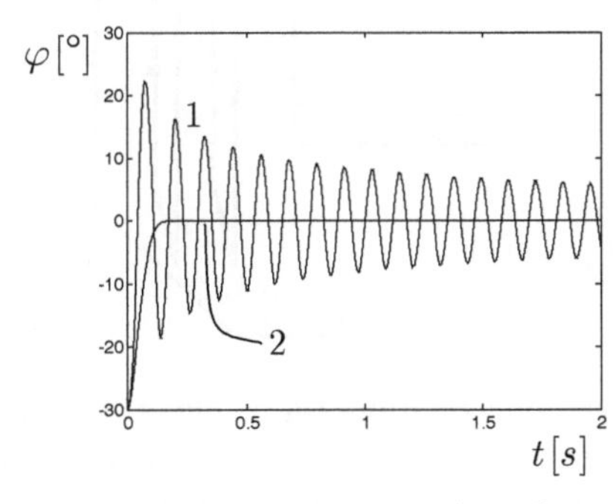

Fig. 8.29. Reluctance stepper motor: transient dynamics for one step of a single-phase drive ($\varphi(0) = -30°$); Curve 1: impedance feedback with $R^{\max} = 2\ \Omega$, Curve 2: mechanical damping $b = 10^{-3}\ \mathrm{Nms}$

■

8.4 Generic ED Transducer: LORENTZ Force

8.4.1 System configuration

Coil moving in a magnetic field Electrodynamic (ED) transducers exploit the LORENTZ force introduced in Eq. (8.22), i.e. the force on an electrical conductor in a magnetic field. Fig. 8.30 shows a schematic implementation where the conductor takes the form of a *coil* rigidly coupled to an elastically-suspended load mass m (here with one mechanical degree of freedom, i.e. one-dimensional translation). It is characteristic of an electrodynamic (ED) transducer to have a (generally constant) *magnetic field* $\vec{B}_0$ acting on the coil, which is induced either electrically or with a permanent magnet.

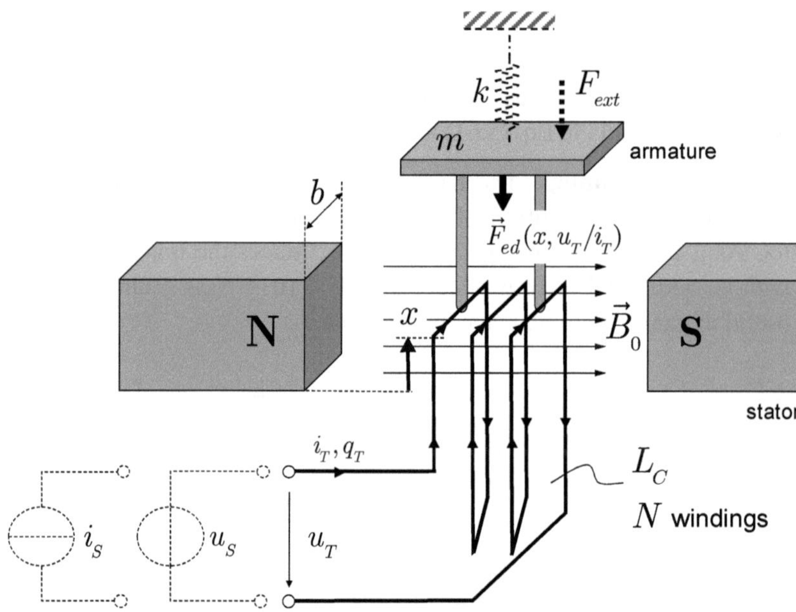

Fig. 8.30. Schematic configuration for a generic translational electrodynamic (ED) transducer with one mechanical degree of freedom (armature moves in one dimension, rigidly connected to a moving rectangular coil in a homogeneous, stationary magnetic field: electrodynamic—LORENTZ—force generation); dashed lines indicate the external loading with voltage/current source and elastic suspension

8.4.2 Constitutive electrodynamic transducer equations

Electrodynamic constitutive basic equations The fundamental physical quantity in electromagnetic transduction is the flux linkage of the reference configuration depicted in Fig. 8.30. According to Eq. (8.17), the flux linkage has two components[16]

$$\psi_T(x, i_T) = \psi_{ed}(x) + \psi_{\mathfrak{F}}(i_T). \tag{8.96}$$

The first component $\psi_{ed}(x)$ describes the magnetic flux linked to the coil via the *externally* induced *magnetic field* $\vec{B}_0$. It is obvious that this flux depends on the relative geometry between the coil and magnetic field. This variable geometry is parameterized here with the variable location x of the upper coil edge.

The second component, $\psi_{\mathfrak{F}}(i)$, describes the magnetic flux generated in the magnetic circuit of the *coil* by the magnetomotive force $\mathfrak{F} = Ni_T$ following Eq. (8.18). This coil-induced magnetic field practically only propagates outside the iron core, and is thus largely independent of coil position, as Eqs. (8.13) and (8.20) give a *coil inductance*

$$L_C = \frac{N^2}{R_{m\Sigma}} \approx \frac{N^2 \mu_0 A_\delta}{\delta} = const. \; .$$

For an electrically linear magnetic circuit and the reference configuration shown in Fig. 8.30, Eq. (8.96) gives the *basic electrodynamic constitutive equation* in *Q-coordinates*

$$\psi_T(x, i_T) = K_{ED} \cdot x + L_C \cdot i_T, \tag{8.97}$$

with coil inductance L_C and *ED force constant* defined by

$$K_{ED} := N \cdot B_0 \cdot b \quad [\mathrm{N/A}], \tag{8.98}$$

where N is the turns count of the coil, B_0 is the magnetic flux density of the externally induced magnetic field, and b is the width of the assumed homogeneous magnetic flux field.

The *electrodynamic constitutive basic equation* in *PSI-coordinates* equivalent to Eq. (8.97) is obtained via a simple rearranging of terms

$$i_T(x, \psi_T) = -\frac{K_{ED}}{L_C} x + \frac{1}{L_C} \psi_T. \tag{8.99}$$

---

16 The flux component ψ_0 in Eq. (8.17) can be interpreted as $\psi_{ed}(x = const.)$.

The remaining derivation is now presented in somewhat more detail than for the electromagnetic (EM) transducer, as the constitutive relations (8.97) and (8.99) were only marginally considered in the foundational Chapter 5.

ELM energy functions As previously discussed, integration of the basic constitutive equations (8.97), (8.99) gives the ELM energy functions in

- *Q-coordinates*

$$T^*_{ed}(x, i_T) = K_{ED} \cdot x \cdot i_T + \frac{1}{2} L_C \cdot i_T^{\,2}, \tag{8.100}$$

- *PSI-coordinates*[17]

$$T_{ed}(x, \psi_C) = \frac{1}{2L_C}\left(\psi_T - K_{ED} \cdot x\right)^2. \tag{8.101}$$

Constitutive ELM transducer equations Differentiation of the ELM energy functions (8.100), (8.101) via the procedure presented in detail in Ch. 5 gives the *constitutive ELM transducer equations* for the *electrodynamic (ED) transducer* in the various coordinate representations:

- *Q-coordinates*

$$\begin{aligned} F_{ed,Q}(i_T) &= K_{ED} \cdot i_T, \\ u_T(\dot{x}, \frac{di_T}{dt}) &= K_{ED} \cdot \dot{x} + L_C \cdot \frac{di_T}{dt}, \end{aligned} \tag{8.102}$$

- *PSI-coordinates*

$$\begin{aligned} F_{ed,\Psi}(x, \psi_T) &= -\frac{K_{ED}^{\,2}}{L_C} x + \frac{K_{ED}}{L_C} \psi_T, \\ i_T(x, \psi_T) &= -\frac{K_{ED}}{L_C} x + \frac{1}{L_C} \psi_T. \end{aligned} \tag{8.103}$$

Eqs. (8.102) and (8.103) describe the *electrodynamic (ED) force laws* and the dynamics at the terminals for the lossless unloaded transducer as a function of either of the electrical terminal variables i_T or $u_T = \dot{\psi}_T$.

---

[17] The integration is non- trivial. It is easiest to compute the kinetic energy function from the kinetic co-energy function in Eq. (8.100) using the LEGENDRE transform $T(\psi) = i \cdot \psi - T^*(i)$ discussed in Ch. 2.

Electrodynamic force or LORENTZ force The force equation in Eq. (8.102) with imposed coil current i_T represents precisely the LORENTZ force from Eq. (8.22) acting on the current-carrying *upper* coil wires

$$F_{ed,Q}(i_T) = Nb \cdot \left(i_T \cdot B_0\right). \tag{8.104}$$

The applicable length of conductor in the magnetic field is $l = N \cdot b$ and the force direction follows the cross product in Eq. (8.22).

The LORENTZ force has two notable properties differing from the reluctance force discussed in previous sections: $F_{ed,Q}(i_T)$ is independent of the coil displacement x, and is linearly dependent on the coil current i_T. The configuration in Fig. 8.30 thus permits *feedback-free, linear, bipolar force generation* at the mechanically attached armature.

Cancellation of LORENTZ force components The fact that only a vertical electrodynamic force $\vec{F}_{ed}$ results for the coil configuration in Fig. 8.30 can be easily illustrated with the representation depicted in Fig. 8.31. Naturally, LORENTZ forces $\vec{F}_{ed,1}$, $\vec{F}_{ed,3}$ also act on the vertical wire sections. However, due to the antiparallel current direction, these forces cancel each other out, so that only the force $\vec{F}_{ed,2}$ acting on the upper wire contributes to the electrodynamic force on the coil.

Energy storage vs. energy transformation The electrodynamic transducer has one interesting and not readily apparent property with respect to energy storage. Closer inspection of the energy functions reveals that only the coil inductance L_C stores energy; the external magnetic field B_0 does not contribute to storage and only serves to transform the energy.

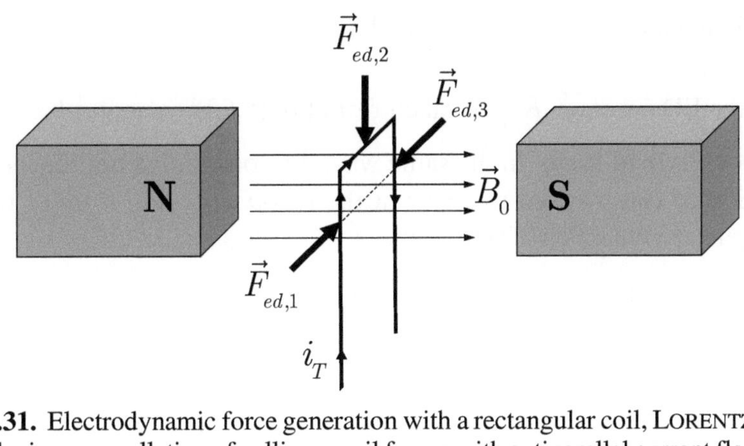

Fig. 8.31. Electrodynamic force generation with a rectangular coil, LORENTZ forces on coil wires: cancellation of collinear coil forces with antiparallel current flows

8.4.3 ELM two-port model

Linear dynamics From the constitutive transducer equations (8.102) and (8.103), the inherent linearity of the dynamics of an electrodynamic (ED) transducer can already be seen. Thus, no further linearization is required and these equations can be directly drawn upon for the two-port representation.

Two-port admittance form The constitutive transducer equations (8.103) in PSI-coordinates directly give the *two-port admittance form* of the unloaded electrodynamic (ED) transducer

$$\begin{pmatrix} F_{ed,\Psi}(s) \\ I_T(s) \end{pmatrix} = \mathbf{Y}_{ed}(s) \cdot \begin{pmatrix} X(s) \\ U_T(s) \end{pmatrix} = \begin{pmatrix} k_{ed,U} & \dfrac{K_{ed,U}}{s} \\ -K_{ed,U} & \dfrac{1}{s}\dfrac{1}{L_C} \end{pmatrix} \begin{pmatrix} X(s) \\ U_T(s) \end{pmatrix}, \tag{8.105}$$

with the defining relations for the constant (drive-independent) *transducer parameters*

- *electrodynamic voltage stiffness* $k_{ed,U} := -\dfrac{K_{ED}^{\;2}}{L_C} \quad \left[\dfrac{\mathrm{N}}{\mathrm{m}}\right],$

- *voltage coefficient* $K_{ed,U} := \dfrac{K_{ED}}{L_C} \quad \left[\dfrac{\mathrm{N}}{\mathrm{Vs}}\right], \left[\dfrac{\mathrm{A}}{\mathrm{m}}\right],$ (8.106)

- *coil inductance* $L_C \; [\mathrm{H}],$

where the ED constant K_{ED} is taken from Eq. (8.98).

Two-port hybrid form In the same way, the constitutive transducer equations (8.102) can by used to arrive at the *two-port hybrid form* of the unloaded electrodynamic (ED) transducer

$$\begin{pmatrix} F_{ed,Q}(s) \\ U_T(s) \end{pmatrix} = \mathbf{H}_{ed}(s) \cdot \begin{pmatrix} X(s) \\ I_T(s) \end{pmatrix} = \begin{pmatrix} k_{ed,I} & K_{ed,I} \\ sK_{ed,I} & sL_C \end{pmatrix} \begin{pmatrix} X(s) \\ I_T(s) \end{pmatrix}, \tag{8.107}$$

with the defining relations for the constant (drive-independent) *transducer parameters*

$$\bullet\ \textit{electrodynamic current stiffness} \qquad k_{ed,I} := 0 \quad \left[\frac{\mathrm{N}}{\mathrm{m}}\right],$$

$$\bullet\ \textit{current coefficient} \qquad K_{ed,I} := K_{ED} \quad \left[\frac{\mathrm{Vs}}{\mathrm{m}}\right],\left[\frac{\mathrm{N}}{\mathrm{A}}\right], \tag{8.108}$$

with the *coil inductance* L_{sp} and *ED force constant* K_{ED} from Eq. (8.98).

8.4.4 Loaded electrodynamic (ED) transducer

Transfer matrix for current drive Eq. (8.107) and Table 5.8 give the *transfer matrix* $\mathbf{G}(s)$, along with the signal-flow diagram in Fig. 8.32,

$$\begin{pmatrix} X(s) \\ U_T(s) \end{pmatrix} = \mathbf{G}(s)\begin{pmatrix} F_{ext}(s) \\ I_S(s) \end{pmatrix} = \begin{pmatrix} V_{x/F,I}\dfrac{1}{\{\Omega_0\}} & V_{x/i}\dfrac{1}{\{\Omega_0\}} \\ V_{u/F}\dfrac{s}{\{\Omega_0\}} & V_{u/i}\cdot s\dfrac{\{\Omega_U\}}{\{\Omega_0\}} \end{pmatrix}\begin{pmatrix} F_{ext}(s) \\ I_S(s) \end{pmatrix}, \tag{8.109}$$

with the parameters

$$\Omega_0^2 = \frac{k}{m}, \quad k_{T,U} = k + K_{ED}^2/L_C, \quad \Omega_U^2 = \frac{k_{T,U}}{m}, \quad \text{where } \Omega_0 < \Omega_U, \tag{8.110}$$

$$V_{x/F,I} = \frac{1}{k}, \quad V_{x/i} = V_{u/F} = \frac{K_{ED}}{k}, \quad V_{u/i} = L_C + \frac{K_{ED}^2}{k}.$$

The *eigenfrequency* of the transfer function at the electrical terminals (electrical impedance) is given by the mechanical eigenfrequency Ω_0 independent of the type of drive, and, as usual, corresponds to the anti-resonance given voltage drive.

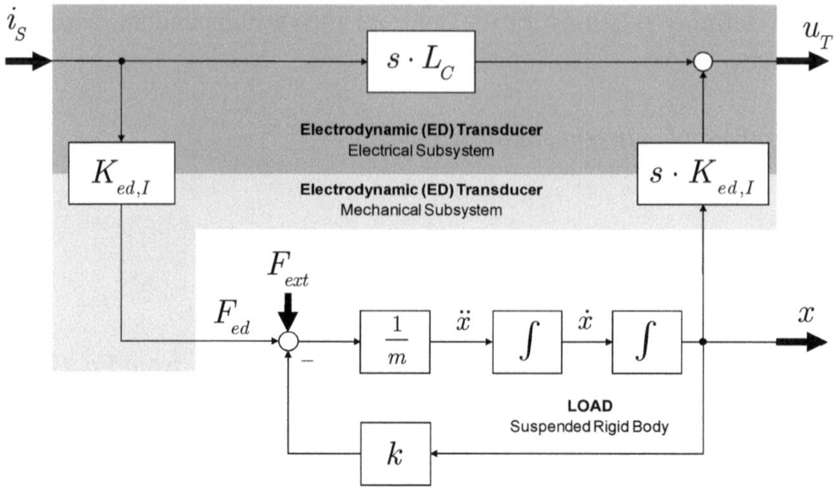

Fig. 8.32. Signal-flow diagram for a loaded *electrodynamic* (ED) transducer with *current drive* (lossless, ideal current source, one translational degree of freedom, mechanical load: elastically suspended rigid body, cf. Fig. 8.30); note the electrically linear generation of force $F_{ed} = K_{ed,I} \cdot i_Q$ which is mechanically decoupled from the armature motion and independent of coil inductance (back EMF)

Transfer matrix with voltage drive Eq. (8.105) and Table 5.8 give the *transfer matrix* $\mathbf{G}(s)$, along with the signal-flow diagram in Fig. 8.33,

$$\begin{pmatrix} X(s) \\ I_W(s) \end{pmatrix} = \mathbf{G}(s) \begin{pmatrix} F_{ext}(s) \\ U_Q(s) \end{pmatrix} = \begin{pmatrix} V_{x/F,U} \dfrac{1}{\{\Omega_U\}} & V_{x/u} \dfrac{1}{s \cdot \{\Omega_U\}} \\ V_{i/F} \dfrac{1}{\{\Omega_U\}} & V_{i/u} \cdot \dfrac{\{\Omega_0\}}{s \cdot \{\Omega_U\}} \end{pmatrix} \begin{pmatrix} F_{ext}(s) \\ U_Q(s) \end{pmatrix} \tag{8.111}$$

with the parameters

$$\Omega_0^{\ 2} = \frac{k}{m}, \quad k_{T,U} = k + K_{ED}^{\ 2}/L_C, \quad \Omega_U^{\ 2} = \frac{k_{T,U}}{m}, \quad \text{where } \Omega_0 < \Omega_U$$

$$V_{x/F,U} = \frac{1}{k_{T,U}}, \quad V_{x/u} = \frac{K_{ED}}{k_{T,U}\, L_C}, \quad V_{i/F} = -\frac{K_{ED}}{k_{T,U}\, L_C}, \quad V_{i/u} = \frac{1}{L_C + \dfrac{K_{ED}^{\ 2}}{k}}. \tag{8.112}$$

For voltage drive, it can be seen that the mechanical stiffness becomes larger than the electrodynamic stiffness, so that the former determines the eigenfrequency of the transducer.

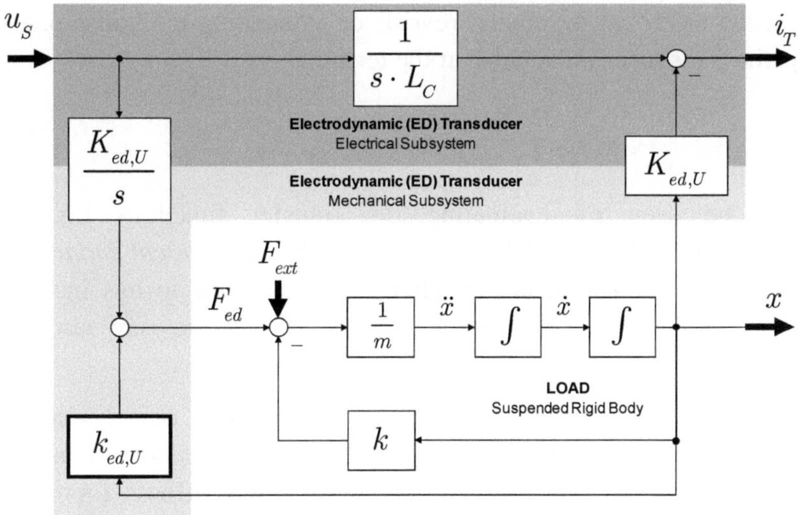

Fig. 8.33. Signal-flow diagram for a loaded *electrodynamic* (ED) transducer with *voltage drive* (lossless, ideal voltage source, one translational degree of freedom, mechanical load: elastically suspended rigid body, cf. Fig. 8.30)

Dynamic analysis

ELM coupling factor Applying the relations from Sec. 5.6 results in the general formula for the *ELM coupling factor* of an *electrodynamic* (ED) transducer

$$\kappa_{ed}^2 = \frac{L_C \cdot K_{ed,U}^2}{k - k_{ed,U}} = \frac{1}{1 + \dfrac{L_C}{K_{ed,I}^2}\left(k - k_{ed,I}\right)} = \frac{1}{1 + \dfrac{k \cdot L_C}{K_{ED}^2}}. \tag{8.113}$$

Eq. (8.113) demonstrates the previously mentioned nature of the energy conversion process in an electrodynamic (ED) transducer. Given negligible mechanical ($k \to 0$) and electrical ($L_C \to 0$) energy storage, the (lossless) electrodynamic (ED) transducer acts as an *ideal* electromechanical *energy transducer* ($\kappa_{ed} = 1$). Regardless, however, the *greatest possible ED force constant* K_{ED} ensures the best possible electromechanical energy conversion ($\kappa_{ed} \approx 1$).

Mechanical inductance An examination of the static inductance of the elastically suspended armature (gain factors $V_{u/i}$, $V_{i/u}$ of the transfer functions at the electrical terminals) reveals an interesting relationship. Effectively, the transducer inductance at the terminals is

$$L_T = L_C + \frac{K_{ED}^{\;2}}{k} = L_C + L_{ED,mech}\,. \tag{8.114}$$

As can be seen by evaluating the transfer functions Eq. (8.109) through (8.112), the added term $L_{ED,mech}$ —the *mechanical inductance*— represents the inductance active when $L_C = 0$ [18]. Due to this increase in transducer inductance, induced transducer voltages u_T are also correspondingly higher.

Fundamental electrodynamic stiffness Analogously, on the mechanical side, there is an added term k_{EDF} —the *fundamental electrodynamic stiffness*. This describes the effective electrodynamic stiffness for the non-suspended armature (with $k = 0$), i.e.

$$k_{T,U} = k + K_{ED}^{\;2}/L_C = k + k_{EDF}\,. \tag{8.115}$$

Lossy transducer: impedance feedback All of the properties relating to impedance feedback and current vs. voltage drive previously discussed in detail for the electromagnetic transducer (e.g. the optimal parameter settings in Fig. 8.17, see Sec. 8.3.6) hold analogously for the electrodynamic transducer (see also Example 8.4).

Under *current drive*, both the *coil resistance* and motion-induced electrical voltage have very small effects. For this reason, power amplifiers with current output are often employed to drive electrodynamic transducers. The transfer matrix in Eq. (8.109) then fully describes the transducer dynamics. If however, additional mechanical damping is to be applied under current drive, a shunt must be employed. In this case, the transfer function change can be estimated as previously described.

When driven by a *voltage source* (power amplifier with voltage output), the finite coil resistance makes stable operation possible, so that for a realistic dynamic analysis, transfer functions incorporating impedance feedback must always be considered. If, as in the case of very small external mechanical damping, additional electrodynamic damping is required, a se-

[18] For the limits $L_C \to 0$ or $k \to 0$, the simultaneous dependence of the gains and of Ω_0, Ω_U on L_C, k must be taken into consideration.

ries resistance or complex impedance can again be employed as a targeted design degree of freedom.

8.4.5 Structural principles

Rotational electrodynamic (ED) transducer

Electrical machines In general, rotational transducers allow large rotational motion amplitudes to be achieved. These can also be used for translational degrees of freedom by employing a variety of transmissions and kinematic machines. All classical *electrical machines* (electric motors, electric generators) are based on the electrodynamic principle. The relevant theory of and the countless implementational types within this domain belong to the standard repertoire of electrical and mechanical engineering curricula. For this reason, a more thorough discussion of electrical machine theory is dispensed with here; the reader is instead referred to the extensive technical literature, e.g. (Hughes 2006).

However, in order to put this theory into the context of the dynamic models developed in this book, a few general, fundamental aspects of simple implementations are somewhat further illuminated in this section.

Homogeneous magnetic field The schematic configuration of a rotational transducer with an air core coil inside a homogeneous magnetic field is sketched in Fig. 8.34a. As previously explained, it is the coil flux linkage which controls the electromechanical conversion. Using the parameters seen in Fig. 8.34a and following Eq. (8.97), this is given as a function of the mechanical degree of freedom φ by

$$\psi_T(\varphi, i_T) = \widehat{K}_{ED} \cdot \sin\varphi + L_C \cdot i_T \tag{8.116}$$

with the *ED torque constant*

$$\widehat{K}_{ED} := N \cdot B_0 \cdot A_C \quad [\mathrm{Nm/A}]. \tag{8.117}$$

The *constitutive ELM transducer equations* in *Q-coordinates* are

$$\begin{aligned} \tau_{ed,Q}(\varphi, i_T) &= \widehat{K}_{ED} \cdot \cos\varphi \cdot i_T\,, \\ u_T(\varphi, \dot{\varphi}, \frac{di_T}{dt}) &= \widehat{K}_{ED} \cdot \cos\varphi \cdot \dot{\varphi} + L_C \cdot \frac{di_T}{dt}\,. \end{aligned} \tag{8.118}$$

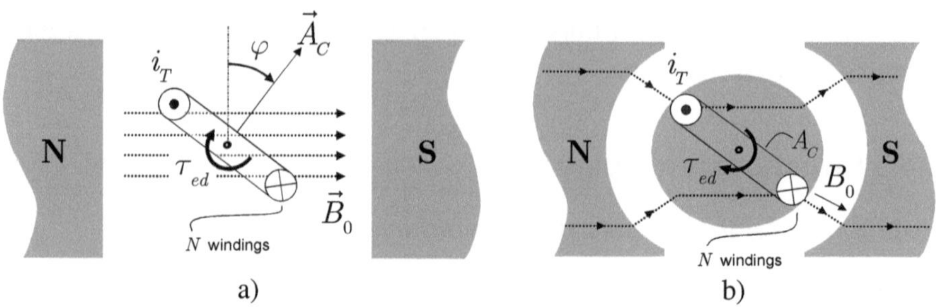

Fig. 8.34. Electrodynamic (ED) rotational transducer: a) planar pole shoes with air core coil (rotating coil), b) radial field with iron armature (dynamo)

Transducer parameters, hybrid form With steady-state values $\varphi = \varphi_R$, $i_T = I_0$, Eq. (8.118) gives the operating-point-dependent transducer parameters in hybrid form (current drive)[19]

- *electrodynamic current stiffness*

$$k_{ed,I}(\varphi_R, I_0) := -\widehat{K}_{ED} \cdot \sin\varphi_R \cdot I_0 \qquad \left[\frac{\text{Nm}}{\text{rad}}\right], \tag{8.119}$$

- *current coefficient* $$K_{ed,I}(\varphi_R) := \widehat{K}_{ED} \cdot \cos\varphi_R \qquad \left[\frac{\text{Vs}}{\text{rad}}\right], \left[\frac{\text{Nm}}{\text{A}}\right].$$

Transfer matrix As usual, Eq. (8.119) and Table 5.8 give the transfer matrix $\mathbf{G}(s)$ for the suspended ($k \neq 0$) and non-suspended ($k = 0$) cases.

Transducer stability As stable operation requires $k_{ed,I} < 0$, it follows directly from Eq. (8.119) that in the angle range $180° < \varphi < 360°$, stable operation is only possible with a reversal of the current ($I_0 < 0$). In a dynamo, this is provided by the commutator (Hughes 2006).

Maximum torque From the torque equation (8.118), it can be seen that the maximum transducer moment is generated at the angle $\varphi = 0$, i.e. when the coil windings lie parallel to the magnetic field. With increasing displacement, the transducer moment decreases; at $\varphi = 90°$ with the coil windings normal to the field, no torque is generated.

19 The transducer parameters in *admittance form* (voltage drive) are obtained analogously.

Radial magnetic field One variant with advantageous transducer dynamics and space utilization is shown in Fig. 8.34b. The flux-guiding iron armature induces a radial magnetic field across the air gap and thus always ensures a relative angle $\sphericalangle(\vec{A}_C, \vec{B}_0) = 90°$, i.e. maximum torque independent of displacement (and constant transducer parameters with $\varphi_R = 0$ in Eq. (8.119)). Furthermore, the significantly smaller air gap as compared to Fig. 8.34a results in a noticeably greater flux density B_0, enabling higher torque for the same electrical and magnetic parameters.

Electrodynamic voice coil transducer

Translational transducers, configuration One of the most widely distributed electrodynamic transducer types is the *voice coil transducer*. Fig. 8.35 shows a schematic configuration with vertical motion. Compared to the reference configuration in Fig. 8.30, the voice coil configuration makes much better use of space. The cylindrical magnetic field ensures maximum flux linkage with the coil. The minimum air gap is limited only by the winding wire width and the need for a small amount of clearance. This results in a high magnetic air gap flux density B_0 and thus a large *ED force constant*

$$K_{ED} = N_{B_0} \cdot B_0 \cdot 2\pi r_0, \tag{8.120}$$

where N_{B_0} represents the number of coil windings linked to the magnetic flux of the pole shoes. The transducer model is given precisely by Eqs. (8.102) through (8.115).

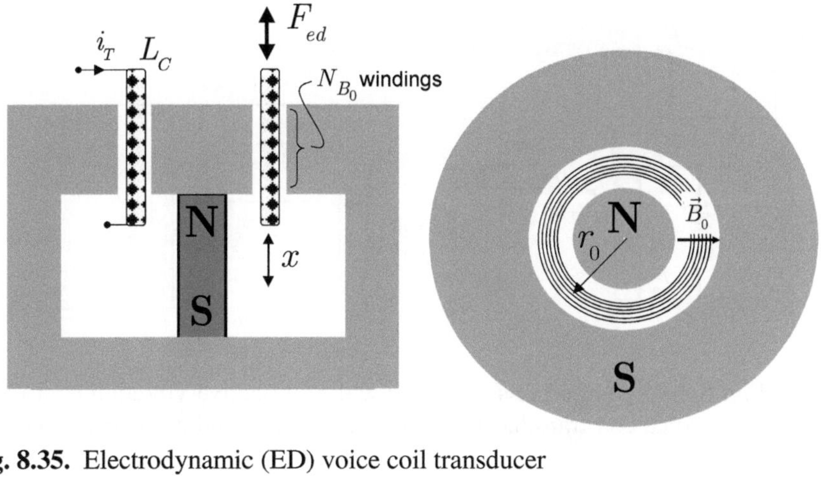

Fig. 8.35. Electrodynamic (ED) voice coil transducer

Applications Voice coil transducers can be operated both in suspended and non-suspended configurations, and are particularly preferred for to their linear operating dynamics. Some classic applications are in speakers and other acoustic transducers, scanning and autofocus systems for disc drives, and shakers for calibration and material testing (Lenk et al. 2011). An additional important area of application is passive and active vibration control in flexible mechanical structures (multibody systems). This mechatronic application is examined in more detail in Example 8.4 below.

Example 8.4 *Vibration control with a voice coil transducer.*

Problem statement The oscillatory mechanical system with negligible damping depicted in Fig. 8.36a is to be insulated from the effects of disturbance forces F_{ext}. The bandwidth of the disturbance forces is assumed to be much greater than the eigenfrequency $\Omega_0 = \sqrt{k_L/m_L}$ of the single-mass oscillator. An electrodynamic voice coil transducer is available as the damping actuator. The implemented solution is to avoid auxiliary energy input and specialized sensors.

Solution approach For a solution without sensors or auxiliary energy input, impedance feedback offers itself as a possibility. Just as in the case of the piezoelectric transducer, the electrodynamic transducer can operate using motion-induced currents and voltages without external auxiliary energy. By connecting the terminals with a suitable electrical load impedance $Z_{FB}(s)$, mechanical energy can be converted into electrical energy and dissipated in the resistive component of the load (semi-active vibration control, see Secs. 5.5.6 and 5.8).

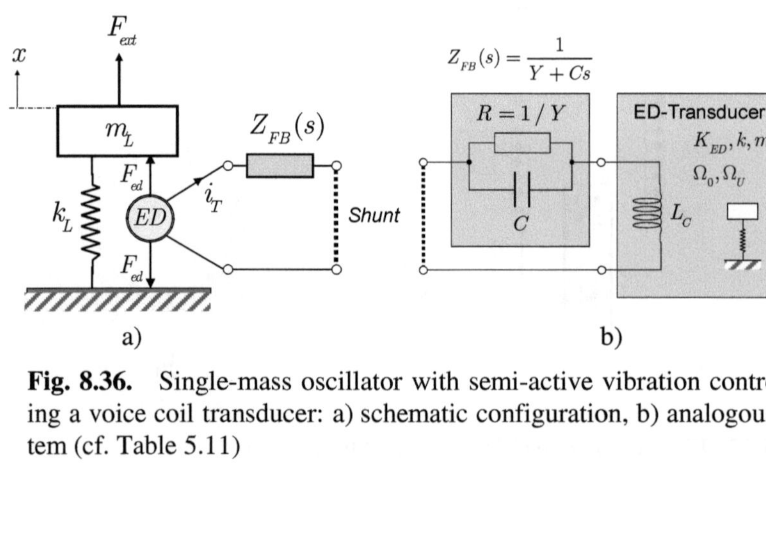

Fig. 8.36. Single-mass oscillator with semi-active vibration control using a voice coil transducer: a) schematic configuration, b) analogous system (cf. Table 5.11)

System parameters This example is based on an implemented, real system (Behrens et al. 2003) with the following parameters:
mechanical system $m_L = 0.15$ kg, $k_L = 56$ kN/m;
ED transducer $K_{ED} = 3.4$ N/A, $L_C = 1$ mH, $R_C = 3.3\ \Omega$.

Solution design To arrive at a solution, the derivations and design formulae of Sections 5.5.6 and 5.8 can be directly applied. As greater damping can generally be achieved with a complex impedance than with purely resistive impedance feedback, the impedance approach using the RC circuit shown in Fig. 8.36b is applied. This corresponds precisely to the standard structure in the right-hand column of Table 5.11. Without loss of generality, the model containing a voltage source with shunting—Table 5.11, upper right—is considered below (there being no difference from a current source with open terminals). For the design formula, Eq. (5.64) in Example 5.1 is adapted following the guidelines presented in Table 5.11 giving the *analytic design equations*

$$C^{opt} = \frac{1}{L_T}\frac{\Omega_I^2}{\Omega_U^4}, \quad Y^{opt} = \frac{1}{R^{opt}} = \frac{2}{L_T}\frac{\Omega_I^2}{\Omega_U^3}\sqrt{\frac{\Omega_U^2}{\Omega_I^2} - 1}\,, \tag{8.121}$$

with a pair of double complex conjugate poles with magnitude Ω_U and *maximum* achievable *damping*

$$d^{opt} = \frac{1}{2}\sqrt{\frac{\Omega_U^2}{\Omega_I^2} - 1}\,. \tag{8.122}$$

Due to the impedance feedback, the resulting disturbance transfer function between the force and mass displacement is (see transfer matrix Eq. (8.111))

$$\tilde{G}_{x/F}(s) = \frac{X(s)}{F_{ext}(s)} = \tilde{G}_{x/F} - Z_{FB} \cdot G_{x/u} \cdot G_{i/F} \frac{1}{1 + Z_{FB} \cdot G_{i/u}}\,. \tag{8.123}$$

Numerical solution With the given system parameters, the characteristic *resonant frequencies* can be computed to be $\Omega_I = \Omega_0 = 611$ rad/s, $\Omega_U = 671$ rad/s. With $L_T = L_C$, Eq. (8.121) gives the optimal component values $C^{opt} = 1.8$ mF, and $R^{opt} = 0.9\ \Omega$. The disturbance transfer function follows from Eq. (8.123):

$$G_{x/F}^{opt}(s) = 6.67\frac{737^2 + 610s + s^2}{\left(671^2 + 305s + s^2\right)^2}\,. \tag{8.124}$$

The maximum *damping* of the double poles of $G_{x/F}^{opt}$ obtained from Eq. (8.122) is $d^{opt} = 0.23$ (as expected, this agrees with Eq. (8.124)).

The magnitude response of $s \cdot G_{x/F}^{opt}$ (the velocity) and the disturbance step response are shown in Fig. 8.37, Curves 1. The desired damping effect can be easily seen (for comparison, the magnitude response Curve 0 is *without* impedance feedback).

Alternate solution In (Behrens et al. 2003), a different design strategy was employed based on the method of (Hagood and Flotow 1991). There, a series RC circuit was employed, i.e. $Z_{FB}(s) = R + 1/Cs$. In that case, the resulting optimal parameters are $C^* = 2.7\ \text{mF}$, $R^* = 0.29\ \Omega$, where R^* was found using a numerical $\mathcal{H}_2$-norm minimization. Again using Eq. (8.123), the disturbance transfer function is then

$$G_{x/F}^{*}(s) = 6.67 \frac{608^2 + 290s + s^2}{\left(502^2 + 119s + s^2\right)\left(740^2 + 171s + s^2\right)}. \tag{8.125}$$

The magnitude response of $s \cdot G_{x/F}^{*}$ (the velocity) and the disturbance step response are shown in Fig. 8.37, Curves 2.

Discussion The two design approaches lead to very similar system dynamics. The comparison design (Behrens et al. 2003) demonstrates somewhat better disturbance rejection in the frequency domain (the peak in the disturbance frequency response is ca. 3 dB lower, though somewhat broader).

In the time domain, both designs are comparable. Due to the larger damping of the poles, the design proposed here using Eq. (8.121) demonstrates faster decay of transients.

Realizing feedback impedances In the simplest case, realization of the feedback impedance is possible using discrete, passive components. However, in some applications, for pragmatic or situational reasons, it is better to resort to electronic realization of the impedance using operational amplifier circuits (Fleming et al. 2000), (Behrens et al. 2003), (Paulitsch et al. 2006).

Pragmatically, electronic implementation enables easier tuning in real systems. The example considered here also demonstrates a situation requiring electronic realization of the impedance. The optimal resistance is significantly less than the inherent coil resistance. This resistance can however be easily realized with suitable electronic means.

Design variation: increased damping If the given maximum damping is insufficient, the *analytic damping formula* (8.122) offers deeper insight into possible design variants. Given the underlying impedance feedback above, it follows from Eq. (8.122) that the maximum achievable damping depends only on the ratio (spread) of the transducer eigenfrequencies Ω_U, Ω_I. In the current case, $\Omega_I = \Omega_0 = \sqrt{k_L/m_L}$. Only Ω_U can be affected by the transducer parameters; from Eq. (8.112),

$$\Omega_U^{\ 2} = \frac{1}{m}\left(k + \frac{K_{ED}^{\ 2}}{L_C}\right). \tag{8.126}$$

An increase in Ω_U can be achieved either by *increasing* the *ED force constant* K_{ED}, or *decreasing* the effective *coil inductance* L_{sp}. In the former case, a more powerful transducer must be employed. The latter case can be realized with electronic inductance compensation, e.g. (Paulitsch et al. 2006).

Using the analytic damping formula (8.122), the order of magnitude of the increase in damping can also be directly deduced as a function of the design degrees of freedom K_{ED} and L_C. Thus, an *increase* of K_{ED} by a factor of $\sqrt{10} = 3.33$, or alternatively, a *decrease* of L_C by a factor of 10, both lead to an *increase* in *damping* to $\tilde{d}^{opt} = 0.72$. The corresponding frequency responses are depicted in Fig. 8.37 in Curves 3.

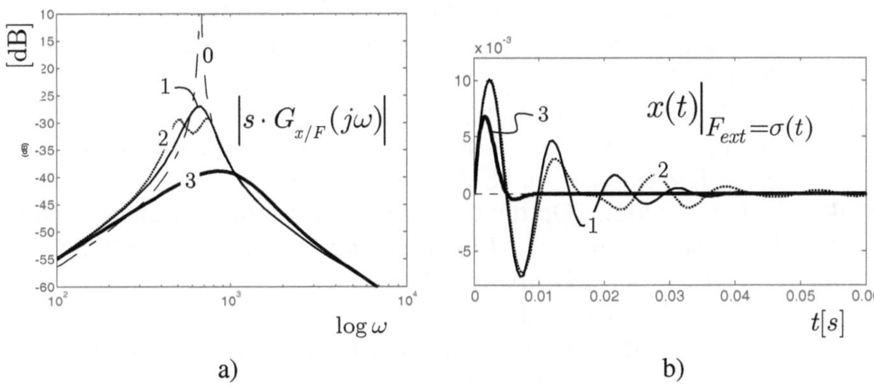

Fig. 8.37. Vibration control with an ED voice coil transducer and impedance feedback: a) magnitude response of the disturbance transfer function $s \cdot G_{x/F}$ (velocity to disturbance force), b) disturbance step response for the velocity of the load mass; legend: 0: $Z_{FB}=0$, 1: Z_{FB} from Eq. (8.121), 2: Z_{FB} from (Behrens et al. 2003), 3: Z_{FB} from Eq. (8.121) at $L_C/10$ or $3.33 \cdot K_{ED}$

■

Bibliography for Chapter 8

Behrens, S., A. J. Fleming and S. O. R. Moheimani (2003). "Electrodynamic vibration supression." *Proc. SPIE Smart Structures and Materials 2003: Damping and Isolation* **5052**(344): 344–355.

Fleming, A. J., S. Behrens and S. O. R. Moheimani (2000). "Synthetic impedance for implementation of piezoelectric shunt-damping circuits." *Electronic Letters* **36**(18): 1525–1526.

Hagood, N. W. and A. v. Flotow (1991). "Damping of structural vibrations with piezoelectric materials and passive electrical networks." *Journal of Sound and Vibration* **146**(2): 243-268.

Hughes, A. (2006). *Electric Motors and Drives: Fundamentals, Types and Applications*. Elsevier.

Hughes, A. and P. J. Lawrenson (1975). "Electromagnetic damping in stepper motors." *Proc. Inst. Elec. Eng.* **122**(8): 819-824.

Jackson, J. D. (1999). *Classical Electrodynamics, Third Edition*. John Wiley and Sons, Inc.

Kallenbach, E., R. Eick, P. Quendt, T. Ströhla, et al. (2008). *Elektromagnete*. Vieweg+Teubner.

Kreuth, H. P. (1988). *Schrittmotoren*. Oldenbourg Verlag.

Krishnan, R. (2001). *Switched Reluctance Motor Drives:Modeling, Simulation, Analysis, Design, and Applications*. CRC Press.

Lenk, A., R. G. Ballas, R. Werthschützky and G. Pfeifer (2011). *Electromechanical Systems in Microtechnology and Mechatronics*. Springer.

Middleton, R. H. and A. Cantoni (1986). "Electromagnetic Damping for Stepper Motors with Chopper Drives." *Industrial Electronics, IEEE Transactions on* **IE-33**(3): 241-246.

Miller, T. J. E. (2002). "Optimal design of switched reluctance motors." *Industrial Electronics, IEEE Transactions on* **49**(1): 15-27.

Ogata, K. (1992). *System Dynamics*. Prentice Hall.

Paulitsch, C., P. Gardonio and S. J. Elliott (2006). "Active vibration damping using self-sensing, electrodynamic actuators." *Smart Materials and Structures* **15**: 499–508.

Russell, A. P. and I. E. D. Pickup (1996). "Analysis of single-step damping in a multistack variable reluctance stepping motor." *Electric Power Applications, IEE Proceedings -* **143**(1): 95-107.

Schweitzer, G. and E. H. Maslen, Eds. (2009). *Magnetic Bearings*, Springer.

Thomas, R. E., A. J. Rosa and G. J. Toussaint (2009). *The Analysis and Design of Linear Circuits*. John Wiley and Sons, Inc.

9 Functional Realization: Digital Information Processing

Background The functionality of a mechatronic product fundamentally depends on its information processing capabilities—popularly referred to as the *"product intelligence"*—in the form of *operating software*. This refers to the ability to suitably control hardware-based system elements such that user commands, formulated at a highly abstract level, are carried out using the best possible (i.e. most robust and exact) motions. In the extreme case, using the same hardware configuration, differing software variants can even be used to solve completely different product tasks ("intelligent mechanics"). From a systems theory point of view, the control and regulation algorithms implemented in operating software close the control loop and determine the system dynamics, creating the "product intelligence".

Contents of Chapter 9 This chapter discusses suitable approaches for modeling phenomena affecting system behavior arising from the *digital realization* of control and regulation algorithms and data communications in a mechatronic system. *Frequency domain models* are chosen as the general descriptive form, enabling a clear depiction of system behaviors using *frequency responses*, while remaining compatible with continuous-time models. Building upon a detailed discussion of *sampling processes*, *signal aliasing*, and the smoothing action of *hold processes*, properties of discrete, sampled *transfer function models* are thoroughly examined. In addition to the s-domain, the presentation draws upon a conformal frequency transformation (the q-transform in the *transformed frequency domain*), allowing BODE diagrams to be employed directly and accurately, and giving a smooth transition to continuous models with very small sampling times. Particular attention is paid to *frequency response aliasing* in *oscillatory* systems (multibody systems). Further, the important model attributes of *quantization* during signal conversion (analog-to-digital and digital-to-analog), *time delays*, and *real-time aspects* are discussed in the context of *signal conversion*, *data communications*, and *controller realization*.

■

9.1 Systems Engineering Context

Product functionality: control and regulation algorithms Execution of the particular tasks of a mechatronic product (the product purpose) is fundamentally shaped by the processing of information in the form of sensor signals and manual user input (Fig. 9.1). The generation of appropriate control signals for the actuators closes the action chain (the control algorithms create a control loop), establishing a fundamental level of robustness to external disturbances and ever-present model uncertainties. Manual user signals can be suitably transformed into control commands or control sequences, and thus initiate automatic execution of operating sequences (forming control algorithms). In layman's terms, information processing capabilities represents the "intelligence" of a mechatronic product.

Systems engineering significance From a systems engineering point of view, the function *"process information"* realizable using digital information processing represents the control and regulation algorithms of a mechatronic system (see Fig. 9.1).

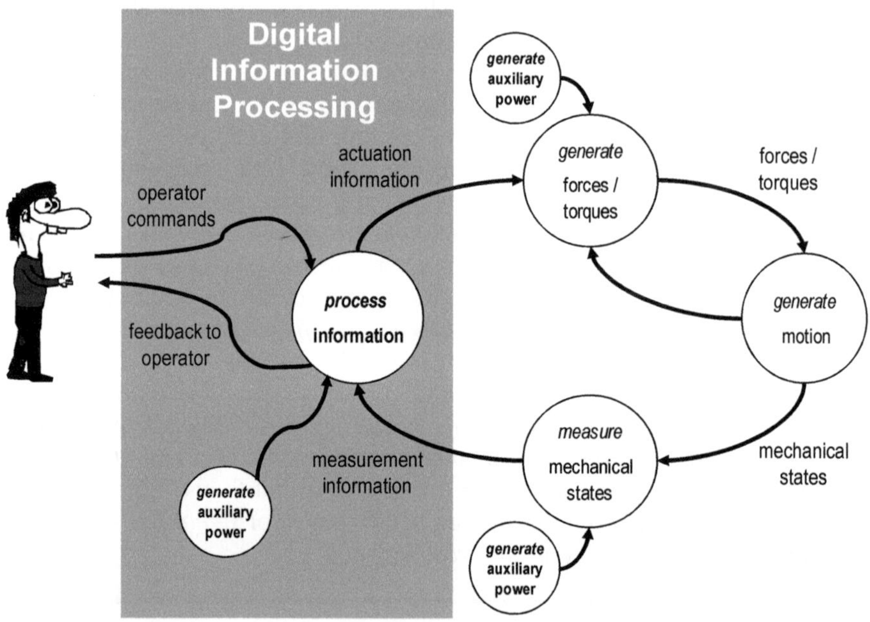

Fig. 9.1. Functional decomposition of a mechatronic system: functional realization using *digital information processing*

In the design of a mechatronic system, the mathematical descriptions of the relationship between inputs (sensor signals, user signals) and outputs (control signals) and of the behavior of the closed action loop are naturally of primary interest. Ultimately, all of the physical behaviors of the functional chain *"generate forces/torques"*, *"generate motion"*, *"measure mechanical states"* are to be produced by the control and regulation algorithms in a way which creates a system which conforms to specifications. The design of such control and regulation algorithms is based on physical models of the components of the above-mentioned functional chain (see Chs. 4 through 8), and is not the subject of considerations in this chapter. A few selected design considerations for control problems are dealt with in the next chapter (Ch. 10).

Rather, the focus of this chapter is the distinctive and mostly parasitic phenomena which appear when using digital information processing realizations.

Analog vs. digital information processing One elementary form of analog information processing at the local level was already discussed in the context of impedance feedback. The advantages of this type of processing are quite evident: at the analog electrical level, there is immediate action—as a rule not requiring specialized sensors—along with minimal signal delays and correspondingly large bandwidths for the feedback system. Against such desirable properties stand the disadvantages of limited functionality and low modifiability of analog implementations. Digital information processing with sufficiently powerful microprocessors offers wide-ranging options for the implementation of complex control and regulation algorithms, along with easy modifiability of the control system. This allows the functionality of a mechatronic product to be adapted and modified with substantially unchanged hardware (structure, sensors, and actuators) using only a modified program (software). This accounts for the fact that modern realization of information processing functions takes place almost exclusively using microprocessors and other digital processing elements (e.g. FPGAs, ASICs).

Mechatronically relevant phenomena In a closed loop, the digital information processing phenomena schematically depicted in Fig. 9.2 are particularly influential on the resulting system behaviors:

- *time discretization:* the processor deals with samples of continuous-time system quantities taken at specified points in time, mathematically described by a *sampling process* (a linear phenomenon);

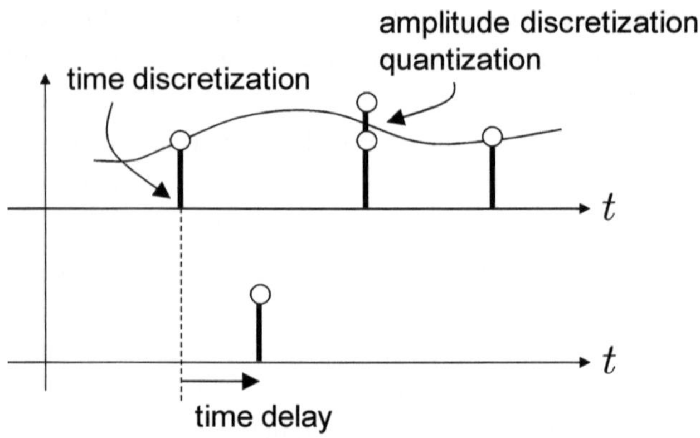

Fig. 9.2. Mechatronically relevant digital information processing phenomena: time discretization, time delay, amplitude discretization (quantization)

- *time delay:* during the generally sequential processing of information, there fundamentally exist time delays between the reading of sensor signals and the active outputting of control signals (a linear phenomenon);
- *amplitude discretization (quantization):* the internal computer representation of discrete-time signals can only take place with a finite word length, resulting in a *magnitude discretization* of the signal values (quantization, a nonlinear phenomenon).

9.2 Definitions

9.2.1 Reference configuration

Digital control loop Fig. 9.3 depicts a signal-oriented representation of the functional model of a mechatronic system shown in Fig. 9.1 as a digital control loop, shown with physically-oriented system elements.

The interfaces between the information processing system—the embedded processor—and the "analog" surroundings are represented by *analog-to-digital converters* (*ADCs*) and *digital-to-analog converters* (*DACs*).

Operator signals can optionally also have analog sources (e.g. joysticks), or be directly generated in the processor. A specialized low-pass filter—the *anti-aliasing filter*—for the suppression of signal ambiguities arising in the sampling process can be seen between the sensor output and the analog-to-digital converter.

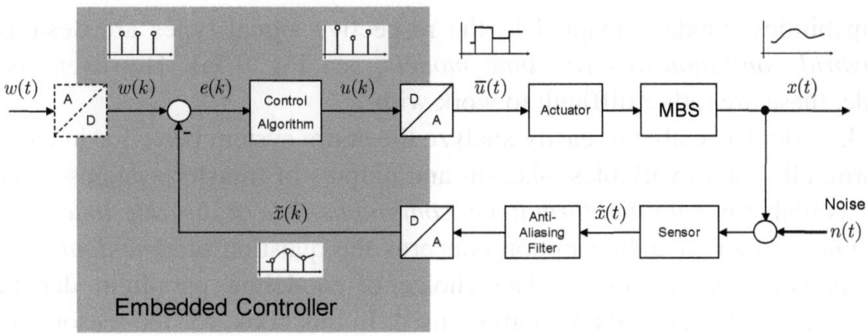

Fig. 9.3. Mechatronic system as a signal-oriented model forming a digital control loop

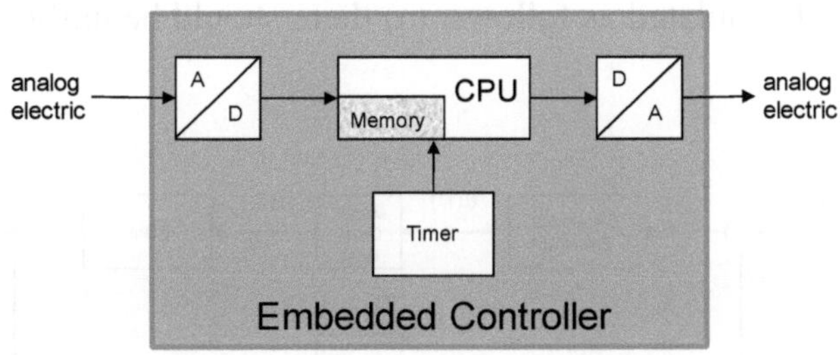

Fig. 9.4. Schematic realization of the function *"process information"* with an embedded microprocessor

All of the information processing functions shown are available in modern *embedded microprocessors* in minimal volumes (see the schematic depiction in Fig. 9.4).

Hybrid continuous/discrete-time system variables Fig. 9.3 also shows the different types of signals within the action loop: *continuous-time* signals for the description of physically "analog" quantities (defined at all points in time), and *discrete-time* signals for the description of internal processor variables. Note the particular stepwise character of the continuous-time output signal $\overline{u}(t)$ of the digital-to-analog converter.

9.2.2 Modeling approaches

Hybrid system character vs. unified model Due to the hybrid compositions of system variables in digital control loops, there is a fundamental question of the appropriate type of model. For the best-possible predictive capabilities, models adapted to the respective signal types are desirable (*hybrid continuous/discrete-time models*, see Fig. 9.5a). However, as a rule, these are rather difficult to work with.

In order to be able to easily analyze the entire system behavior in closed form, all system variables—i.e. in- and outputs of transfer systems—must be available in a *unified format*, i.e. *continuous-time* or *discrete-time*.

One further modeling option concerns the question of *time-domain vs. frequency-domain models*. The choice of modeling paradigm depends strongly on the methods of analysis used. In this book, for the reasons previously discussed, frequency-domain models in the form of *frequency responses* are preferred.

Thus, the requirements on a model type suitable to systems design can be succinctly formulated as follows: (a) there should be unified signal cha-

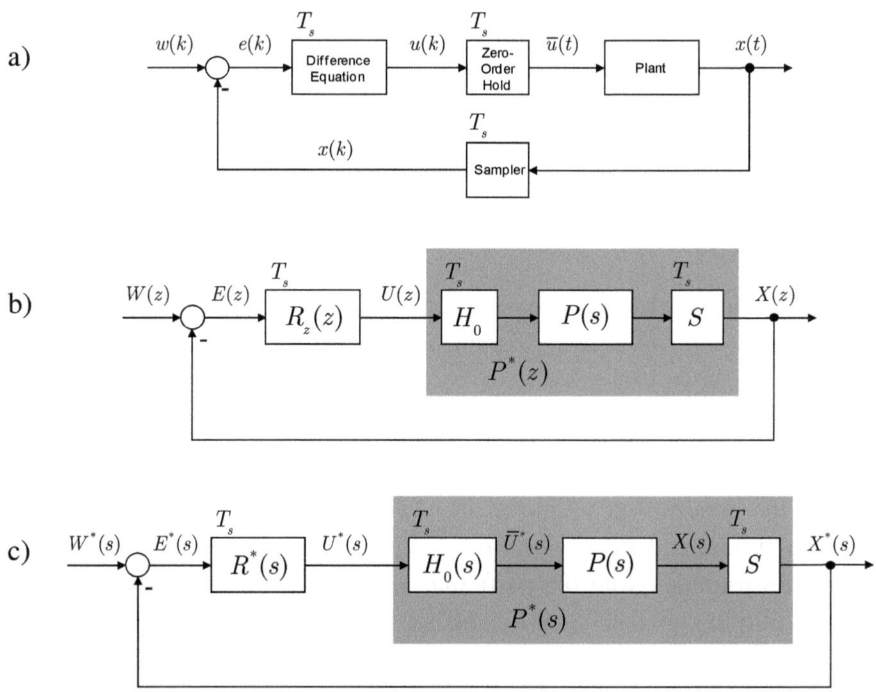

Fig. 9.5. Signal-oriented descriptions of a digital control loop: a) hybrid continuous/discrete-time model, b) discrete-time model (z-transform), c) sampled continuous-time model (LAPLACE transform, preferred model in this book), symbols described in main text

racteristics within the model, (b) it should be easy to manipulate and interpret frequency responses.

Signal-oriented discrete-time model Often, a *unified discrete-time treatment* of signals is selected, in which computation is performed on numerical sequences and the z-transform in the complex plane is employed (see Fig. 9.5b). Many powerful methods of design and analysis are based on this approach (Franklin et al. 1998), (Kuo 1997). There is however one deciding disadvantage if frequency responses—and particularly BODE diagrams—are to be employed. In this case, the frequency response is no longer a rational function in $j\omega$ but rather a rational function in $e^{j\omega}$, i.e. a transcendental function in $j\omega$. This causes frequency response treatments to lose much of their attraction.

Signal-oriented, sampled continuous-time model The alternative treatment is based on sampled continuous-time system variables (Tou 1959). Such variables remain formally within the continuous-time modeling paradigm and permit the LAPLACE transform to be applied as usual. However, the sampling process does result in frequency responses which are periodic at the sampling frequency, and which are also relatively difficult to work with. However, given certain preconditions, many practical problems can make use of sufficiently accurate approximate relationships, allowing common continuous-time frequency response methods to also be applied to this representation with minimal changes. This chapter thus pursues this path, allowing straightforward application of all the frequency domain models derived so far.

9.3 Sampling

Time discretization A digital computer cannot process continuous signals, only sequences of numbers. Thus, continuous-time, analog signals must generally be sampled, i.e. *time-discretized*, at equidistant times with sampling period T_s. This process is implemented in an analog-to-digital converter, the abstract mathematical model of which is a *sampler* A (see Figs. 9.5 and 9.6).

Strictly speaking, the sampled values are also subject to a magnitude discretization corresponding to the bit width of the converter (i.e. quantized, e.g. 8 bits $\Rightarrow$ 2^8=256 values). This is neglected in the current discussion.

Definition 9.1 *Sampler, numerical sequence* A *sampler* assigns to the function $x(t)$ the numerical sequence $\left(x(k)\right)$, where $x(k) := x(kT_a)$, $k = 0, \pm 1, \pm 2, \ldots$ (see Fig. 9.6).

Definition 9.2 δ *-sampler, sampled function* A δ *-sampler* assigns to a function $x(t)$ the *sampled function*

$$x^*(t) := \sum_{n=-\infty}^{+\infty} x(t) \cdot \delta(t - nT_s),$$

see Figs. 9.6 and 9.7.

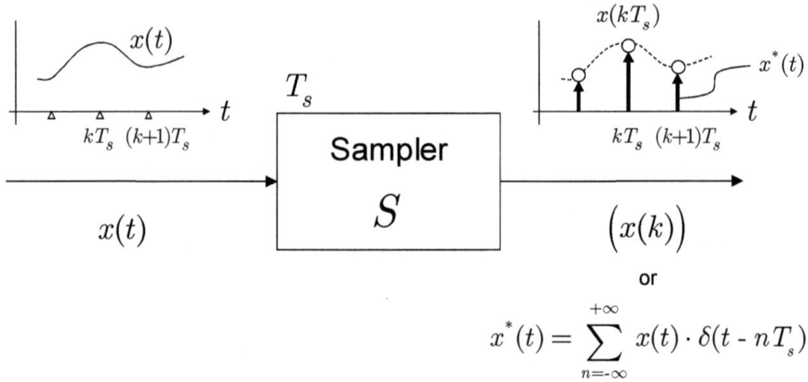

Fig. 9.6. Mode of operation of a sampler *A*: sequence $\left(x(k)\right)$, sampled function $x^*(t)$

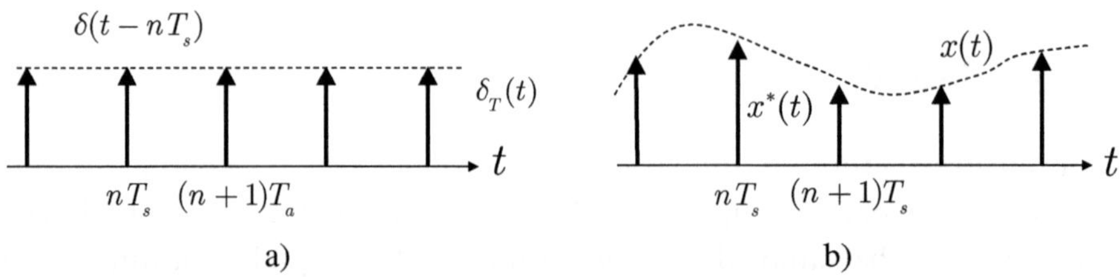

Fig. 9.7. Mode of operation of a δ -sampler: a) δ -impulse series, b) sampled function $x^*(t)$

Sampled function in the complex domain Due to the periodicity seen in Fig. 9.7b, the δ*-impulse series* $\delta_T(t)$ can be transformed into a complex FOURIER series

$$\delta_T(t) := \sum_{n=-\infty}^{+\infty} \delta(t - nT_s) = \sum_{n=-\infty}^{+\infty} c_n e^{jn\omega_s t} \quad , \ \omega_s := \frac{2\pi}{T_s} ,$$

$$c_n = \frac{1}{T_s} \int_{-\frac{T_s}{2}}^{+\frac{T_s}{2}} \delta(t) \cdot e^{-jn\omega_s t} \cdot dt = \frac{1}{T_s} \int_{0^-}^{0^+} \delta(t) \cdot dt = \frac{1}{T_s} ,$$

$$\Rightarrow \quad \delta_T(t) = \frac{1}{T_s} \sum_{n=-\infty}^{+\infty} e^{jn\omega_s t} \ .$$

This gives the sampled function

$$x^*(t) = \sum_{n=-\infty}^{+\infty} x(t) \cdot \delta(t - nT_s) = \frac{1}{T_s} \sum_{n=-\infty}^{+\infty} x(t) \cdot e^{jn\omega_s t} ,$$

and, applying the LAPLACE transform frequency shift $\mathcal{L}\left\{f(t) \cdot e^{\alpha t}\right\} = F(s - \alpha)$,

$$X^*(s) = \mathcal{L}\left\{x^*(t)\right\} = \frac{1}{T_s} \sum_{n=-\infty}^{+\infty} X(s + jn\omega_s) \tag{9.1}$$

or, with $s = j\omega$, the *frequency spectrum* of $x^*(t)$

$$X^*(j\omega) = \underbrace{\frac{1}{T_s} X(j\omega)}_{\text{base spectrum}} + \underbrace{\frac{1}{T_s} \sum_{m=1}^{+\infty} X\big(j(\omega \pm m\omega_s)\big)}_{\text{mirror frequency spectra "harmonics"}} \ . \tag{9.2}$$

> Periodic functions (here, the δ-impulse series) have discrete frequency spectra.
>
> Functions sampled with period T_a (here, $x^*(t)$) have periodic frequency spectra (mirror spectra) with (frequency) period $\omega_s = 2\pi/T_s$ and magnitude reduced by a factor of $1/T_s$ relative to the spectrum of the input function (here, $x(t)$).

Example 9.1 *Sampling a harmonic oscillation.*

Consider the harmonic oscillation

$$x(t) = \cos\omega_o t = \frac{1}{2}\left(e^{j\omega_o t} + e^{-j\omega_o t}\right).$$

Sampling with a period T_a, or sampling frequency $\omega_s = 2\pi/T_s > 2\omega_o$, results in the sampled signal

$$x^*(t) = \frac{1}{2T_s}\sum_{n=-\infty}^{+\infty}\left(e^{j(\omega_o+n\omega_s)t} + e^{-j(\omega_o+n\omega_s)t}\right) =$$

$$= \frac{1}{2T_s}\left(e^{j\omega_o t} + e^{-j\omega_o t}\right) + \frac{1}{2T_s}\sum_{m=1}^{\infty}\left(e^{j(\omega_o \pm m\omega_s)t} + e^{-j(\omega_o \pm m\omega_s)t}\right).$$

Fig. 9.8 depicts the signal $x(t)$, its spectrum, and the spectrum of the sampled signal $x^*(t)$ for the values $\omega_o = 1$ and $\omega_s = 3$.

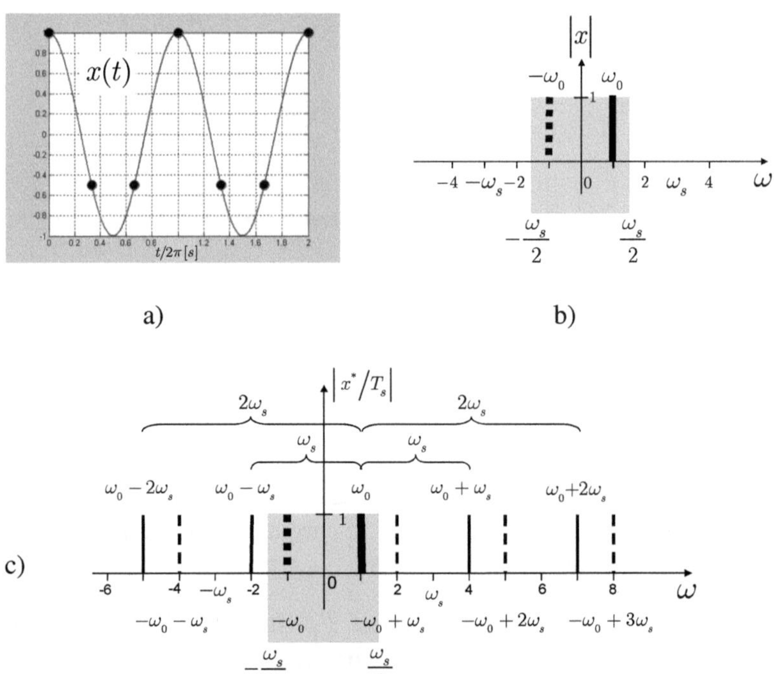

Fig. 9.8. Sampling a harmonic signal: a) continuous signal, b) spectrum of the continuous signal, c) spectrum of the sampled signal

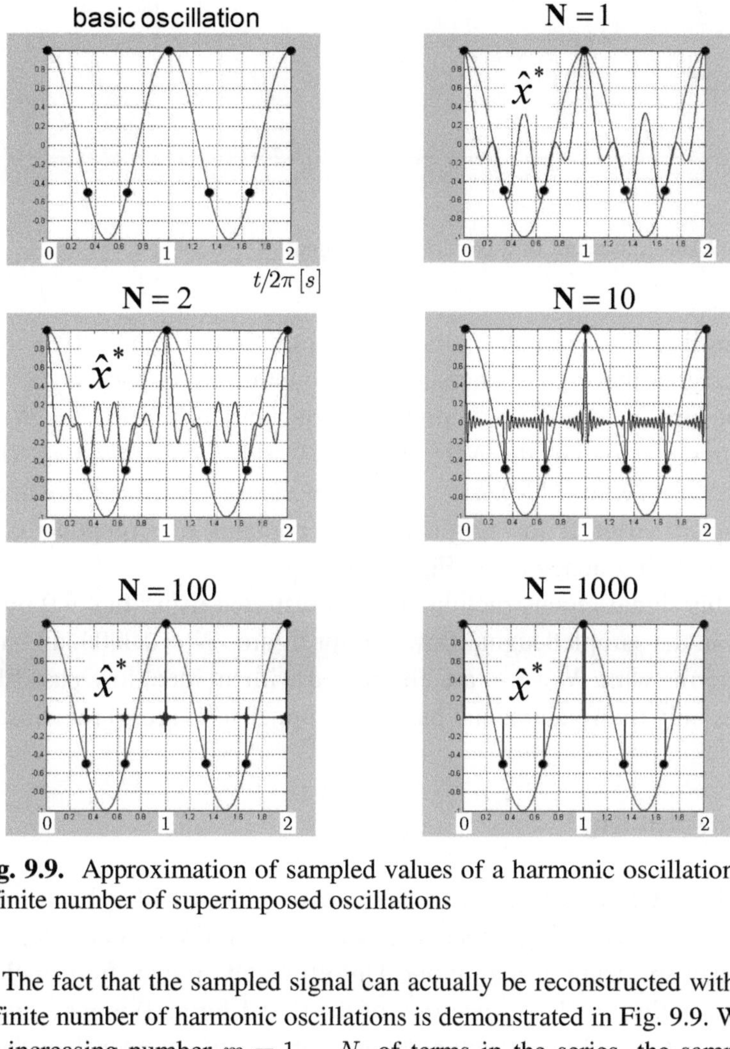

Fig. 9.9. Approximation of sampled values of a harmonic oscillation by a finite number of superimposed oscillations

The fact that the sampled signal can actually be reconstructed with an infinite number of harmonic oscillations is demonstrated in Fig. 9.9. With an increasing number $m = 1, ..., N$ of terms in the series, the sampled points $x(kT_a)$ become more and more pronounced.

■

Band-limited signals For a band-limited signal $x(t)$, i.e. one whose spectrum is zero above some bounding frequency ω_B, sampling results in a periodic spectrum as shown in Fig. 9.9. If $\omega_B < \omega_s/2$, the spectra do not overlap.

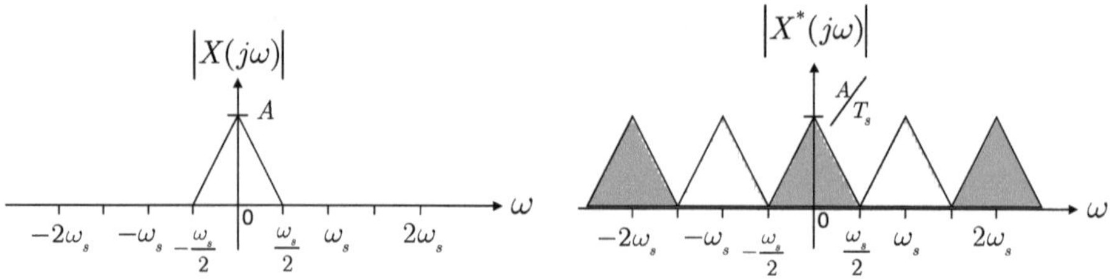

Fig. 9.10. Spectrum of a band-limited signal

Sampling theorem, NYQUIST frequency In communications engineering and signal theory, the characteristic bounding frequency $\omega_s/2$ of the sampling process (half the sampling frequency) is termed the *NYQUIST frequency*. Following the *NYQUIST-SHANNON sampling theorem* (Franklin et al. 1998), (Kuo 1997), signals can be completely reconstructed from the sampled series if the original signal contains no components above the NYQUIST frequency (as in Fig. 9.10).

This signal reconstruction can be illustrated using Fig. 9.9 by imagining that in the sampled signal (lower-right figure, $N = 1000$), all harmonics—i.e. terms with $N \geq 1$—are eliminated with an ideal low-pass filter. In this case, only the fundamental harmonic (upper-left figure) remains.

9.4 Aliasing

Ambiguities due to mirror frequencies Even with a band-limited input signal, the sampling process generates an infinite (periodic) mirror frequency spectrum in the output. Unfortunately, mirror frequencies can also occur *within* the *base spectrum* of the (band-limited) input signal, as when the input signal frequency contains components greater than the NYQUIST frequency $\omega_s/2$ (see Fig. 9.11). This is the normal case for control loops, as at a minimum, the measurement noise will always contain arbitrarily large frequencies. Such mirror frequency components in the base frequency band lead to a corruption of information in the impulse series, which is termed *aliasing*. The mirror frequency components are interpreted as seemingly low-frequency signal components, or rather, they can not be distinguished from true low-frequency components.

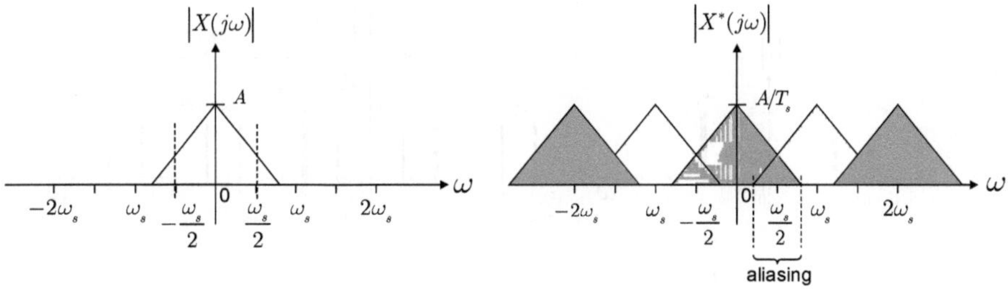

Fig. 9.11. *Aliasing*: ambiguities due to mirror frequencies in the base band when sampling frequency is too low

> A harmonic oscillation with frequency $\omega_2 = \omega_1 \pm n\omega_s$, sampled at T_s, or $\omega_s = 2\pi/T_s$, generates the same numerical sequence as a sampled harmonic oscillation with the base frequency ω_1.
>
> Signals with high mirror frequencies appear in the output of the sampler as low-frequency numerical/impulse series with the corresponding base frequency. Downstream discrete-time transfer systems (digital controllers, digital filters) are excited by external high-frequency signals as if by low-frequency ones.

Example 9.2 *Aliasing in the sampling of a harmonic oscillation.*

Aliasing Consider the harmonic oscillation from Example 9.1, though this time sampled at a lower sampling frequency $\omega_s/2 < \omega_o$. Figs. 9.12a through c show spectra and time histories for $\omega_o = 2$ and $\omega_s = 3$.

The lowest possible aliasing frequency is $\omega_{alias,\min} = \omega_o - \omega_s = -1$, i.e. $|\omega_{alias,\min}| < \omega_s/2$. There is thus an alias signal (harmonic carrier wave of the sampled values) at a lower frequency than the actual signal frequency. The original signal x cannot be reconstructed from the sampled values by applying an ideal low-pass filter (with cutoff frequency $\omega_s/2$): only the alias frequency $\omega_{alias,\min}$ occurs in the filter passband, and an alias signal x_{alias} results as the reconstructed harmonic carrier wave (Fig. 9.12c). Thus, a downstream controller would mistakenly interpret this sampled series as a low-frequency signal with $\omega_{alias,\min}$.

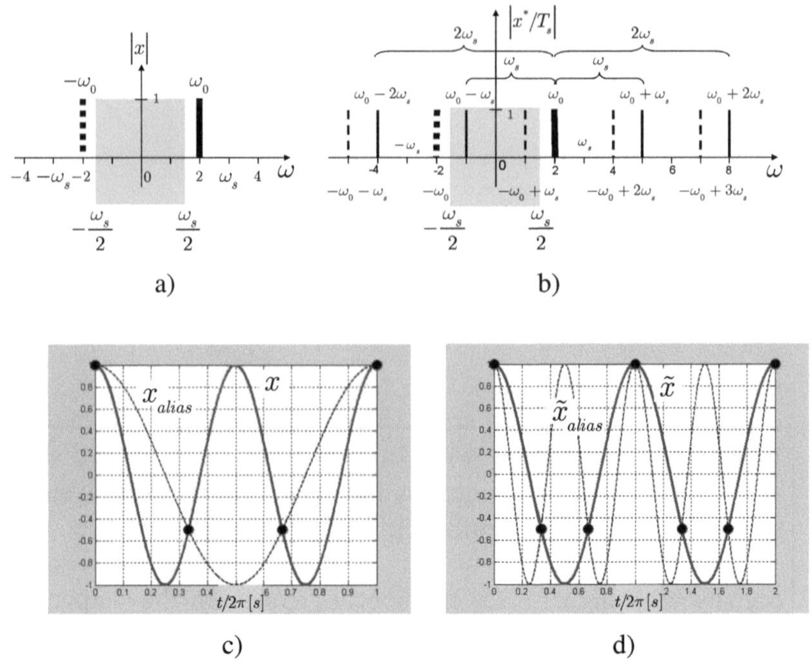

Fig. 9.12. Sampling a harmonic signal with aliasing: a) spectrum of the continuous signal, b) spectrum of the sampled signal, c) signal and aliased signal with least-possible alias frequency for $\omega_s/2 < \omega_o$, d) signal and aliased signal with least-possible alias frequency for $\omega_s/2 > \omega_o$, see main text

No aliasing For comparison, Fig. 9.12d depicts a signal *without* aliasing, with $\omega_o = 1$ and $\omega_s = 3$ (cf. Example 9.1). The least-possible alias frequency is now $\omega_{alias,\min} = \omega_o - \omega_s = -2$, i.e. $\left|\omega_{alias,\min}\right| > \omega_s/2 > \omega_o$, and is thus greater than the signal frequency. The carrier wave with the least-possible frequency is now the signal frequency; the base signal x can be reconstructed via an ideal low-pass filter.

■

Anti-aliasing filter The disruptive, yet ever-present, phenomenon of aliasing in sampling processes can be mitigated to a limited extent by placing a low-pass filter—an *anti-aliasing filter*—in front of the sampler. It is convenient to select the filter cutoff frequency to be equal to half the sampling frequency. For the greatest possible damping of amplitudes above the cutoff frequency, the filter order should be as large as possible. However,

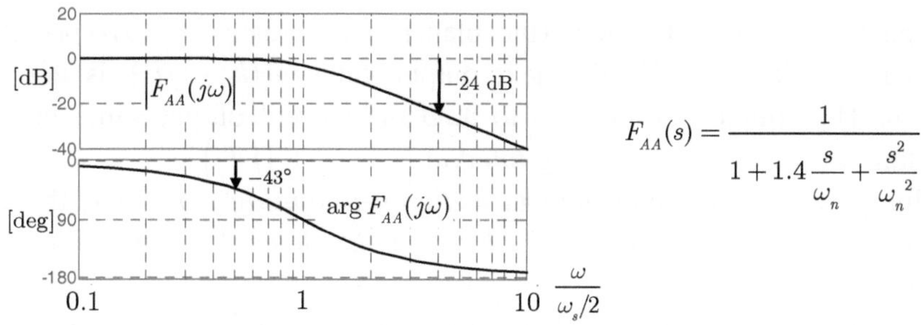

Fig. 9.13. Second-order low-pass filter as anti-aliasing filter

as part of a *closed* action chain (the control loop), this technique is subject to strict limitations, due to the resulting negative filter phase shift which disturbs the *control loop stability*. Fig. 9.13 illustrates the frequency response of a second-order low-pass filter. In the passband, there is already a negative phase shift of $\varphi = -43°$ at $\omega = \omega_s/4$, while in the stopband, at $\omega = 2\omega_s$, the magnitude damping is only $-24\text{ dB} \triangleq 0.06$. These two points demonstrate the limitations of available design options.

An *anti-aliasing filter* with a cutoff frequency of $\omega_n = \omega_s/2$ results in the following signal distortions:

- in the *passband* $\omega_n < \omega_s/2$, there is no amplitude damping, though there is a significant *negative phase shift* (➔ stability problems in the control loop!);
- in the *stopband* $\omega_n \geq \omega_s/2$, there is *finite amplitude damping*, i.e. the high-frequency signal components still impinge on the sampler and are transformed into low-frequency series, though with reduced magnitude.

Anti-aliasing filters can only be realized as *analog filters*, i.e. digital filters *cannot* be used for this purpose, as these latter are naturally also subject to aliasing.

9.5 Hold Element

Signal transformation, signal extrapolation The numerical sequence of the control variable must be transformed into a continuous-time function (in general, an electrical voltage) to ensure a continuous effect of the control on the analog subsystem (the plant). This creates an *extrapolation problem*, as at time kT_s, the next output value $u((k+1)T_s)$ is not yet known. The simplest solution is to hold the current output value $u(kT_s)$ constant for the current sampling interval $\left[kT_s,(k+1)\,T_s\right)$.

This process is implemented in a *digital-to-analog converter*, the abstract mathematical model of which is a *hold element* H_0 (see Figs. 9.5 and 9.14).

Strictly speaking, the sampled values are also subject to a magnitude discretization corresponding to the bit width of the converter (i.e. quantized, e.g. 10 bits $\Rightarrow 2^{10}=1024$ values). This is neglected in the current discussion.

Definition 9.3 *Zero-order hold (ZOH), numerical sequence* A zero-order hold assigns to a numerical sequence $\left(u(k)\right)$ the function $\overline{u}(t)$, where $\overline{u}(t) := u(kT_s)$ for $kT_s \le t < (k+1)\,T_s$, see Fig. 9.14.

Definition 9.4 *Zero-order hold (ZOH), sampled function* A zero-order hold assigns to a sampled function $u^*(t)$ the function $\overline{u}(t)$, where $\overline{u}(t) := u^*(kT_s)$ for $kT_s \le t < (k+1)\,T_s$, see Fig. 9.14.

Order of the hold element The operation of a hold element as an *extrapolator* is clearly illustrated in Fig. 9.14. Generalizing the hold action, the continuous-time function $\overline{u}(t)$ active during the current sampling interval $\left[kT_s,(k+1)\,T_s\right)$ can be described by an Nth-order polynomial. In the case shown, this is a 0th-order polynomial—requiring only one interpolation value, here $\overline{u}(kT_s) = u(kT_s)$—creating a *zero-order hold* (*ZOH*).

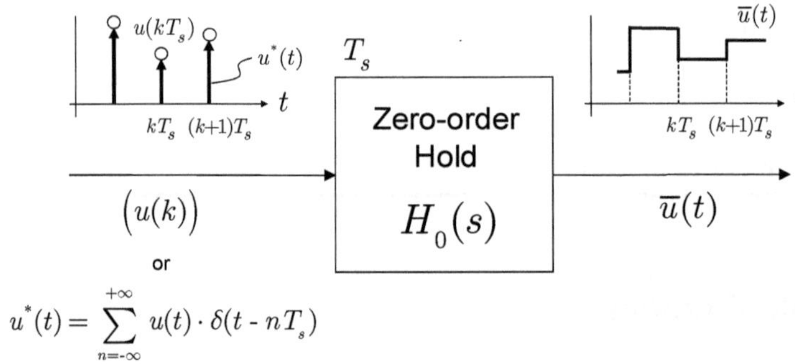

Fig. 9.14. Operation of a zero-order hold H_0: numerical sequence $\left(u(k)\right)$, sampled function $u^*(t)$, filtered output signal $\overline{u}(t)$

If preceding elements of the series are included, a higher-order polynomial can also be constructed, creating an *N th-order hold.*

The practical use of higher-order hold elements in closed action chains (control loops) is limited, so that such extrapolators are not further considered here. A more extensive system-theoretical discussion can be found in (Reinschke 2006).

Hold elements as transfer systems If the input series $\big(u(k)\big)$ is interpreted as a sampled series of impulses $u^*(t)$, then the hold element can be interpreted as a *linear continuous-time transfer system* with transfer function $H_0(s)$ (representing a pulse-amplitude modulation).

From Fig. 9.14, using the unit step function $\sigma(t)$, the output a hold element is then

$$\overline{u}(t) = u(0)\sigma(t) + \big[u(1)-u(0)\big]\sigma(t-T_s) + \big[u(2)-u(1)\big]\sigma(t-2T_s) + \ldots,$$

and further, applying the LAPLACE transform,

$$\overline{U}(s) = u(0)\frac{1}{s} + \big[u(1)-u(0)\big]\frac{e^{-T_s \cdot s}}{s} + \big[u(2)-u(1)\big]\frac{e^{-2T_s \cdot s}}{s} + \ldots$$

$$\overline{U}(s) = \frac{1-e^{-T_s \cdot s}}{s}\Big\{u(0) + u(1)\cdot e^{-T_s \cdot s} + u(2)\cdot e^{-2T_s \cdot s} + \ldots\Big\}$$

$$\overline{Y}(s) = \frac{1-e^{-T_s \cdot s}}{s} Y^*(s) = H_0(s)\cdot Y^*(s)\,,$$

thus giving the *transfer function* for a *zero-order hold element*

$$H_0(s) = \frac{1-e^{-T_s s}}{s}\,. \tag{9.3}$$

Linearity, time variance A zero-order hold is a *linear* transfer system, though in its general form, it is *time-varying.* In (Reinschke 2006), it is shown that interchanging the order of sampling and hold operations results in different output signals in the case when the input signal is not displaced along the time axis by an integer multiple of the sampling period. This effect does not appear if the output series $\big(u(k)\big)$ is always tightly synchronized with the hold operation in the digital-to-analog converter.

Frequency response Subsequent discussion of system dynamics is based on the frequency response of the zero-order hold. A quick calculation applied to Eq. (9.3) gives this *frequency response*

$$H_0(j\omega) = \frac{2 \cdot \sin\left(\frac{\omega T_s}{2}\right)}{\omega} e^{-j\omega\frac{T_s}{2}} = H_0(\omega) \cdot e^{-j\omega\frac{T_s}{2}}, \tag{9.4}$$

$$H_0(j0) = T_s \,.$$

The frequency response equation (9.4) reveals two quite interesting properties of the zero-order hold. The frequency-dependent magnitude response has a low-pass character (see Fig. 9.15), and the frequency-dependent *phase response* is that of a *time-delay element*, with a delay equal to *half the sampling period.*

Low-pass nature Fig. 9.15a clearly shows the low-pass nature of the hold. This can also be illustrated heuristically. At the input of the hold, there is a series of impulses with an infinitely wide-band spectrum, while at the output, a smooth step function appears. The high-frequency mirror spectra are thus more-or-less filtered out of the signal.

Equivalent system: time delay element To easily evaluate the system behavior as part of a transfer chain for sufficiently low frequencies $\omega < \omega_s/4$ where $H_0(\omega) \approx T_s$, the transfer function of the zero-order hold can be approximated by a time delay of half the sampling period (Fig. 9.15b):

$$H_0(s) \approx T_s \cdot e^{-\frac{T_s}{2}s} \quad \text{for } \omega < \omega_s/4 \,. \tag{9.5}$$

Ideal smoothing The dotted line in Fig. 9.15 shows the amplitude response for (non-realizable) ideal smoothing. If such a filter were in fact realizable, its effect would be well-illustrated by Fig. 9.9. The series of impulses at the input of the hold corresponds approximately to the lower-right figure with $N = 1000$ terms. An ideal smoothing filter would remove all terms with $N \geq 1$, leaving only the base oscillation at the output of the ideal smoothing filter, i.e. the upper-left figure[1].

[1] This corresponds conceptually to the reconstruction of a band-limited continuous carrier signal from a sampled series, as described at the end of Sec. 10.3 in the discussion of the NYQUIST-SHANNON sampling theorem.

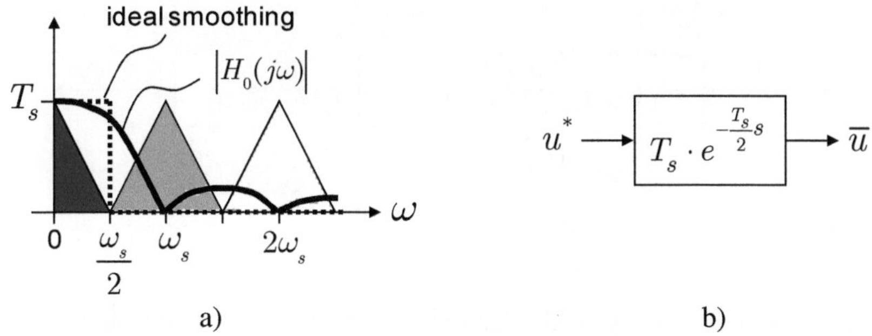

Fig. 9.15. Zero-order hold: a) magnitude of the frequency response (shown schematically), b) equivalent system for sufficiently small frequencies $\omega < \omega_s/4$

9.6 Sampled Plant Frequency Response

Sampled plant The *sampled plant frequency response* designates the frequency response from the impulse series of the control variable to the impulse series of the sampled measured variable (see Fig. 9.16). This response encompasses the entire transfer chain < hold element – actuator – multibody system – sensor – anti-aliasing filter – sampler >, see Fig. 9.3. At the input, there is first a smoothing of the mirror frequencies of the input impulse series by the hold element, then the individual transfer properties of the continuous plant (beware resonances in multibody systems!), and finally, the sampling process with its newly-generated mirror frequency spectra.

At first glance, as it consists of an infinite number of mirror frequency spectra, the sampled plant frequency response may be very difficult to work with. Happily, given very general conditions, for many practical cases, simplified approximate relations can be used for rough calculations at low- to mid-frequency ranges relative to the half sampling frequency. In any case, frequency responses can be easily computed using computer-aided tools (e.g. MATLAB).

Sampled plant transfer function From the signal flow in Fig. 9.16, it follows that

$$
\begin{aligned}
X(s) &= H_0(s) \cdot P(s) \cdot U^*(s), \\
X^*(s) &= \frac{1}{T_s} \sum_{m=1}^{+\infty} X(s \pm jm\omega_s) = \\
&= \frac{1}{T_s} \sum_{m=1}^{+\infty} H_0(s \pm jm\omega_s) \cdot P(s \pm jm\omega_s) \cdot U^*(s \pm jm\omega_s).
\end{aligned}
$$

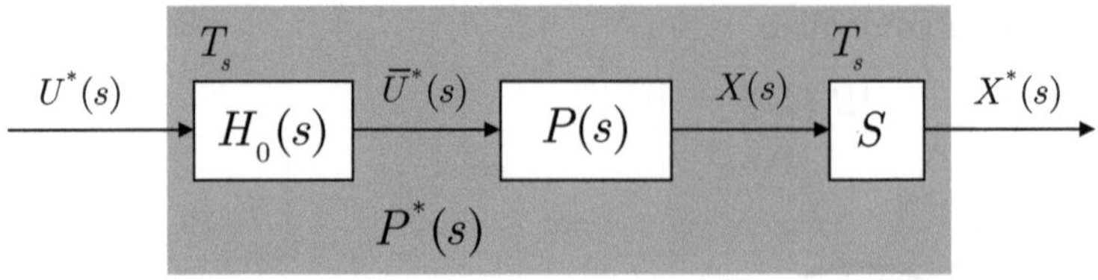

Fig. 9.16. Sampled plant transfer function

Due to the periodicity of the frequency spectrum of a sampled function, $U^*(s) = U^*(s \pm jm\omega_s)$, giving the *sampled plant transfer function*[2]

$$P^*(s) := \frac{X^*(s)}{U^*(s)} = \frac{1}{T_s} \sum_{m=1}^{+\infty} H_0(s \pm jm\omega_s) \cdot P(s \pm jm\omega_s) . \tag{9.6}$$

Sampled plant frequency response As usual, substituting $s = j\omega$ into the transfer function Eq. (9.6) results in the frequency response—the *sampled plant frequency response*—

$$\boxed{P^*(j\omega) = \frac{1}{T_s} \sum_{m=1}^{+\infty} H_0\left(j(\omega \pm m\omega_s)\right) \cdot P\left(j(\omega \pm m\omega_s)\right)} . \tag{9.7}$$

Properties of P^* The frequency response (9.7) is a *transcendental function*[3], as, according to Eq. (9.3), $H_0(s)$ contains an exponential term. Due to its complex argument, the frequency response is periodic, with period ω_s. For $\omega = \omega_s/2$, $P^*(j\omega)$ is real[4], and its magnitude is an even function of frequency, resulting in the schematic representation shown in Fig. 9.17. Thus, for evaluation purposes, only the *baseband* $0 \le \omega \le \omega_s/2$ need be considered (the gray domain in Fig. 9.17).

---

2 Using this transfer function, given a known input $u^*(t)$, the time evolution of the output $x(t)$ at the sampling times $t = kT_a$ can also be calculated. However, this is not further pursued here, as a frequency-domain analysis is preferred. In (Reinschke 2006), a further interesting extension is presented, in which the behavior *between* sampling times can also be computed analytically.

3 Incidentally, the same result is found via the more common derivation using the z-transfer function and then the substitution $z = e^{j\omega T_s}$. A detailed discussion of the properties of $P^*(j\omega)$ can be found e.g. in (Tou 1959).

4 Terms of the following form always appear: ... $H_0 P(j(2n+1)\,\omega_a/2) +$
$+ H_0 P(-j(2n+1)\,\omega_a/2) + ... = 2\,\mathrm{Re}[... + H_0 P(j(2n+1)\,\omega_a/2) + ...], n = 1, ...$

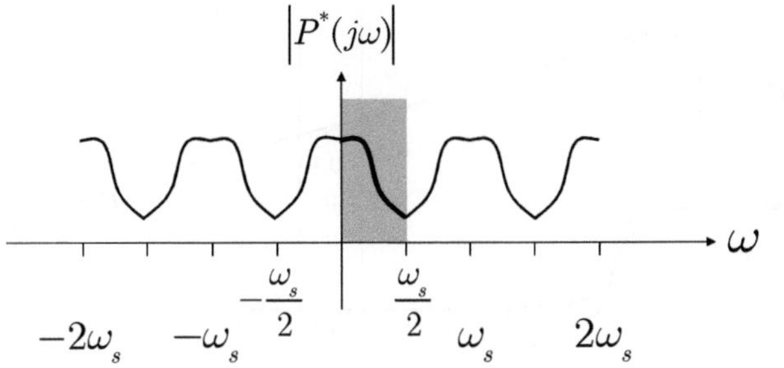

Fig. 9.17. Sampled plant frequency response (schematically depicted)

Approximation of low-pass behavior For plants $P(s)$ of a low-pass nature, the seemingly complicated form of Eq. (9.7) can be significantly simplified for approximating investigations.

In Eq. (9.7), given a low-pass plant, only a few terms have any appreciable contribution:

$$P^*(j\omega) = \frac{1}{T_s} H_0\left(j\omega\right) P\left(j\omega\right) + \frac{1}{T_s} H_0\left(j(\omega - \omega_s)\right) P\left(j(\omega - \omega_s)\right) + \dots \quad (9.8)$$

The first term in Eq. (9.8) represents the dominant component, and the second term is the most significant correction term. It is easy to see that for a given frequency ω, the remaining terms contain the mirror frequencies $\omega \pm m\omega_s$, which tend to very high magnitudes, but, due to the low-pass nature of $H_0(j\omega)$ and an assumed low-pass nature of $P^*(j\omega)$, make no appreciable contribution. Eq. (9.8) thus already represents a rather good approximation of $P^*(j\omega)$.

For rough computations, the dominant term in Eq. (9.8) can also be used alone, which, along with Eq. (9.5), gives the approximation

$$P^*(j\omega) \approx \frac{1}{T_s} H_0\left(j\omega\right) P\left(j\omega\right) = P\left(j\omega\right) \cdot e^{-j\omega\frac{T_s}{2}}, \quad (9.9)$$

which is quite valid in the range $\omega < \omega_s/4$.

Thus, in the low-frequency domain (relative to the NYQUIST frequency), the sampled plant frequency response can be rather well approximated by the continuous frequency response of the plant, with a correction coming from a time delay element. This implies that, primarily, only the phase response need have a *negative phase correction* applied.

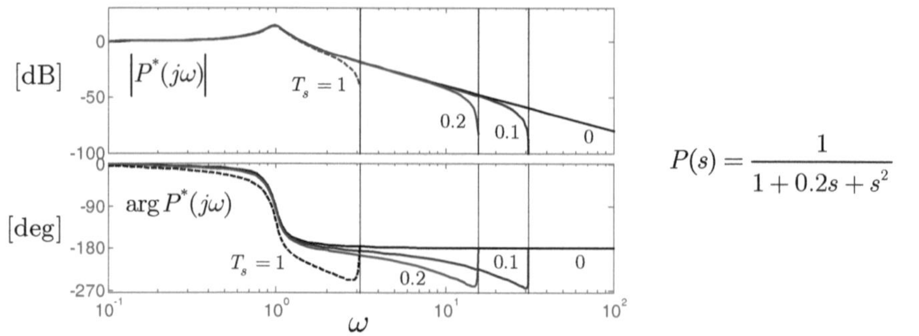

Fig. 9.18. Sampled plant frequency responses $P^*(j\omega)$ for various sampling times

Fig. 9.18 depicts sampled plant frequency responses for a well-damped oscillatory system with various sampling times (only the respective base-bands $0 \leq \omega \leq \omega_s/2$ are shown).

> Given a *low-pass* continuous plant consisting of the chain < hold element – actuator – multibody system – sensor – anti-aliasing filter >, the *sampled plant frequency response* can be approximated at sufficiently small frequencies $\omega < \omega_s/4$ by the continuous frequency response of the plant $P(j\omega)$, corrected with a time delay $= T_s / 2$.
>
> Note: The negative phase shift of the hold element fundamentally degrades closed loop stability in comparison to a continuous control loop. Given an equivalent loop gain, the phase margin is always reduced by the negative phase shift of the time delay component.

9.7 Aliasing in Oscillatory Systems

Oscillatory systems: multibody systems As discussed in previous chapters, in mechatronic systems, the plant $P(s)$ very often contains weakly-damped oscillatory components of the form

$$P(s) = \frac{\cdots}{\cdots \left(1 + 2d_0 \frac{s}{\omega_0} + \frac{s^2}{\omega_0^2}\right)}, \quad d_0 \ll 1,$$

arising from the elastic suspension of moving armature components in electromechanical transducers, or from multibody systems. In such systems, given low mechanical damping, large amplitudes $\left|P(j\omega)\right|$ can appear even at high frequencies at such resonances.

Fig. 9.19 shows a typical case, where the eigenfrequency $\omega_0 > \omega_s/2$ is greater than the NYQUIST frequency. In this case, the low-pass approximation of Eq. (9.8) naturally no longer applies, and the frequency response $P^*(j\omega)$ also contains significant contributions from selected additional terms. For the case represented in Fig. 9.19, it is quite apparent that the term

$$\begin{aligned} P^*(j\omega^*) &= \dots \frac{1}{T_s} H_0\left(j(\omega^* + \omega_s)\right) P\left(j(\omega^* + \omega_s)\right) + \dots = \\ &= \dots \frac{1}{T_s} H_0\left(j\omega_0\right) P\left(j\omega_0\right) + \dots \end{aligned}$$

has a significant contribution at the frequency $\omega^* = \omega_0 - \omega_s$. The low-pass property of the hold element already has a rather large effect at the frequency ω_0. However, since $\left|P\left(j\omega_0\right)\right| \gg 1$, the frequency response has a (very) large contribution, which is also apparent in $P^*(j\omega^*)$ as a resonant peak, though at the *mirror frequency* $\omega^* = \omega_0 - \omega_s$. The high-frequency resonance ω_0 is thus mirrored in the (low-frequency) base spectrum of the

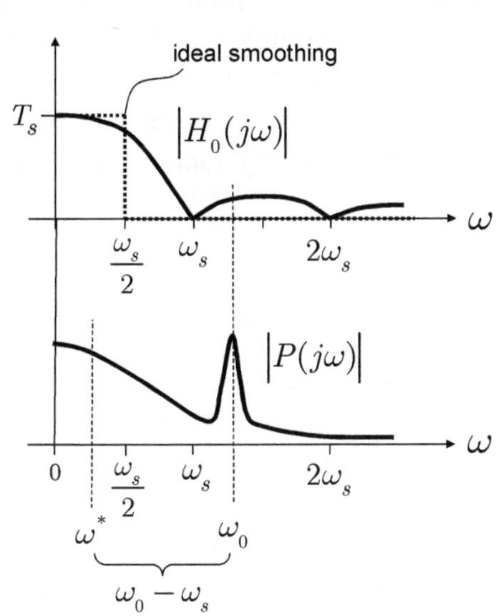

Fig. 9.19. Weakly-damped oscillatory plant with resonant frequency $\omega_0 > \omega_s/2$

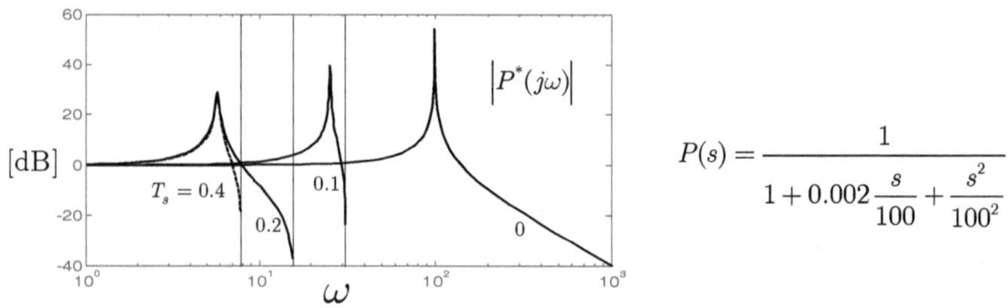

Fig. 9.20. Aliasing of a high-frequency resonance in the sampled plant frequency response

sampled plant frequency response. Fig. 9.20 shows this effect for a variety of sampling times.

Aliasing resonant frequencies From the geometric relationships between frequencies in Fig. 9.19, it is easy to derive a condition for the aliasing frequencies of the base frequency band. For the *aliasing frequencies* ω_i^* of a resonant frequency ω_i, it must obviously hold that

$$\omega_i^* := \left\{ \omega \;\middle|\; \omega = \omega_i \pm m\omega_s, \quad m \in \mathbb{N} \text{ and } |\omega| \leq \frac{\omega_s}{2} \right\}. \tag{9.10}$$

Aliasing condition in the complex s-plane The aliasing issue can also be quite nicely visualized in the complex s-plane with the distribution of poles of the continuous plant transfer function $P(s)$. Fig. 9.21 presents a few typical pole locations. Ultimately, it is the *imaginary component* of the plant poles which is the determining factor for aliasing, with the non-aliasing condition

$$\left|\mathrm{Im}(p_i)\right| < \frac{\omega_s}{2}. \tag{9.11}$$

Condition (9.11) describes a horizontal strip arranged symmetrically about the real axis—the *base strip*, see Fig. 9.21. Plant poles in this region only influence mirror frequencies of $P^*(j\omega)$ outside the base frequency band (the NYQUIST band).

Plant poles outside the base strip—i.e. inside the *complementary strip* shown in Fig. 9.21—lead to aliasing frequencies inside the base frequency band of $P^*(j\omega)$.

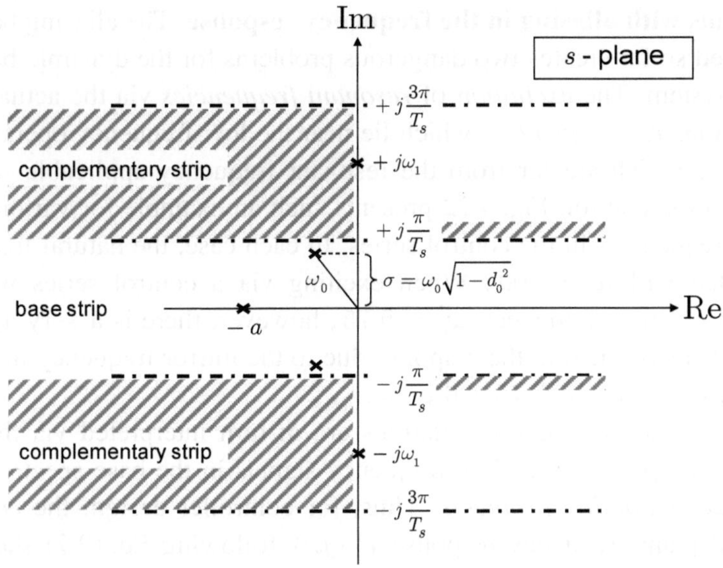

Fig. 9.21. Complex s-plane with poles of the continuous plant transfer function $P(s)$

Fig. 9.21 also shows the influence of damping on resonant frequencies: with increasing damping, high-frequency resonances become represented in $P^*(j\omega)$ without aliasing. The case shown in Fig. 9.18 corresponds to the pair of poles with ω_0 in Fig. 9.21, and the case shown in Fig. 9.19, the pair of poles with ω_1 in Fig. 9.21. For *real poles* $p = -a$, there can fundamentally be *no aliasing*.

Analytical formula for aliasing resonances From the transformation of poles of the sampled plant frequency response—which is relatively simple to calculate analytically—using a representation equivalent to a z-transfer function or a q-transfer function in the *transformed frequency domain* (Gausch et al. 1993), a simpler analytical relation than Eq. (9.10) can be found for the undamped *aliasing resonant frequencies* (Janschek 1978):

$$\boxed{\omega_i^* = \frac{2}{T_s}\arctan\frac{\sin\omega_i T_s}{1+\cos\omega_i T_s}}\,. \tag{9.12}$$

It is easy to verify the formula in Eq. (9.12) using the examples in Fig. 9.20. It is noteworthy—though not surprising considering the relationships represented—that identical aliasing resonances $\omega_0^* = 5.75$ exist for both sampling periods $T_s = 0.2$ and $T_s = 0.4$.

Problems with aliasing in the frequency response The aliasing behavior discussed so far creates two dangerous problems for the dynamic behavior of the system. The *excitation* of *resonant frequencies* via the actuator can occur at *mirror frequencies* which lie near the low-frequency operating regime and which are far from the resonant frequency applicable for continuous-time control. Fig. 9.22 presents such an example for various excitation frequencies in the control series. In each case, the natural frequency is excited with $\omega_0 = 100$. When exciting via a control series with the aliasing resonant frequency $\omega_0^* = 5.75$, however, there is a very apparent build up (resonance) in the response due to the mirror frequency in the excitation series at $\omega = 5.75 + 6\omega_s = \omega_0$.

Such excited resonant oscillations are in turn interpreted via the sampling in the processor as low-frequency signals in the base spectrum, and can lead to *stability problems*. Thus, for controller design, the *complete* sampled plant frequency response $P^*(j\omega)$ following Eq. (9.7) should be used in all cases, for which computer-aided calculation is recommended.

Avoiding aliasing in the frequency response The aliasing issue always appears when there are natural frequencies of the plant $\omega_i > \omega_s/2$. However, the influence of such resonances can be significantly reduced if the resonant frequencies are sufficiently well filtered by the *anti-aliasing fil-*

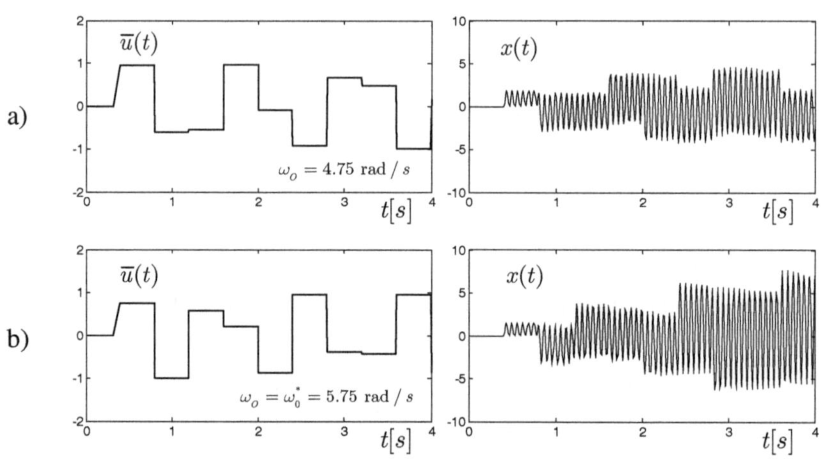

Fig. 9.22. Excitation of aliasing resonant frequencies for the example in Fig. 9.20 with $T_s = 0.4$ and a harmonic excitation carrier function: a) $\omega_o = 4.75 \neq$ aliasing resonant frequency, b) $\omega_o = 5.75 =$ aliasing resonant frequency

ter (as an integral component of $P(s)$). In the best case, this results in $P(j\omega) \approx 0$ for $\omega > \omega_s/2$ (corresponding to the ideal smoothing in Fig. 9.19). Note, however, that in this case, the excited resonant frequencies are rendered unobservable, and can no longer be affected by the control.

However, in all cases it is critical to consider resonant frequencies in the neighborhood of the cutoff frequency of the anti-aliasing filter, as the filter is not yet effective in this range.

9.8 Digital Controllers

Realization Digital controllers are realized as computer programs. This brings two important limitations into play. First, though in principal, arbitrarily complex algorithms can be implemented, the transfer properties of the control algorithm must (a) fit the plant, and (b) be analytically representable in such a way that a theoretical treatment of the system is possible (in particular, for meaningful stability analyses). For this reason, linear discrete-time transfer systems, which can generally be described using *linear difference equations*, are often chosen for mechatronic systems. In many applications, using a limited number of standard linear components (e.g. PID controllers, lead-lag elements, or low-pass, band-stop, and band-pass filters), substantive preliminary designs for baseline investigations of the system can already be made, as will be shown in Ch. 10.

The second important limitation concerns *real-time capability*. The model assumptions made up to now require strict periodic sampling and synchronous output of control signal series. This means that all computational operations must be completed within one sampling interval. In particular for embedded microprocessors, limited computational power requires limiting implementations to "simple" control algorithms.

Control algorithm: linear difference equation As is well known, a linear control algorithm with one input and one output can generally be represented by an *n-th-order difference equation* (Franklin et al. 1998), (Kuo 1997) (see Fig. 9.23a)

$$u(k) + a_1 u(k-1) + \ldots + a_n u(k-n) = b_0 e(k) + \ldots + b_m e(k-m) . \quad (9.13)$$

This representation also gives one possible implementation[5], as for known $e(k)$, the element $u(k)$ can be recursively calculated from preceding values $u(k-i),\ e(k-i)$.

z-transfer function For design and analysis, the difference equation (9.13) can be represented via the z-transform as a *z-transfer function* (Franklin et al. 1998), (Kuo 1997)

$$R_z(z) = \frac{U(z)}{E(z)} = \frac{b_0 + \ldots + b_m z^{-m}}{1 + a_1 z^{-1} + \ldots a_n z^{-n}} = \frac{b_0 z^m + \ldots + b_m}{z^n + a_1 z^{n-1} + \ldots a_n}. \tag{9.14}$$

Discrete controller frequency response Using the transformation function $z = e^{T_s s}$ and the usual $s = j\omega$, $R_z(z)$ gives the *discrete controller frequency response* $R_z(e^{j\omega T_s})$. However, this is a *transcendental function* in ω and thus difficult to compute or represent graphically.

Sampled controller transfer function In order to eliminate the transcendental exponential functions, these can be replaced by a rational function approximation of the exponential, e.g. a first-order *PADÉ approximation*[6]

$$z = e^{sT_s} \approx \frac{1 + \frac{T_s}{2} s}{1 - \frac{T_s}{2} s}. \tag{9.15}$$

Using the substitution in Eq. (9.15), the z-transfer function Eq. (9.14) can then be written as a function of the variable s, and formally represented as a *sampled transfer function* $R^*(s)$. This is now a rational function of s, and substituting $s = j\omega$, gives the *sampled controller frequency response* $R^*(s)$ also as a rational function in $j\omega$, in a form equivalent to the continuous-time transfer system. However, this clearer representational form comes at the cost of the approximation in Eq. (9.15). For sufficiently low frequencies, however, it also holds approximately that

$$R^*(j\omega) \approx R_z(e^{j\omega T_s}) \quad \text{for } \omega < \frac{\omega_s}{4}. \tag{9.16}$$

5 A variety of equivalent state space representations can be directly derived from the difference equation (Lunze 2008), exhibiting different numerical properties for computational implementation (e.g. with limited word lengths).

6 Also known as the *TUSTIN transform* in the literature.

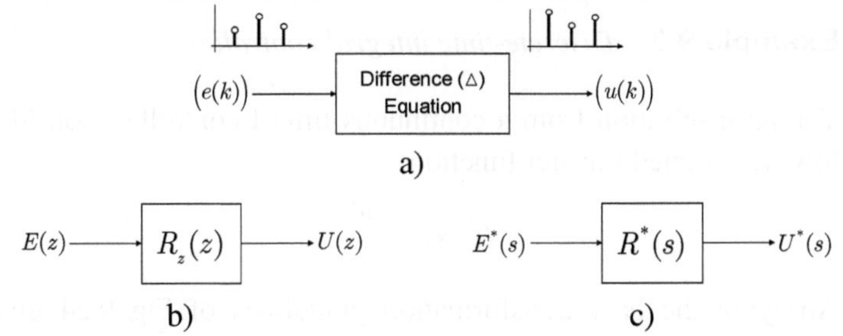

Fig. 9.23. Equivalent representations for a linear control algorithm: a) *difference equation* (input/output (I/O) model: numerical sequences), b) z-transfer function (I/O-model: numerical sequences), c) *sampled s-transfer function* (I/O-model: series of impulses)

Employing to the approximate relation (9.16) and taking advantage of the rational structure of $R^*(s)$, given a known frequency response $P^*(j\omega)$, controller design can proceed as usual for continuous-time systems. In particular, well-known controller structures from the world of continuous-time design, such as e.g. PID controllers, with their well-known, clear relationships between controller parameters and the system frequency response, can be used.

Realization of $R^*(s)$ as a difference equation To interpret and realize a sampled controller transfer function $R^*(s)$ as a difference equation, all that is required is to execute the steps carried out so far in reverse order, as shown in Fig. 9.24.

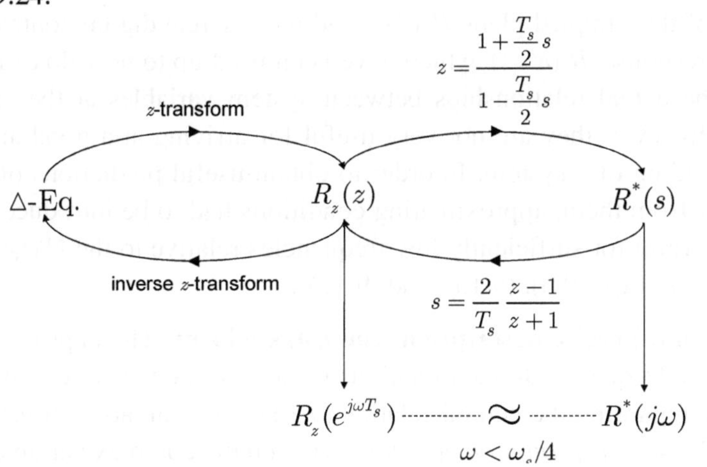

Fig. 9.24. Schema for the transformations "difference (Δ)-equation $\Leftrightarrow R_z(z) \Leftrightarrow R^*(s)$"

Example 9.3 *Discrete-time integral controller.*

Taking inspiration from a continuous-time I-controller, consider the following sampled transfer function:

$$R^*(s) = \frac{K}{s}.$$

Applying the back-transformation guidelines of Fig. 9.24 gives as the first step the z-transfer function

$$R_z(z) = K \cdot \frac{T_s}{2} \frac{z+1}{z-1} = K \cdot \frac{T_s}{2} \frac{1+z^{-1}}{1-z^{-1}},$$

and further, the first-order difference equation

$$y(k) = K \frac{T_s}{2} \cdot \left[e(k) + e(k-1) \right] + y(k-1)$$

in which the well-known *trapezoid rule* for numerical integration can be recognized.

■

9.9 Transformed Frequency Domain

Representational difficulties for discrete ω-frequency responses The representations of discrete-time behaviors of the discrete frequency response of the sampled plant $P^*(j\omega)$ and the discrete digital controller frequency response $R_z(e^{j\omega T_s})$ which have been used up to now do exactly describe the actual relationships between system variables at the sampling times. However, they are not very useful for arriving at a good analytical understanding of a system. In order to obtain useful predictions of system behavior from them, approximating conditions had to be introduced which are only valid for sufficiently low frequencies relative to the NYQUIST frequency (e.g. Eqs. (9.9), (9.15), and (9.16)).

A clear and precise description: the *q*-transform The approximate nature of the frequency domain relations used up to now can be easily made exact by regarding the central relation (9.15) not as an approximation, but as *defining* a *transformation equation* between the complex variables s or z and a new complex variable *q*. The resulting *q-transform*, introduced by (Schneider 1977), differs from the long-used *bilinear transform* (e.g. (Tou

1959)) in, among other things, the fact that a direct relation to sampled continuous-time systems can be established. Thus, descriptions of both continuous-time and sampled discrete-time systems can be smoothly merged, in both their structure and parameters and as both a function of sampling time and of frequency. Clear and predominantly simple analytical relations between continuous and discrete system parameters are made available, and above all, unmodified application of familiar continuous-time frequency domain methods is possible (BODE diagrams, frequency response curve methods, etc.). The short introduction to the *q-transform* and the *transformed frequency domain* presented below is based on (Janschek 1978); for a more thorough presentation and applications of this modeling approach for analysis and design, the reader is referred to (Gausch et al. 1993), (Horn and Dourdoumas 2006).

Definition 9.5 *q-transform* Let the q-transform be defined as the conformal map

$$q := \Omega_0 \tanh\left(\frac{s}{\Omega_0}\right) = \Omega_0 \frac{e^{\frac{s}{\Omega_0}} - e^{-\frac{s}{\Omega_0}}}{e^{\frac{s}{\Omega_0}} + e^{-\frac{s}{\Omega_0}}}, \quad \Omega_0 = \frac{2}{T_s},$$
$$s := \Omega_0 \operatorname{arctanh}\left(\frac{q}{\Omega_0}\right) = \frac{1}{T_s} \ln \frac{1 + \frac{q}{\Omega_0}}{1 - \frac{q}{\Omega_0}}, \tag{9.17}$$

where $s = \delta + j\omega,\ q = \Delta + j\Omega\ \in \mathbb{C}$.

Relation to the z-transform From the defining equation (9.17) and the previously-discussed relation between the z- and s-domains, it follows that

$$z = e^{sT_s} = \frac{1 + \frac{q}{\Omega_0}}{1 - \frac{q}{\Omega_0}}. \tag{9.18}$$

Eq. (9.18) describes a *bilinear transformation* between the variables z and q and a conformal map from the interior of the complex z unit circle onto the complex left q-half-plane[7].

[7] This property is exploited in the commonly-used bilinear map, e.g. (Tou 1959).

From the defining equation (9.18), it is also easy to infer the applicability of the q-transform to the description of numerical sequences by using the calculation rules of the z-transform with z or z^{-1} following Eq. (9.18). Based on this equivalence, all concepts of the z-transform for the description of discrete-time systems (transfer functions, back transformation of a q-transfer function into a difference equation, etc.) can be straightforwardly transferred to the q-domain.

q-transfer function The *q-transfer function* $G_q(q)$ from a discrete-time input signal $u(k)$ to a discrete-time output signal $x(k)$ can be obtained from a corresponding z-transfer function $G_z(z)$ by substituting Eq. (9.18)

$$G_q(q) = G_z(z)\Big|_{z=\frac{1+\frac{q}{\Omega_0}}{1-\frac{q}{\Omega_0}}} . \tag{9.19}$$

Image of the imaginary axis: the transformed frequency For $s = j\omega$, the following relation can be derived:

$$q = j\Omega = \Omega_0 \frac{e^{j\frac{\omega}{\Omega_0}} - e^{-j\frac{\omega}{\Omega_0}}}{e^{j\frac{\omega}{\Omega_0}} + e^{-j\frac{\omega}{\Omega_0}}} = j\Omega_0 \tan\frac{\omega}{\Omega_0},$$

with the *transformed frequency*

$$\boxed{\Omega = \Omega_0 \tan\frac{\omega}{\Omega_0}} . \tag{9.20}$$

Eq. (9.20) represents the central relationship between the “real” continuous-time frequency ω, and the transformed frequency in the q complex plane.

Using Eq. (9.20), the finite ω base frequency band $[0, \omega_s/2]$ is mapped onto the entire (infinite) Ω-axis, see Fig. 9.25.

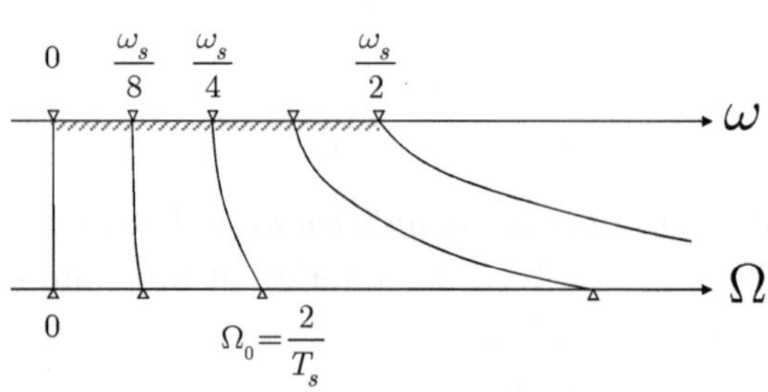

Fig. 9.25. Mapping of ω-axis onto the Ω-axis

Transformed frequency response The transformed frequency response of a q-transfer function $G_q(q)$ is obtained via the substitution $q = j\Omega$, i.e.

$$G_q(j\Omega) = G_q(q)\Big|_{q=j\Omega} . \tag{9.21}$$

Eq. (9.21), together with Eq. (9.19), describes a noteworthy property of the q-transfer function which renders it so attractive for practical use (Fig. 9.26).

Use in systems design As $G_q(q)$ is a *rational function* of the variable q, and—due to the frequency transformation Eq. (9.20)—is defined on an *infinite* interval of the transformed frequency, the frequency response can be treated formally in the same manner as in a *continuous-time system* described by an s-transfer function. In particular, *BODE diagrams* and their simple construction rules (linear/quadratic factors, cutoff frequencies, asymptotic characteristics, etc.) can be adopted *without change.*

Thus, not only can *exact* frequency responses (in contrast to the approximations of previous sections) be easily constructed over the entire frequency domain (even by hand), all familiar structural and parameter relationships relevant for design (gain, phase margin, etc.) can also be employed as usual. Due to the nonlinear spreading of frequencies (Fig. 9.25), there are, as would be expected, equivalencies between both time evolutions and frequency responses in the different domains for "sufficiently low" frequencies $\Omega < \Omega_0$.

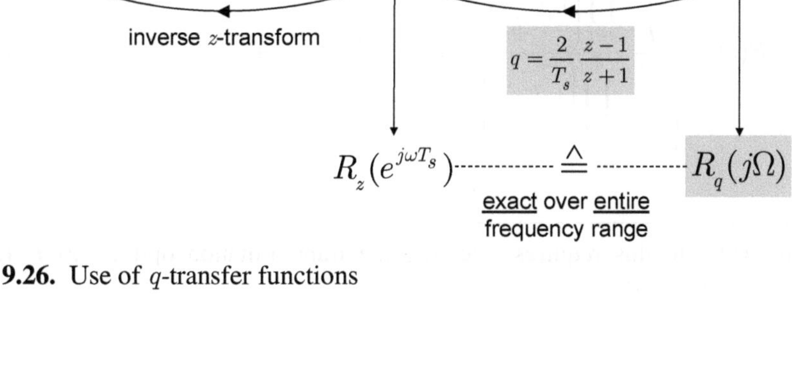

Fig. 9.26. Use of q-transfer functions

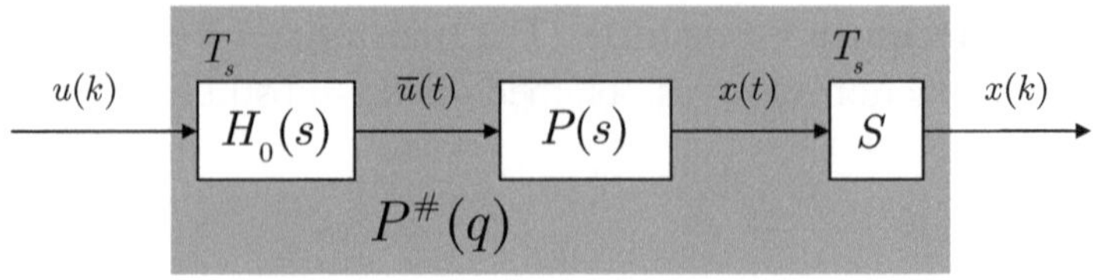

Fig. 9.27. q-transfer function for a sampled continuous-time system with a hold element

Sampled continuous systems with input holds The q-transform proves to be of particular utility in the calculation of transfer functions of sampled continuous-time systems with a hold element at the input, as shown in Fig. 9.27 (cf. the sampled plant transfer function, Fig. 9.16). The customary calculation using *z-transfer functions* (Franklin et al. 1998), (Kuo 1997) only results in muddled correspondences between z-transfer function parameters and those of the continuous model $P(s)$, or if a computer-aided tool is used[8], only a numerical model with fixed parameters. The precise calculation of $P^*(j\omega)$, on the other hand, only results in a *non-parametric* frequency response; at best, using the previously-discussed approximations, $P^*(s)$ can be represented analytically by $P(s)$ at the lower frequency ranges.

Using the *q-transform*, it is shown in (Janschek 1978) that without much calculation, the transfer function and all important parameters can be approximated *analytically* and highly accurately over the *entire frequency domain* $0 \leq \Omega < \infty$, even by hand. In all cases, the exact transfer function can be straightforwardly calculated via the *z-transfer function* and subsequent substitution of Eq. (9.18) (Fig. 9.28).

Proposition 9.1 *Properties of* $P^{\#}(q)$ (Proof in (Janschek 1978)):
For a sampled continuous-time system (with direct feedthrough)

$$P(s) = \left[\frac{K}{s^r} \frac{\prod_{j=1}^{m} \left(1 + \frac{s}{b_j}\right)}{\prod_{i=1}^{l} \left(1 + \frac{s}{a_i}\right)} + d \right] \cdot e^{-sT_T} \qquad r + l = n > m \tag{9.22}$$

[8] In MATLAB, this requires selecting the transformation option *"ZOH (zero-order hold)"* for $P(s)$.

with an *input hold* as in Fig. 9.27 and a *time delay*

$$T_T = \lambda T_s, \quad \lambda = 0, 1, 2, \ldots,$$

the corresponding *q-transfer function*[9] $P^{\#}(q)$ has the following properties:

P1
$$P^{\#}(q) = \left[\frac{K}{q^r} \frac{\prod_{k=1}^{n-1}\left(1+\frac{q}{B_k}\right)\left(1-\frac{q}{\Omega_0}\right)}{\prod_{i=1}^{l}\left(1+\frac{q}{A_i}\right)} + d \right] \cdot \left(\frac{1-\frac{q}{\Omega_0}}{1+\frac{q}{\Omega_0}} \right)^{\lambda}, \tag{9.23}$$

$$\Omega_0 = \frac{2}{T_s}.$$

P2
$$A_i = \Omega_0 \tanh\left(\frac{a_i}{\Omega_0}\right). \tag{9.24}$$

P3 The gain K and number r of poles in the original are preserved.

P4 The zero $(1 - q/\Omega_0)$ in the *right* half-plane always exists and describes the hold process.

P5 Independent of the particular form of $P(s)$, there are always $n-1$ zeros B_k.

P6 $A_i \to a_i$ and $B_j \to b_j$ as $T_s \to 0$, see upper and lower loops with limits in Fig. 9.28.

P7 For the zeros B_k, there is no analytical relation equivalent to that for the poles (Eq. (9.24)); the location of the zeros B_k can, however, be estimated as described below.

P8 A zero is a *significant zero (SZ)* if $B_j \to b_j$ for $T_s \to 0$, otherwise, it is a *negligible zero (NZ)*.

P9 SZs and NZs can not coincide with poles A_i for any T_s.

P10 Real SZs and NZs retain their *relative* position to real poles; as such: for each of the left and right q-half-planes, if poles and zeros are ordered by magnitude, then no two zeros may appear in immediate succession.

P11 No SZ or NZ may lie at $q = 0$ if $P(s)$ has no zero at $s = 0$.

P12 Provided they fulfill property P10, NZs can lie only *outside* the interval $[-\Omega_0, \Omega_0]$.

---

9 The *superscript #* indicates the special system structure with an input hold as shown in Fig. 9.27. A q-transfer function for a discrete transfer system (e.g. a difference equation) without this property is given the *subscript* q (cf. Eq. (9.19)).

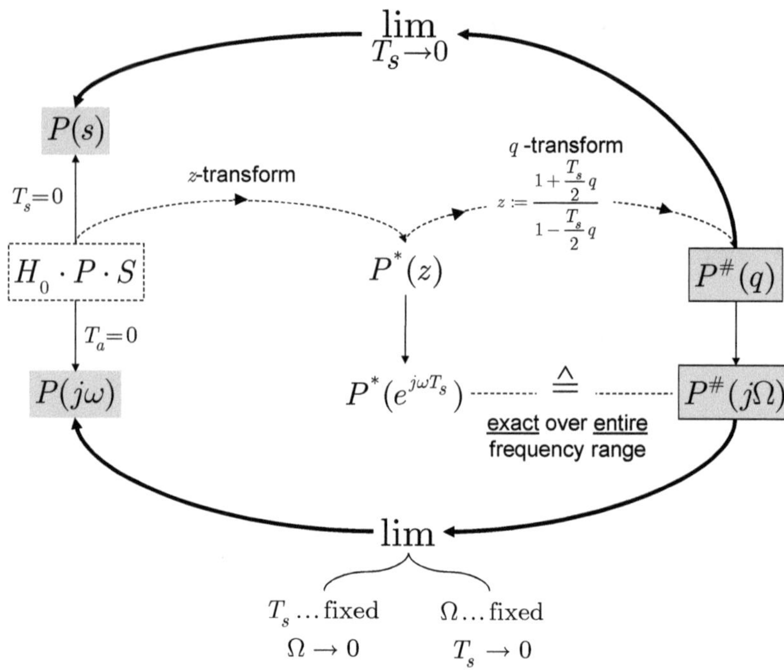

Fig. 9.28. Using the q-transfer function $P^\#(q)$ of a sampled continuous-time system $P(s)$ with an input hold

Using q-transfer functions with input holds Fig. 9.28 presents an overview diagram for the use of q-transfer functions $P^\#(q)$ with input hold elements. The dashed path indicates the exact calculation of $P^\#(q)$ via the *z-transform*; this path is also used for *numerical* calculations with computer-aided tools (e.g. MATLAB). The actual analysis and design steps are then carried out using the frequency response $P^\#(j\Omega)$, which can be manipulated in the usual manner using controller frequency responses $R(j\Omega)$ (e.g. stabilization via the NYQUIST criterion, see Ch. 10). The resulting controller transfer function $R(j\Omega)$ can then be converted into a difference equation by following Fig. 9.26.

The continuity properties of $P^\#(q)$ also permit deeper *analytical* insight into system behavior various sampling periods. For practical cases, using Proposition 9.1, important structural properties and system parameters can be estimated quite accurately to create an analytical representation of $P^\#(q)$ as a function of the sampling period. In the process, the *physical*

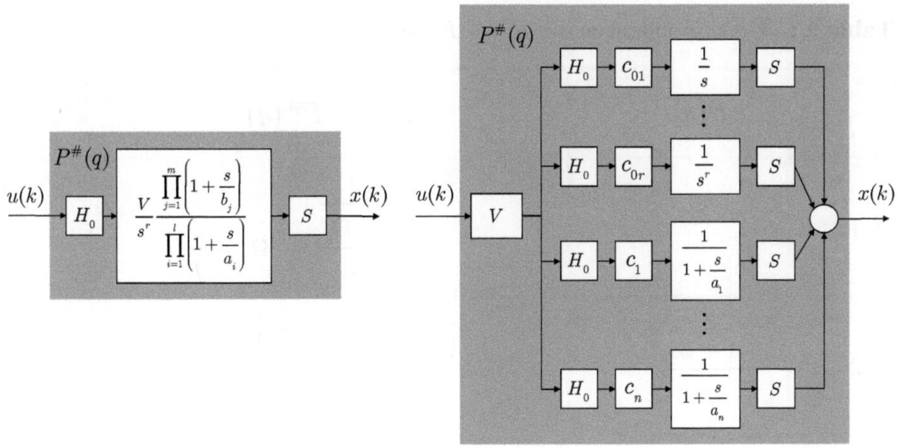

Fig. 9.29. Construction of $P^{\#}(q)$ via elementary transfer functions and partial fraction decomposition

dependencies of $P(s)$ (see Chs. 4 through 8) also remain clear in the discrete transfer function $P^{\#}(q)$ and can be explicitly taken into consideration for design[10].

Exact analytical construction In some cases, the exact analytical derivation of $P^{\#}(q)$ from $P(s)$ is desired. Due to the linearity of the transforms involved, this can proceed following the schema shown in Fig. 9.29. $P(s)$ is decomposed into partial fractions, and analytical correspondences for the resulting elementary terms are applied. Using modern computer algebra programs, $P^{\#}(q)$ can then also be easily reconstructed from the substituted partial fraction terms.

Relevant correspondences for elementary terms as well as several common transfer functions are presented in Table 9.1. The properties of $P^{\#}(q)$ in Proposition 9.1 are also clearly represented there.

---

10 In the experience and estimation of the author, this property forms the *deciding advantage* of the use of the q-transform compared to other descriptions.

Table 9.1. Correspondences for $P(s),\ P^{\#}(q)$

	$P(s)$	$P^{\#}(q)$	
C1	$\dfrac{1}{s}$	$\dfrac{1-\dfrac{q}{\Omega_0}}{q}$	$\Omega_0 = \dfrac{2}{T_s}$
C2	$\dfrac{1+\beta s}{s^2}$	$\dfrac{\left(1+\beta^{\#}q\right)\left(1-\dfrac{q}{\Omega_0}\right)}{q^2}$	$\beta^{\#} = \beta$
C3	$\dfrac{1}{1+\dfrac{s}{a}}$	$\dfrac{1-\dfrac{q}{\Omega_0}}{1+\dfrac{q}{A}}$	$A = \Omega_0 \tanh\left(\dfrac{a}{\Omega_0}\right)$
C4	$\dfrac{1+\beta s}{\left(1+\dfrac{s}{a}\right)^2}$	$\dfrac{\left(1+\beta^{\#}q\right)\left(1-\dfrac{q}{\Omega_0}\right)}{\left(1+\dfrac{q}{A}\right)^2}$	$\beta^{\#} = \beta\left[\left(\dfrac{a}{A}\right)^2 - \left(\dfrac{a}{\Omega_0}\right)^2\right] + {}$ $+\dfrac{1}{A} + a\left(\dfrac{1}{\Omega_0^2} - \dfrac{1}{A^2}\right)$
C5	$\dfrac{1+\beta s}{s\left(1+\dfrac{s}{a}\right)}$	$\dfrac{\left(1+\beta^{\#}q\right)\left(1-\dfrac{q}{\Omega_0}\right)}{q\left(1+\dfrac{q}{A}\right)}$	$\beta^{\#} = \beta + \dfrac{1}{A} - \dfrac{1}{a}$
C6	$\dfrac{1+\beta s}{1+2d_N\dfrac{s}{\omega_N}+\dfrac{s^2}{\omega_N^2}}$	$\dfrac{\left(1+\beta^{\#}q\right)\left(1-\dfrac{q}{\Omega_0}\right)}{1+2D_N\dfrac{q}{\Omega_N}+\dfrac{q^2}{\Omega_N^2}}$	$D_N, \Omega_N, \beta^{\#}$ see next page
C7	$\dfrac{\left(1+\beta_1 s\right)\left(1+\beta_2 s\right)}{s\left(1+2d_N\dfrac{s}{\omega_N}+\dfrac{s^2}{\omega_N^2}\right)}$	$\dfrac{\left(1+\beta_1^{\#}q+\beta_2^{\#}q^2\right)\left(1-\dfrac{q}{\Omega_0}\right)}{q\left(1+2D_N\dfrac{q}{\Omega_N}+\dfrac{q^2}{\Omega_N^2}\right)}$	$D_N, \Omega_N, \beta_1^{\#}, \beta_2^{\#}$ see next page

Table 9.1. Correspondences for $P(s)$, $P^{\#}(q)$, continued

C6, cont.	$A_N = \Omega_0 \dfrac{\sinh\left(2d_N \dfrac{\omega_N}{\Omega_0}\right)}{\cosh\left(2d_N \dfrac{\omega_N}{\Omega_0}\right) + \cos\left(2\dfrac{\omega_N}{\Omega_0}\sqrt{1-d_N^2}\right)}$ $B_N = \Omega_0 \dfrac{\sin\left(2\dfrac{\omega_N}{\Omega_0}\sqrt{1-d_N^2}\right)}{\cosh\left(2d_N \dfrac{\omega_N}{\Omega_0}\right) + \cos\left(2\dfrac{\omega_N}{\Omega_0}\sqrt{1-d_N^2}\right)}$ $\Omega_N = \sqrt{A_N^2 + B_N^2}, \quad D_N = \dfrac{A_N}{\Omega_N}, \quad \beta^{\#} = \dfrac{D_N}{\Omega_N} + \dfrac{\omega_N}{\Omega_N}\sqrt{\dfrac{1-D_N^2}{1-d_N^2}}\left(\beta - \dfrac{d_N}{\omega_N}\right)$
C7, cont.	$\beta_1^{\#} = \beta_1 + \beta_2 + 2\left(\dfrac{D_N}{\Omega_N} - \dfrac{d_N}{\omega_N}\right)$ $\beta_2^{\#} = \dfrac{1}{\Omega_N^2} + \dfrac{D_N}{\Omega_N}\left(\beta_1 + \beta_2 - 2\dfrac{d_N}{\omega_N}\right) +$ $+ \dfrac{\omega_N}{\Omega_N}\sqrt{\dfrac{1-D_N^2}{1-d_N^2}}\left(\beta_1\beta_2 - \dfrac{d_N}{\omega_N}\left(\beta_1 + \beta_2\right) + 2\dfrac{d_N^2}{\omega_N^2} - \dfrac{1}{\omega_N^2}\right)$

Example 9.4 *q-transfer function for an oscillatory system.*

The plant q-transfer function and the BODE diagram of the damped oscillatory system depicted in Fig. 9.18 are to be constructed for $T_s = 1\ \text{s}$.

Solution From Table 9.1, C6, with $d_N = 0.1$, $\omega_N = 1$, $\beta = 0$, it follows that

$$P^{\#}(q) = \frac{\left(1+\dfrac{q}{58}\right)\left(1-\dfrac{q}{2}\right)}{1+2\cdot 0.12\dfrac{q}{1.1}+\dfrac{q^2}{1.1^2}} = \frac{[58][-2]}{\{0.12;\,1.1\}}.$$

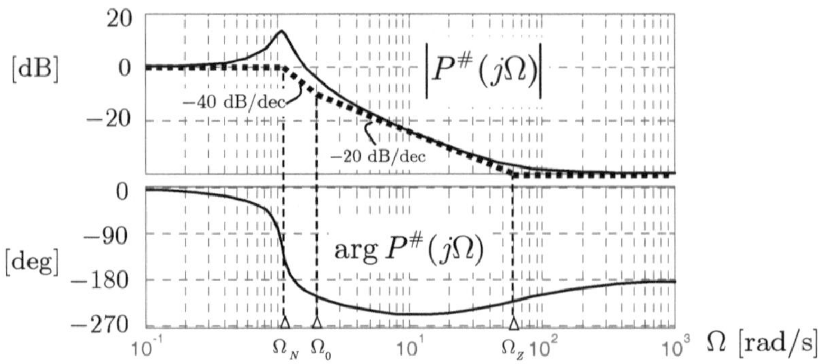

Fig. 9.30. BODE diagram for Example 9.4 (cf. Fig. 9.18) ■

9.10 Signal Conversion

Amplitude discretization: quantization System variables outside of an embedded processor are abstracted as continuous-time and continuous-valued quantities $\in \mathbb{R}$, i.e. they are all quantities of a physical nature, such as forces/moments or analog electrical signals. However, inside the processor, such quantities are generally only available in a *discrete-valued* form with a *finite representational precision.* In the numerical representation for internal computer operations, real numbers are approximated using suitably long word lengths (floating-point numbers), and for many applications, these can be considered *quasi-continuous-valued.*

However, at the computer interfaces consisting of analog-to-digital and digital-to-analog converters, the situation is quite different. There, fundamentally, only a greater or lesser number of *discretization steps*—the *quantization*—can be used, which are determined by the so-called converter word length.

Converter word length The converter word length is understood to be the number of binary digits (bits) available for the discrete-value coding of an analog, continuous-valued signal, e.g. *8-bit* $\triangleq 2^8 = 256$ *discretization* or *quantization steps*; with a 12-bit length, there are $2^{12} = 4096$ steps, and for 16-bit words, there are $2^{16} = 65536$ steps.

Quantization steps The absolute size of discretization steps depends on the numerical range of the original variable being represented. In a signal converter, an *interval*

$$\left[x\right]=\left[\underline{x},\overline{x}\right]=\left\{x\middle|\underline{x}\leq x\leq\overline{x};\quad \underline{x},x,\overline{x}\in\mathbb{R}\right\}$$

with width $\Delta[x]=\overline{x}-\underline{x}$ is mapped to an integer interval

$$\left[g\right]=\left[\underline{g},\overline{g}\right]=\left\{g\middle|\underline{g}\leq g\leq\overline{g};\quad \underline{g},g,\overline{g}\in\mathbb{Z}\right\}$$

with width $\Delta[g]=\overline{g}-\underline{g}=2^N$, where N represents the converter word length.

The *physical quantity* x describes an electrical signal (e.g. for a bipolar voltage, $\underline{x}=-10$ V, $\overline{x}=+10$ V), which itself is a representation of a physical quantity which has been created in the measuring component.

The *internal processor variable* g can be interpreted as a fixed-point number with a word length N, which can be coded as either a unipolar (N data bits) or bipolar (1 sign bit, $N-1$ data bits) number.

The *physical quantization step*, important for the analysis of system behavior, is defined as

$$Q_x=\frac{\Delta[x]}{2^N}. \tag{9.25}$$

Quantization curve Quantization can be represented in the form of a nonlinear characteristic curve. Both of the different curves shown in Fig. 9.31 are important for signal conversion.

An *analog-to-digital converter* has a *symmetric* characteristic curve corresponding to the mathematical operation of *rounding* (Fig. 9.31a), while a *digital-to-analog converter* is described by an *asymmetric* characteristic curve, corresponding to the mathematical operation of *truncation*, (Fig. 9.31b).

Properties of quantization curves For small signals, the deadband about the origin is of particular importance. In an ADC, very small input signals $\left|x\right|<Q_x/2$ are not detected and thus evaluated as $x=0$ in the processor in a symmetric fashion.

For a DAC, the situation is somewhat more complicated. Assuming a numerical representation in the processor with greater precision than the converter word length $N_{D/A}$ (e.g. a fixed-point number with $N_x>N_{D/A}$), then for bipolar conversion, the quantization process implies that at even a

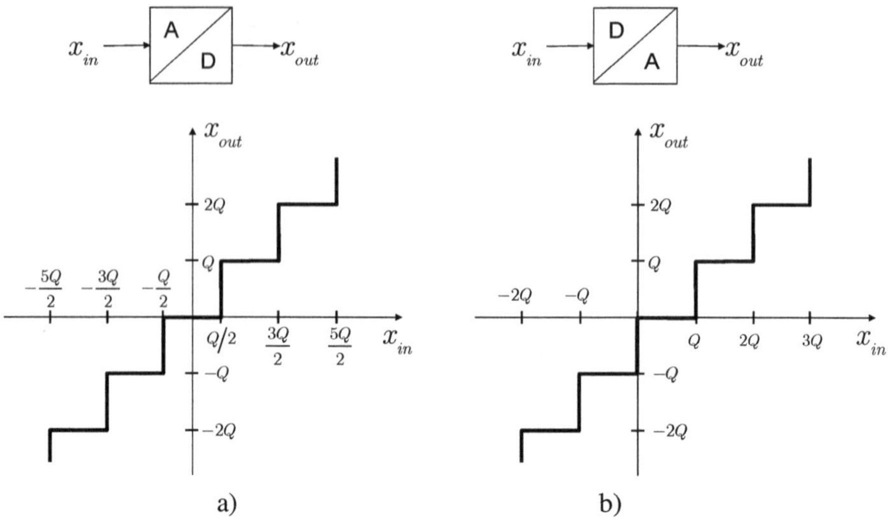

Fig. 9.31. Quantization curves for signal converters: a) symmetric, rounding operation, e.g. an analog-to-digital converter, b) asymmetric, truncation operation, e.g. a digital-to-analog converter ($Q \triangleq Q_x$)

negligible negative value of x_E (i.e. when the sign bit equals one), the smallest negative converter output $x_A = -Q_x$ is generated. In a closed loop, this can lead to unsteady flutter or even undesirable limit cycling.

The DAC quantization curve can be made *symmetric* by adding an offset of $Q_x/2$ before outputting the control variable. This is always possible if a longer word length is used in the processor.

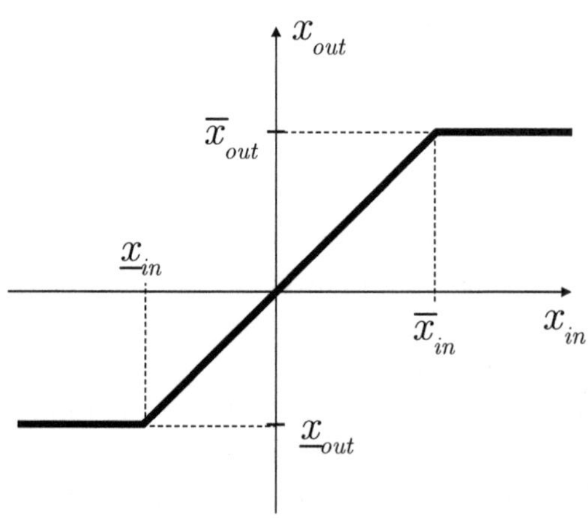

Fig. 9.32. Saturation curve

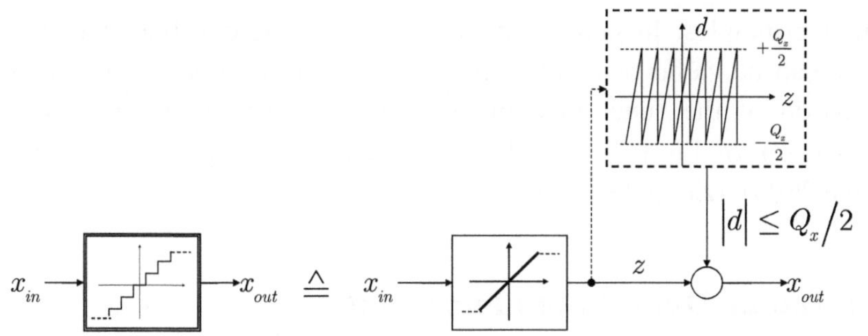

Fig. 9.33. Equivalent piecewise-linear quantization model (dashed lines at right describe nonlinear effects)

Saturation curve Due to their interval mapping, signal converters have a characteristic *saturation curve* (Franklin et al. 1998) limited by the interval bounds $\underline{x}, \overline{x}$, see Fig. 9.32. This additional nonlinear property can be neglected when analyzing small-signal behaviors, if it is assumed that applicable system variables stay within the linear domain $x_E \in \left[\underline{x}_E, \overline{x}_E\right]$[11].

Equivalent linear model A useful tool for dynamic analysis is the equivalent linear quantization model depicted in Fig. 9.33, consisting of a linear component with a bounded-amplitude output disturbance $|d| \leq Q_x/2$ —the *quantization noise*. It is customary to assume a *uniform distribution* for the quantization noise.

For completeness, Fig. 9.33 (right) also indicates the nonlinear properties of the saturation curve and the generating mechanism for the quantization noise (dashed lines).

Time domain dynamics The signal conversion from analog to digital domains and vice versa fundamentally requires a certain amount of time. For dynamic analysis, this implies a *time delay* $\tau_{A/D}, \tau_{D/A}$ between the input signal and output signal. For ADCs, processes for value conversion often use counters or integrators (averaging converters) employing a voltage-time conversion (*dual-slope* converters). Both types of converter require a certain amount of time to evaluate the signal and interpret it as a digital value. In DACs, circuits consisting of weighted resistance networks and operational amplifiers are often used, in which it is primarily the settling time which determines the maximum cycle frequency. Due to their func-

[11] This applies generally for all system variables under consideration.

tioning principles, in general $\tau_{A/D} > \tau_{D/A}$; concrete values specific to a device can be gathered from respective data sheets. For a more extensive discussion of the functional and dynamic properties of converters, the reader is referred to related technical literature, e.g. (Hristu-Varsakelis and Levine 2005), (Zurawski 2005).

9.11 Digital Data Communications

Analog information networking In the overview Fig. 9.1, the information chain *measurement – information processing – force generation* is schematically represented along with associated information flows. In simple mechatronic systems, each of these functional units is realized in exactly one device unit: a sensor, embedded microprocessor, and actuator. In such cases, the information flow in the form of electrical measurement and control signals is also very simple in structure. In particular, for spatially compact solutions, analog electrical signals (voltages and currents) with point-to-point wiring are generally used for this flow, resulting in no noteworthy aberrations in the dynamics (Hristu-Varsakelis and Levine 2005), (Zurawski 2005).

Spatially distributed systems More complex mechatronic systems, in particular those of a macroscopic nature, are often spatially distributed and composed of a large number of sensors, actuators, and processing units—called *nodes* in this context—e.g. in automobiles, CNC machine tools, and industrial machines. In such cases, many nodes with a variety of assigned purposes must be interconnected. Such arrangements are then termed *sensor-processor-actuator networks*[12].

Communication topologies Fig. 9.34 shows a few commonly used topologies for digital data communications. The indicated nodes N_i all represent named devices. In the case of sensors and actuators, integrated signal converters are assumed, so that all data connections can be assumed digital. The topologies depicted in Fig. 9.34a-c each use *point-to-point* connections, whereas the *linear topology* shown in Fig. 9.34d employs a common

[12] Often only the imprecise short form *sensor-actuator network* is used, with which either the link to separate processing units is implicitly understood, or local information processing in the sensors and actuators is assumed.

transport medium—a *communications bus*. With point-to-point connections, the transmission medium can be used exclusively, conflict-free, and with minimal data delays; however, for multiple connections between nodes, wiring and connector costs become large (e.g. the point-to-point topology, Fig. 9.34a). For the star and ring topologies (Fig. 9.34b,c) a smaller number of wires is achieved at the cost of limited connectivity and larger data delays when information is transported between nodes.

A particularly popular and economical topology is the linear topology in Fig. 9.34d. Using a single transmission medium (e.g. a twisted pair or bus line), all nodes can, in principle, be connected to each other as desired with much less wiring than the point-to-point topology in Fig. 9.34a.

Serial data communications For reasons of transmission security and communication economy (or the complexity/use ratio), as a rule, serial data communications are used outside of processing units and for any appreciable transmission length. In such a setup, digital data are serially transferred on the transmission medium as binary-coded data packets or *datagrams*. Fig. 9.35 shows a schematic depiction of the data flow between two serially connected nodes N_i, N_{i+1} for the reading of a sampled value from the ADC (Node N_i) into the microprocessor (Node N_{i+1}).

Serial bus systems A particularly economical arrangement is serial data communications combined with a linear bus topology as in Fig. 9.34c; these are termed a *serial bus systems* or, if particular real-time properties are fulfilled, *field busses* (Hristu-Varsakelis and Levine 2005), (Mahalik 2003), (Mackay et al. 2004), (Pfeiffer et al. 2008), (Etschberger 2001).

Table 9.2 lists three examples of widely distributed serial bus systems, where PROFIBUS and CAN-Bus can be counted as field busses.

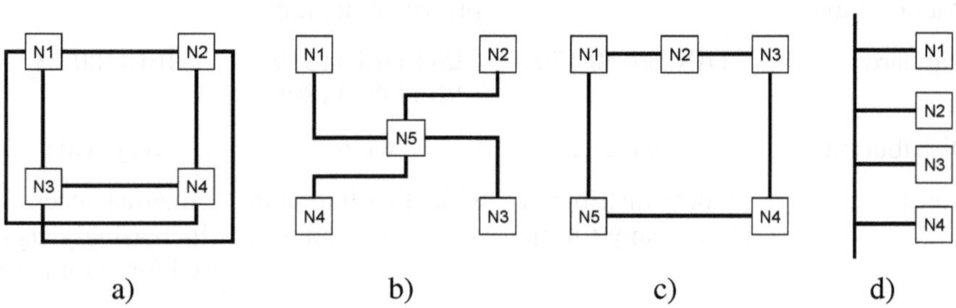

Fig. 9.34. Common topologies for data communications: a) point-to-point topology, b) star topology, c) ring topology, d) linear topology

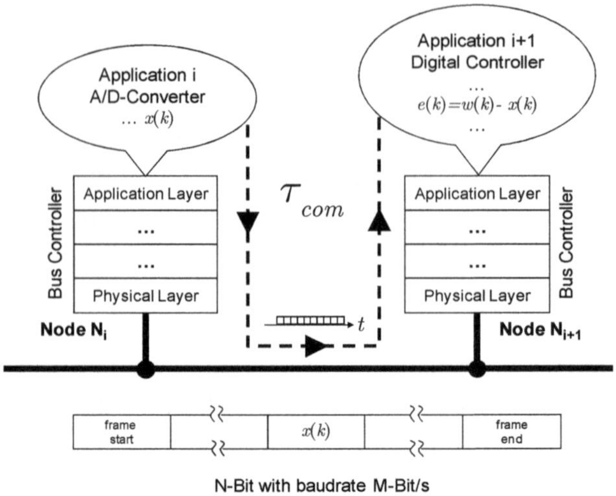

Fig. 9.35. Data flow in a serial bus system: schematic example of two nodes with individual controllers

Table 9.2. Examples of serial bus systems

Property	Profibus	CAN-Bus[13]	Ethernet
Nodes	max. 12	max. 64	many
Speed	12 Mbit/s	1 Mbit/s	10–1000 Mbit/s
Bus access	mostly master/slave	CSMA/CA (multi-master)	CSMA/CD (multi-master)
User data / datagram	0–32 bytes	0–64 bytes	46–1500 bytes
Guaranteed reaction time	yes	only for top-priority datagrams	no
Standard	DIN EN 50 170	ISO DIS 11519 ISO DIS 11898	IEEE 802.3
Distribution	wide	wide	very wide
Notes	widely distributed in Germany/Europe	wide distribution in automotive industry	internet access, increasingly also used for automation

[13] CAN = Controller Area Network.

Communications protocols Constructing datagrams for serial data communications generally follows the *OSI network layer model*[14] (Mahalik 2003). This data model sets up a framework of requirements for the structured processing of datagrams (the communications protocol) in the form of layers. The sequential processing hierarchy is indicated in Fig. 9.35. Depending on the implementation, for each communications action, more or less complex and time-consuming processing steps must thus be performed.

Bus access modes The primary disadvantage of bus-based data communications consists of the shared use of the transmission medium. However, this leaves communications fundamentally subject to conflicts. To avoid such conflicts and ensure defined timing requirements, bus access must be coordinated using suitable access rules.

There is a general distinction between *controlled* and *random* bus access. Under *controlled, deterministic bus access* following the master-slave paradigm, a master node cyclically accesses its slave nodes, or cyclically permits each slave node to transmit for a given time (the flying master paradigm). This guarantees that within a defined wait time, each slave node can send its data at least once, allowing *defined reaction times* (real-time conditions) to be maintained (e.g. Table 9.2, Profibus). Disadvantages include the fact that slave nodes are queried even if there is no data to send, and that in the case of a fault in the master, no notifications are sent.

Under *random bus access*—specifically *carrier sense multiple access (CSMA)*—each node desiring to transmit listens on the shared bus and begins to send only when the bus is available. The sequencing of transmitters is not pre-assigned, but rather evolves as necessary. If a second participant begins sending at the same time, a collision of datagrams on the transfer medium ensues. In a protocol *with collision avoidance (CSMA/CA)*, a priority system is used to allow the higher-priority datagram to be handled preferentially (Table 9.2, CAN-Bus). In a protocol *with collision detection (CSMA/CD)*, both datagrams are terminated, and both transmitters attempt to re-access the bus at a later time (Table 9.2, Ethernet).

In random CSMA/CA access protocols, guaranteed reaction times can only be achieved for datagrams with high priority, while for CSMA/CD protocols, no defined reaction times can be guaranteed.

[14] Open Systems Interconnection (OSI) Reference Model, standard of the International Standard Organization (ISO).

Time delays Digital data communications, and in particular serial protocols, have inherent *communications time delays* τ_{com}. Fig. 9.35 shows these using the example of data transport from an ADC (sensor measurement) to a processor (control algorithm). The communications delay τ_{com} comprises the execution times for the data processing in the communications protocol, the actual transport time on the bus medium, and also wait times for bus access. An accurate estimate of such delays is indispensible for the dynamic analysis of a mechatronic system. The delay properties of the information flow between sensors, control computers, and actuators play a decisive role in determining closed-loop stability.

9.12 Real-Time Aspects

Embedded controller realization In an embedded microprocessor, a control algorithm is realized as a sequential program following the scheme set out below (cf. Fig. 9.3).

At each *sampling time* $t = kT_s$, the following generic *program sequence* (in pseudocode) is executed:

τ_A	`get_ADC(x)`	A/D conversion, data transfer ADC→ CPU
	`get_reference(w)`	Reference value generation
τ_B	`e:=w-x`	Forming the controller error
	`u:=control(e)`	Control algorithm, e.g. difference equation
τ_C	`put_DAC(u)`	Data transfer CPU→ DAC, D/A conversion

The operations `get_ADC`, `get_reference`, and `control` represent calls to language-specific program modules which service peripheral devices, interfaces, and data connections, as well as activate control algorithms.

Temporal processing scheme Due to sequential execution and finite processing speeds, these operations are carried out with the indicated processing times τ_A, τ_B, τ_C[15]. These include the processing times for signal conversion $\tau_{A/D}, \tau_{D/A}$ and data communications τ_{com} already discussed in previous sections. The temporal processing scheme is shown in Fig. 9.36.

[15] The computational time for a control algorithm is primarily determined by the number of multiplications; addition/subtraction times are negligibly small in comparison.

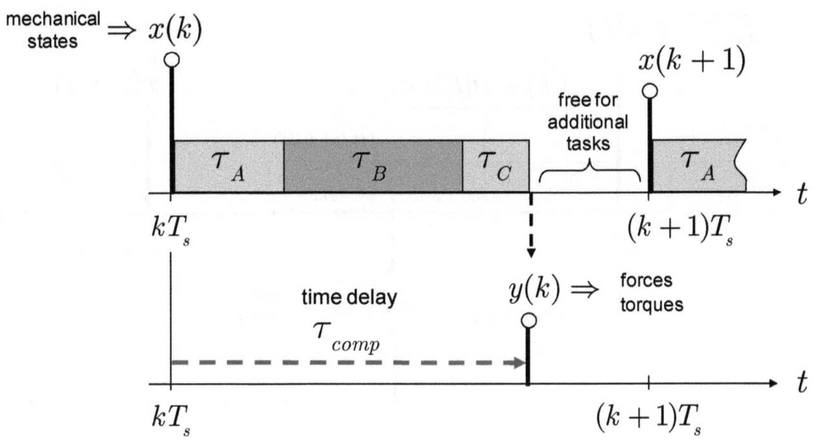

Fig. 9.36. Temporal processing scheme for a digital control algorithm

Computational time delay Fig. 9.36 demonstrates the fundamental computational delay $\tau_{comp} = \tau_A + \tau_B + \tau_C$ between the observation time $t = kT_s$ or sampling of measured variables (here, motion quantities) and the action of control variables (here, forces and moments) on the plant, which only occurs at time $t = kT_s + \tau_{comp}$. To this extent, the synchronous action of $x(k)$, $e(k)$ and $u(k)$ assumed up to now in dynamic models must be suitably corrected. In frequency domain models, this can be easily described using a computational delay τ_{comp} implemented as a *time delay element* $e^{-s\tau_{comp}}$.

Real-time condition As can be seen from Fig. 9.36, all operations required for time step k must be finished before the next time step $k+1$ can begin, i.e. for all time steps, the so-called *real-time condition* must be met:

$$\tau_{comp} = \tau_A + \tau_B + \tau_C < T_s . \tag{9.26}$$

From a system theoretical point of view, the computational delay τ_{comp} should be kept a small as possible so as to enable the *smallest possible sampling period* and minimize dynamic influences of the sample-and-hold process. In general, this design goal is limited by the cost and complexity of solutions, so that here suitable design compromises must also be found.

In all cases, due to Eq. (9.26), the *worst-case* estimate is

$$\tau_{comp,\max} = T_s . \tag{9.27}$$

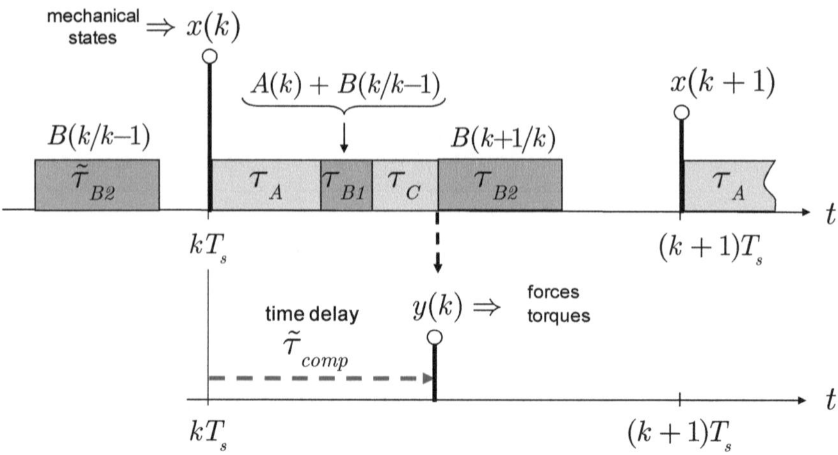

Fig. 9.37. Temporal processing scheme for minimum computational delay

Minimizing computational delays Due to the destabilizing action of computational delays, it is incumbent upon the designer to minimize them. Within certain limits, for a given hardware realization, they can be reduced with a clever program structure compared to the execution scheme shown in Fig. 9.36.

In the controller difference equation (9.13) for the control variable $u(k)$, all values $u(k-i)$, $e(k-i)$, $i = 1,2,...$ are in fact already known at the end of step $k-1$, so that a partial sum $B(k/k\text{–}1)$ can already be precomputed at the end of step $k-1$ according to the following scheme:

$$u(k) = \underbrace{b_0 e(k)}_{A(k)} + \underbrace{b_1 e(k\text{–}1) + \ldots + b_m e(k\text{–}m) - a_1 u(k\text{–}1) - \ldots - a_n u(k\text{–}n)}_{B(k/k\text{–}1)} .$$

In time step k, all that remains to compute is the term $A(k)$ and its addition to $B(k/k\text{–}1)$, resulting in the time-optimal computational scheme with minimum delay shown in Fig. 9.37.

9.13 Design Considerations

Relevant phenomena For systems design, the closed-loop dynamic behavior of a mechatronic system is of particular interest. Previous sections

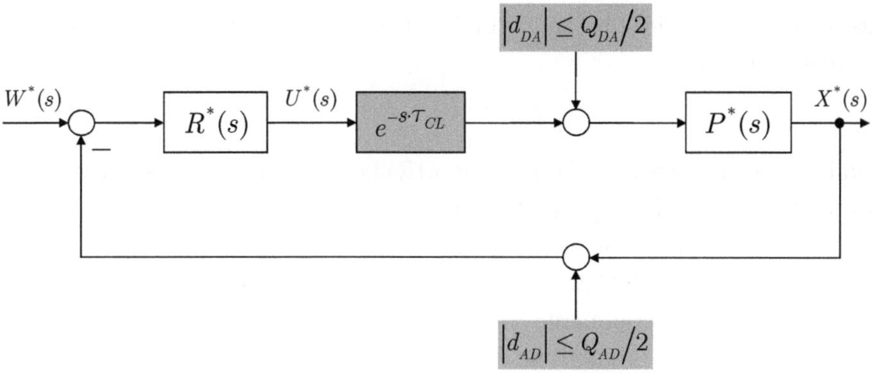

Fig. 9.38. Equivalent linear model for dynamic analysis of a digital control loop (an anti-aliasing filter is included in $P^*(s)$)

discussed important phenomena of *digital information processing*, which usually have a negative influence on the system dynamics. These include *harmonic spectra* and *ambiguities* (aliasing) due to the sampling process, *time delays* due to finite processing and data transport speeds, and the nonlinear effects of *quantization* and *amplitude saturation* during signal conversion.

Equivalent linear model For predictive analytical evaluation of the dynamic behavior of a digitally controlled mechatronic system, the equivalent linear frequency domain model depicted in Fig. 9.38 can be used. Using the transfer functions $P^*(s)$, $R^*(s)$, the sampling process can be modeled as previously discussed: $P^*(s)$ contains all continuous system elements including the anti-aliasing filter. The time delay element $e^{-sT_{CL}}$ describes applicable delay effects and the bounded-amplitude disturbances representing the quantization noise.

Effective control loop delay Depending and the model accuracy, the effective *control loop delay* τ_{CL} accounts for different effects. In all cases, this includes the computational delay τ_R discussed in Sec. 9.12, which is the sum of all computer and information processing delays. To ensure the real-time condition is met, following Eq. (9.27), a sensible design goal is $\tau_{C,\max} = T_s$.

There is a certain degree of flexibility in modeling $P^*(s)$. Here, either the exact sampled plant frequency response $P^*(j\omega)$ from Eq. (9.7) or the approximate frequency response of Eq. (9.9) can be used. For both these cases, holding to the real-time condition results in the maximum *effective control loop delay*

$$\tau_{CL} = T_s \,. \tag{9.28}$$

Analysis requirements Time delay effects from the delay element $e^{-s\tau_{CL}}$ dominate the stability properties of the closed loop. The resulting frequency-dependent phase delay is extremely critical for stability, particularly in the context of MBS eigenfrequencies of the mechanical subsystem. For these reasons, applicable analysis procedures must especially be in a position to provide predictions of the dynamics of *control loops* with *time delays*. In this context, the advantages of the clear and easy-to-use *frequency domain* model representation become distinct.

In the subsequent Ch. 10, it will be shown that the well-known *NYQUIST stability criterion* combined with a particular frequency response representation in the form of *NICHOLS diagrams* is eminently suited to the clear analysis of all discussed phenomena of mechatronic systems.

Variable delays Time delays are particularly critical for closed-loop stability. Delays from algorithm execution, signal conversion, and digital data communications under controlled bus access methods can be rather well estimated with maximum (*worst case*) values $\tau_{CL,\max}$.

For systems with *random bus access* however, often no guaranteed upper bounds can be determined. Particularly critical cases are datagram losses due to collisions, resulting in highly variable latencies. Such phenomena can have unpleasant or even catastrophic repercussions on the closed-loop system dynamics. The analysis and targeted mitigation of such disadvantageous properties using control system measures have lately been a focus of scientific interest and, under the title of "Networked Control Systems", are the subject of current research (Wang and Liu 2010), (at 2008).

Bibliography for Chapter 9

at (2008). "Schwerpunktheft: Digital vernetzte Regelungssysteme." *at-Automatisierungstechnik* **56**(1): 1-57.

Etschberger, K. (2001). *Controller Area Network*. IXXAT Automation GmbH.

Franklin, G. F., J. D. Powell and M. L. Workman (1998). *Digital Control of Dynamic Systems*. Addison-Wesley.

Gausch, F., A. Hofer and K. Schlacher (1993). *Digitale Regelkreise.* Oldenbourg-Verlag.

Horn, M. and N. Dourdoumas (2006). *Regelungstechnik.* Pearson Studium.

Hristu-Varsakelis, D. and W. S. Levine, Eds. (2005). *Handbook of Networked and Embedded Control Systems (Control Engineering)* Birkhäuser Boston.

Janschek, K. (1978). *Über die Behandlung von Abtastsystemen im transformierten Frequenzbereich.* Diplomarbeit. Institut für Regelungstechnik, Technische Universität Graz.

Kuo, B. C. (1997). *Digital Control Systems.* Oxford University Press.

Mackay, S., E. Wright, D. Reynders and J. Park (2004). *Practical Industrial Data Networks: Design, Installation and Troubleshooting (IDC Technology).* Newnes.

Mahalik, N. P., Ed. (2003). *Fieldbus Technology: Industrial Network Standards for Real-Time Distributed Control* Springer.

Pfeiffer, O., A. Ayre and C. Keydel (2008). *Embedded Networking with CAN and CANopen.* Copperhill Media Corporation.

Reinschke, K. (2006). *Lineare Regelungs- und Steuerungstheorie.* Springer.

Schneider, G. (1977). "Über die Beschreibung von Abtastsystemen im transformierten Frequenzbereich." *Regelungstechnik* **25**(9): A26-A28.

Tou, J. T. (1959). *Digital and Sampled-data Control Systems.* McGraw-Hill Book Company.

Wang, F.-Y. and D. Liu, Eds. (2010). *Networked Control Systems: Theory and Applications*, Springer.

Zurawski, R., Ed. (2005). *Embedded Systems Handbook*, CRC Press.

10 Control Theoretical Aspects

Background The behavior of a mechatronic system is fundamentally determined by the closed action chain which composes it. Here, *all* involved *physical phenomena* are regulated by *all* involved *system components* at a common, abstract modeling level. The feedback interactions of the entire set of system components can thus be considered at the level of a *control loop*, and analyzed and systematically assembled using well-known methods of control theory. Thus, control plays a central role in systems design: "*Mechatronics is much more than control, but there is no mechatronics without control*".

Contents of Chapter 10 This chapter deals with certain aspects of control which are particularly applicable to the design of mechatronic systems, and which cannot be found in the standard literature in this thematic combination. For the model class employed here—*linear models in the frequency domain*—the dynamics of weakly-damped, elastically-coupled *multibody systems* particularly characterize the design problem. Starting with an introductory consideration of a *control loop structure* with *two design degrees of freedom*, the discussion then moves to the most influential *modeling uncertainties*. To evaluate *robust stability properties*, the *NYQUIST criterion* is introduced in the otherwise oft-ignored *intersection/frequency response curve formulation*. Along with the representation of the frequency response in the form of a *NICHOLS diagram*, this formulation provides *the* fundamental tool for clear, manually accomplishable controller design. On this basis, *robust* and generalizable *control strategies* are presented for particular classes of systems: *single-mass oscillators*, *collocated* and *non-collocated multibody systems*, and *active vibration isolation*. Particular consideration is also given to *observability and controllability problems* for *control in relative coordinates*, and for measurement and actuation locations placed at *oscillation nodes*. Finally, specialized problems which can appear in the context of the realization of a *digital controller* for a multibody system are addressed. ■

10.1 Systems Engineering Context

Robustness requirements The top-level product task for a mechatronic system consists of realizing purposeful motions in the face of varying or unknown operating and environmental conditions. As a rule, this task can only be satisfactorily addressed with a closed action chain—a control loop. The important role of control in the context of systems design was already alluded to in the introduction: *"Mechatronics is much more than control, but there is no Mechatronics without control"* (Janschek 2008).

The following primary specifications are central to any design considerations:

- *robust stability* in the presence of parameter uncertainties and variations,
- *disturbance rejection* of undesirable disturbance inputs,
- *command following* relative to the desired motion.

Here, the stability of the entire feedback system is a fundamental and necessary prerequisite for the two dynamic properties of controller response and disturbance rejection (i.e., the transient and the steady-state dynamics), which are externally much more visible.

Mechatronic problem statement At its core, this is a problem of *robust control loop design.* In the last decades, numerous approaches to this problem have been developed in the control theory community, and have been successfully deployed in a wide variety of industrial applications (Åström 2000), (Zhou and Doyle 1998).

For the mechatronic systems of interest here, and in the context of systems design, the following requirements for *usable* procedures for designing control loops present themselves:

1. A procedure should predict the dynamics of *multibody systems (MBSs)* (e.g. resonances, anti-resonances, collocation, uncertainties in damping, and eigenfrequencies) in a clear and transparent manner.
2. A procedure should predict system behaviors related to *parasitic dynamic properties* (delays) in all applicable system components (sensors, actuators, and information processing) in a clear and transparent manner.
3. *Higher-order systems* should be easy to deal with (i.e. MBSs with multiple eigenfrequencies and parasitic dynamics).

4. The dynamic properties of digital information processing and *digital control* should be considered in a clear and transparent manner.
5. A straightforward, *manual parameterization of the controller* should be possible, in order to be able to explicitly and directly incorporate the analytical relations based on physical system parameters presented in Chs. 4 to 9.
6. It should be simple to employ *experimentally* determined *plant models* (and particularly, multibody system attributes).

10.2 General Design Considerations

Methods for design Today, many powerful methods and procedures—some of which are computer-aided—are available for the design of control loops, and in particular also address *robustness* to uncertainties. A well-prepared overview of these can be found in (Åström 2000); the reader is referred to (Åström and Murray 2008) for a deeper, quite readable introduction to these topics, and to (Zhou and Doyle 1998), (Weinmann 1991) for a more detailed mathematical description of common methods and procedures, including well-known catchphrases such as *μ-synthesis*, *$\mathcal{H}_2$ optimal control*, and *$\mathcal{H}_\infty$ control*.

However, this book does not further delve into these methods for the following reasons: their understanding and use requires a significant mathematical foundation, which cannot necessarily be assumed of the reader, and which cannot be provided in this book for reasons of space. For systems of a somewhat realistic order, such methods can, in fact, only be applied with the aid of computers[1]. However, this implies a loss of clarity in the model (see requirements above), which is somewhat disadvantageous in the context of systems design—particularly for initial, rough investigations.

In order to avoid these disadvantageous aspects, this chapter presents control loop design in the *frequency domain* based on the *frequency response*. This allows all of the requirements set out above to be met quite satisfactorily, and the required mathematical and systems theoretical foundations may be assumed known by the reader. Only in addressing the prop-

[1] For example, MATLAB offers excellent toolboxes for this purpose.

erties of MBSs with weakly-damped eigenmodes is an extension of the customary frequency response representation undertaken (see Sec. 10.4):

- the otherwise seldom-used *NICHOLS diagram* (*gain-phase plot*) is introduced as a complement to BODE diagrams (logarithmic frequency response curves),
- the *NYQUIST criterion* is introduced in the seldom-used *intersection formulation.*

It will be shown that using these additional representations, it is possible to design control loops with an easily understandable, manual procedure, which distinctly includes important system attributes relating to robustness (the MBS eigenfrequencies and damping). As a result, fundamental controller configurations can be devised and feasibility predictions made, supporting the system design. The controller models obtained in this manner then serve as feasible initial solutions for more refined designs using the specialized methods listed above.

Control loop configuration: design degrees of freedom It will be shown that the three primary tasks introduced above—robust stability, disturbance rejection, and controller response—cannot be solved in a completely independent manner. In this chapter, a *control loop configuration* (which has proven itself in practice and which is easy-to-use for design) with *two design degrees of freedom* serves as the foundation for further considerations (see Fig. 10.1). The two design degrees of freedom can be illustrated in the following sequential design steps:

- In the *first step*, using the *control algorithm* in the feedback loop—here a linear controller with the transfer function $H(s)$—both the robust stability and disturbance rejection tasks are addressed (though only by making certain reciprocal compromises).
- In the *second step*, based on the closed control loop, the controller response can then be corrected with a special pre-processing of the reference inputs using *filter algorithms*—here linear filters $F(s)$, $A(s)$—as desired.

The *design task* for the configuration shown here thus consists of finding suitable *structures* and *parameters* for the dynamic transfer functions $H(s)$, $F(s)$, and $A(s)$.

Control loop with two degrees of freedom The control loop configuration shown in Fig. 10.1 possesses two correcting mechanisms which can be

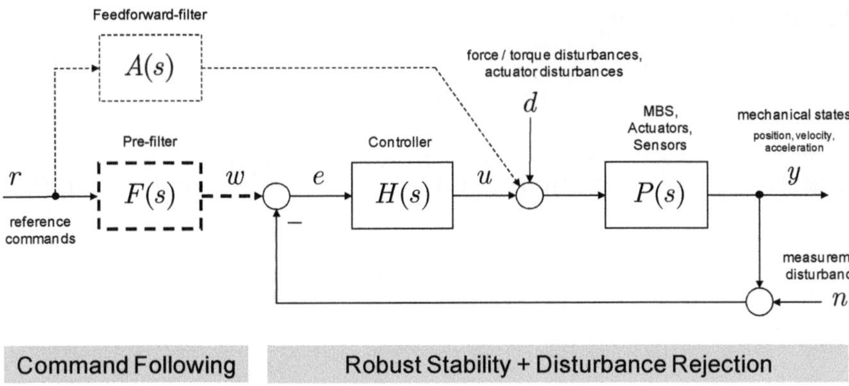

Fig. 10.1. Standard control loop with two design degrees of freedom (controller and pre-filter/feedforward-filter)

freely chosen during design: the actual *controller* with transfer function $H(s)$, and a *pre-filter* with transfer function $F(s)$. The plant $P(s)$ is considered given, and encompasses the dynamics of the mechatronic system components, i.e. the mechanical structure (e.g. a multibody system), actuators, and sensors. The inputs considered are the *reference input* $r(t)$; *disturbance forces/torques*, and *actuator disturbances* $d(t)$ at the input to the plant; and *measurement disturbances (noise)* $n(t)$ in the feedback path.

System transfer functions for design The system properties important for design can be represented using the following transfer functions.

Open-loop transfer function $L(s) := H(s)\,P(s)$ (10.1)

Sensitivity $$S(s) := \frac{1}{1 + L(s)} \tag{10.2}$$

Complementary sensitivity $$T(s) := \frac{L(s)}{1 + L(s)} \tag{10.3}$$

Reference input transfer function

$$T_r(s) := \frac{Y(s)}{R(s)} = \frac{F(s)\,L(s)}{1 + L(s)} = F(s)\,T(s)$$
$$\tilde{T}_r(s) := \frac{Y(s)}{R(s)} = \frac{A(s)\,P(s) + L(s)}{1 + L(s)} = \left(\frac{A(s)}{H(s)} + 1\right)T(s) \tag{10.4}$$

Disturbance rejection

$$T_d(s) := \frac{Y(s)}{D(s)} = \frac{P(s)}{1+L(s)} = P(s)\,S(s) \tag{10.5}$$

Measurement noise rejection $T_n(s) := \dfrac{Y(s)}{N(s)} = \dfrac{L(s)}{1+L(s)} = T(s)$ (10.6)

Actuation penetration

$$\left[\begin{aligned} T_{u,r}(s) &:= \frac{U(s)}{R(s)} = \frac{F(s)\,H(s)}{1+L(s)} \\ T_{u,n}(s) &:= \frac{U(s)}{N(s)} = \frac{H(s)}{1+L(s)} \\ T_{u,d}(s) &:= \frac{U(s)}{D(s)} = \frac{L(s)}{1+L(s)} = T(s) \end{aligned}\right. \tag{10.7}$$

Control loop classification The fundamental idea behind the control loop configuration in Fig. 10.1 is easy to understand: there is a division of labor between (i) *feedback control* to address parameter uncertainties and exogenous disturbances, which are considered given, and (ii) *feedforward control* for the transfer of the directly modifiable reference inputs.

In (Reinschke and Lindert 2006), a nice historical overview of two-degree-of-freedom structures is presented. These extend back to (Graham 1946) and have successfully proven themselves to this day in a wide variety of implementations, e.g. (Reinschke 2006), (Hagenmeyer and Zeitz 2004), (Kreisselmeier 1999).

The configuration considered here (Fig. 10.1) presents no loss of generality, rather *"Some structures have been presented as fundamentally different from the others. It has been suggested that they have virtues not possessed by others, and have been given special names...all 2DOF configurations have basically the same properties and potentials...."* (Horowitz 1963).

Pre-filter vs. feedforward filter Two possible realizations are sketched for the open-loop control segment in Fig. 10.1: a *pre-filter* $F(s)$ and a *feedforward filter* $A(s)$. Systems theoretically, both are in fact equivalent, as can be seen by equating $T_r(s)$ and $\tilde{T}_r(s)$ in Eq. (10.4). For

$$F(s) = \frac{A(s)}{H(s)} + 1, \tag{10.8}$$

the identical controller response results.

Differences in the two formulations only result for the layout and realization of the system. For the *feedforward filter* $A(s)$, access to the actuator input or controller output is required, whereas the pre-filter $F(s)$ involves modification of the reference value. Both courses of action are equally easy to implement.

For the system layout, the *pre-regulation variant* is somewhat more popular since it is conceptually easier to implement. For the most *ideal* controller response $\tilde{T}_r(s) = 1$, the feedforward filter

$$A(s) = \frac{1}{P(s)} \tag{10.9}$$

would be chosen, independent of the dynamic properties of the feedback loop (independent of $L(s) = H(s)\,P(s)$). In reality, however, the problem occurs that, for low-pass behavior in $P(s)$, an improper transfer function (degree of the numerator > degree of the denominator) results for $A(s)$. The ideal design condition can thus only be realized over a finite frequency range. However, in certain cases with sufficiently good knowledge of the model, the condition in Eq. (10.9) can be realized via a reference value generator, e.g. robot controllers (Siciliano et al. 2009) and flatness-based control approaches (Rudolph 2003).

However, for *MBS plants* which are the focus here and which have uncertain eigenmodes and structural damping, the approach of Eq. (10.9) loses much of its attraction. It will be shown that the repercussions on the controller response $\tilde{T}_r(s)$ of inexact compensation of $P(s)$ by $A(s)$ are more sensitive than those of the pre-filter variant[2]. As can be seen from Eq. (10.4), $F(s)$ need only compensate "unevenness" in $T_r(j\omega)$ in the working frequency range. In the process, it is convenient to note that due to the feedback in $T(s)$, parameter variations in $P(s)$ are already suppressed or reduced.

For these reasons, the discussion below employs *pre-filtering* as the *basis for design* (thick dashed branch including $F(s)$ in Fig. 10.1). Due to the equivalence relation (10.8), designs can naturally also be converted into the pre-regulation variant and vice versa at any time.

---

[2] Displacement of weakly-damped eigenfrequencies has a drastic effect on the frequency response, which cannot be compared to parameter variations in the (typically low-pass) plant.

Separating design tasks The system transfer functions quite clearly demonstrate the separation of design tasks discussed above. Using the *controller* $H(s)$, disturbances (10.5), (10.6), and stability (via the identical characteristic polynomials of all transfer functions which include the feedback) can be addressed, though not independently of each other. Using the *pre-filter* $F(s)$, the controller response (10.4) can, within certain limits, be independently adjusted. Critical compromises in the design are thus required first and foremost for the rejection of disturbances $(d(t),\, n(t))$ as a function of *inputs*, and the stability as a function of *parameters*.

Systems theoretical design limitations The fundamental role of the sensitivity and complementary sensitivity of a system with feedback, introduced in Eqs. (10.2) and (10.3), can be seen in the transfer functions (10.5) and (10.6) for disturbance rejection. In the ideal case, $S(j\omega) = 0$ or $T(j\omega) = 0$ ensure complete rejection of the undesirable disturbances $d(t)$ and $n(t)$ from the controlled variable $y(t)$. In reality, sufficiently small values $\left|S(j\omega)\right| \ll 1$ and $\left|T(j\omega)\right| \ll 1$ over as wide a frequency range as possible must suffice. Unfortunately, however, $S(j\omega)$ and $T(j\omega)$ cannot be independently—much less simultaneously—kept arbitrarily small. As can be easily verified, in general

$$S(s) + T(s) = 1\,,$$

i.e. $S(j\omega)$ and $T(j\omega)$ can fundamentally only take small values over complementary frequency ranges. As a result, input disturbances and measurement noise can also only be suppressed in complementary frequency ranges. Furthermore, $S(j\omega)$ and $T(j\omega)$ can only be kept small over a limited frequency range, as, according to *BODE's sensitivity integral*, for an $L(s)$ with a pole surplus[3] ≥ 2,

$$\int_0^\infty \log\left|S(j\omega)\right| d\omega = \pi \sum p_k\,, \quad \int_0^\infty \frac{\log\left|T(j\omega)\right|}{\omega^2}\, d\omega = \pi \sum \frac{1}{z_i}\,, \qquad (10.10)$$

where p_k are the poles and z_k are the zeros of $L(s)$ in the right half-plane (Åström and Murray 2008). Small magnitudes of the sensitivity within a frequency range thus fundamentally imply a concomitant increase within a different, complementary range (the "waterbed effect").

[3] This is equivalent to the requirement that $\lim_{s\to\infty} sL(s) = 0$. For a strictly proper open-loop transfer function $L(s)$ with relative degree ≥ 1, Eq. (10.10) requires a minor extension, see (Reinschke 2006).

Naturally, additional design limitations on $H(s)$ exist due to the closed-loop stability requirement, for which see Sec. 10.4.

Realization-related design limitations Up to this point, only exogenous system quantities (inputs or outputs) have been considered. However, important limitations in the realization of systems arise in the form of implementation- and device-related limitations on internal variables. Most importantly, these involve the control variable $u(t)$ and limitations on achievable forces/torques in the actuators. This condition is reflected in the various transfer functions of the *actuation penetration* in Eq. (10.7) for the respective input variables. Most often, there is a limit on the magnitude of these transfer functions.

Key role played by the open-loop transfer function All of the system transfer functions introduced here depend fundamentally on the open-loop transfer function $L(s) := H(s)\,P(s)$ in Eq. (10.1). The controller transfer function $H(s)$ does not explicitly affect the transfer behavior between exogenous variables, but does so implicitly via its product with the plant transfer function $P(s)$. Using this formulation, many further design considerations can be conducted in a generalized manner without particular knowledge of the plant. In this section, it is thus possible to consider only the frequency responses of applicable transfer functions.

In general, frequency ranges having large and small $|L(j\omega)|$ can be distinguished, allowing descriptive approximations of system transfer functions to be derived (see Table 10.1). Due to both systems-theoretically-determined stability difficulties and certain physical attributes (namely, their low-pass behavior), for plants of a low-pass nature, these ranges can be more precisely identified as the low-frequency (= large-magnitude $|L(j\omega)|$) and high-frequency (= small-magnitude $|L(j\omega)|$) ranges.

General design rules From Table 10.1, the previously mentioned complementary character of the system properties under consideration is clearly recognizable, so that the following generally valid *design rules* can be derived.

- A *large magnitude* $|L(j\omega)|$ automatically ensures low sensitivity and a good controller response ➔ for low frequencies, this can be realized via an integrator in the plant or controller, or with a selectively large controller gain based on frequency.
- A *large magnitude* $|L(j\omega)|$ implies a high complementary sensitivity and thus poor rejection of measurement noise; low-frequency measure-

ment noise can generally only be suppressed with frequency-selective (complex) controller zeros (however, see next point).

- For good rejection of input disturbances, a large magnitude $\left|L(j\omega)\right|$ with *simultaneously* large magnitude $\left|H(j\omega)\right|$ of the controller frequency response is required ➔ compensating controllers with frequency-selective low controller gain (complex controller zeros) are thus counterproductive in this regard, and to be avoided.
- A *small magnitude* $\left|L(j\omega)\right|$ ensures low complementary sensitivity and thus good rejection of measurement noise ➔ the controller gain should only be as high as necessary; in all cases, low-pass behavior should be implemented at high frequencies (the degree of the denominator of $H(s)$ should be greater than the degree of its numerator).
- Shortcomings of (compromises in) the closed-loop controller response can be compensated using the pre-filter $F(s)$ in such a way that $F(j\omega)\cdot T(j\omega)=1$ (i.e. $F(j\omega)=1/T(j\omega)$). However, in the process, especially at higher frequencies, the product $\left|F(j\omega)\right|\cdot\left|H(j\omega)\right|$ should be kept sufficiently small, in order to avoid actuator saturation ➔ $F(s)$ and $H(s)$ should exhibit low-pass behavior at high frequencies.

Table 10.1. System properties for large/small magnitudes of the open-loop transfer function (highlighted fields indicate positive influence of the controller $H(s)$ or pre-filter $F(s)$)

	$\left\|L(j\omega)\right\| \gg 1$	$\left\|L(j\omega)\right\| \ll 1$
$\left\|S(j\omega)\right\|$	$1/\left\|L(j\omega)\right\| \approx 0$	≈ 1
$\left\|T(j\omega)\right\|, \left\|T_n(j\omega)\right\|, \left\|T_{u,d}(j\omega)\right\|$	≈ 1	$\left\|L(j\omega)\right\| \approx 0$
$\left\|T_r(j\omega)\right\|$	$\left\|F(j\omega)\right\|$	$\left\|F(j\omega)\right\| \cdot \left\|L(j\omega)\right\|$
$\left\|T_d(j\omega)\right\|$	$\left\|P(j\omega)\right\| / \left\|L(j\omega)\right\| = 1/\left\|H(j\omega)\right\|$	$\left\|P(j\omega)\right\|$
$\left\|T_{u,r}(j\omega)\right\|$	$\left\|F(j\omega)\right\| / \left\|P(j\omega)\right\|$	$\left\|F(j\omega)\right\| \cdot \left\|H(j\omega)\right\|$
$\left\|T_{u,n}(j\omega)\right\|$	$1/\left\|P(j\omega)\right\|$	$\left\|H(j\omega)\right\|$

Particularities for multibody systems In mechatronic systems, elastic coupling in the mechanical structure (i.e., an MBS) introduces distinct difficulties when compared to plants with pronounced low-pass dynamics. As an illustrative example, consider a *weakly-damped single-mass oscillator* (Fig. 10.2a). In order to avoid possible instability (a detailed treatment of robust stability is presented in Secs. 10.4 and 10.5), *ideal* actuators and sensors without additional parasitic dynamics are assumed.

For control, consider a real *PID controller* $H(s)$ (see Sec. 10.5.2). Let the controller be realistically parameterized as a *compensating controller*, i.e. the mechanical eigenfrequency ω_0 is compensated in the numerator of the controller transfer function. This assumes the ideal case of exact knowledge of the mechanical parameters. No pre-filter is included here. This gives the following generic system transfer functions:

$$P(s) = K_P \frac{1}{1 + 2d_0 \dfrac{s}{\omega_0} + \dfrac{s^2}{\omega_0^2}}, \qquad \omega_0 := \sqrt{\frac{k}{m}}, \; d_0 := \frac{b}{2}\sqrt{\frac{1}{mk}}, \tag{10.11}$$

$$H(s) = K_H \frac{1 + 2d_N \dfrac{s}{\omega_N} + \dfrac{s^2}{\omega_N^2}}{s\left(1 + \dfrac{s}{\omega_D}\right)}, \qquad d_N = d_0, \; \omega_N = \omega_0, \; \omega_D = 5\omega_0,$$

so that given positive controller gain K_H, all applicable closed-loop transfer functions are inherently stable (all transfer functions are consistently proper: they posses a common second order characteristic polynomial with positive coefficients).

The resulting frequency responses are depicted in Fig. 10.2b,c; the corresponding closed-loop step responses can be seen in Fig. 10.2d.

Controller response A thoroughly satisfactory controller response can be seen: a flat frequency response $|T(j\omega)| \approx 0\,\mathrm{dB}$ over a wide frequency range, and a well-damped step response $y_r(t)$ consistent with a crossover frequency ω_C and phase margin Φ_M; the mechanical resonance ω_0 can be seen in neither the frequency responses $L(j\omega)$, $T(j\omega)$ nor the reference input step response $y_r(t)$. Is the compensating PID controller thus a "perfect" controller? By no means: this question must unfortunately be answered in the negative, as will be shown below.

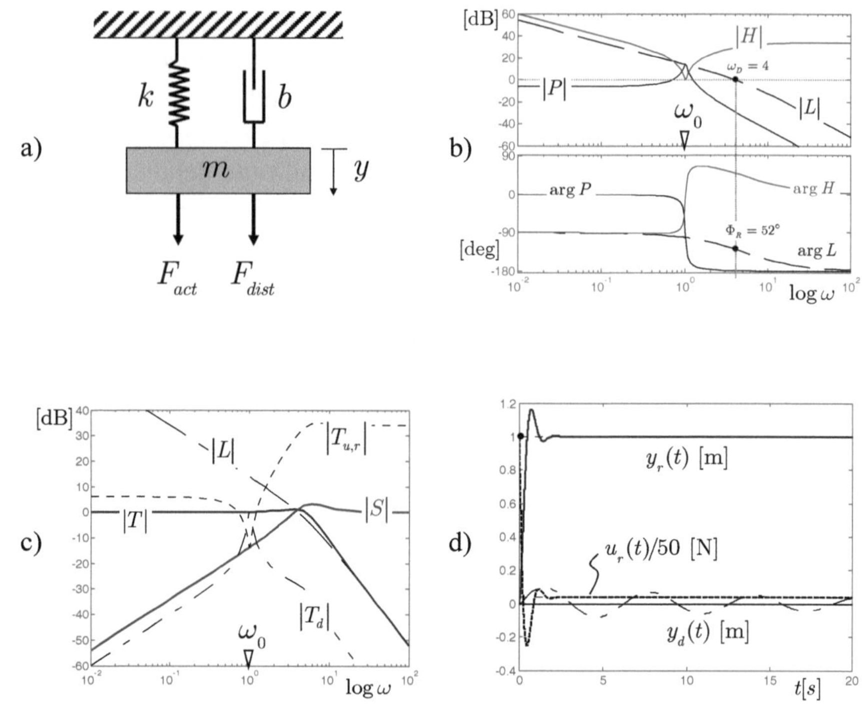

Fig. 10.2. Position control of a weakly-damped single-mass oscillator using a compensating PID controller: a) equivalent mechanical model, b) open-loop frequency responses, c) closed-loop frequency responses, d) closed-loop step responses for reference inputs and disturbance forces

Disturbances The *response to disturbances* at the plant input is absolutely unsatisfactory. Natural oscillations ω_0 excited by disturbance forces are not actively damped by the controller $H(s)$, and decay only as a result of the inherent mechanical damping (see $y_d(t)$ in Fig. 10.2d). Indeed, at the eigenfrequency ω_0, the controller has a complex zero resulting in a "gain gap" masking out this frequency band of the feedback signal. In this band, the controller is thus ineffective and blind. For reference inputs, on the other hand, this gap acts as a band-stop filter in the forward loop: the eigenfrequency ω_0 is filtered out of the reference input signal and thus cannot even excite the natural oscillation. The resonant behavior in response to disturbance forces can be recognized in the disturbance frequency response $T_d(j\omega)$ (Fig. 10.2c).

Control variable A further drawback of the compensating PID controller presented here is the resulting behavior of the control variable. It can be seen from the actuation penetration $T_{u,r}(j\omega)$ of the reference input that, due to the (very high) finite gain of the PID controller $H(s)$ at high frequencies (resulting from the degree of the numerator equaling the degree of the denominator), high-frequency components in the reference input result in large amplitudes of the control variable (Fig. 10.2c). Compare also the evolution of the control variable $u_r(t)$ for a step reference input (Fig. 10.2d). The same response incidentally also results for the actuation penetration of the measurement noise (see Eq. (10.7)), i.e. high-frequency measurement noise can superfluously task the actuator and lead to unnecessary "flutter".

Challenges for MBS controller design This elementary introductory example of the simplest imaginable controlled multibody system has already demonstrated important differences compared to customary design strategies applicable to plants with low-pass behavior. Conventional compensation approaches are completely unusable here. In addition to the problems highlighted above, realistically, *parameter uncertainties*—particularly in the mechanical parameters of masses, stiffnesses, and structural damping—must be expected; many of these parameters can only be insufficiently determined, or are subject to change over time. In addition, the real (parasitic) dynamics of measurement and actuation devices should be considered, and thought should be given to the targeted insertion of low-pass elements in the feedback loop to ensure high-frequency disturbance rejection. Finally, additional undesirable dynamic influences arise from the use of digital control concepts (e.g. sampling and aliasing).

With the necessary consideration of additional parasitic dynamics in the control loop, however, the problem of stability immediately becomes the focus of design activity. Above all, this task is made difficult by the fact that—in addition to the previously mentioned system *parameter uncertainties* and *parasitic dynamics*—the *high order* of the multibody system especially renders necessary a robust and comprehensive consideration of stability. Design solutions which do not take these aspects into account are doomed to failure from the start if applied to a real mechatronic system, and serve at most as didactic, academic examples.

Transparent manual design of controllers Subsequent sections attempt to offer practical strategies and solutions for all of the challenges raised above, presenting a transparent and largely manual approach to controller design. Given sufficiently representative models for design, the resulting controllers can be used early in the design process to assess achievable system dynamics and the implications of these for system components. The developed (generally low-order) controller structures and accompanying parameters serve as a solid basis (initial values) for further, detailed analyses, and possibly more involved, computer-aided controller design and optimization procedures.

10.3 Modeling Uncertainties

10.3.1 MBS parameter uncertainties

Uncertain system parameters Flexible multibody systems with lumped elements are characterized both by their structural properties (the topology of the mechanical system) and the individual element parameters of *mass* m_i, *stiffnesses* k_i, and *damping* b_i (Fig. 10.3). For one, these model quantities can only be precisely determined to a limited extent, and additionally fluctuate during operation due to environmental factors (e.g. temperature, humidity, etc.), deterioration, or changing operating conditions.

In terms of the design task, such difficult-to-determine material properties should thus be interpreted as *uncertain system parameters*. For controller design, the primary concern is the transfer characteristics between various input and output quantities, described by transfer functions of the form

$$P_{MBS}(s) = K_{MBS} \frac{\prod_{j=1}^{M} \{d_{z,j}; \omega_{z,j}\}}{\prod_{k=1}^{N} \{d_{p,k}; \omega_{p,k}\}}, \tag{10.12}$$

where, in turn, the parameters of $P_{MBS}(s)$ are complicated functions of the MBS parameters, i.e.

$$\begin{aligned} K_{MBS} &= K_{MBS}(m_i, k_i), \\ d_{z,j} &= d_{z,j}(m_i, k_i, b_i), \quad d_{p,k} = d_{p,k}(m_i, k_i, b_i), \\ \omega_{z,j} &= \omega_{z,j}(m_i, k_i, b_i), \quad \omega_{p,k} = \omega_{p,k}(m_i, k_i, b_i). \end{aligned}$$

For the *MBS frequencies* $\omega_{p,k}, \omega_{z,j}$, the general relation

$$\omega_{p/z} \sim \sqrt{\frac{k^+}{m^+}} \tag{10.13}$$

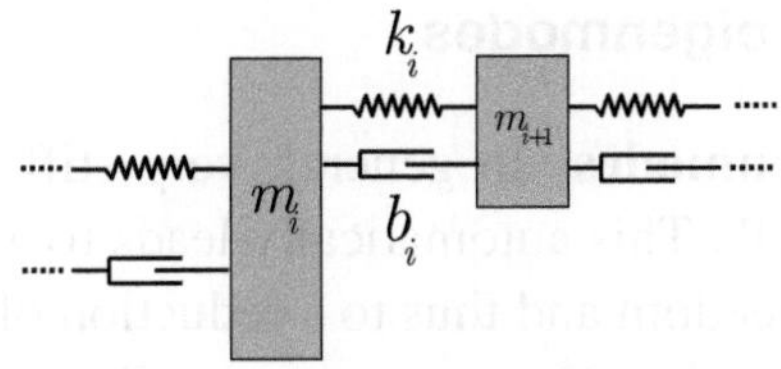

Fig. 10.3. Typical MBS structure with lumped mechanical elements

holds, where k^+ and m^+ represent linear combinations of the individual elements k_i and m_i. The smaller and lighter the masses involved (or the stiffer the elastic ties), the higher the resulting MBS frequencies.

For one, the *damping constants* $d_{p,k}$, $d_{z,j}$ are quite difficult to determine, and, in addition, they vary over a wide range. Typical values are: for steel (structure, joints, bearings, etc.) $d \simeq 0.01–0.02$, for light-weight structures (composites) $d \simeq 0.001–0.005$, and for dissipative elastic links $d \simeq 0.05–0.2$. The resulting actual magnitude peaks of the frequency response can thus be estimated only approximately. In all cases, it holds for damping coefficients $d_{p,k}$ or $d_{z,j}$ that

$$\lim_{b_i \to 0} d_{p,k/z,j} = 0, \quad \forall i, j, k \,,$$

so that in the limit, mechanical damping vanishes, and, for *maximum robustness*, infinite magnitude peaks must be taken into consideration.

An overview of possible variations in MBS frequency responses which must be considered in a design is presented in Fig. 10.4.

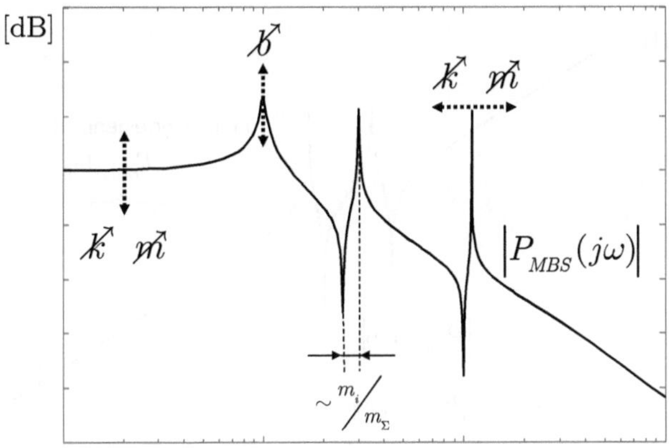

Fig. 10.4. Variations in MBS frequency responses due to MBS parameter variations

10.3.2 Unmodeled eigenmodes

High-frequency eigenmodes In general, very stiff linking elements are assumed ideally "rigid". This automatically leads to a reduction of the mechanical degrees of freedom and thus to a reduction of the model order and the number of eigenmodes. However, in actuality, such cases present eigenmodes with very high eigenfrequencies corresponding to Eq. (10.13). In FEM models, high-order eigenmodes are generally characterized by small effective masses, resulting in high eigenfrequencies. In any case, in real systems, high-frequency structural dynamics are always afflicted with large uncertainties in the model order and in the parameters

Model reduction: simplified design model For controller design, it is advantageous to use an MBS model of suitably low order, which nevertheless reproduces *all relevant* behaviors of the real system, via a process termed *model reduction*. Procedures implementing this concept employ various measures to distinguish the relevant and less relevant eigenmodes. The resulting low-order design models thus neglect real, existing eigenmodes; as a rule, it is high-frequency eigenmodes which are neglected (Fig. 10.5).

Unmodeled eigenmodes: spillover The dynamics of eigenmodes missing (unmodeled) in a design model naturally still influence the stability of the closed control loop in the real system. In unfavorable cases, this can lead to undesirable and dangerously unstable behaviors. Such undesirable interactions of unmodeled eigenmodes in a control loop are known as *control*

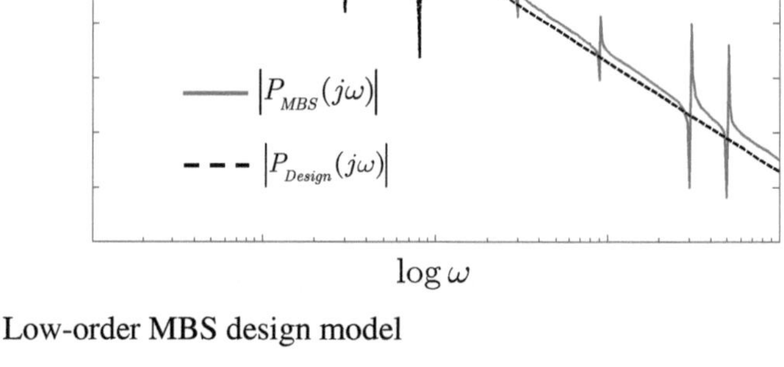

Fig. 10.5. Low-order MBS design model

spillover. The interacting eigenmodes are also often termed *spillover* eigenmodes. Sensible, reality-based controller design must thus necessarily take these unmodeled eigenmodes into account in a suitable manner so as to avoid potential stability problems.

10.3.3 Parasitic dynamics

Negative phase shift Mechatronic systems consist of a multiplicity of system components, whose individual dynamics can often only be modeled with difficulty or unwarrantable effort. Particularly disturbing to closed-loop stability, however, are parasitic dynamic effects arising from low-pass components and delays, and inducing negative phase shifts in the system. Of particular importance are any *delay effects*, modeled by a transfer element

$$G(s) = e^{-\tau s}.$$

Since the negative phase shift

$$\arg G(j\omega) = -\tau\omega \tag{10.14}$$

acts without a simultaneous reduction in magnitude, the deleterious effect even increases linearly with frequency.

As a result, high-frequency (spillover) eigenmodes are prime candidates for engendering stability problems. Sensible, reality-based controller design must thus also necessarily take *parasitic delay effects* into account in a suitable manner so as to avoid potential stability problems.

10.4 Robust Stability for Multibody Systems

10.4.1 NYQUIST criterion in intersection formulation

Closed control loop As is well known, evaluation of the stability of a *closed* control loop (as in Fig. 10.6) is possible knowing only the frequency response $L(j\omega)$ of the *open loop* by employing the NYQUIST criterion (Ogata 2010), (Horowitz 1963).

Advantages of the NYQUIST criterion The great attraction and popularity of this classical stability criterion is due to several factors. The open-loop frequency response (of the chain consisting of the controller, actuator, plant, and sensors) is easy to determine both in principle and in practice

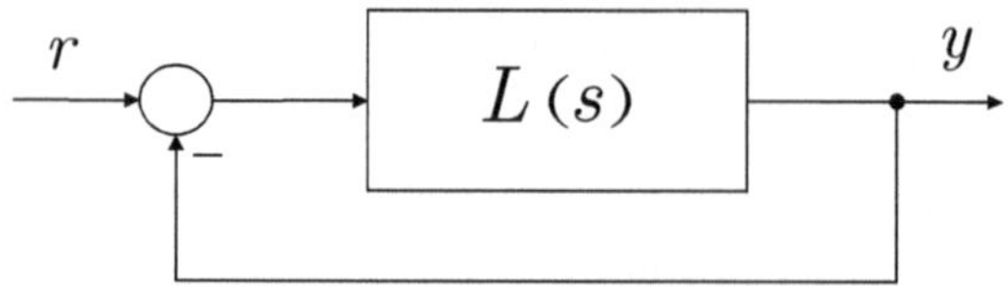

Fig. 10.6. Standard control loop for the NYQUIST criterion

using analytical or experimental modeling. In addition, changes in the component transfer functions and their effects on the open-loop frequency response are more or less easily visible. In its general form, the NYQUIST criterion also holds for systems with transcendental components in their transfer functions, and is thus eminently suited to the analysis of systems with delay components.

Graphical interpretation The graphical interpretation of the NYQUIST criterion additionally provides descriptive, comprehensive insight into a system, simplifying systems design. However, for practical use, a suitable representation and interpretation of the frequency response adapted to the class of problem being considered is of great importance.

The commonly-used formulation of the NYQUIST criterion using the continuous *phase change* of the NYQUIST plot of $L(j\omega)$ can also be stated in a completely equivalent form considering special *intersections* of the plot with the real axis (Föllinger 1994). This relatively seldom-used interpretation, termed the *intersection formulation*, in combination with frequency response curves, enables very clear application of the NYQUIST criterion to multibody systems.

System prerequisites Consider a control loop as in Fig. 10.6 with a linear open-loop transfer function

$$L(s) = \frac{K_L}{s^r}\frac{N(s)}{D(s)} \cdot e^{-T_t s} \quad , \quad N(0) = D(0) = 1$$

with the following properties:

1. there are no poles on the imaginary axis, except
2. at most *three* poles at $s = 0$ (numbered $r = 0,1,2,3$), (10.15)
3. the number of poles in the right half-plane is n_R,
4. the gain K_L is *positive*,
5. degree $N(s) <$ degree $D(s) + r$ without common roots in $N(s)$ and $D(s)$, and
6. a delay $T_t \geq 0$.

The limiting conditions set out above are applied here only as a means of simplifying the criterion while allowing practically all important cases of multibody systems to be addressed. In particular, *complex conjugate poles* with *arbitrarily small* finite *damping* are permitted, corresponding to (very) weakly-damped MBS eigenmodes[4]. As a result, multiple intersections of $L(j\omega)$ with the unit circle (the 0-dB line of the magnitude curve) are explicitly allowed[5]. With (at most) three integrators in the open loop, an integral component to compensate constant disturbance forces/torques can be included in the controller, even in the case of freely moving masses (double integrators).

Definition 10.1 *Positive/negative intersections* A *positive* or *negative* intersection of the frequency response $L(j\omega)$ in the range $+0 \leq \omega < +\infty$ is understood to be a penetration of the negative real axis (sign change of the imaginary part) by $L(j\omega^*)$, for which

$$\left|L(j\omega^*)\right| > 1 \ (\triangleq 0\,\mathrm{dB}) \quad \text{and} \quad \arg L\left(j\omega^*\right) = (2q+1)\cdot 180^\circ, \qquad (10.16)$$

with $q \in \mathbb{Z}$,

where the *phase curve* either *rises* (i.e. positive gradient) or *falls* (i.e. negative gradient), see Fig. 10.7. Points at which the NYQUIST plot only touches the real axis without penetrating it are not counted.

In the case $r = 2$, the starting point $\omega^* = 0$ is counted as half an intersection (sign corresponding to the gradient, i.e. $+1/2$ for a rising phase and $-1/2$ for a falling phase).

---

4 In real problems, purely passive multibody systems always have finite mechanical damping, so that Properties 1 and 2 are practically always fulfilled. The variously-used simplifying assumption of negligibly small damping of eigenmodes (purely imaginary poles) is a hypothetical limiting case which can be used to advantage for simplified analysis (approximating formulae for eigenfrequencies) of multibody systems.

5 In the commonly-used simplified formulation of the NYQUIST criterion (the "positive phase margin" test), only *one* intersection with the unit circle is permitted (Föllinger 1994). This condition is not a significant limitation in plants with low-pass behavior (e.g. chemical engineering). However, as a rule, it is violated in multibody systems.

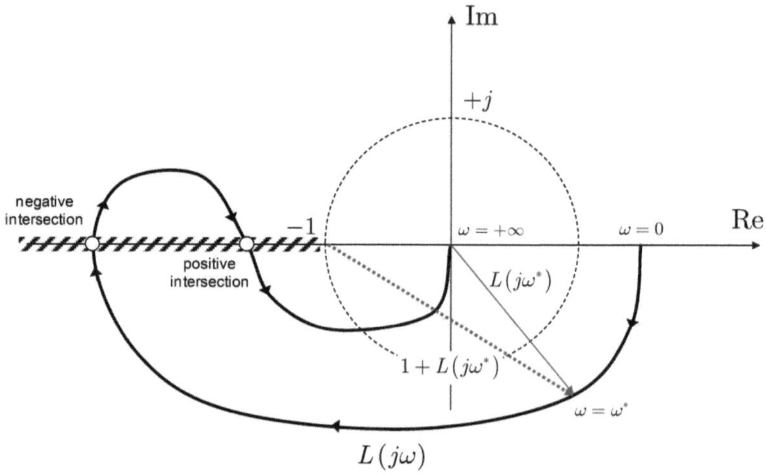

Fig. 10.7. Intersections of the NYQUIST plot

Proposition 10.1. *NYQUIST criterion in intersection formulation*[6] (Föllinger 1994)

The *closed control loop* $T(s) = L(s)/(1 + L(s))$ is *stable* if the following hold for the open-loop transfer function $L(s)$ of Eq. (10.16):

(C1) $\arg L(j\omega) \neq (2q+1) \cdot 180°$ for all $\left|L(j\omega)\right| = 1$ ($\hat{=}$ 0 dB),

where $q \in \mathbb{Z}$, i.e. there is no intersection with the "critical" point -1.

(C2) The *difference D* in the *number* of positive and negative intersections is

$$D = \frac{n_R}{2} \qquad \text{for } r = 0, 1,$$

$$D = \frac{n_R}{2} + \frac{1}{2} \qquad \text{for } r = 2,$$

$$D = \frac{n_R}{2} + 1 \qquad \text{for } r = 3.$$

If any of the conditions (C1), (C2) *is not* met, the control loop is *unstable*.

---

6 A more detailed proof can be found in (Föllinger 1994).

Proposition 10.1 can also be used to derive the following

Proposition 10.2. (Föllinger 1994)

The *closed control loop* $T(s) = L(s)/(1 + L(s))$ is always *unstable* when the open-loop transfer function $L(s)$ possesses an *odd* number of poles in the right s-half-plane.

The important points are thus only the intersections of the NYQUIST plot $L(j\omega)$ with the negative real axis (frequency response curves: $\arg L(j\omega^*) = (2q+1) \cdot 180°$) *outside* the unit circle (frequency response curves: $\left|L(j\omega)\right|_{\mathrm{dB}} > 0$). For smooth frequency responses, these can be easily determined from BODE diagrams, as is shown in the following example.

Example 10.1 *Double integrator system.*

The frequency response (BODE diagram) of a double integrator transfer function as in Eq. (10.15), with $r = 2$ and $n_R = 0$ (no poles in the right s-half-plane) and depicted in Fig. 10.8, describes the open-loop response of a standard control loop (Fig. 10.6).

To evaluate closed-loop stability, following Proposition 10.1, only the cross-hatched frequency range, where $|L(j\omega)| > 1$ ($\left|L(j\omega)\right|_{\mathrm{dB}} > 0\mathrm{dB}$), need be considered. There exist two positive intersections (one of which is at $-\infty$ on the abscissa) and one negative intersection, so that the difference in the number of intersections is

$$D = \frac{1}{2} - 1 + 1 = \frac{1}{2},$$

and the closed control loop is *stable*.

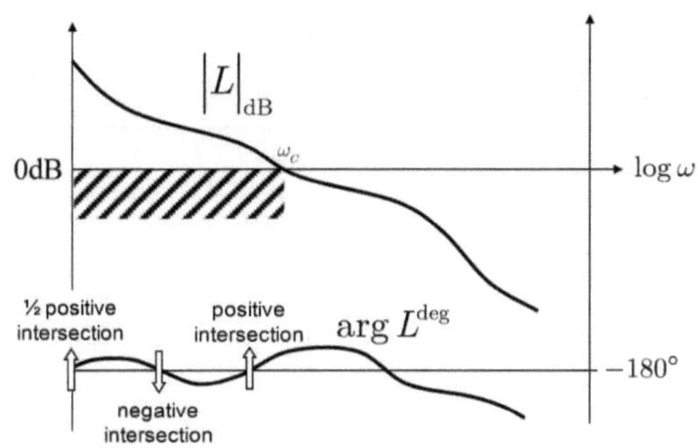

Fig. 10.8. BODE diagram for a double integrator system (qualitative dynamics without concrete numerical values) ■

10.4.2 Stability analysis with the NICHOLS diagram

Nichols diagram For multibody systems, due to the weakly-damped complex conjugate zeros and poles of their transfer functions, performing any type of stability analysis using the graphical representation of the frequency response in the form of a BODE diagram is very unwieldy. A significantly clearer representation is offered by the orthogonal representation of *magnitude* (*dB*, decibels) and *phase* (° or *deg*, degrees) in the so-called *NICHOLS diagram* (*gain-phase plot*), see Fig. 10.9.

Critical stability regions The region of interest for closed-loop stability is the -1 point and the negative real axis to its left. In the NICHOLS diagram, these domains map to an infinite number of so-called *critical stability regions*, i.e. to the points ($\left|L(j\omega)\right|_{\mathrm{dB}} = 0\,\mathrm{dB}$, $\arg L(j\omega) = (2q+1)\cdot 180°$) or the upper half-lines at $(2q+1)\cdot 180°$, for $q \in \mathbb{Z}$ (see Fig. 10.9).

Since, for a stable system, the -1 point cannot be touched, and since there should always be a safety margin, an exclusion zone can be placed around -1 (a *gain-phase stability margin box*). For a stable control loop, the frequency response $L(j\omega)$ should never intersect this box.

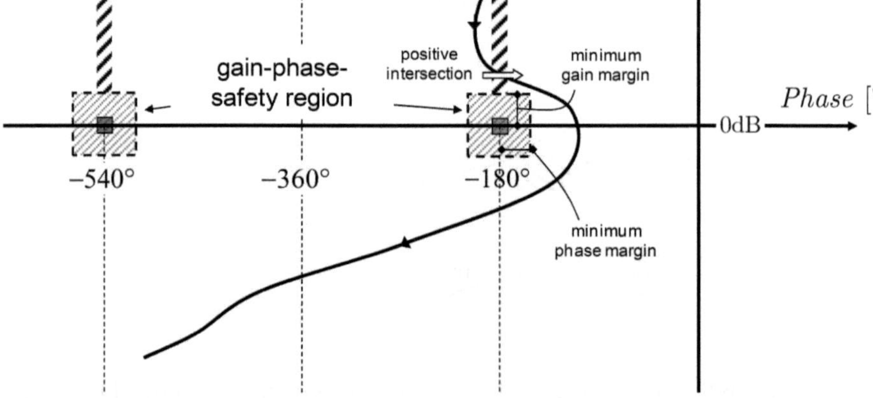

Fig. 10.9. NICHOLS diagram for a stable control loop (cf. Fig. 10.8)

The intersections of $L(j\omega)$ with the critical stability regions determine the intersections for the NYQUIST criterion, where a passage from left to right is counted as a positive intersection (negative intersections are correspondingly a passage from right to left).

Elastic eigenmodes Every *collocated eigenmode* (i.e. collocated sensor/ actuator arrangement) in the form of complex conjugate pairs of zeros and poles

$$G_1(s) = \frac{\left(1 + 2d_z \dfrac{s}{\omega_z} + \dfrac{s^2}{\omega_z^2}\right)}{\left(1 + 2d_p \dfrac{s}{\omega_p} + \dfrac{s^2}{\omega_p^2}\right)}$$

is represented in the NICHOLS diagram by a *loop* (Fig. 10.10a,c).

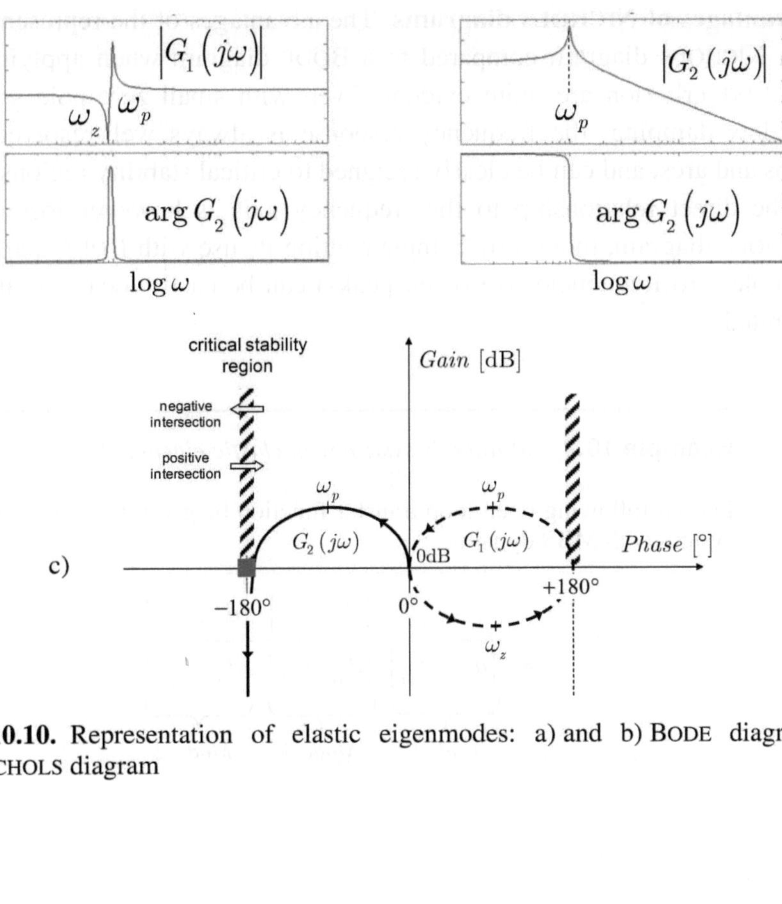

Fig. 10.10. Representation of elastic eigenmodes: a) and b) BODE diagrams, c) NICHOLS diagram

All *non-collocated eigenmodes* in the form of complex conjugate pairs of poles

$$G_2(s) = \frac{1}{\left(1 + 2d_p \dfrac{s}{\omega_p} + \dfrac{s^2}{\omega_p^2}\right)}$$

are represented as *arcs* (Fig. 10.10b,c).

The heights of the loops and arcs are determined by the corresponding damping d_z, d_p (of the resonant peaks). For vanishingly small damping, the vertical extents become infinite. The magnitude maxima and minima lie at the frequencies ω_p and ω_z, respectively.

The width of the arcs is typically 180° (the phase change of a second-order delay). The width of loops can be at most 180° (for very low damping); it is reduced when zeros ω_z and poles ω_p lie very close together and for higher damping.

Advantages of NICHOLS diagrams The advantages of the representation as a NICHOLS diagram compared to a BODE diagram when applying the NYQUIST criterion are quite evident. Even with small zero-pole spacing and low damping, the frequency response is always well-resolved into loops and arcs, and can be clearly assigned to critical stability regions.

The direct relationship to the frequency axis is however lost in the NICHOLS diagram, though by complementing its use with BODE diagrams, the pole/zero frequencies (resonant peaks) can be easily localized and interpreted.

Example 10.2 *Multibody system with elastic eigenmodes.*

Let the following open-loop transfer function be given for a mechatronic system with MBS dynamics:

$$L(s) = \frac{K_L}{s} \underbrace{\frac{1}{\{d_{p0}, \omega_{p0}\}}}_{Mode\ 0} \underbrace{\frac{\{d_{z1}, \omega_{z1}\}}{\{d_{p1}, \omega_{p1}\}}}_{Mode\ 1} \underbrace{\frac{\{d_{z2}, \omega_{z2}\}}{\{d_{p2}, \omega_{p2}\}}}_{Mode\ 2} \cdots.$$

Figs. 10.11 and 10.12 contrast the frequency response $L(j\omega)$ represented as a BODE diagram (magnitude curve) and a NICHOLS diagram. The elastic eigenmodes (common Mode 0, elastic Modes 1, 2, 3) can be clearly seen, as well as an additional negative phase shift due to parasitic dynamics or delays in actuators, sensors, and the information processing (not explicitly represented in the transfer function).

Stability analysis with the NICHOLS diagram gives:

- no intersections with the critical points (0 dB, -180°), (0 dB, -540°), etc.;
- one positive and one negative intersection with the critical stability region $\left|L\left(j\omega\right)\right| \geq 1$ and $\arg L\left(j\omega\right) = -180°$;
- since $r = 1$ and $D = 1 - 1 = 0$, it follows that the closed control loop is *stable*.

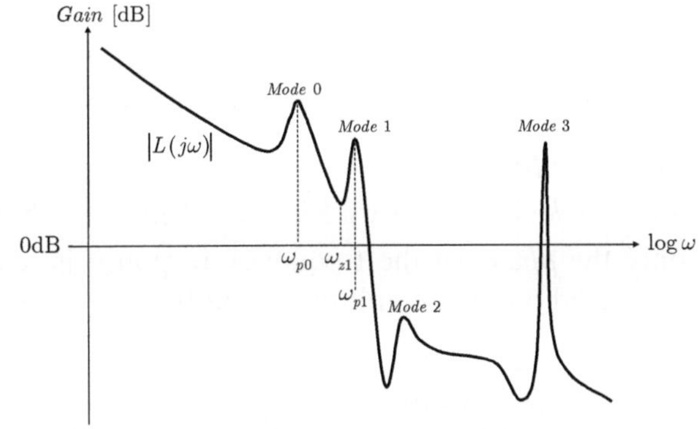

Fig. 10.11. Magnitude curve for MBS with elastic eigenmodes

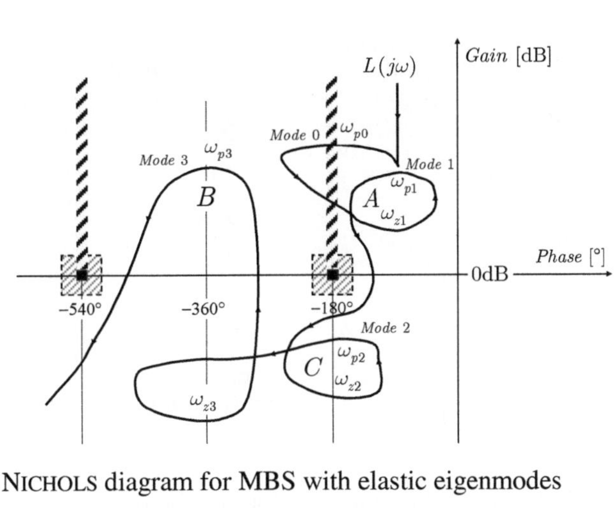

Fig. 10.12. NICHOLS diagram for MBS with elastic eigenmodes ■

10.4.3 Robust stability of elastic eigenmodes

If the *open-loop* frequency response of the chain *"controller × MBS"* is such that the eigenmodes lie within certain regions of the gain-phase plain, and that they do not leave these regions even given MBS parameter uncertainties (in stiffness k, mass m, or damping b), then the *closed control loop* is *robustly stable* to these parameter uncertainties.

Definition 10.2 *Robust stability regions* Let the following *robust stability regions* be defined for the frequency response $L(j\omega)$ of the open-loop transfer function (Fig. 10.13).

- *Phase-lead stability region*

  $-180° < \arg L(j\omega) < 180°,$

  only the *phase* of the frequency response is relevant, there are no conditions placed on the magnitude of the frequency response.

- *Phase-lag stability region*

  $-540° - k \cdot 360° < \arg L(j\omega) < -180° - k \cdot 360°, \quad k \in \mathbb{N},$

  only the *phase* of the frequency response is relevant, there are no conditions placed on the magnitude of the frequency response.

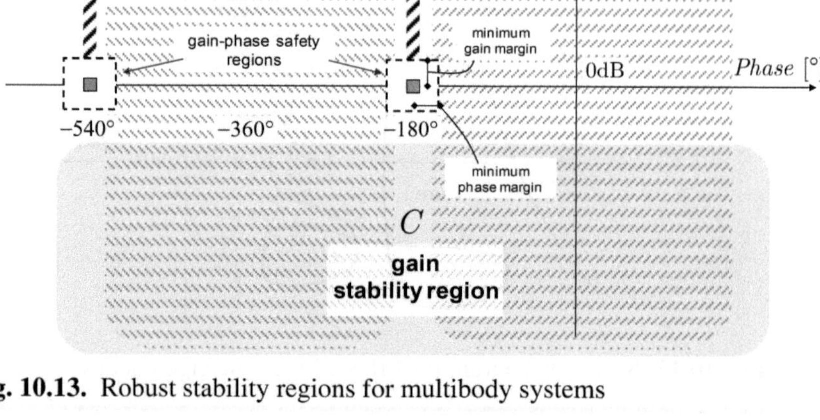

Fig. 10.13. Robust stability regions for multibody systems

- *Gain stability region*

  $$\left|L\left(j\omega\right)\right| < 1,$$

  only the *magnitude* of the frequency response is relevant, there are no conditions placed on the phase of the frequency response.

- *Critical stability regions*

  $$\left|L\left(j\omega\right)\right| \geq 1 \quad \text{and} \quad \arg L\left(j\omega\right) = 180° \cdot (2q+1), \qquad q \in \mathbb{Z},$$

  the number and type of allowed intersections corresponds to the NYQUIST criterion.

- *Gain-phase stability margin box*

  $$\Delta_{dB} \leq \left|L\left(j\omega\right)\right|_{dB} \leq -\Delta_{dB} \quad \text{and}$$

  $$-\Phi_M \leq \arg L\left(j\omega\right) + 180° \cdot (2q+1) \leq \Phi_M, \qquad q \in \mathbb{Z}$$

  where Δ_{dB} is the *gain margin* magnitude and Φ_M is the *phase margin* magnitude.

Compare these to the frequency response from Example 10.2 shown in Fig. 10.12. The eigenmodes there are assigned to the robust stability regions (A,B,C). Eigenmodes 1 and 3, for example, are robustly stable with respect to arbitrary variations in damping, even in the extreme case of vanishingly small damping. For approximately constant damping, Eigenmode 2 is robustly stable to variations in the eigenfrequency and arbitrary phase shifts (delays).

Research context Interestingly, this form of robust stability analysis, in particular, the graphical NYQUIST curve representation in the NICHOLS diagram, is hardly represented in the standard control theory literature. However, this approach is widely distributed and time-tested in the aerospace industry, e.g. for the control of spacecraft with flexible structures (Bittner et al. 1982), (Janschek and Surauer 1987). In such applications, which absolutely require the most accurate design verification before operational use, the strengths of the frequency domain approach also become evident: the relationships between parameters in the design are clear, parasitic dynamic effects can be included, and even nonlinear phenomena can be addressed (Bittner et al. 1982). Due to the nature of its existing applications, it is also not remarkable that the few available references on the subject come from authors with precisely this background, e.g. (Lurie and Enright 2000), (Sidi 1997).

10.5 Manual Controller Design in the Frequency Domain

10.5.1 Robust control strategies

Goal This section considers the structuring of the open-loop frequency response $L(j\omega)$ for a standard control loop as shown in Fig. 10.14. The objective pursued here is to form the frequency response $L(j\omega)$ using dynamic corrective elements such that desired regions of robust stability in the gain-phase plain are occupied for particular frequency ranges (termed *loop shaping*). If the frequency response $L(j\omega)$ can be successfully kept inside these regions despite variations in the plant parameters (including the multibody system, actuators, and sensors), then a *robustly stable* complete system can be ensured.

Model uncertainties and manual design For mechatronic systems, model parameters representing structural elasticity are particularly subject to uncertainties and changes over time. Luckily, the resulting variations in eigenfrequencies and damping of eigenmodes can be relatively well decoupled in the NICHOLS diagram representation. This offers a practical method to account for particular physical phenomena in a targeted manner as part of manual controller design. In contrast to purely formal procedures, using this method, the order of the controller can also be limited to a reasonable size from the beginning.

Standard control loop configuration Without loss of generality, this section considers unity feedback of a controller and plant connected in series $L(s) = H(s)\,P(s)$ (Fig. 10.14). Other configurations (see Sec. 10.2) differ only in their time responses to reference and disturbance inputs. In mechatronic systems, the plant $P(s)$ describes the mechanical structure (the multibody system), actuators, and sensors; disturbances at the plant input are generally considered in the form of forces and torques.

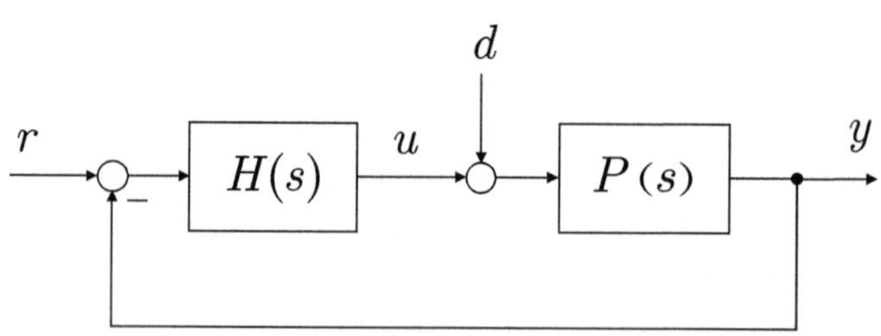

Fig. 10.14. Standard control loop with one degree of freedom

Active stabilization of low-frequency eigenmodes

Frequency response condition: $L(j\omega) \in$ *phase-lead stability region* with $|L(j\omega)| \gg 1$.

In the lower frequency range, it is, in principle, simple to set up a phase lead relative to the critical -180° stability region. With the additional magnitude condition $\left|L\left(j\omega\right)\right| \gg 0\,\mathrm{dB}$ (e.g. for an integrating component), stable closed-loop dynamics generally result, and within this frequency range (the controller bandwidth $0 \leq \omega \leq \omega_B$), $\left|T\left(j\omega\right)\right| = \left|L\left(j\omega\right)\right| / \left|1 + L\left(j\omega\right)\right| \approx 1$. Even intersections with the critical -180° stability region are acceptable, as long as they do not violate the stability condition.

Dynamic properties The magnitude condition $|L(j\omega)| \gg 1$ is particularly easily met for all eigenmodes in this frequency range. The resonant peaks at the open-loop eigenfrequencies and their associated natural oscillations are *actively suppressed* in the closed loop (a process termed *active stabilization*). In this frequency range, the multibody system follows the reference input signal nearly ideally; as a rule, this is also associated with good rejection of disturbance forces/torques.

Robustness properties All eigenmodes within the controller bandwidth are robustly stable to *arbitrarily low* damping $d > 0$, and *sufficiently tight frequency bounds* on the eigenfrequencies, as long as the critical stability bounds (phasing) are not violated (see Fig. 10.12, Mode 0 and 1).

Active damping of high-frequency eigenmodes

Frequency response condition: $L\left(j\omega_{p,i}\right) \in$ *phase-lag stability region* with $\left|L\left(j\omega_{p,i}\right)\right| \gg 1$.

Outside the controller bandwidth, it is fundamentally no longer possible to realize the magnitude condition $|L(j\omega)| \gg 1$ over a large frequency range. However, for high-frequency eigenmodes, $\left|L\left(j\omega_{p,i}\right)\right| \gg 1$ is often naturally satisfied (especially with marginal damping). With a phase response that suitably conforms to stability requirements (i.e. it is in the *phase-lag stability region* or has an acceptable number of intersections with the critical stability regions), it then also holds for such high-frequency eigenmodes that $\left|T\left(j\omega_{p,i}\right)\right| = L\left(j\omega_{p,i}\right) / \left(1 + L\left(j\omega_{p,i}\right)\right) \approx 1$.

Dynamic properties The motions of eigenmodes $\omega_{p,i}$ in the phase-lag stability region with $\left|L\left(j\omega_{p,i}\right)\right| \gg 1$ are actively affected (suppressed) in the same manner as with active control. However, since this active influence occurs only locally relative to the frequency range, this process is termed *active damping*. In any case, however, eigenmotions excited by disturbance forces/torques or by nonlinearities are countered by the feedback, and any oscillations are actively damped.

Robustness properties All eigenmodes within the indicated frequency range are robustly stable to *arbitrarily low* damping $d > 0$, and *sufficiently tight frequency bounds* on the eigenfrequencies, as long as the critical stability bounds (phasing) are not violated. This is an ideal measure for the handling of eigenmodes above the controller bandwidth (mid-range frequencies) which have *sufficiently well-determined* eigenfrequencies and *arbitrary* damping (see Fig. 10.12, Mode 3).

Gain stabilization with eigenfrequency parameter uncertainties

Frequency response condition: $L\left(j\omega_{p,i}\right) \in$ *gain stability region.*

Due to low-pass components which are always present in real systems, a reduction in the open-loop magnitude naturally occurs, particularly at high frequencies. For the case $\left|L\left(j\omega\right)\right| < 0\,\text{dB}$ or $\left|L\left(j\omega\right)\right| < -\Delta_{dB}$ there are no limitations on the phase of $L(j\omega)$, as no critical stability bounds are exceeded.

Dynamic properties However, any eigenmodes $\omega_{p,i}$ with $\left|L\left(j\omega_{p,i}\right)\right| < 1$ and which have been excited cannot be actively influenced, since $\left|T\left(j\omega_{p,i}\right)\right| = \left|L\left(j\omega_{p,i}\right)\right| / \left|1 + L\left(j\omega_{p,i}\right)\right| \approx \left|L\left(j\omega_{p,i}\right)\right|$. The feedback has no effect on such natural oscillations; they can only be damped by inherent mechanical damping, if such exists.

Robustness properties If the magnitude condition $\left|L\left(j\omega_{p,i}\right)\right| < 0\,\text{dB}$ or $\left|L\left(j\omega_{p,i}\right)\right| < -\Delta_{dB}$ is not violated within the tolerance of the mechanical damping, then there is robust stability to *arbitrary* eigenfrequency locations. This is an ideal measure against *spillover* eigenfrequencies with *poorly-known* frequency range but *assessable* damping (see Fig. 10.12, Mode 2).

Phase stabilization with uncertain damping

Frequency response condition: $L\left(j\omega_{p,i}\right) \in$ *phase-lead stability region* or *phase-lag stability region.*

Given unknown damping of eigenmodes $\omega_{p,i}$, and thus theoretically very large magnitude peaks, only a phase which conforms to the stability requirements can guarantee robust stability. Depending on the frequency range, either a phase lead or a phase lag is preferred. At low frequencies, phase lead can be used, while in the mid-to-high frequency range, fundamentally only phase lag is possible.

Dynamic properties Depending on the magnitude $\left|L\left(j\omega_{p,i}\right)\right| <> 1$, passive inherent damping or active damping of natural oscillations results.

Robustness properties All eigenmodes within the phase-lead or phase-lag frequency range are robustly stable to *arbitrarily low* damping $d > 0$ and *sufficiently tight tolerances* on the eigenfrequencies, as long as the critical stability bounds (phasing) are not exceeded. This is an ideal measure against *spillover* eigenfrequencies with *assessable* frequency range and *unknown* damping (see Fig. 10.12, phase-lead: Mode 0 and 1, phase-lag: Mode 3).

10.5.2 Generic controller types for multibody systems

This section will show that for mechatronic systems with pronounced structural elasticity (flexible multibody systems), a manageable number of generic transfer elements suffice for robust control. In light of the plant complexity, it is perhaps somewhat surprising at first glance that in most cases a *PID controller* can serve as the *backbone* for robust control. Admittedly, the controller must be parameterized from a somewhat different standpoint than is customary for low-pass plants (cf. the introductory example in Sec. 10.2). In addition, it is somewhat surprising that for further correction of the magnitude and phase behavior of the system within certain frequency ranges, *low-pass elements* (customarily avoided due to their parasitic phase delays) prove to be exceedingly useful elements. For frequency-selective corrections, *notch filters* (band stops) and *anti-notch filters* (band passes) are also particularly suitable.

Complementing these, other common standard elements—such as PD elements (lead compensators), lag compensators, etc.—are naturally also put to use. A basic understanding of their dynamic properties and design rules is assumed, or may be drawn from the corresponding standard control theory literature, e.g. (Ogata 2010), (Horowitz 1963).

PID controllers

Consider the following *real PID controller* with parallel structure (see Fig. 10.15):

$$H_{PID}(s) = K_P + K_I \frac{1}{s} + K_D \frac{s}{1+T_N s} = K_{PID} \frac{1 + 2d_N \dfrac{s}{\omega_N} + \dfrac{s^2}{\omega_N^2}}{s\left(1 + \dfrac{s}{\omega_D}\right)}, \tag{10.17}$$

$$\text{where}\quad K_{PID} = K_I;\; \omega_D = \alpha\omega_N;$$

$$\frac{K_P}{K_I} = \frac{1}{\omega_N}\left(2d_N - \frac{1}{\alpha}\right);\; \frac{K_D}{K_I} = \frac{1}{\omega_N^2}\left[1 - \left(2d_N - \frac{1}{\alpha}\right)\frac{1}{\alpha}\right].$$

Steady-state accuracy The integrator ensures steady-state accuracy given proportional dynamics in the plant, or can be inserted as a disturbance torque estimator for constant disturbances (infinite gain at zero frequency → ideal disturbance rejection, cf. Eq. (10.5) and T_d in Table 10.1).
Positive phase shifter With the quadratic numerator, a *positive phase shift* can be realized for $\omega > 0.1\omega_N$. With a sufficiently large $d_N \simeq 0.5–1$, this enables *broad-band positive phase correction* in the lower frequency range, shifting eigenmodes into the phase-lead stability region.

Active damping in the phase-lead stability region For eigenmodes in the phase-lead stability region, the increasing magnitude of the numerator

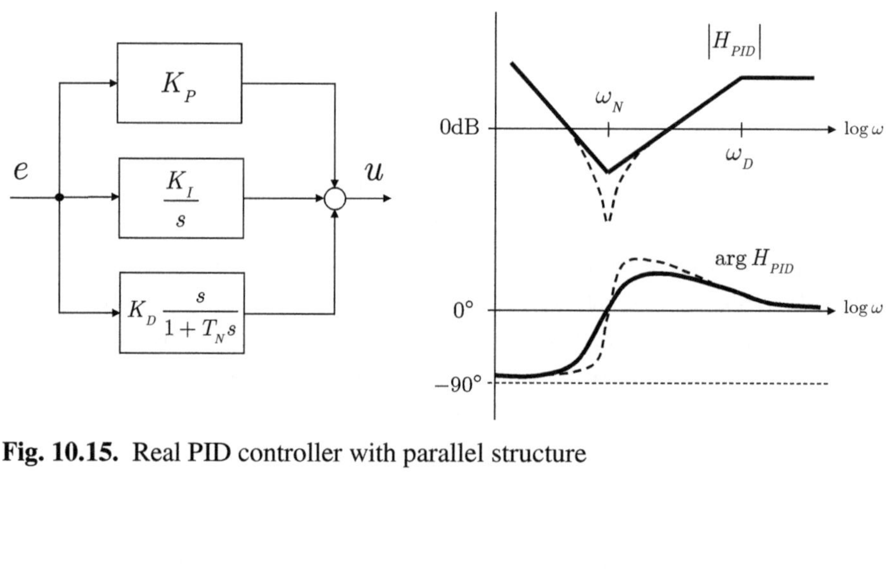

Fig. 10.15. Real PID controller with parallel structure

at frequencies $\omega \ll \omega_N$ and $\omega \gg \omega_N$ allows for equally broad-band generation of high controller gains, and thus good disturbance rejection and active damping of eigenmodes in this frequency range (see Examples 10.4, 10.5, 10.6, 10.7, 10.10, 10.11).

Phase stabilization with low-pass term The additional first-order denominator term (delay term) is required for realization (degree of the numerator = degree of the denominator). Note that with a clever choice of this delay term in the denominator, the phase of high-frequency eigenmodes can be shifted into the phase-lag stability region, ensuring robustness even in the face of vanishingly low damping. Thus, in favorable cases, using a *single PID controller*, both *active control* of the lowest eigenfrequency and simultaneous *phase stabilization* of higher eigenfrequencies (i.e. a robust low-order controller) can be achieved.

PI parameterization In certain cases, the full potency of the positive phase shift of the quadratic numerator is not required, while the sometimes-bothersome increase in magnitude above the cutoff frequency ω_N is to be avoided. In such cases, a slight change in the parameterization using $K_D = 0$ gives the well-known PI controller structure (see Example 10.9)

$$H_{PI}(s) = K_{PI} \frac{1 + \frac{s}{\omega_N}}{s}. \tag{10.18}$$

PI/PD parameterization It is easy to see that a wider bandwidth with a positive phase shift can be achieved by choosing $d_N > 1$ in Eq. (10.17). In this case, it is convenient to split the quadratic numerator into two linear terms, resulting in the series connection of a PI controller and a PD element (lead compensator)

$$H_{PI/PD}(s) = K_{PI/PD} \frac{1 + \frac{s}{\omega_{N1}}}{s} \frac{1 + \frac{s}{\omega_{N2}}}{1 + \frac{s}{\omega_D}}, \quad \omega_{N1} < \omega_{N2} < \omega_D, \tag{10.19}$$

cf. also Example 10.5.

Compensating eigenmodes If needed, the common mode (lowest eigenfrequency) can be approximately compensated using the quadratic PID numerator. Note, however, that, as a rule, exact compensation is not possible. The damping d_N should thus not be made too small, to avoid steep phase changes and failed control of the system. Sec. 10.2 already gave an account of the dangers and shortcomings of such a compensating controller (see also Example 10.3). For these reasons, this type of parameterization should only be applied with *extreme caution* and *restraint.*

Universal PID structure Thus, in all of the examples presented here, the generalized PID structure of Eq. (10.17) will be employed, whereby the special PI (10.18) and PI/PD (10.19) structures simply represent particular parameterizations of Eq. (10.17). This again highlights the fundamental importance and utility of a basic PID controller for mechatronic systems, even and especially those with pronounced elastic multibody properties.

Low-pass element

$$H_{LP1}(s) = \frac{1}{\left(1 + \dfrac{s}{\omega_{LP}}\right)^i}, \quad i = 1,2,... \tag{10.20}$$

$$H_{LP2}(s) = \frac{1}{\left(1 + 2d_{LP}\dfrac{s}{\omega_{LP}} + \dfrac{s^2}{\omega_{LP}^2}\right)^i}, \quad i = 1,2,... \tag{10.21}$$

Gain stabilization For $\omega > \omega_{LP}$, eigenfrequencies whose damping can be estimated can be kept sufficiently far below the 0-dB-line (see the *gain stability region*). This is perhaps the most common use for a low-pass element in a system with structural elasticity, and always produces gain stabilization at sufficiently high eigenfrequencies. However, such a low-pass element holds many more possibilities for design.

Negative phase shifter Already far below the cutoff frequency ω_{LP}, there is a noticeable negative phase shift. As a result, a low-pass element can be used to shift the phase of certain eigenfrequencies (particularly those with poorly-assessable damping) into desired phase regions ☞ the *phase-lag stability region.* Note that, due to its destabilizing ramifications (the reduction in phase margin), this negative phase shift is customarily avoided. However, for multibody systems, it can be useful and expedient.

Phase separation with complex conjugate poles Often, a clear separation of eigenmodes into the phase-lead and phase-lag regions is desired in order to avoid possible intrusion into the critical stability regions given uncertain damping. In Fig. 10.12, for example, given very low damping, Mode 2 would lead to an unstable closed control loop. On the other hand, however, the phase of the eigenmodes in the *phase-lead stability regions* should not be affected. In such a case, a low-pass element with complex conjugate poles, i.e. $d_{LP} \ll 1$, can be chosen. If ω_{LP} is also placed in the phase-lead stability region, but above the highest phase-lead eigenfrequency (Mode 1 in Fig. 10.12), then there are two consequences. First, the parasitic resonant peak of the low-pass is absorbed by the feedback, i.e. $\left|T\left(j\omega_{LP}\right)\right| \approx 1$ since $\left|L\left(j\omega_{LP}\right)\right| \gg 1$. Then, due to the low damping $d_{LP} \ll 1$, the desired negative phase shift results, affecting only frequencies $\omega > \omega_{LP}$. In this way, all higher eigenmodes can be shifted to the left in the gain-phase plain (cf. Example 10.6). However, this is only truly possible when the eigenfrequencies are sufficiently separated. If needed, additional corrective measures must be applied to alleviate possible stability problems in the critical -540° stability region.

Band stop / notch filter

$$H_{Notch}(s) = \frac{1 + 2d_N \dfrac{s}{\omega_0} + \dfrac{s^2}{\omega_0^2}}{1 + 2d_D \dfrac{s}{\omega_0} + \dfrac{s^2}{\omega_0^2}} \quad \text{where } d_N \ll d_D \tag{10.22}$$

Compensation of eigenfrequencies In the frequency range $\omega \approx \omega_0$, H_{Notch} acts as a *band stop* with $\left|H_{Notch}\right| \ll 1$. This allows eigenfrequencies $\omega_{pi} \approx \omega_0$ to be approximately compensated or damped in their amplitude. This property can be used as needed for gain stabilization of eigenmodes by minimizing the excitation of such natural oscillations using the feedback loop. Note however, that as a rule, exact compensation is not possible, especially when the location of the eigenfrequencies is uncertain. For this reason, the damping d_N should not be too low, so as to decrease the magnitude over as wide a frequency band as possible (compare to the quadratic numerator of the PID controller). Thus, a notch filter should be employed for such purposes only with great care.

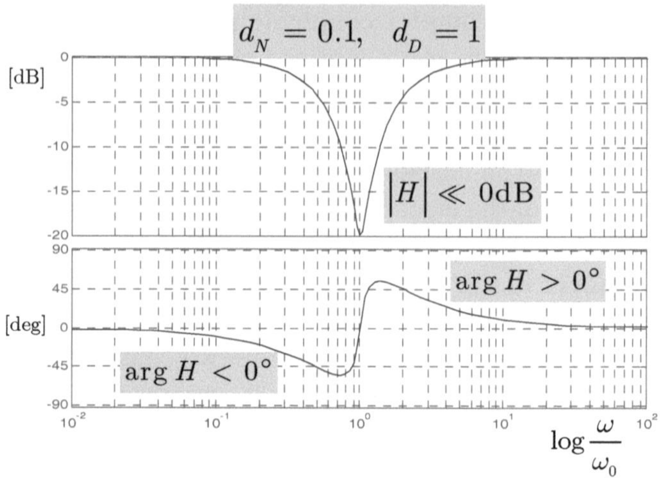

Fig. 10.16. Normalized frequency response of a *notch filter* for $d_N = 0.1, \ d_D = 1$

Positive phase shifter A significantly more attractive property of a notch filter is its interesting phase response. For $\omega > \omega_0$, a *positive phase shift* can be seen in Fig. 10.16, without an increase in magnitude[7]. This property can now be profitably exploited, in particular, to actively stabilize low-frequency eigenmodes, i.e. to shift them into the *phase-lead stability region* of the gain-phase plain. The ever-present parasitic negative phase shift for $\omega < \omega_0$ is then of no consequence, provided that $|L(j\omega)| > 1$, and only intersections with the critical stability region -180° which conform to stability requirements occur (for example, see Mode 0 in Fig. 10.12).

If a notch filter is used for such a purpose, then a limited amount of variation in the eigenfrequencies will play no determining role, as it is not specific eigenfrequencies which are compensated, but rather the focus is on the positive phase shift within a finite frequency band.

---

7 In a customary phase lead corrective element (e.g. PD controller or lead element), the positive phase shift comes at the cost of an increase in magnitude at high frequencies. In order to avoid this destabilizing effect at high frequencies, the magnitude is decreased at lower frequencies using a lag element (with accompanying negative phase shift). If the frequency ranges of such a lead-lag compensator are allowed to merge, then a frequency response similar to Fig. 10.16 results. Thus, a notch filter can be interpreted as a generalized lead-lag compensator with complex conjugate zeros.

Phase contraction Using a notch filter, frequency-dependent phase contraction in the gain-phase plain can be achieved. Relative to the notch frequency ω_0, the phase is shifted towards negative values (left in the NICHOLS diagram) for lower frequencies and towards positive values (right in the NICHOLS diagram) for higher frequencies. As a result, eigenmodes below and above ω_0 are brought closer together in phase (a *phase contraction*, cf. Examples 10.8 and 10.9). This can be used to advantage to place eigenmodes with very low damping in the *phase-lag stability region* in a robustly stable manner.

Band pass/ anti-notch filter

$$H_{Anti-Notch}(s) = \frac{1 + 2d_N \dfrac{s}{\omega_0} + \dfrac{s^2}{\omega_0^2}}{1 + 2d_D \dfrac{s}{\omega_0} + \dfrac{s^2}{\omega_0^2}} \quad \text{where } d_N \gg d_D \tag{10.23}$$

Frequency response The anti-notch filter has precisely the inverse frequency response of the notch filter, i.e. a *band-selective amplification* at ω_0, and a *positive phase shift* below ω_0 and *negative phase shift* above ω_0 (cf. Fig. 10.16).

Band-selective amplification The apparent purpose of an anti-notch filter lies in its band-selective amplification around the anti-notch frequency ω_0. Recall that for good rejection of input disturbances (forces/torques), a high controller gain is required (cf. Eq. (10.5) and T_d in Table 10.1). To actively damp natural oscillations excited by external disturbance forces/torques, it is precisely a large, frequency-selective amplification which is called for. By the same token, harmonic input disturbances with known frequency (e.g. unbalanced rotating masses) can be suppressed with this type of frequency-selective controller component. In both cases, the phase response (see next point) should naturally be included in considerations of stability.

Phase separation A further, oft-ignored application of the anti-notch filter lies in its phase separation properties. This fundamentally represents the inverse behavior of the notch filter in the gain-phase plain. Relative to the anti-notch frequency ω_0, the phase is shifted towards positive values (right in the NICHOLS diagram) for lower frequencies and towards negative values (left in the NICHOLS diagram) for higher frequencies. As a result, eigenmodes below and above ω_0 are shifted apart in phase (a *phase separation*). This can be used to advantage to separate eigenmodes with very small damping into different stable phase regions (e.g. phase-lead vs. phase-lag) in a robustly stable manner (cf. Example 10.7). In comparison to the equally phase-separating second-order low-pass with complex poles, the anti-notch filter enables band-limited phase correction.

10.5.3 Transient dynamics under unity feedback

Approximating relation The response of a closed control loop with unity feedback (see Fig. 10.14 or the feedback portion of the generic two-degree-of-freedom configuration in Fig. 10.1) can, under certain assumptions, be quite nicely described using characteristic values of the open-loop frequency response $L(j\omega)$. If the dynamics of the closed loop can be characterized by a *dominant pair of poles* or if the *magnitude curve* $|L(j\omega)|$ fulfills the following conditions:

$$\begin{aligned} &|L(j\omega)| \gg 1 && \text{for} \quad \omega < \omega_C\,, \\ &|L(j\omega)| \approx -20\,\text{dB/dec} && \text{for} \quad \omega \approx \omega_C\,, \\ &|L(j\omega)| \ll 1 && \text{for} \quad \omega > \omega_C\,, \end{aligned}$$

where ω_C is the *crossover frequency* at which $|L(j\omega_C)| = 1$, then the following approximate relations hold:

$$\boxed{\begin{aligned} \omega_C \cdot t_r &\approx 1.5\,, \\ \Phi_M[\text{deg}] + o[\%] &\approx 70\,. \end{aligned}}\,. \tag{10.24}$$

For the definition and meaning of the variables in Eq. (10.24) see Fig. 10.17.

Model control loop with dominant pair of poles The given approximate relations (10.24) were derived for the second-order model control loop depicted in Fig. 10.18. The resulting system is also second-order and can be represented as a second-order delay element. The following *heuristic* interpretation is meant to illustrate that these approximations hold not only for this specific model control loop, but also can be applied to a large class of *"similar" open-loop transfer functions*.

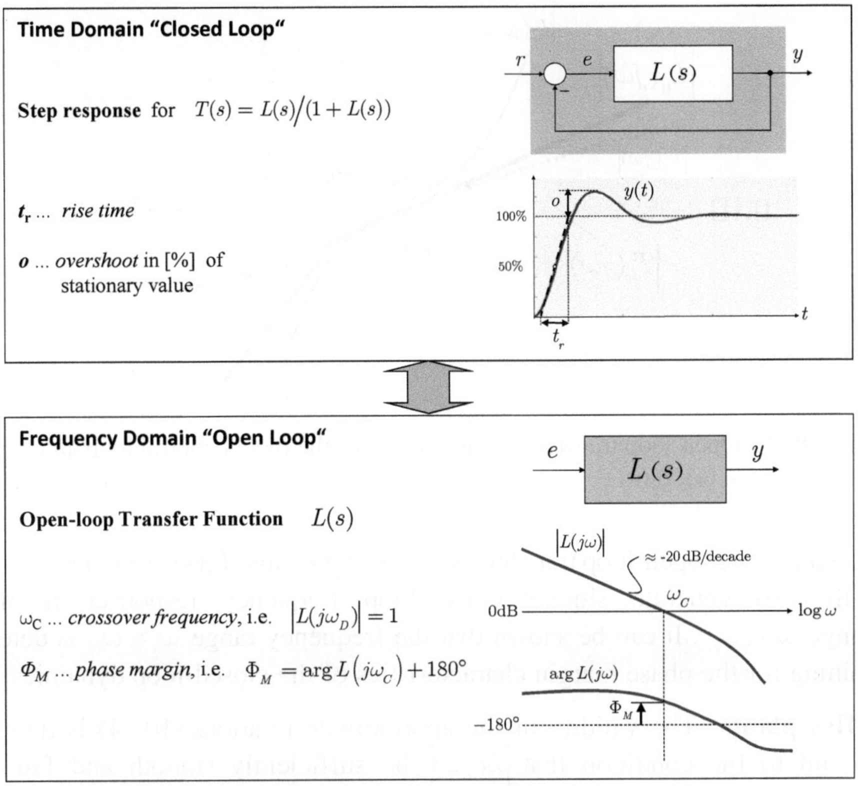

Fig. 10.17. Relationship between characteristic parameters of the open and closed standard control loop describing the transient closed-loop dynamics

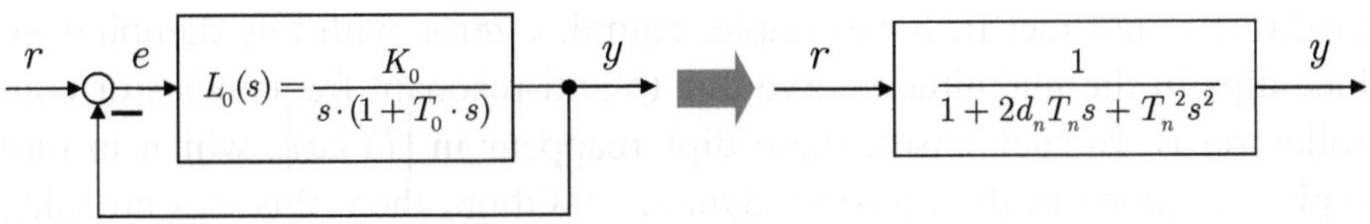

Fig. 10.18. Model second order control loop (dominant pair of poles in the closed loop)

The magnitude curve $|L_0(j\omega)|$ of the model control loop is sketched in Fig. 10.19. For practically sensible phase margins, the crossover frequency lies in the frequency range where the magnitude decrease is -20 dB/decade. Due to the properties of feedback, the exact evolution of $|L_0(j\omega)|$ is irrelevant for the closed-loop frequency response $T(j\omega)$, as long as $|L_0(j\omega)| \gg 1$, since, in the frequency range $\omega < \omega_C$, it always holds that $|T(j\omega)| \approx 1$.

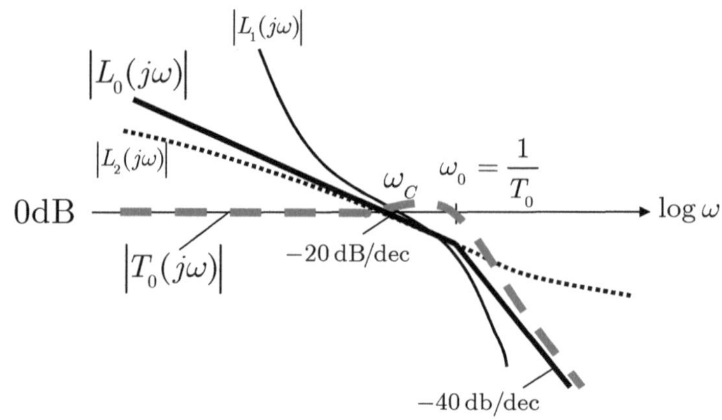

Fig. 10.19. Open-loop transfer function $L_0(s)$ of the model control loop and similar systems $L_1(s), L_2(s)$

Thus, other open-loop transfer functions $L_1(s)$ and $L_2(s)$ (see Fig. 10.19) will also generate similar closed-loop frequency responses in the range $\omega < \omega_C$. It can be shown that the frequency range $\omega \approx \omega_C$ is determining for the phase margin characteristics of the closed-loop dynamics.

MBS plants The validity of the approximate relations (10.24) is tightly bound to the condition that $|L(j\omega)|$ be sufficiently smooth and fall at -20 dB/decade near the crossover frequency. This is equivalent to the condition that $|T(j\omega)|$ also be sufficiently smooth and possess a magnitude peak due only to the phase margin (see Fig. 10.19). In MBS plants, this condition is not met if, for example, complex zeros with low damping induce dips in the magnitude curve due to anti-resonant frequencies or controller zeros. In such cases, these dips reappear in $|T(j\omega)|$, which in turn implies changes in the transient dynamics. Often, then, this is a creeping behavior, systems theoretically brought about by complex pole and zero pairs having marginal damping. Such dynamics are illustrated in Example 10.4.

10.5.4 Control of a single-mass oscillator

Model considerations As the most elementary and illustrative multibody system, Sec. 10.2 introduced a single-mass oscillator in Eq. (10.11) (Fig. 10.2a). In the context of the considerations there—ignoring parasitic dynamics and assuming exact compensation of the eigenfrequency—

some of the principal deficits of a compensating controller were already spelled out.

In Example 10.3 in this section, these deficits are once more clarified from the point of view of stability analysis on the basis of more realistic assumptions (including parasitic dynamics and compensation errors), corroborating the *impracticality* of such compensation concepts.

Robust control strategy Further below, Example 10.4 presents an alternative *robust* controller layout with *phase-lead stabilization* and a *pre-filter*, avoiding all disadvantages of a compensating controller. The control approach presented there can categorically be taken as the basis for control of higher-order multibody systems, as will be shown in subsequent sections.

Example 10.3 *Compensating control of a single-mass oscillator.*

System configuration Consider the single-mass oscillator in Fig. 10.2a. Let the plant—including parasitic dynamics (e.g. of actuators or sensors)—be given by the following transfer function:

$$P(s) = K_P \frac{1}{\left\{d_0; \omega_0\right\}} \frac{1}{\left[\omega_{par}\right]}, \text{ where } \begin{aligned} &K_P = 0.5;\ d_0 = 0.01;\ \omega_0 = 1; \\ &\omega_{par} = 10\omega_0 . \end{aligned}$$

To enable a comparison, consider the following two compensating PID controllers:

$$H_1(s) = 4\frac{\left\{0.01; \omega_0\right\}}{s\left[10\omega_0\right]}, \quad H_2(s) = 4\frac{\left\{0.1; \omega_0\right\}}{s\left[10\omega_0\right]} .$$

H_1 compensates the nominal eigenmode “exactly”, while in H_2, a somewhat higher numerator damping is assumed. The achievable system performance is depicted in Fig. 10.20.

Discussion of dynamics As expected, the stability of the system differs distinctly from the idealized case in Sec. 10.2. Due to the parasitic dynamics, the control loop is now only conditionally stable—though in any case, there are sufficient stability margins (Fig. 10.20b). It is also easy to verify the validity of the approximate relations (10.24) for the transient dynamics by applying a step reference input. It is interesting to note the differing disturbance responses for the two controllers. With the controller H_2, considerably better damping of force disturbances results (Fig. 10.21d). This is due to the higher controller gain at the eigenfrequency ω_0, i.e. higher damping at the controller zero (Fig. 10.21c).

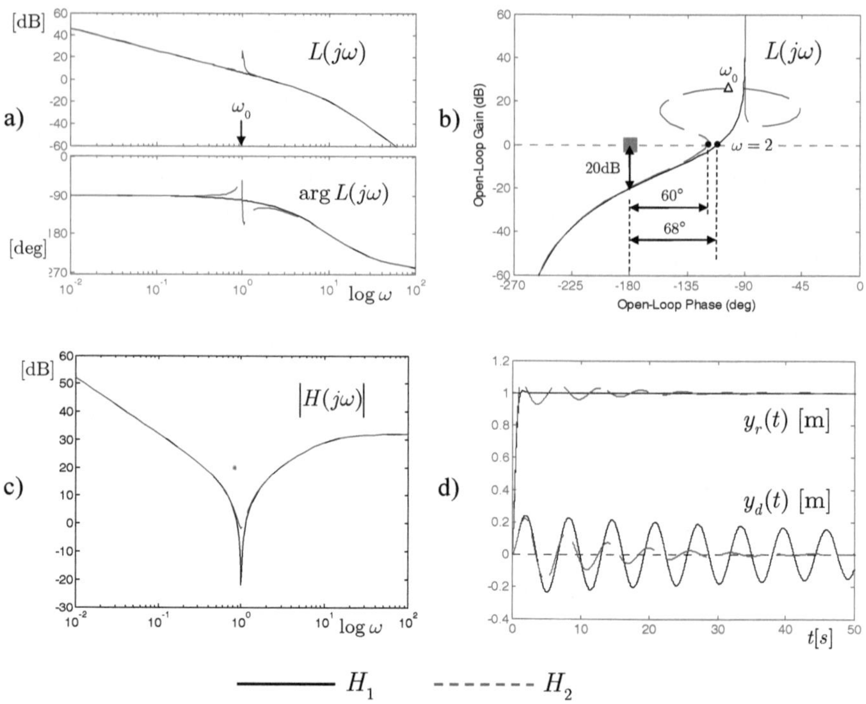

Fig. 10.20. System dynamics with a compensating PID controller: a) BODE diagram (open-loop), b) NICHOLS diagram (open-loop), c) controller frequency response, d) step responses (y_r is reference input step response, y_d is response to step in force disturbance)

MBS parameter variations Realistically, it must be assumed that the MBS eigenfrequency can not be exactly compensated by the controller zero. A particularly critical case is if the MBS eigenfrequency ω_p is *lower* than the PID controller zero ω_N. This implies that a leftwards loop with a phase shift of up to -180° appears in the NICHOLS diagram, potentially leading to intersections with the critical stability region. With even small deviations, this can have catastrophic consequences, as shown in Fig. 10.21. If the MBS eigenfrequency is only 5% lower than assumed, PID compensation with the controller H_1 leads to an *unstable* control loop. In this case—no doubt mostly serendipitously—the control loop with controller H_2 remains stable despite intersections with the critical stability region (conforming to the NYQUIST criterion in intersection formulation, cf. Sec. 10.4.1). In the case of H_2, due to the higher damping, the effect of the controller zero is spread over a wider frequency band, compensating the narrowband negative phase shift of the MBS eigenfrequency to some degree. This favorable behavior points the way towards a robust parameterization of the PID controller, as presented in Example 10.4.

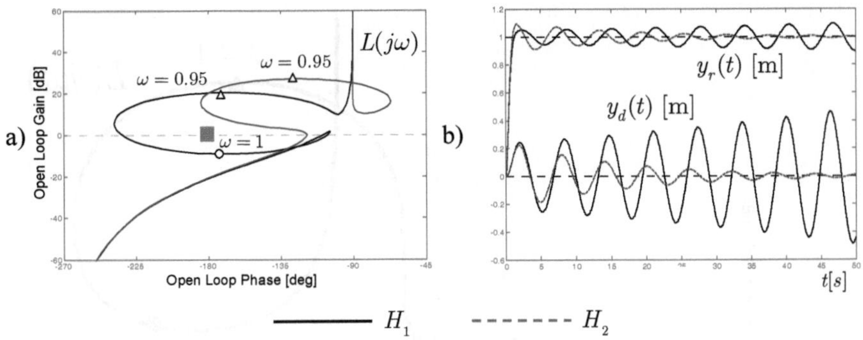

Fig. 10.21. System dynamics given parameter variation of the MBS eigenfrequency $\omega_p = 0.95 \cdot \omega_0$: a) NICHOLS diagram (open-loop), b) step responses (y_r to reference input step, y_d to disturbance force step) ■

Example 10.4 *Robust control of a single-mass oscillator.*

System configuration Consider once more the single-mass oscillator from Example 10.3 with a PID controller. In this case, however, the PID controller is parameterized as follows:

$$H(s) = 1.2 \frac{\left\{0.4;\ 0.5\omega_0\right\}}{s\left[10\omega_0\right]} \tag{10.25}$$

Phase-lead stabilization Comparing the above parameterization of the PID controller to that in Example 10.3, it can be seen that, due to the controller zero which is closer to lower frequencies *near* the MBS eigenfrequency, a broad-band positive phase shift is effected, moving the arc of the MBS eigenfrequency to the right in the NICHOLS diagram (positive phase) (Fig. 10.22). At the same time, here, the MBS eigenfrequency already lies in the rising (differentiating) leg of the magnitude curve of the PID controller (cf. Fig. 10.15), implying an increase in the magnitude of $\left|L(j\omega)\right|$ (Fig. 10.22). The MBS eigenmode is thus placed in the phase-lead stability region, with all the advantages of insensitivity to arbitrarily low damping and (sufficiently limited) variations in the MBS eigenfrequency.

Disturbances As a result of the relatively high gain of the PID controller at the MBS eigenfrequency, force disturbances at the plant input are very well damped and very quickly controlled (Fig. 10.23b).

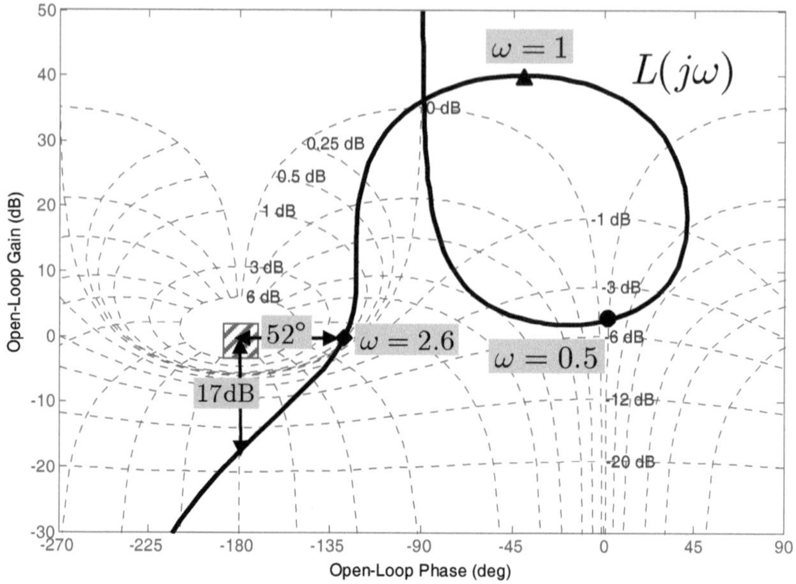

Fig. 10.22. NICHOLS diagram for single-mass oscillator with robust PID controller (phase-lead stabilization)

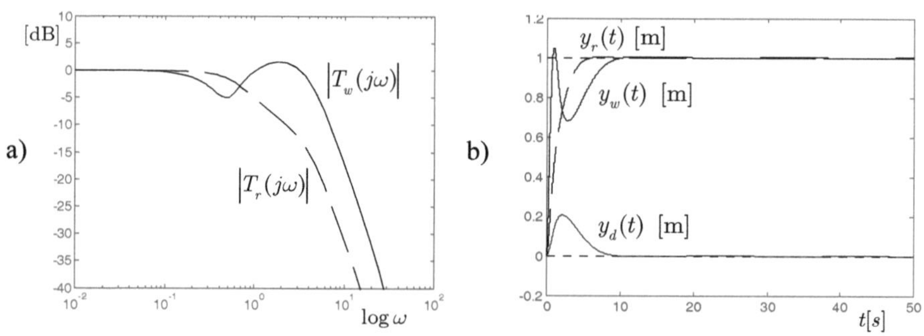

Fig. 10.23. System dynamics with robust PID controller: a) magnitude curves for controller response relative to r (with pre-filter) and w (without pre-filter), b) step responses (y_r for the reference input r, y_w for the reference input w, y_d for the force disturbance)

Controller response The frequency response $T_w(j\omega)$ from the input w to the outputs y resulting from the feedback is represented in Fig. 10.23a. A dip in the magnitude can be seen inside the passband of $T_w(j\omega)$, which in turn results in a curious transient response in the output $y_w(t)$ (Fig. 10.23b). This is a very salient and *inevitable* phenomenon in multibody systems. The cause is nicely illustrated in the open-loop NICHOLS diagram (Fig. 10.22). In this figure, an additional grid is plotted for the complementary sensitivity $T(j\omega)$ (Eq. (10.3)) (here, identical to $T_w(j\omega)$). Due to the loops in the frequency response $L(j\omega)$, the lower arcs fundamentally lead to values $|T(j\omega)| < 1$. These dips in $L(j\omega)$ practically always exist in MBS problems, either due to the controller frequency response (as here), or to the MBS zeros (anti-resonant frequencies). In the current case, it is easy to verify that the dip in the magnitude is not a result of the relatively low numerator damping of the PID controller, but of the bend in the magnitude at the numerator corner frequency (cf. Fig. 10.15). Due to the functional coupling of $L(j\omega)$ and $T(j\omega)$ in Eq. (10.3), these dips in magnitude of $T(j\omega)$ are thus inevitable in MBS plants, so that *pure feedback* can fundamentally only realize a limited-quality controller response.

Pre-filter design Exploiting the second design degree of freedom, taking the reference input frequency response $T_w(j\omega)$ as fixed, and using a suitable parameterization of a pre-filter $F(s)$, the frequency response of the output y to an external reference input r can be kept as *flat* as possible at *0 dB* over the *widest* possible frequency band (cf. Fig. 10.1). In the current case, this can be accomplished with a band-limited increase in magnitude at $\omega \approx 0.4$ and a decrease in magnitude around $\omega \approx 2$. One possible compatible filter transfer function is an anti-notch/lag combination:

$$F(s) = \frac{\left\{1;\,0.5\right\}}{\left\{0.4;\,0.5\right\}} \frac{\left[1.8\right]}{\left[0.35\right]}. \tag{10.26}$$

The now satisfactory controller response is depicted in Fig. 10.23.

MBS parameter variations In the current case, all the advantages of phase-lead stabilization already discussed in Sec. 10.5.1 come to bear quite clearly. In any case, there is robust stability in the presence of arbitrarily low MBS damping. In addition, it is to be expected that a limited variation in the MBS eigenfrequency will also only lead to a limited left-right shift of the frequency response loop of $L(j\omega)$.

Due to the large phase margin resulting from the phase lead of the PID controller, there is high phase robustness. With unchanged controller/filter parameters (10.25), (10.26), the MBS eigenfrequency can be drastically changed to $\omega_p = 0.8\omega_0$ and $\omega_p = 0.5\omega_0$. The still satisfactory, and above all *stable*, dynamics can be seen in Fig. 10.24 (cf. compensating PID controller Fig. 10.21).

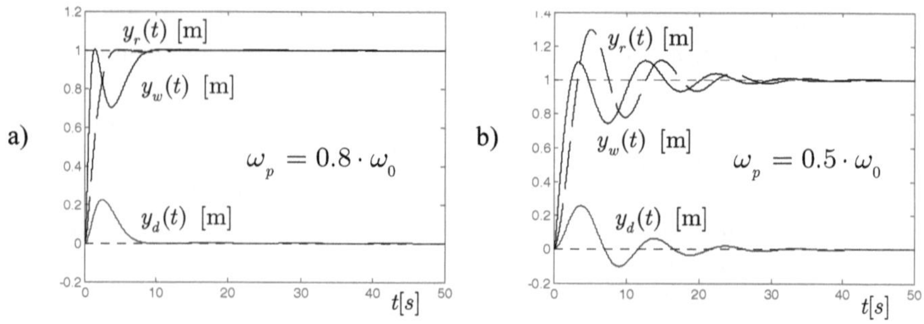

Fig. 10.24. System dynamics with robust PID controller given parameter variation in the MBS eigenfrequency (step responses y_r to reference input, r, y_w to reference input, w, y_d to force disturbance): a) $\omega_p = 0.8 \cdot \omega_0$, b) $\omega_p = 0.5 \cdot \omega_0$

■

10.5.5 Collocated MBS control

Collocated sensor-actuator arrangement In high-order multibody systems the issue of the type of sensor and actuator arrangement immediately arises. As previously discussed, in a *collocated* setup, the resulting sensor/actuator transfer functions always have alternating pole/zero distributions of the following kind:

$$P_{MBS}(s) = \cdots \frac{\left\{\omega_{z,i-1}\right\}\left\{\omega_{z,i}\right\}\left\{\omega_{z,i+1}\right\}}{\left\{\omega_{p,i-1}\right\}\left\{\omega_{p,i}\right\}\left\{\omega_{p,i+1}\right\}}\cdots,$$

where

$$\omega_{z,i-1} < \omega_{p,i-1} < \omega_{z,i} < \omega_{p,i} < \omega_{z,i+1} < \omega_{p,i+1} .$$

For stability, this is favorable to the extent that the *negative* 180° phase shifts of the *eigenfrequencies* $\omega_{p,j}$ are each absorbed by *positive* 180° phase shifts of the anti-resonant frequencies $\omega_{z,j}$. Naturally, the phase shift of the lowest eigenmode (the common mode) also comes to bear, as well as any phase lag due to parasitic dynamics. However, the relative phase of any two neighboring eigenmodes is limited, and no large phase shifts can appear. Roughly similar conditions as in a single-mass oscillator result, with the addition of a few MBS eigenfrequency and anti-resonance pairs.

Robust control strategy The robust control strategy is based on the results found for the single-mass oscillator:

- *2 degrees of freedom* with *controller + pre-filter*: to balance deficiencies in the controller response due to dips in the open-loop magnitude response or limited feedback bandwidth;
- *PID controller* with *phase-lead* parameterization: in favorable cases, enables robust phase-lead stabilization of whole clusters of sufficiently closely-spaced eigenmodes in the lower frequency range;
- *phase-shifting transfer elements* (optional): to separate the phases of eigenmodes or clusters of eigenmodes for robust phase-lag stabilization.

In the following Examples 10.5 and 10.6, the application of this control strategy is demonstrated in two typical and easily generalizable MBS configurations.

Example 10.5 *Collocated control of a two-mass oscillator.*

System configuration Consider the elastically-suspended two-mass oscillator depicted in Fig. 10.25.
For the *collocated* sensor-actuator layout y_1/F_1, accounting for additional *parasitic dynamics*, the following transfer function results:

$$P_{MBS}(s) = \frac{Y_1(s)}{F(s)} = K_P \frac{1}{\{d_{p0}, \omega_{p0}\}} \frac{\{d_z, \omega_z\}}{\{d_{p1}, \omega_{p1}\}} \frac{1}{[\omega_{par}]},$$

$$K_P = \frac{2}{3k}, \; \omega_{p0} = \sqrt{\frac{k}{m}}, \; \omega_{p1} = \sqrt{\frac{3k}{m}}, \; \omega_z = \sqrt{\frac{2k}{m}},$$

$$m = 1 \text{ kg}, \quad k = 10 \text{ N/m},$$
$$\omega_{p0} = 3.2 \text{ rad/s}, \; \omega_{p1} = 5.5 \text{ rad/s}, \; \omega_z = 4.5 \text{ rad/s},$$
$$d_{p0} = 0.05, \; d_{p1} = d_z = 0.0005,$$
$$\omega_{par} = 55 \text{ rad/s} .$$

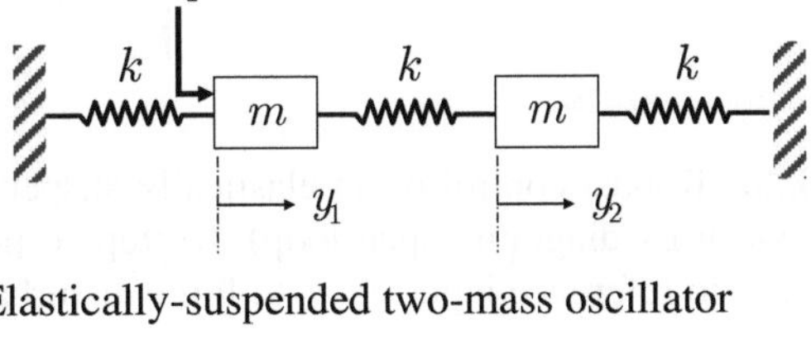

Fig. 10.25. Elastically-suspended two-mass oscillator

Control approach Controller design following the same steps as in Example 10.4 gives the following transfer functions for the *controller* and *pre-filter*:

$$H(s) = 10\frac{\left[0.32\right]\left[2\right]}{s\left[50\right]}, \tag{10.27}$$

$$F(s) = \frac{\left[0.2\right]}{\left[0.3\right]}\frac{1}{\left[4.5\right]^2}\frac{\left\{0.1;\ 4.5\right\}}{\left\{1;\ 4.5\right\}}. \tag{10.28}$$

The dynamics of the controlled system are depicted in Fig. 10.26.

Discussion In the current case, it is possible to place the two relatively *closely-spaced* eigenmodes *together* into the phase-lead stability region, resulting in robust stability for both eigenmodes.

The need for a pre-filter for a satisfactory controller response is based on the localized dip in the magnitude of $L(j\omega)$ due to the PID controller and the MBS zero (anti-resonant frequency).

The parameterization of the dynamic corrective elements presents two peculiarities. As a *PI/PD controller*, the controller is parameterized according to Eq. (10.19) to affect both eigenmodes with a positive phase shift over a wider frequency band.

In the pre-filter, a *notch filter* with $\omega_{Notch} \approx \omega_z$ can be seen. The reason for taking this measure can be described as follows: as can be easily recognized from the *root locus* (Ogata 2010) of this control loop depicted in Fig. 10.27, the closed loop possesses a weakly-damped pair of poles quite near the anti-resonant frequency ω_z. The weakly-damped oscillation accompanying this pair of poles is excited by input variables (the ref-

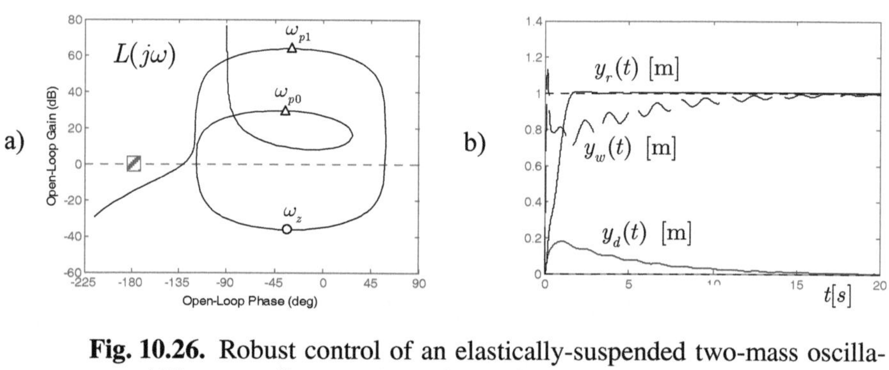

Fig. 10.26. Robust control of an elastically-suspended two-mass oscillator: a) NICHOLS diagram (open-loop), b) step responses (y_r to reference input, r, y_w to reference input, w, y_d to force disturbance)

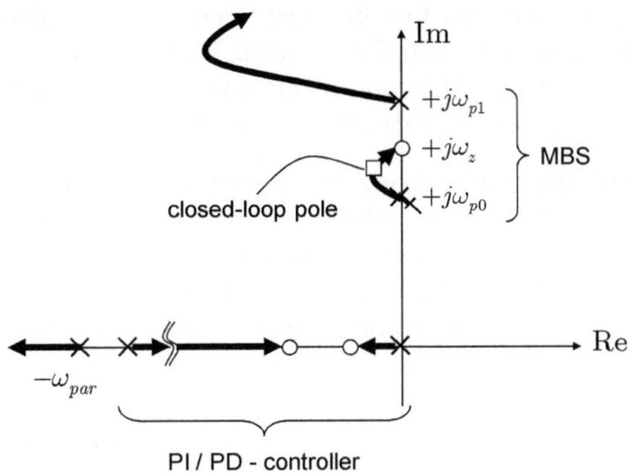

Fig. 10.27. Root locus for robust control of an elastically-suspended two-mass oscillator (not to scale)

erence and disturbances), and, due to the differing gains in the reference input and force disturbance channels, can be more clearly seen in the reference input step response. For this reason, this excitation frequency is masked out in the reference channel using a notch filter.

The weakly-damped pair of closed-loop poles at the anti-resonant frequency of P_{MBS} incidentally also has a very clear physical meaning. Due to the control, the left mass is so-to-speak "held fixed", so that the rest of the MBS structure oscillates against this virtually restrained mass at precisely the anti-resonant frequency.

■

Example 10.6 *Robust control of a free mass with attached flexible structure.*

System configuration In this example, a *higher-order* multibody system is to be considered. The chosen configuration consists of a free (unsuspended) mass with an attached flexible structure. Examples of physical implementations include elastic drive trains, flexible manipulators, and satellites with flexible booms. Let the MBS plant model with lumped parameters including *parasitic dynamics* have the following form:

$$P_{MBS}(s) = \frac{1}{s^2}\frac{\{3\}}{\{4\}}\frac{\{8\}}{\{10\}}\frac{\{90\}}{\{100\}}\frac{\{500\}}{\{510\}}\frac{1}{[50]}\,.$$

Let the transfer function be experimentally determined (e.g. Sec. 2.7.3). For the structural damping, assume all that is known is that it is very small throughout. The control is to ensure that constant disturbance forces are compensated and that structural oscillations are sufficiently damped.

Control approach As all that is required here is disturbance rejection, implementing constant reference control with a standard control loop as in Fig. 10.14 suffices. The requirement for the compensation of constant disturbance forces necessitates an integrator in the controller, the implementation of which is not completely trivial, as the MBS plant already contains a double integrator (the common mode of the free mass). As there are no further assumptions regarding structural damping, applying a *worst-case* design approach, *vanishingly small damping* should be anticipated for all flexible eigenmodes. This is evidently a collocated configuration, as can be easily discerned from the arrangement of poles and zeros. In addition, two clusters of eigenmodes can be recognized, separated by a somewhat larger frequency gap. Thus, the following control strategy presents itself. The low-frequency cluster with the eigenfrequencies $\omega = (4, 10)$ is to be placed along with the common mode $\omega = 0$ in the phase-lead stability region using a PID controller. Due to the parasitic dynamics, it will not be possible to also keep the high-frequency eigenmode cluster $\omega = (100, 510)$ inside the *phase-lead stability region*. Thus, this cluster is to be shifted phase robustly into the phase-lag stability, region requiring a suitable phase shifter. This control strategy can be realized using the following robust *controller*:

$$H(s) = 0.003 \underbrace{\frac{\{1;\ 0.05\}}{s\,[40]}}_{PID} \underbrace{\frac{1}{\{0.1;\ 40\}}}_{LP2} .$$

Discussion Due to the I-component in the controller, triple-integrating dynamics result for the open loop. According to the NYQUIST criterion, this requires a positive intersection with the critical stability region at -180°. This is effected by the numerator of the PID controller (Fig. 10.28b). The choice of an additional low-pass element for phase separation of the two eigenmode clusters can be discerned in Fig. 10.28a,b. For a narrow-band phase separation, the cutoff frequency of the low-pass element is put in the frequency gap, and a relatively low damping is chosen. Thus, the high-frequency eigenmode cluster is now placed phase robustly in the *phase-lag stability region.* The controller gain should then be increased as far as possible so there remains just sufficient gain margin relative to the critical point (0 dB, -180°)—in the current case, $\Delta = 15\text{dB}$ at $\omega = 30$.

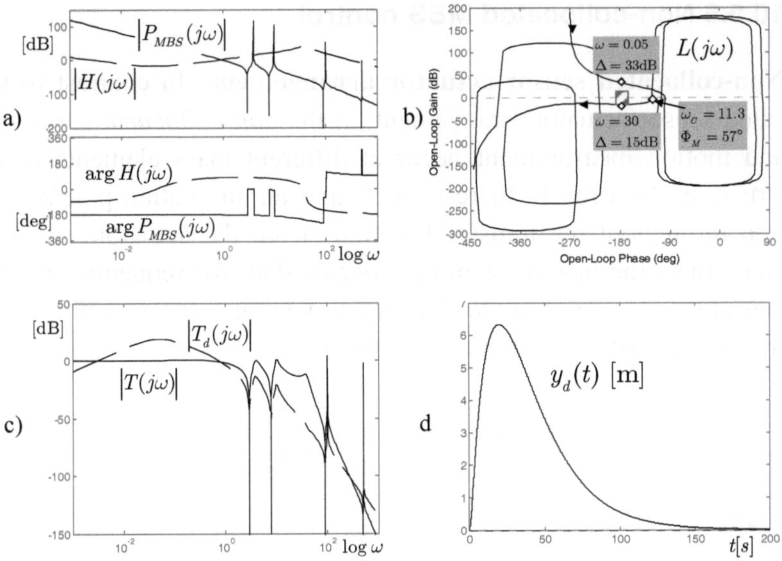

Fig. 10.28. System dynamics of an MBS with free mass and flexible structure: a) BODE diagram (open- and closed-loop), b) NICHOLS diagram (open-loop), c) complementary sensitivity and disturbance frequency response, d) force disturbance step response

The result for closed-loop control is shown in Fig. 10.28c,d. The active damping of structural eigenmodes and the desired time evolution with constant disturbance torques can be seen.

Unmodeled eigenmodes The strategy presented above provides a very good approach in general for the stabilization of unmodeled eigenmodes (*spillover*). Often, all that is known of such eigenmodes is the approximate frequency range without further information about structural damping. If such an uncertain frequency range can be successfully placed well within the center of the *phase-lag stability regions*, high robustness is ensured.

Role of the low-pass element Note that, though the stabilization of such modes is achieved here using a low-pass element, its effectiveness is *not* due to its *magnitude drop*, but rather, its phase lag. It is mistakenly often assumed that all high-frequency eigenmodes must be damped with a low-pass element to guarantee $|L(j\omega)| < 0\,\text{dB}$ at high frequencies. This requirement is in no way necessary—as this example shows—and cannot even be guaranteed in the extreme case of very low structural damping. ■

10.5.6 Non-collocated MBS control

Non-collocated sensor-actuator arrangement In contrast to the collocated sensor-actuator arrangement, in the *non-collocated* case, force input and motion measurement occur at different mass elements of the MBS structure. As a result, there is no longer an alternating pole/zero distribution throughout the transfer functions from the actuators to the sensors. Here, then, the inconvenient case occurs that two sequential eigenfrequencies are *no longer* separated by an intervening anti-resonant frequency, i.e. the plant transfer function has the form

$$P_{MBS}(s) = \cdots \frac{\left\{\omega_{z,i-1}\right\}}{\left\{\omega_{p,i-1}\right\}} \frac{1}{\left\{\omega^*_{p,i}\right\}} \frac{\left\{\omega_{z,i+1}\right\}}{\left\{\omega_{p,i+1}\right\}} \cdots , \qquad (10.29)$$

where

$$\omega_{z,i-1} < \omega_{p,i-1} < \omega^*_{p,i} < \omega_{z,i+1} < \omega_{p,i+1} .$$

This is unfavorable for stability to the extent that *negative* 180° phase shifts of the *eigenfrequency* $\omega^*_{p,i}$ are no longer absorbed by the now-missing anti-resonant frequency $\omega_{z,i}$, and due to the subsequent eigenfrequency $\omega_{p,i+1}$, a total *negative phase shift* of *-360°* occurs.

As a result, it is thus fundamentally no longer possible to place eigenmodes $\omega_{p,j}$, $j > i$ into the *phase-lead stability region*, as the requisite phase shift would lead to improper transfer functions or unusable controller gains incompatible with stability. Thus, in non-collocated arrangements, performance penalties and compromises in bandwidth, disturbance rejection, and active damping of high-frequency eigenmodes must generally be reckoned with.

Robust control strategy The robust control strategy for non-collocated sensor-actuator arrangements is largely based on the collocated case with a few specific highlights:

- *2 degrees of freedom* with *controller + pre-filter*: to balance deficiencies in the controller response from dips in the open-loop magnitude or limited feedback bandwidth;
- *PID controller* with *phase-lead* parameterization: in favorable cases, enables robust phase-lead stabilization of *at most one* cluster of sufficiently

closely-spaced eigenmodes below the particular eigenfrequency $\omega^*_{p,i}$ in Eq. (10.29);

- *phase-shifting transfer elements*: to separate the phases of eigenmodes $\omega_{p,j} > \omega^*_{p,i}$ for robust phase-lag stabilization.

The following Examples 10.7 and 10.8 deal with two typical non-collocated system configurations and present additional easily generalizable control strategies.

Example 10.7 *Non-collocated control of a two-mass oscillator.*

System configuration Consider the singly-suspended two-mass oscillator depicted in Fig. 10.29.

For the *non-collocated* sensor-actuator arrangement y_2/F_1, the MBS parameters $m_1 = 50$ kg, $m_2 = 5$ kg, $k_1 = k_2 = 400$ N/m, $b_1 = b_2 = 0.05$ Ns/m, and additional *parasitic dynamics* result in the following transfer function:

$$P_{MBS}(s) = \frac{Y_2(s)}{F(s)} = K_P \frac{1}{\{d_{p0}, \omega_{p0}\}} \frac{1}{\{d_{p1}, \omega_{p1}\}} \frac{1}{[\omega_{par}]},$$

$$K_P = 0.025 \text{ m/N}, \ \omega_{p0} = 2.7 \text{ rad/s}, \ \omega_{p1} = 9.5 \text{ rad/s},$$

$$\omega_{par} = 40 \text{ rad/s}.$$

For the given configuration, a robust control approach is to be developed, ensuring a sensible controller response having no steady-state error and sufficient disturbance rejection.

Control approach Given the requirements, a control loop configuration with two degrees of freedom is again envisaged here. The layout of the feedback controller is based on the points laid out above; the layout of the

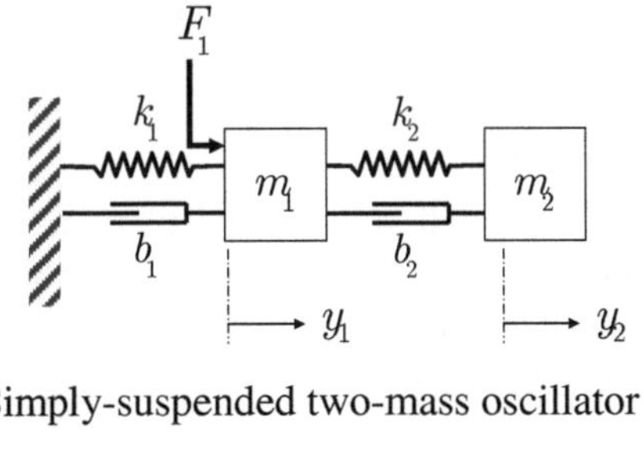

Fig. 10.29. Simply-suspended two-mass oscillator

pre-filter proceeds as previously discussed. This results in the following transfer functions for the *controller* and *pre-filter*:

$$H(s) = 20 \underbrace{\frac{\{2;\ 2.7\}}{s[13.5]}}_{PID} \underbrace{\frac{\{1;\ 6.5\}}{\{0.2;\ 6.5\}}}_{Anti-Notch},$$

$$F(s) = \frac{[0.05]}{[0.5]} \frac{1}{\{1;\ 0.7\}} \frac{\{0.2;\ 2.7\}}{\{1;\ 2.7\}}.$$

Discussion The NICHOLS diagram of the open loop $L_0(j\omega)$ including only the PID controller is shown in Fig. 10.30a. It can be seen that only the first eigenmode can be shifted into the *phase-lead stability region* using the controller zero. For the high-frequency eigenmode, at best, phase-lag stabilization can be achieved. In the critical stability region, *phase spreading* is necessary to separate the two frequency response loops. Ideally suited to this purpose is an *anti-notch filter* whose characteristic frequency $\omega_{Anti-Notch}$ is properly placed in the gap between the two eigenfrequencies (see $L(j\omega)$ in Fig. 10.30a). Due to the particularities of the magnitude and phase in this case, the controller gain can only be increased to a limited extent, so that only limited active damping of the first eigenmode $\omega_{p,0}$ can be realized ($y_w(t)$, $y_d(t)$ in Fig. 10.30b). The layout of the *pre-filter* takes this into account with a *notch filter*, so that frequency components $\omega_{p,0}$ are filtered out of the reference input ($y_r(t)$ in Fig. 10.30b).

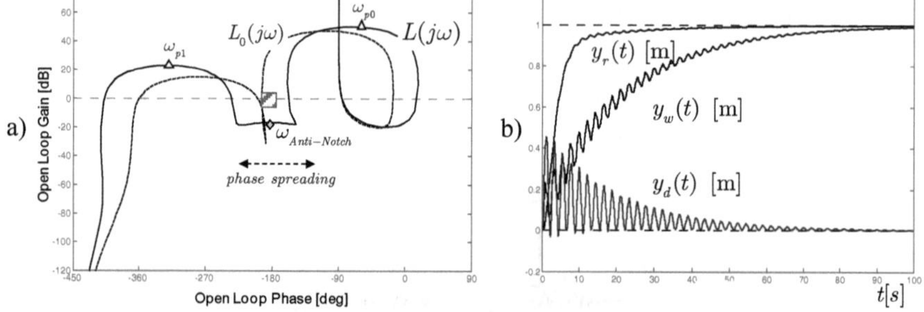

Fig. 10.30. Non-collocated robust control of a simply-suspended two-mass oscillator: a) NICHOLS diagram (open-loop, L_0: PID controller only, L: PID + anti-notch), b) step responses (y_r to reference input r, y_w to reference input w, y_d to force disturbance)

Overall, even for non-collocated arrangements, thoroughly acceptable system dynamics can be achieved, though with considerable compromises and performance penalties compared to a collocated arrangement. ■

Example 10.8 *Non-collocated control of a multibody system.*

System configuration Let a higher-order multibody system possess the following experimentally determined transfer function from actuator to sensor output:

$$P_{MBS}(s) = \frac{1}{\{0.7;\ 1\}} \frac{1}{\{0.2;\ 2\}} \frac{\{0;2.5\}}{\{0;3\}} \frac{1}{\{0.2;\ 10\}} \frac{\{0;15\}}{\{0;16\}} \frac{1}{[40]} .$$

For a few eigenmodes, there is very low structural damping (here assumed equal to zero). A robust controller is to ensure that constant disturbance forces are handled, that structural oscillations are sufficiently damped, and that the response to input steps has neither steady-state error nor transient overshoot.

Control approach Due to requirement for steady-state accuracy, an I-component should once again be envisioned here. A glance at the MBS plant shows that it is evidently a *non-collocated* arrangement: the first two eigenmodes are not separated by a zero. At $\omega = 10$, there is additionally a weakly-damped oscillation, which could also originate from actuator or sensor dynamics. In any case, this induces another destabilizing phase shift of -180°. It is easy to verify here that phase-lead stabilization of the first eigenmodes can be accomplished only with difficulty, as the remaining eigenmodes lie very close together. There is no gap in the frequency as there was in Example 10.7. Thus, the control strategy of phase-lag stabilization of the *entire MBS structure* is selected here for further investigation. The controller crossover frequency must thus lie below the first eigenfrequency, giving a lower bound on the achievable rise time. For this case, in order to achieve a phase margin of more than 70°, the phase lead of a *PI controller* suffices. One challenge here, however, is the requirement that eigenmodes be robustly placed in the *phase-lag stability region.* Due to the expected smooth behavior of $\left|L(j\omega)\right|$ near the crossover frequency, smooth behavior of $\left|T(j\omega)\right|$ in the passband is also to be expected, so that magnitude correction with a pre-filter can be dispensed with. Account for these limiting conditions, the following controller can be found:

$$H_1(s) = \underbrace{0.05 \frac{[4]}{s}}_{PI} \underbrace{\frac{\{0.2;\ 10\}}{\{1;\ 10\}}}_{Notch}.$$

Discussion The open-loop NICHOLS diagram of $L_0(j\omega)$ including only the PI controller is shown in Fig. 10.31a. As expected, the rigid body dynamics (below the first eigenmode) are sufficiently stabilized with the phase lead of the PI controller. However, at both flanks of the *phase-lag stability region*, the eigenmodes extend toward the critical stability regions or intersect them inadmissibly.

For this reason, a *notch filter* is placed approximately in the center of the eigenmode cluster (here coincidentally at one of the eigenfrequencies—limited variation of which does not, however, change the system dynamics). Due to the *phase response* of the notch filter, the phase near the notch frequency is now *contracted* in precisely such a way that both eigenmode cluster portions (left and right of the notch frequency) are pulled away from the critical stability regions (see $L_1(j\omega)$ in Fig. 10.31a).

Controller response As expected, the controller response is smooth. Due to the magnitude/phase response below the critical stability region (0 dB, -180°), a shorter rise time is not possible. The gain can not be increased without inadmissibly decreasing the gain margin.

Disturbances On the whole, the disturbance response is acceptable, only the undamped natural oscillation at $\omega = 3$ is disturbing. The reason for this can be seen in the controller frequency response (Fig. 10.32). Though the open-loop transfer function $L_1(j\omega)$ has a very high magnitude at this frequency, the vanishing plant damping means that the controller has low gain in relative terms, and thus poor disturbance rejection (cf. Eq. (10.5) and $T_d(j\omega)$ in Table 10.1).

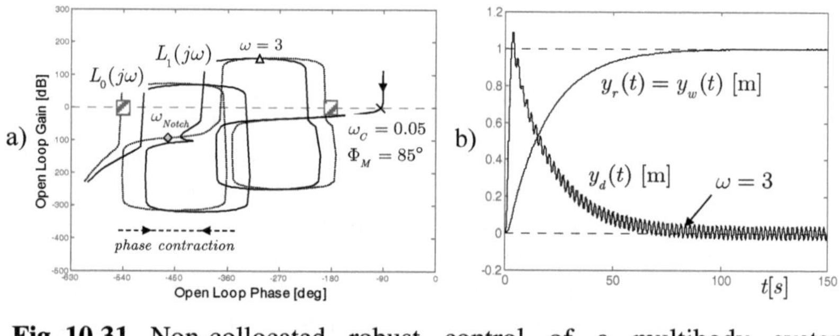

Fig. 10.31. Non-collocated robust control of a multibody system: a) NICHOLS diagram (open-loop, L_0: only PI controller, L_1: PI + Notch), b) step responses (y_r/y_w to reference input r/w, y_d to force disturbance)

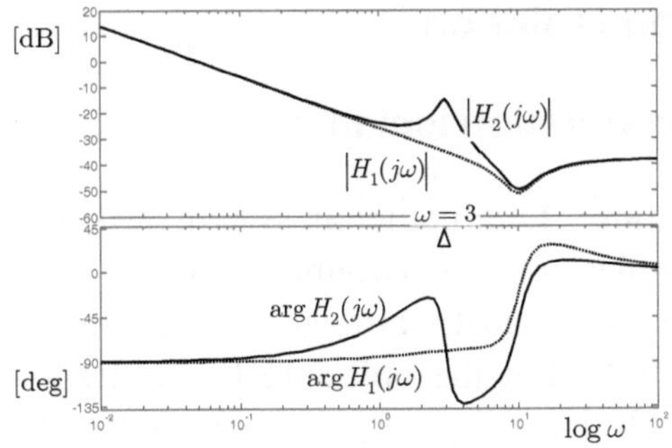

Fig. 10.32. Controller frequency response: H_1: PI + notch, H_2: H1 + anti-notch

Anti-notch filter for disturbance rejection If this disturbance response is unacceptable, then, using an additional *anti-notch filter*, a larger controller gain could be locally brought to bear at this eigenfrequency (+*20 dB* at $\omega = 3$, see Fig. 10.32):

$$H_2(s) = \underbrace{0.05 \frac{[4]}{s}}_{PI} \underbrace{\frac{\{0.2;\ 10\}}{\{1;\ 10\}}}_{Notch} \underbrace{\frac{\{1;\ 3\}}{\{0.1;\ 3\}}}_{Anti-Notch}.$$

Note however, that the phase separation of the anti-notch filter (see also Example 10.7) partially counters the action of the notch filter (Fig. 10.33). Despite this, using the extended controller, there is distinct active damping of the eigenmode at $\omega = 3$, so that all design requirements are fulfilled in a comparatively simple and clear, robust controller. In practice, the effects of parameter variations in individual cases would naturally still need to be investigated.

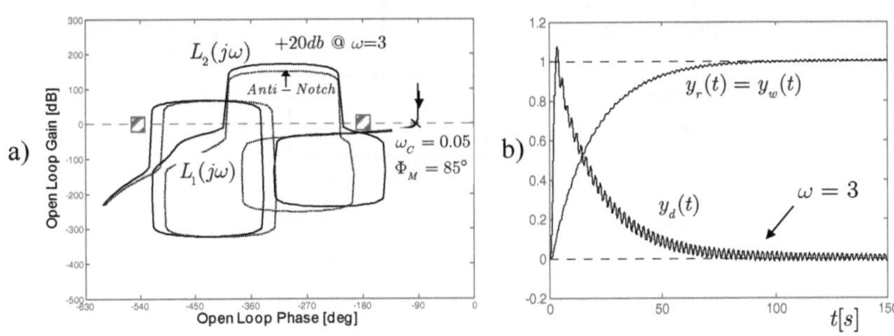

Fig. 10.33. Improved disturbance rejection with additional anti-notch filter: a) NICHOLS diagram (open-loop, L_1: controller H_1, L_2: controller H_2), b) step responses (y_r/y_w of reference input r/w, y_d of force disturbance) ■

10.6 Vibration Isolation

10.6.1 Passive vibration isolation

Vibration isolation One common system task consists of the motion isolation of an object from environmental disturbances (Preumont 2002), (Karnopp 1995). A schematic arrangement for such vibration isolation is shown in Fig. 10.34. Let the object to be isolated—here assumed to be a rigid body with mass m_0—be subject to the following disturbance sources:

- *displacement excitation*: this consists of *imposed motions* $z(t)$ of the base bearing the object. For the case where the mass M_B of the base is significantly larger than the mass of the object (e.g. a building foundation, the ground for seismic studies), exogenous forces acting on the moving base play no role in the motion of the object to be isolated. Similarly, the back effect of the (small) object mass is negligible.
- *force excitation*: this consists of imposed disturbance forces, acting directly on the object, e.g. unbalanced rotating masses (automobile engine, flywheels).

In both cases, the object position y_0 and its time derivatives $\dot{y}_0$, $\ddot{y}_0$ are to be affected as little as possible by the disturbance sources (for example, acceleration is highly relevant for the ride comfort of an automobile).

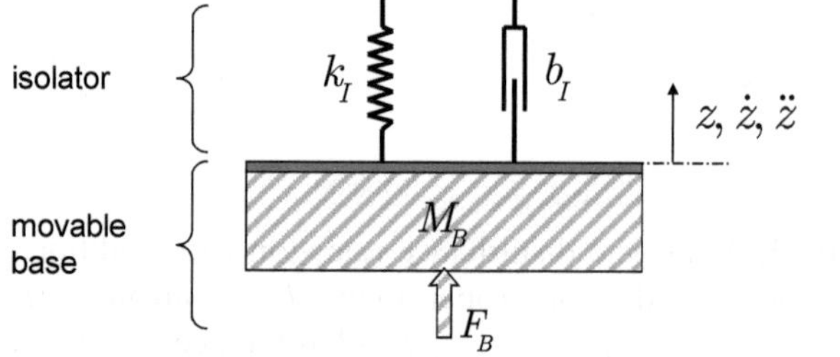

Fig. 10.34. Passive vibration isolator

Passive vibration isolator In the configuration in Fig. 10.34, vibration isolation is achieved with an elastic bearing of stiffness k_I and a velocity-dependent damper b_I (forming a passive spring-damper system).

The action of an isolator is expressed by its so-called *transmissivity*

$$T_{isol}(s) = \frac{L\{y_0(t)\}}{L\{z(t)\}} = \frac{L\{\dot{y}_0(t)\}}{L\{\dot{z}(t)\}} = \frac{L\{\ddot{y}_0(t)\}}{L\{\ddot{z}(t)\}} . \tag{10.30}$$

Eq. (10.30) defines the general relationship between response and excitation for the paired motion quantities of position, velocity, and acceleration.

For the configuration of a passive vibration isolator shown, the resulting transmissivity is

$$T_{isol}(s) = \frac{k_I + b_I s}{k_I + b_I s + m_0 s^2} = \frac{1 + 2d_0 \dfrac{s}{\omega_0}}{1 + 2d_0 \dfrac{s}{\omega_0} + \dfrac{s^2}{\omega_0^{\,2}}} ,$$

$$\text{where } \omega_0 = \sqrt{\frac{k_I}{m_0}}, \quad \frac{2d_0}{\omega_0} = \frac{b_I}{k_I} . \tag{10.31}$$

A fundamentally low-pass behavior can be seen in Eq. (10.31), i.e. disturbances with frequency content sufficiently above the natural frequency ω_0 are damped, while below the natural frequency, no disturbance rejection is possible. One nuisance, however, is presented by the numerator with cutoff frequency $\omega_0 / 2d_0$, which is both differentiating and dependent on the mechanical damping *b*.

This numerator term fundamentally implies a magnitude peak around the resonant frequency, even with arbitrarily high damping *b*, and for high frequencies, a magnitude descent of only *-20 dB/decade*. A greater magnitude descent above the natural frequency (and then also only in a limited frequency band) can only be achieved for very low damping, though at the cost of a significant increase in magnitude near the natural frequency (Fig. 10.35).

Ideally, by eliminating the numerator term, a magnitude descent of *-40 dB/decade* at high frequencies, and an optimal frequency band separation between passband and stopband could be achieved at a damping of $d_0 = 0.7$ (Fig. 10.35). For such an ideal or quasi-ideal configuration, in this section, let the characteristic frequency ω_B be termed the *transmissiv-*

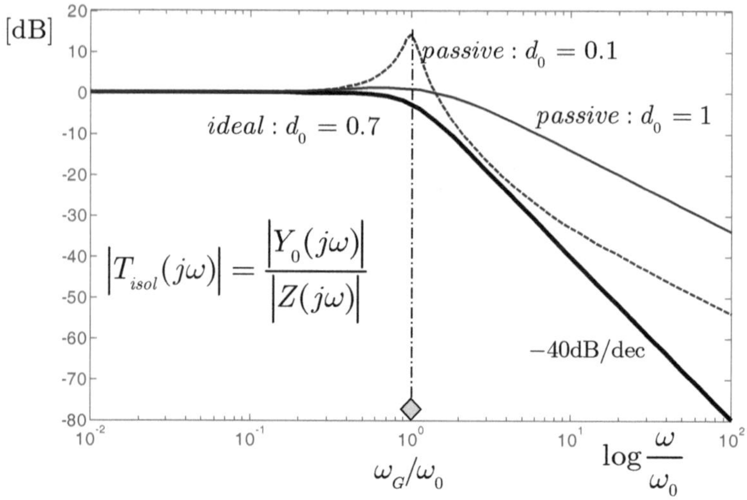

Fig. 10.35. Transmissivity of a passive vibration damper

ity bounding frequency[8]. Such ideal filtering, however, is fundamentally *impossible* using the passive configuration shown, so that designing a passive vibration isolator always requires compromises.

10.6.2 Active vibration isolation: the skyhook principle

Relative velocity The physical explanation for the insufficiencies of a passive vibration damper lies in its principle of operation. The velocity-dependent damping force generally results from the *relative motion* of the two attachment points of the damper. Here, physically, only the relative velocity between the base and object can be used for damping, whereas the motion of the mass is considered relative to inertial space (i.e. the inertial or absolute velocity). Were the damper to be fixed to inertial space, the numerator term in the transmissivity (10.31) would not even appear, and the damper would act as a pure second-order delay (cf. the ideal behavior in Fig. 10.35 and introductory example in Sec. 10.2).

Active isolator: skyhook principle The ideal damping discussed above can be realized by means of the following technique. By measuring the *inertial (absolute)* velocity of the object to be isolated, a force proportional

[8] See Fig. 10.35: at ω_B there is typically a decrease in magnitude of -3 dB indicating the separation of passband and stopband.

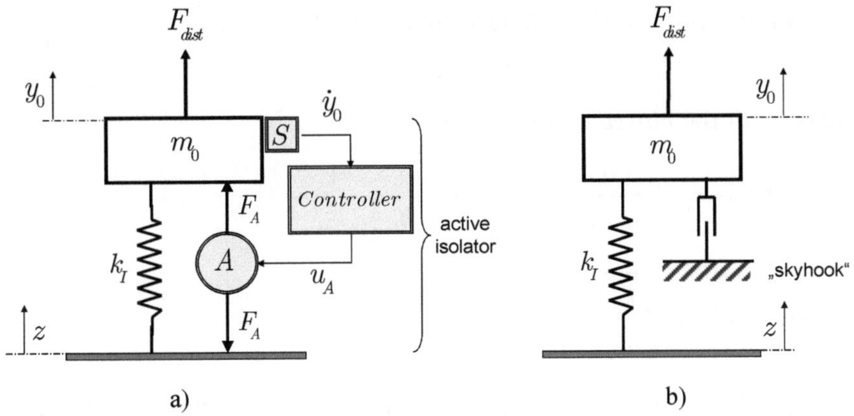

Fig. 10.36. Active vibration damper, skyhook principle: a) inertial velocity feedback, b) equivalent damper configuration

to it can be accurately generated in a suitable actuator and applied to the object. The object is thus virtually "hooked" into inertial space (see Fig. 10.36b) and damped as a function of absolute velocity.

A schematic configuration for a physical realization of this *skyhook principle* is shown in Fig. 10.36a. For velocity sensing, inertial sensors (inertial masses) are suitable. Using a control configuration, a suitable correction signal is generated; the realization of damping forces can occur via electromechanical or hydraulic actuators. Hallmarks of the configuration presented in Fig. 10.36 are that no passive damper is required so that a *soft* spring (with a low eigenfrequency) can be selected for the elastic support, and that the actuator force is applied to the base (foundation). For a sufficiently large foundation mass, there is then also no further back effect.

Velocity feedback

Ideal velocity feedback The dynamic effect of velocity feedback for the skyhook arrangement in Fig. 10.36 can be described—given ideal assumptions (no parasitic dynamics)—by the following mathematical model:

$$\text{MBS:} \qquad Y_0(s) = \frac{1}{k_I + m_0 s^2}\left[F_{dist}(s) + F_A(s) + k_I Z(s)\right],$$

$$\text{Actuator:} \quad F_A(s) = K_A \cdot u_A(s)$$

$$\text{Controller:} \quad u_A(s) = -H(s) \cdot sY_0(s)\,.$$

For purely *proportional feedback* of the absolute velocity

$$H(s) = K_H ,$$

this gives for the *closed control loop*

$$Y_0(s) = \frac{1}{\{\tilde{d}_0;\omega_0\}} Z(s) \ + \ \frac{1}{k_I}\frac{1}{\{\tilde{d}_0;\omega_0\}} F_{dist}(s) \ ,$$

$$\text{where } \omega_0 = \sqrt{\frac{k_I}{m_0}}, \ \ \tilde{d}_0 = \frac{\omega_0}{2k_I} K_A K_H \ \Rightarrow \ \mathrm{K}_H = \tilde{d}_0 \frac{2k_I}{\omega_0 K_A} . \tag{10.32}$$

It can be seen from Eq. (10.32) that with a suitable choice of the feedback gain K_H, arbitrary damping $\tilde{d}_0$—and thus the *ideal transmissivity* of Fig. 10.35—can be achieved. The selection of the eigenfrequency is realized via the support stiffness k_I.

Control loop design for velocity feedback The system dynamics (10.32) derived above hold only for idealized assumptions. In order to estimate the dynamics under realistic conditions, consider the equivalent control loop configuration depicted in Fig. 10.37 and the NICHOLS diagram of the open loop in Fig. 10.38a.

The ideal case of Eq. (10.32) is represented by $L_0(j\omega)$. For positive controller gains, it can be seen that the stability is not limited. However, if additional parasitic dynamics (sensors, actuators, MBS structures) are considered, then stability problems and the limits of purely proportional feedback become clearly apparent. Thus, in such cases, a more complex controller design following the robust control strategies presented in the previous sections is required.

Improved stability margin It is worth noting here that, due to the differentiating nature of the velocity feedback, the NICHOLS curve begins at +90° and is significantly further from the critical -180° stability region than for non-differentiating plants. This implies that in this case, MBS eigenmodes of an attached flexible structure can be shifted (for example, using a PID controller) into the phase-lead stability region (and thus actively damped) over a significantly greater bandwidth than, for comparison, using position feedback. Indeed, in the latter case, the curve starts at 0° (for proportional controllers) or -90° (for integral controllers), so that for higher MBS eigenfrequencies, insufficient positive phase shift is available for phase-lead stabilization. Such a velocity feedback combined with position

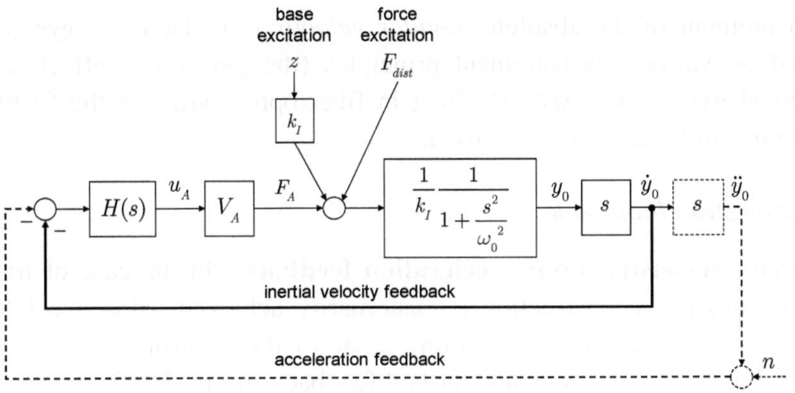

Fig. 10.37. Skyhook principle: block diagram of inertial velocity feedback vs. acceleration feedback (dashed line)

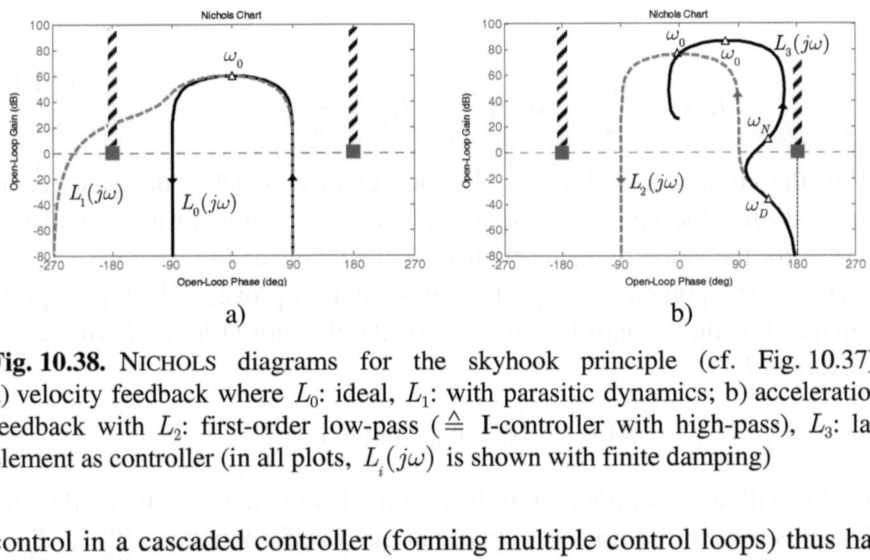

Fig. 10.38. NICHOLS diagrams for the skyhook principle (cf. Fig. 10.37): a) velocity feedback where L_0: ideal, L_1: with parasitic dynamics; b) acceleration feedback with L_2: first-order low-pass ($\triangleq$ I-controller with high-pass), L_3: lag element as controller (in all plots, $L_i(j\omega)$ is shown with finite damping)

control in a cascaded controller (forming multiple control loops) thus has clear dynamic advantages compared to single-loop position control. However, these advantages come at the cost of additional device complexity in the form of a *velocity sensor*.

Absolute velocity measurement A critical aspect for implementation is the physical measurement of absolute velocity, which requires so-called *inertial sensors*. For indirect measurement of the *translational* absolute velocity, motions (displacements) of a test mass which result from inertial forces are measured. This principle is realized in *accelerometers* or, in a specialized feedback structure, as a *geophone* (Preumont 2006). For direct measurement of the absolute *angular velocity* (angular rate), *gyroscopes* based on various measurement principles (the gyroscopic effect in mechanical gyros, the SAGNAC effect in fiber-optic gyros, or the CORIOLIS force in solid-state gyros) are used.

Acceleration feedback

Velocity reconstruction: acceleration feedback In the case of translations, velocity reconstruction is customarily achieved using accelerometers, thus feeding an acceleration signal to the controller. This is thus termed *acceleration feedback* (dashed feedback in Fig. 10.37, implementational configuration in Fig. 10.39).

Stability problem for I-controllers Intuitively, the following obvious controller with *integrator* could now be applied:

$$H(s) = \frac{K_H}{s} \cdots \text{ or } H(s) = K_H \frac{1 + \dfrac{s}{\omega_D}}{s} \cdots . \tag{10.33}$$

With the controller in Eq. (10.33), the velocity signal is de facto reconstructed *after* the integrator. At first glance, this results in the same situation as for the velocity feedback discussed above (see Fig. 10.38a).

However, upon closer inspection, this solution proves to be highly problematic. For the I-controller in Eq. (10.33), the closed-loop *characteristic polynomial*

$$\Delta(s) = s\left(k + K_H s + m s^2\right)$$

results, with an *unstable* root at the origin. However, as can be easily verified, this root $s = 0$ does *not* appear in the transfer functions $T_{\ddot{y}_0/z}$, $T_{u_A/z}$. The reason for this is that it *cancels* an equal term in the numerator (see observability problems in Sec. 10.7).

The stability problem only becomes apparent when considering the transfer functions $T_{u_A/n}$ and $T_{y_0/n}$ from the measurement noise n in the acceleration channel to the control variable u_A or the mass position y_0 (Fig. 10.37), where this unstable pole actually appears. A constant measurement error (offset, bias), as is practically always present, would induce an actuation force and mass displacement growing without bound. Due to the differentiating nature of the system relative to the velocity meas-

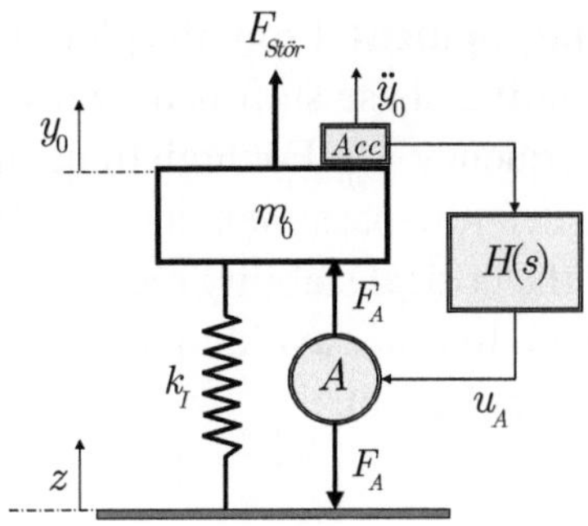

Fig. 10.39. Active vibration isolator with acceleration feedback

urement, these unstable dynamics are not visible in the control loop—this represents a case of hidden *instability* of internal system quantities[9].

Stable acceleration feedback One simple possibility to avoid this instability lies in the use of a *high-pass filter* at the output of the integrator to filter out constant sensor signal components:

$$H(s) = \frac{K_H}{s} \frac{\frac{s}{\omega_{HP}}}{1 + \frac{s}{\omega_{HP}}} = \frac{K_H}{\omega_{HP}} \frac{1}{1 + \frac{s}{\omega_{HP}}}. \tag{10.34}$$

Essentially, this series connection acts as a low-pass filter (here first-order) on the fed-back acceleration signal (Li and Goodall 1999).

An accessible justification for the controller approach in Eq. (10.34) can be found directly in the NICHOLS diagram for acceleration feedback in Fig. 10.38b. In order to avoid stability problems with the critical stability region at +180°, a *negative* phase shift must be applied sufficiently *below* the eigenfrequency ω_0—in the current case, using the low-pass component in Eq. (10.34) (see $L_2(j\omega)$ in Fig. 10.38b). For the ideal case without parasitic dynamics, the control loop is then stable without limitation. There is also no pole-zero cancellation, as can be easily verified. Given parasitic dynamics, however, the same problems as in the velocity feedback case appear (cf. Fig. 10.38a,b).

9 In some representations, this problematic state of affairs is unfortunately obscured, e.g. (Preumont 2002).

Robust stability with lag element From the plot of $L_2(j\omega)$ in Fig. 10.38b, it can be seen that a negative phase shift is only necessary in the frequency range below the eigenfrequency ω_0. For high frequencies, however, the -90° phase shift of a first-order low-pass element is rather troublesome, as the curve is pushed toward the critical stability region at -180° (this is a problem given parasitic dynamics). In order avoid this shift, in place of the low-pass component, the use of a lag element[10]

$$H_{lag}(s) = K_H \frac{1+\frac{s}{\omega_N}}{1+\frac{s}{\omega_D}}, \quad \omega_D < \omega_N, \tag{10.35}$$

suggests itself, for which the phase shift at high frequencies is 0°. This favorable influence is demonstrated in Fig. 10.38b insomuch as the NICHOLS curve in the frequency range of the pole-zero pair $\omega_D - \omega_N$ avoids the critical point $\left(0\,\text{dB}, +180°\right)$ (negative phase shift, bulging to the left) and then once more runs approximately along the +180° line. Of equal importance is the more positive phase at high frequencies, so that for high-frequency MBS eigenmodes, there is significantly higher phase margin available in the presence of parasitic dynamics than for the low-pass controller[11] (see also Example 10.9).

Design considerations for the skyhook principle

Design givens Often, the following task statement is given: there exists a load mass m_0 $[\text{kg}]$ to be isolated, possibly with an attached flexible structure (MBS), and a *bounding frequency* f_B $[\text{Hz}]$, beyond which exciting disturbances $z(t)$ or $F_{dist}(t)$ in Fig. 10.36 are to be suppressed with a magnitude descent of -40 dB/decade.

---

10 Here, however, the numerator is used, inducing a positive phase shift relative to the negative phase shift of the denominator (the low-pass behavior). For very high frequencies relative to its corner frequency, the lag element has almost no phase shift. The lag element thus also represents a PI controller where a low-pass term is employed instead of the integrator).

11 Note: for $\omega \to \infty$, since *degree of the numerator of* H_{lag} = *degree of the denominator of* H_{lag}, the NICHOLS curve $L_3(j\omega)$ has a finite magnitude.

System configuration In addition to the sensors and a suitable actuator, the bearing stiffness k_I and the controller $H(s)$ (its structure and parameters) are to be determined. Fundamentally, the bounding frequency f_B naturally determines the bearing stiffness, and thus the eigenfrequency of the isolator

$$2\pi f_B \approx \omega_0 = \sqrt{k_I/m_0}\,. \tag{10.36}$$

From Eq. (10.36), the order of magnitude of k_I can be estimated. Due to the possibility of electromechanical tuning via the acceleration feedback, the bearing stiffness can, however, be kept variable within certain regions, as will be shown.

A determining factor for design of the controller is the lowest eigenfrequency; this is generally the common mode at $\omega_0 = \sqrt{k_I/m_0}$. From the NICHOLS curve in Fig. 10.38, it is clear that the closed-loop bandwidth—and thus the corner frequency for the transmissivity (desired value of $\omega_B = 2\pi f_B$)—must be lower than the eigenfrequency ω_0 (approximately equal to the frequency at which $L(j\omega)$ cuts the 0 dB line). Thus, in general, $\omega_0 > \omega_B$, so that physically, a scaling

$$\omega_B = \alpha \cdot \omega_0 = \alpha \cdot \sqrt{k_I/m_0}\,,\quad 0.1 \le \alpha \le 0.7 \tag{10.37}$$

can be used. Inside this α-range, for a *given* ω_0 (i.e. fixed k_I, m_0), the corner frequency ω_B of the transmissivity can be *freely* selected using suitable controller parameters, as will be shown below.

Thus, by taking the "electronic" path, compromises in the mechanical configuration—e.g. structural and material bounding conditions on the bearing, variations in the load mass—can be elegantly compensated, as should be expected of a mechatronic design

Skyhook-controller parameterization By using the simple first-order controller structures introduced above, sensible control loop dynamics can be achieved, even for higher-order systems—though a lag controller (10.35) does always exhibit significant advantages compared to a low-pass controller (10.34).

The following *design rules* for setting up linear first-order corrective elements (Janschek 2009) ensure robust stability and, to a good approximation, a transmissivity as in Fig. 10.35 with $d_0 \approx 0.7$ (typically -3 dB magnitude descent at ω_B) for a *lag controller* with a ratio $0.1 \le \omega_B/\omega_0 \le 0.7$:

$$\text{lag element:} \qquad \boxed{H_3(s) = K_H \frac{1+\dfrac{s}{100\,\omega_D}}{1+\dfrac{s}{\omega_D}}} \tag{10.38}$$

with the controller parameters

$$\boxed{K_H = \frac{k_I}{K_A\,\omega_0^{\;2}} \frac{25.8}{\left(\omega_B/\omega_0\right)^{2.7}}, \qquad \omega_D = 0.05\,\omega_0 \left(\omega_B/\omega_0\right)^{1.7}}. \tag{10.39}$$

The comparable, though not recommended, *low-pass controller* is

$$\text{low-pass (I-controller + high-pass):} \quad H_2(s) = K_H \frac{1}{1+\dfrac{s}{\omega_D}}. \tag{10.40}$$

Using the design rules in (10.38), (10.39), a quick estimation of realistically achievable system dynamics can be comfortably made. These rules can be equally applied to higher-order systems, as will be shown in Example 10.9, below.

For further considerations regarding controller optimization, the reader is referred to the applicable literature, e.g. (Li and Goodall 1999).

Dynamic properties of lag controller vs. low-pass controller For comparison, various dynamic properties of a lag and a low-pass controller are shown in Fig. 10.38b and Fig. 10.40 for the case $\alpha = \omega_B/\omega_0 = 0.2$. The two controller variants display a thoroughly similar time response to a disturbance step $z(t) = \sigma(t)$ of the base (Fig. 10.40b). However, according to Fig. 10.38b, the *lag* controller $H_3(s)$ possesses clear advantages in its *stability margin* in the face of parasitic dynamics and variations in the eigenfrequency ω_0 (e.g. due to variation of the load mass m_0) and in the *isolation* from high-frequency excitation (Fig. 10.40a).

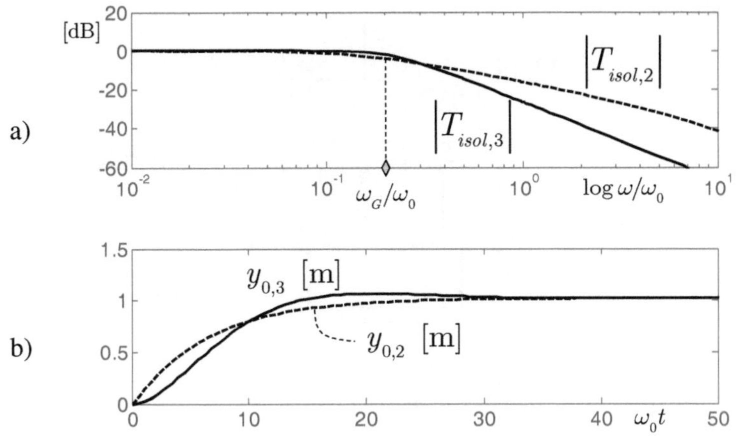

Fig. 10.40. Acceleration feedback—skyhook principle: a) transmissivity, b) step response to excitation $z(t)$ in base; for both figures, subscript 2 = low-pass controller $H_2(s)$, subscript 3 = lag controller $H_3(s)$

Example 10.9 *Skyhook damper for two-mass system.*

System configuration The multibody system shown in Fig. 10.41 is to be decoupled from displacement excitations $z(t)$ of the foundation with a *bounding frequency* $f_B = 1$ Hz using active vibration isolation based on the *skyhook principle*. For this purpose, an accelerometer and a high-bandwidth actuator are available.

The configuration has the following physical parameters:

MBS: $m_0 = 50$ kg, $m_1 = 5$ kg, $k_1 = 40000$ N/m, $b_1 = 0.01$ Ns/m,

Parasitic dynamics: $\tau_{par} = 1/\omega_{par} = 0.025$ s.

Solution approach For the choice of bearing stiffness, let the ratio $\alpha = \omega_B/\omega_0 = 0.5$ be chosen as an example. Eq. (10.37) then gives the natural frequency (lowest eigenfrequency)

$$\omega_0 \approx \sqrt{k_I / (m_0 + m_1)} \approx 2 \cdot 2\pi f_B = 12.1 \text{ rad/s}, \tag{10.41}$$

and further, using Eq. (10.41), the required bearing stiffness $k_I = 8000$ N/m.

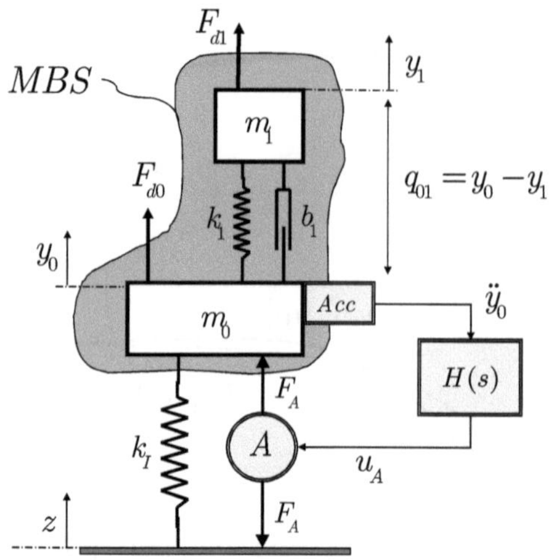

Fig. 10.41. Two-mass system with active vibration isolator: skyhook principle with acceleration feedback

This gives the *design model* for the plant transfer function

$$P(s) = \frac{1}{k_I} \frac{s^2}{\{d_{p0}; \omega_{p0}\}} \frac{\{d_z; \omega_z\}}{\{d_{p1}; \omega_{p1}\}} \frac{1}{[\omega_{par}]}$$

where

$$\omega_{p0} = 12.1 \text{ rad/s}, \ \omega_{p1} = 93.8 \text{ rad/s}, \ \omega_z = 89.4 \text{ rad/s}.$$

For the *controllers*, consider the two following variations with controller parameterizations following Eqs. (10.38) through (10.40):

$$\text{I-controller with high-pass: } H_1(s) = \frac{9160 \cdot 0.19}{\cancel{s}} \frac{\frac{\cancel{s}}{0.19}}{1 + \frac{s}{0.19}}, \quad (10.42)$$

$$\text{lag element: } \quad H_2(s) = 9160 \frac{1 + \frac{s}{19}}{1 + \frac{s}{0.19}}. \quad (10.43)$$

Discussion The *low-pass* controller (10.42) reveals two fundamental weaknesses. First, due to the parasitic dynamics, a stability problem appears in the critical -180° stability region (Fig. 10.42a). Any additional phase lag or spillover eigenmodes would lead to an unstable control loop. Second, the eigenmode ω_{p1} is weakly damped, resulting in a clear resonance in the actual stopband of the transmissivity (Fig. 10.42b). Though this weakly-damped resonance is not visible in the step response Fig. 10.42c due to its small residual, it is all the more evident in the *relative motion* of the two masses (Fig. 10.42d).

A practical solution to both problems is offered by the *lag controller* (10.43). Here, the favorable phase response of the lag element for the eigenmode ω_{p1} can be seen; there is now a comfortable stability margin to the critical stability region at -180°. Due to the more favorable phase and the somewhat larger gain at the eigenfrequency ω_{p1}, this eigenmode is now also significantly better damped, as demonstrated by the relative motion in Fig. 10.42d.

Otherwise, the predictions for the dynamics of this higher-order system agree quite well with the predictions of the design rules. In particular, the transmissivity demonstrates the desired behavior (Fig. 10.42b).

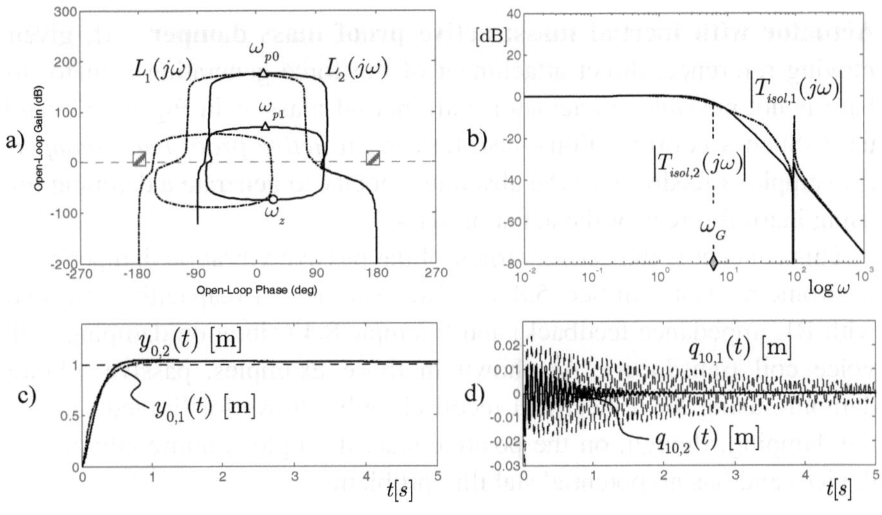

Fig. 10.42. System dynamics of the two-mass system with active vibration isolator (skyhook principle): a) NICHOLS diagram (open-loop), b) transmissivity , c) step response to excitation of base, d) step response in the *relative position* $q_{10} = q_1 - q_0$ to excitation of base $z(t)$; for all figures, subscript 1 = low-pass controller $H_1(s)$, subscript 2 = lag controller $H_2(s)$

■

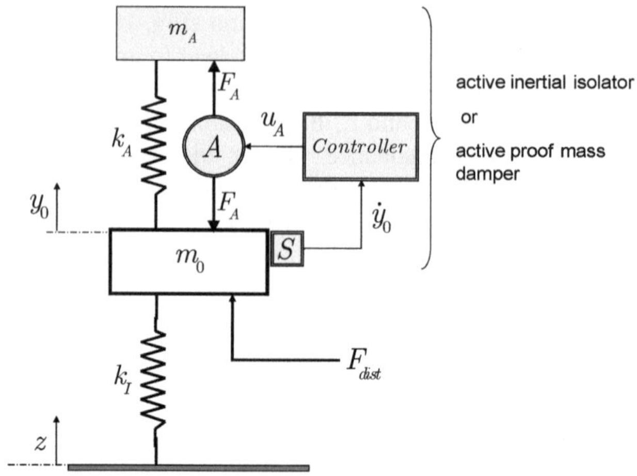

Fig. 10.43. Active proof mass damper

10.6.3 Active proof mass damper

Actuator with inertial mass: active proof mass damper If, given a moving reference, direct attachment of the force-generating actuator to a base is not possible, an actuator with inertial mass as in Fig. 10.43 can be used. Such a configuration—also termed an *active proof mass damper*—also employs feedback of the absolute velocity to generate a damping force using inertial forces of the actuator mass.

This concept is the active variant of the passive vibration damper or mechatronic resonator in Sec. 5.8, cf. also Example 5.1 (capacitive transducer with RL impedance feedback) and Example 8.4 (vibration damping with a voice coil transducer). As shown in those examples, passive vibration damping using local electrical feedback only allows for limited setting of the damping, though, on the positive side, it employs minimally complex devices and has no potential stability problems.

10.7 Observability and Controllability Aspects

10.7.1 General properties

The well-studied control theoretical concepts of *state controllability* and *state observability* have an important and simultaneously illustrative significance for multibody systems. Specifically, in certain cases, some eigenmodes cannot be externally affected, i.e. using the available sensor-actuator arrangement. In such cases, natural oscillations which have been excited (e.g. by external disturbance forces) cannot be actively damped. Such eigenmotions then take place "out of sight" of the control loop and only decay with their own mechanical damping. The considerations in this section are intended to somewhat further illuminate and give access to modeling these processes.

Multibody system in state-space representation A multibody system with n degrees of freedom

$$\mathbf{M}\,\ddot{\mathbf{y}} + \mathbf{D}\,\dot{\mathbf{y}} + \mathbf{K}\,\mathbf{y} = \mathbf{P}\,\mathbf{f} \tag{10.44}$$

with the coordinates $\mathbf{y} \in \mathbb{R}^{n}$, mass matrix $\mathbf{M} \in \mathbb{R}^{n \times n}$, damping matrix $\mathbf{D} \in \mathbb{R}^{n \times n}$, stiffness matrix $\mathbf{K} \in \mathbb{R}^{n \times n}$, generalized forces $\mathbf{f} \in \mathbb{R}^{m}$, and input force matrix $\mathbf{P} \in \mathbb{R}^{n \times m}$ can, by defining the *state vector*[12]

$$\mathbf{x} := \begin{pmatrix} \mathbf{y} & \dot{\mathbf{y}} \end{pmatrix}^{T}, \quad \mathbf{x} \in \mathbb{R}^{2n}, \tag{10.45}$$

be transformed into the following *MBS state-space representation*:

$$\dot{\mathbf{x}} = \begin{pmatrix} \mathbf{0} & \mathbf{I} \\ \text{-}\mathbf{M}^{-1}\mathbf{K} & \text{-}\mathbf{M}^{-1}\mathbf{D} \end{pmatrix} \mathbf{x} + \begin{pmatrix} \mathbf{0} \\ \mathbf{M}^{-1}\mathbf{P} \end{pmatrix} \mathbf{f} = \mathbf{A}_{MBS} \cdot \mathbf{x} + \mathbf{B}_{MBS} \cdot \mathbf{f}\,, \tag{10.46}$$

where $\mathbf{A}_{MBS} \in \mathbb{R}^{2n \times 2n}$ represents the system matrix, and $\mathbf{B}_{MBS} \in \mathbb{R}^{2n \times n}$, the input matrix.

Defining *measurements* $\mathbf{z} \in \mathbb{R}^{m}$ as a linear combination of the states gives the measurement equation

$$\mathbf{z} = \mathbf{C}_{MBS} \cdot \mathbf{x} = \begin{pmatrix} \mathbf{C}_{p} & \mathbf{C}_{v} \end{pmatrix} \mathbf{x} \tag{10.47}$$

with the measurement matrix (output matrix) $\mathbf{C}_{MBS} \in \mathbb{R}^{m \times 2n}$, and the weights $\mathbf{C}_{p}$, $\mathbf{C}_{v}$ for the positions and velocity sensors, respectively.

---

12 As is well known, the definition of states is not unique, i.e. infinitely many state representations can be created from linear combinations of available states. In motion processes, however, the definition of *position* and *velocity coordinates* as the state turns out to be particularly appropriate.

Without loss of generality, no further actuator or sensor dynamics are taken into consideration in the system equations (10.46) and (10.47), as these generally do not affect controllability or observability.

Definition 10.3. *State controllability* The multibody system (10.46) is termed *state-controllable* if the state vector $\mathbf{x}$ can be taken to any desired final state $\mathbf{x}(t_1)$ from an arbitrary initial state $\mathbf{x}(0)$ in some finite time interval $(0, t_1)$ using a bounded input $\mathbf{f}(t)$.

Criterion for state controllability To test for state controllability, the well-known *KALMAN criterion* on the *KALMAN controllability matrix* $\mathbf{S}$ can be employed. The multibody system (10.46) is state-controllable if and only if

$$\mathbf{S} = \left(\mathbf{B}_{MBS}, \mathbf{A}_{MBS}\mathbf{B}_{MBS}, \ldots, \mathbf{A}_{MBS}^{2n-1}\mathbf{B}_{MBS}\right) \in \mathbb{R}^{2n \times 2n^2}$$

has *full row rank*, i.e. (10.48)

$$\operatorname{rank}\mathbf{S} = \operatorname{rank}\left(\mathbf{B}_{MBS}, \mathbf{A}_{MBS}\mathbf{B}_{MBS}, \ldots, \mathbf{A}_{MBS}^{2n-1}\mathbf{B}_{MBS}\right) = 2n\,.$$

Definition 10.4. *State observability* The multibody system (10.46) with measurement equation (10.47) is termed *state-observable* if knowing the evolution of the output vector $\mathbf{z}(t)$ over some finite time interval $0 \le t \le t_1$ allows the initial state $\mathbf{x}(0)$ to be computed.

Criterion for state observability To test for state observability, the well-known *KALMAN criterion* on the *KALMAN observability matrix* $\mathbf{W}$ can be employed. The multibody system (10.46) is state-observable if and only if

$$\mathbf{W} = \begin{pmatrix} \mathbf{C}_{MBS} \\ \mathbf{C}_{MBS}\mathbf{A}_{MBS} \\ \vdots \\ \mathbf{C}_{MBS}\mathbf{A}_{MBS}^{2n-1} \end{pmatrix} \in \mathbb{R}^{m \cdot 2n \times 2n}$$

has *full column rank*, i.e. (10.49)

$$\operatorname{rank}\mathbf{W} = \operatorname{rank}\begin{pmatrix} \mathbf{C}_{MBS} \\ \mathbf{C}_{MBS}\mathbf{A}_{MBS} \\ \vdots \\ \mathbf{C}_{MBS}\mathbf{A}_{MBS}^{2n-1} \end{pmatrix} = 2n\,.$$

Decomposition of the state space If the controllability (10.48) or observability (10.49) conditions are violated, the state space spanned by the states (10.45) can be partitioned into complementary subspaces (Reinschke 2006):

- the controllable and observable subspace $\mathfrak{R}_A$,
- the controllable and unobservable subspace $\mathfrak{R}_B$,
- the uncontrollable and observable subspace $\mathfrak{R}_C$,
- the uncontrollable and unobservable subspace $\mathfrak{R}_D$.

Using a suitable regular state transformation

$$\mathbf{x} = \mathbf{T}\tilde{\mathbf{x}} = \mathbf{T}\begin{pmatrix}\tilde{\mathbf{x}}_A \\ \tilde{\mathbf{x}}_B \\ \tilde{\mathbf{x}}_C \\ \tilde{\mathbf{x}}_D\end{pmatrix}, \tag{10.50}$$

the states can also be clearly partitioned according to the subspaces presented above (Reinschke 2006). The corresponding state-space model then has the form

$$\begin{aligned}\dot{\tilde{\mathbf{x}}} &= \tilde{\mathbf{A}}\tilde{\mathbf{x}} + \tilde{\mathbf{B}}\mathbf{f} \\ \mathbf{z} &= \tilde{\mathbf{C}}\tilde{\mathbf{x}}\ .\end{aligned} \tag{10.51}$$

It is well known that both the eigenvalues of the state-space model (here, the MBS eigenfrequencies) and the input/output dynamics are invariant to a regular state transformation (10.50).

Transfer matrix The only subspace accessible for control with sensor-actuator pairs proves to be the both controllable *and* observable subspace $\mathfrak{R}_A$. This can be seen clearly by computing the transfer matrix $\mathbf{G}_{\mathbf{z}/\mathbf{f}}(s)$ of the multibody system. Applying the partitioned state-space model (10.51) gives[13]

[13] The matrices $\tilde{\mathbf{A}}_i$, $\tilde{\mathbf{B}}_i$, $\tilde{\mathbf{C}}_i$, $i = A, B, C, D$ are obtained as expected from the partition.

$$\mathbf{G}_{\mathbf{z}/\mathbf{u}}(s) = \tilde{\mathbf{C}} \cdot \left(s\mathbf{I}_{2n} - \tilde{\mathbf{A}}\right)^{-1} \cdot \tilde{\mathbf{B}} =$$

$$\left(\tilde{\mathbf{C}}_A \; \mathbf{0} \; \tilde{\mathbf{C}}_C \; \mathbf{0}\right) \begin{pmatrix} \left(s\mathbf{I}_{nA} - \tilde{\mathbf{A}}_A\right)^{-1} & * & * & * \\ \mathbf{0} & \left(s\mathbf{I}_{nB} - \tilde{\mathbf{A}}_B\right)^{-1} & * & * \\ \mathbf{0} & \mathbf{0} & \left(s\mathbf{I}_{nC} - \tilde{\mathbf{A}}_C\right)^{-1} & * \\ \mathbf{0} & \mathbf{0} & \mathbf{0} & \left(s\mathbf{I}_{nD} - \tilde{\mathbf{A}}_D\right)^{-1} \end{pmatrix} \begin{pmatrix} \tilde{\mathbf{B}}_A \\ \tilde{\mathbf{B}}_B \\ \mathbf{0} \\ \mathbf{0} \end{pmatrix}$$

$$\Rightarrow \quad \mathbf{G}_{\mathbf{z}/\mathbf{u}}(s) = \tilde{\mathbf{C}}_A \cdot \left(s\mathbf{I}_{nA} - \tilde{\mathbf{A}}_A\right)^{-1} \cdot \tilde{\mathbf{B}}_A . \qquad (10.52)$$

Thus, only those states belonging to the both controllable and observable subspace $\Re_A$ are still visible in the transfer dynamics. For a multibody system, this would imply that not all eigenmodes are visible in the transfer function between a sensor-actuator pair. The number of complex conjugate pairs of poles, and thus the order of the transfer function, would therefore be less than the number of degrees of freedom n—in particular, equal to $n_A/2$ if n_A represents the dimension of the controllable and observable subspace $\Re_A$. The eigenmodes belonging to the complementary subspaces $\Re_B$, $\Re_C$, $\Re_D$ are *not visible* in the corresponding transfer function, and accordingly *cannot be affected* by the associated sensor-actuator pair[14]. Mathematically, this reduction in order of the transfer function results from a cancellation of the uncontrollable or unobservable complex conjugate pairs of poles by corresponding zeros.

10.7.2 MBS control in relative coordinates

Application One potential case giving rise to controllability and observability problems is in the control of multibody systems in relative coordinates. This is always the case if the relative positions of two mass elements are used as measured or controlled variables, or if an actuator is connected to two moving mass elements.

---

14 The state transformation (10.50) does not need to be explicitly carried out; this occurred here simply for demonstration purposes. When forming the transfer matrix, only the *observable and controllable* subsystems are automatically retained.

Example two-mass oscillator As an illustrative example of control in relative coordinates, consider the elastically suspended two-mass oscillator with $n = 2$ degrees of freedom depicted in Fig. 10.44. In the configuration shown, the separation of the two masses is to be controlled via a relative measurement $z = y_2 - y_1$ and an actuator connected between these two masses having the actuation force F_A.

With the states $\mathbf{x} = \begin{pmatrix} y_1 & y_2 & \dot{y}_1 & \dot{y}_2 \end{pmatrix}^T$, the following state-space model results for the MBS plant:

$$\begin{aligned} \dot{\mathbf{x}} &= \mathbf{A}\mathbf{x} + \mathbf{b}F_A \\ z &= \mathbf{c}^T\mathbf{x} \end{aligned} \qquad (10.53)$$

$$\text{where } \mathbf{A} = \begin{pmatrix} 0 & 0 & 1 & 0 \\ 0 & 0 & 0 & 1 \\ -\dfrac{k_1 + k_2}{m_1} & \dfrac{k_2}{m_1} & 0 & 0 \\ \dfrac{k_2}{m_1} & -\dfrac{k_2 + k_3}{m_1} & 0 & 0 \end{pmatrix}, \quad \mathbf{b} = \begin{pmatrix} 0 \\ 0 \\ -\dfrac{1}{m_1} \\ \dfrac{1}{m_2} \end{pmatrix}, \quad \mathbf{c} = \begin{pmatrix} -1 \\ 1 \\ 0 \\ 0 \end{pmatrix}.$$

Observability analysis The *KALMAN observability matrix* (10.49) is the determining factor in the observability of the MBS plant (10.53). In the case considered here, this must be composed for the output $z = y_2 - y_1$, i.e.

$$\mathbf{W} = \begin{pmatrix} \mathbf{c}^T \\ \mathbf{c}^T\mathbf{A} \\ \mathbf{c}^T\mathbf{A}^2 \\ \mathbf{c}^T\mathbf{A}^3 \end{pmatrix} =$$

$$\begin{pmatrix} -1 & 1 & 0 & 0 \\ 0 & 0 & -1 & 1 \\ \dfrac{k_1}{m_1} + \dfrac{k_2}{m_1} + \dfrac{k_2}{m_2} & -\dfrac{k_3}{m_2} - \dfrac{k_2}{m_1} - \dfrac{k_2}{m_2} & 0 & 0 \\ 0 & 0 & \dfrac{k_1}{m_1} + \dfrac{k_2}{m_1} + \dfrac{k_2}{m_2} & -\dfrac{k_3}{m_2} - \dfrac{k_2}{m_1} - \dfrac{k_2}{m_2} \end{pmatrix} \qquad (10.54)$$

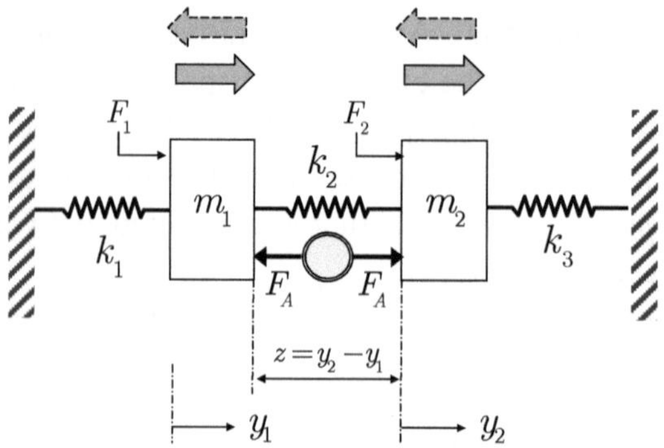

Fig. 10.44. Elastically suspended two-mass oscillator

If can be seen from (10.54) that in the case

$$\frac{k_1}{m_1} = \frac{k_3}{m_2}, \tag{10.55}$$

the third and fourth row of the matrix **W** become linearly dependent on the first and second row, respectively, and thus a total rank deficiency of two appears, i.e.

$$\operatorname{rank} \mathbf{W} = 2,$$

violating the observability condition.

Controllability analysis Equivalently, it can be verified that in this case, the *controllability condition* (10.48) is also violated; the controllability matrix also has a rank deficiency of two, i.e.

$$\operatorname{rank} \mathbf{S} = 2.$$

Physical interpretation: rigid-body eigenmode The lack of observability when condition (10.55) holds can be very clearly recognized in the eigenvector analysis of the algebraic MBS problem (cf. Eq. (10.44) and Ch. 4)

$$\left(-\omega^2 \cdot \mathbf{M} + \mathbf{K}\right) \cdot \mathbf{v} = \mathbf{0}. \tag{10.56}$$

For the case of a *rigid-body eigenmode*, elements of the associated eigenvector are equal, i.e. the eigenvector has the form

$$\mathbf{v}_{\text{rigid-body}} = \begin{pmatrix} \alpha & \alpha & \dots & \alpha \end{pmatrix}^T. \tag{10.57}$$

The associated eigenmotion with eigenfrequency $\omega_{\text{rigid-body}}$ is thus characterized by the fact that all mass elements oscillate in phase with the same amplitude, so that the relative motion between all masses is zero (see the motion indicated in Fig. 10.44). Thus, this eigenmode cannot be detected by measuring relative motion quantities.

It is easy to verify that in the two-mass oscillator under consideration, precisely for condition (10.55), there is a *rigid-body eigenmode* with the eigenfrequency

$$\omega_{\text{rigid-body}} = \sqrt{\frac{k_1}{m_1}} = \sqrt{\frac{k_3}{m_2}}.$$

This also clearly explains the unobservability of the system.

The *uncontrollability* can be explained in a similar manner. In this case, due to the anti-symmetric force generation of F_A, no in-phase control forces can be applied to the masses at the rigid body frequency to suppress rigid body motion.

In-phase eigenmode components The dynamics discussed above also apply in general when a relative measurement takes place between MBS coordinates which have *equal elements* in an *eigenvector* of the algebraic MBS (10.56). If, for an eigenfrequency ω_i with eigenvector

$$\mathbf{v}_i = \begin{pmatrix} \ldots & \underset{j}{\alpha} & \ldots & \underset{k}{\alpha} & \ldots \end{pmatrix}^T,$$

it holds for some two components that $v_{i,j} = v_{i,k}$, then this eigenfrequency is *unobservable* given relative measurement of the associated coordinates $(y_j - y_k)$ and *uncontrollable* given relative actuation of the associated mass elements (m_j, m_k).

> **Uncontrollability and unobservability for control in relative coordinates** *Rigid-body eigenmodes* or *in-phase eigenmode components* are neither controllable nor observable via measurement or actuation of relative coordinates of the associated components.

Transfer dynamics In the case of uncontrollability of unobservability, the order of the MBS transfer function between measurement and actuation locations is less than the order of the multibody system by at least two. This reduced order comes about due to a cancellation of the unobservable and uncontrollable modes with complex conjugate numerator zeros.

For the case of the two-mass oscillator in Fig. 10.44, the transfer functions from the forces F_1, F_2 acting on the inertial space or the relative force F_A to the relative position $z = y_2 - y_1$ can be represented as follows:

$$Z(s) = K_1 \frac{\{\omega_1\}}{\Delta(s)} F_1(s) + K_2 \frac{\{\omega_2\}}{\Delta(s)} F_2(s) + K_A \frac{\{\omega_A\}}{\Delta(s)} F_A(s),$$

$$\omega_1 = \sqrt{\frac{k_3}{m_2}}, \quad \omega_2 = \sqrt{\frac{k_1}{m_1}}, \quad \omega_A = \sqrt{\frac{k_1 + k_3}{m_1 + m_2}}, \tag{10.58}$$

$$\Delta(s) = \frac{m_1 m_2}{\tilde{k}} s^4 + \frac{m_1(k_2 + k_3) + m_2(k_1 + k_2)}{\tilde{k}} s^2 + 1 = \{\omega_{p1}\}\{\omega_{p2}\}$$

$$\text{where } \tilde{k} = k_1 k_2 + k_1 k_3 + k_2 k_3 \,.$$

It can be seen from Eq. (10.58) that when the rigid-body eigenmode condition (10.55) holds, the zeros of all three transfer functions (for absolute as well as relative force application) take on the same value. For this case, it is also easy to verify that one of the two eigenfrequencies ω_{p1} or ω_{p2} also possesses this value, and thus the expected pole-zero cancellation appears. The common mode is no longer visible to the controller, and cannot be affected by the relative force F_A nor the absolute forces F_1, F_2. If the rigid body condition (10.55) is *not* met, then numerator and denominator of all three transfer functions share no common factors, leaving both eigenfrequencies ω_{p1}, ω_{p2} visible in the transfer dynamics. In particular, active damping of both eigenmodes using control in relative coordinates is then also possible. This will be illustrated in the following example.

Example 10.10 *Control of a two-mass oscillator in relative coordinates.*

System configuration Consider the two-mass oscillator configuration in Fig. 10.44 with the specific parameters:

$$m_1 = 10 \text{ kg}, \ m_2 = 2 \text{ kg}, \ k_1 = 1000 \text{ N/m}, \ k_2 = 800 \text{ N/m}\,.$$

Case A: $k_3 = 200$ N/m ➔ $P_1(s) = \dfrac{Z(s)}{F_A(s)} = 10^{-3} \dfrac{\cancel{\{0;10\}}}{\cancel{\{0;10\}}\,\{0;24,1\}}$

Case B: $k_3 = 1000$N/m ➔ $P_2(s) = \dfrac{Z(s)}{F_A(s)} = 7.7 \cdot 10^{-4} \dfrac{\{0;12.9\}}{\{0;11.7\}\{0;30.7\}}$.

In Case A, there is a rigid-body eigenmode with the rigid-body eigenfrequency $\omega_{\text{rigid-body}} = 10$, creating a pole-zero cancellation. In Case B, controllability and observability are fully preserved.

The results for control in relative coordinates including *parasitic dynamics* with the time constant $T_{par} = 1/\omega_{par} = 0.01\,\text{sec}$, and a robust *PID controller*

$$R_{PID} = 700\frac{\{1;5\}}{s\,[50]}$$

are presented in Fig. 10.45.

Discussion of the system dynamics Initial inspection of the NICHOLS diagram in Fig. 10.45a shows the robust PID controller apparently guaranteeing robust stability for both Cases A and B. In fact, however, for Case A, dangerously unstable dynamics are present.

These can be seen in the disturbance force excitation of Mass 1. Fig. 10.45b depicts the response of the position coordinate y_1 to a disturbance step. In *Case A*, the common mode is naturally also excited, but oscillates undamped (or with its own mechanical damping), as it is not visible to the control loop ($y_{1,A}(t)$). If the disturbance force contains harmonic components at the unobservable eigenfrequency, then the multibody system will mechanically resonate, without this being visible to the controller. In *Case B*, on the other hand, both eigenmodes are controllable and observable, and are thus actively damped ($y_{1,B}(t)$).

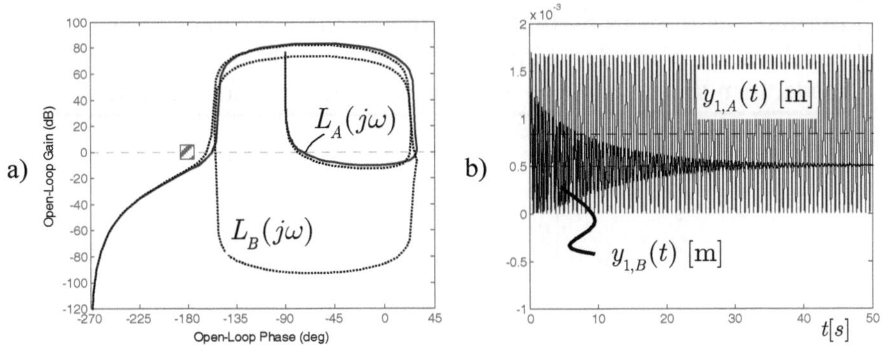

Fig. 10.45. Control of a two-mass oscillator in relative coordinates: a) NICHOLS diagram (open-loop), b) step response $y_1(t)$ to a disturbance force step F_1; subscript A: unobservable common mode, subscript B: observable common mode.

■

10.7.3 Measurement and actuation locations at oscillation nodes

A careful choice of measurement and actuation locations is particularly important for higher-order or infinite-dimensional multibody systems. If a measurement location is placed at an *oscillation node*, then the local displacement is equal to zero, and cannot be detected by a motion sensor (Fig. 10.46). Control theoretically, this case again presents an *unobservable system*. Equivalently, the system is *uncontrollable* if an actuator is placed at the oscillation node. Due to the virtual fixture at the oscillation node, the mass element can not be influenced in this motion mode.

Both effects are again expressed in a reduced transfer function, where the corresponding eigenmode is not visible due to the pole-zero cancellation. Mathematically, in a multibody system with lumped parameters, this can be very easily seen in the eigenvectors of the algebraic MBS eigenvalue problem (10.56).

If, for an eigenfrequency ω_i with the eigenvector

$$\mathbf{v}_i = \begin{pmatrix} \dots & \underset{j}{0} & \dots \end{pmatrix}^T ,$$

it holds for a component that $v_{i,j} = 0$, then this eigenfrequency is *unobservable* via measurement of the associated coordinate y_j and *uncontrollable* via actuation at the associated mass element m_j.

> **Uncontrollability and unobservability for measurement and actuation locations in oscillation nodes** If the measurement or actuation location coincides with the oscillation node of an MBS eigenmode, then this eigenmode is uncontrollable and unobservable.

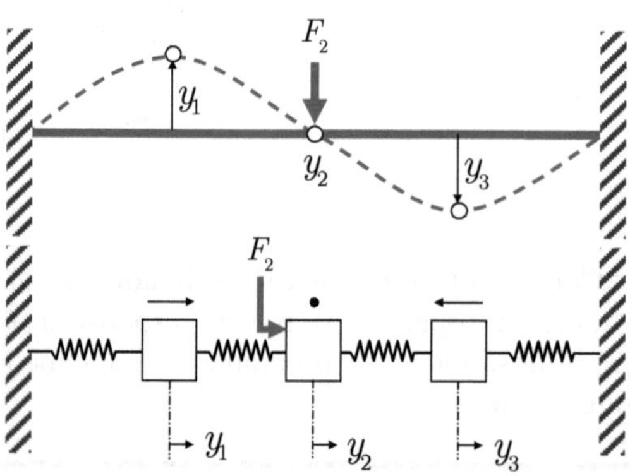

Fig. 10.46. Eigenmode of a multibody system with oscillation node

10.8 Digital Control

10.8.1 General design process

The implementation of control for a mechatronic system generally takes place using digital means with a microprocessor as the core element. As a result, all relevant system quantities must be modeled as discrete-time variables, and treated correspondingly in the design model of the control loop. If the description using sampled signals introduced in Ch. 9 is employed for this purpose, then all previously discussed principles for controller design can also be straightforwardly applied with a few extensions for digital control. In addition, due to the chosen frequency-domain description, a seamless transition between continuous and digital solutions is available; the effect of digital signal processing is immediately and maximally visible in the frequency responses[15].

Design model Let the control loop configuration depicted in Fig. 10.47 serve here as the design model. In addition to the system components of actuator, multibody system, sensors, and controller considered so far, the analog/digital interfaces and the obligatory anti-aliasing filter can be seen.

The following two system transfer functions relevant to design can be abstracted from this model:

- the discrete MBS plant $P^*(s)$,
- the discrete controller $H^*(s)$.

Note that the discrete *MBS plant* $P^*(s)$ comprises all continuous-time modeled system elements outside of the microprocessor, in particular, also the anti-aliasing filter. Let these continuous plant dynamics be encapsulated in the transfer function $P(s)$.

---

15 In the estimation of the author, only the frequency-domain description offers this clear approach, which is why this description is regarded here as the optimal approach for manually viable and mentally penetrable system designs. The basic solutions found in this way can then be employed as eminently suitable, feasible initial solutions for more strongly mathematically formalized design and optimization processes in the frequency and time domains (including state-space methods) (Zhou u. Doyle 1998).

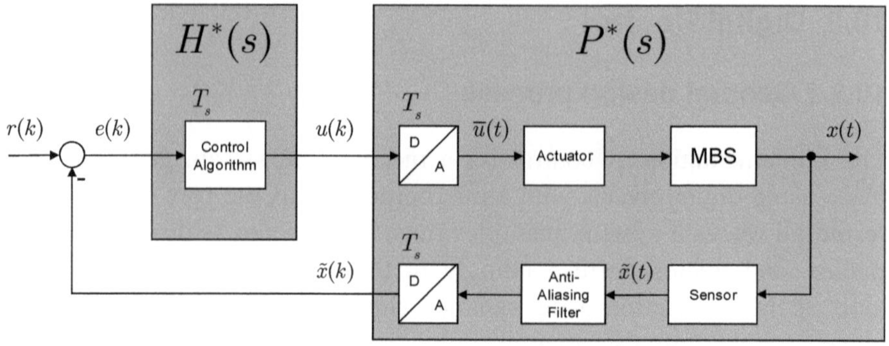

Fig. 10.47. Digital control loop with mechatronic system components

As previously discussed, assuming a *zero-order hold* having transfer function $H_0(s)$ (realized in the digital-to-analog converter), the discrete transfer function $P^*(s)$ is then (cf. Ch. 9)

$$P*(s) = \frac{1}{T_s}\sum_{n=-\infty}^{+\infty} H_0(s + jn\omega_s)\cdot P(s + jn\omega_s), \tag{10.59}$$

where T_s is the *sampling period*, and $\omega_s = 2\pi/T_s$, the *sampling frequency*.

The *discrete controller* $H^*(s)$ represents a linear difference equation implemented in the microprocessor, which can in turn be represented as a z-transfer function $H_D(z)$. The following well-known bijective correspondence exists between these two representations:

$$H_D(z) = H^*(s)\,\Big|_{s=\frac{2}{T_s}\frac{z-1}{z+1}} \quad \text{and} \quad H^*(s) = H_D(z)\,\Big|_{z=\frac{1+\frac{T_s}{2}s}{1-\frac{T_s}{2}s}}. \tag{10.60}$$

The *frequency responses* of the design transfer functions are obtained from the substitutions

$$P^*(j\omega) = P^*(s)\,|_{s=j\omega},$$

$$H^*(j\omega) = H^*(s)\,|_{s=j\omega}, \quad H_D(e^{j\omega T_s}) = H_D(z)\,|_{z=e^{j\omega T_s}} \tag{10.61}$$

where $0 \le \omega \le \omega_s/2$.

Design procedure The frequency response $P^*(j\omega)$ fundamentally exhibits the same MBS properties as its continuous counterpart. The characteristic complex conjugate pole-zero pairs are essentially unchanged, so that all discussed robust control strategies employing NICHOLS diagrams can be applied without modification.

Design steps for digital control

1. Calculate $P^*(j\omega)$: (a) *exactly* via Eq. (10.59), or (b) *approximately* via $P^*(j\omega) \approx P(j\omega)\cdot e^{-j\omega\frac{T_s}{2}}$, $0 \le \omega \le \omega_s/4$.
2. Determine a suitable *controller* $H^*(s)$ by shaping the frequency response $L^*(j\omega) = H^*(j\omega)\, P^*(j\omega)$ in the NICHOLS diagram using the robust control strategies of Secs. 10.5 through 10.7. A continuous $H(s)$ can be used for approximate calculations if the critical frequency ranges are sufficiently less than $\omega_s/4$.
3. Calculate the *controller difference equation* from $H(z)$ and Eq. (10.60)[16].
4. *Verify* the design via *simulation* of the closed control loop in the time domain with $H(z)$ from Step 3 and the MBS plant using one of the following options: (a) *discrete-time*: $P^*(s)$ is represented as a *difference equation* with sampling period T_s[17], (b) *continuous-time*: $P(s)$ is simulated via *numerical integration* with sufficiently small step size[18], (c) *continuous-time*: $P(s)$ is simulated via the *transition matrix* with an adaptive step size[19].

---

16 In MATLAB, this is accomplished with the transformation option *"Tustin"*.

17 In MATLAB, this is accomplished with the transformation option *"ZOH"* for $P(s)$.

18 Note the fundamental problems existing for explicit integration methods when applied to marginally stable systems (Sec. 3.3). Such cases occur here when the eigenmodes are very weakly damped or undamped—it is precisely in *worst-case* investigations that such marginal cases are intentionally played out.

19 Cf. Sec. 3.4. Particularly for higher-order systems having weakly-damped eigenmodes, this allows for the numerically most stable and efficient simulation solution.

Practice-based procedure In the experience of the author, in the course of controller design, it pays to set up an initial conception using a continuous system, if necessary including the delay term of Step 1(b). This already allows the *best-possible* performance to be rather realistically estimated. In comparison to the continuous case, less favorable dynamic conditions always result for the discrete-time case, due to the additional anti-aliasing filter and the mirror frequencies of $P(s)$. These can be compensated for to a certain degree given some skill and intuition in shaping the frequency response. However, in all cases, there will be trade-offs and compromises in the control performance (bandwidths, loop gain, and stability margins).

10.8.2 Rigid-body-dominated systems

System characteristics Let the term rigid-body-dominated system apply to such multibody systems in which—in the frequency range of interest—there is only one eigenmode or a closely-spaced cluster of collocated eigenmodes. These eigenmodes can then be easily placed in the phase-lead stability region and the sampling time can be appropriately selected such that $\omega_{p,i} \ll \omega_s/2$ holds for all eigenfrequencies. In this case, near the crossover frequency ω_C (which determines stability), $P^*(j\omega)$ and thus also $L^*(j\omega)$ will display the behavior shown in Fig. 10.48. The eigenmodes will thus all be located below the crossover frequency (i.e. $\omega_{p,i} < \omega_D$), above which there is a smooth frequency response $P^*(j\omega)$ with low-pass behavior not affecting closed-loop stability.

Control precision as a function of sampling time Given the conditions above, the degradation of control precision when employing a digital controller and the maximum possible sampling time can be easily estimated. The effect of *digital signal processing* can be approximately characterized with a *delay term* (see Ch. 9).

$$\Delta_s(s) = e^{-s\frac{T_s}{2}}.$$

For the obligatory *anti-aliasing filter*, assume a *second-order low-pass* filter with a cutoff frequency equal to half the sampling frequency

$$F_{AA}(s) = \frac{1}{\left\{0.7; \frac{\omega_s}{2}\right\}}. \tag{10.62}$$

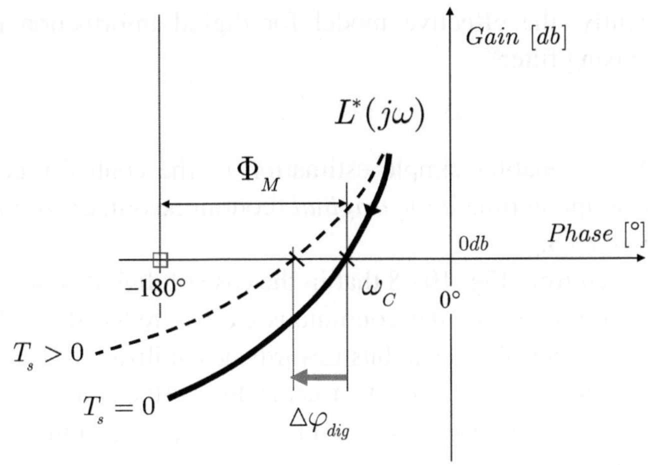

Fig. 10.48. Phase loss in the sample-and-hold process and anti-aliasing filter

This gives the effect of *digital control* at sufficiently low frequencies to be approximately

$$\Delta_{dig}(s) = \Delta_s(s) \cdot F_{AA}(s) .$$

As there is no decrease in magnitude below the anti-aliasing filter cutoff frequency, the crossover frequency ω_C valid for $T_s = 0$ will also remain constant for $T_s > 0$. In the case of digital control, the frequency response in this frequency range is simply shifted left in phase (negative phase) in the NICHOLS diagram. This comes from

$$\Delta\varphi_{dig} = \arg \Delta_{dig}(j\omega_C) = -\frac{T_s}{2}\omega_C - \arctan \frac{2 \cdot 0.7 \dfrac{\omega_C}{\omega_s/2}}{1 - \left(\dfrac{\omega_C}{\omega_s/2}\right)^2} . \qquad (10.63)$$

For cases relevant in practice, $\omega_C \ll \omega_s/2$, so that Eq. (10.63) gives the simple *approximate relation* for the sampling-time-dependent *phase loss*[20]

$$\boxed{\Delta\varphi_{dig}^{[°]} \approx -\omega_C^{[1/s]} \cdot T_s^{[s]} \cdot \frac{180}{\pi}} , \qquad (10.64)$$

[20] This relation naturally holds in general in the considered frequency range.

or, equivalently, the effective model for digital information processing with anti-aliasing filter[21]

$$\Delta_{dig}(s) \approx e^{-sT_s} .$$

Eq. (10.64) now enables simple estimation of the control precision as a function of sampling time if the *original* (continuous-time) *controller* is retained *without change.*

It can be seen from Fig. 10.48 that in the case of digital control, the original phase margin Φ_M of the continuous case is reduced by $\Delta\varphi_{dig}$. The tendency of the control loop is thus towards destabilization. If the assumptions made in Sec. 10.5.3 hold, then using Eqs. (10.64) and (10.24), even the *transient dynamics* (*overshoot*) can be very simply quantified.

Maximum possible sampling time The *maximum possible sampling time* given an unchanged controller can also be estimated from Eq. (10.64) if the maximum allowable phase loss is taken to be the phase margin Φ_M, i.e.

$$\boxed{T_{s,\max}^{[s]} \approx \Phi_M^{[°]} \frac{\pi}{180} \frac{1}{\omega_C^{[1/s]}}} . \tag{10.65}$$

Example 10.11 *Digital control of a single-mass oscillator.*

Consider the single-mass oscillator from Example 10.4 with the robust PID controller (10.25). The open-loop frequency response for various sampling times including the anti-aliasing filter (10.62) is shown in Fig. 10.49. The increasingly destabilizing action of increasing sampling time can be seen. Applying the phase margin $\Phi_M = 52°$ and crossover frequency $\omega_C = 2.6 \; 1/s$ of the continuous case in the approximate relation (10.65) gives a critical sampling time

$$T_{s,crit} = 52 \frac{\pi}{180} \frac{1}{2.6} = 3.5 \; s ,$$

which agrees very well with the exact frequency responses of Fig. 10.49. Note also the individual frequency bounds as a function of sampling time. In the current case, the approximate relation (10.24) for the transient dynamics of the closed control loop can also be applied quite nicely. Verification of these relationships is left as an exercise to the reader.

---

21 Below its cutoff frequency, the anti-aliasing filter thus acts approximately as a delay of $T_s/2$.

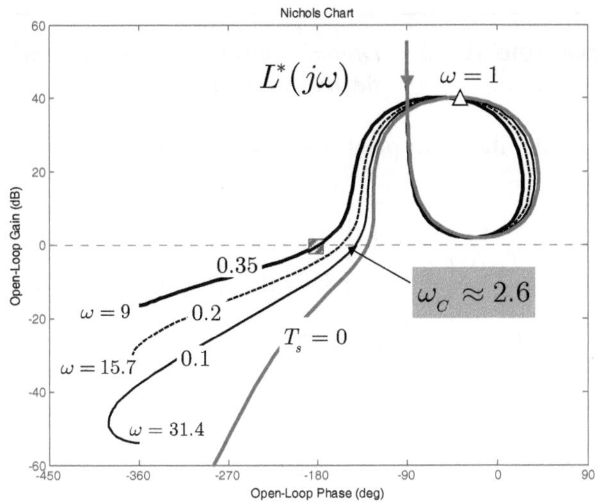

Fig. 10.49. Digital control of a single-mass oscillator: open-loop NICHOLS diagram with PID controller for various sampling times

■

10.8.3 Systems with unmodeled eigenmodes (spillover)

High-frequency eigenmodes In systems with high-frequency eigenmodes—particularly those with non-negligible unmodeled eigenmodes—sometimes rather appreciable changes in the frequency response $P^*(j\omega)$ must be reckoned with as compared to the continuous case. Indeed, here the *mirror frequencies* $\omega = \omega_i \pm n\omega_s$ come to full prominence.

Aliasing of eigenfrequencies As previously discussed, eigenfrequencies $\omega_i > \omega_s/2$ are mirrored in a subharmonic frequency ω_i^* lying within the base frequency band $0 < \omega < \omega_s/2$ (see Sec. 9.7). As a result, the frequency response in the low-frequency range fundamentally changes, with all accompanying negative consequences for closed-loop stability. The location of pole-zero loops in the NICHOLS diagram can now not be predicted without further effort.

In addition, the negative phase shift due to digital information processing and the anti-aliasing filter naturally also applies here, though this can be rather well estimated using Eq. (10.64).

The possible consequences for controller design are illustrated in the following example.

Example 10.12 *Digital control of a free mass with attached flexible structure.*

Consider the MBS plant from Example 10.6 including an anti-aliasing filter:

$$P_{MBS}(s) = \frac{1}{s^2}\frac{\{3\}}{\{4\}}\frac{\{8\}}{\{10\}}\frac{1}{[50]}\frac{1}{\{0.7;62.8\}}\cdot\frac{\{90\}}{\{100\}}\frac{\{500\}}{\{510\}}.$$

Selecting a sampling time of $T_s = 0.05$ s, then—since $\omega_s/2 = \pi/T_s = 62.83$—the two eigenmodes at $\omega_{p,3} = 100$ and $\omega_{p,4} = 510$ lie outside the base frequency band of the discrete frequency response $P^*_{MBS}(j\omega)$. For the eigenfrequency $\omega_{p,3} = 100$, for example, the resulting subharmonic mirror frequency is

$$\omega^*_{p,3} = \omega_0 - \omega_s = 100 - 125.67 = -25.7$$

and for the eigenfrequency $\omega_{p,4} = 510$, it is

$$\omega^*_{p,4} = 100 - 4\cdot 125.67 = -7.3\,.$$

For illustration, the discrete open-loop frequency response $L^*(j\omega)$ including the *controller*

$$H^*(s) = 0.003\,\frac{\{1;\ 0.05\}}{s[40]}\,\frac{1}{\{0.1;\ 40\}} \tag{10.66}$$

is depicted in Fig. 10.28. The z-transfer function for $T_s = 0.05$ s associated with Eq. (10.66) is

$$H_D(z) = \frac{10.88 - 0.0545z - 21.82z^2 + 0.0546z^3 + 10.94z^4}{-0.8182z + 0.8182z^2 - z^3 + z^4}.$$

Discussion Comparing with the continuous case (see Fig. 10.28b) again clearly and predictably reveals the effect of digital information processing and the anti-aliasing filter. The two low-frequency eigenmodes remain almost unchanged—only the phase margin at the crossover frequency $\omega_C = 11$ reduces from $\Phi_M = 57°$ to $\Phi_M = 28°$, completely in accord with the approximate relation (10.64) for phase loss. The subharmonic eigenfrequency $\omega^*_{p,3} = 25.7$ can be nicely recognized in $L^*(j\omega)$, though with reduced stability margin. The second, high-frequency eigenfrequency $\omega^*_{p,4} = 7.3$, however, is not visible in $L^*(j\omega)$. This is obviously due to the fact that for the eigenfrequency $\omega_{p,4} = 510$, the low-pass components in $P_{MBS}(s)$ and the hold element already act so strongly that

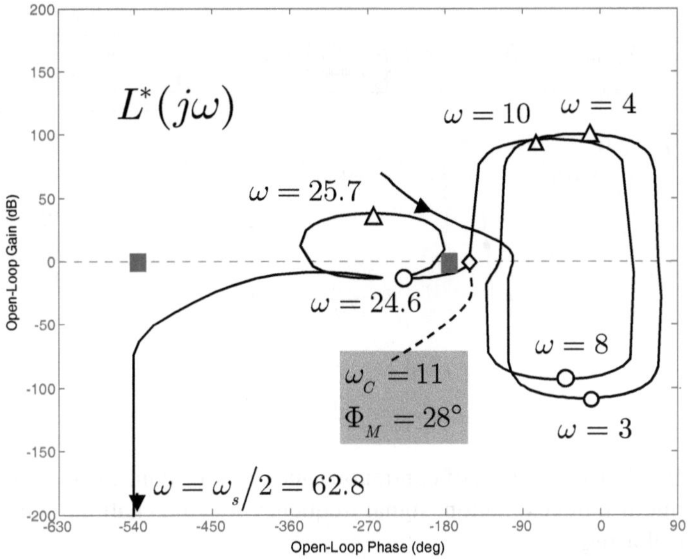

Fig. 10.50. Digital control of a multibody system with high-frequency eigenmodes: open-loop NICHOLS diagram for $T_s = 50\text{ms}$

in the discrete frequency response $P^*(j\omega)$, no appreciable mirror frequency terms of this eigenfrequency have any effect. Overall, there is thus less robust stability and reduced phase margin. If larger phase margins are desired, an attempt could be made to insert an additional phase-separating corrective element (anti-notch filter) to spread the frequency response near $\omega \approx 15$. In any case, with the available controller (10.66), time-domain dynamics in response to force disturbances are almost unchanged from the continuous case. Verification of this is left as an exercise to the interested reader.

■

11.8.4 Aliasing in digital controllers

Filtering of analog vs. digital controllers When transitioning from a continuous controller to a digital one, the following previously-discussed phenomenon should be considered. A *continuous controller* of low-pass character inherently suppresses high-frequency components, e.g. sensor noise in the feedback channel. In a *digital controller* with equivalent frequency characteristics, however, high-frequency continuous input variables are sampled in the feedback channel, resulting in subharmonic mirror

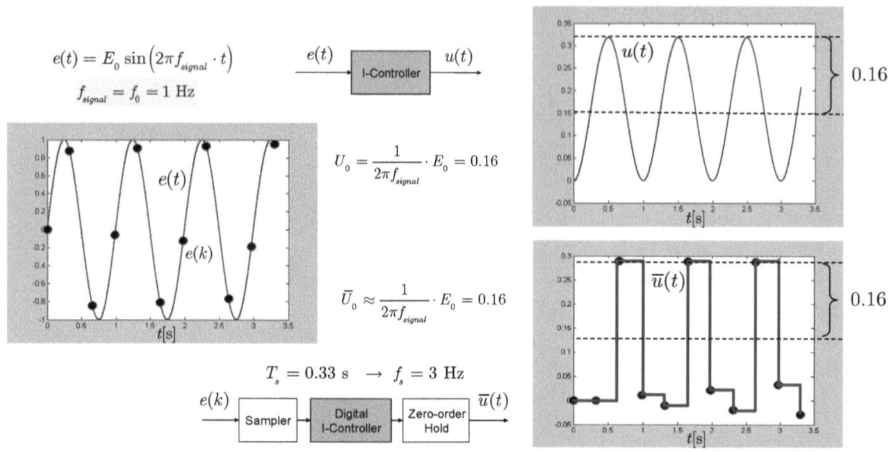

Fig. 10.51. Comparison of continuous and digital I-controllers with low-frequency continuous input excitation: signal frequency *less* than half the sampling frequency ⇒ no aliasing

frequencies due to aliasing effects. High-frequency sensor noise can thus give rise to a low-frequency controller input signal, which is then processed by the low-frequency component of the controller. An illustrative example of this type of aliasing is shown in Figs. 10.51 and 10.52 for an *integral controller* (*I-controller*). In this type of controller, a large gain is applied to low-frequency input signals. This can be clearly seen in Fig. 10.52, where in the continuous case, the I-controller ensures good disturbance rejection, while the *digital I-controller* evaluates the input signal unacceptably high and maintains it practically undiminished in the control loop. This effect is incidentally also present when employing an anti-aliasing filter, though with reduced amplitude.

10.8.5 Aliasing of measurement noise

High-frequency measurement noise In the case of digital control, particularly of weakly-damped multibody systems, and in contrast to continuous (analog) control, aliasing of high-frequency measurement noise can lead to disagreeable effects, as will be shown in the following.

Fig. 10.53 once again shows the fundamental configuration of a digital control loop, where those system elements having particular importance in the following considerations are highlighted in gray. As a rule, measure-

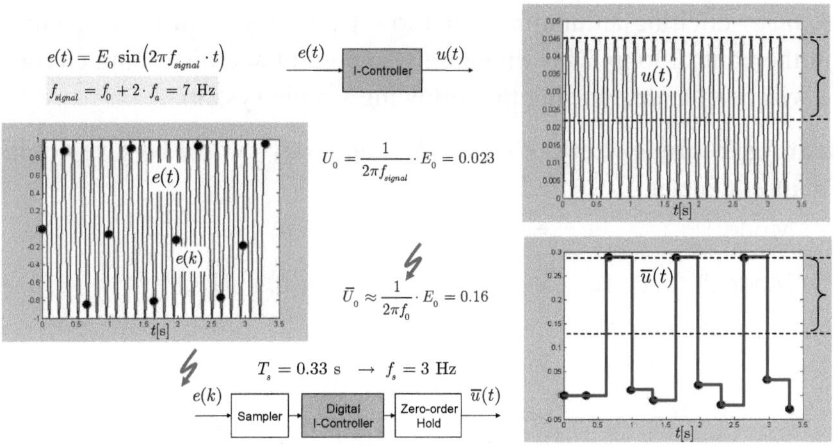

Fig. 10.52. Comparison of continuous and digital I-controllers with high-frequency continuous input excitation: signal frequency *greater* than half the sampling frequency ⇒ aliasing

ment noise is a more or less broadband signal, whose high-frequency components are subject to fundamental aliasing effects. The measured quantity $\tilde{x}(k)$ sampled by the analog-to-digital converter always contains subharmonic components of the noise signal $n(t)$; the anti-aliasing filter only limits their amplitudes. Via the control algorithm and the hold element (digital-to-analog converter), these subharmonic signals are relayed on to the MBS plant. If by chance, subharmonic signal components coincide with weakly-damped MBS eigenfrequencies, then these natural oscillations excited by high-frequency measurement noise will become apparent in the controlled variable $x(t)$.

As laid out in Sec. 10.2, it is precisely in the low-frequency range that, as a rule, there is poor disturbance rejection. Following the robust design strategy, due to $|L^*(j\omega)|$ and Eq. (10.6), it holds that the *measurement noise rejection* $|T_n^*(j\omega)| \approx 1$. This presents the unusual case of *high-frequency*

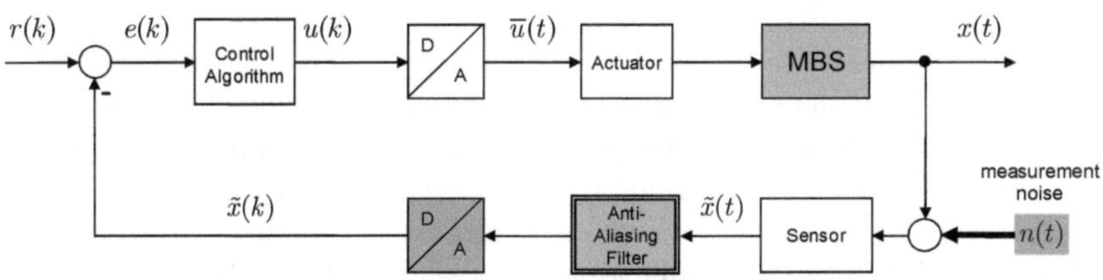

Fig. 10.53. Digital control loop with measurement noise

disturbances eliciting an undiminished low-frequency disturbance component in the control loop. This immensely important and disturbing circumstance will be elucidated using the following simple system.

Illustrating example Let the following *multibody system* be given, with the eigenfrequencies

$$\text{Mode 1:} \qquad \omega_{p,1} = 10 \ \text{rad/s} \ \triangleq \ 1.6 \ \text{Hz},$$

$$\text{Mode 2:} \qquad \omega_{p,2} = 52.8 \ \text{rad/s} \ \triangleq \ 8.4 \ \text{Hz},$$

the *plant transfer function*

$$P_{MBS}(s) = \frac{1}{\{0.1;10\}} \frac{\{0.001;40\}}{\{0.001;52.8\}},$$

and *high-frequency harmonic measurement noise*[22]

$$n(t) = N_0 \cdot \sin\left(\omega_n \cdot t\right), \ \omega_n = 115.6 \ \text{rad/sec}. \tag{10.67}$$

For the controller, consider the following analog and digital *I-controllers*

$$H(s) = H^*(s) = \frac{1.6}{s}.$$

Analog controller In the continuous-time case, the high-frequency disturbance component lies far *above the controller bandwidth.* Due to the low-pass nature of the analog controller and the MBS plant, the controlled variable is only marginally affected—the effect of *measurement noise* in this case is thus *inconsequential* (Fig. 10.54a).

Aliasing As for any digital control, two aliasing effects should be distinguished here: *signal aliasing* of the high-frequency sensor noise, and *frequency response aliasing* via high-frequency MBS eigenmodes. Selecting a *sampling time* $T_s = 0.1$ s gives the *sampling frequency*

$$\omega_s = 2\pi \,/\, T_s = 62.8 \ \text{rad/s} \ \text{ or } \ \frac{\omega_s}{2} = 31.4 \ \text{rad/s}$$

---

[22] For clarity, only *one* particular *harmonic* signal is considered here. In the general case, the coincidence of frequency components found further below in this example is of course quite improbable. Similar dynamics exist, however, for *broad-band noise signals*, which do contain all frequencies—i.e. also those which can lead to critical subharmonic excitation. To this extent, the signals considered here do have *general validity*.

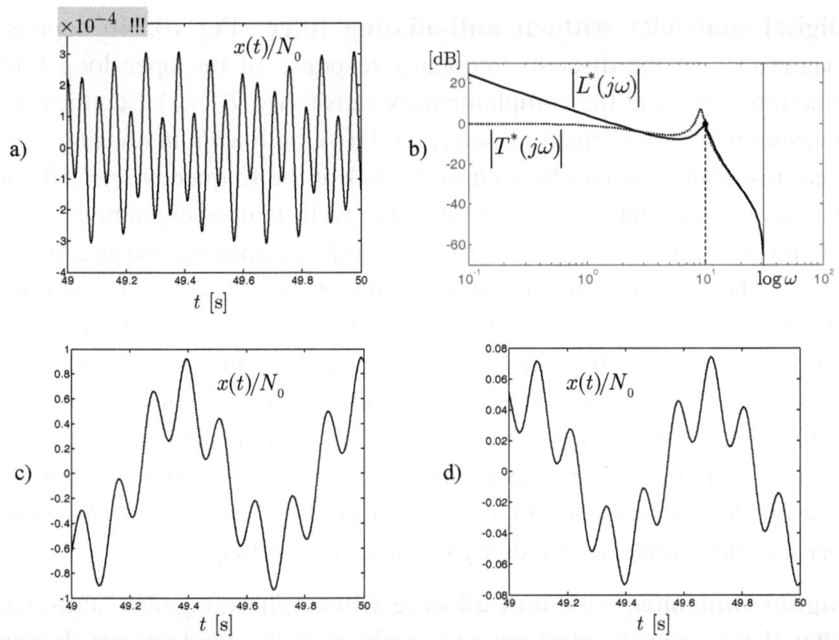

Fig. 10.54. Excitation of eigenfrequencies by measurement noise: a) controlled variable, *analog* I-controller; b) magnitude plot of discrete frequency response for $T_a = 0.1$ s ; c) controlled variable, *digital* I-controller ($T_a = 0.1$ s) *without* anti-aliasing filter; d) controlled variable, *digital* I-controller ($T_a = 0.1$ s) *with* anti-aliasing filter

and for signal aliasing, the *subharmonic mirror frequency* of the measurement noise within the frequency baseband

$$\omega_n^* := \omega_n - 2\omega_s = -10 \text{ rad / s}. \tag{10.68}$$

The digital controller interprets the measurement noise as a sampled sequence with precisely the frequency ω_n^*.

The second eigenfrequency $\omega_{p,2}$ in turn appears in the discrete frequency response as a mirrored eigenfrequency $\omega_{p,2}^* = 10$ (see Sec. 9.7), which happens to coincide with the lowest eigenfrequency $\omega_{p,1}$. This further implies that a signal frequency $\omega = 10$ simultaneously excites both MBS eigenfrequencies. Since the measurement noise coincidentally[23] also has a subharmonic mirror frequency at $\omega = \omega_n^* = 10$, a multiply unfavorable situation exists.

[23] In this example, naturally selected intentionally for illustration; see previous footnote.

Digital controller without anti-aliasing filter Fig. 10.54b depicts the magnitudes of the discrete frequency response of the open-loop transfer function $L^*(s)$ and the complementary sensitivity $T^*(j\omega)$ (identical to the measurement noise rejection, see Eq. (10.6)), both *without* aliasing filter. A clear resonant peak can be seen in $L^*(j\omega)$ at the frequency $\omega = 10$, originating from the superposition of the rather well-damped eigenmode $\omega_{p,1}$ and the mirror frequency $\omega^*_{p,2} = 10$ of the weakly-damped second eigenmode.

In the measurement noise rejection $T^*(j\omega)$, a magnitude $\left|T^*(j1)\right| \approx 0\,\mathrm{dB} \triangleq 1$ results at $\omega = 10$. This is also precisely the amplification factor with which the harmonic measurement noise (10.67) appears at the plant output (see Fig. 10.54c). The high-frequency disturbance is transferred by the control loop practically undiminished to the controlled variable y in the low-frequency mirror frequency. As expected, both MBS eigenmodes are excited. In the digital control loop, the *high-frequency measurement noise* has suddenly become *very consequential.*

Digital controller with anti-aliasing filter The obligatory anti-aliasing filter (here, second-order) changes nothing in the fundamental dynamics. Only the *repercussions* of the frequency response aliasing and signal aliasing are *reduced* (in the lower resonant peak and lower signal amplitude at the sampler, respectively). Fig. 10.54d presents the controlled variable for this case; a decrease in magnitude of the measurement noise by a factor of $0.06 \triangleq -24\,\mathrm{dB}$ results. Compared to the analog I-controller, this disturbance rejection is particularly large. This is due to the fact that the disturbance frequency still lies within the first decade of the stopband of the anti-aliasing filter.

Bibliography for Chapter 10

Åström, K. J. (2000). Model uncertainty and robust control. *Lecture Notes on Iterative Identification and Control Design.* Lund, Sweden. Lund Institute of Technology: 63–100.

Åström, K. J. and R. M. Murray (2008). *Feedback systems: an introduction for scientists and engineers.* Princeton University Press.

Bittner, H., H. D. Fischer and M. Surauer (1982). Design of Reaction Jet Attitude Control Systems for Flexible Spacecraft. *Proceedings of 9-th IFAC/ESA Symposium "Automatic Control in Space 1982",* Noordwijkerhout, The Netherlands.

Föllinger, O. (1994). *Regelungstechnik, Einführung in die Methoden und ihre Anwendung*. Hüthig Verlag.

Graham, R. E. (1946). "Linear Servo Theory." *Bell System Technical Journal*: 616-651.

Hagenmeyer, V. and M. Zeitz (2004). "Flachheitsbasierter Entwurf von linearen und nichtlinearen Vorsteuerungen." *at-Automatisierungs-technik* **52**(1): 3-12.

Horowitz, I. M. (1963). *Synthesis of Feedback Systems*. New York. Academic Press.

Janschek, K. (2008). "Optimized system performances through balanced control strategies (Editorial)." *Mechatronics* **18**(5-6): 262-263.

Janschek, K. (2009). Skyhook-Einstellregeln für lineare Korrekturglieder erster Ordnung. Interner Bericht, Institut für Automatisierungs-technik, Technische Universität Dresden.

Janschek, K. and M. Surauer (1987). Decentralized/hierarchical control for large flexible spacecraft *10th IFAC World Congress*, Munich, Federal Republic of Germany. pp.53-60.

Karnopp, D. (1995). "Active and Semi-Active Vibration Isolation." *Transactions of the ASME* **117**(June): 177-185.

Kreisselmeier, G. (1999). "Struktur mit zwei Freiheitsgraden." *at-Automatisie-rungstechnik* **47**(6): 266–269.

Li, H. and R. M. Goodall (1999). "Linear and non-linear skyhook damping control laws for active railway suspensions." *Control Engineering Practice* **7**(7): 843-850

Lurie, B. J. and P. J. Enright (2000). *Classical feedback control with MATLAB*. New York. Marcel Dekker.

Ogata, K. (2010). *Modern Control Engineering*. Prentice Hall.

Preumont, A. (2002). *Vibration Control of Active Structures - An Introduction*. Kluwer Academic Publishers.

Preumont, A. (2006). *Mechatronics, Dynamics of Electromechanical and Piezoelectric Systems*. Springer.

Reinschke, K. (2006). *Lineare Regelungs- und Steuerungstheorie*. Springer.

Reinschke, K. and S.-O. Lindert (2006). Anmerkungen zu regelungstechnischen Konzepten, insbesondere zur Stabilisierung von Regelkreisen mit instabilen Reglern. *Workshop GMA-Fachausschuss 1.40 Theoretische Verfahren der Regelungstechnik*, Bostalsee, Deutschland, Lehrstuhl für Systemtheorie und Regelungstechnik (Prof. Kugi), Universität des Saarlandes. pp.124-142.

Rudolph, J. (2003). *Beiträge zur flachheitsbasierten Folgeregelung linearer und nichtlinearer Systeme endlicher und unendlicher Dimension*. Aachen. Shaker Verlag.

Siciliano, B., L. Sciavicco, L. Villani and G. Oriolo (2009). *Robotics: Modelling, Planning and Control*. Springer.
Sidi, M. J. (1997). *Spacecraft Dynamics and Control: A Practical Engineering Approach*. New York. Cambridge University Press.
Weinmann, A. (1991). *Uncertain Models and Robust Control*. Springer.
Zhou, K. and J. C. Doyle (1998). *Essentials of Robust Control*. New Jersey. Prentice Hall.

11 Stochastic Dynamic Analysis

Background Mechatronic products are usually designed for high-precision control of mechanical state variables. Thus, small-signal and stochastic input responses take on a particular significance in the overall behavior of the system. Fundamentally, the noise levels induced by sensors, actuators, and environmental disturbances determine the achievable system accuracy. Starting from the conceptual design model, meaningful, quantitative calculations which can estimate the effects of stochastic inputs are essential to enabling robust predictions of the behavior of the system.

Contents of Chapter 11 This chapter discusses fundamental methods for the description and modeling of system behaviors induced by dynamic stochastic inputs. The first, longer portion is devoted to the characterization and modeling of noise processes. Following a short introduction to important concepts of stochastic systems theory (*random variables, random processes*), the focus turns to *spectral representations* and *propagation behavior* in LTI systems. In addition, *white* and *colored noise processes* are introduced as important modeling concepts. Building on this foundation, models for *real noise sources* are discussed, with reference to typical noise specifications in data sheets.

In the second part, disturbance propagation in compound systems is investigated, and appropriate computational methods in the form of analytical, numerical, and simulation-based *covariance analysis* are presented.

■

11.1 Systems Engineering Context

Non-deterministic operational conditions One of the primary tasks of systems design is to predict the behavior of a system under representative operational conditions. In real systems, many of the influencing factors are inherently stochastic[1], explaining the need for corresponding methods to describe and analyze them. The dynamic properties of a system which determine its accuracy depend fundamentally (especially in high-precision applications) on its stochastic inputs and the degree to which these are amplified or suppressed by the system. In contrast to deterministic inputs, the time history of random influences cannot be predicted, with the result that deterministic concepts such as command and disturbance feedforward control are fundamentally not applicable to this type of input. Good rejection of stochastic disturbance inputs can be easily achieved using an appropriate controller structure and parameterization (including specialized filter algorithms, e.g. the KALMAN filter). However, design of the controller requires predictive, quantitative computational methods in order to estimate the effects of stochastic inputs, starting with the initial conceptual model. This requirement is further addressed in the remainder of this chapter.

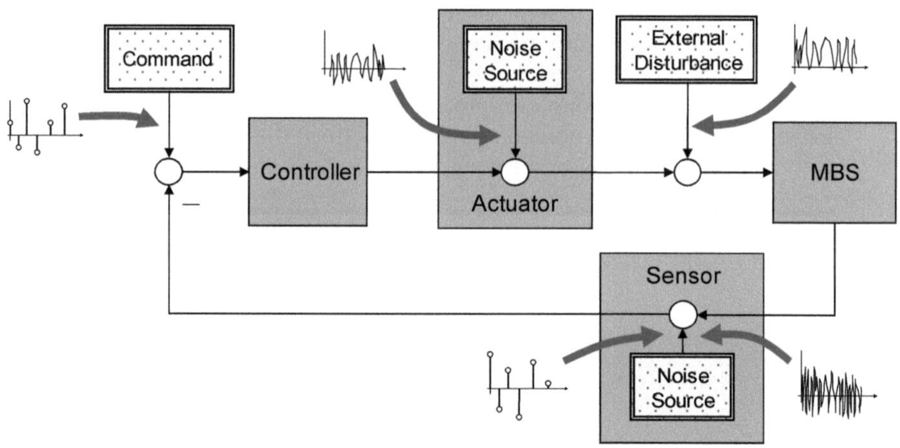

Fig. 11.1. Mechatronic system with stochastic inputs

[1] Also termed *random*.

Stochastic inputs A schematic overview of important stochastic inputs to a mechatronic system is shown in Fig. 11.1:

- *plant disturbances*: e.g. external disturbances, inputs of coupled systems;
- *sensor noise*: noise in sensors and sensor amplifiers—the most common noise source;
- *actuator noise*: inherent noise in auxiliary energy sources, actuator components (e.g. a hydraulic servo valve), and power amplifiers;
- *operator commands*: unpredictable operator actions, e.g. *pilot-in-the-loop*.

Depending on their generating source, these disturbances can be modeled as *continuous signals* or as *discrete random events*. In both cases, their behavior is described using *static* parameters for the *amplitude distribution*, and *spectral* parameters for the *dynamic* properties.

Disturbance modeling and disturbance propagation In the context of this book, two fundamental areas of stochastic dynamic analysis are of interest. First, an essential prerequisite is an accurate *description of the disturbances* as stochastic processes (the *disturbance modeling* shown on the left of Fig. 11.2). The information required for this modeling can be determined from data sheets of the components employed, or via experimental investigation. Second, the analysis requires a description of how such stochastic inputs *propagate* through the mechatronic system, and what their effect is on system variables (the *disturbance propagation* shown on the right of Fig. 11.2). This task can be very efficiently handled via analytical means using the *covariance analysis* presented in Sec. 11.6.

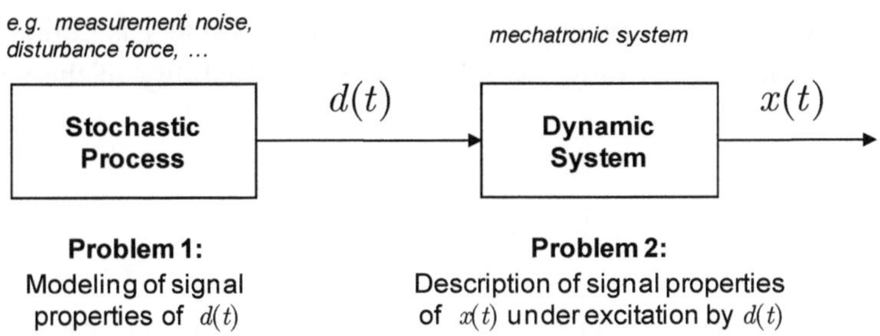

Fig. 11.2. Basic tasks of stochastic dynamic analysis

11.2 Elements of Stochastic Systems Theory

11.2.1 Random variables

In order to allow quantitative description of the amplitude characteristics of random system quantities, well-known results of probability theory for the statistical description of random events are presented here (Hsu 2011), (Cooper and McGillem 1999).

Random variables Representing the result of a random event, a *random variable* takes on a value from a certain range according to chance.

Probability distribution The *cumulative distribution function* (*CDF*)

$$F(x) = P\left\{\xi \leq x\right\} \tag{11.1}$$

over a random variable x describes the probability that $\xi \leq x$.

Probability density Given certain conditions, a probability distribution will have a *probability density function* (*PDF*)

$$f(x) \;=\; \frac{dF(x)}{dx}, \tag{11.2}$$

so that Eq. (11.1) can be equivalently expressed as

$$F(x) = \int\limits_{-\infty}^{x} f(x') \, dx' = P\left\{\xi \leq x\right\}.$$

Statistical parameters The shape of a distribution is described mathematically using *statistical parameters* (numerical values).

Expected value The *expected value* of a random value x is defined as the sum of all values in its range weighted by the probability of the value occurring:

$$E[x] := \int\limits_{-\infty}^{+\infty} x \, f(x) \;\; dx \,.$$

This represents an "averaging" (mean) of all possible realizations of the random variable x.

Similarly, the expected value of a *function* $g(x)$ is defined by

$$E[g(x)] := \int_{-\infty}^{+\infty} g(x)\, f(x)\, dx$$

The expectation operator is a *linear operator*, i.e. the *principle of superposition* holds:

$$E[c_1 x_1 + c_2 x_2] = c_1 \cdot E[x_1] + c_2 \cdot E[x_2], \tag{11.3}$$

for random variables x_1, x_2 and constants $c_1, c_2 \in \mathbb{R}$.

Moments of a distribution The elementary statistical properties of a probability distribution are defined by a particular set of expected values—the *moments* of the distribution:

- nth *ordinary* moment:

$$E\left[x^n\right] := \int_{-\infty}^{+\infty} x^n\, f(x)\, dx\ , \tag{11.4}$$

- nth *central* moment:

$$E\left[\left(x - E[x]\right)^n\right] := \int_{-\infty}^{+\infty} \left(x - E[x]\right)^n f(x)\, dx\ . \tag{11.5}$$

Mean of a distribution

$$\mu_x := E\left[x\right] := \int_{-\infty}^{+\infty} x \cdot f(x)\, dx\ . \tag{11.6}$$

Physical dimension: $\dim \mu_x = \dim x$

The *mean* is the first ordinary moment (see Eq. (11.4)) and describes the location of the distribution.

Variance of a distribution

$$\operatorname{var}(x) := \sigma_x^{\ 2} := E\left[\left(x - \mu_x\right)^2\right] := \int_{-\infty}^{+\infty} \left(x - \mu_x\right)^2 f(x)\, dx\ . \tag{11.7}$$

Physical dimension: $\dim \operatorname{var}(x) = \left(\dim x\right)^2$

The *variance* is the 2nd central moment (see Eq. (11.5)), and describes the *spread* of the distribution. Note that

$$\mathrm{var}(x) = E\left[x^2\right] - \mu_x^{\,2}\,. \tag{11.8}$$

Covariance Given two random variables x_1 and x_2, their *covariance*[2] is defined by

$$\begin{aligned}\mathrm{cov}\left(x_1, x_2\right) &:= \sigma_{x1x2} := E\left[\left(x_1 - \mu_{x1}\right)\left(y - \mu_{x2}\right)\right] := \\ &= \int_{-\infty}^{+\infty}\int_{-\infty}^{+\infty} \left(x_1 - \mu_{x1}\right)\left(x_2 - \mu_{x2}\right)\, f\left(x_1, x_2\right)\, dx_1\, dx_2\,.\end{aligned} \tag{11.9}$$

Physical dimension: $\dim \mathrm{cov}(x_1, x_2) = \dim x_1 \times \dim x_2$

It directly follows that $\mathrm{cov}\left(x_1, x_1\right) = \mathrm{var}(x_1)$. As a result, the variance (11.7) is often (imprecisely) referred to as the *covariance* of the random variable x_1 (see Sec. 11.6: Covariance analysis).

Correlation coefficient The normalized covariance

$$\rho_{x1x2} := \frac{\mathrm{cov}(x_1, x_2)}{\sigma_{x1}\sigma_{x2}} = \frac{\sigma_{x1x2}}{\sigma_{x1}\sigma_{x2}} \tag{11.10}$$

is termed the *correlation coefficient* of the random variables x_1, x_2.

CAUCHY-SCHWARZ inequality

$$\sigma_{x1x2}^2 \le \sigma_{x1}^2 \cdot \sigma_{x2}^2 \quad \text{or} \quad \left|\sigma_{x1x2}\right| \le \sigma_{x1} \cdot \sigma_{x2}\,. \tag{11.11}$$

Statistical independence Following Eqs. (11.10) and (11.11), it holds in general that

$$-1 \le \rho_{x1x2} \le 1\,.$$

If $\rho_{x1x2} = 0$, the two random variables x_1 and x_2 are termed *statistically independent* or *uncorrelated.*

---

2 Note the differing notation between *covariance* variables σ_{xy} (not $\sigma_{xy}^{\,2}$!) and *variance* variables $\sigma_x^{\,2}$. This notation is important for two reasons: (i) σ_{xy} can also be negative, (ii) $\dim \sigma_{xy} = \dim x \times \dim y$ just as $\dim \sigma_x = \dim x$.

Covariance matrix For a vector of random variables $\mathbf{x} = \begin{pmatrix} x_1 & x_2 & \dots & x_n \end{pmatrix}^T$, the *covariance matrix* is defined to be

$$\mathbf{P} := E\left[\mathbf{x}\mathbf{x}^T\right] = \begin{pmatrix} \sigma_{x1}^2 & \sigma_{x1x2} & \cdots & \sigma_{x1xn} \\ \sigma_{x1x2} & \sigma_{x2}^2 & \cdots & \vdots \\ \vdots & & \ddots & \vdots \\ \sigma_{x1xn} & \cdots & \cdots & \sigma_{xn}^2 \end{pmatrix}. \tag{11.12}$$

The variances of the individual random variables $x_1, x_2, \dots, x_n$ are located along the diagonal of $\mathbf{P}$, while the covariances measuring the variables' mutual statistical dependences, are located on the off-diagonals.

Standard deviation

$$\sigma_x := \sqrt{\mathrm{var}(x)}\,. \tag{11.13}$$

Physical dimension: $\dim \sigma = \dim x$

The *standard deviation* is the square root of the variance, and directly describes the *spread* of the distribution.

Mean square

$$E\left[x^2\right] = \sigma_x^{\;2} + \mu_x^{\;2} = \int_{-\infty}^{+\infty} x^2\ f(x)\ dx\ . \tag{11.14}$$

Physical dimension: $\dim E\left[x^2\right] = \left(\dim x\right)^2$

The *mean square* is the ordinary 2nd moment (see Eq. (11.4)). Note the relation to the variance (Eq. (11.8)).

Root mean square

$$x_{rms} = \sqrt{E\left[x^2\right]}\,. \tag{11.15}$$

Physical dimension: $\dim x_{rms} = \dim x$

The *root mean square* (*RMS*) is the square root of the mean square. Note the special case of a *zero-mean* random variable:

$$\mu_x = 0 \ \ \Rightarrow \ \ x_{rms} = \sigma_x\,. \tag{11.16}$$

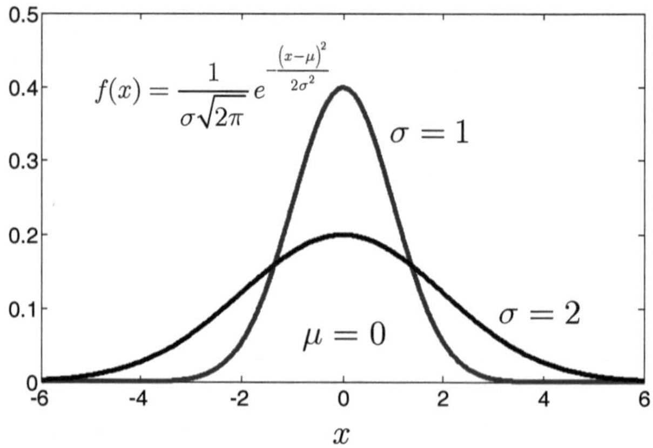

Fig. 11.3. Normal or GAUSSian distribution

Mean of a sum The following relation holds for the mean of a sum of random variables:

$$E[x_1 + x_2] = E[x_1] + E[x_2] = \mu_{x1} + \mu_{x2}. \tag{11.17}$$

Variance of a sum The following relation holds for the variance of the sum $x = x_1 + x_2$ of random variables:

$$\sigma_x^{\ 2} = \sigma_{x1}^{\ 2} + \sigma_{x2}^{\ 2} + 2\sigma_{x1x2}. \tag{11.18}$$

Normal distribution Both in theory and in practice, the most important probability distribution is the *normal* or *GAUSSian distribution*[3]. This distribution is completely defined by two statistical parameters: the *mean* μ and *standard deviation* σ in the *GAUSSian* or *bell curve* density function (see Fig. 11.3):

$$N(\mu, \sigma): \quad f(x) = \frac{1}{\sigma\sqrt{2\pi}} e^{-\frac{(x-\mu)^2}{2\sigma^2}} \quad . \tag{11.19}$$

[3] Johann Carl Friedrich GAUSS, (1777-1855), German mathematician, astronomer, geographer, and physicist.

Table 11.1. Statistical certainty and confidence intervals for the normal distribution

Confidence interval	Statistical certainty [%]
$n = 1: \quad \mu \pm \sigma$	68.2
$n = 2: \quad \mu \pm 2\sigma$	95.4
$n = 3: \quad \mu \pm 3\sigma$	99.7

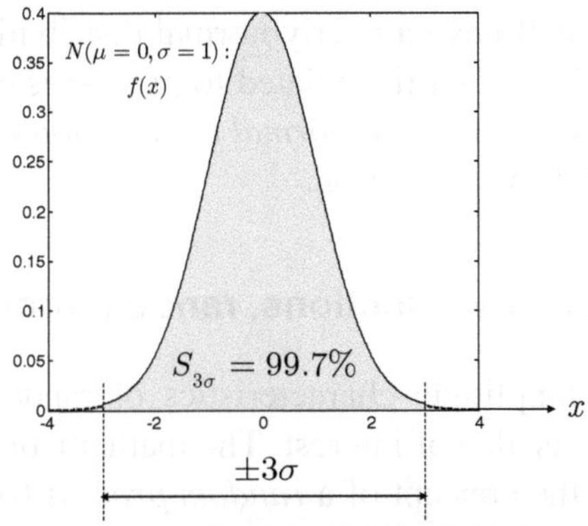

Fig. 11.4. Statistical certainty and confidence interval of the normal distribution

Statistical certainty One attribute of a probability distribution which is important in practical applications is the *statistical certainty*

$$S(n) = \int_{\mu - n\sigma}^{\mu + n\sigma} f(u)\,\mathrm{d}u = P\left\{\mu - n\sigma < x \leq \mu + n\sigma\right\},$$

i.e. $S(n)$ represents the probability with which observations of the random variable x fall in the interval $(\mu - n\sigma, \mu + n\sigma]$. This interval is also called the *confidence interval.* Characteristic confidence values obtained for the GAUSSian bell curve are shown in Table 11.1.

In practice, the confidence interval $n = 3: \quad \mu \pm 3\sigma$ is often used to approximately describe observed *minimal* and *maximal values* of the random variable (the "3σ value"). In this way, a direct correspondence between observed extrema and the standard deviation can be made[4] (Fig. 11.4).

[4] Strictly seen, this is only permissible if it has already been established that the observed random values actually follow a normal distribution. This is often assumed without comment on physical grounds (see *Central Limit Theorem*), but in certain cases, can be completely in error.

Central Limit Theorem The fundamental significance of the normal distribution in technical and physical phenomena results from an interesting, generally applicable mathematical property of independent random variables. The *Central Limit Theorem* states that, under certain very general conditions, every sum of independent random variables is *asymptotically normally distributed.* In particular, the sum of a very large number of independent random variables, each with a small standard deviation relative to that of the sum, will have a nearly normal distribution. Using this property, when calculating quantities related to processes consisting of the *sum* of many *independent events*, the *normal distribution* can be employed as a representative model for estimation.

11.2.2 Stochastic time functions, random processes

In addition to the amplitude characteristics of random system variables, their time evolution is also of interest. The mathematical description of this evolution employs the concept of a *random process* (or stochastic process) (Hsu 2011), (Cooper and McGillem 1999).

Random process When a random variable depends on time, it is referred to as a *stochastic time function*, and a collection or ensemble of such variables is termed a *random process* (Fig. 11.5). In this case, the statistical parameters belonging to the stochastic time functions are naturally also time-dependent.

Stationary random process If the static parameters of a random process are constant and independent of starting time, the process is termed *stationary.*

Ergodic random process If all statistics (expected values) calculated at any point in time (i.e. *all* members of the ensemble viewed at the same time) are *identical* to the statistics obtained from taking the *expectation over time* for *one* representative member of the ensemble, the random process is termed *ergodic.* This means that *any member* of an ergodic ensemble is representative of the entire ensemble (all members) (Fig. 11.5). An important practical implication is that the statistics of all members can be obtained solely by *time-averaging any member* of the ensemble[5]. That

[5] An example of a stationary ensemble that is *not ergodic*: an ensemble of constant variables, as none of the members is representative of the others. Arbitrarily long observation of any particular member of the ensemble can not be applied to any other particular member.

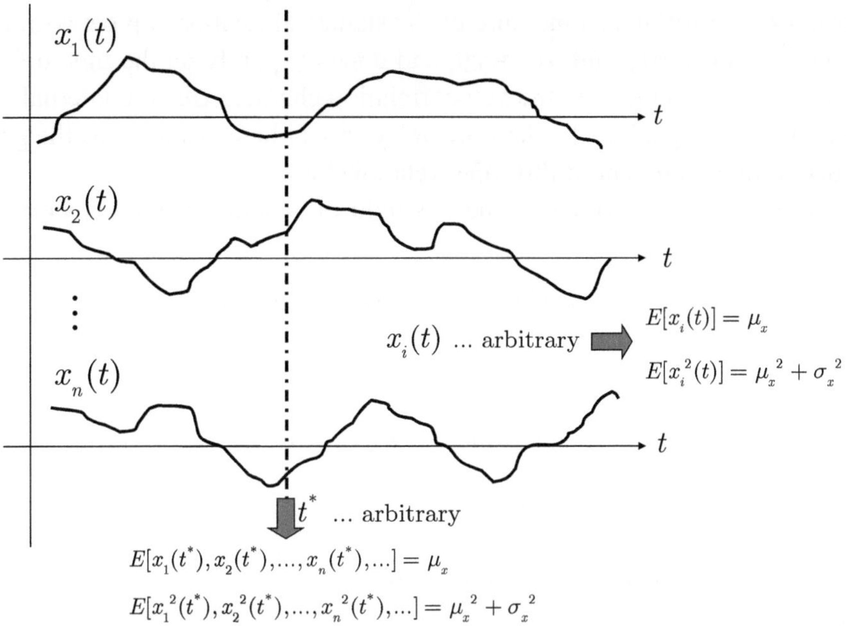

Fig. 11.5. Stochastic time functions as members of an ensemble forming an ergodic random process

is, *no* knowledge of the probability distributions is required; the statistical *expectations* can be replaced with *time averages (means)*. This opens the way for direct experimentation as a way to determine the statistics of measurable random system variables.

Temporal mean

$$\mu = E\left[x(t)\right] = \lim_{T\to\infty} \frac{1}{2T} \int_0^T x(t)\, dt \tag{11.20}$$

Quadratic temporal mean

$$E\left[x(t)^2\right] = \sigma^2 + \mu^2 = \lim_{T\to\infty} \frac{1}{2T} \int_0^T x(t)^2\, dt \tag{11.21}$$

Autocorrelation

$$r_{xx}(\tau) := E\left[x(t)\, x(t+\tau)\right] = \lim_{T\to\infty} \frac{1}{T} \int_0^{+T} x(t)\, x(t+\tau)\, dt \tag{11.22}$$

Physical dimension: $\dim r_{xx} = \left(\dim x\right)^2$

The *autocorrelation* is a measure of the statistical relationship between the function values $x(t)$ and $x(t+\tau)$, and depends purely on the time difference τ. In a certain way, the autocorrelation characterizes the internal *dynamics* of the signal $x(t)$. The time delay τ can be considered as the generalized time argument of this inner relationship.

The following important properties hold in general for the autocorrelation:

$$\begin{aligned}
&\text{(a)} \quad r_{xx}(\tau) = r_{xx}(-\tau) \quad \text{(an even function)},\\
&\text{(b)} \quad r_{xx}(0) = E\left[x(t)^2\right] = \sigma^2 + \mu^2,\\
&\text{(c)} \quad \mu = 0 \quad \text{(zero-mean): } r_{xx}(0) = \sigma^2,\\
&\text{(d)} \quad \max_{\tau} r_{xx} = r_{xx}(0),\\
&\text{(e)} \quad r_{xx}(\infty) = E\left[x(t)\right]^2 = \mu^2,\\
&\text{(f)} \quad \mu = 0 \quad \text{(zero-mean): } r_{xx}(\infty) = 0.
\end{aligned} \tag{11.23}$$

Cross-correlation

$$r_{xy}(\tau) := E\left[x(t)\,y(t+\tau)\right] = \lim_{T\to\infty} \frac{1}{T}\int_0^{+T} x(t)\,y(t+\tau)\,dt. \tag{11.24}$$

Physical dimension: $\dim r_{xy} = \dim x \times \dim y$

The *cross-correlation* is a measure of the statistical relationship between two stochastic time functions $x(t)$ and $y(t)$.

Spectral representation The spectral representation of functions using FOURIER and LAPLACE transforms requires an analytical description of the functions. In the case of stochastic signals, such a description is not available, so that a spectral representation is not immediately available. For a stochastic signal, at most the distribution of probabilities with which values occur, or even just the statistics (moments) of this distribution, are known. To solve this dilemma, the *autocorrelation* $r_{xx}(\tau)$ with a generalized time argument "τ" can be used as the representative *analytical* description of the stochastic function. As shown above, the autocorrelation contains information about both the amplitude distribution—via the moments μ and

σ^2—as well as information about internal interactions—i.e. the dynamics of the stochastic signal $x(t)$. Thus, as a practical matter, $r_{xx}(\tau)$ fulfills all the formal prerequisites for applying a FOURIER transform.

Power spectral density Formal application of the FOURIER transform to the autocorrelation function (11.22) gives the *power spectral density* (*PSD*)

$$\boxed{S_{xx}(\omega) = \int_{-\infty}^{+\infty} r_{xx}(\tau)\, e^{-j\omega\tau}\, d\tau}. \tag{11.25}$$

Physical dimension:

$$\dim S_{xx} = \left(\dim x\right)^2 / Hz \quad \text{or} \quad \left(\dim x\right)^2 \times \text{s}.$$

Inverting this transformation and applying to the power spectral density gives, in turn, the autocorrelation

$$\boxed{r_{xx}(\tau) = \frac{1}{2\pi}\int_{-\infty}^{+\infty} S_{xx}(\omega)\, e^{j\omega\tau}\, d\omega}. \tag{11.26}$$

Due to the properties of the autocorrelation (a real, even function), the power spectral density (11.25) is also a real, even function.

Eqs. (11.25) and (11.26) are known in the literature as the *WIENER-KHINCHIN relations* (Hsu 2011), (Cooper and McGillem 1999), and are ultimately the key to a compact description of stochastic signals in the frequency domain. In addition, they enable straightforward calculation of disturbance propagation, given knowledge of the relevant transfer functions.

Physical interpretation In the same way as the FOURIER integral, the power spectral density $S(\omega)$ represents the power at each frequency, whence the name. The autocorrelation $r_{xx}(\tau)$ itself represents simple *power*, i.e. *energy (work) per time*. Thus, for deterministic time functions, PARSEVAL's Theorem (Föllinger 2007) directly gives

$$\int_{-\infty}^{+\infty} x(t)^2\, \mathrm{d}t = \frac{1}{2\pi}\int_{-\infty}^{+\infty} \left|X(\omega)\right|^2\, d\omega\,.$$

In other words, the total energy of the time function equals the total energy contained in the spectral density. This can be verified with the WIENER-KHINCHIN relations by setting $\tau = 0$.

11.2.3 LTI systems with stochastic inputs

Fig. 11.6 shows a linear time-invariant (LTI) system with the impulse response $g(t)$ (or equivalently the transfer function $G(s)$) and a stochastic input $u(t)$ which can be described as a stationary random process. A description of the output $y(t)$ is sought. Due to the stationary character of the input and the linearity of the system, the output $y(t)$ can also be described as a stationary random process (Giloi 1970). For this task, the *autocorrelation*

$$r_{yy}(\tau) = \lim_{T\to\infty} \frac{1}{2T} \int_{-T}^{+T} y(t)\, y(t+\tau)\; dt \tag{11.27}$$

naturally presents itself as a descriptive measure. In order to obtain the output behavior, the *convolution integral*

$$y(t) = g(t) * u(t) = \int_{0}^{\infty} g(q)\, u(t-q)\; dq \tag{11.28}$$

can be used to calculate $y(t)$ given a known input $u(t)$, even in the case of stochastic signals.

Substituting for $y(t)$ in Eq. (11.27) with Eq. (11.28), and switching the order of integration gives

$$\begin{aligned} r_{yy}(\tau) = &\int_0^\infty \int_0^\infty g(q_1)\, g(q_2) \cdot \\ &\cdot \lim_{T\to\infty} \frac{1}{2T} \int_{-T}^{+T} u\left(t - q_1\right)\, u\left(t - q_2 + \tau\right)\; dt\; dq_1\; dq_2 \end{aligned} \tag{11.29}$$

and, following Eq. (11.22) defining the autocorrelation,

$$r_{yy}(\tau) = \int_0^\infty \int_0^\infty g(q_1)\; g(q_2)\; r_{uu}(q_1 - q_2 + \tau)\; dq_1\; dq_2 . \tag{11.30}$$

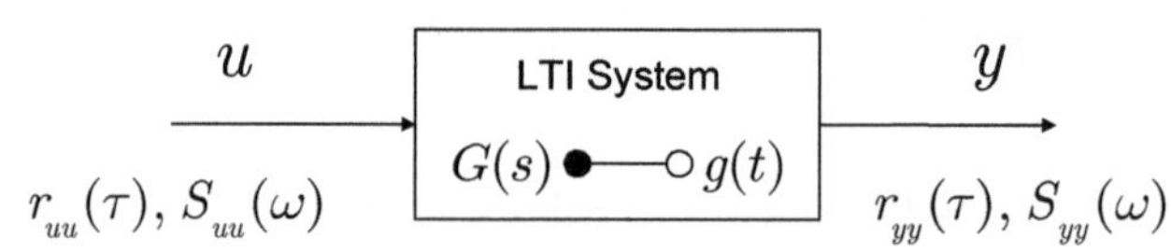

Fig. 11.6. Linear time-invariant system with stochastic inputs

Replacing the autocorrelation $r_{uu}(\tau)$ in (11.30) with the FOURIER integral

$$r_{uu}(\tau) = \frac{1}{2\pi} \int_{-\infty}^{+\infty} S_{uu}(\omega)\, e^{j\omega\tau}\, d\omega$$

results in[6]

$$r_{yy}(\tau) = \frac{1}{2\pi} \int_{-\infty}^{+\infty} \int_{-\infty}^{\infty} \int_{-\infty}^{\infty} g(q_1)\, g(q_2)\, S_{uu}(\omega)\, e^{j\omega(q_1 - q_2 + \tau)}\, dq_1\, dq_2\, d\omega\,. \quad (11.31)$$

Noting the FOURIER integrals

$$G(-j\omega) = \int_{-\infty}^{+\infty} g(\tau)\, e^{j\omega q_1}\, dq_1 \quad \text{and} \quad G(j\omega) = \int_{-\infty}^{+\infty} g(\tau)\, e^{-j\omega q_2}\, dq_2\,,$$

and equating Eq. (11.26) and Eq. (11.31) gives the following relation:

$$\begin{aligned} r_{yy}(\tau) &= \frac{1}{2\pi} \int_{-\infty}^{+\infty} S_{yy}(\omega)\, e^{j\omega\tau}\, d\omega = \\ &= \frac{1}{2\pi} \int_{-\infty}^{+\infty} G(j\omega) G(-j\omega)\, S_{uu}(\omega)\, e^{j\omega\tau}\, d\omega\,. \end{aligned}$$

Finally, comparing integrands gives the fundamental result

$$\boxed{S_{yy}(\omega) = \left|G(j\omega)\right|^2 \cdot S_{uu}(\omega)}\,. \quad (11.32)$$

Eq. (11.32) thus encapsulates the relationship between the power spectral densities of stochastic inputs and resulting outputs. Note that the phase of the system plays no role in this relationship[7].

The *variance* $\sigma_y^{\,2}$ of the *output* $y(t)$ can be derived from Eqs. (11.23)(c) and (11.26):

$$\boxed{\sigma_y^{\,2} = r_{yy}(0) = \frac{1}{2\pi} \int_{-\infty}^{+\infty} \left|G(j\omega)\right|^2 S_{uu}(\omega)\, d\omega}\,. \quad (11.33)$$

---

6 Since the impulse response $g(t) = 0$ for $t < 0$, the lower bounds of the integrals can be readily extended to $-\infty$.

7 The relation in (11.32) should not be confused with the seemingly similar relation for the *cross-spectral density* $S_{uy}(j\omega) = G(j\omega) \cdot S_{uu}(j\omega)$ (see Sec. 2.7.3), in which the phase of the system *does* play a role and can thus be used to determine the transfer function $G(s)$.

Determining the probability distribution It is well known that linear operations on random variables do not change their probability distributions. As a result, the distribution of amplitudes of the input $u(t)$ remains unchanged after passing through an LTI system, i.e. the output signal $y(t)$ possesses the same type of amplitude distribution as the input signal (a normal input distribution would thus remain normal).

11.3 White Noise

Model-based treatment of noise sources as random processes is enabled at a fundamental level by the mathematical concept of *white noise*. In a white noise signal, signal values arbitrarily close in time are assumed to be statistically independent. As will be shown, such signals are not physically realizable, but, in the form of fictitious signals, can indeed be formulated mathematically.

Fictitious continuous white noise *White noise* $n(t)$ can be thought of as a series of closely-spaced, statistically independent DIRAC impulses with a given probability distribution for their amplitudes (e.g. a normal distribution). Due to the assumption of statistical independence, it holds for the *autocorrelation* of $n(t)$ that

$$r_{nn}(\tau) = S_{n0} \cdot \delta(\tau)\,, \tag{11.34}$$

i.e. $r_{nn}(\tau) \neq 0$ only at $\tau = 0$, when the signal $n(t)$ coincides with itself in time (Fig. 11.7b).

The associated *power spectral density* of white noise can then be calculated from Eq. (11.25):

$$S_{nn}(\omega) = S_{n0} = const. \tag{11.35}$$

with constant power density over an infinitely wide frequency band (Fig. 11.7c). Naturally, this is only possible in the ideal case, explaining the non-realizability of such a signal.

Continuous white noise is thus specified using *one single* parameter: the constant *power spectral density* S_{n0} with physical dimension $\dim S_{nn} = \left(\dim n\right)^2 / Hz$.

Note that continuous *white noise* thus defined represents at most a purely *computational quantity* with the practical properties (11.34) and (11.35). In this sense, such a signal should be considered as *fictitious* continuous white noise. Despite this limitation, this abstract model is of fundamental importance in the analysis of real noise processes, as every real noise source can be represented as resulting from excitation by a fictitious continuous white noise source.

Quasi-continuous white noise One commonly used approximation of real noise processes is so-called *quasi-continuous white noise* $\hat{n}(t)$. This type of noise can always be used when the source has constant power density over a finite frequency range $[0, \omega_b]$ and its bandwidth ω_b is significantly greater than the bandwidth of the affected dynamical system. In such a case, the power spectral density $S_{\hat{n}\hat{n}}(\omega)$ disappears at large frequencies, and the total power thus remains finite (Fig. 11.8).

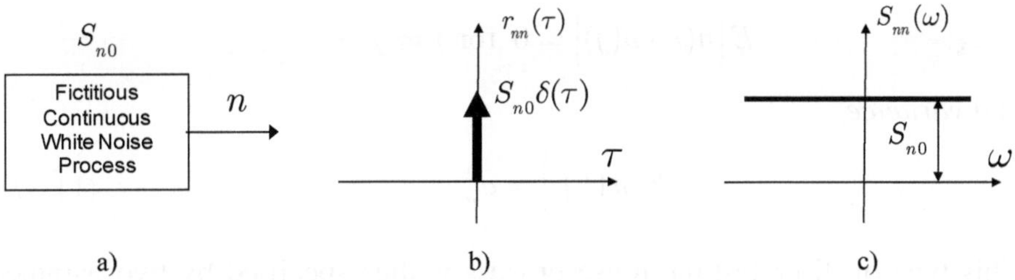

Fig. 11.7. Model representation for fictitious continuous white noise: a) noise source, b) autocorrelation, c) power spectral density

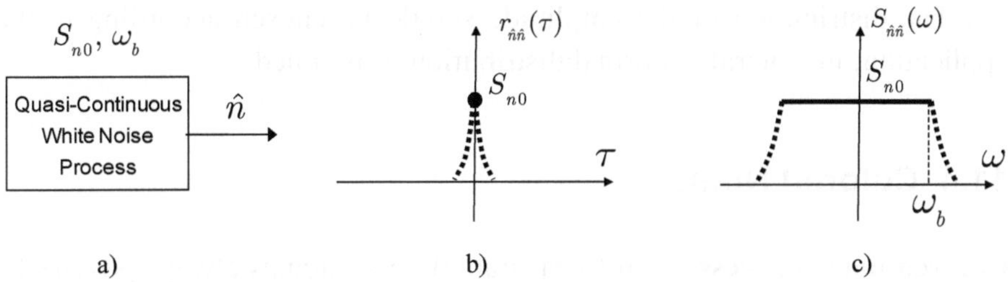

Fig. 11.8. Quasi-continuous white noise: a) noise source, b) autocorrelation, c) power spectral density

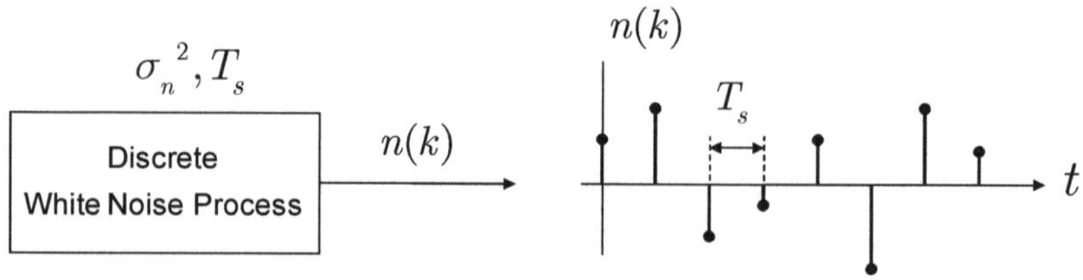

Fig. 11.9. Discrete white noise as discrete-time random process

Discrete white noise A direct conceptual extension of the continuous case is *discrete white noise.* This type of noise is represented as a zero-mean *numerical series*

$$n(k) = \{n(0),\ n(T_s),\ n(2T_s),\ \ldots\ n(kT_s),\ \ldots\} \tag{11.36}$$

defined at *discrete times* $t_k = kT_s$, with *statistically independent* sequential elements, i.e.

$$E\left[n(i) \cdot n(j)\right] = 0 \text{ for } i \neq j$$

and *variance*

$$E\left[n(k)^2\right] = \sigma_n^2 . \tag{11.37}$$

This type of discrete-time noise process is thus specified by two parameters: the *variance* $\sigma_n^{\ 2}$ (with $\dim \sigma_n^{\ 2} = (\dim n)^2$) or *standard deviation* σ_n (with $\dim \sigma_n = \dim n$), and the *sampling time* T_s.

A realizable system model for such a process can be conceptualized as a zero-mean random number generator with variance $\sigma_n^{\ 2}$ and sampling time T_s. The distribution of the amplitude should be chosen according to the application; in general, a normal distribution is assumed.

11.4 Colored Noise

Colored noise process Real (measurable) noise signals always possess finite power. As a result of Eq. (11.26), this implies that the *power spectral density* must be *band-limited.* Thus, all physically realizable systems exhibit a low-pass behavior at sufficiently large frequencies, with bandwidth ω_b. In some cases, the passband also contains particular smaller stop- or passbands.

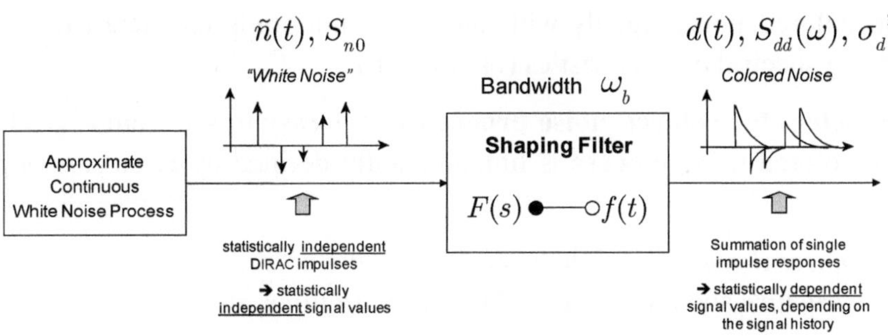

Fig. 11.10. Colored noise process generated from a (zero-mean) band-limited white noise process

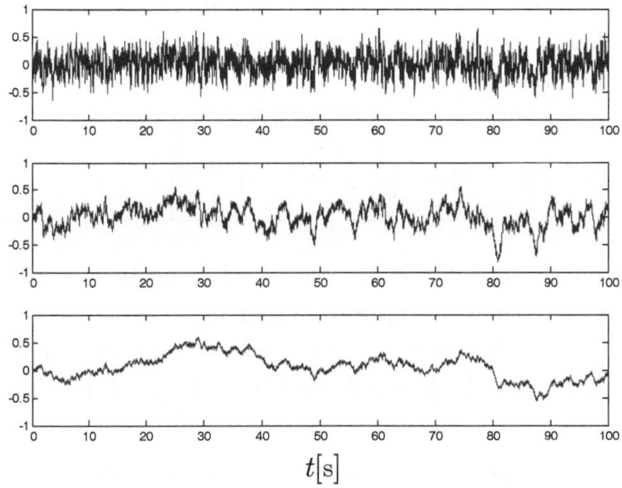

Fig. 11.11. Colored noise signals with *equal* amplitude characteristics $N(\mu = 0, \sigma = 0.2)$ yet *differing* spectral characteristics (bandwidth)

This introduces the concept of employing appropriate *filtering* of the homogeneous, infinitely wide power spectral density of white noise to obtain more complex behaviors (Fig. 11.10). Combining a fictitious white noise source with a *shaping filter* $F(s)$ in this manner results in *colored noise* or a colored noise process[8]. A fictitious white noise process is illustrated in Fig. 11.10 in an approximate realization in the form of a series of statistically independent DIRAC impulses. Fig. 11.11 shows three zero-

8 This terminology results from the analogy to white and band-limited light.

mean colored noise signals with the same amplitude characteristics, but differing spectral characteristics (bandwidth).

Parameters for colored noise processes It is easy to see from Fig. 11.10 that a colored noise process is fundamentally defined by *three sets of parameters*:

- the *power density* S_{n0} of the generating *white noise*,
- the *structure* and *parameters* of the *shaping filter*, and
- the *power spectral density* $S_{dd}(\omega)$ of the *output signal* $d(t)$ or its *variance* σ_d^2.

Note, however, that of these three sets of parameters, only *two* are *independent*. Indeed, the previously noted relation (11.33) links the inputs and outputs of a linear system, i.e. for an arrangement as shown in Fig. 11.10, it holds in general that

$$\boxed{\sigma_d^2 = S_{n0} \cdot \left[\frac{1}{2\pi} \int_{-\infty}^{+\infty} \left| F(j\omega) \right|^2 d\omega \right]}. \tag{11.38}$$

Shaping filters Commonly used types of shaping filter are first- and second-order *lag elements*. In digital components (particularly sensors), discrete white noise is also often "filtered" with a *zero-order hold* element. In the latter case, the sampling time T_s also becomes an important parameter.

Table 11.2. Covariance relations for common shaping filters (S_{n0} is the power spectral density of the fictitious white noise source)

	Shaping filter	Covariance of the output signal
1st-order:	$F(s) = \dfrac{K}{1 + T_0 s}$	$\sigma_d^2 = \dfrac{K^2}{2T_0} S_{n0}$
2nd-order:	$F(s) = \dfrac{K}{1 + 2d_0 T_0 s + T_0^2 s^2}$	$\sigma_d^2 = \left(\dfrac{K}{2}\right)^2 \dfrac{1}{d_0 T_0} S_{n0}$
Sample-and-hold:	$F(s) = \dfrac{1 - e^{-T_s s}}{T_s\ s}$	$\sigma_d^2 = \dfrac{S_{n0}}{T_s}$

Covariance relations for common shaping filters Relation (11.38) allows the noise model to be easily constructed if the integral is known. As this function is defined solely by the shaping filter, it can be fairly easily determined analytically for several standard types. A few important types of filter are presented below (see also Table 11.2); for additional types, see *Appendix A*.

First-order shaping filter The simplest form of signal shaping, often sufficient by itself, is obtained with a first-order lag element (low-pass filter)[9]

$$F(s) = \frac{K}{1+T_0 s}. \tag{11.39}$$

For this filter, the covariance relation (11.38) gives

$$\sigma_d^{\,2} = S_{n0} \cdot \left[\frac{1}{2\pi} \int_{-\infty}^{+\infty} \frac{K^2}{1+T_0^{\,2}\omega^2}\, d\omega \right] = S_{n0} \frac{K^2}{T_0} \frac{1}{\pi} \arctan \omega T_0 \Big|_{\omega=0}^{\infty},$$

and it follows that

$$\boxed{\sigma_d^{\,2} = S_{n0} \frac{K^2}{2T_0}}. \tag{11.40}$$

Sample-and-hold shaping filter A wide-band noise process arising in a digital, discrete-time system and subsequently affecting a continuous-time system via a hold element can be modeled as continuous white noise with a downstream *sample-and-hold element* (Fig. 11.12). In this case, the covariance relation (11.38) using a zero-order hold, and taking into account the sampling process (see Ch. 10), gives

$$\sigma_{\bar{d}}^{\,2} = S_{n0} \cdot \left[\frac{1}{2\pi} \int_{-\infty}^{+\infty} \left(\frac{1}{T_s} \frac{2 \sin\left(\frac{\omega T_s}{2}\right)}{\omega} \right)^2 d\,\omega \right]$$

---

9 The resulting colored noise process is generally also termed a first-order MARKOV process, and is used when working in the time domain with state space models.

and thence, the fundamental relation

$$\boxed{\sigma_{\bar{d}}^{\ 2} = S_{n0} \frac{1}{T_s}}. \tag{11.41}$$

In particular, the covariance relation (11.41) can always be used when a *discrete-time noise source* following Eqs. (11.36) and (11.37) is combined with a hold element, and the equivalent white noise source is required for analysis.

White noise simulation model The sample-and-hold shaping filter approach presented above suggests an intuitive heuristic simulation model for a white noise process. If the sampling period is kept sufficiently small with respect to the time constants of the system to be excited by the white noise process, then the model shown in Fig. 11.13 produces a continuous staircase function $\bar{n}(t)$ which approximates a wide-band process with approximately constant power spectral density S_n. The random number generator can employ any probability density available in the simulation tool (normal, uniform, etc.). This simulation white noise generator can be combined with any shaping filter to realize simulation models of colored noise.

Correlation time Colored noise is often also referred to as (time-) *correlated* noise. This terminology can be explained via the previously mentioned conceptualization of white noise as a series of closely-spaced, statistically independent DIRAC impulses. Given this model, the output

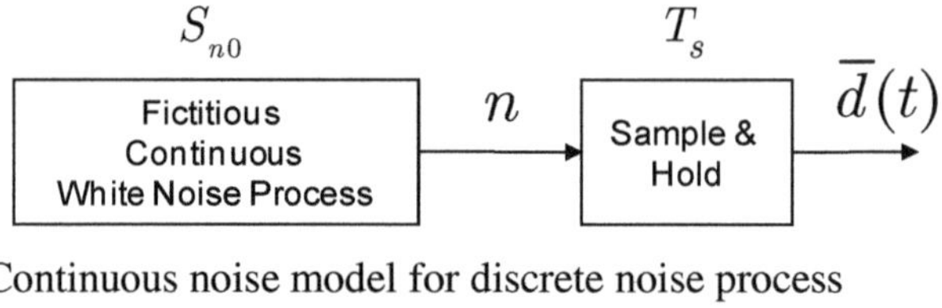

Fig. 11.12. Continuous noise model for discrete noise process

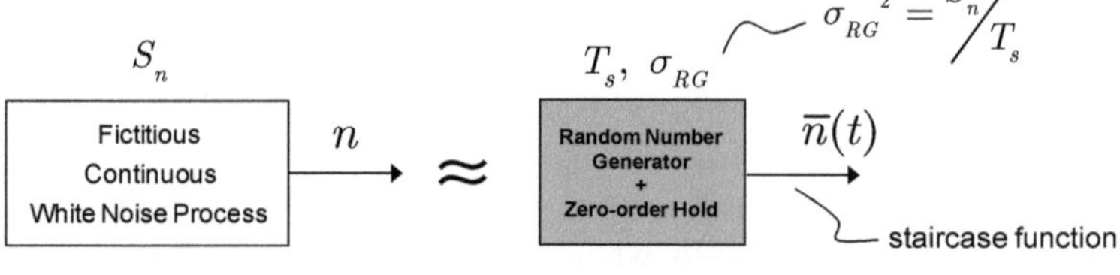

Fig. 11.13. Simulation model for a continuous white noise process

signal $d(t)$ in the arrangement shown in Fig. 11.10 can be seen to be composed of a series of superimposed impulse responses. Following the principle of superposition, the output $d(t^*)$ at a given time t^* will thus contain contributions due to all previous input impulses $\delta(t^* - \tau)$. As a result, at every instant of time, the previous history (time correlation) can be seen in $d(t)$—a result of the "memory" of the shaping filter. In this context, the term *correlation time* T_{corr} is generally used to designate the time span after which an impulse response has sufficiently decayed as to make no further contributions to later times. The common *shaping filters* presented above result in the following correlation times:

- 1st-order: $T_{corr} = T_0$, time constant T_0, i.e. output has decayed to $1/e$;
- 2nd-order: $T_{corr} = 2.15 \cdot T_0$, time constant T_0, damping $d_0 = 1$;
- sample-and-hold: $T_{corr} = T_s$, sampling time T_s.

11.5 Modeling Noise Sources

Physical noise sources Physical noise sources should be modeled as colored noise sources, as shown in Fig. 11.10. The constitutive parameters to be specified in a given model are the *amplitude characteristics* (the distribution, statistical parameters) and the *spectral characteristics* (the shaping filter). As a rule, in an actual design process, *data sheets* or *empirical data* are available for the system components under consideration. These should be used to derive the information required for modeling. This section presents several typical configurations commonly seen in practice.

Spectral characteristics: shaping filter Only in very rare cases are the spectral characteristics of a noise source available as an explicit, frequency-dependent power spectral density. Data sheets often refer to the *bandwidth* $\omega_b = 2\pi f_b$ of the noise signal, but the exact shape of the spectrum remains obscure. A reasonable assumption in such cases is a first- or second-order *shaping filter*, with a filter time constant of $T_0 = 1/\omega_b$. If the dynamics of the electronics of the component are roughly known, they can be used to guide the form of the shaping filter to a good approximation. By and large, the actual physical noise signal arises in precisely this portion of the component. For a digital component (primarily sensors), which generally lack significant dynamics in the creation of the signal, a zero-order hold should be assumed to describe the smoothing of the noise impulse series.

Amplitude characteristics In many cases, the amplitude characteristics of a stochastic signal are well-approximated by a *normal distribution* $N(\mu_d, \sigma_d)$. As long as the component description (data sheet) makes no explicit statements to the contrary, a normal distribution is implicitly implied. Though the mean μ_d is a random variable for a stationary random process, it remains constant within some finite observation interval. For this reason, the mean is often removed from any stochastic analysis and is treated as a *systematic*, constant quantity (*bias* or *offset*). All that remains as a stochastic input, then, is the zero-mean, normally-distributed random process $N(0, \sigma_d)$. Thus, the amplitude characteristics are solely defined by the *standard deviation* σ_d or *variance* σ_d^2 of the input signal $d(t)$.

In cases where the datasheet does not specify parameters in the above terms, there are generally three typical, more indirect specifications for the amplitude characteristics. These are explained below.

Peak-to-peak amplitude A specification of the amplitude in the form

$$d_{\mathrm{pp}} \quad [\dim d] \tag{11.42}$$

implies a *min-max evaluation* of the signal amplitudes occurring in the noise *output* signal $d(t)$ giving a *peak-to-peak (PP)* value, as shown in Fig. 11.14. This value is also straightforward to measure. If a normal distribution $N(0, \sigma_d)$ is assumed, then, with a statistical certainty of $S = 99.7\%$, the amplitude values of $d(t)$ lie within the interval

$$[-3\sigma_d, 3\sigma_d] \approx [d_{\min}, d_{\max}] = d_{pp},$$

from which it immediately follows that

$$\boxed{\sigma_d \approx \frac{d_{pp}}{6}}. \tag{11.43}$$

Note that, as a result of relation (11.38), σ_d depends on the shaping filter used. If the shaping filter changes (e.g. due to electrical loading of the amplifier), the signal variance also changes.

RMS value An amplitude specification of the form

$$d_{\mathrm{rms}} \quad [\dim d] \tag{11.44}$$

indicates the *root mean square (RMS)* value of the noise *output* signal $d(t)$, following Eq. (11.15) with the *implicit assumption* of a zero-mean normal distribution $N(0,\sigma_d)$. As a result, Eq. (11.16) gives

$$\boxed{\sigma_d = d_{\mathrm{rms}}} \tag{11.45}$$

It should again be noted that due to Eq. (11.38), σ_d here also depends on the shaping filter used.

Power density At first glance, an amplitude specification of the following type (often without a name/symbol, but rather a simple numerical value with physical dimension) appears unusual:

$$\begin{aligned} &S_{n0} \quad \left[(\dim d)^2 / Hz\right] \quad \text{or} \quad \left[(\dim d)^2 \times \mathrm{s}\right] \\ &\sqrt{S_{n0}} \quad \left[\dim d / \sqrt{Hz}\right] \quad \text{or} \quad \left[\dim d \times \sqrt{\mathrm{s}}\right]. \end{aligned} \tag{11.46}$$

Such a specification obviously does *not* indicate the amplitude characteristics of the *output* noise signal $d(t)$, but rather, explicitly gives the *power spectral density* of the *input white noise* of the shaping filter (see Fig. 11.10). If no other distribution is specified, a normal distribution is assumed here as well. One *advantage* of this amplitude specification lies in the fact that it is *independent* of the shaping filter employed. For this reason, this specification is preferred for sensors in which external circuitry can be used to set the signal bandwidth (and thus in inverse proportionality, the noise amplitude).

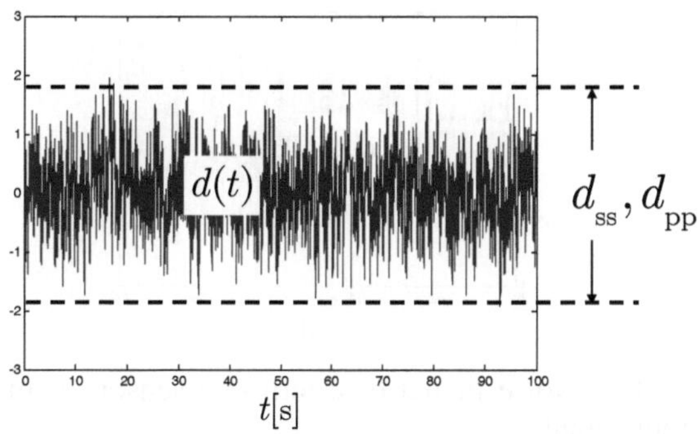

Fig. 11.14. Amplitude specification via peak-to-peak values

Example 11.1 *Noise model for a turn rate sensor.*

The following information is gathered from the data sheet of a turn rate sensor (gyroscope):
- Noise: 0.25 $\mathrm{deg/s_{rms}}$
  measured with bandwidth 0.1…30 Hz

This noise specification clearly indicates the *RMS value (standard deviation)* of normally distributed, zero-mean colored noise, i.e.

$$\sigma_d = 0.25 \ \ \mathrm{deg/s} \,.$$

To complete the noise model, it remains to specify the white noise source and the shaping filter. A reference point for the noise bandwidth is provided by the specified measurement bandwidth $f_b = 30 \ Hz$. The order of the shaping filter remains unspecified. As an example, a *second-order lag shaping filter* following Table 11.2, with parameters

$$T_0 = \frac{1}{2\pi f_b} = 0.0053 \ \mathrm{s}\,, \ d_0 = 0.7, \ K = 1\,,$$

is assumed.

The *power spectral density* of the white noise source is then obtained using Table 11.2, giving

$$S_{n0} = \sigma_d^{\ 2} 4 D_0 T_0 = \sigma_d^{\ 2} \frac{4D_0}{\omega_0} = 9.4 \cdot 10^{-4} \frac{\mathrm{deg}^2}{\mathrm{s}} \,. \tag{11.47}$$

The complete analytical noise model along with a typical output signal is shown in Fig. 11.15.

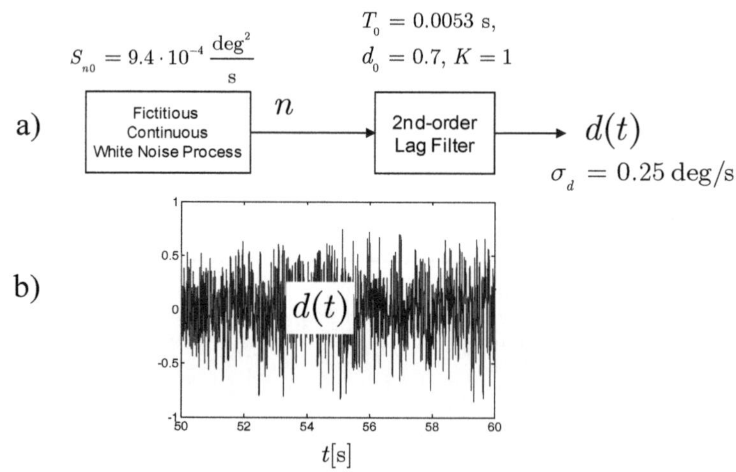

Fig. 11.15. Noise model of a turn rate sensor: a) mathematical model, b) output signal

■

Example 11.2 *Noise model for a magnetic field sensor.*

The following information is gathered from the data sheet of a magnetic field sensor:

- Noise: $30\ \mathrm{nT}/\sqrt{\mathrm{Hz}}$
- Maximum frequency: 100 Hz

This noise specification clearly indicates the *square root of the power spectral density* of the equivalent white noise, i.e.

$$\sqrt{S_{n0}} = 30\ \ \mathrm{nT} \,/\, \sqrt{\mathrm{Hz}} \quad \Rightarrow \quad S_{n0} = 900\ \mathrm{nT}^2/\mathrm{Hz}\,. \qquad (11.48)$$

Shaping filter The bandwidth of an appropriate shaping filter cannot be gathered directly from the specification. The indicated maximum frequency is clearly the *maximum physical* bandwidth of the sensor element. In a concrete application, the actual bandwidth is chosen to be just as large as is necessary for the rest of the system, so as to keep the noise at the sensor output as small as possible (cf. the reciprocal relationship of noise covariance and bandwidth in Table 11.2). Bandwidth matching can be achieved either by placing the sensor element in an appropriate circuit, or using a measurement amplifier. In the case presented here, as an example, a band-limited amplifier with $f_b = 10$ Hz and a first-order lag behavior is assumed, so that for the *shaping filter*, it follows that (cf. Table 11.2)

$$T_0 = 1/_{2\pi f_b} = 0.016\ \mathrm{s}\,,\ \ K = 1\,.$$

Output signal The a*mplitude characteristics* of the output signal of the measurement amplifier are obtained from the covariance relation

$$\sigma_d = \sqrt{\frac{S_{n0}}{2T_0}} = 168\ \ \mathrm{nT}\,.$$

The complete analytical noise model of the magnetic field sensor, including amplifier, is shown in Fig. 11.16.

Changing the measurement electronics It is worth noting the following advantageous circumstances in this particular noise specification. Clearly, it is the *general* noise properties of the sensor which have been characterized above. In a concrete application, if the sensor is outfitted with measurement electronics exhibiting other frequency characteristics, then a shaping filter with these particular characteristics should be used, while maintaining *the same* power spectral density (11.48).

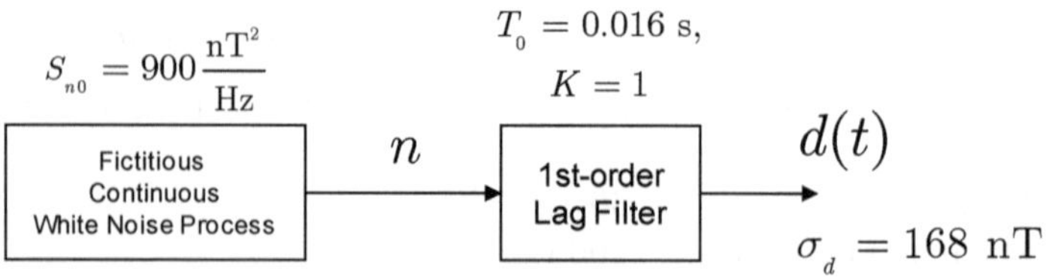

Fig. 11.16. Noise model for a magnetic field sensor

11.6 Covariance Analysis

Disturbance propagation The mathematical key to describing disturbance propagation is provided by the covariance relation (11.33). Using this relation, the variance of an output can be calculated for an LTI system given a known transfer function and stochastic input with known power spectral density (cf. Fig. 11.6). This variance describes the amplitude distribution of the output, and thus characterizes the effects of a stochastic disturbance input. A particularly simple relation results for the special case of white noise stochastic input. In this case, the power spectral density degenerates to a constant, and the simple relation (11.38) results. Note that the transfer function in the integrand must encompass the complete signal path from the white noise source to the system variable under examination.

Compound systems Fig. 11.17 shows the sequential arrangement of two LTI systems with a white noise source as input. As per the covariance relation (11.38), it holds for the variance of the output variable $x_2(t)$ that

$$\begin{aligned} \sigma_{x_2}^{\ 2} &= S_{n0} \frac{1}{\pi} \int_0^\infty \left| G_1(j\omega) \cdot G_2(j\omega) \right|^2 d\omega = \\ &= S_{n0} \frac{1}{\pi} \int_0^\infty \left| G_1(j\omega) \right|^2 \cdot \left| G_2(j\omega) \right|^2 d\omega \end{aligned} \tag{11.49}$$

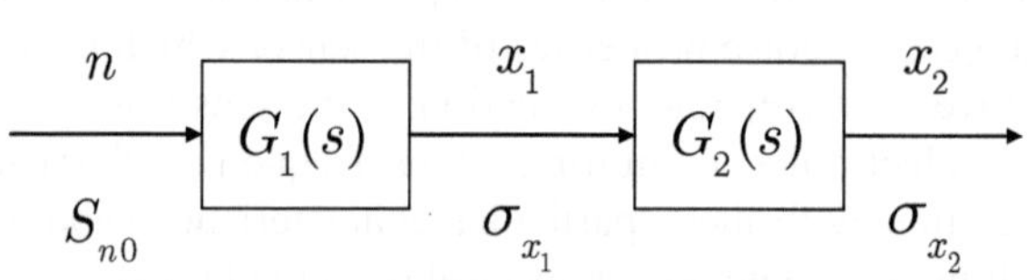

Fig. 11.17. Covariance calculation for a compound system

Application: colored noise source and LTI system The arrangement shown in in Fig. 11.17 represents the standard case for an LTI system $G_2(s)$ with a stochastic disturbance in the form of a colored noise source with shaping filter $G_1(s)$. The colored noise signal $x_1(t)$ has variance

$$\sigma_{x_1}^{\ 2} = S_{n0} \frac{1}{\pi} \int_0^\infty \left|G_1(j\omega)\right|^2 d\omega , \tag{11.50}$$

whereas the output $x_2(t)$ propagated through the LTI system possesses the variance defined by Eq. (11.49).

Incorrect method of computation Note that, in general,

$$\sigma_{x_2}^{\ 2} = S_{n0} \frac{1}{\pi} \int_0^\infty \left|G_1(j\omega)\right|^2 \left|G_2(j\omega)\right|^2 d\omega \ \neq \ \sigma_{x_1}^{\ 2} \cdot \frac{1}{\pi} \int_0^\infty \left|G_2(j\omega)\right|^2 d\omega . \tag{11.51}$$

If the right side of (11.51) were to hold, this would imply that the standard deviation σ_{x_2} (or variance $\sigma_{x_2}^{\ 2}$) of the system variable x_2 could be calculated locally using only the variance of the input x_1. Unfortunately, this is *not* the case. Rather, in the integrand, the entire chain from the white noise source n up through the variable of interest x_2 must be taken into account. This increases the order of the resulting transfer function and complicates analytical computation of the integral.

Simplification for wide-band noise source In the system configuration shown in Fig. 11.18a, $F(s)$ represents the shaping filter for the noise source, and $T(s)$ the system transfer function (e.g. that of a mechatronic system) from the noise input to the output variable $y(t)$. The bandwidth configuration can be seen in Fig. 11.18b.

Here, the bandwidth of the noise source is significantly greater than the bandwidth of $T(s)$, so that Eq. (11.49) simplifies to

$$\sigma_y^{\ 2} = \frac{S_{n0}}{\pi} \int_0^{\omega^*} \underbrace{\left|F(j\omega)\right|^2}_{\approx 1} \cdot \left|T(j\omega)\right|^2 d\omega \ + \ \frac{S_{n0}}{\pi} \int_{\omega^*}^\infty \left|F(j\omega)\right|^2 \cdot \underbrace{\left|T(j\omega)\right|^2}_{\approx 0} d\omega$$

$$\sigma_y^{\ 2} \approx \frac{S_{n0}}{\pi} \int_0^{\omega^*} \left|T(j\omega)\right|^2 d\omega = \frac{S_{n0}}{\pi} \int_0^\infty \left|T(j\omega)\right|^2 d\omega . \tag{11.52}$$

In this case, then, the shaping filter can be completely ignored, and the LTI system $T(s)$ can be taken to be directly acted upon by the white noise (Fig. 11.18c).

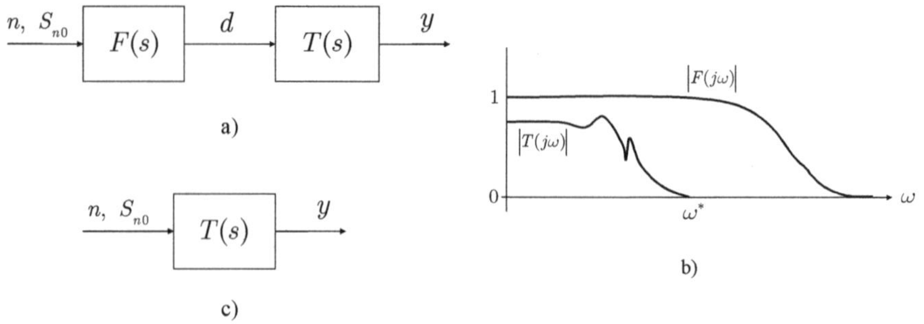

Fig. 11.18. Wide-band noise source for a narrow-band LTI system: a) shaping filter and LTI system, b) frequency responses of shaping filter and LTI system, c) replacement system given the wide-band shaping filter

Covariance analysis The stochastic dynamic analysis presented here is generally known under the name *covariance analysis*[10]. Using this type of analysis, the *variances* of system variables are calculated for a given *LTI system* under *stationary* conditions and the assumption of *white noise sources* as the system inputs (cf. Eq. (11.38)). In this way, important information is gained concerning the probabilities associated with the amplitudes of these system variables.

Using this method, it is possible to consider the entirety of all stochastic ensembles of a particular input variable in *one single* computation. If the analytical formulas in Table 11.2 or *Appendix A* are used for simple, typical transfer functions, important analytical relationships between system parameters and signal variances can be derived.

During the course of system design, the relevant transfer functions can often be approximated with second- to third-order systems for rough design considerations. In this case, the analytical formulas presented above can be used directly.

Numerical covariance computation To compute the covariance for higher-order systems, it is advantageous to evaluate the covariance relations (11.33) or (11.38) numerically. A straightforward method for this computation is to numerically integrate the frequency response, i.e. the area under the squared frequency response graph is simply approximated. In practice, however, computer-aided tools often employ a different method based on a *state-space representation* of the system dynamics (Gelb 1974).

[10] "Covariance" is used here as a general term for the "variance" of the output signal (cf. note on "covariance" in Sec. 11.2.1).

The LTI system shown in Fig. 11.6 can be equivalently represented as a state-space model

$$\begin{aligned}\dot{\mathbf{x}} &= \mathbf{A}\mathbf{x} + \mathbf{B}\mathbf{n} \\ \mathbf{y} &= \mathbf{C}\mathbf{x} \\ E[\mathbf{n}(t)\mathbf{n}(\tau)^T] &= \mathbf{S}_{n0}\delta(t-\tau) = \text{diag}\left\{S_{n1,0} \quad \dots \quad S_{nn,0}\right\}\delta(t-\tau),\end{aligned} \tag{11.53}$$

where Eq. (11.53) represents a generalized multivariable system with m outputs $\mathbf{y}(t) \in \mathbb{R}^m$, and where the inputs have been chosen to be n statistically independent *white noise processes* $\mathbf{n}(t) \in \mathbb{R}^n$. It can be shown that the time-varying *covariance matrix* $\mathbf{P}_x(t)$ of the state vector can be obtained via a system of matrix differential equations—the *matrix RICCATI equations*—giving

$$\dot{\mathbf{P}}_x(t) = \mathbf{A}\mathbf{P}_x(t) + \mathbf{P}_x(t)\mathbf{A} + \mathbf{B}\mathbf{S}_{n0}\mathbf{B}^{\mathrm{T}}. \tag{11.54}$$

In a comparable manner to Eq. (11.38) the *steady-state* solution $\mathbf{P}_{x,\infty}$ to Eq. (11.54) is the solution to an algebraic system of equations, the *LYAPUNOV equation*

$$\mathbf{A}\mathbf{P}_{x,\infty} + \mathbf{P}_{x,\infty}\mathbf{A} + \mathbf{B}\mathbf{S}_{n0}\mathbf{B}^{\mathrm{T}} = \mathbf{0}, \tag{11.55}$$

whence it follows for the *steady-state covariance matrix* $\mathbf{P}_{y,\infty}$ of the outputs $\mathbf{y}(t)$ that

$$\mathbf{P}_{y,\infty} = \mathbf{C}\mathbf{P}_{x,\infty}\mathbf{C}^T. \tag{11.56}$$

Eq. (11.56) gives equivalent values to those obtained from the spectral covariance relation (11.38), but it possesses one great advantage. In the LYAPUNOV equation (11.55), not only are the *autocovariances* of the state variables calculated (the elements of $\mathbf{P}_{x,\infty}$ on the main diagonal), but automatically also the cross-covariances of all states (on the off-diagonals) which account for the internal system coupling. The same also holds for the output covariances in Eq. (11.56). This significantly increases the efficiency of numerical calculation, particularly in the multi-variable case, as compared to the spectral relation.

In MATLAB, for example, numerical calculation of the covariance can be performed with the function

[Py,Px]=covar(sys,Sn0),

where the LTI system is represented by sys (the remaining variables are self-explanatory).

Covariance calculation via simulation Due to the assumed ergodicity of the noise process, the covariance of the system output variables can naturally also be calculated using a time average (expectation over time, see Eq. (11.21)). This can be approximated with a simulation experiment in which the averaging takes place over a sufficiently long period and suitable approximate white noise generators are employed (see Ch. 3). In general, the computation required for numerical integration over a long time interval is greater than for a spectral covariance calculation.

Monte-Carlo simulation Clear fundamental advantages of simulation-based covariance calculations appear in *nonlinear* systems. In this case, analytical covariance calculation is only approximately possibly using linearized system models. When combined with a sufficiently *large number* of simulation experiments conducted on a *nonlinear simulation model*, characteristic statistical parameters for the system variables of interest can be calculated via an *ensemble average*, given carefully chosen varying stochastic parameters. Such a process is termed a *Monte-Carlo simulation*. However, attention should be paid to the fact that, due to the nonlinearities of the system model, the probability distributions of the outputs will generally *not* be normally distributed. The estimated standard deviation σ_y thus provides only limited information about the actual amplitudes of $y(t)$. In such cases, higher statistical moments should thus also be estimated (Lin 1991).

Noise excitation in multibody systems Elastic coupling in multibody systems leads to oscillatory system components with the structure shown in Fig. 11.19. Wide-band noise excitation results in the output signal variance

$$\sigma_y^{\,2} = S_{n0}\left(\frac{V_{MBS}}{2}\right)^2 \frac{\omega_0}{d_0} \tag{11.57}$$

(cf. Eq. (11.52), also valid for higher-order systems in an equivalent manner, see *Appendix A*). As expected, there is a reciprocal relationship between the variance and the damping d_0 of the resonant mode, but there is also an additional direct proportional dependence on the natural frequency ω_0. As a consequence, with wide-band noise excitation, lightly-damped spillover eigenmodes especially lead to a significant increase in the noise level. This adverse effect can be compensated for with suitable bandwidth limiting in the control loop. Note, however, the potential negative impact on robust stability (see Ch. 10). It should be further noted that

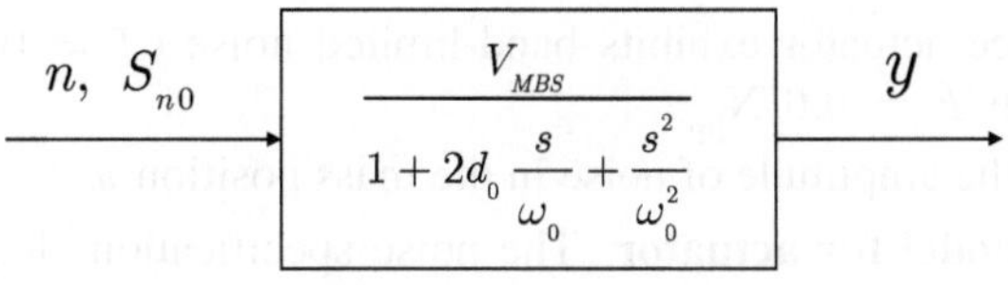

Fig. 11.19. Multibody subsystem with noise excitation

since the damping d_0 is often poorly known, predictions obtained from Eq. (11.57) can exhibit significant indeterminacy, particularly in the case of very light damping.

Example 11.3 *Covariance analysis for a mechatronic system.*

Problem statement Consider an elastically-suspended single-mass system with an integral controller (Fig. 11.20)

$$P_{MBS}(s) = \frac{Y(s)}{U(s)} = V_{MBS} \frac{1}{1 + 2d_0 \dfrac{s}{\omega_0} + \dfrac{s^2}{\omega_0^2}}$$

$$H(s) = \frac{V_I}{s}; \qquad V_I = 4; \; d_0 = 0.2$$

$$V_{MBS} = \frac{1}{2k}, \; \omega_0 = \sqrt{\frac{k}{m}}; \qquad k = 10 \text{ N/sec}; \; m = 1 \text{ kg} .$$

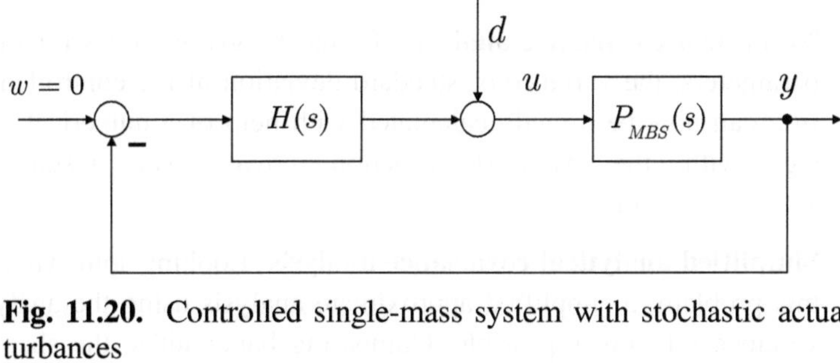

Fig. 11.20. Controlled single-mass system with stochastic actuator disturbances

The force actuator exhibits band-limited noise ($f_b = 10$ Hz, first-order lag) with $F = 0.6\ \mathrm{N_{pp}}$.

Find the amplitude of noise in the mass position y.

Noise model for actuator The noise specification clearly indicates the peak-to-peak value for a colored noise process. Under the implicit assumption of a normal amplitude distribution $N(0,\sigma_d)$, it follows that the standard deviation is

$$\sigma_d = \frac{0.6\ \mathrm{N_{pp}}}{6} = 0.1\ \mathrm{N}\,.$$

The resulting shaping filter is

$$F(s) = \frac{1}{1+T_0 s} \quad \text{where } T_0 = \frac{1}{2\pi f_b} = 0.016\ \mathrm{s}\,.$$

Using the corresponding formula from Table 11.2, it then follows that the power spectral density of the white noise source is

$$S_u = \sigma_u^{\ 2} \cdot 2T_0 = 3.2 \cdot 10^{-4}\ \mathrm{N^2/Hz}\,.$$

Noise transfer function The relevant transfer function between the white noise source n_u and the controlled variable y can be computed as (see also Fig. 11.21a)

$$T_{y/n} = F(s) \cdot T_{y/d} = F(s) \frac{P(s)}{1+H(s)P(s)}\,,$$

$$T_{y/n} = F(s) \frac{\frac{1}{V_I} s}{1 + \frac{1}{V_L} s + \frac{2d_0}{V_L \omega_0} s^2 + \frac{1}{V_L \omega_0^{\ 2}} s^3} \quad \text{where } V_L = V_I \cdot V_{MBS}\,. \tag{11.58}$$

Numerical covariance analysis Using the given values for the system parameters, the variance or standard deviation of the controlled variable $y(t)$ can now be directly calculated with help of a numerical computer, e.g. with the MATLAB function covar. The result is then $\sigma_{y,num} = 1.9\ \mathrm{mm}$.

Simplified analytical covariance analysis Looking somewhat closer at this problem, a simplified approximate analysis using the analytical covariance relation is possible. Comparing bandwidths, the shaping filter bandwidth $\omega_{b,F} = 62,8 \text{ rad/s}$ can be seen to be more than a decade greater than the relevant dynamics of the control loop

$$T_{y/d} = \frac{\frac{1}{V_I} s}{\left(1 + T_1 s\right)\left(1 + 2\frac{\tilde{d}_0}{\tilde{\omega}_0} s + \frac{1}{\tilde{\omega}_0^2} s^2\right)} \approx$$
$$\approx \frac{\frac{1}{V_I} s}{\left(1 + \frac{1}{V_L} s\right)\left(1 + 2\frac{d_0}{\omega_0} s + \frac{1}{\omega_0^2} s^2\right)}. \tag{11.59}$$

The exact values of T_1, $\tilde{d}_0$, and $\tilde{\omega}_0$ can be easily determined numerically. For a rough calculation, the simple approximations $T_1 \approx 1/V_L = 5 \text{ rad/s}$, $\tilde{d}_0 \approx d_0$, and $\tilde{\omega}_0 \approx \omega_0 = 3.16 \text{ rad/s}$ can be used (the justification for this assumption comes from, e.g., a qualitative examination using root locus curves).

Thus, the control loop is excited by *quasi-continuous white* noise, i.e. the shaping filter can be removed from consideration for the analysis (cf. Fig. 11.18c).

The *analytical* covariance relation of the simplified noise transfer function (11.59) can be taken from *Appendix A*:

$$T(s) = \frac{Ks}{\left(1 + 2d_0 T_0 s + T_0^2 s^2\right)\left(1 + T_1 s\right)} \quad \Rightarrow$$
$$\Rightarrow \quad \sigma_d^2 = \left(\frac{K}{2}\right)^2 \frac{\omega_0^3}{d_0} \frac{1}{1 + 2d_0 \omega_0 T_1 + \left(\omega_0 T\right)^2}, \tag{11.60}$$

resulting, with the given numerical parameters, in the *analytical* predication for the standard deviation of the output variable $y(t)$ $\sigma_{y,analyt} = 1.8 \text{ mm}$.

Simulation-based covariance analysis If the control loop structure in Fig. 11.20 is implemented in a simulation tool (e.g. SIMULINK) along with a model for a colored noise process, according to the methods presented in Ch. 3, the result—given a simulation time of 2000 s and a step size of 0.01 s in a RUNGE-KUTTA-4 integration—is a simulation-based estimated value of the standard deviation of the output variable $y(t)$: $\sigma_{y,sim} = 1.9 \text{ mm}$ (see Fig. 11.21b).

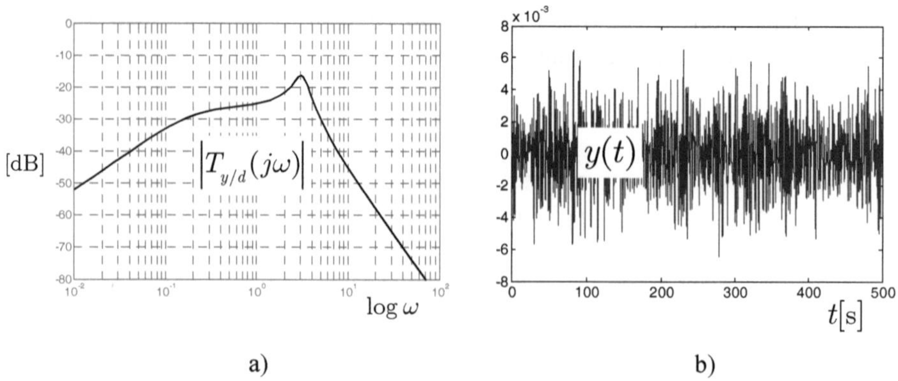

Fig. 11.21. Disturbance propagation for actuator noise: a) frequency response graph of $\left|T_{y/d}(j\omega)\right|$, b) output variable $y(t)$

Discussion As expected, the solutions obtained using different methods agree. Note, however, the differences in computational requirements and transparency of the models. Via relation (11.60), the *analytically*-based result enables the type of insight important to *system design* into structural relationships and parameter sensitivities. To a certain extent, this capability remains when relation (11.59) is used to obtain a rougher approximation of some of the complex system dynamics. Even in this latter case, system tendencies and orders of magnitude can generally be rather well assessed.

In principle, *numerical-* or *simulation-based* covariance calculation is—when aided by a computer—also easily applied to arbitrarily complex system structures. With this method, the applicability of analytical results (obtained from simplified models) can be justifiably confirmed. In addition, simulation can also be employed in the case of non-linear systems. Thus, the three methods presented here should always be used complementarily.

■

Bibliography for Chapter 11

Cooper, G. and C. McGillem (1999). *Probabilistic Methods of Signal and System Analysis*. Oxford University Press.

Föllinger, O. (2007). *Laplace-, Fourier- und z-Transformation.* Hüthig Verlag.

Gelb, A., Ed. (1974). *Applied Optimal Estimation*, M.I.T. Press.

Giloi, W. (1970). *Simulation und Analyse stochastischer Vorgänge* Oldenbourg Verlag.

Hsu, H. (2011). *Schaum's Outline of Probability, Random Variables, and Random Processes, Second Edition (Schaum's Outline Series).* McGraw-Hill.

Lin, C.-F. (1991). *Modern Navigation, Guidance, and Control Processing.* Prentice Hall.

12 Design Evaluation: System Budgets

Background The ultimate goal of systems design is to ensure that an engineered system—here a mechatronic system—achieves a desired level of performance relative to specified product demands. The degree to which these demands are fulfilled should, if possible, be evaluated via quantifiable metrics. These metrics should describe the parameters of time-dependent system variables in a compact form (information compression) and, in the best case, have analytical connections to important design parameters. Ultimately, the performance of particular aspects of a system (mechanical state variable evolution, energy use, thermal behavior, etc.) depends on many, generally time-varying inputs and disturbances. For this reason, component performance metrics must be appropriately combined. In this context, we speak of setting up a *budget* or of a *budgeting* of metrics.

Contents of Chapter 12 This chapter discusses fundamental methods for the *quantitative evaluation of designs* and for the establishing of *system budgets*. *Performance metrics* are introduced as the evaluative measure. These make use of the mathematical concept of a metric as a generalized distance. Working from a statistical perspective, general *budgeting rules* are presented for both *linear* and *nonlinear* combinations (*quadratic* vs. *max-sum*). An important metric for mechatronic systems—*product accuracy*—is introduced as the measure of the correctness of system motions as compared to a reference motion. Given the variable time evolutions of inputs and disturbances (constant, harmonic, or stochastic), budgeting approaches for *heterogeneous metrics* are discussed along with relevant application-specific metrics in the domains of metrology and aerospace. Approaches to *design optimization* are presented as an important application of system budgets in systems design. Two fully worked-out design examples are used to demonstrate the calculation and use of system budgets. ■

12.1 Systems Engineering Context

System specification vs. design evaluation How "good" a design is, is not a subjective question, but should be objectively measurable. The yardstick by which this may be measured is the *system specifications*, which clearly state the required system capabilities (see Fig. 1.9).

A golden rule in setting specifications is that every *functional specification must be "testable"*, i.e. it must be verifiable by experiment. Obviously, such experiments should not have their initial trial conducted on the finished product. This suggests the concept of conducting experiments using system models, which arise naturally as part of systems design. Such models allow the quality of a design solution to be evaluated in a concrete manner, even in the early stages of the design process.

Measurable specifications: performance metrics In the simplest case, a functional specification can be verified with a binary yes-no decision, e.g. the machine automatically turns off or it doesn't. However, many specifications target performance of a steady-state or dynamic nature, e.g. positioning accuracy, attitude stability, rise time for generating a force, etc.

These performance properties generally cannot be determined using single, instantaneous values. Rather, they are composed of various individual components of heterogeneous origin. In the case of time-varying, often stochastic, disturbances and system variables, neither instantaneous values nor exact time histories are specifiable, and thus also not verifiable. As a result, in order to measure the performance properties of a system, appropriate *performance parameters* must be defined. These should be experimentally ascertainable using both the system models and the realized system. This type of variable—also called a *performance metric*—suitable for the description of the dynamic behavior of mechatronic systems is a primary subject of this chapter.

An experienced system designer will begin by defining system requirements on the basis of performance metrics, so as to ensure their best possible verifiability. Thus, a knowledge of performance metrics is essential, starting right from the definition of system requirements.

Budgeting of performance metrics For the systematic analysis of various, generally heterogeneous, inputs and disturbances, the *principle of superposition* may be applied. Using this principle, inputs, disturbances, and their effects are analyzed separately, and combined in the end. This is termed the *budgeting* of performance metrics or the drawing up of *system budgets*. Complicating factors in this type of analysis are the heterogeneous nature of the inputs and disturbances, and the generally *nonlinear* nature of system dynamics. Both factors can be managed with appropriate mathematical methods, as is shown below.

Design evaluation, design optimization When performance characteristics can be *quantitatively* evaluated, then two competing design solutions can be objectively compared, and the solution chosen which has the capability to better fit the system requirements. This forms the basis of *design evaluation*.

A further result of this approach is that the performance metrics can also be a helpful tool for obtaining optimized design solutions. This forms the basis of *design optimization*. System parameters can be modified in a targeted, iterative fashion, and the modified performance properties can be evaluated with the help of the system model. For such model-aided optimization, *analytical performance models* are particularly helpful, as relations between the performance metrics can often be constructed using them, promoting transparent and systematic design optimization. The performance models presented in this book are particularly suited to such a design process, as described below.

12.2 Performance Metrics: Performance Parameters

Performance models In order to describe specific performance properties of a mechatronic system, suitable *performance models* which capture the properties of interest with sufficient fidelity are used (see Sec. 2.1). In addition to the time evolution of motion variables (position, velocity, and acceleration), the design engineer is also interested in variables internal to the machine (electrical voltages and currents, etc.) and system properties such as resource usage (auxiliary energy, or information processing), thermal behavior, mass distribution, and much more. All of these generally time-varying properties can be described with appropriately defined *system variables* $y_i(t)$. The behavior of these system variables depends on the actual *operational and environmental conditions* $u_j(t)$ as well as *configuration parameters* $p_k(t)$ (Fig. 12.1).

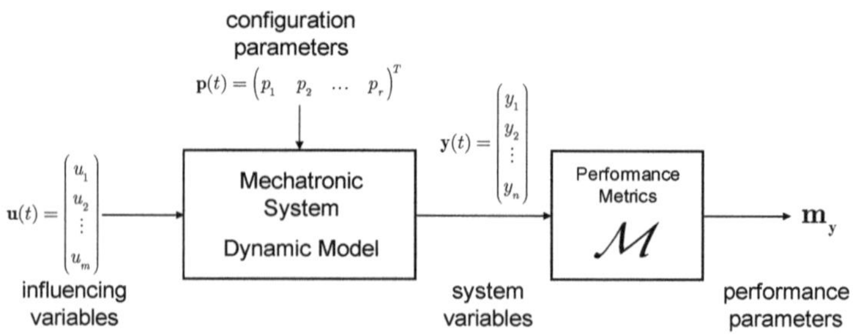

Fig. 12.1. Mechatronic system with performance parameters

Stationary performance parameters As a practical expedience, the behavior of time-varying system variables is often specified using derived parameters which describe generalized *constant* properties such as averages, extreme values, variances, etc., over a defined time horizon. Using this "averaging" viewpoint, large classes of operational and environmental conditions and parameter variations can be concisely represented. In order to maintain a high capacity for the prediction of *complete* time histories, appropriate mathematical concepts must be used when deriving such parameters. In the case of *systems design*, the evaluation of these parameters proceeds *computationally* using dynamic models. If appropriately chosen, the parameters can be *experimentally* determined during *system validation* with appropriate procedures applied to the finished design product.

Performance metrics When creating predictive performance parameters, the mathematical concept of a *metric* as a generalized distance may be employed. In general, a metric is understood to be a mathematical function which assigns a non-negative value to all pairs of elements in a metric space. This value represents the separation of the two elements within the space. A stricter axiomatic definition is as follows:

Definition 12.1 *performance metric*: Let Ω be any set of dynamic properties of a system. A performance metric is a map $\mathcal{M}$: $\Omega \times \Omega \to \mathbb{R}$ with the following properties for any elements $a, b, c \in \Omega$:

$$\begin{aligned} &\text{Identity of indiscernibles:} && \mathcal{M}(a,b) = 0 \Leftrightarrow a = b\,, \\ &\text{Symmetry:} && \mathcal{M}(a,b) = \mathcal{M}(b,a)\,, \\ &\text{Triangle inequality:} && \mathcal{M}(a,b) \le \mathcal{M}(a,c) + \mathcal{M}(c,b)\,. \end{aligned} \tag{12.1}$$

From the axiomatic relations (12.1), it follows directly that

$$\mathcal{M}(a,b) \geq 0 . \tag{12.2}$$

Whenever (12.1) holds, $(\Omega, \mathcal{M})$ is termed a *performance metric space.*

Metrics for deterministic system variables A natural approach to determining performance metrics for deterministic system variables (described by real values) is offered by the well-known *p-norms*:

$$\left\| m_y \right\|_p = \underset{t \in [t_0, t_f]}{\mathcal{M}} (y(t), y_{ref}) := \left(\sum_{j=1}^{n} \left| y(t_j) - y_{ref} \right|^p \right)^{1/p} \tag{12.3}$$

where $t_j \in [t_0, t_f]$, $p \geq 1$.

The *1-norm*—i.e. the *absolute value* (Fig. 12.2a,b)—is well suited to *steady-state quantities* and *system variables* constant over an observation interval $[t_0, t_f]$. In this case, a single element is sufficient for computation, i.e. $p = 1$.

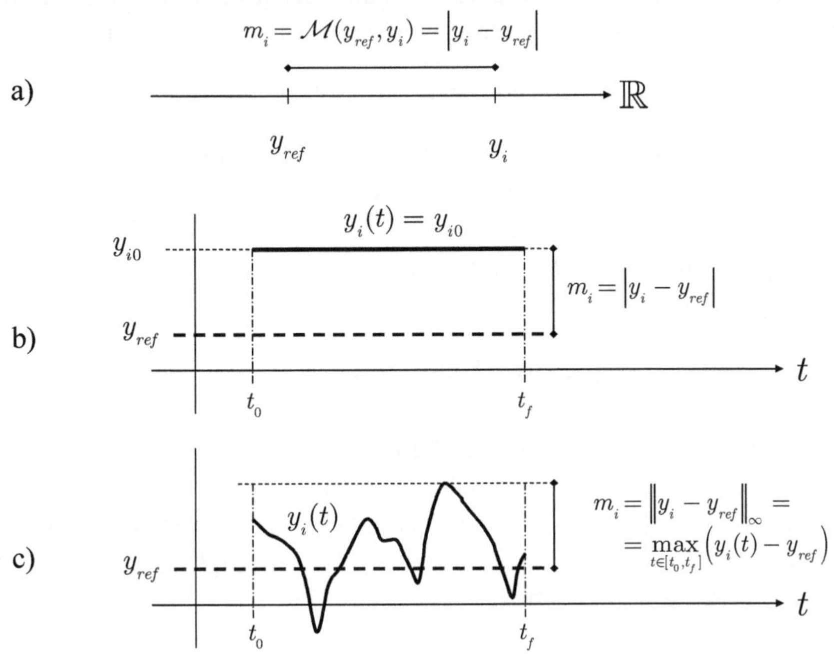

Fig. 12.2. Performance metrics for deterministic signals: a) *absolute value* for a stationary quantity, b) *absolute value* for a constant quantity, c) *max norm* for a time-varying quantity

Time-varying quantities require a sufficient number of observations $y(t_j)$, $j = 1,...n$, $n \gg 1$ to describe. Appropriate metrics are then either the *Euclidian norm*—a *power-proportional* metric—with $p = 2$, or the *max norm* (or *infinity norm)*—the maximum distance to a reference value—with $p = \infty$ (see Fig. 12.2c).

Metrics for stochastic system variables Stochastic system variables generally require evaluating *expected values* in order to define a metric. *Variances* can be used directly as a metric (as they are non-negative). For *average values*, as with any constant quantity (Fig. 12.2b), the absolute value (1-norm) of the distance to a reference value should be used.

Max-norm vs. 3σ-values Experimental determination of the *max norm* of a stochastic system variable is subject to great difficulty. In fact, given the generally applicable assumption of a normal distribution of values, the max norm is rather meaningless, as amplitudes of any particular size will appear, given sufficient time.

Thus, for practical reasons, "approximate maxima" in the form of so-called "*3σ values*" of the variables are generally used. Assuming a normal distribution $N(0, \sigma_y)$, the amplitudes of $y(t)$ fall in the interval

$$[-3\sigma_y, 3\sigma_y] \approx [y_{\min}, y_{\max}]$$

with a statistical certainty of $S = 99.7\%$ (see. Sec. 11.2.1). Such a metric is specified with the addendum "(3σ)", e.g. 9 mm (3σ). This metric, based on standard deviation or covariance, is both computationally and experimentally easy to determine (see Sec. 11.6).

12.3 Linear Budgeting of Metrics

Linear superposition As a rule, the behavior of a system variable is determined by excitations from multiple inputs. Let the causal relationship between inputs $u_j(t)$ and a system variable $y_i(t)$ be modeled by a *linear function* $y_i = \mathcal{G}(u_1, u_2, ..., u_n)$. Following the principle of superposition, the combined excitation is then

$$\begin{aligned} y_i &= \mathcal{G}(u_1, u_2, ..., u_n) = \\ &= \mathcal{G}(u_1, 0, ..., 0) + \mathcal{G}(0, u_2, ..., 0) + ... + \mathcal{G}(0, 0, ..., u_n), \end{aligned} \tag{12.4}$$

i.e. the total excitation can be calculated via a linear combination of individual elements $y_{i/uj}$:

$$y_{i/\Sigma} = y_{i/u1} + y_{i/u2} + \ldots + y_{i/un} \,. \tag{12.5}$$

Let each element be assigned and individual metric (Fig. 12.3)

$$m_{yi/uj} = \mathcal{M}(y_{i/uj}, y_{ref}) \,. \tag{12.6}$$

In order to express the metric of the summed signal using the individual metrics $m_{yi/uj}$, these quantities must be *appropriately combined* (Fig.12.3)

$$m_{yi/\Sigma} = m_{yi/u1} \oplus m_{yi/u2} \oplus \ldots \oplus m_{yi/un} \,, \tag{12.7}$$

where the symbol $\oplus$ describes an *appropriate summation rule*.

Note that the metrics do not describe coincident events, but rather time-varying behaviors at different instants *or* quantities which are averaged over a finite time horizon.

Combining metrics: budgeting For notational convenience in the following discussion, consider the combination of individual metrics m_j in the form

$$m_\Sigma = m_1 \oplus m_2 \oplus \ldots \oplus m_n \,. \tag{12.8}$$

Some type of calculation rule following (12.8) and enabling accurate determination of the metric m_Σ of the summed signal $y_{i/\Sigma}$ using the individual metrics m_j is required. This type of enumeration and calculation is termed a *budget* or *budgeting* of the metrics m_j.

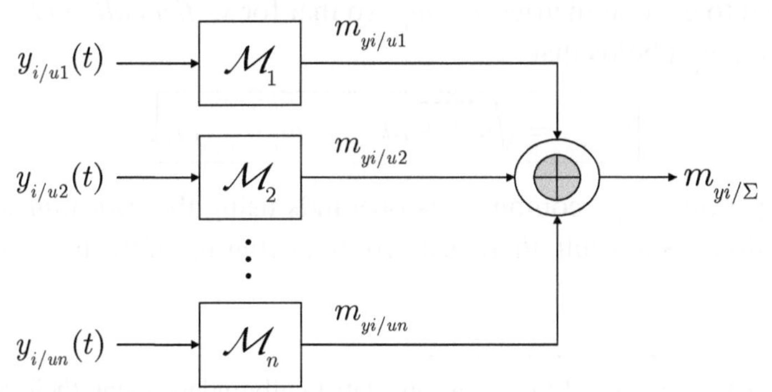

Fig. 12.3. Superposition of metrics

Depending on the contents of the metrics, the budget can be an *uncertainty* budget, *error* budget, *energy* budget, *mass* budget, *computational processing load* budget, *bus usage* budget, etc.

Statistical viewpoint If the system variables $y_i(t)$ are considered to be stationary random processes, then the amplitude characteristics of the individual elements $y_{i/uj}, y_{i/uk}$ can be modeled using their *variances* $\sigma^2_{yi/uj}, \sigma^2_{yi/uk}$. The *statistical dependence* of two elements $y_{i/uj}$ and $y_{i/uk}$ is in turn described by the *covariance* $\sigma_{yi/uj,\ yi/uk}$. The variance of the sum of two elements $y_{i/uj}, y_{i/uk}$ is then given following Eq. (11.18)

$$\sigma^2_{yi/\Sigma} = \sigma^2_{yi/uj} + \sigma^2_{yi/uk} + 2\sigma_{yi/uj,\ yi/uk}\,. \tag{12.9}$$

In general, the covariance $\sigma_{yi/uj,\ yi/uk}$ is not known, or is difficult to obtain. In practical applications, generally only the two extreme cases are examined: the *minimal (best-case)* and *maximal (worst-case) statistical* dependences[1]. If more exact results are required, $\sigma_{yi/uj,\ yi/uk}$ must be determined using the underlying probability density functions and applied in Eq. (12.9).

Quadratic summation Assuming that two component quantities $y_{i/uj}, y_{i/uk}$ are *statistically independent*, it follows that $\sigma_{yi/uj,\ yi/uk} = 0$, so that Eq. (12.9) can be simplified to

$$\sigma^2_{yi/\Sigma} = \sigma^2_{yi/uj} + \sigma^2_{yi/uk} \quad \text{or} \quad \boxed{\sigma_{yi/\Sigma} = \sqrt{\sigma^2_{yi/uj} + \sigma^2_{yi/uk}}}\,. \tag{12.10}$$

In this case, then, *variances* are combined *linearly*. In addition to the special case of variance as the metric considered here, Eq. (12.10) can also be applied to general metrics m_i, m_k, so that for *statistically independent* metrics m_i, m_k it holds that

$$\boxed{m_{\Sigma,\text{rss}} = \sqrt{m_i^2 + m_k^2} =: m_i \oplus_{\text{rss}} m_k}\,. \tag{12.11}$$

The *operator* $\oplus_{rss}$ combines its operands using the *root sum square (RSS)* operation. As a result, there is a certain averaging of the two operands.

---

1 As shown below, if the inputs are statistically independent, their metrics can be added *quadratically*, otherwise they must be added *linearly*. Quadratic addition always results in a smaller sum (*best case*) than linear addition (*worst case*).

Thus, given *statistical independence* of the individual metrics, the budget equation (12.8) can be written as follows:

$$m_{\Sigma,\mathrm{rss}} = m_1 \oplus_{\mathrm{rss}} m_2 \oplus_{\mathrm{rss}} \ldots \oplus_{\mathrm{rss}} m_n \,. \tag{12.12}$$

Max-sum If the two component quantities $y_{i/uj}, y_{i/uk}$ are *statistically dependent*, then $\sigma_{yi/uj,\, yi/uk} \neq 0$. In this case, the CAUCHY-SCHWARZ inequality (11.11) can be used to establish an *upper bound* on $\sigma_{yi/uj,\, yi/uk}$. In this case, Eq. (12.9) is used to define an upper bound on the variance of the combined value

$$\sigma^2_{yi/\Sigma} \leq \sigma^2_{yi/uj} + \sigma^2_{yi/uk} + 2\sigma_{yi/uj}\sigma_{yi/uk} = \left(\sigma_{yi/uj} + \sigma_{yi/uk}\right)^2$$

$$\boxed{\sigma_{yi/\Sigma} \leq \sigma_{yi/uj} + \sigma_{yi/uk}} \,. \tag{12.13}$$

Thus, for statistically dependent quantities, the *standard deviations* are *linearly* combined.

Again, Eq. (12.13) can be extended to general metrics m_i, m_k, so that for *statistically dependent* metrics m_i, m_k

$$\boxed{m_{\Sigma,\max} = m_i + m_k =: m_i \oplus_{\max} m_k} \tag{12.14}$$

The *operator* $\oplus_{\max}$ combines its operands using the unsigned addition of positive values. The two operands are always added in the manner giving the maximum possible result.

Thus, under conditions of *statistical dependence* of the individual metrics, the budget equation (12.8) can be written as follows:

$$m_{\Sigma,\max} = m_1 \oplus_{\max} m_2 \oplus_{\max} \ldots \oplus_{\max} m_n \,. \tag{12.15}$$

Quadratic vs. max sum: best case vs. worst case It can be easily shown that in general

$$m_{\Sigma,rss} < m_{\Sigma,\max} \,. \tag{12.16}$$

If a quadratic summation is permissible, the summed value will be smaller than for linear summation. Thus, *quadratic (RSS) summation* is considered the *best case* assessment and the *max sum* is considered the *worst case* assessment.

The max sum represents a *conservative overestimate* of the actual summed value, i.e. actual values of the function $y_{i,\Sigma}(t)$ can lie significantly below the bound described by the metric $m_{i,\Sigma,\max}$. On the other hand, however, $m_{i,\Sigma,\max}$ gives a *safe estimate* of the maximum possible combination of the component values (a safe upper bound). Depending on the problem statement at hand, either variant may be chosen.

12.4 Nonlinear Budgeting of Metrics

Nonlinear performance models In many cases, system variables must be related using nonlinear models. In the simplest case, this is a *nonlinear algebraic* relation

$$y_\Sigma = g(y_1, y_2, \ldots, y_n). \tag{12.17}$$

This section describes how metrics of these system variables can be appropriately combined (Taylor 1982).

Statistical viewpoint Once again, consider the system variables $y_i(t)$ as stationary random processes with statistical parameters

$$\mu_i = E[y_i] = \overline{y}_i, \quad \sigma_i = \sqrt{\operatorname{var}(y_i)}. \tag{12.18}$$

Then each system variable $y_i(t)$ can be modeled as

$$y_i(t) = \overline{y}_i + \delta y_i(t), \quad E[\delta y_i] = 0, \quad \operatorname{var}(\delta y_i) = \sigma_i^{\,2}. \tag{12.19}$$

Assuming sufficiently small standard deviations σ_i, Eq. (12.17) can be locally linearized, and, neglecting higher-order terms, a linear approximation obtained (shown here for two component variables)

$$y_\Sigma = \overline{y}_\Sigma + \delta y_\Sigma \approx g(\overline{y}_1, \overline{y}_2) + \left.\frac{\partial g}{\partial y_1}\right|_{\overline{y}_1, \overline{y}_2} \cdot \delta y_1 + \left.\frac{\partial g}{\partial y_2}\right|_{\overline{y}_1, \overline{y}_2} \cdot \delta y_2 \tag{12.20}$$

with *sensitivity coefficients*

$$\frac{\partial g}{\partial y_1}, \frac{\partial g}{\partial y_2}. \tag{12.21}$$

The variance σ_Σ^2 of the combined variable y_Σ can then be computed using statistical combination, as in Eq. (12.9), giving

$$\sigma_\Sigma^2 = \mathrm{var}(\delta y_\Sigma) = \left(\frac{\partial g}{\partial y_1}\right)^2 \sigma_1^2 + \left(\frac{\partial g}{\partial y_2}\right)^2 \sigma_2^2 + 2\frac{\partial g}{\partial y_1}\frac{\partial g}{\partial y_2}\sigma_{12}\,. \tag{12.22}$$

Given *statistical independence* of δy_1 and δy_2, it follows analogously to Eq. (12.10) that

$$\sigma_\Sigma = \sqrt{\left(\frac{\partial g}{\partial y_1}\right)^2 \sigma_1^2 + \left(\frac{\partial g}{\partial y_2}\right)^2 \sigma_2^2}\,. \tag{12.23}$$

Similarly, for *correlated* variables δy_1, δy_2 an estimated upper bound can be derived:

$$\sigma_\Sigma \le \left|\frac{\partial g}{\partial y_1}\right| \sigma_1 + \left|\frac{\partial g}{\partial y_2}\right| \sigma_2\,. \tag{12.24}$$

Nonlinear combination of metrics Relations (12.23) and (12.24), derived for the special metric of variance, can again be extended to general metrics, giving for the two summation approaches:

- *Quadratic summation*

$$\boxed{m_\Sigma = \sqrt{\left(\frac{\partial g}{\partial y_1}\right)^2 m_1^2 + \left(\frac{\partial g}{\partial y_2}\right)^2 m_2^2} = \left(\frac{\partial g}{\partial y_1} m_1\right) \oplus_{\mathrm{rss}} \left(\frac{\partial g}{\partial y_2} m_2\right)}\,, \tag{12.25}$$

- *Max-summation*

$$\boxed{m_\Sigma = \left|\frac{\partial g}{\partial y_1}\right| m_1 + \left|\frac{\partial g}{\partial y_2}\right| m_2 = \left(\left|\frac{\partial g}{\partial y_1}\right| m_1\right) \oplus_{\max} \left(\left|\frac{\partial g}{\partial y_2}\right| m_2\right)}\,. \tag{12.26}$$

Example 12.1 *Planar manipulator with one degree of freedom.*

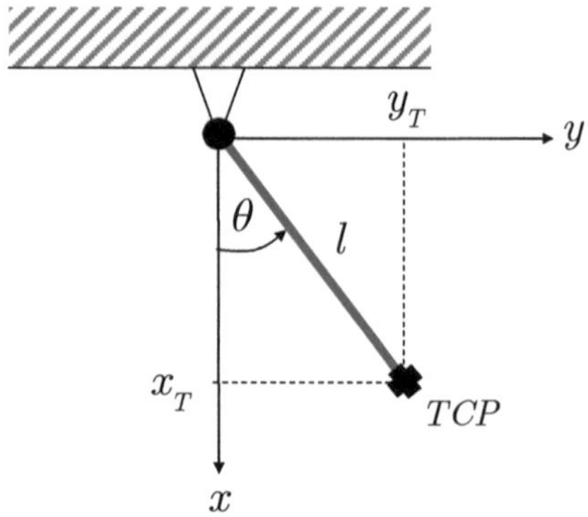

Fig. 12.4. Planar manipulator with one degree of freedom

Problem statement The planar manipulator shown in Fig. 12.4, with parameters $l = 80$ cm, $\theta_{\max} = \pm 45°$, has one rotational degree of freedom θ. For a certain class of command signals, a joint controller ensures a control accuracy of $\delta\theta = 0.9° \ (3\sigma)$.

Find the positioning accuracy of the *tool center point (TCP)* in Cartesian coordinates.

Statistical performance model The Cartesian coordinates of the *tool center point* are related to the controlled manipulator axis θ via the following nonlinear geometric equations

$$\begin{aligned} x_T &= l \cdot \cos\theta \\ y_T &= l \cdot \sin\theta \end{aligned} .$$

The Cartesian position uncertainties can be determined using sensitivity coefficients

$$\delta x_T = \left| \frac{\partial x_T(\theta)}{\partial \theta} \right| \cdot \delta\theta = \left(l \cdot \sin\theta \right) \cdot \delta\theta ,$$

$$\delta y_T = \left| \frac{\partial y_T(\theta)}{\partial \theta} \right| \cdot \delta\theta = \left(l \cdot \cos\theta \right) \cdot \delta\theta .$$

Both components δx_T and δy_T depend on the angle θ. It is easy to verify that the respective maximum position uncertainties occur at the angles θ^* as follows:[2]

$$\left(\delta x_T\right)_{\max} = \left(l \cdot \sin\theta\right)\Big|_{\theta^*=45^\circ} \cdot \delta\theta =$$

$$= \left(80 \cdot 0,707\right)\frac{\text{cm}}{\text{rad}} \cdot 0.9^\circ \cdot \frac{\pi}{180^\circ} = 0,89\ \text{cm}\ (3\sigma)$$

$$\left(\delta y_T\right)_{\max} = \left(l \cdot \cos\theta\right)\Big|_{\theta^*=0^\circ} \cdot \delta\theta =$$

$$= \left(80 \cdot 1\right)\frac{\text{cm}}{\text{rad}} \cdot 0.9^\circ \cdot \frac{\pi}{180^\circ} = 1,26\ \text{cm}\ (3\sigma)$$

The computed Cartesian TCP uncertainties retain the same "3σ" metric as the underlying base quantity $\delta\theta$, i.e. the computed positioning accuracy *approximates* the *maximum possible* Cartesian positioning deviations with respect to a given nominal value (with an approximate statistical certainty of $S = 99.7\%$).

■

Example 12.2 *Planar manipulator with two degrees of freedom.*

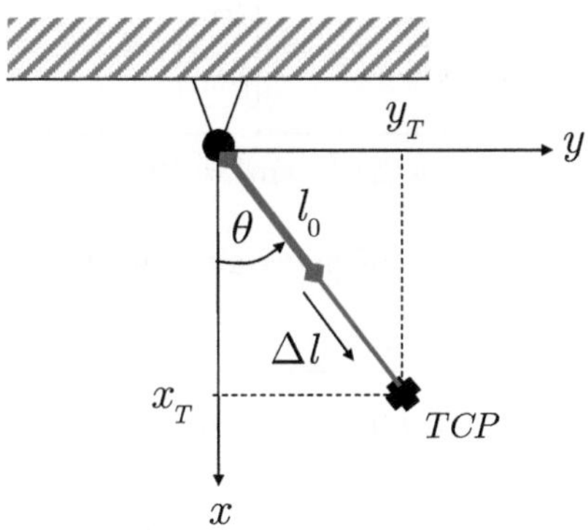

Fig. 12.5. Planar manipulator with two degrees of freedom

---

2 Note that the sensitivity coefficients have the physical dimension [cm/rad], so that the angular uncertainty must be converted to [rad].

Problem statement Fig. 12.5 shows a planar manipulator modified from that of Example 12.1 with an additional controlled sliding joint with degree of freedom Δl. For a certain class of command signals, a controller ensures a control accuracy of $\delta\theta = 0.9°\ (3\sigma)$ and $\delta\Delta l = 2\ \text{mm}\ (3\sigma)$. The system parameters are: $l_0 = 40\ \text{cm}$, $\Delta l \in [0, 40\ \text{cm}]$, $\theta_{max} = \pm 45°$.

Find the positioning accuracy of the *tool center point (TCP)* in Cartesian coordinates.

Statistical performance model The Cartesian coordinates of the *tool center point* are determined by the two controlled manipulator axes θ and Δl via the following nonlinear geometric relation, shown here for the x-coordinate:

$$x_T = (l_0 + \Delta l) \cdot \cos\theta \, .$$

The sensitivity coefficients are then

$$\left| \frac{\partial x_T(\theta, \Delta l)}{\partial \theta} \right| = (l_0 + \Delta l) \cdot \sin\theta \, ,$$

$$\left| \frac{\partial x_T(\theta, \Delta l)}{\partial \Delta l} \right| = \cos\theta \, .$$

Since it can be assumed that both manipulator axes $\theta, \Delta l$ are uncorrelated, a *quadratic summation* of the two uncertainty components can be conducted

$$\begin{aligned} \delta x_{T,rss} &= \sqrt{\left|\frac{\partial x_T}{\partial \theta}\right|^2 \cdot \delta\theta^2 + \left|\frac{\partial x_T}{\partial \Delta l}\right|^2 \cdot \delta\Delta l^2} = \\ &= \sqrt{(l_0 + \Delta l)^2 \cdot \sin^2\theta \cdot \delta\theta^2 + \cos^2\theta \cdot \delta\Delta l^2} \end{aligned} .$$

The maximum Cartesian uncertainty results at $\Delta l^* = 40$ cm, $\theta^* = 45°$, so that

$$\left(\delta x_{T,rss}\right)_{max} = \sqrt{\left(0{,}89\ \text{cm}\right)^2 + \left(0.14\ \text{cm}\right)^2} = 0{,}90\ \text{cm}\ (3\sigma) \, .$$

It can be seen that the positioning uncertainty of the linear axis Δl is negligible compared to that of the angular uncertainty, and has practically no influence on the Cartesian uncertainty.

■

12.5 Product Accuracy

Product task, product accuracy The overarching product task of a mechatronic system consists of realizing a purposeful motion behavior (see Sec. 1.1). Thus, the accuracy with which the *motion behavior* is realized relative to a *reference motion* (command value) is a key property of the product. This property will be termed the *product accuracy.*

Fig. 12.6 shows an example of a single-loop controller configuration with unity feedback, highlighting the relevant system variables. The controlled variable y can be most easily interpreted as the mechanical state (*position* and *velocity*), but can also represent any other internal variables of a mechatronic system.

Linear superposition of dynamic inputs The relevant dynamic inputs can be seen to be the command variable w, as well as the force/torque disturbances d and measurement noise n. The *product accuracy* δy is obtained by considering the corresponding transfer functions:

$$\begin{aligned} \delta y &= T_{\delta y/w} \cdot w + T_{\delta y/d} \cdot d + T_{\delta y/n} \cdot n = \\ &= \frac{1}{1+L} w - \frac{P}{1+L} d + \frac{L}{1+L} n \,. \end{aligned} \tag{12.27}$$

Thus, the product accuracy can be budgeted directly as a linear superposition of the individual contributions δy_i of the inputs w, d and n (Fig. 12.7).

Constant inputs Computation of the individual contributions of constant inputs $u_i(t) = U_{i0} = const.$ (e.g. offset, bias) can be easy accomplished using the *final value theorem* of the LAPLACE transform[3]

$$\delta y_{\infty,ui} = \lim_{s \to 0} s \cdot T_{y/ui}(s) \frac{U_{i0}}{s} = \lim_{s \to 0} T_{y/ui}(s) \cdot U_{i0} \,. \tag{12.28}$$

Harmonic inputs For harmonic inputs $u_i(t) = U_{i0} \sin \omega_* t$, the amplitudes of the components are obtained using the *frequency response*

$$\delta y_{h,ui} = U_{i0} \cdot T_{y/ui}(j\omega_*) \,. \tag{12.29}$$

[3] Naturally under the condition that the transfer function $T_{y/ui}(s)$ is stable.

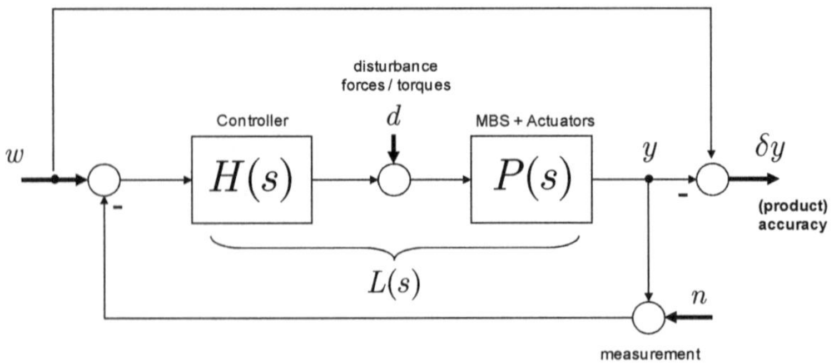

Fig. 12.6. Product accuracy for a mechatronic system including dynamic inputs

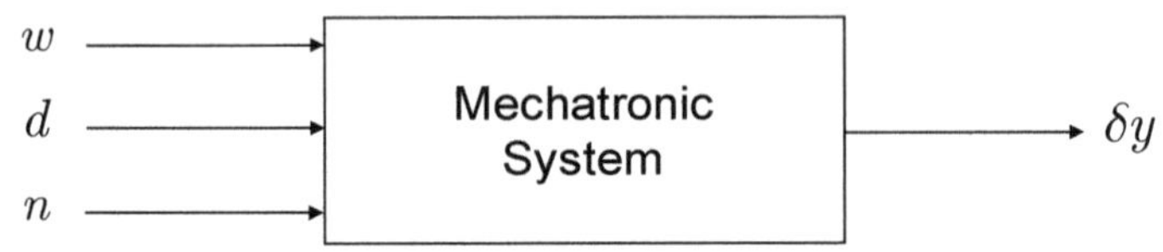

Fig. 12.7. "Product accuracy" budgeting model for a mechatronic system

Transient amplitudes The determination of transient amplitudes caused by step-wise changes in the inputs can often be accomplished for the standard configuration shown in Fig. 12.6 by means of the well-known approximating equation (10.24) using the *crossover frequency* and *phase margin* of the closed-loop transfer function $L(j\omega)$ (see Sec. 10.4.3). For more complicated circumstances, an evaluation via simulation is recommended.

Stochastic inputs The component values for stochastic inputs can be obtained using the *covariance analysis* presented in Ch. 11.

12.6 Budgeting Heterogeneous Metrics

Heterogeneous behavior of inputs The various inputs to a system generally have differing temporal properties. There are three classes of time histories which are usually sufficient for evaluating the overall system behavior: *constant*, *harmonic*, *random*. Each of these classes has appropriate metrics (see Sec. 12.2) to describe the desired properties. The superposition of these component metrics, then, should describe the overall effect of the underlying inputs on system variables of interest (Fig. 12.8).

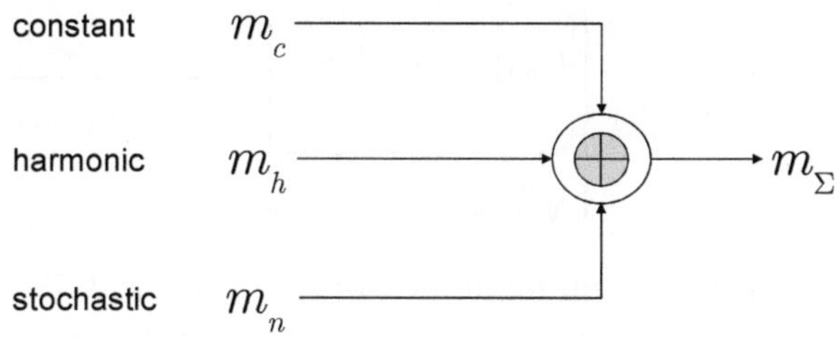

Fig. 12.8. Superposition of heterogeneous metrics

Table 12.1. Max-metrics suitable for superposition given various classes of time histories

Time history	Metric
Constant	m_c : 1-Norm
Harmonic	m_h : $\lvert Amplitude \rvert$
Random (normally distributed)	m_n : 3σ

In the case of product accuracy (see Sec. 12.5), each of the inputs $w(t)$, $d(t)$, and $n(t)$ can contain both constant, harmonic, and random components. However, what is of interest is *one single* value for the achievable motion accuracy δy which is to be estimated via a suitable superposition.

Principle of superposition As a result, a fundamental requirement for the superposition operations appears: the component metrics should describe *equivalent* properties, i.e. they should either all describe maximum values, or all describe variances or standard deviations. In practice, the metrics shown in Table 12.1 have proven to be valuable in describing magnitude values[4].

Practical budgeting approach The following approach to budgeting has shown practical utility when computing a combined metric m_Σ from constant, harmonic, and random component input metrics (see Fig. 12.8):

$$m_\Sigma = m_{c,\mathrm{rss}} + \begin{Bmatrix} m_{h,\max} \\ m_{h,\mathrm{rss}}^{\;1)} \end{Bmatrix} + m_{n,\mathrm{rss}} \,, \tag{12.30}$$

1) valid only for equal frequencies and statistically independent phases

4 In the arena of industry, project-specific metrics must be taken into account in certain cases. These must be clearly and unambiguously stated as part of the definition of system requirements, see Sec. 1.2.

$$m_\Sigma = \sqrt{m_{c,1}^{\;2} + \ldots m_{c,n}^{\;2}} + \left\{ \begin{matrix} \left|m_{h,1}\right| + \ldots + \left|m_{h,k}\right| \\ \\ \sqrt{m_{h,1}^{\;2} + \ldots m_{h,k}^{\;2}} \end{matrix} \right\} + \sqrt{m_{n,1}^{\;2} + \ldots m_{n,r}^{\;2}} \;. \quad (12.31)$$

In the case of a quadratic summation (*root sum square*, *RSS*), the statistical independence of the component metrics must naturally be verified for each particular case. If independence cannot be verified, the components must be added linearly (with a max summation).

Fig. 12.9 shows a few typical time histories and their superposition, and allows the budgeting approach of Eqs. (12.30) and (12.31) to be easily verified. In the case of a harmonic variable, a quadratic summation is only permissible when either purely power-proportional metrics are of interest, or when time histories with the same frequency and statistically independent phase are combined. In all other cases, the oscillatory nature of such signals requires linear addition of the component amplitudes to establish a maximum combined amplitude (Fig. 12.10).

Application-specific metrics The choice of variables and manner of combining them into a budget depends on the specific application and the parties involved. There is no global standard covering every case. Since a performance budget is intended to quantify a product requirement in a compact form, budgeting rules must always be explicitly agreed upon for every project. Often, this results in a clash of interests of the various groups. After all, in the case of an *"RSS"* summation, large component contributions have a smaller effect on the total than is the case for a max-summation. Thus, there is often an effort by the suppliers of subsystems and components to define the summed parameter (e.g. positioning accuracy) quadratically. In this way, a component requirement can be more easily met and exceeded. The judgment of whether individual component contributions are in fact statistically independent is sometimes more a question of belief than objectively determinable. Whenever possible, one should thus strive to adhere to any applicable standards, two of which are discussed here as examples.

Metrological budgeting approach Rules from the domain of metrology for the determination of *measurement uncertainty*[5] can be applied directly

5 The term measurement *error* has been replaced with the conceptually more correct *uncertainty* which afflicts a measurement.

to accuracy budgets of mechatronic systems in an obvious way. After all, the process of "measurement" is one of the elementary system functions of a mechatronic product. The applicable international standardized framework is the so-called *Guide to the expression of uncertainty in measurement (GUM)* (ISO 1995), (Kirkup and Frenkel 2006).

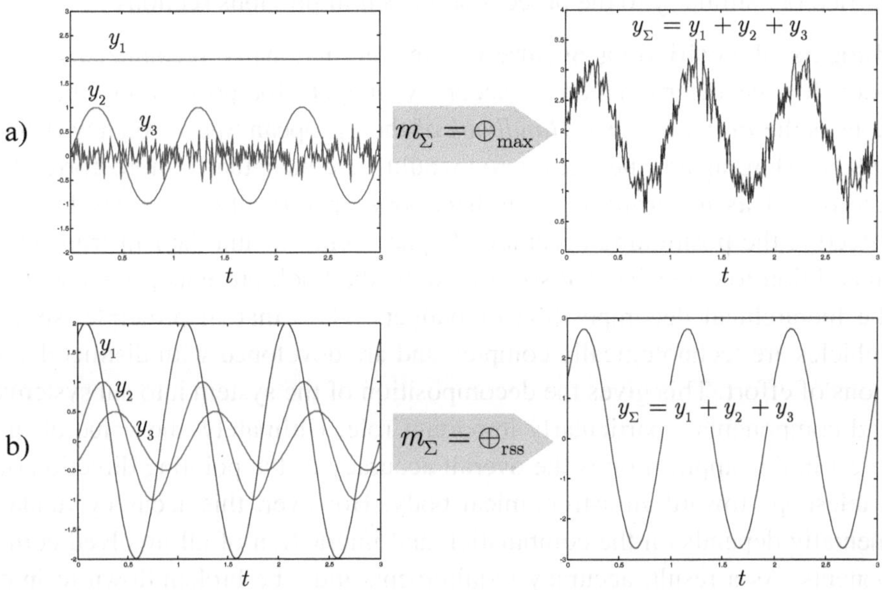

Fig. 12.9. Superposition of time histories and associated metrics: a) constant, harmonic and random variables, b) harmonic oscillations with the same frequency

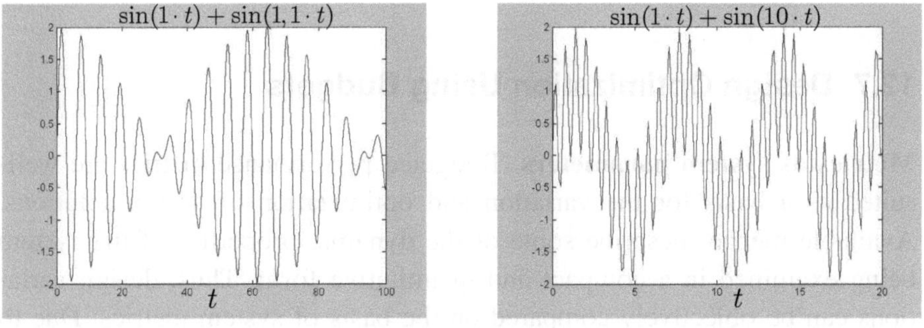

Fig. 12.10. Combination of two harmonic oscillations with differing frequencies

Roughly, the GUM distinguishes between two types of evaluation:

- **Type A evaluation**: evaluating uncertainty by the *statistical analysis* of series of observations,
- **Type B evaluation**: evaluating uncertainty by *means other than* the statistical analysis of series of observations.

The budgeting of component contributions of such uncertainties is then carried out similarly to the process described in previous sections.

Budgets of positioning accuracy An internationally recognized handbook for the determination of accuracy budgets for positioning applications is the *Pointing Error Handbook* of the European Space Agency (ESA 1993). (Here again, the term *error* would be conceptually more correctly interpreted as *uncertainty*.) This handbook specifically examines factors affecting the positioning accuracy of space vehicles and their instruments. In addition to extensive statistical models, the book presents principles for the hierarchical decomposition of budgets. Note that in principle, space vehicles are technologically complex and are developed with distinct divisions of effort. This gives the decomposition of the system into subsystems and components a particularly important role. Ultimately, the value of interest in this application is the overall accuracy of the pointing direction of a telescope toward an astronomical body. However, this accuracy fundamentally depends on the combination and interaction of all involved components. As a result, accuracy requirements must be broken down to into subsystem and component levels. In addition, inevitable uncertainties must be accounted for using appropriate mathematical models and budgeting. In this respect, this reference forms an outstanding basis for the design of mechatronic systems aimed at *high-precision* positioning tasks.

12.7 Design Optimization Using Budgets

Metrics as system parameters Budgeted performance metrics are well-suited as a basis for the variation and optimization of design solutions. Available metrics describe some of the dynamic properties of the system being examined in a compact and quantitative form. Thus, design variations can be objectively compared on the basis of system metrics. Due to their special algebraic properties, budgeting using the metrics in (12.1) and (12.2) ensures that the combined system variables always consist of cumulatively summed quantities. These in turn can represent complex system behaviors in an extremely concise form, e.g. the "product accuracy".

Design task The summed quantities depend on both given, fixed system parameters, and on freely selectable design parameters. A "favorable" choice of these design parameters will lead to a "good" design solution, an "unfavorable" choice instead leads to an undesirable "bad" design solution. The fine art of systems design, then, consists of determining appropriate "favorable" design parameters.

Design optimization The design task described above can be formulated and solved as a mathematical optimization with the aid of performance models and metrics, and clear budgeting rules, see Fig. 12.11 (compare Fig. 1.8).

Consider a set of free *design parameters* $\mathbf{par} = \{par_i\}, \ i = 1,...,N$, taken from a *definition set* Ω_{par}. Examples of design parameter are parameters governing the forms of signals for command variables (e.g. a ramp), device properties (dynamics, bias, noise parameters, etc.), structural properties (mass, geometry, etc.). The larger definition set Ω_{par} also represents parameters such as device type, volume and mass limitations, and so on.

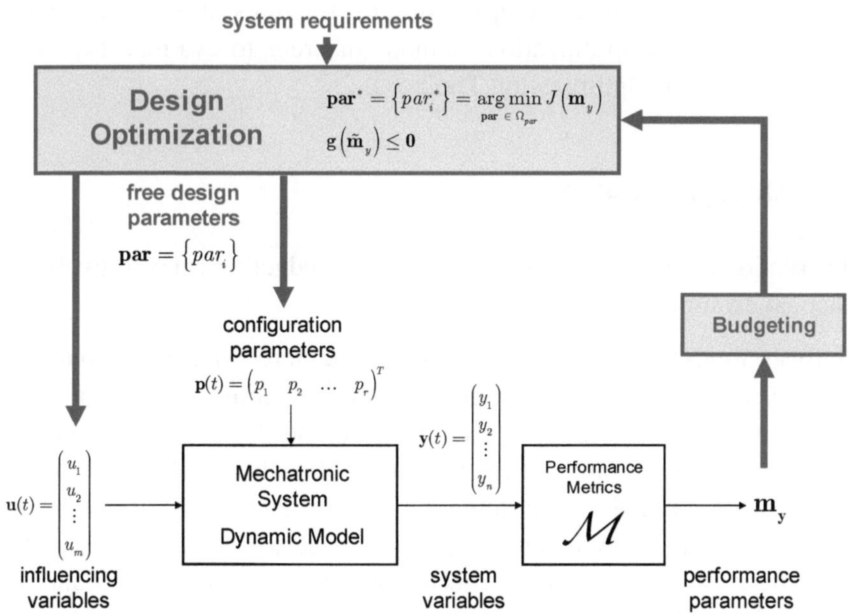

Fig. 12.11. Design optimization using dynamic models and budgeted performance metrics

The goal is to find the *optimal design parameters* $\mathbf{par}^* = \{par_i^*\}$ minimizing an *objective function* $J(\mathbf{m}_y)$ composed of performance metrics $\mathbf{m}_y$. Suitable objective functions include, for example, the product accuracy, total mass, total volume, or power usage.

Often there are also additional system-specific *secondary conditions*, which can generally be described with an algebraic inequality constraint on certain metrics $\tilde{\mathbf{m}}_y$, e.g. maximum forces, accelerations, or velocities.

The design task can thus be compactly formulated as the following *constrained optimization*:

$$\begin{aligned} \mathbf{par}^* &= \{par_i^*\} = \underset{\mathbf{par} \in \Omega_{par}}{\arg\min}\, J(\mathbf{m}_y) \\ \mathbf{g}(\tilde{\mathbf{m}}_y) &\leq \mathbf{0}\,. \end{aligned} \tag{12.32}$$

The optimization (12.32) can be solved either *analytically* or *numerically*, depending on the presence of suitable metrics. Analytical approaches should be preferred, as they allow for better insight into the interrelationships of the system. However, this is generally only possible with simplified models. At a minimum, though, a simplified analytical approach allows reliable tendencies to be derived and high-level operational principles to be evaluated. For more complex system structures, it is advantageous to employ numerical optimization methods in order to evaluate Eq. (12.32) (e.g. the MATLAB Optimization Toolbox).

12.8 Design Examples

This section demonstrates the application of budgeting rules to two simple, practical examples.

Analysis models The system models used here—*analysis models*—are typically dynamic models for preliminary, orienting assessments of the system at the beginning of a design project, and describe its principal dynamic behaviors. Because of the low order of such models, *analytical* formulae can be applied to determine the metrics, and as a result, it becomes possible to analytically identify dependences between free system parameters. These can then be used directly in evaluating design variants on a quantitative, objective basis.

Refined budgets In the advanced stages of a design, the initial, simplified analysis models can be made as detailed as desired (e.g. higher-order models incorporating parasitic dynamics, MBS eigenmodes, more complex controller architectures) and the corresponding metrics must then naturally be determined numerically. However, the budgeting rules used with the original analysis model can be adopted unchanged. In this way, a consistent refinement of the analysis of the system capability is possible using this method.

Example 12.3 *Positioning budget for satellite attitude control.*

Given an attitude-controlled satellite, the achievable *orientation accuracy* and *stability* for a given system configuration are to be evaluated using the simplified analysis model shown in Fig. 12.12.

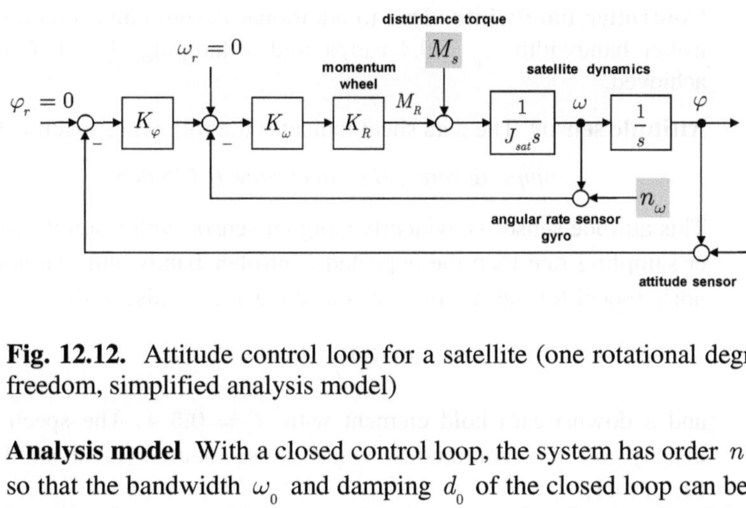

Fig. 12.12. Attitude control loop for a satellite (one rotational degree of freedom, simplified analysis model)

Analysis model With a closed control loop, the system has order $n = 2$, so that the bandwidth ω_0 and damping d_0 of the closed loop can be freely selected over a wide range using the proportional gains K_φ, K_ω in the relations

$$K_\varphi = \frac{\omega_0}{2d_0}, \quad K_\omega = \frac{2d_0\omega_0 J_{sat}}{K_R}.$$

The system task is *set point control* ($\varphi_r = \omega_r = 0$), so that the input variables for the attitude accuracy are the *disturbance torque* M_s, and the *noises* n_φ, n_ω in the *attitude sensor* and the *orientation rate sensor*.

Using the previously mentioned abbreviated notation,

$$\left\{d_0; \omega_0\right\} := 1 + 2d_0 \frac{s}{\omega_0} + \frac{s^2}{\omega_0^2},$$

the relevant transfer functions are determined to be

$$\begin{aligned}
T_{\varphi/Ms}(s) &= \frac{K_1}{\left\{d_0; \omega_0\right\}}, & T_{\omega/Ms}(s) &= \frac{K_1 \cdot s}{\left\{d_0; \omega_0\right\}}, & K_1 &= \frac{1}{J_{sat}\omega_0^2}, \\
T_{\varphi/n\varphi}(s) &= -\frac{1}{\left\{d_0; \omega_0\right\}}, & T_{\omega/n\varphi}(s) &= -\frac{s}{\left\{d_0; \omega_0\right\}}, & & \\
T_{\varphi/n\omega}(s) &= -\frac{K_2}{\left\{d_0; \omega_0\right\}}, & T_{\omega/n\omega}(s) &= -\frac{K_2 \cdot s}{\left\{d_0; \omega_0\right\}}, & K_2 &= \frac{2d_0}{\omega_0}.
\end{aligned} \tag{12.33}$$

Satellite parameter The *moment of inertia* of the rigid satellite is $J_{sat} = 4.2\ \mathrm{kgm}^2$.

Controller bandwidth Due to additional design considerations, a controller bandwidth $\omega_0 = 0.4\ \mathrm{rad/s}$ and a damping $d_0 = 0.7$ are to be achieved.

Attitude sensor The data sheet indicates the following specifications:

sampling rate 2 Hz, accuracy 0.135 deg.

This attitude sensor is evidently a digital sensor with a significantly higher sampling rate than the expected controller bandwidth. An appropriate noise model for the sensor is thus discrete white noise with

$$\sigma_\varphi = \mathrm{accuracy}(3\sigma)/3 = 0.045\ \ \mathrm{deg}$$

and a downstream hold element with $T_a = 0.5\ \mathrm{s}$. The spectral power density of the continuous white noise source is then, following Eq. (11.41),

$$S_\varphi = \sigma_\varphi^{\ 2} \cdot T_a = 0.001\ \mathrm{deg}^2/\mathrm{Hz}\,. \tag{12.34}$$

Orientation rate sensor The data sheet indicates the following specifications

update rate 100 Hz, noise 1 deg/√hr.

Due to the very high sample rate, this noise source can be modeled directly as continuous white noise. Observe, however, that the square root of the noise power density[6] is specified with a particular time basis (*hour*), i.e.

$$S_\omega = \left(1\ \text{deg}/\sqrt{\text{hr}}\right)^2 = 2.8 \cdot 10^{-4}\ \text{deg}^2/\text{s}\,. \tag{12.35}$$

Disturbance torques A typical disturbance torque profile, caused by aerodynamic effects and other disturbances varying periodically with the orbit frequency Ω_0, can be defined as follows:

$$\begin{gathered} M_s(t) = M_{s0} + M_{s1} \sin \Omega_0 t\,, \\ M_{s0} = 5 \cdot 10^{-4}\ \text{Nm},\ M_{s1} = 5 \cdot 10^{-4}\ \text{Nm},\ \Omega_0 = 0.001\ \text{rad/s}\,. \end{gathered} \tag{12.36}$$

Noise calculations As this is a low-order system, the effects of sensor noise can be examined using the transfer functions (12.33) and noise power densities (12.34), (12.35) via an analytical *covariance relation* (see Sec. 11.6) as follows:

Variance of the attitude error

$$\begin{aligned} \sigma_{\varphi/\varphi}^2 &= \frac{1}{4}\frac{\omega_0}{d_0} S_\varphi && [\text{deg}^2] \\ \sigma_{\varphi/\omega}^2 &= \frac{d_0}{\omega_0} S_\omega && [\text{deg}^2] \end{aligned} \tag{12.37}$$

Variance of the orientation rate error

$$\begin{aligned} \sigma_{\omega/\varphi}^2 &= \frac{1}{4}\frac{\omega_0^3}{d_0} S_\varphi && [\text{deg}^2/\text{s}^2] \\ \sigma_{\omega/\omega}^2 &= d_0 \omega_0 S_\omega && [\text{deg}^2/\text{s}^2] \end{aligned} \tag{12.38}$$

Disturbance torque calculations The effects of the *constant disturbance torque* can be obtained using the final value theorem of the LAPLACE transform (Eq. (12.28)):

$$\begin{aligned} \delta\varphi_c &= \frac{M_{s0}}{J_{sat} \cdot \omega_0^2} \cdot \frac{180}{\pi} && [\text{deg}], \\ \delta\omega_c &= 0 && [\text{deg/s}]. \end{aligned} \tag{12.39}$$

6 One should not be confused by the physical units, as $[(\text{deg/s})^2/\text{Hz}] = [\text{deg}^2/\text{s}]$.

Evaluating the relevant frequency responses (12.33) at the location $s = j\Omega_0$ while noting that $\Omega_0 \ll \omega_0$ gives the effect of the *harmonic disturbance*:

$$\begin{aligned}\delta\varphi_h &\approx \frac{M_{s1}}{J_{sat}\cdot\omega_0^2}\cdot\frac{180}{\pi} \quad [\text{deg}],\\ \delta\omega_h &\approx \frac{M_{s1}\cdot\Omega_0}{J_{sat}\cdot\omega_0^2}\cdot\frac{180}{\pi} \quad [\text{deg/s}].\end{aligned} \tag{12.40}$$

Budgeting When calculating the overall budget for orientation accuracy and stability, the noise sources of the attitude sensor and the orientation rate sensor can be considered to be independent, so that budgeting statement (12.31) may be applied:

Orientation accuracy (3σ)

$$\delta\varphi_\Sigma = \delta\varphi_c + \delta\varphi_h + \sqrt{\left(3\sigma_{\varphi/\varphi}\right)^2 + \left(3\sigma_{\varphi/\omega}\right)^2} \quad [\text{deg}], \tag{12.41}$$

Orientation stability (3σ)

$$\delta\omega_\Sigma = \delta\omega_c + \delta\omega_h + \sqrt{\left(3\sigma_{\omega/\varphi}\right)^2 + \left(3\sigma_{\omega/\omega}\right)^2} \quad [\text{deg/s}]. \tag{12.42}$$

The numerical evaluation of Eqs. (12.41), (12.42) for the given controller configuration with $\omega_0 = 0.4\ \text{rad/s}$ and $d_0 = 0.7$ is presented in Table 12.2. A typical time history with all combined disturbance inputs is shown in the simulation results in Fig. 12.13.

Table 12.2. Accuracy budget for attitude control loop with controller bandwidth $\omega_0 = 0.4\ \text{rad/s}$ and damping coefficient $d_0 = 0.7$

Disturbance	Orientation [deg]	Orientation rate [deg/s]
constant disturbance torque	0.043	0
harmonic disturbance torque	0.043	$4.3\cdot10^{-5}$
attitude sensor noise	0.036 (3σ)	0.014 (3σ)
orientation rate sensor noise	0.073 (3σ)	0.025 (3σ)
Total	$\delta\varphi_\Sigma = 0.16$	$\delta\omega_\Sigma = 0.03$

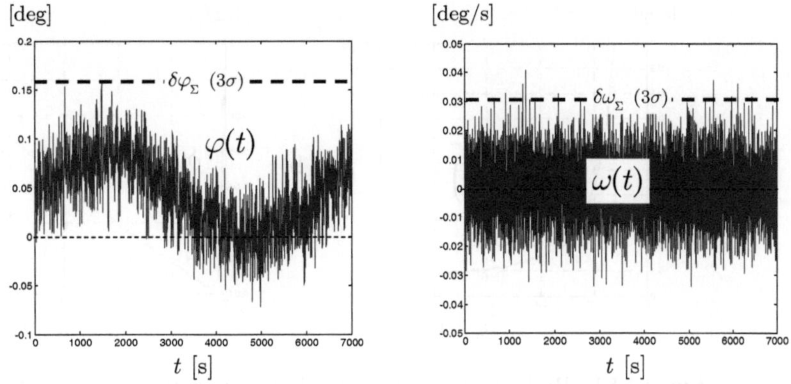

Fig. 12.13. Simulated system calculations with combined deterministic and stochastic excitation (controller bandwidth $\omega_0 = 0.4\ \mathrm{rad/s}$, damping $d_0 = 0.7$)

Design optimization Observe that Eqs. (12.41) and (12.42) in combination with the metrics (12.37) through (12.40) represent *analytically* analyzable budgeting equations which depend on the design parameters, i.e.

$$\begin{aligned} \delta\varphi_\Sigma &= \delta\varphi_\Sigma\left(M_{s0}, M_{s1}, \Omega_0, S_\varphi, S_\omega, J_{sat}, \omega_0, d_0\right), \\ \delta\omega_\Sigma &= \delta\omega_\Sigma\left(M_{s0}, M_{s1}, \Omega_0, S_\varphi, S_\omega, J_{sat}, \omega_0, d_0\right). \end{aligned} \tag{12.43}$$

Depending on which of the parameters are set by the problem specifications, the analytical budgeting equations (12.43) offer an outstanding design tool, with which "favorable" (optimal) settings for the remaining *free* design parameters can be determined.

If the satellite configuration and the devices used are taken as given in this example, then for a particular choice of the control loop damping $d_0 = 0.7$, only the *controller bandwidth* ω_0 remains as a *free* design parameter. A graphical evaluation of the budgeting equations $\delta\varphi_\Sigma(\omega_0)$, $\delta\omega_\Sigma(\omega_0)$ is depicted in Fig. 12.14.

Obviously, both the attitude deviation $\delta\varphi_\Sigma$ and the orientation rate deviation $\delta\omega_\Sigma$ are to be simultaneously kept as small as possible. From Fig. 12.14, it is possible to tell that these objectives are inherently contradictory as concerns the controller bandwidth. An *optimal* bandwidth ω_0^* which minimizes both deviations does not exist. A favorable choice of ω_0^* will thus result in a compromise solution. Depending on the tasks of the satellite, the final design will be geared more towards either orientation accuracy or towards stability.

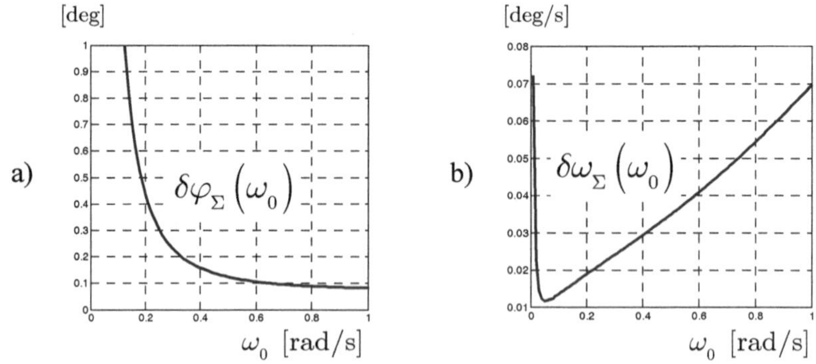

Fig. 12.14. Budget sums (3σ) as a function of the controller bandwidth: a) orientation accuracy, b) orientation stability (orientation rate) ■

Example 12.4 *Controlled tool axis.*

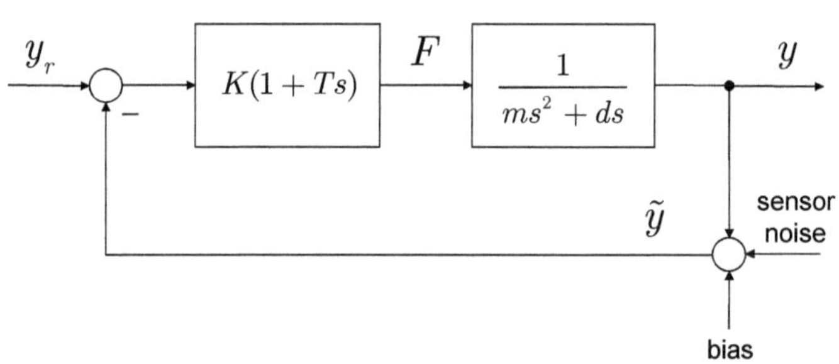

Fig. 12.15. Analysis model for a controlled tool axis

Consider the controlled linear tool axis shown in Fig. 12.15, with the following configuration parameters:

Linear axis: $m = 1$ kg, $d = 10$ Ns/m

Controller: $K = 500$ N/m, $T = 0.2$ s

Sensor: Noise $5 \cdot 10^{-4}$ mm/$\sqrt{\text{Hz}}$, 3db-bandwidth 500 Hz ,
Bias 0.01 mm .

Assume that the command inputs are harmonic signals of the form $y_r(t) = Y_r \sin(\omega_r t), |Y_r| \le 3$ mm . What is the maximum frequency of harmonic command signal that will still ensure a *position accuracy* of $\delta y = 0.03$ mm (3σ) is maintained?

Bandwidth relations As a first step, it is useful to carry out an initial overview of the bandwidth relations in the system. It is easy to verify the following relations for the control loop:

$$L(s) = K\frac{1+Ts}{ms^2+ds} \, ,$$

$$T_{y/yr}(s) = \frac{L(s)}{1+L(s)} = \frac{1+Ts}{1+2d_0\dfrac{s}{\omega_0}+\dfrac{s^2}{\omega_0}} \, ,$$

$$\omega_0 = \sqrt{\frac{K}{m}} = 22.4 \text{ rad/s}, \quad d_0 = \left(\frac{d}{K}+T\right) = 2.46 \, .$$

From here, it can be seen that the bandwidth of the sensor noise is significantly higher than the controller bandwidth. Thus, the sensor noise can be interpreted as *continuous white noise* with noise power density

$$S_n = \left(5\cdot 10^{-4} \text{ mm}/\sqrt{\text{Hz}}\right)^2 = 2.5\cdot 10^{-2} \text{ mm}^2/\text{Hz}\, .$$

(A PSD shaping filter is not required in this case).

Positioning budget The *positioning accuracy* can now be budgeted according to Eq. (12.27)

$$\delta y = \delta y_{bias} + \delta y_{noise} + \delta y_{command} \, .$$

Sensor Bias

$$\delta y_{bias} = \lim_{s\to 0}\left|-\frac{L(s)}{1+L(s)}\right| Bias = 0.01 \text{ mm}\, . \tag{12.44}$$

Sensor noise (Covariance relation, see *Appendix A*)

$$\delta y_{noise} = 3\sigma_{y/sensor} = 3\frac{S_n}{2}\sqrt{\frac{\omega_0}{d_0}\left(1+\left(\omega_0 T\right)^2\right)} = 0.01 \text{ mm}\, . \tag{12.45}$$

Command signal The control error of a harmonic command signal is determined by the frequency response

$$T_{\delta y/yr}(j\omega_r) = \frac{1}{1+L(j\omega_r)}$$

giving

$$\delta y_{command} = \left| \frac{1}{1 + L(j\omega_r)} \right| Y_r . \qquad (12.46)$$

A simple side calculation shows that frequencies ω_r which satisfy these requirements must be sufficiently smaller than unity. At the same time, it remains the case that $\left|L(j\omega)\right| \gg 1$, so that Eq. (12.46) simplifies to

$$\delta y_{command} \approx \left| \frac{1}{L(j\omega_r)} \right| Y_r = \frac{\omega_r}{K} \sqrt{\frac{d^2 + m^2 \omega_r^{\,2}}{1 + T^2 \omega_r^{\,2}}} Y_r \approx \frac{\omega_r}{K} d Y_r . \qquad (12.47)$$

Design solution The design requirement

$$\delta y = \delta y_{bias} + \delta y_{noise} + \delta y_{command} \leq 0.03 \text{ mm}$$

can then be combined with the intermediate results (12.44) through (12.47) to obtain the *design solution*

$$\boxed{\omega_r \leq 0.17 \text{ rad/s}} .$$

■

Bibliography for Chapter 12

ESA (1993). *ESA Pointing Error Handbook.* European Space Agency.

ISO (1995). Guide to the expression of uncertainty in measurement (GUM), dtsch: "Leitfaden zur Angabe der Unsicherheit beim Messen", 1. Auflage 1995, Deutsches Institut für Normung - Beuth-Verlag. I. O. f. S. (ISO). Geneva, Switzerland.

Kirkup, L. and R. B. Frenkel (2006). *An Introduction to Uncertainty in Measurement Using the GUM (Guide to the Expression of Uncertainty in Measurement).* Cambridge University Press.

Taylor, J. R. (1982). *An Introduction to Error Analysis.* University Science Books.

Appendix A

Covariance Formulae

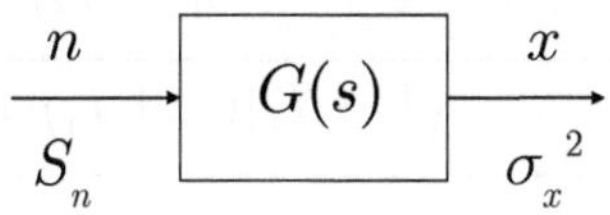

n : white noise $\qquad$ x : colored noise

spectral density $S_n \left[(\dim n)^2/\mathrm{Hz}\right]$ $\qquad$ variance $\sigma_x^2 \left[(\dim x)^2\right]$

P-T_1	$G(s) = \dfrac{K}{1+T_1 s}$	$\sigma_x^2 = \dfrac{K^2}{2T_1} S_n$
PD-T_2	$G(s) = K \dfrac{1+T_D s}{(1+T_1 s)(1+T_2 s)}$	$\sigma_x^2 = \dfrac{K^2}{2} \cdot \dfrac{1+\dfrac{T_D^2}{T_1 T_2}}{T_1+T_2} S_n$
	$G(s) = K \dfrac{1+T_D s}{1+\dfrac{2D}{\omega_0}s+\dfrac{1}{\omega_0^2}s^2}$	$\sigma_x^2 = \left(\dfrac{K}{2}\right)^2 \dfrac{\omega_0}{D}\left(1+(\omega_0 T_D)^2\right) S_n$
P-T_2	$G(s) = \dfrac{K}{(1+T_1 s)(1+T_2 s)}$	$\sigma_x^2 = \dfrac{K^2}{2(T_1+T_2)} S_n$
	$G(s) = \dfrac{K}{1+\dfrac{2D}{\omega_0}s+\dfrac{1}{\omega_0^2}s^2}$	$\sigma_x^2 = \left(\dfrac{K}{2}\right)^2 \dfrac{\omega_0}{D} S_n$
D-T_2	$G(s) = \dfrac{K_D \cdot s}{(1+T_1 s)(1+T_2 s)}$	$\sigma_x^2 = \dfrac{K_D^2}{2} \cdot \dfrac{1}{T_1 T_2 (T_1+T_2)} S_n$
	$G(s) = \dfrac{K_D s}{1+\dfrac{2D}{\omega_0}s+\dfrac{1}{\omega_0^2}s^2}$	$\sigma_x^2 = \left(\dfrac{K_D}{2}\right)^2 \dfrac{\omega_0^3}{D} S_n$

$$G(s)=\frac{K\left(1+T_{D1}s\right)\left(1+T_{D2}s\right)}{\left(1+T_1s\right)\left(1+T_2s\right)\left(1+T_3s\right)}$$

$$\sigma_x^2=\frac{K^2}{2}\cdot\frac{T_1T_2+T_1T_3+T_2T_3+\left(T_{D1}^2+T_{D2}^2\right)+T_{D1}^2T_{D2}^2\dfrac{T_1+T_2+T_3}{T_1T_2T_3}}{\left(T_1+T_2\right)\left(T_1+T_3\right)\left(T_2+T_3\right)}S_n$$

$$G(s)=\frac{K\left(1+T_{D1}s\right)\left(1+T_{D2}s\right)}{\left(1+\dfrac{2D}{\omega_0}s+\dfrac{1}{\omega_0^2}s^2\right)\left(1+T_1s\right)}$$

$$\sigma_x^2=\left(\frac{K}{2}\right)^2\frac{\omega_0}{D}\frac{1+2D\omega_0T_1+\omega_0^2\left(T_{D1}^2+T_{D2}^2\right)+\left(T_{D1}T_{D2}\omega_0^2\right)^2\left(1+\dfrac{2D_3}{\omega_0T_1}\right)}{1+2D\omega_0T_1+\left(\omega_0T_1\right)^2}S_n$$

PD₂-T₃

$$G(s)=\frac{K\left(1+\dfrac{2D_z}{\omega_{0z}}s+\dfrac{1}{\omega_{0z}^2}s^2\right)}{\left(1+T_1s\right)\left(1+T_2s\right)\left(1+T_3s\right)}$$

$$\sigma_x^2=\frac{K^2}{2}\cdot\frac{T_1T_2+T_1T_3+T_2T_3+\left(\dfrac{2\left(2D_z^2-1\right)}{\omega_{0z}^2}\right)+\dfrac{T_1+T_2+T_3}{T_1T_2T_3\omega_{0z}^4}}{\left(T_1+T_2\right)\left(T_1+T_3\right)\left(T_2+T_3\right)}S_n$$

$$G(s)=\frac{K\left(1+\dfrac{2D_z}{\omega_{0z}}s+\dfrac{1}{\omega_{0z}^2}s^2\right)}{\left(1+\dfrac{2D}{\omega_0}s+\dfrac{1}{\omega_0^2}s^2\right)\left(1+T_1s\right)}$$

$$\sigma_x^2=\left(\frac{K}{2}\right)^2\frac{\omega_0}{D}\cdot\frac{1+2D\omega_0T_1+2\left(\dfrac{\omega_0}{\omega_{0z}}\right)^2\left(2D_z^2-1\right)+\left(\dfrac{\omega_0}{\omega_{0z}}\right)^4\left(1+\dfrac{2D}{\omega_0T_1}\right)}{1+2D\omega_0T_1+\left(\omega_0T_1\right)^2}S_n$$

$$G(s)=\frac{K}{\left(1+T_1 s\right)\left(1+T_2 s\right)\left(1+T_3 s\right)}$$

$$\sigma_x^2=\frac{K^2}{2}\cdot\frac{T_1T_2+T_1T_3+T_2T_3}{\left(T_1+T_2\right)\left(T_1+T_3\right)\left(T_2+T_3\right)}S_n$$

P-T_3

$$G(s)=\frac{K}{\left(1+\frac{2D}{\omega_0}s+\frac{1}{\omega_0^2}s^2\right)\left(1+T_1 s\right)}$$

$$\sigma_x^2=\left(\frac{K}{2}\right)^2\frac{\omega_0}{D}\cdot\frac{1+2D\omega_0T_1}{1+2D\omega_0T_1+\left(\omega_0T_1\right)^2}S_n$$

■

Index

MIX
Papier aus verantwortungsvollen Quellen
Paper from responsible sources
FSC® C105338

If you have any concerns about our products,
you can contact us on
ProductSafety@springernature.com

In case Publisher is established outside the EU,
the EU authorized representative is:
Springer Nature Customer Service Center GmbH
Europaplatz 3, 69115 Heidelberg, Germany

Printed by Libri Plureos GmbH
in Hamburg, Germany